Mastering CAD/CAM

Ibrahim Zeid

Northeastern University

Higher Education

Boston   Burr Ridge, IL   Dubuque, IA   Madison, WI   New York   San Francisco   St. Louis
Bangkok   Bogotá   Caracas   Kuala Lumpur   Lisbon   London   Madrid   Mexico City
Milan   Montreal   New Delhi   Santiago   Seoul   Singapore   Sydney   Taipei   Toronto

Higher Education

MASTERING CAD/CAM

Published by McGraw-Hill, a business unit of The McGraw-Hill Companies, Inc., 1221 Avenue of the Americas, New York, NY 10020. Copyright © 2005 by The McGraw-Hill Companies, Inc. All rights reserved. No part of this publication may be reproduced or distributed in any form or by any means, or stored in a database or retrieval system, without the prior written consent of The McGraw-Hill Companies, Inc., including, but not limited to, in any network or other electronic storage or transmission, or broadcast for distance learning.

Some ancillaries, including electronic and print components, may not be available to customers outside the United States.

This book is printed on acid-free paper.

2 3 4 5 6 7 8 9 0 DOC/DOC 0 9 8

ISBN 0–07–286845–7

Publisher: *Elizabeth A. Jones*
Senior sponsoring editor: *Suzanne Jeans*
Developmental editor: *Amanda J. Green*
Marketing manager: *Dawn R. Bercier*
Senior project manager: *Kay J. Brimeyer*
Senior production supervisor: *Laura Fuller*
Lead media project manager: *Audrey A. Reiter*
Media technology producer: *Eric A. Weber*
Designer: *Rick D. Noel*
Cover design and illustration: *Rokusek Design*
Compositor: *Interactive Composition Corporation*
Typeface: *9/13 Times Roman*
Printer: *R. R. Donnelley Crawfordsville, IN*

Trademarks

AutoCAD is a registered trademark of Autodesk, Inc. CADKEY is a registered trademark of Kubotech Corporation. CATIA is a registered trademark of Dassault Systèmes SA. DirectX is a registered trademark of Microsoft Corporation. Java 3D is a registered trademark of Sun Micorsystems, Inc. Maple is a registered trademark of Waterloo Maple Inc. Mathcad is a registered trademark of Mathsoft Engineering and Education, Inc., http://www.mathsoft.com. Mathematica is a registered trademark of Wolfram Research, Inc. MATLAB is a registered trademark of the Math Works, Inc. OpenGL is a registered trademark of Silicon Graphics, Inc. (SGI). Pro/Engineer is a registered trademark of Parametric Technology Corporation (PTC). SolidWorks is a registered trademark of SolidWorks Corporation. Teamcenter, Unigraphics, Parasolid, Solid Edge, and I-deas are registered trademarks of EDS, Inc.

Library of Congress Cataloging-in-Publication Data

Zeid, Ibrahim.
    Mastering CAD/CAM / Ibrahim Zeid. — 1st ed.
        p.   cm.
    Includes bibliographical references and index.
    ISBN 0–07–286845–7
    1. CAD/CAM systems. I. Title.

TS155.6.Z455   2005
670'.285—dc22   2004042608
                                           CIP

www.mhhe.com

ABOUT THE AUTHOR

Dr. Abe Zeid is a professor of Mechanical and Industrial Engineering at Northeastern University. He is the recipient of both the Northeastern Excellence in Teaching Award and the SAE Ralph R. Teetor Educational Award. Professor Zeid is the author or co-author of many publications in journals and conference proceedings. He has been a guest of foreign governments and a keynote speaker to many organizations throughout the world. Professor Zeid teaches a variety of subjects in different fields including CAD/CAM, object-oriented design and concepts, and Internet and information systems fundamentals. Dr. Zeid has written four books including this one. The other three are CAD/CAM Theory and Practice (McGraw-Hill), Mastering the Internet and HTML (Prentice Hall), and Mastering the Internet, XHTML, and JavaScript (Prentice Hall). He is a fellow of the American Society of Mechanical Engineers.

To the memory of my parents
who taught me life by doing

Brief Contents

Part III: Computer Graphics

Part IV: Product Design and Development

Part V: Product Manufacturing and Management

Contents

Part IV: Product Design and Development

Chapter 15 **Mass Properties**   **577**

Chapter 16 **Assembly Modeling**   **615**

Part V: Product Manufacturing and Management

Preface

OVERVIEW. This new book follows in the footsteps of my *CAD/CAM Theory and Practice* (McGraw-Hill, 1991), which has been widely adopted and used throughout the world. This new book, *Mastering CAD/CAM*, has been written to respond to many suggestions made since the first book was published. It is also written to include the new concepts that have been developed since 1991. Designed to meet the demands of both practice-oriented and theoretically based courses, the book meets these conflicting demands by making CAD/CAM systems its focus. It explains the functionalities of these systems in a generic and syntax-independent fashion so that students can use their system of choice, and it relates the mathematical developments and concepts to these systems to unravel their secrets.

The book achieves a mastery level in CAD/CAM by carefully balancing the breadth and depth of topic coverage with the syntax-independent use of CAD/CAM systems. The book has 23 chapters, over 904 equations, over 328 figures, 562 screenshots from using different CAD/CAM systems, 97 examples, 36 tutorials done on different CAD/CAM systems, 75 geometric models of parts and assemblies, 31 real-life objects modeled (including a golf ball, razor, slipper, household fan, AC duct, universal joint, a telephone, a glass of wine, a spiral, a spring, and a candy dish), 383 end-of-chapter problems, a comprehensive index, and three appendices.

The book covers CAD/CAM systems and related software in a syntax-independent fashion by focusing on their semantics, and not their syntax. The semantics of these systems are the same although their syntax differs. The book uses and discusses three sets of software: (1) CAD/CAM systems: SOLIDWORKS®, Pro/ENGINEER®, CATIA®, Unigraphics, I-DEAS®, CADKEY®, and AutoCAD®; (2) Programming and graphics: Java™, Java 3D™, C, C++, OpenGL®, DirectX®, VRML, FTP, Telnet, and Web browsers; (3) Symbolic equation solvers: MATLAB®, Maple®, Mathcad®, and Mathematica®.

The rationale behind this book is simple but effective. Students need a comprehensive and complete source of CAD/CAM knowledge in order to become proficient in using any CAD/CAM system. This knowledge includes understanding 3D modeling and viewing, geometric modeling, computer graphics, and product design and manufacturing. Both students and instructors should find this book useful, as it provides "one-stop shopping" for all their learning and teaching needs.

The purpose and goal of this book is to present the fundamental concepts of CAD/CAM and its tools in a generic framework. These concepts and tools are supplemented with examples, tutorials, and problems to provide students with hands-on experience so that they can master the concepts. The book strikes a delicate balance between subject depth and breadth, and between generic and practical aspects of CAD/CAM. Regarding depth and breadth, the book covers the basic topics about CAD/CAM. Regarding generic and practical aspects, the book relates the generic concepts to their use in technology, software, and practical applications.

FEATURES. This book has many pedagogical and content features:

- Eye-catching page design, and graphics design for section and example headings.
- Abundant figures and screen captures to illustrate concepts.
- Goal, objectives, and headlines at the start of each chapter.
- Tutorials and problems at the end of each chapter.
- Each example and tutorial has three pedagogical elements to allow interactivity and deep understanding. *Solution strategy* describes the thinking behind the solution. *Discussion* provides insight into the solution. *Hands-on* exercises ask the reader to extend or modify the solution. This is a confidence builder during the learning process.
- Each example focuses on one chapter concept only, while each tutorial combines a few chapter concepts together to provide a more practical application. Many tutorials use real-life objects to stimulate and motivate students.
- All topics are covered with depth and breadth.

AUDIENCE. This book fills an important need in the market. **Students** in Mechanical and Industrial Engineering need a book that explains the subject matter in a simple, yet comprehensive and coherent way with enough examples and hands-on tutorials. This book offers concentrated knowledge to its readers so that they can find what they need very quickly.

Instructors need a book that provides them with enough topics, examples, tutorials, problems, and pedagogy. For example, the instructor may use the examples and tutorials in a lab setting. The instructor can also access the book's companion website, located at **http://www.mhhe.com/zeid1**. A solutions manual accompanies the book as well.

Professionals, usually pressed for time, need a book that they can use for self-teaching purposes. They also need a book that provides them with answers to specific questions they may have while using a CAD/CAM system. With this book, professionals can tap into its many examples, tutorials, and concepts for answers.

ORGANIZATION. This book can be used in a series of semester courses—for example, in a two undergraduate course sequence, or an undergraduate course followed by a graduate course. The undergraduate course(s) can use Chapters 1 to 10, 16, 18, and 20. The course can use two or three additional chapters, depending on the course's focus and philosophy. For example, CAD courses can use Chapters 15, 17, and 19. CAM courses can use Chapters 21, 22, and 23. The coverage of Chapters 6 to 9 may exclude the mathematical rigor of these chapters and focus more on the practice-oriented concepts. The graduate course can use chapters 1 to 14 with the mathematical rigor. In addition, courses with a CAD focus can use three more chapters from Part IV, and CAM courses can use three more chapters from Part V.

The book is organized into five related parts. Part I covers the effective use of CAD/CAM systems. This part develops the basic skills required for using any commercial system. Part II discusses, in detail, geometric modeling. Part III covers computer graphics concepts. Part IV focuses on product design and development. Part V covers the manufacturing and management concepts of products. This organization is beneficial in accommodating different course requirements and readers' backgrounds.

The problems section is divided into three parts: theory, lab, and programming. The theory and lab parts are ideal for practice-oriented courses. The theory and programming parts are ideal for theoretically based courses.

The book is written in such a way that some chapters stand on their own; that is, the chapters need not be taught sequentially. Such an appraoch accommodates different teaching styles. For example, Chapter 12 may be covered before the chapters in Part II. Chapter 20 may be covered following Chapter 4. Chapter 16 may be covered following Chapter 6, 7, 8, 9, or 10. Similarly, the chapters in Parts IV and V can be covered in different order than that of the book.

ACKNOWLEDGEMENTS. I am indebted to all of the people who helped directly or indirectly to write this book. I would like to thank the following reviewers for their valuable comments, suggestions, and advice throughout the project. There is no doubt that their suggestions have influenced and enhanced this book.

Holly K. Ault, *Worcester Polytechnic Institute*

David Ben-Arieh, *Kansas State University*

Jan Helge Bøhn, *Virginia Technological University*

Rajesh N. Dave, *New Jersey Institute of Technology*

Hazim El-Mounayri, *Indiana University-Purdue University, Indianapolis*

Louis J. Everett, *University of Texas–El Paso*

Georges Fadel, *Clemson University*

Mustapha S. Fofana, *Worcester Polytechnic Institute*

Jaime Hernandez-Mijangos, *University of Wisconsin–Platteville*

Kent L. Lawrence, *University of Texas at Arlington*

Sang-Joon Lee, *San Jose State University*

Jami J. Shah, *Arizona State University*

Yin-Lin Shen, *George Washington University*

Dušan Šormaz, *Ohio University, Athens*

Silvanus J. Udoka, *North Carolina A&T State University*

Shih-Liang (Sid) Wang, *North Carolina A&T State University*

Chun Zhang, *Florida A&M University*

Jack G. Zhou, *Drexel University*

Thanks are due to the McGraw-Hill staff for their patience and professional help. The valuable experience and vision of Jonathan Plant, Senior Editor, have permitted the successful launch of the project. Thank you, Jonathan, for starting the project! His e-mail messages, phone calls, and visits were crucial to getting the project started. I would also like to thank Amanda Green, Developmental Editor, for handling the many facets of the production process. Our many phone and face-to-face meetings were so important in keeping the project on schedule. Her many communications with the production team were invaluable in getting a head start on the project. Thank you, Amanda, for a job well done! The flexibility and vision of Jill Peter, Lead Project Manager, allowed a very smooth and productive schedule. Thank you, Jill, for your support, and best of luck in your new position! I would also like to thank Kay Brimeyer, Senior Project Manager, for taking over the project and keeping us on track.

Many other individuals and organizations have been involved with the book in one form or another. The copyeditor has done a superb job catching all my typos that even escaped my spell checker. Interactive Composition Corporation has done an excellent job rendering the book art. Rokusek Design, Inc. has done a great job designing the book cover. All my students, graduate and undergraduate, and my assistants (TAs and RAs) have changed my views on the CAD/CAM subject and the way I teach it over the years. Each time I think I got it, I only discover that I am still missing something. The help of all of my students is reflected in this book, be it an idea, an example, a tutorial, or a screen capture.

Last, but not least, very special thanks are due to my family and friends who supported me from start to finish with their love, support, and encouragement, which are greatly appreciated.

Ibrahim Zeid
zeid@coe.neu.edu
Northeastern Univeristy
Boston, Massachusetts

Using CAD/CAM Systems

This part covers all the plumbing of CAD/CAM systems, an important goal that students must achieve. This part helps put Parts II - V in context. The goal of this part is to provide a solid and clear understanding of the basic concepts of using CAD/CAM systems for 3D modeling and viewing. To achieve this goal, we accomplish the following objectives:

1. Understand the intricate nature of CAD/CAM systems **(Chapter 1)**
2. Understand 3D modeling and viewing and the differences between the two **(Chapter 2)**
3. Understand the productivity tools offered by CAD/CAM systems **(Chapter 3)**
4. Understand the content and associativity of engineering drawings **(Chapter 4)**
5. Understand customizing CAD/CAM systems via macros/API programming **(Chapter 5)**

Introduction

GOAL

Understand and master the nature of CAD/CAM systems, their basic structure, their use in engineering tasks, and their use to create geometric models of simple parts.

OBJECTIVES

After reading this chapter, you should understand the following concepts:

- Product life cycle
- Scope of CAD/CAM
- CAD/CAM applications
- Acquiring CAD/CAM systems
- Installing CAD/CAM systems
- Becoming familiar with CAD/CAM systems
- GUIs and `Help` menus
- Demo parts

CHAPTER HEADLINES

1.1 Introduction

CAD/CAM has been utilized in engineering practice in many ways including drafting, design, simulation, analysis, and manufacturing. CAD/CAM systems are commonly used in daily engineering tasks. Engineering companies large and small acquire CAD/CAM systems and train their engineers, either in-house or on a vendor's site, on how to use them. Sometimes engineers use the brute-force approach to learn how to use a system. It is not uncommon for different groups within the same company to use different CAD/CAM systems.

Experience and wisdom have it that CAD/CAM users become very inefficient in using CAD/CAM systems unless they understand the fundamental concepts on which these systems are built. This book covers all the concepts you need to know, without focusing on the syntax of any single CAD/CAM system. However, we always guide you on how to apply these concepts to your systems.

CAD/CAM is diverse subject, and the courses that teach it are more diverse. What should we cover and how can we cover it? The answer to the first question defines the scope ofCAD/CAM that we discuss in Section 1.3. This scope is primarily an outcome of the product life cycle that we cover in Section 1.2. As these two sections reveal, CAD/CAM spans four major areas: geometric modeling, computer graphics, design applications, and manufacturing applications. A CAD/CAM course typically focuses on one or two of these areas.

The answer to the second question defines the method of subject delivery. Some courses use CAD/CAM systems to illustrate the concepts, while others use more generic techniques such as programming languages and graphics libraries.

This book covers the four major areas of CAD/CAM as it is used by professionals and engineers in industry. It uses CAD/CAM systems as tools to explain and apply the concepts. This approach provides the best of two worlds: concepts and practice. The book does not cover the syntax of one system or another. It leaves the choice of a system to the reader. Some readers may decide not to use a system at all.

1.2 Product Life Cycle

Figure 1.1 shows the life cycle of a typical product. The product begins with a need which is identified based on customers' and markets' demands. The product goes through two main processes from inception to a finished product: the design process and the manufacturing process. Synthesis and analysis are the two main subprocesses of the design process. The philosophy, functionality, and uniqueness of the product are all determined during synthesis. During synthesis, a design takes the form of sketches and layout drawings that show the relationship among the various product parts. These sketches and drawings can be created using a CAD/CAM system or simply hand-drawn on paper. They are used during brainstorming discussions among various design teams and for presentation purposes.

The analysis subprocess begins with an attempt to put the conceptual design into the context of engineering sciences to evaluate the performance of the expected product. This requires design modeling and simulation. An important aspect of analysis is the "what if" questions that help us to eliminate multiple design choices and find the best solution to each design problem. The outcome of analysis is the design documentation in the form of engineering drawings (also known as blueprints).

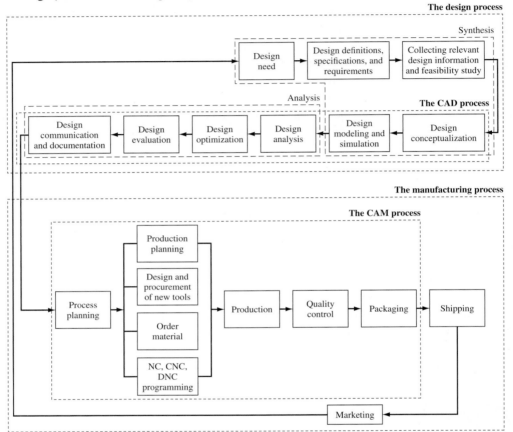

Figure 1.1   A typical product cycle.

The manufacturing process begins with the process planning and ends with the actual product. Process planning is considered the backbone of the manufacturing process since it attempts to determine the most efficient sequence in which to produce the product. A process planner must be aware of the various aspects of manufacturing to plan properly. The planner typically works with the blueprints and may communicate with the design team to clarify or request changes in the design to fit manufacturing requirements. The outcome of the process planning is a production plan, tools procurement, material order, and machine programming. Other special manufacturing needs such as design of jigs and fixtures or inspection gages are planned.

Once the process planning phase is complete, the actual production of the product begins. The manufactured parts are inspected and usually must pass certain standard quality control (assurance) requirements. Parts that survive inspection are assembled, packaged, labeled, and shipped to customers. Market feedback is usually incorporated into the design process. With this feedback, a closed-loop product cycle results, as Figure 1.1 shows.

1.3 Scope of CAD/CAM

The product cycle shown in Figure 1.1 serves as the basis upon which to define the scope of CAD/CAM. The CAD process is a subset of the design process and the CAM process is a subset of the manufacturing process. Engineers involved in the design process are usually themselves the CAD designers who execute the CAD process. Similarly, manufacturing engineers execute the CAM process.

At the core of the CAD and CAM processes is a geometric model of the product under design. Activities of the CAD process include mass properties, finite element analysis, dimensioning, tolerancing, assembly modeling, generating shaded images, and documentation and drafting. Activities of the CAM process include CAPP (computer aided process planning), NC (numerical control) programming, design of injection molds, CMM (coordinate measuring machines) verifications, inspection, assembly via robots, and packaging.

The CAD process and its tools utilize three disciplines: geometric modeling, computer graphics, and design. The CAM process utilizes the disciplines of CAD itself, manufacturing, and automation. Figure 1.2 shows these disciplines.

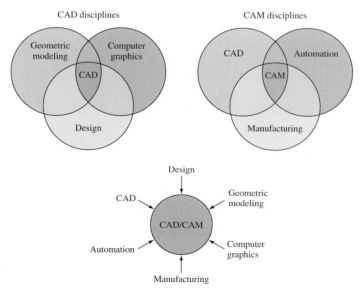

Figure 1.2   CAD and CAM disciplines.

1.4 CAD/CAM Systems

A **CAD/CAM system** is a complex application that requires both hardware and software. The **hardware** is a computer (either a PC or a workstation) with a mouse and a keyboard. The **software** is a computer program written in C or C++ as the primary language. Some other languages such as Scheme and Java are also used. For example, ACIS is written in C++ and uses Scheme. The software has a multilayered GUI (graphical user interface) that provides users with menus and icons that enables them to perform CAD/CAM activities from creating geometry to running analyses and computations.

CAD/CAM software can run as a client/server or standalone application. A network is required for the client/server mode. The software is installed on a central computer, the server, and any computer, the client, connected to the network can run the software. The software supports concurrent users (seats). Multiple users can access the software at the same time. This is the typical setup used in companies, laboratories, and academic institutions. The single seat (standalone) model, while less common, requires installing the software on each computer that is intended to run it. Thus, the computer of each seat acts as client and server at the same time.

Installing CAD/CAM software is fairly easy. The software is a self-extracting file (file installs itself) that comes on a CD. Simply insert the CD in a CD-ROM drive and follow instructions. Many CAD/CAM vendors offer student versions of their software for a reasonable price. The student version is usually a subset of the full-blown commercial version.

CAD/CAM software utilizes a data structure to save the geometry and topology (discussed later in the book) of geometric models. We refer to these models as solid models or parts. Throughout the book, we use the words *geometric model*, *solid model*, and *part* interchangeably. We also use CAD/CAM and CAD interchangeably throughout the book. The **data structure** is a well-defined storage scheme that stores model data. The physical file that stores the model data in the format offered by the data-structure scheme is called a database. Thus, a CAD **database** is the file that stores the model information. Each file has a name and an extension. The CAD user selects a meaningful name that reflects the part stored in the file — for example, *block* or *piston*. The file extension type is assigned by the CAD system; the CAD user has no control over it. The file extension depends on the type of information it stores. Many CAD systems use the *.prt*, *.assem*, and *.dwg* for part, assembly, and drawing files, respectively.

CAD systems have a steep learning curve associated with them. Some systems are easier to learn than others. Nevertheless, new CAD users are faced with two challenging problems. First, they must learn the structure of the software GUI and where to find commands when needed. Second, they must understand the concepts of 3D modeling and viewing, and how to control geometric construction—a subject we cover in Chapter 2. Before using a CAD system, users should learn the theoretical concepts of geometric modeling and computer graphics that we cover in this book. Without these concepts, users tend to use the brute-force approach by attempting to learn from the software GUI — a very counterproductive approach that costs users and their companies large sums of money that are nevertheless hard to measure. The loss is usually expressed in terms of delays and projects that take a lot longer to finish.

Learning and using one CAD system should help accelerate learning and using other systems. The initial time investment pays dividends when you migrate from one system to another. The training on one system should ease the transition to another system. The initial learning experience includes both the concepts and the system syntax. Migrating from one system to another does not change the concepts because they are independent of any system. Only the syntax changes, which makes it more manageable for users to adapt.

CAD/CAM software is designed to run on all platforms and operating systems (OSs), including the four major ones: Unix, Linux, Windows, and Macintosh. However, SolidWorks is a native Windows application, that is, it only runs on the Windows OS. You buy the software that suits your existing OS. The software GUI is independent of the OS to prevent confusing the users. However, the look and feel of the GUI will differ somewhat between operating systems. For example, the look and feel of a GUI running on Unix is different than its look and feel when it runs on Windows. This is due to the fact that the rendering and display functions of a GUI depend on libraries provided by each OS.

1.5 CAD/CAM Applications

There are considerable numbers of applications for the various existing CAD/CAM systems. Each application has its own strengths and is usually targeted toward a specific market and group of users. For example, there are mechanical, electrical, and architectural CAD and CAM products. An inspection of these various systems reveals that they have a generic structure and common modules. Awareness of such structure and modules enables users to better understand system functions for both evaluation and training purposes. The major available modules are discussed here.

The **geometric engine** (module) is the heart of a CAD/CAM system. It provides users with functions to perform geometric modeling and construction, editing and manipulation of existing geometry, drafting and documentation. The typical modeling operations that users can engage in are model creation, cleanup, documentation, and printing/plotting. Shaded images can be generated as part of model documentation.

The creation of a geometric model of an object represents a means and not a goal for engineers. Their ultimate goal is to be able to utilize the model for design and manufacturing purposes. The **applications module** achieves this goal. This module varies from one software system to another. However, there are common applications shared by most CAD/CAM systems. Mechanical applications include mass property calculations, assembly analysis, tolerance analysis and synthesis, sheet metal design, finite element modeling and analysis, mechanisms analysis, animation techniques, and simulation and analysis of plastic injection molding. Manufacturing applications include CAPP, NC, CIM, robot simulation, and group technology.

The **programming module** allows users to customize systems by programming them to fit certain design and manufacturing tasks. Chapter 5 covers this module in detail. Programming a CAD/CAM system requires advanced knowledge of the system architecture, its database format, and a high-level programming language such as C, C++, Java, Scheme, or others.

The **communications module** is crucial if integration is to be achieved between the CAD/CAM system, other computer systems, and manufacturing facilities. It is common to network the system to transfer the CAD database of a model for analysis purposes or to transfer its CAM database to the shop floor for production. This module also serves the purpose of translating databases between CAD/CAM systems using graphics standards such as IGES and STEP.

The **collaborative module** is emerging as an outcome of the widespread of the World Wide Web and the Internet. This module supports collaborative design. Various design teams in different geographical locations can work concurrently on the same part, assembly, or drawing file in real time over the Web. One team can make changes that other teams can view and accept or reject.

The various CAD/CAM systems support engineering design, analysis, and manufacturing applications. They all have a flexible pricing structure that allows customers to add on applications as needed. A basic license for a system includes its geometric engine and mass properties calculations. The majority of the vendors bundle their applications with the geometric engine. SolidWorks, however, uses a different approach. It does not get involved with applications. It redirects its customers to contact SolidWorks partners directly to acquire applications from them.

1.6 Acquiring a CAD/CAM System

Many CAD/CAM systems exist today. We classify them into four groups depending on the market they serve and the tools, functionalities, and flexibility they provide. The four groups are low end, midrange, high end, and specialized. We must admit that we run the risk of misclassification as all CAD/CAM systems have been getting better over the years to survive competition. The classification is only a guide; it carries with it the history of the evolution of CAD/CAM systems. Low-end systems tend to target users who are not sophisticated and whose products are not complex; a typical product consists of a small number of parts whose geometry is not complicated. These users tend to focus on basic geometric modeling and drafting. Examples include AutoCAD, Autodesk Inventor, and CADKEY.

Midrange systems target users who have complex modeling needs. The number of parts per product is large enough for midrange applications. Unlike low-end systems, midrange systems support design and manufacturing applications. They either bundle them with the geometric modeling engine or work with partners. Examples include SolidWorks, Pro/E, and MasterCAM.

High-end systems are the legacy CAD/CAM systems that appeared in the 1970s and evolved over the years. These systems support the modeling, analysis, and manufacturing of complex products such as airplanes, cars, and others. Examples include Unigraphics, Parasolid,

SDRC I-DEAS, and CATIA. Unigraphics, Parasolid, and I-DEAS are offered by EDS. CATIA and SolidWorks are offered by Dassault Systems.

Specialized systems include ACIS (Spatial Corp.) and Parasolid (EDS). Each provides a very robust and universal geometric modeling and graphics kernel (software) that companies can license to build fully functional CAD/CAM systems. ACIS software serves as the core kernel of these systems. SolidWorks and Unigraphics SolidEdge are ACIS-enabled systems; they use ACIS as their kernel.

Which of these systems should a company use? The wide variety of systems makes the selection process rather difficult. CAD/CAM selection committees find themselves developing long lists of guidelines to screen available choices. These lists typically begin with cost criteria and end with sample models or benchmarks chosen to test system performance and capabilities. In between comes other factors such as compatibility requirements with existing in-house computers, prospective departments that plan to use the system, and the credibility of CAD/CAM systems' suppliers. In addition, the background, skill level, and aptitude of employees for new technology and its associated training must be considered carefully.

The selection of a CAD/CAM system is greatly influenced by the size and the complexity of the company seeking a system. The smaller the size of the company, the more specialized its CAD/CAM needs and the less CAD/CAM power it needs. For example, a small consulting agency tends to select an easy-to-use system such as AutoCAD, CADKEY, SolidEdge, or SolidWorks.

A midsize company would select from a number of systems such as SolidWorks, Pro/E, Unigraphics, I-DEAS, and CATIA, but often not with the PDM (product data management) or other high-end software modules.

Large companies tend to buy the same CAD/CAM systems as the midsize entities, but they acquire most, if not all, of the high-end applications including, PDM. Large organizations can easily justify the up-front cost of these systems because they have complex products that span many departments that must share and modify product data concurrently. This is where PDM becomes a crucial CAD/CAM software module to have.

In contrast to many selection guidelines that may vary sharply from one organization to another, the technical evaluation criteria are by and large the same. They are usually based on and are limited by the existing CAD/CAM theory and technology. These criteria are ease of using the system, the quality of documentation, software maintenance, support and service, geometric modeling capabilities, design and manufacturing applications, and programming languages available for system customizing.

Once a CAD/CAM system is selected and acquired, it needs installation, training, use, support, and maintenance. A system administrator installs it. Intended designers and users attend either in-house or vendor training courses. CAD/CAM managers develop policies, guidelines, and standards that designers and engineers use daily to achieve their CAD/CAM tasks. The system administrator is always in charge of maintaining the system, backing it up, and updating it by installing new revisions that are released by the vendors.

Ongoing system support is important to the success and effective use of a newly acquired CAD/CAM system. Users typically need help to adapt a CAD system to their company-specific needs such as setting rules and conventions of filenames, part templates, and so forth.

1.7 Getting Started

When we acquire and install a CAD/CAM system, we are faced with the question of where to start. The first goal of using a system should be simple; we would like to create a simple part such as a box. Regardless of the specific CAD/CAM system that is used, follow the steps shown in Table 1.1 to get started. Refer to the online users' and reference manuals for your CAD/CAM system for specific instructions. You should also consult the online help system during the daily use. We also recommend that readers apply the generic concepts covered throughout this book to their particular CAD/CAM systems.

Table 1.1   Steps for getting started on a CAD/CAM system.

Goal: log in, understand the system GUI, use the `Help` menu, and try to use the system

Step	Description
1. Get an account	Contact the system administrator or manager to obtain an account: a username and a password.
2. Log in	Find a computer that can run the software; here we assume a client/server configuration to run the software, and not a single user installation. This computer is the client. Type the username and the password to log in.
3. Start software	Ask the system administrator or manager how to start the CAD software. Or ask for the OS-level command that is required to start up the CAD software (program). We essentially run the executable, like any other program. For example type `solidworks`, `proe`, or `ug` in an OS window (`Terminal` in Unix) to run SolidWorks, Pro/E or Unigraphics, respectively. Windows (or Linux) provides a shortcut to the CAD program, like all other programs. Simply double-click it, or click this command sequence: `Start` (Windows menu on the bottom left corner of computer screen) `=> All Programs =>` select CAD program name.
4. Use GUI	Take time and browse through the user interface, and become familiar with it. It is similar to other programs such as Word or PowerPoint. There should be a menu bar at the top of the software window with menus such as `File`, `Edit`, `View`, and so forth. Each menu has menu items — for example, `File` has `New`, `Open`, and so forth. Try to locate the menu or buttons (icons) that create geometry.
5. Get help	Become familiar with the online `Help` menu located at the extreme right of the menu bar. `Help` comes in handy for consultation while using the CAD system. It is the online documentation of the software.

Table 1.1   Steps for getting started on a CAD/CAM system. (Continued)

Goal: log in, understand the system GUI, use the `Help` menu, and try to use the system	
Step	Description
6. Change database units	CAD/CAM systems use the English (inch) or metric (centimeter) system as measurement units for model databases. Each CAD/CAM system uses one measurement system as the default unit for its databases. Make sure to change the database measurement unit before beginning construction. If we forget to change the units before construction, there is no problem. CAD/CAM systems allow users to change the database units even after a model has been created. A user simply selects the desired measurement unit and updates the model database.
7. Use system	Use the brute-force approach to create a box (cube). Get help from someone who knows how to drive (use) the system.
8. Print a copy	Locate the `Print` function and print a hardcopy of the box model. Try this sequence: `File => Print`.
9. Save part file	Click this sequence: `File => Save As =>` provide a part filename. Make sure you specify the directory where you wish to save the part file. We can also use buttons (icons) provided by the GUI to save.
10. Log out	Log out, by using an `Exit` function on the GUI (`File => Exit`), or simply kill the window by clicking its x icon on the top right corner.

1.8   Tutorials

Throughout this book, tutorials serve as hands-on activities to illustrate the concepts we cover in each chapter. The reader should do them using a CAD/CAM system of choice. Coverage of the tutorials is done in a generic way that is independent of any system syntax. We offer planning strategies and steps for how to create the geometric models of parts and assemblies. The reader can translate these steps into commands and activities on a particular CAD/CAM system. We have tested this approach extensively, and it works. The tutorials of this book have been tested over the years using four major CAD/CAM systems: SolidWorks, Pro/E, Unigraphics, and CATIA. The version included in this book has been created using SolidWorks.

1.8.1   Getting Started

This tutorial shows how to implement some of the steps shown in Table 1.1. Let us assume that you have logged in and have the CAD/CAM software up and running. Apply the following concepts to your CAD/CAM system:

1. **Two ways to execute commands.** We can either use the menubar approach or the toolbar approach as shown in the following screenshots. Each of these approaches has advantages and disadvantages as shown below.

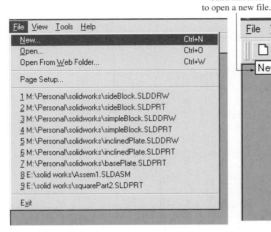

Click this icon
to open a new file.

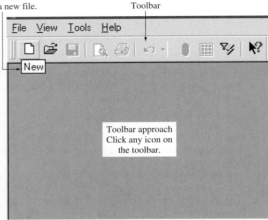

Toolbar

Menubar Approach:

Advantages:
1. Commands are grouped and placed under menus.
2. It does not occupy a lot of the modeling window space.
3. Always available.

Disadvantages:
1. One must remember how the commands are grouped together.

Toolbar Approach:

Advantages:
1. Provides visual interaction through icons.
2. One-click access to commands.

Disadvantages:
1. One must activate all the necessary toolbars.
2. Makes the screen clumsy and reduces the modeling window space

2. **System-level customization.** We can change system settings to fit our modeling needs and environment, and save them. These settings affect the CAD/CAM system properties. They apply to all models. System-level options that we can set include database colors, drawing properties, and others as shown in the upcoming screenshot.

3. **Document-level customization.** Document-level settings allow users to independently control the properties of each document. Users can access and change these properties only when the document is open. Document-level options that we can set include units, grid, image attributes, and others as shown in the forthcoming screenshot.

4. **Modes of CAD/CAM systems.** When we create a new file in a CAD system, the system usually prompts us to select the mode in which we wish to create the new file, as shown in the upcoming screenshot. The most common modes are `Part` mode, `Drawing` mode, and `Assembly` mode. Other modes may exist, depending on the CAD system, such as

Simulation mode, Animation mode, Analysis mode, and Machining mode. The screenshot on the right shows the common modes.

The relevance of selecting the right mode is that the CAD system will be configured accordingly to select the right CAD module and make available the commands that are applicable to that mode. For example, if you select the Part mode, the Assembly module and thus the assembly commands will not appear because you do not need them. Thus, mode selection increases the efficiency of CAD systems and reduces the time needed to create the model.

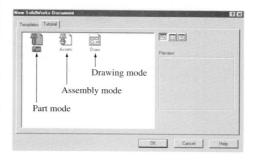

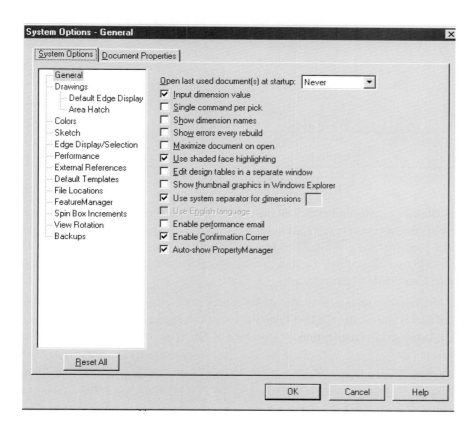

System-level customization

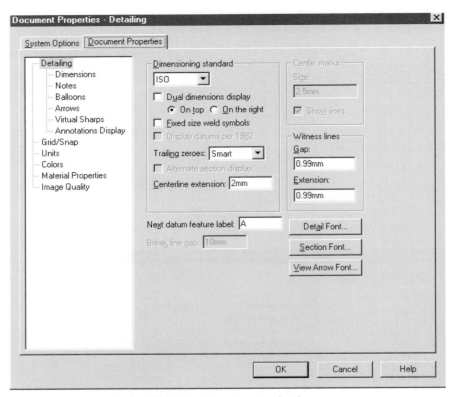

Document-level customization

5. **Coordinate systems and sketch planes.** Coordinate systems and sketch planes are the most important concepts in a CAD system. They are used to input, store, and display model geometry and graphics. The three coordinate systems that achieve these tasks are Model (Master) Coordinate System (MCS), Working Coordinate System (WCS), and Screen Coordinate System (SCS).

An **MCS** is the reference space of the model with respect to which all the model's geometrical data is stored. In a CAD system MCS is generally shown by displaying the X, Y, and Z axes. The orthogonal planes formed by the axes of MCS are datum planes, sketch planes, or construction planes.

A **WCS** can be thought of as a portable coordinate system. It is often used when the desired plane of sketching is not easily defined as one of the MCS planes. The user can define a WCS (and thus a sketch plane) for such a condition and relate it to MCS using a transformation matrix that allows geometric data to be stored with respect to MCS. In SolidWorks or Pro/E there is no physical display of WCS; however when you are creating/selecting sketch planes you are essentially creating WCSs.

An **SCS** is a 2D device-dependent coordinate system whose origin is located at the lower left corner of the screen. It is mostly used in view-related clicks such as definitions of view origin and select views for graphics operations.

Sketch planes, datum planes, or construction planes, as different CAD systems call them, is also one of the most important concepts in CAD that complements coordinate systems. **Sketch planes** are the orthogonal planes created by the axes of MCS or WCS. They can also be other planes in any orientation. Creating/selecting a sketch plane is the very first step toward creating a CAD model. Moreover, one has to repetitively create/select sketch planes throughout modeling activities. How your CAD model will be oriented in 3D space depends on which sketch plan you use for sketching. As one may have figured out by now, a **sketch plane** is a plane that we sketch on. The following screenshot shows MCS and sketch planes of a typical CAD/CAM system.

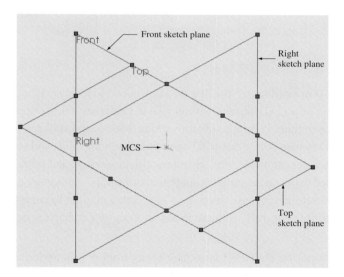

6. **Sketching (2D operations).** In item 5 we said that the first step in CAD modeling is creating/selecting the sketch plane; we now proceed to the next step in CAD modeling — sketching. CAD systems offer many sketching commands/tools, which allow us to create various 2D shapes using geometrical entities such as lines, circles, arcs, fillets, rectangles, and chamfers. All CAD systems offer these sketching tools. The upcoming screenshot shows SolidWorks sketching tools as an example. They can be accessed by clicking the `Sketch` icon (identified in the following screenshot) after selecting a sketch plane.

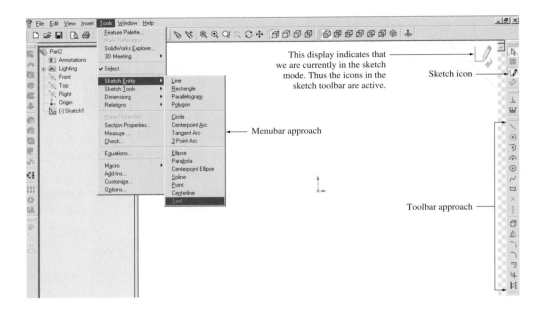

7. **Modeling (3D operations).** Now that we know how to create sketches using the sketch tools provided by a CAD system, we can take the sketches that we create and apply various 3D operations to them to obtain various features and surfaces which we call CAD models. CAD systems offer many 3D operations for both solids and surfaces such as extrusion, revolve, cut, holes, ribs, chamfers, fillets, sweeps, and lofts. These commands can be selected using the menubar or toolbar approach. It is important to know at this stage that 3D operations may require more than one sketch in order to completely define the feature or surface. We shall see these in the tutorials of the chapters that will follow.

CAD models can be classified into three types from a geometric construction point of view: 2½D models, 3D models, and a combination of both.

2½D models are the ones that have uniform cross section and thickness in a direction perpendicular to the plane of the cross section. Axisymmetric models also fall under this category. Models that are made up of many 2½D features also fall into this category; however just to distinguish them we may call them composite 2½D models.

2½D models are simple to create and use simple commands such as extrude and revolve. The creation of a 2½D model follows two steps. First, we create the model cross section in a sketch plane. Second, we extrude the cross section if the model has uniform thickness, or we revolve it if it is an axisymmetric model. The upcoming screenshots show sample 2½D models.

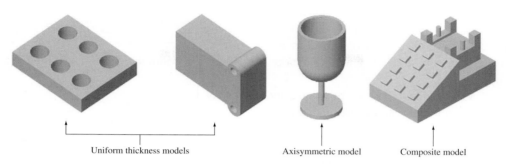

Uniform thickness models   Axisymmetric model   Composite model

Sample 2 1/2D models

3D models are the ones that do not have a uniform cross section and/or do not have a constant thickness. Such models usually require more than one sketch in different sketch planes and use advanced commands such as sweep and loft. The screenshots on the right show two 3D models.

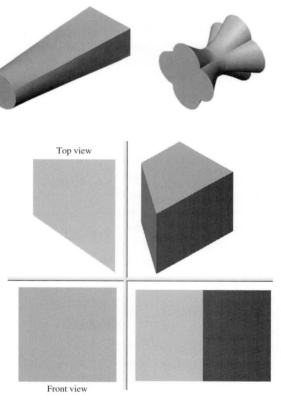

As we have seen, creating a 2½D model is usually much easier than creating a 3D model. It is important to visualize/interpret if a model can be created as a 2½D model because this may affect the ease with which a model can be created—whether we should use simple/fewer commands or advanced/more commands. Consider the model shown in the screenshot on the right. If we realize that cross section is the one in Top view then it is a simple extrude command (2½D), but if one creates a rectangle in Front view and extrudes it, then a cut is also required for the tapered face, making the model creation more complex.

8. **Viewing.** Once we create a model using 3D operations applied to sketches, or while in the process of creating models and sketches, CAD systems allow us to view those models in many different ways. Most of the viewing operations on a CAD system can be classified into three groups: view orientation, view modes, and view manipulations.

View orientation includes standard views such as front, top, right, and isometric. They basically align the model on-screen in a way

that displays the selected view. SolidWorks, for example, provides the view orientation toolbar shown here.

View modes allow us to change the display of the model to different types such as wireframe, hidden, or shaded. In SolidWorks one can access these through the toolbar shown on the right.

View manipulations allow us to dynamically rotate, pan, and zoom the model to gain better control over its viewing. In

SolidWorks one can access these through the toolbar shown here.

9. **Productivity tools.** All CAD systems provide certain productivity tools/features that reduce the time required to create a model and also increase the accuracy of the model. Examples of such productivity tools are geometric modifiers, geometric arrays, and grids.

Geometric modifiers allow us to use existing information about a model or a sketch without having to calculate that information. To put this in simple words, if we want to create a new line starting from the end point of an existing line, CAD systems highlight the desired end point when the cursor is moved close to that point. Three common geometric modifiers are end, origin (center or mid), and intersection. Some CAD systems also offer other modifiers such as tangent, quadrant, near, perpendicu-

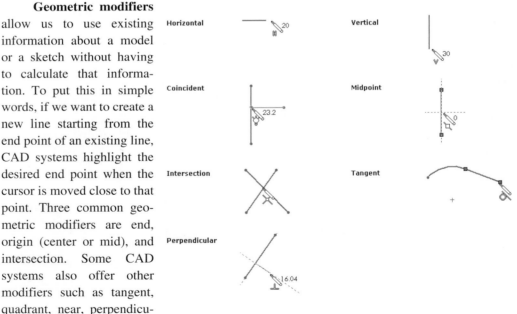

lar, and parallel. The terminology used for modifiers in different CAD systems may be different. The modifiers also enforce a concept of adding relationship between entities. Some typical modifiers are shown in the screenshot on this page. Example modifiers are end, midpoint (center or origin), intersection, horizontal, vertical, coincident, tangent, and perpendicular.

Geometric arrays is a very useful productivity tool offered by all CAD systems. It is a number of identical entities placed uniformly at specified locations — that is, in a pattern. The two types of geometric arrays are rectangular and circular. Geometric arrays are offered at both the sketching stage and modeling stage. At the sketching stage they allow the user to repeat the sketches in a pattern, and at the modeling stage they allow the user to repeat the features in a pattern. The screenshots below show sample geometric arrays.

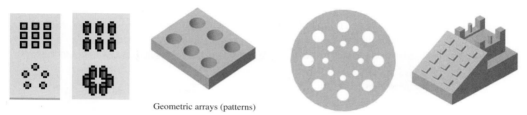

Geometric arrays (patterns)

Grids are useful sketching aids. A **grid** is a network of uniformly spaced points. There are two types of grids available: rectangular and radial. A grid is displayed as a series of dots. Different CAD systems allow many settings for grid properties such as grid spacing, grid activation, and grid appearance. In SolidWorks grid properties are considered such as document properties — that is, document-level settings we discussed in item 3 of this tutorial. The screenshot below shows how one can customize the grid in SolidWorks.

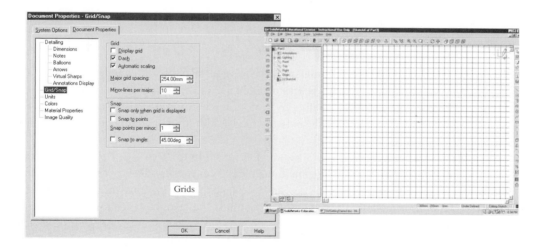

Grids

10. **Management tools.** CAD systems provide management tools so that a user can group together the entities of similar nature, which in turn assists in editing, visualization, and efficient access to CAD models. Some management tools that CAD systems offer are layers and colors.

The primary use of **layers** is to group together CAD entities of similar nature. Layers can be thought of as sheets of transparencies, each layer holding some information about the model such as sketches, features, dimensions, or notes. Layers also have independent properties such as visibility, color, and selectability. Layers in CAD systems may exist for all modes such as modeling, drawing, assembly, or other modes. In SolidWorks the concept of layers is applicable only to drawing mode.

Colors are helpful in distinguishing the entities of geometric models from each other. In SolidWorks, colors is one of the properties that can be customized at both system level and document level. The following screenshots show color customization in SolidWorks.

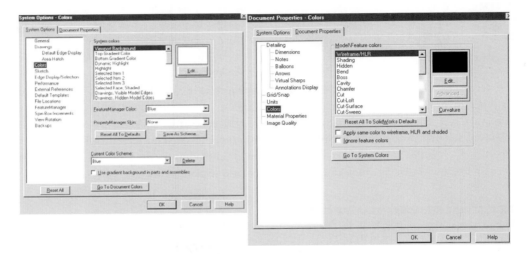

11. **Storing and retrieving CAD models.** Make sure you save the CAD model you are working on at the end of each CAD session. You need to retrieve the model later by opening its file to resume its modeling by performing new CAD/CAM operations. These operations include either editing the existing model definition or adding new entities to it.

Tutorial 1.8.1 discussion:

In this tutorial we have looked at some fundamental concepts of CAD systems in general. We have also seen different functionalities and tools that CAD systems offer to aid in the step-by-step creation of models. We have briefly outlined the steps to be followed to create a CAD model — open a new file, select the mode, change the software settings, select the sketch plane, create sketches, create models, and view models. Finally, we have applied all the concepts we discussed to a CAD software program, SolidWorks.

This tutorial has also covered many concepts that will be explained throughout the book in more detail. More importantly, we will relate these concepts to their theoretical and mathematical background. If the tutorial feels overwhelming, that is to be expected. While it is not intentional, it should make you appreciate the theoretical aspects of geometric modeling.

Tutorial 1.8.1 hands-on exercise:

Apply this tutorial to your CAD/CAM system. Do you see a lot of differences between the material described here and that which your CAD system offers? Can you find all of the items easily? Try to construct some of the models shown in the screenshots. Are they easy to create? Make up your own dimensions.

1.8.2   Use FTP and Telnet Sessions Together

How do we access files anytime, anywhere? The Internet and the Web have made the answer to this question simple. Use Telnet to access computer accounts remotely, and FTP to copy files from one computer to another. This is useful in working with CAD/CAM files.

This tutorial shows the use of both FTP (WS_FTP program) and Telnet (Windows Telnet) sessions together. On the client side (local computer) we use the Windows OS, and on the server side (remote computer) we use the Unix OS. We need to create a new directory, called *testDir*, on the Unix computer, upload a file, called *myFile.txt,* from the local computer to it, and list the contents of the *testDir* directory to make sure the file transfer was successful. Follow these steps:

1. **Start a Telnet session** on the Windows computer. Click: `Start` (Windows button) => Run => `telnet` => OK. Type o *gateway.coe.neu.edu* in the DOS window that pops up, as shown in the following screenshot.

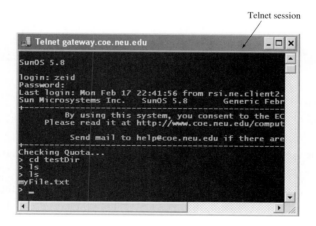

2. **Log in** to the Unix computer remotely by typing the username and password.
3. **Create the directory** using this Unix command: `mkdir` *testDir.*
4. **Go to this directory** using the command `cd` *testDir.*
5. **List the contents** of *testDir* using the command `ls`. Nothing should be there.
6. **Start an FTP session** on the Windows computer. Use a standalone FTP program. Log in. The forthcoming screenshot shows the client window of the WS_FTP program.

7. **Upload the file**, *myFile.txt*, to the *testDir* directory. Select it in the FTP local system window and click the arrow pointing to the remote system window. Make sure the *testDir* directory is selected in the remote system window.

8. **Check the contents** of *testDir*. Type `ls` in the Telnet session DOS window. *myFile.txt* should be there, as the Telnet screenshot in the previous page shows.

9. **Close both sessions**. Type `logout` to close Telnet; click `Close` to close the WS_FTP program.

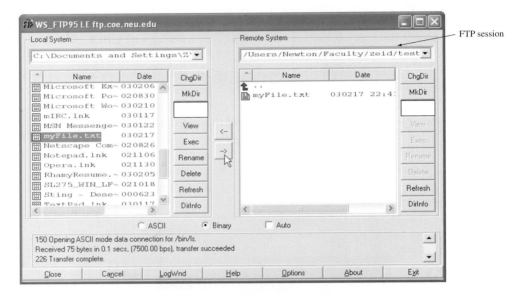

Tutorial 1.8.2 discussion:

Let us assume that you are using a CAD system at home and you want to move CAD files to your account at work or school. Or, let us assume that you are using a CAD system in a library or in a public lab where you may not have an account; how do you save your CAD files? Obviously you can use a floppy or a zip disk, but more often than not these corrupt data files.

The best solution is to use FTP to upload the CAD files from home, the library, or the lab computer to your account in your department. Assume that you have created a folder (directory) called *CADFiles*. You move the files there; no need for a disk that you may have forgotten at home.

Tutorial 1.8.2 hands-on exercise:

Download WS_FTP shareware (or CuteFTP or other FTP shareware) from the Web. It is a self-extracting file. Double-click it to install. Run a test by uploading or downloading a file from one computer to another. Use Telnet to make sure the upload or download is successful. Use this idea all the time in your daily tasks.

Problems

Part I: Theory

1.1   Perform a literature and/or a Web search on the history of CAD/CAM. Hint: We identify four generations. First generation is 2D CAD focusing on drafting (early 1970s). Second generation is 3D wireframes and surfaces running on proprietary systems focusing on engineering design and detailed drafting (late 1970s through the 1990s). Third generation is 3D CAD, parametrics, feature-based design, associative modeling and drafting, and digital product (late 1980s through the 1990s). Fourth generation is digital product, value chain and product lifecycle management (early 2000s to present).

1.2   CAD/CAM is an interdisciplinary field. Many magazines and journals exist. Go to your library, and check the Web and make a list of available journals, magazines, and newsletters in CAD, CAM, CAE (Computer Aided Engineering), and CIM (Computer Integrated Manufacturing). Use Appendix A as a start.

1.3   Make a list of some of the design projects that you were involved with. Apply the CAD tools defined in Section 1.3 to them.

1.4   Some design applications would be better with CAD than others. Why? What characteristics make an application a good candidate for CAD?

1.5   What are the three modeling modes offered by CAD/CAM systems? What do they do?

1.6   CAD/CAM systems have one of the following orientations for MCS: XY plane horizontal or XY plane vertical. Sketch each system in 3D space.

1.7   How do you control the 3D modeling/construction of your geometric modeling activities on a CAD/CAM system?

1.8   What is a 2 ½D model? Sketch some examples.

Part II: Laboratory

1.9   Familiarize yourself with the CAD/CAM system you will use to do the laboratory assignments at the end of each chapter. Make sure you can use the user interface (GUI) and online help (documentation) provided by the system. Use Tutorial 1.8.1 as a guide.

1.10 Let us assume that you are in an office or a computer lab. If you look around you, there are many objects that serve as good CAD models. Classify each of the following objects as 2½D, composite 2½D, or 3D models: keyboard, mouse, mouse cable, computer monitor, printer, chair, computer tables, a cell phone, a printer cable, your backpack, trash paper basket, and computer system box.

1.11 Which objects of problem 1.10 have geometric arrays (patterns)? Which of these use rectangular patterns and which use circular ones?

1.12 Develop a good layering scheme for CAD models. Hint: Think of the model data and use a layer for each type, for example, a layer for geometry, a layer for dimensions, a layer for tolerances, and so forth.

1.13 How can you make a good use of grids for sketching? Hint: set up the grid spacing to a value, and use the grid as a measurement tool if you need it.

1.14 Sketch the following cross sections of two of the parts shown in Tutorial 1.8.1. Submit screenshots of the sketches. All dimensions are in inches. Make sure to switch database units from mm (millimeters) to inches before you begin construction.

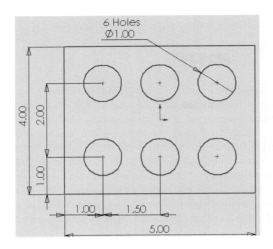

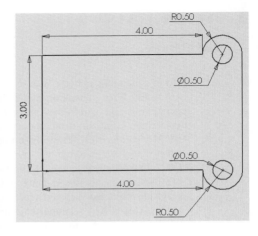

Part III: Programming

1.15 Familiarize yourself with a high-level programming language such as C, C++, or Java.

CHAPTER 2

3D Modeling and Viewing

GOAL

Understand and master the basic concepts of 3D modeling and viewing, the difference between them, the control of modeling via the sketch planes, and the effective use of CAD/CAM systems in geometric modeling.

OBJECTIVES

After reading this chapter, you should understand the following concepts:

- Types of geometric models
- Coordinate systems
- Sketching and sketch planes
- Modeling entities and features
- Modeling operations
- Modeling strategies
- Model viewing
- System modes

CHAPTER HEADLINES

2.1 Introduction

When we use a CAD/CAM system, our first goal is to create a geometric model of an object. Such a model serves as a digital representation, in a computer, that we can use later for a variety of engineering activities such as analysis and manufacturing. The representation is well structured in the model database, and the database structured content is stored in the part file of the model.

What geometric model do we really create and what kind of digital representation does a CAD/CAM system use? Solid models are what CAD/CAM systems use. We create a solid model of an object. A **solid model** is a complete, unique, and unambiguous representation of an object. The model resembles the object. An object, such as a cube, has sides (6), edges (12), and corners (8). Its corresponding solid model has faces, edges, and vertices to represent its sides, edges, and corners, respectively. Figure 2.1 shows the terminology that CAD/CAM systems use when sketching and creating solid models. Such terminology is universal among all systems.

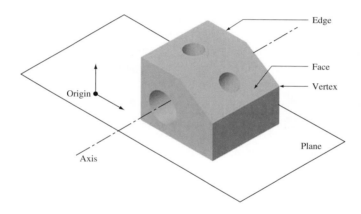

Figure 2.1   Terminology of a solid model.

We use 3D modeling concepts during the process of creating a solid model of an object. We typically need to define and use different WCSs to control the modeling. As we construct geometric entities (items), they reside in the sketch plane (*XY* plane) defined by the current WCS. The creation of a solid model begins with the creation of points, curves, and surfaces. The combination of these entities together with 3D operations such as extrusion and revolution create solid models.

We also utilize the concept of parametrics to create solid models. We simply sketch geometry freely, similar to freehand sketching; we can assign dimensions at the time of creation or later. This is a very useful concept that frees us from worrying about dimensions while we are designing objects and models.

CAD/CAM databases are built to store the full 3D definition of a solid model. As such, the database is fully 3D, associative, centralized, and integrated. The associativity concept implies that new information can be derived from users' input. For example, if the two end points of a line are input, the line length and its orientation in space can be calculated. The centralized concept means that any change in or addition to a geometric model in one of its views is automatically reflected in its existing views or any views that may be defined later. The integrated concept implies that a geometric model of an object can be utilized in all phases of a product cycle (refer to Chapter 1).

We utilize 3D viewing concepts during solid modeling. We can view the model from different angles and orientations in 3D space. We create 3D and/or 2D views when we need them. This chapter is dedicated to the full coverage of the 3D modeling and viewing concepts.

2.2   The Three Modeling Approaches

While all CAD/CAM systems create parametrics solid models, CAD designers can create the models in different ways. We identify three modeling approaches that designers can choose from to create solid models. They are primitives, features, and sketching. The existence of the three approaches is attributed to the evolution of solid modelers over the years. The early solid modelers provided only the primitives approach. As science and technology advanced, features, along with primitives, became available. Further developments have led to sketching, the third approach.

The primitives approach views a solid model as a combination of simple, generic, and standard shapes that can be combined. These shapes are the primitives. Primitives include a block (box), cylinder, sphere, cone, wedge, and torus. These primitives are combined via the Boolean operations union, subtraction, and intersection. We cover primitives and Boolean operations in detail later in the book.

The features approach is similar to the primitives approach; it replaces primitives with features and embeds Boolean operations in the feature definition. For example, let us consider creating a hole in a block. Using the primitives approach, we follow these steps:

1. Create the block using a block primitive.
2. Create a cylinder in the right location and orientation relative to the block.
3. Subtract the cylinder from the block.

Using the features approach, we follow these steps:

1. Create the block using a block feature.
2. Create the hole in the block by creating a hole feature in the right location and orientation relative to the block.

In the features approach, step 2 combines steps 2 and 3 of the primitives approach.

The sketching approach is similar to the features approach, with one change. Instead of using predefined shapes only, such as holes and ribs, it allows CAD designers to create much more elaborate and more general features starting from a sketch. Examples include extrusion, revolution, linear and nonlinear sweep, loft, spirals, and helicals. A CAD designer uses 2D operations to sketch a region or a cross section, followed by 3D operations, such as extrusion, to create the feature.

Which of the three approaches is best for creating solid models? The sketching approach is the most commonly used approach because it combines the best of the three of them. It allows the creation of many more features than the features approach. It also enables CAD/CAM vendors to increase the modeling domain of their systems without having to predefine and store many features in system libraries. We can think of the sketching approach as a way to create features on demand. Whenever we need a feature, we create it beginning with a sketch.

The primitives approach is the least-used approach. CAD designers are not fond of this approach because its terminology is rooted in the set theory where Boolean operations come from. Moreover, editing solids created this way is inefficient. For example, the deletion of a hole requires adding a cylinder to plug the hole, unless we edit the solid tree and prune the hole branch, and regenerate the solid.

Which CAD/CAM systems support which approaches? Legacy systems such as CATIA, Unigraphics, and I-DEAS support all three approaches, by the virtue of their existence since solid modeling started. They must support the three approaches to stay competitive. Newer systems such as Pro/E and SolidWorks support only the sketching approach because it is all that is needed to create any solid model. They do not support primitives and do not offer Boolean operations.

The sketching approach utilizes the following steps to create any feature:

1. **Select or define a sketch plane.** We need a sketch plane to sketch geometry on. The sketch plane controls the orientation of the sketched geometry in the 3D modeling space.
2. **Sketch 2D profile.** This profile is typically a cross section of the 3D model that we wish to construct. The profile typically consists of curves such as lines, arcs, chamfers, and splines. The sketched geometry resides in the currently active sketch plane.
3. **Modify sketch dimensions and update sketch.** We modify the dimensions of the profile to reflect the final dimensions of the solid model under construction. Step 2 provides a rough sketch that we refine in this step. After changing the dimensions, update the sketch to reflect the new dimensions.
4. **Create the feature.** Apply one of the 3D operations, such as extrusions, to create the feature.

The sketch plane we select or define in step 1 sets the orientation of the finished model in its 3D space. For example, if we select the Front plane, the 2D profile we sketch in step 2 becomes the front view of the resulting 3D model. If we select the Top plane instead, the same profile becomes the top view of the model. This is equivalent to rotating the model 90 degrees.

EXAMPLE 2.1   **Create a feature using the sketching approach.**
Create a block using a CAD/CAM system. Start with a sketch.

SOLUTION   We apply the four steps just
described to create the chamfered block, shown on
the right, that has the cross section shown on the
right and a thickness of 60 mm (millimeters). All
dimensions are in mm.

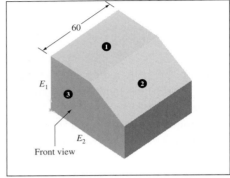

Front view

Modeling strategy:

1. Log in and start the CAD/CAM system.
2. Open a new part file.
3. Select the `Front` sketch plane.
4. Sketch the profile shown on the right. Sketch
   freely; ignore the dimensions for now.
5. Use the sketch dimension icon to modify
   dimensions to their final values shown on the
   right.
6. Use an extrusion feature icon or command
   and extrude the sketch by a distance of 60
   mm in one direction from the profile plane.
7. Save the file using `File => Save As`
   `=> example21`.
8. Print the model using `File => Print`.
9. Capture the screen by hitting the `Print`
   `Screen` key on the keyboard. Then open a
   Word document and paste the buffer content by using `Edit => Paste`.

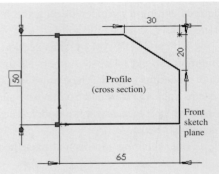

Example 2.1 discussion:

We can create the profile in two different ways. We create a rectangle and use a chamfer
command to chamfer its top right corner. Or we can sketch five contiguous lines. To modify the
dimension of any line, activate the sketcher dimension mode, click on any dimension value, type
the new value, and click `OK` or its equivalent. As the screenshots show, the profile sketched in
the `Front` sketch plane becomes the `Front` view of the `Isometric` view.

We will not use steps 1, 2, 7, 8, and 9 after this example to eliminate redundancy.

Example 2.1 hands-on exercise:

Re-create the chamfered block by using the `Top` sketch plane instead of the `Front` plane.
What happens to the orientation of the resulting `Isometric` view? Explain your answer.

Create the three through holes shown above. Hole ❶ ($d = 15$ mm) is perpendicular to the
top face and centered. Hole ❷ (d = 15 mm) is perpendicular to the inclined plane and centered.
Hole ❸ ($d = 25$ mm) is perpendicular to the front view and is located at distances of 20 mm from
both edges E_1 and E_2.

2.3   Types of Geometric Models

We have discussed the types of geometric models briefly in Tutorial 1.8.1. Objects and their geometric models can be classified into three types from a geometric construction point of view. These are 2½D, 3D, or a combination of both. 2½D objects are further classified into three subtypes: extrusions, axisymmetric, and composite. An extrusion has a constant cross section and a thickness in a direction perpendicular to the plane of the cross section. Figure 2.2 shows an example. An axisymmetric object has a constant cross section that is revolved about an axis of revolution through an angle $0 < \theta \le 360$. The composite 2½D object is a combination of multiple 2½D objects, as shown in Figure 2.2.

A 3D object is one that does not have any geometric uniformity in any direction, as shown in Figure 2.2. In such a case, we can start with 2½D objects and modify them, or find other strategies. The 3D object shown in Figure 2.2 is a block trimmed by two inclined planes.

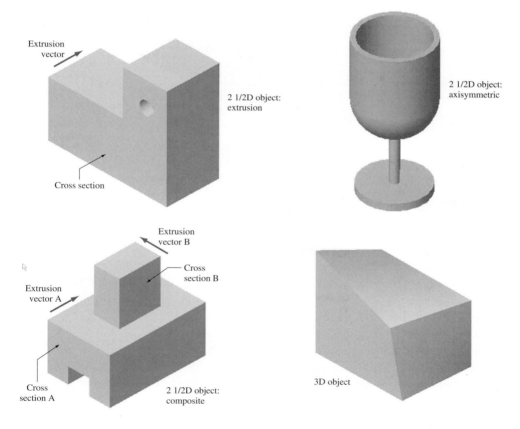

Figure 2.2   Types of geometric models.

Constructing 2½D objects requires only constructing the proper cross sections and projecting them along the proper direction by the thickness value. This is a much more efficient way of construction than calculating and inputting the coordinates of all of the corner points of the model. As complex as it might look, the phone model shown in Figure 2.3 is a composite 2½D. We extrude the three cross sections in the directions shown, cut out the excess material between the handset support pins, and use a rectangular array to create the buttons.

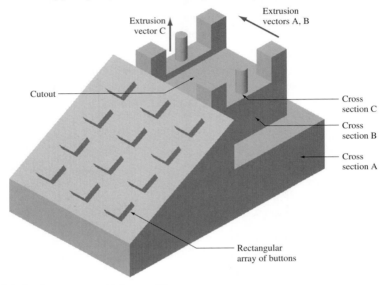

Figure 2.3   A telephone geometric model.

EXAMPLE 2.2	**Create an axisymmetric object.**
	Create a pin.

SOLUTION   We create a pin, shown on the right, that has the cross section shown on the right. All dimensions are in inches.

Modeling strategy:

1. Select the Front sketch plane.
2. Sketch the profile shown on the right. Sketch freely; ignore the dimensions for now.
3. Use the sketch dimension icon to modify dimensions to their final values shown on the right.
4. Create axis A, to be used as an axis of revolution.

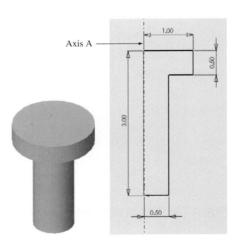

5. Use a revolve feature icon or command and revolve the profile geometry 360 degrees around the axis A.

Example 2.2 discussion:

The axis is not a line. It is an axis, a special type of geometry known as datum features or reference geometry. We discuss this subject in Section 2.8. If you simply create a line, the revolve command will fail.

Example 2.2 hands-on exercise:

Change the pin to a stepped pin along its length. Use a small radius of 0.375 inch starting at a distance of one inch from the top of its head.

2.4 Coordinate Systems

Three types of coordinate systems are needed in order to input, store, and display model geometry and graphics. These are Model Coordinate System (MCS), Working Coordinate System (WCS), and Screen Coordinate System (SCS), respectively. Other names for MCS are database, master, or world coordinate system. Another name for SCS is device coordinate system. Throughout this book, MCS, WCS, and SCS are used. We have covered each briefly in Tutorial 1.8.1.

2.4.1 Model Coordinate System

The **model coordinate system** is defined as the reference space of the model with respect to which all the model geometrical data is stored. It is a cartesian system which forms the default coordinate system used by a particular software program. The X, Y, and Z axes of the MCS can be displayed on the computer screen. The origin of the MCS can be arbitrarily chosen by the user while its orientation is established by the software. The three default sketch planes of a CAD/CAM system define the three planes of the MCS, and their intersection point is the MCS origin. When a CAD designer begins sketching, the origin becomes a corner point of the profile being sketched. The sketch plane defines the orientation of the profile in the model 3D space. This is how we attach the MCS to a geometric model.

In order for the user to communicate properly and effectively with a model database, the relationships between the MCS orthogonal (sketch) planes and the model views must be understood by the user. Typically, the software chooses one of two possible orientations of the MCS in space. As shown in Figure 2.4a, the XY plane is the horizontal plane and defines the model top view. The front and right side views are consequently defined by the XZ and YZ planes, respectively. Figure 2.4b shows the other possible orientation of the MCS where the XY plane is vertical and defines the model front view. As a result, the XZ and the YZ planes define the top and the right side views, respectively.

Existing CAD/CAM software uses the MCS as the default WCS (see Section 2.4.2). In both orientations, the XY plane is the default construction (sketch) plane. If the user utilizes such a plane, the first face of a model to be constructed becomes the top or front view, depending on which MCS is used.

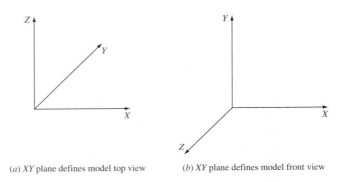

(a) XY plane defines model top view   (b) XY plane defines model front view

Figure 2.4   Possible orientations of MCS in 3D space.

The MCS is the only coordinate system that the software recognizes when storing or retrieving graphical information in or from a model database. Many existing software packages allow the user to input coordinate information in cartesian (x, y, z) and cylindrical (r, θ, z) coordinates, This input information is transformed to (x, y, z) coordinates relative to the MCS before being stored in the database.

Obtaining views is a form of retrieving graphical information relative to the MCS. If the MCS orientation does not match the desired orientation of the object being modeled, users become puzzled and confused. Another form of retrieving information is entity verification. Coordinates of points defining the entity are given relative to MCS by default. However, existing software allows users to obtain the coordinates relative to another system (WCS) by using the proper commands or modifiers.

EXAMPLE 2.3   **Mismatch of model orientation in modeling space.**
Show how to control a model orientation in 3D space.

SOLUTION   The screenshot on the right shows two possible orientations of the chamfered block of Example 2.1. Each orientation is decided by the first sketch plane chosen for construction. In orientation ❶, we use the Front (*XY*) sketch

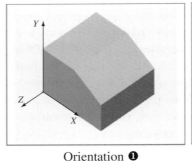

Orientation ❶

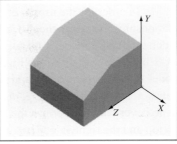

Orientation ❷

plane to create the model profile and extrude it in the −*Z* direction. In orientation ❷, we use the Right (*YZ*) sketch plane to create the model profile and extrude it in the −*X* direction.

Example 2.3 discussion:

In both cases the orientation of the MCS is the same in space; however, the orientation of the model, and therefore its views, is different. This orientation problem becomes apparent when we generate views for an engineering drawing or for other purposes. Some CAD designers do not view this as an important issue, but it is. Designers usually tend to patch the problem (and their lack of understanding of model orientation in the 3D modeling space) by using the CAD software to rotate the model to generate the right views for engineering drawings. However, rotating the model in its space does not change its initial orientation, which is decided by the first sketch plane used in model construction.

Example 2.3 hands-on exercise:

Create the chamfered block in orientation ❷. Generate the three standard views: top, front, and right. Now rotate the model to create orientation ❶. Generate the same three views. What do you get? Why?

2.4.2   Working Coordinate System

It is often convenient in the development of geometric models and the input of geometric data to refer to an auxiliary coordinate system instead of the MCS. This is usually useful when a desired plane (face) of construction is not easily defined as one of the MCS orthogonal planes, as in the case of inclined faces of a model (see Example 2.1). The user can define a cartesian coordinate system whose *XY* plane is coincident with the desired plane of construction. That system is the Working Coordinate System, WCS. It is a convenient user-defined system that facilitates geometric construction. It can be established at any position and orientation in space that the user desires. While the user can input data in reference to the WCS, the CAD software performs the necessary transformations to the MCS before storing the data. The ability to use two separate coordinate systems within the same model database in relation to one another gives the user great flexibility. Some commercial software refers to the WCS as is; Unigraphics offers an example. Other software refers to it as a sketch plane (Pro/E and SolidWorks) or construction plane.

A WCS requires three noncollinear points to define its *XY* plane. The first defines the origin, the first and the second define the *X* axis, and the third point with the first define the *Y* axis. The *Z* axis is determined as the cross product of the two unit vectors in the directions defined by the lines connecting the first and the second (the *X* axis), and the first and the third points (*Y* axis). We will use the subscript *w* to distinguish the WCS axes from those of the MCS. The $X_w Y_w$ plane becomes the active sketch (working) or construction plane if the user defines a WCS. In this case, the WCS and its corresponding $X_w Y_w$ plane override the MCS and the default sketch plane, respectively. As a matter of fact, the MCS with its default sketch plane is the default WCS with its $X_w Y_w$ plane. All CAD/CAM software packages provide users with three standard WCSs (sketch planes) that correspond to the three standard views: Front, Top, and Right sides. The user can define other WCSs or sketch planes.

There is only one active WCS (sketch plane) at any one time. If the user defines multiple WCSs in one session during a model construction, the software recognizes only the last one and stores it with the model database if the user stores the model. The model tree displayed by the CAD software shows the last selected (activated) sketch plane.

How is the WCS related to the MCS, and vice versa? Once a WCS is defined, user coordinate inputs are interpreted by the software in reference to this system. The software calculates the corresponding homogeneous transformation matrix between the WCS and the MCS to convert these input values into coordinates relative to the MCS before storing them in the database. The transformation equation can be written as:

$$\mathbf{P}_m = [T]\mathbf{P}_w \tag{2.1}$$

where $\mathbf{P}_m$ is the position vector of a point relative to the MCS and $\mathbf{P}_w$ is the vector of the point relative to the active WCS. Each vector is given by:

$$\mathbf{P} = \begin{bmatrix} x & y & z & 1 \end{bmatrix}^T \tag{2.2}$$

The matrix $[T]$ is the homogeneous transformation matrix. It is a 4×4 matrix and is given by:

$$[T] = \begin{bmatrix} t_{11} & t_{12} & t_{13} & t_{14} \\ t_{21} & t_{22} & t_{23} & t_{24} \\ t_{31} & t_{32} & t_{33} & t_{34} \\ 0 & 0 & 0 & 1 \end{bmatrix} = \begin{bmatrix} [R]_w^m & \mathbf{P}_{w,org}^m \\ 000 & 1 \end{bmatrix} \tag{2.3}$$

where $[R]_w^m$ is the rotation matrix that defines the orientation of the WCS relative to the MCS and $\mathbf{P}_{w,org}^m$ is the position vector that describes the origin of the WCS relative to the MCS. The columns of $[R]_w^m$ give the direction cosines of the unit vectors in the X_w, Y_w and Z_w directions relative to the MCS as shown in Figure 2.5. These direction cosines are the components of the unit vectors along the axes of the MCS.

If the WCS axes are along the MCS axes, then the direction cosines become 1, -1, or 0, and Figure 2.5 is greatly simplified. In such a case, we can write the columns of $[T]$ by inspection. Also, if we know two unit vectors, we cross multiply them to find the third vector. For example, if X_w and Y_w are aligned along the Z and X axes of the MCS, respectively, as shown at right, the transformation between the WCS and MCS is given by:

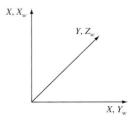

$$[T]_w^m = \begin{bmatrix} 0 & 1 & 0 & 0 \\ 0 & 0 & 1 & 0 \\ 1 & 0 & 0 & 0 \\ 0 & 0 & 0 & 1 \end{bmatrix}, [T]_m^w = \begin{bmatrix} 0 & 0 & 1 & 0 \\ 1 & 0 & 0 & 0 \\ 0 & 1 & 0 & 0 \\ 0 & 0 & 0 & 1 \end{bmatrix} \tag{2.4}$$

Observe that one of the matrices in Eq. (2.4) is the inverse of the other; their multiplication produces the identity matrix, $[I]$.

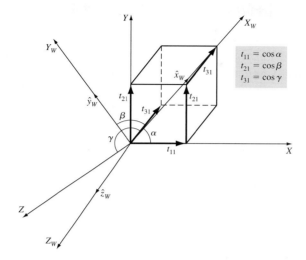

$$t_{11} = \cos \alpha$$
$$t_{21} = \cos \beta$$
$$t_{31} = \cos \gamma$$

Figure 2.5   Direction cosines of WCS relative to MCS.

EXAMPLE 2.4   **Relate WCS and MCS.**
Find the centers of holes ❶ and ❷ relative to the MCS. The holes are centered in their respective planes.

SOLUTION   The screenshot on the right shows the two holes. We use the dimensions of Example 2.1. The sketch plane required to create hole ❸ is the Front plane of the MCS. Thus the WCS for this hole is identical to the MCS. Therefore, $[T] = [I]$. Thus, the center of hole ❸ is (20, 20, 0) relative to the MCS.

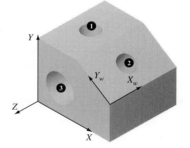

For hole ❷, we need to define a sketch $(X_w Y_w)$ plane on the inclined plane to create the hole. Let us assume that the hole center is P. Using the geometry of Example 2.1, we write:

$$P_w = \begin{bmatrix} 30 & 5\sqrt{13} & 0 & 1 \end{bmatrix}$$

$$\hat{x}_w = \begin{bmatrix} 0 & 0 & -1 & 1 \end{bmatrix}, \hat{y}_w = \begin{bmatrix} -3/\sqrt{13} & 2/\sqrt{13} & 0 & 1 \end{bmatrix}$$

Using the orthogonal axes of the WCS, and using the right-hand rule, we write:

$$\hat{z}_w = \hat{x}_w \times \hat{y}_w = \begin{vmatrix} \hat{i} & \hat{j} & \hat{k} \\ 0 & 0 & -1 \\ \dfrac{-3}{\sqrt{13}} & \dfrac{2}{\sqrt{13}} & 0 \end{vmatrix} = 5/\sqrt{13}\hat{k} = \begin{bmatrix} 0 \\ 0 \\ \dfrac{5}{\sqrt{13}} \end{bmatrix}$$

The transformation matrix becomes:

$$[T] = \begin{bmatrix} 0 & -3/\sqrt{13} & 0 & 65 \\ 0 & 2/\sqrt{13} & 0 & 30 \\ -1 & 0 & 5/(\sqrt{13}) & 0 \\ 0 & 0 & 0 & 1 \end{bmatrix}$$

The coordinates of the center P relative to the MCS becomes:

$$P_m = [T]_w^m \ P_w = \begin{bmatrix} 63.85 \\ 30.77 \\ -30 \\ 1 \end{bmatrix}$$

Example 2.4 discussion:

This example illustrates what goes on behind the scenes with the CAD/CAM software when a CAD designer selects different sketch planes to create geometric models. Equations (2.1) to (2.3) are programmed into the software and are called repeatedly for each construction the designer makes in the sketch plane.

The coordinates of the center, P_m, make sense given the dimensions of the problem. For example, $z_m = -30$ makes sense because the hole center is located in the negative direction of the Z axis.

Example 2.4 hands-on exercise:

Repeat the above calculations to find the coordinates of the center of hole ❶ relative to the MCS. Do the calculations match the hole location by inspection?

2.4.3   Screen Coordinate System

In contrast to the MCS and WCS, the **screen coordinate system** (SCS) is defined as a 2D device-dependent coordinate system whose origin is usually located at the lower left corner of the graphics display, as shown in Figure 2.6. The physical dimensions of a device screen (aspect ratio) and the type of device (raster) determine the range of the SCS. The SCS is mostly used in view-related clicks such as definitions of view origin and window or clicking a view to select it for graphics operations.

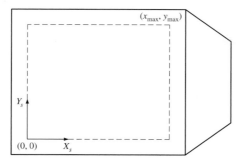

Figure 2.6   Typical SCS.

A 1024×1024 display has an SCS with a range of $(0,0)$ to $(1024,1024)$. The center of the screen has coordinates of $(512,512)$. This SCS is used by the CAD/CAM software to display relevant graphics by converting directly from MCS coordinates to SCS (physical device) coordinates. A normalized SCS can also be utilized. The range of the SCS can be chosen from $(0,0)$ to $(1,1)$. Such representation can be translated by device-dependent codes to the appropriate physical device coordinates. The third method of defining an SCS is by using the drawing size that the user chooses. If a size A drawing is chosen, the range of the SCS becomes $(0,0)$ to $(11,8.5)$ while size B produces the range $(0,0)$ to $(17,11)$. The rationale behind this method stems from the conventional drawing board so that the drafting paper is represented by the device screen.

A transformation operation from MCS coordinates to SCS coordinates is performed by the software before displaying the model views and graphics. Typically, for a geometric model, there is a data structure to store its geometric data (relative to MCS), and a display file to store its display data (relative to SCS).

2.5 Sketching and Sketch Planes

When a CAD designer defines a sketch plane, he or she invokes the sketcher of the CAD system. Some CAD systems like SolidWorks and Pro/E have only the sketcher to create geometry and eventually solid models. Legacy systems such as Unigraphics and CATIA require CAD designers to invoke the sketcher specifically.

Sketchers provide CAD designers with various sketching entities and tools such as lines, circles and so forth as shown in the forthcoming screenshot. Sketchers have been designed to read the CAD designer's mind. They understand design intent to a degree. For example, when a designer moves the mouse cursor near a line, the sketcher can flash the end point or midpoint, depending on which point it is closer to. It can also predict when two lines should be perpendicular, as in the case of defining two datum axes. Another example is that the sketcher can use an existing point in the vicinity of sketching as a center for a circle. Pro/E has an Intent

Manager (IM) that designers can turn on and off. SolidWorks simplifies this issue a great deal; no IM, just a sketch plane.

Line   Axis   Circular array

   Sketchers also provide geometric constraints and relationships. A sketcher provides relations icons that a designer clicks, and then selects the entities to apply relations to. Based on the selected entities, the sketcher makes available what it thinks will be the correct set of constraints to choose from. The following screenshot shows an example.

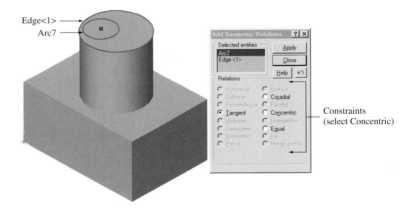

   If we need sketch planes beyond the standard Top, Front, and Right offered by a sketcher, we must create them. Because a new sketch plane is considered geometry that is not part of the model under construction (but it is part of the model's geometric definition), it is referred to as a datum plane, which is part of the model reference geometry. There are various ways to create datum planes. The following screenshots show some of them.

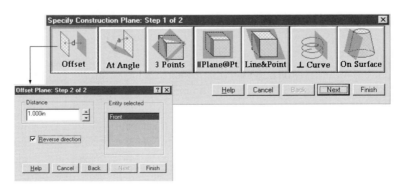

A sketch status is a very important issue to the sketcher. If a sketch (profile) is not correctly defined, a valid solid cannot be generated from it. Sketchers use color codes to display sketch geometry to alert designers. After a sketch is created, it can exist in one of three states. An **underdefined (underconstrained) sketch** is displayed in blue; additional dimensions or relations are required. A **fully-defined (fully-constrained) sketch** is displayed in black; no additional dimensions or relations are required. An **overdefined (overconstrained) sketch** is displayed in red; the sketch contains conflicting dimensions or relations or both. They must be removed. The following screenshots show the three sketch states.

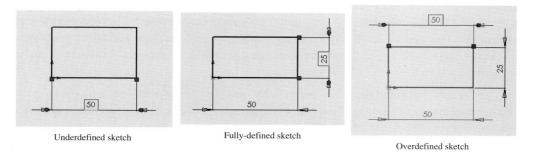

Underdefined sketch   Fully-defined sketch

Overdefined sketch

2.6 Parameters and Dimensions

Modifying solid models easily is a key to testing "what-if" scenarios during product design in search of the best design. The ease of such modification is rooted in the concept of parametrics, where a model is parametrized automatically by the CAD/CAM system during construction, at no (mental) cost or effort to the designer. A CAD designer can simply change the values of parameters and ask the CAD/CAM system to regenerate or re-create the model. Figure 2.7 shows a simple parametric solid model. The CAD/CAM system keeps track of all the geometric constraints that are required to maintain the validity of the model topology.

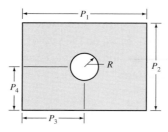

Figure 2.7   Parameters of a solid model.

During the construction of a sketch, a solid modeler derives the requires set of parameters that are required to fully define the sketch. The dimensions that the designer inputs for these parameters make up the values of these parameters. Consider the part shown in Figure 2.7. It is a block with a hole in it. The five parameters, P_1 to P_4 and R, fully define the sketch. We can assign any values to these parameters and redraw the sketch by using a regeneration function of a CAD system.

The sketch parameters and dimensions work together. Parameters are the generalization of the sketch definition; they create a template of the sketch. Dimensions are the specification of the sketch definition; they create a specific instance of it. This is analogous to the concept objects in OOP (object-oriented programming). The parametrized sketch is analogous to a *class*, and the dimensioned sketch is similar to an *instance* of the class. Similar to the relationship between a class and an instance, the dimensioned sketch derives its definition from the parametrized sketch.

A sketch parametrization is not usually unique. There may be multiple ways to parametrize the same sketch. If we think of the geometric and topological constraints or relations between the sketch entities, in addition to its geometry, the choices of parametrization could be many. However, regardless of the parametrization way we choose, there is always a minimum set of parameters that we must use to fully define the sketch geometry and topology. Consider Figure 2.7. The five parameters shown are the minimum required set. However, we can add relations such as $P_3 = 0.5P_1$, $P_4 = 0.5P_2$, and $R = 0.25P_1$. In addition, we may add perpendicular or parallelism constraints between the sketch entities.

CAD/CAM systems present sketch parameters and constraints to designers at different levels of sophistication. Legacy systems, such as Unigraphics and CATIA, allow designers to get very involved in defining relations and constraints manually. This could become very involved and confusing for complex sketches, such as those found in the automotive industry. The parametrization of the sketch could fail many times before a correct set of parameters is found.

Other systems such as Pro/E and SolidWorks use the automation approach. After the designer creates a sketch, the designer adds dimensions to it. Pro/E uses a two-step approach. First, the designer clicks a `Dimension` icon on the sketcher and dimensions the sketch. In this step, Pro/E generates parameters and displays them. When the designer regenerates the sketch, Pro/E assigns values to the parameters based on the size of the initial sketch. In the second step, the designer manually selects one dimension at a time (from the set of step 1), modifies it using a `Modify` icon, and then regenerates the entire sketch for the final dimensions. The forthcoming screenshots show the two steps.

SolidWorks uses a simpler approach. It does not display the sketch parameters to the designer because the designer does not care too much about them. SolidWorks displays only the sketch dimensions. It provides the designer with a `Dimension` icon on the sketcher toolbar. Upon clicking this icon, a dimension `Modify` window pops up. The window displays the current value of the dimension; this is similar to the Pro/E first step. The designer inputs the new desired dimension value and clicks a tick mark, ☑; this similar to the Pro/E second step.

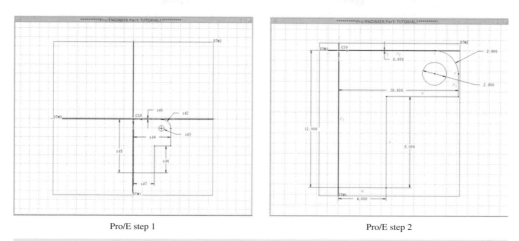

| Pro/E step 1 | Pro/E step 2 |

Note: Designers may skip step 1. They also may use Pro/E IM (Intent Manager), which simplifies construction and dimensioning.

The sketch parametrization offers a great design tool for CAD designers. They can freely sketch their ideas without having to worry about dimensions. While sketching, they try to keep the proportions of the sketch reasonable. After a sketch is complete, they can look at the dimensions and accept them or modify them. For example, a designer may adjust a value of 3.45 inches (or 35.79 mm) to 3.5 inches (36 mm).

We summarize the parameterics solid modeling approaches into these three steps:

Generate parameters (via sketching) => Assign dimensions => Regenerate

We follow this approach throughout the book. Sometimes, to avoid clutter, we do not provide dimensions for a geometric model. Interested readers can simply generate sketches on their CAD systems that look similar to the model presented in the book. The sketcher in turn generates the dimensions.

2.7   Basic Features

An analysis of existing CAD/CAM systems reveals that they all offer a basic set of features. This set is universal among them, although each system may use different names to designate them. This basic set includes the following features: extrusion, revolution, hole, cut, sweep, loft, fillet, chamfer, rib, shell, draft, patterns, spiral, and helix. The forthcoming screenshot shows typical features of a CAD/CAM system. We cover how each feature is defined and created.

SolidWorks features toolbar

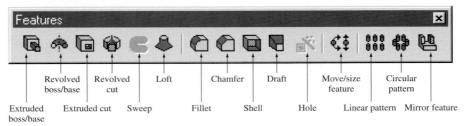

Extruded   Extruded cut   Sweep   Fillet   Shell   Hole   Linear pattern Mirror feature
boss/base

Revolved   Revolved   Loft   Chamfer   Draft   Move/size   Circular
boss/base   cut   feature   pattern

An **extrusion** feature is used to create a uniform-thickness model. It requires a cross section and an extrusion vector (direction and a distance) as shown below. We have also already covered it in Section 2.3 and Tutorial 1.8.1.

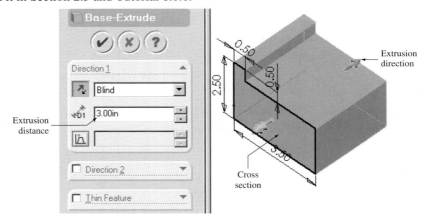

A **revolution** feature is used to create axisymmetric models. It requires a cross section, an axis of revolution, and an angle of revolution, as shown below.

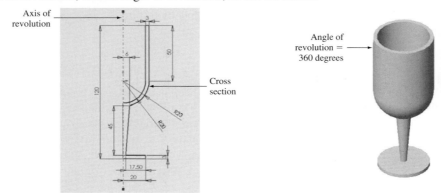

A **hole** feature is used to create holes in models. It requires a hole diameter or radius (size), a length, and an axis to define orientation.

A **cut** feature is used to cut material from another existing feature. Different CAD systems use different names for the cut feature. Different types of cuts can be made such as extruded or revolved cuts. A slot is also a form of a cut feature.

A **sweep** feature is used to create a model with a constant cross section along a nonlinear axis. It is a generalization of the extrusion feature. It requires a cross section and a sweep curve, as shown below. If the sweep curve is linear, the sweep becomes an extrusion.

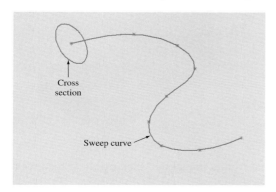

A **loft** feature is used to create a model with a variant cross section along a linear/nonlinear axis. It is a generalization of the sweep feature. It requires a set of cross sections as shown below. A guide curve may be used to blend the cross sections. If no guide curve is specified, a linear blending is assumed.

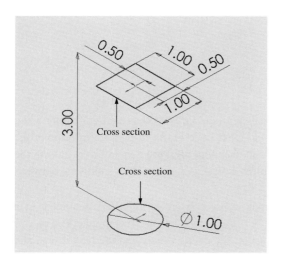

The **fillet** and **chamfer** features are used to change the corners of a model. A fillet rounds the corner, thus eliminating its sharpness. This is a good engineering practice because it reduces the stresses around the corners. A chamfer creates a transition between two edges of a model.

A **rib** feature is used as a stiffener for models. It can be viewed as an extrusion. It requires a line, the faces to stiffen, and a thickness as shown below.

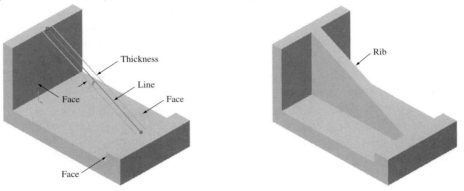

A **shell** feature is used to carve out material from a model by "shelling" it. It requires a face to shell and thickness to keep as shown below.

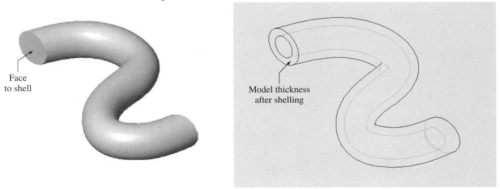

A **draft** feature is used to create a slant (taper) in model faces. It is a required feature when we create models of injection molds. These molds must have drafts to be able to remove the molded parts from the molds. These drafts release the vacuum pressure generated by pulling the plastic parts out of the molds. A draft feature requires a draft angle (usually a small one between five and ten degrees) and the faces to draft.

A **pattern** feature is a geometric array that is used to create repeated geometry in a specific order, as we have shown in Tutorial 1.8.1. A pattern can be rectangular (linear) or circular. A linear pattern requires the feature to repeat, the distance between the repetitions, and the number of repetitions. A circular pattern requires the feature to repeat, an angle between the repetitions, and the number of repetitions. Or, we can specify the total angle that the pattern sweeps and the number of repetitions. The CAD system calculates the angle between repetitions from this data, by dividing the total angle by the number of repetitions.

A **spiral** feature is used to create spirals. The spiral is considered a sweep along a helix curve as shown below. It requires a cross section and a helix curve. The helix curve requires a pitch and a number of revolutions.

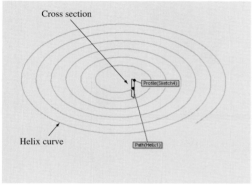

A **helix** feature is used to create helical springs. The helix is considered a sweep along a helix curve as shown below. It requires a (circular) cross section and a helix curve. The helix curve requires a pitch and a number of revolutions. The helix can be created tapered if needed.

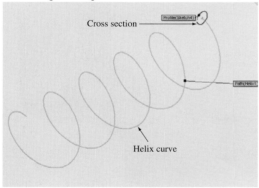

Featured-based modeling uses these features and others, depending on what a CAD system offers, to create geometric models of imaginative objects or engineering parts, products, and assemblies. Some systems, such as SolidWorks, call the first feature we create or an existing feature a *base* feature. We add or subtract from the base feature until we create the final model.

We should realize that many current CAD/CAM systems, such as Pro/E and SolidWorks, do not allow more than one solid during any geometric construction. These systems do not allow what we call "disjoint solids" or "multiple bodies" during construction. We will discuss this issue later in more detail. Legacy systems, such as Unigraphics and CATIA, do allow multiple solids to exist concurrently in the same database. This is sometimes useful because a CAD designer may generate disjoint solids during intermediate modeling steps before combining them using Boolean operations.

Let us investigate this issue further by using SolidWorks. The creation of any feature of any part begins by creating a cross section that is adjacent to a face of an existing feature; that is, it is constructed off a face. Thus, all existing features and cross sections are joined during any phase of construction. Pro/E and SolidWorks do not allow the creation of disjoint solids. An example of a disjoint solid is two blocks placed apart from each other in the modeling space. The error that appears on the screen when we attempt to create a disjoint solid is shown in the screenshot on the right. To avoid this error, we make sure that we adopt a modeling strategy such that creates cross sections, and hence features, that are adjacent to or penetrate existing faces/planes.

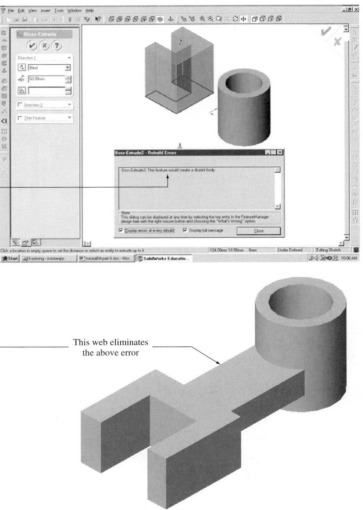

This web eliminates the above error

The part shown here is a disjoint solid, which we have to connect with a web. If a CAD/CAM system does not allow the creation of disjoint solids like this, we must create the web first.

2.8 Datum Features

The creation of a solid model begins with a sketch that is used in conjunction with a feature operation to create features. As we create more and more features, we create models. A CAD system makes the assumption that any geometry created during sketching is part of the profile under construction. More often than not, we need to create geometry to define construction operation. Such geometry is not part of the profile. How can we tell the sketcher and the CAD system not to treat that geometry in a regular way? We define reference geometry.

Reference geometry is a special type of geometry that is used to define other geometry. It is not ordinary geometry. Consider the cross section and the axis that are used to create a revolved feature. The cross section is ordinary geometry while the axis is reference geometry. Reference geometry is part of the feature definition, and it cannot be deleted unless the feature is deleted first.

Reference geometry comes in the form of datum features. CAD systems allow designers to create these types of datum features: planes, axes, curves, points, and coordinate systems. Think of these datum features as ordinary features, with the difference that they are not part of the sketch (profile) geometry. Thus, their creation requires the same definitions as their ordinary counterparts. Pro/E makes extensive use of datum features. It has an extensive `Datum` menu under its `Feature Class` menu. Any of the above-listed datum features are available under the Pro/E `Datum` menu. SolidWorks offers only three datum features under its `Reference Geometry` menu (click `View => Toolbar => Reference Geometry`). They are `Coordinate System`, `Datum Axis`, and `Datum Plane`.

A **datum plane** is used when we need a nonstandard construction plane. We have covered datum planes in Section 2.5. A CAD system offers a rich menu of defining datum planes as shown in Section 2.5.

A **datum axis** is used for various purposes including defining an axis of revolution for revolved features or an axis for a circular pattern. Some CAD systems, such as SolidWorks, offer an axis feature as part of other menus, depending on the context for modeling and whether a feature creation needs it or not.

A **datum curve** or **point** feature is used when we need to create curves or points in the model database that are needed to define a sketch. For example, consider a set of points that define a spline curve. We may create the points explicitly first, before using them to create the spline. In this case, we must create them as datum points, as Pro/E does.

A **datum coordinate system** feature is useful when we need a CAD system to use it temporarily in some calculations, such as measuring entity length or calculating mass properties, instead of using the MCS. It can also be used to export documents to IGES, STL, STEP, or other formats. The required information for creating the coordinate system is X axis, Y axis and origin. For example, we can select an edge as the origin and two faces as the X and Y axes as shown in the forthcoming screenshot. Or we can select a vertex as an origin and two edges as the X and Y axes as shown in the screenshot.

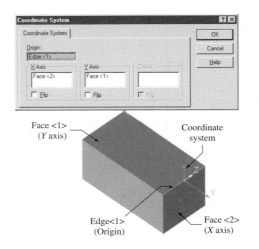 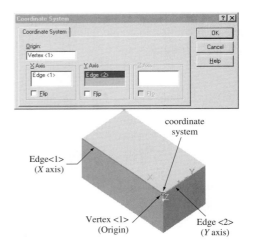

2.9 Geometric Constraints

We use geometric constraints or relations as part of sketch definition. These constraints could be geometrical (such as a length equals another) or topological (such as two lines are always perpendicular). They are useful in relating sketch entities together. For example, the length of a line could be set to be twice the length of another line. When the value of the latter length changes, the length of the former line changes as well, thus maintaining the geometric constraint. Another example includes centering a hole in a rectangle; whenever the rectangle changes size, the hole is always in the center of the rectangle.

We can use equations to express more elaborate constraints. We cover this subject in more detail later in the book.

2.10 Modeling Operations

As we create feature after feature to create a solid model, what happens to all these features? Do they remain individually in the model database? CAD systems combine them as we create them. We begin with the first feature, which represents the total solid at this point. When we create the next feature, the CAD system can add it or subtract it from the first feature, and so forth.

CAD systems utilize Boolean operations to combine features. These operations are union, subtraction, and intersection. For example, when we use an extrusion, the resulting feature is added to the existing feature or solid. When we create a cut, the resulting feature is subtracted from the existing feature or solid. The intersection happens as a supporting operation.

Legacy CAD systems such as Unigraphics, CATIA, and I-DEAS support Boolean operations explicitly. CAD designers can find them under a `Tools` or `Edit` menu. Systems such as SolidWorks or Pro/E do not have Boolean operations explicitly. The interesting question is how do we intersect two solids explicitly on these systems? The answer is that we do not; these systems do not support the existence of disjoint solids.

The details of Boolean operations and the set theory behind them is covered later in the book in more detail.

2.11 Heterogeneous Modeling

The construction of a solid model is a series of sketches and features that are combined to create the final solid. We begin with a sketch that we draw on a sketch plane. We then apply a 3D operation to transform the 2D sketch to a 3D feature. We repeat these two steps to create other features. As features are created, the CAD system combines them to produce the solid.

A sketch is a collection of curves. A curve may be a line, circle, arc, chamfer, or spline. In more complex solids, we may use the curves to create surfaces before creating a feature or the final solid. We cover curves and surfaces in more detail later in the book. Thus, a model database may contain three types of modeling classes: curves, surfaces, and solids.

Some CAD systems, such as Pro/E and SolidWorks, only allow the existence of curves and surfaces as part of a sketch definition. The only valid entity that these systems allow to exist individually is a solid, and only one solid, as we discussed at the end of Section 2.7.

Legacy CAD systems, such as Unigraphics and CATIA, allow the existence of curves, surfaces, and solids at the same time in a model database. These systems allow designers to create curves and surfaces that are not part of a solid definition yet. We can view this coexistence as three modeling modes: curves, surfaces, and solids. We refer to the existence of these three modeling modes as **heterogeneous modeling**. Whether heterogeneous modeling is good or bad is a debatable issue that we do not cover here. We can only point out that it may be good for research purposes and other experimentation activities in geometric modeling.

2.12 Modeling Strategies

CAD/CAM software is complex and requires training and understanding of its philosophy and underlying principles. It should be considered a complex engineering modeling tool, like other software for computational fluid dynamics and finite element analysis. CAD designers should use it with care if they wish to be efficient in using it.

The efficient use of a CAD system begins with developing a modeling strategy. A **modeling (planning) strategy** is a sequence of thoughts about the best, easiest, and fastest way to create the geometric model of the object we wish to model. This modeling strategy is the high-level thoughts of what we need to do on a CAD system. This strategy may change when we execute it.

We view a modeling strategy as a way for CAD designers to organize their thoughts. It is similar to a flowchart or pseudocode in programming activities. The development of an effective modeling strategy requires a CAD designer to answer a series of questions:

1. **Determine model type and subtype.** Is it 2½D or 3D? Within 2½D type, is it extrusion, revolution, or composite 2½D? A large percentage of engineering parts fall into the 2½D group.
2. **Observe geometric characteristics of model.** Is the model symmetric with respect to one plane or more? If so, a CAD designer can construct only half of the model and then use the Mirror command to create the full model. The Translate and Rotate commands can also be used.
3. **Choose model orientation in 3D space.** How is the model oriented in 3D? Which model face or view is aligned with which view? This decides the first sketch plane of the model whether it is Top, Front, or Right.
4. **Choose model origin.** Where is the MCS origin located on the model sketch? Is it the bottom left corner of the sketch (profile) or somewhere else?
5. **Decide on other geometric details.** Do we need geometric modifiers to speed up construction? What color and layer scheme do we need?
6. **Avoid unnecessary calculations.** Can we use the CAD system to perform geometric calculations for us, to save time? For example, we may use an intersection or end modifier instead of explicitly making calculations by hand.

These six questions offer a guide. CAD designers may amend or change them. We should mention that CAD designers tend to naturally jump to a CAD workstation and begin building a model, just like a programmer jumps to writing code. Wisdom has it that the faster we build models or write code, the worse the final product is. We simply ask designers to spend a few minutes to investigate the geometric properties of the model under construction and find the smartest and fastest way to create it.

Once we answer the above six questions, what does a modeling strategy look like? Examples 2.1 and 2.2 offer examples. It consists of steps on how to construct a model. Each step does not have to include deep-level details about each dimension or number used in the step; just a high-level plan as Examples 2.1 and 2.2 show. The best presentation of a planning strategy is to include screenshots as shown in Examples 2.1 and 2.2.

2.13 Master Model

The use of CAD/CAM systems in practice requires a lot of preparation. It is not just an issue of getting CAD designers to create models and assemblies. It is also an issue of consistency and reuse of these models and assemblies. A typical company has many departments (including design, manufacturing, and marketing) and subcontractors. The company may also outsource various CAD activities. In such a case, all these groups and individuals must have a way of knowing how to read a CAD file. This is where the concept of a master model becomes useful.

A **master model** is a template file that has the structure of CAD data in a consistent way for all groups to use and follow. CAD managers in companies usually develop the content and structure of master models and their files. The content of a master model includes layering schemes, modeling and drawing conventions, model views, model coloring schemes, and so forth.

CAD designers learn the structure of the master models and their files. They also learn how to use them. A CAD designer may begin a model. Another designer can then continue the model because they all know where model information is and how to add new information.

2.14 System Modes

Constructing geometric models and producing drawings are two basic and popular functions of CAD/CAM software. The software provides users with two basic modes to enable them to perform the two functions. These are the model and drawing modes. Only one mode can be active at any one time. Both model and drawing modes utilize existing information in the model database. However, if the model mode is active, the result of every modeling operation is recorded in the model database. The drawing mode offers two options: local and associative. In the local option, changes to a drawing are local to the drawing; they do not affect the model database. In the associative option, changes in the drawing mode affect the model geometry and database.

While the main purpose of the model mode is to construct a model geometry, the main purpose of the drawing mode is to generate engineering drawings. The three major activities of the drawing mode are model cleanup, documentation, and plotting. Model cleanup is the most tedious and time-consuming activity a user can engage in. Starting with the three standard orthographic views from the model database, the user should hide or change fonts of entities according to the standard drafting rules. The user is usually faced with two main problems during model cleanup. The first is overlapping entities. This problem appears if the user has to blank (hide) these entities or change their fonts to dashed font. Reliable automatic hidden line removal algorithms are helpful in model cleanup.

The second problem is that cleanup work is usually not recoverable. If the user destroys a view during the cleanup process, the only alternative is to start from scratch by repeating the work on a "fresh" view and discarding any previous cleanup work. It is also possible that the CAD/CAM system might crash in the middle of a drawing session, which results in loss of the drafting work also. The general advice is that the user must save more frequently during an active session.

Once the cleanup work is completed, model documentation includes adding dimensions and text notes, generating drawings, and producing bills of materials. Users are assumed to know the standard rules of dimensioning and tolerancing. CAD/CAM systems provide users with rich menus and commands for dimensioning and tolerance purposes.

Obtaining high-quality plots of the generated drawings is usually achieved by issuing the proper `Plot` command. The following standard drawing sizes are supported by all systems:

SIZE A   8 1/2 × 11
SIZE B   17 × 11
SIZE C   17 × 22
SIZE D   34 × 22
SIZE E   34 × 44

In addition to these standard sizes, user-defined drawing sizes are always provided. There is always one of these sizes that is used as a default by the software. The mapping between the orientation of a drawing on the display screen and on the drafting paper mounted on the plotter must be understood by the user to avoid surprises. There is a modifier that can be used with the plotter command to rotate the drawing orientation on the screen by 90° before it is plotted on the paper.

2.15 Model Viewing

While model construction involves the creation of the model database, viewing affects the way the model is displayed on the screen. Modeling and viewing are interrelated. Views are defined by the various angles from which a model can be observed, as Figure 2.8 shows. Essentially, the observer changes position in the MCS, while the model maintains its original orientation. This effectively lines up the view with the plane of the computer screen. The geometric transformation that generates the orthographic views is covered later in the book. An infinite number of views can be defined for a model. Most software provides commands for the standard views. Figure 2.9 shows the standard six 2D views of a model. Any 3D view of a model can be generated by rotating the model in real time in its modeling space.

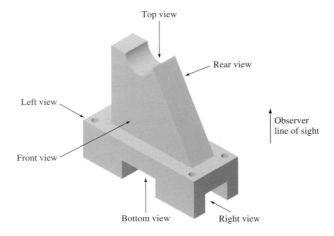

Figure 2.8   Standard viewing angles.

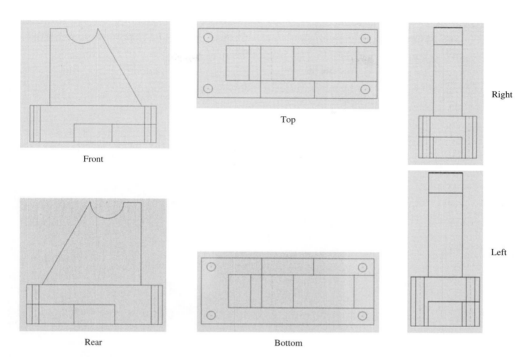

Figure 2.9   Standard 2D views of a typical CAD model.

2.16 VRML Web-Based Viewing

The Internet has affected CAD/CAM as it has many other fields. One noticeable change is that we can view CAD models in a Web browser without having the CAD systems that have created them up and running. This is a significant advantage, especially for marketing engineers and mobile designers. A marketing engineer may need to share product 3D shaded views with a group of customers. Or, a CAD designer may want to present a CAD model to a group of engineers in a meeting.

VRML (Virtual Reality Modeling Language) enables us to display CAD models in a Web browser. Moreover, we can manipulate these models. For example, we can rotate a 3D view in the browser window to look at other invisible sides of it.

We need to download a VRML plug-in and install it in the browser. Search the Web using a search engine, such as http://www.google.com, for VRML plug-in freeware (free software). Silicon Graphics' website provides VRML plug-in. Open a Web browser and type http://www.cai.com/cosmos as the URL. Look for a VRML link and follow instructions. The downloaded file is self-extracting. Double-click it and follow instructions to install it.

We need to use a CAD system to generate a VRML file before we can use the VRML plug-in. While the CAD system is running, save the CAD file in VRML format to generate a

VRML file. Simply click this sequence: File => Save As => select the VRML format and type a filename. The VRML file extension is *.wrl* (for *world*). See the upcoming screenshot.

    We now open a browser window and open the *.wrl* file to view it. Use this sequence to open the file: File (menu on browser menu bar) => Open => select the *.wrl* file => OK. While the VRML file is open in the browser window, we can manipulate it using the buttons offered by the plug-in as shown in the following screenshot.

Browser window

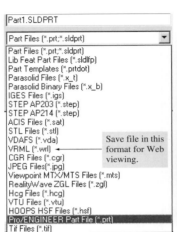
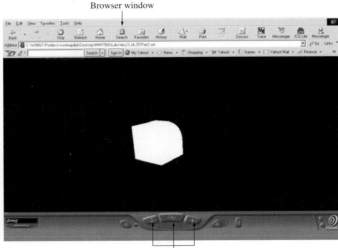

Save file in this format for Web viewing.

Click any of these buttons to rotate model

2.17   Tutorials

2.17.1   Create a Rectangular Plate

    This tutorial shows how to create a 2½D uniform thickness model. We create the rectangular plate, shown on the right. All dimensions are in inches.

Modeling strategy:

1. Select Front sketch plane.
2. Sketch the profile shown in the Front view.
3. Use extrusion with a thickness of 3 inches, to create the plate with no hole.
4. Select the Top sketch plane.

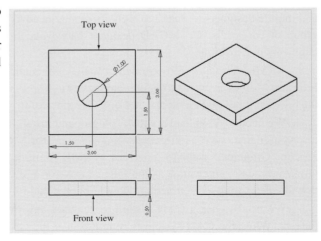

5. Sketch the circle in the center of the plate with a diameter of 1 inch.

6. Use a cut (extrude cut) operation to cut the hole from the plate.

Tutorial 2.17.1 discussion:

If we swap steps 1 and 4 in the modeling strategy, the Front and Top views shown would swap places.

Tutorial 2.17.1 hands-on exercise:

There is an easier way to create this model. Start with the Top sketch plane, create the rectangle and the circle inside it, and extrude the sketch with a 0.5 inch thickness.

2.17.2 Create a Base Plate

This tutorial shows how to create a 2½D uniform thickness model. We create the base plate, shown on the right. All dimensions are in inches.

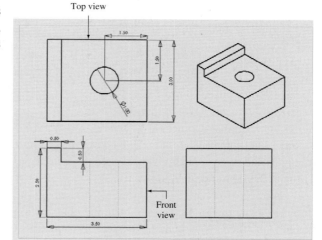

Top view

Modeling strategy:

1. Select the Front sketch plane.

2. Sketch the profile shown in the Front view.

3. Use extrusion with a thickness of 3 inches, to create the plate with no hole.

4. Define the top of the plate as a sketch plane.

5. Sketch the circle in the center of the plate with a diameter of 1 inch.

Front view

6. Use a cut (extrude cut) operation to cut the hole from the plate.

Tutorial 2.17.2 discussion:

If we swap steps 1 and 4 in the modeling strategy, the Front and Top views shown would swap places.

We must define a sketch plane in step 4 although it is in the same orientation as the Top sketch plane. The Top sketch plane of a CAD system passes through the origin of the system MCS. Thus, the sketch plane parallel to it must be defined by selecting the top plane of the plate.

Tutorial 2.17.2 hands-on exercise:

Re-create the model in a different way. Start with the Top sketch plane. Create a 3 × 3 rectangle with a circle, of 1 inch diameter, in the center of it. Extrude the sketch 2 inches up. Use the top plane of the resulting extrusion as a sketch plane, sketch a 3 × 0.5 rectangle, and extrude the rectangle half an inch up to finish the model.

2.17.3 Create a Support Bracket

This tutorial shows how to create a 2½D uniform thickness model. We create a support bracket, shown on the right. All dimensions are in inches.

Modeling strategy:

1. Select the `Front` sketch plane.
2. Sketch the profile shown in the `Front` view.
3. Use extrusion with a thickness of 5 inches to create the bracket without the cutouts shown in the screenshot on the right.

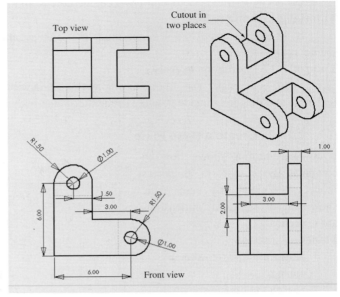

Tutorial 2.17.3 discussion:

The creation of this support bracket in any other way becomes inefficient. The above modeling strategy is the most efficient one. Think of other strategies and compare them with the one presented here.

Tutorial 2.17.3 hands-on exercise:

The above modeling strategy creates the model shown in the accompanying screenshot, no cutouts. Create the cutouts to complete the construction of the support bracket. Hint: Define a sketch plane for each cutout, create a rectangular extrusion, and cut it out from the model shown below. The dimensions of the rectangular extrusion do have to match those of the model on the right. Make the dimensions large enough to make the extrusion stick in the front and the back of the model. It does not really matter because the extrusion acts as a knife to cut material from the model.

2.17.4 Create a Golf Ball

This tutorial shows how to create a 2½D axi-symmetric model with a circular pattern. We create a golf ball, shown on the right. All dimensions are in inches. The golf ball diameter is 1.25 inches. The dent profile diameter is 0.06 inch with a depth of 0.015 inch.

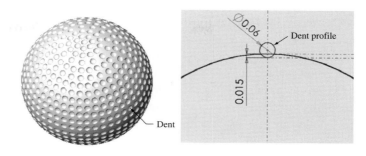

Modeling strategy:

1. Select the Front sketch plane.
2. Sketch a circle with diameter 1.25 inches. Also sketch an axis that passes through the circle center.
3. Revolve the circle about the axis an angle of 180 degrees to create a sphere that makes up the profile of the golf ball.
4. Sketch the dent profile in the Front sketch plane. Revolve cut it to create one dent. Also in the same sketch, create an axis which is 7 degrees inclined to the vertical axis, as shown in the screenshot below.
5. Circular pattern the dent profile with an instance of 2. Use the angular dimension of 7 degrees as the reference direction of the pattern and the angle as 7 degrees, as shown in the screenshot below.
6. Now using the central axis of the sphere as the reference, create the circular pattern of the dent as shown in the screenshot below.

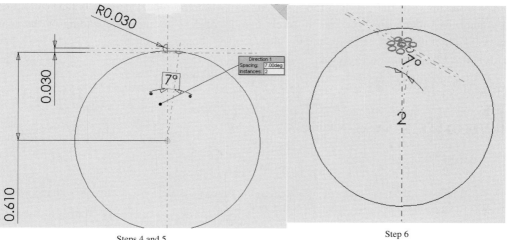

Steps 4 and 5   Step 6

7. Repeat step 5 but now create an instance of the dent at 14 degrees from the central axis. After that, repeat step 6 to create a circular pattern of that instance. Select the number of instances to fit the perimeter.

8. Repeat steps 5 and 6 to create the dent instance up to 90 degrees with increments of 7 degrees (21 degrees, ..., 84 degrees, 90 degrees — last increment is 6 degrees) and create the circular pattern at each level of instance as shown on the screenshot below.

9. Create the mirror image of the top half of the golf ball dent pattern (select all the patterns), by using the Top plane as the mirror plane, to complete the model as shown in the following screenshot below.

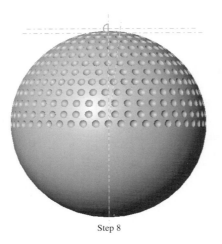
Step 8

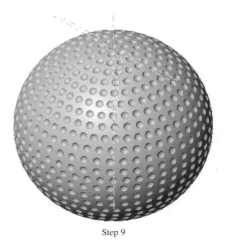
Step 9

Tutorial 2.17.4 discussion:

As complex as the golf ball looks, it is an axisymmetric model with a circular pattern of the dents. The circular pattern has a different number of dent instances, depending on the dent level. Think of the earth's equator and the altitude lines above it and below it.

Tutorial 2.17.4 hands-on exercise:

Can you think of another way to create the golf ball model?

PROBLEMS

Part I: Theory

2.1   Give an example of how the centralized integrated database concept can help with the what-if situations that arise during the design process.

2.2   Discuss the contents of a database for a line, a circle, and an arc.

2.3   Can you define a nonorthogonal WCS? How is three-point definition interpreted by software?

2.4   What is underdefined, fully-defined and overdefined sketching? Explain your answer with sketches.

2.5   What are the different types of geometric relations? Why would you use them in 3D geometric modeling?

2.6   What information do you need to define a coordinate system? Why do CAD/CAM systems allow you to define a coordinate system (i.e., what is the use of it)? Provide an example.

Part II: Laboratory

Use your in-house CAD/CAM system to answer the questions in this part.

2.7   How can you define a sketch plane?

2.8   Create the following sketch FULLY DEFINED, without having to calculate any other dimensions explicitly. Use only geometric modifiers (do not use any trim option), geometric constraints, and the shown dimensions. The parallel lines in the sketch indicate that the lines are of equal length.

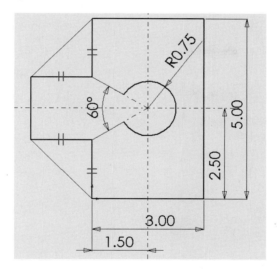

2.9   Provide the minimum required dimensions needed to fully define the following sketch. Is your answer the only option to fully define this sketch? Support your answer by providing at least three different options of fully defining this sketch.

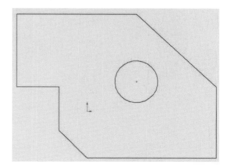

2.10 Create the model that has these two views.

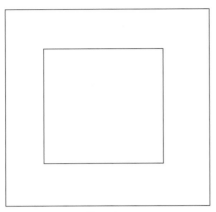

Front view

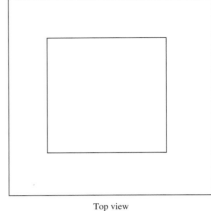

Top view

2.11 Create this axisymmetric model. All dimensions in mm.

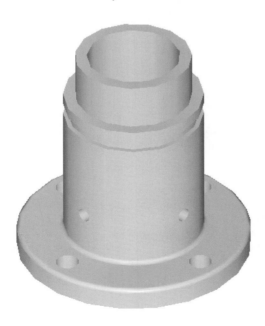

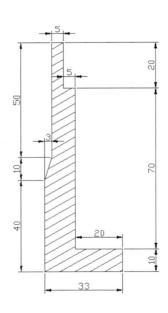

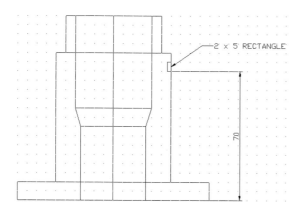

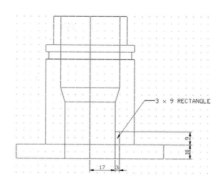

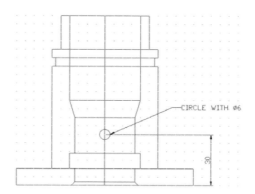

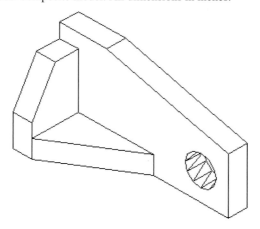

2.12 Create this 2½D composite model. All dimensions in inches.

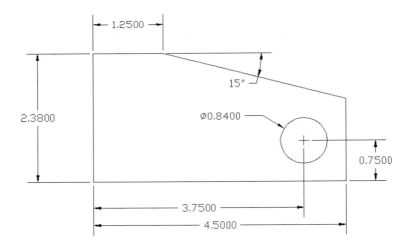

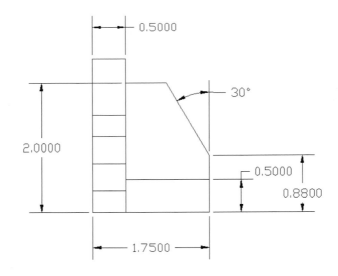

CHAPTER 3

Modeling Aids and Tools

GOAL

Understand and master modeling aids and manipulations, know the available tools offered by CAD/CAM systems, and learn how and when to use the tools to speed up geometric construction.

OBJECTIVES

After reading this chapter, you should understand the following concepts:

- Geometric modifiers

- Layers

- Colors

- Grids

- Groups

- Entity selection

- Entity manipulations

- Entity editing

CHAPTER HEADLINES

3.1 Introduction

We have discussed briefly in Tutorial 1.8.1 some of the productivity tools that CAD/CAM systems offer. Although CAD/CAM systems may seem different in their capabilities due to the syntax and format of user interface offered by each one, they all provide a generic set of modeling aids and tools. It is the intent of this chapter to introduce some of the common aids and tools offered by many of these systems. To best utilize the material presented here, you should refer to system documentation and manuals for your own CAD/CAM system.

The awareness of the available modeling aids provides designers with the ability to think in terms of them when performing design tasks. These aids should help speed up the completion of design tasks with CAD/CAM systems. To effectively use these aids during a design session on a CAD/CAM system, you should kae them part of the modeling (planning) strategy (refer to Chapter 2) when you create the design.

One of the most recognized benefits of using CAD/CAM systems in design and engineering applications is increasing productivity. The principal time-saving factor in CAD as compared with manual geometric construction comes in the manipulation and editing of geometric data and entities. One of the most time-consuming tasks in manual construction is correcting and revising existing geometry.

Good understanding and efficient use of functions and commands on a CAD/CAM system enable users of the system to better correct errors and mistakes. While it is always easiest to simply delete a partially wrong entity and reconstruct it, it is also the most nonproductive way to deal with mistakes. It does not help you to grow in your ability yo use the system. For example, if a correct entity is created in the wrong position, you can move it to the desired one instead of deleting it and re-creating it.

In this chapter we present a set of common aids and tools that are useful for creating designs and geometric models on CAD/CAM systems. The material is presented in a generic form with the goal of introducing the basic concepts. No syntax details are given. Readers can attempt to find the corresponding menus and toolbars for these concepts on their CAD/CAM systems and study their syntax. In essence, we present only the semantics of these aids and tools here.

3.2 Geometric Modifiers

Geometric modifiers are used by any CAD/CAM system to facilitate entering and extracting information to and from the system. A **geometric modifier** is a word that changes the mode of input and output in a command. A major advantage of using geometric modifiers is to be able to deal with specific existing geometric information in a geometric model without having to calculate that information explicitly. These modifiers apply mostly to curves.

The three common geometric modifiers are the `end`, the `origin` (center), and the `intersection` modifiers as shown in Figure 3.1. The `end` modifier signifies an end point of a curve. Each curve has two end points. If the curve is closed such as a circle, the two end points are coincident. The `origin` modifier identifies the center of a curve. The origin of a line is its midpoint, the origin of a point is the point itself, and the origin of a circle is its center. The origin of a general curve such as a B-spline is its center of curvature. The intersection modifier indicates the intersection point of two entities.

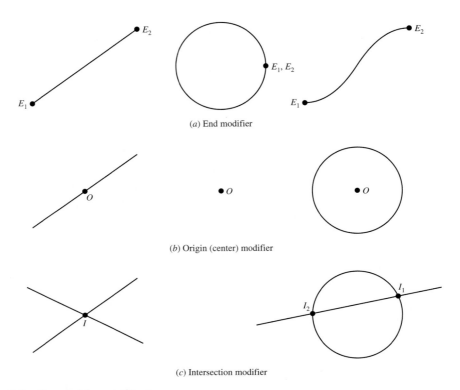

(a) End modifier

(b) Origin (center) modifier

(c) Intersection modifier

Figure 3.1   Geometric modifiers.

A command that uses a geometric modifier requires the user to click a button followed by selecting (clicking) the desired entity. In the case of the end or intersection modifier where more than one solution (point) exists, the user must click the entity close to the desired solution. A modifier may remain in effect once used until another modifier is used.

EXAMPLE 3.1	Use geometric modifiers.

Show how to use geometric modifiers.

SOLUTION   Figure 3.2 shows some examples of using geometric modifiers. The first example in the figure creates two lines between the end points of the line and the circle. The second example shows the use of the origin modifier and the interchange of modifiers. The third example shows the creation of the new circle with a center at I_1. If the clicks d_1 and d_2 were closer to I_2, the dashed circle would have been created instead.

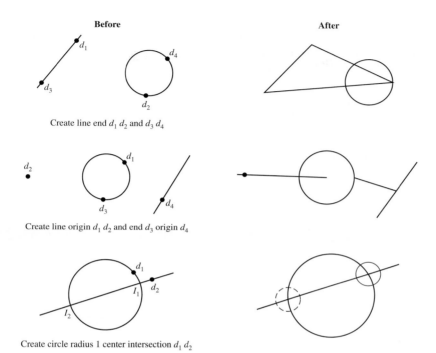

Figure 3.2   Typical uses of geometric modifiers.

Example 3.1 discussion:

If we do not use the geometric modifiers, we would have to re-enter the coordinates of the end or center points. Moreover, if we try to click close to these points thinking we got them (eye-balling approach), we fail because clicking on the screen never works.

Example 3.1 hands-on exercise:

Create a line connecting the centers of the line and circle of the first case shown in Figure 3.2. Do it using a CAD system.

3.3 Layers

Oftentimes, users of CAD/CAM systems may want to group and/or separate certain types of information related to the models or parts they create on these systems. For example, a user may want to separate the dimensions and other drafting information from the geometry of a given model. This can be accomplished by using layers. Most CAD/CAM systems provide their users with a large enough number of layers. Layer numbering begins at 0 or 1 and is incremented by 1. Users, however, can use any layer numbers, within the range, in their CAD models. We usually skip a few — we can use layer 10 for geometric entities, and layer 20 for dimensions.

A **layer** can be thought of as a sheet of transparency. Users can mix and organize these sheets as they desire to deal with and/or present their models effectively. Typically, a user can assign geometric entities to layers, turn layers on/off, and assign colors to layers. Following are the activities that we can do with layers:

1. **Create/delete layers.** To create a new layer, the CAD designer invokes a layer manager on the CAD system. The manager allows the designer to assign a name to the new layer (or use the default name), and change its properties. User-created layers can be deleted. A CAD system provides one default layer (0 or 1). The designer cannot delete or change its name but can change its properties. The following screenshot shows an example of a layer manager (of AutoCAD).

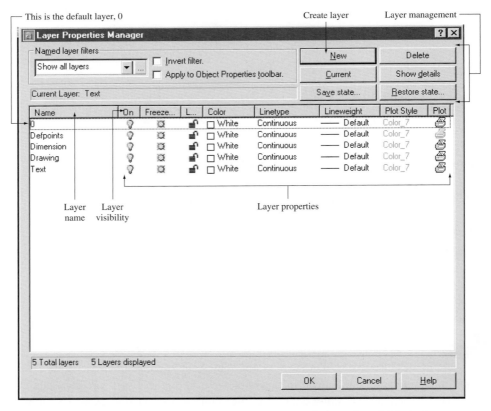

2. **Modify layer properties.** Each layer has properties (attributes) that CAD designers can customize according to their needs. As shown in the above screenshot, layer properties include its name, color, line type, line weight, and its visibility state (on/off). The default names are 1, 2, and so forth. A CAD designer may override these names and use others such as geometry or dimensions. Any geometric construction on a layer assumes its

color (red, green, ...), its line type (solid, dash, ...), and its line weight (line thickness as 0.25 mm, 0.30 mm, ...). Layer visibility controls the number of visible layers that are displayed on the screen at any one time. If a layer state is on, its content is visible on the screen, otherwise it is hidden if it is off.

3. **Use layers in geometric construction.** Any new geometric construction belongs to the layer that is current at the time of construction. We can make any existing layer current by clicking the `Current` button shown in the foregoing screenshot. We can only have one current layer at a time. However, we can have many concurrent visible layers from which we can select entities to aid in geometric construction. Thus, we can have only one current layer, but multiple visible layers.

Figure 3.3 shows an example of using layers. We sketch geometry on layer 0 and create dimensions on layer 100. We can display both layers by making them visible, or we can display any one of them. We may turn off layer 100 that holds dimensions while constructing new geometry. We turn it on when we need to add dimensions.

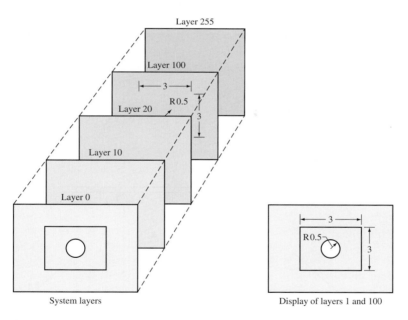

Figure 3.3   Layers.

Assignment of entities to layers can be done either globally or locally. If the user selects a particular layer (makes this layer active or current), any geometric entities created afterwards are assigned to this layer until the user selects another layer. Layer assignment can be done locally to an entity. When the user is creating the entity, a layer modifier can be used. The resulting entity resides on the specified layer regardless of the currently selected one.

In typical industrial CAD/CAM installations, standard layering schemes are usually developed by company management, and system users are required to adhere to these schemes. One example would be to use layer 1 for indexing, the next fifty layers for model geometry, the following fifty for manufacturing notes, another fifty for drafting, another fifty for analysis and technical illustrations, and the last fifty for construction aids and intermediate construction steps.

3.4 Colors

Colors are helpful in distinguishing entities of geometric models from each other. CAD/CAM systems provide users with color palettes to choose colors from. Color is assigned by layer; therefore, all entities assigned to a particular layer are displayed in the color assigned to that layer. If no color assignments are made, the default color is used for all layers. Users can assign a desirable color to a certain layer by using a layer property manager.

3.5 Grids

A **grid** is a network of uniformly spaced points superimposed (overlaid) on the screen, as shown in Figures 3.4 and 3.5. Grids are displayed as a series of dots, one dot for each grid point.

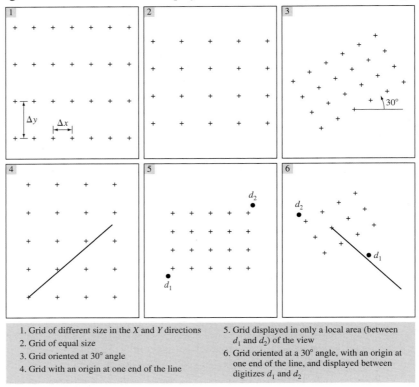

1. Grid of different size in the X and Y directions
2. Grid of equal size
3. Grid oriented at 30° angle
4. Grid with an origin at one end of the line

5. Grid displayed in only a local area (between d_1 and d_2) of the view
6. Grid oriented at a 30° angle, with an origin at one end of the line, and displayed between digitizes d_1 and d_2

Figure 3.4   Rectangular grids.

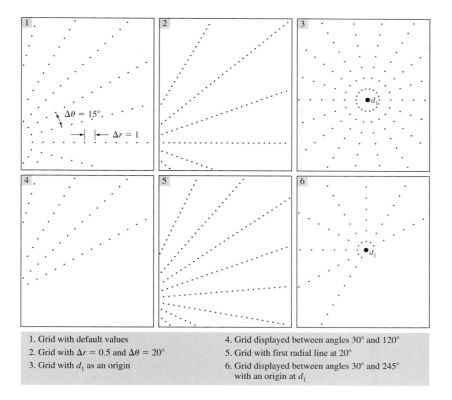

1. Grid with default values
2. Grid with $\Delta r = 0.5$ and $\Delta\theta = 20°$
3. Grid with d_1 as an origin

4. Grid displayed between angles 30° and 120°
5. Grid with first radial line at 20°
6. Grid displayed between angles 30° and 245°
   with an origin at d_1

Figure 3.5　Radial grids.

A grid can be turned on (activated) or off (deactivated) in any view at any time. The default origin for grids is usually the origin of the sketch. If a grid size (spacing between grid points) is specified too small for the grid points to be easily distinguished, the grid is not displayed on the screen; also, the grid is not visible if the grid points are farther apart than the width and/or height of the sketch. Changing the magnification (zooming the sketch in or out) of the view will cause the grid points to become visible. Even if the grid is not visible, it is still active, and mouse clicks are snapped to the closest grid point.

There are two types of grids available: rectangular and radial. A rectangular grid, shown in Figure 3.4, is particularly useful in constructing equally spaced entities, enabling the user to enter exact points by using mouse clicks rather than explicit coordinate entry. Use of rectangular grids also greatly facilitates sketching, moving entities from point to point, measuring between points, and trimming (using grid points as trimming boundaries). The user can specify the distance between grid points, the origin of the grid, and the angle (orientation) of the grid.

A grid command typically activates or deactivates the rectangular grid. The grid manager allows us to control the origin, size, orientation, and/or the local area in which the grid is to be displayed. The default orientation of a rectangular grid is parallel to the X and Y axes of the active WCS. Changing the orientation of the grid can be achieved by using an orientation angle. Figure 3.4 shows some rectangular grids.

A radial grid, shown in Figure 3.5, is useful primarily in the construction of concentric circles and axisymmetric and radial constructions. The radial grid consists of a series of dots (along radial lines) extending from a point of origin. The radial grid has properties similar to those of a rectangular grid. Figure 3.5 shows some radial grids.

Grids are useful for operations such as sketching, planning layouts, placing text at a specific location, and using freehand clicks to indicate geometry location. When the grid is activated in a sketch, clicks in that sketch are snapped (moved) to the nearest grid point, if the grid snap option is on. The following screenshot shows grid properties.

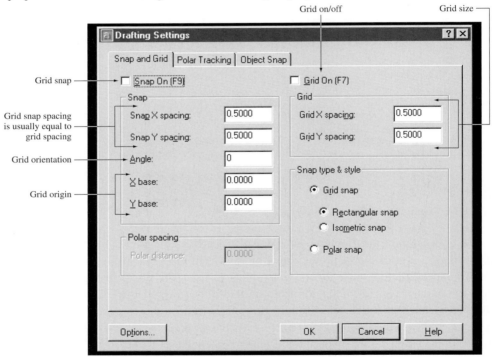

Grid management includes the following activities:

1. **Turn grid on/off.** A grid can be turned on or off with a simple command or a shortcut key such as F9. A grid is active when it is on.

2. **Turn grid snap on/off.** A grid snap can be turned on or off with a simple command or a shortcut key such as F7. When a grid is active, the grid snap takes effect if it is set to on. The grid snap is useful to use during sketching. However, it sometimes becomes an obstacle and slows the free movement of the mouse. In such a case, turn it off.

3. **Set grid size and other properties.** A grid has a size that specifies the spacing between its grid points. We can set the size as we desire to help in fast construction. We use the grid as a ruler, and its points as ticks (marks). Other grid properties that we can set include the grid origin (shown as X base and Y base in the foregoing screenshot), and its orientation (shown as Angle in the foregoing screenshot). We cover these properties in the remainder of the section.

3.6 Groups

Groups are useful when it is required to treat a selected number of entities temporarily as one (compound) entity. A group can be created by selecting the individual entities that are to make the group. A group can be manipulated by selecting any entity that belongs to it. An operation executed on a group affects all the entities in the group. More than one group can be constructed if required.

Groups eliminate the need for continually constructing a window around the entities that are to be manipulated several times, as in the case of animation. Consider the motion of a mechanism (or a robot) consisting of a few links which may rotate and/or translate in space. It will be efficient to combine the entities of each link into a group which can be rotated or translated as one entity from one location to another.

Once a group is no longer needed, the user can delete (or disassociate) it, that is, release its entities so that each can be treated individually again. A user may not be allowed to delete an entity which is a member of a group before deleting the group first.

3.7 Dragging and Rubber Banding

Dragging is a technique of moving an object around with a locating device such as a mouse. Dragging can be used to position objects and symbols. Dragging is achieved by the software's reading the locating device at least once each refresh cycle and using the position of the device to move (translate) the object.

Rubber banding is another technique usually used to construct lines. The push of a button marks the starting point of the line. The coordinates of the point are based on the position of the locating device — a mouse, for example. The line is drawn from this point to wherever the cursor is positioned. The line end point moves as the cursor moves until the button is pushed again or released to disconnect the end point from the locating device. Figure 3.6 shows the use of the rubber-banding technique.

3.8 Clipping

Various projections of an object's geometry can be obtained by defining views. A view requires a view window. If any part of the geometry is not inside the window, it is made invisible by the CAD software through a process known as clipping. Any geometry lying wholly outside the view boundary is not mapped to the screen, and any geometry lying partially inside and partially outside is cut off at the boundary before being mapped. If clipping is not done properly, a CAD system will produce incorrect pictures due to an overflow of internal coordinate registers. This effect is known as wraparound.

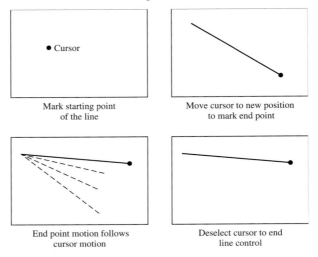

Figure 3.6   Line construction with rubber-banding technique.

3.9 Entity Selection Methods

Manipulation and editing operations usually act on existing entities. Each operation requires the user to identify or select the desired entities and input data. The most obvious method of identifying entities is clicking them. This method, however, is inefficient if many entities are required or if the entities to be clicked happen to be in a dense area (closely displayed entities) of the computer screen. This section discusses the most common methods of selecting entities. These methods can be mixed, that is, more than one method can be used simultaneously in a manipulation/editing command.

3.9.1   Individual Entity

If only one or a few entities are to be selected, the user can simply click them. The user must click close to the entities. The CAD/CAM system usually highlights the successfully selected entities to provide the user with visual feedback before hitting the Return key for final acceptance and processing of the operation. If no entity is highlighted, it means that the system did not recognize the clicks and the user should click again. If the wrong entities are selected, the user can use the Backspace key to deselect them.

3.9.2   All Displayed Entities

In some cases, all entities of a given view displayed on the screen are to be selected. Consider the example of construction of a 2½D model. The user first creates the geometry of the face of the model that is perpendicular to the direction of the uniform thickness. Next, the user must select all the displayed entities to extrude them to create the extruded feature. In these cases, there is a modifier that can be used with the desired manipulation/editing or a select-entity operation to indicate that all entities in a displayed view are desired. This modifier is usually followed by a click within the desired window. In this case, all displayed entities in the window are highlighted. If a substantial number of entities (not all of them) in a view are desired, these entities can be moved to a specific layer and the layer can be selected.

3.9.3 Groups

Groups are easily managed and manipulated. Creating groups themselves requires selecting the group entities. Thus, any of the methods described in this section can be used to create groups. Once groups are created, they can be used as a method of entity selection.

3.9.4 Enclosing Polygon or Window

Entities can be selected by enclosing them within the boundaries of a polygon. A polygon can be any shape as shown in Figure 3.7. A polygon is defined by its corner points, which are input by the user as clicks. A polygon can take any shape to selectively enclose the desired entities. If only two points are given to define a polygon, they are the end points of the

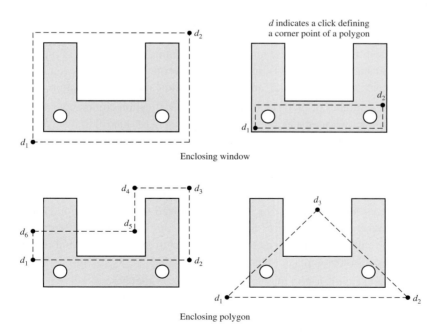

Figure 3.7 Use of polygon to select entities.

diagonal of a rectangle. The enclosing polygon becomes an enclosing window in this case. Entities or end points of entities which lie on the boundary of a polygon are considered to be within the polygon. If both end points of an entity (such as a line or a circle) lie within the polygon, the entire entity is affected by the editing commands. If only one end point is included (entity is partially enclosed), different CAD/CAM systems give different results. Some systems do not allow the editing command to affect the entity at all, while others apply the commands to the included end points only.

3.9.5 Chaining Contiguous Entities

An entity and all entities forming a contiguous path with it can be selected by chaining them. It is enough to select (click) one entity of the contiguous entities to form a chain. Some CAD/CAM systems may require another click to indicate the direction of chaining. Multiple branch points, where several entities are contiguous at the same point, result in multiple chains. All of these chains or a particular one of them can be selected. Chaining is useful for selecting long series of contiguous entities when they are surrounded by other geometry, making selection by a polygon impractical, and when selecting each entity in the chain individually would be tedious and time-consuming.

3.10 Entity Verification and Copying

Information about existing entities in a model database can be obtained by using an entity verification command. Such a command usually provides the entity type, the geometric information of the entity, and its layer number. For example, a verification of a line provides the coordinates for its start and end points, its length, its layer, and its name. Secondary information could include line type, weight, width, and date of creation and/or modification.

There is a number of modifiers related to the verification command. One can verify all the entities of a model without selecting them by using an ALL modifier. We can also verify an entity relative to MCS or WCS. For example, the coordinates of the end points of a line can be obtained relative to one of these coordinate systems. Such flexibility is helpful during geometric construction. Figure 3.8 shows an example.

Existing entities can be copied or duplicated. Two cases may arise. An entity may be copied in its current location or may be copied and moved to a new location. The need for the first case arises when the user desires to copy an existing entity to a different layer than its current one. If the entity is complex and the user is about to experiment with it, it is important to back it up in case of loss or damage to the entity.

The second case is useful in constructing geometry that has repetitive patterns. The user creates one pattern and duplicates it in the proper locations. Consider, for example, how would you construct a model of a gear. You create the geometry of one tooth precisely and duplicate it around the gear blank. In this case, the tooth is copied multiple times in a circular fashion around the blank.

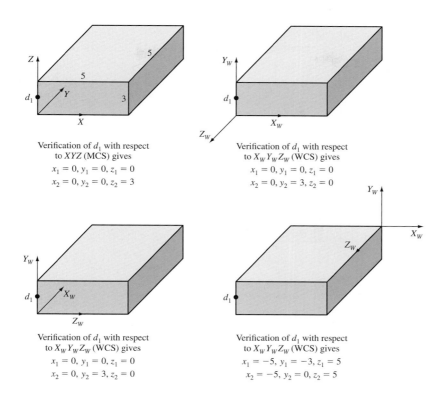

Figure 3.8 Entity verification.

3.11 Geometric Arrays

A **geometric array (pattern)** is a number of identical entities placed uniformly at specified locations. Two types of geometric arrays exist: rectangular and circular. Entities in rectangular arrays are separated by increments along the X, Y, and/or Z axes of the proper coordinate system. Entities in circular arrays are separated by increments in the radial and/or angular directions.

Figures 3.9 and 3.10 show some array examples. The base entities are the ones used to generate the arrays while the arrows show the directions of copying the base entities to form the arrays. The user can specify the number of copies and the spacing between them (Δx, Δy, and/or Δz for rectangular arrays, or Δr and/or $\Delta \theta$ for circular arrays) in each direction. Case D in both figures shows that the resulting arrays do not have to coincide with the base entity.

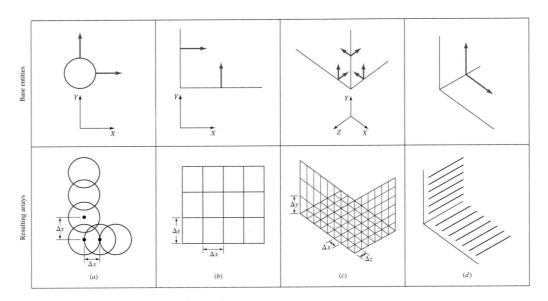

Figure 3.9   Rectangular arrays (patterns).

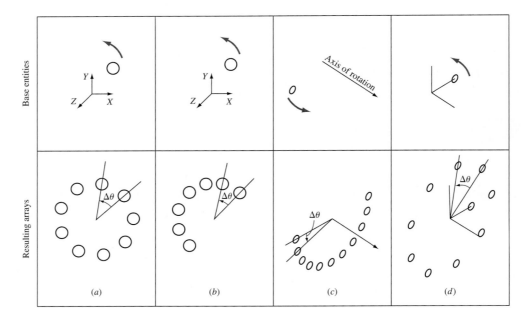

Figure 3.10   Circular arrays (patterns).

3.12 Transformation

Transformation operations are useful CAD/CAM productivity tools for manipulating existing entities. They can be used to translate, rotate, mirror (reflect), and scale entities. These operations and their related input are based on the mathematical background we cover later in the book. A `copy` modifier is usually available with these operations, in which case a copy of the existing entity is made and acted upon by the operation, thus keeping the existing entity unchanged.

The `translation` operation requires a translation vector, that is, a translation distance and a translation direction. If the layer modifier is used, a different layer from that of the existing one may be chosen for the copy. The `rotation` operation requires a rotational vector, that is, an axis of rotation and an angle of rotation. The angle of rotation could be positive or negative. A positive angle is an angle represented by a vector whose direction is in the positive direction of the axis of rotation. Figure 3.11 shows some examples of translating and rotating entities.

The `mirror` operation requires a mirroring axis. The existing entity is always mirrored onto the other side of the mirroring axis in reverse, that is, as if it were being viewed in a mirror. The `mirror` operation moves the existing entity to the other side of the axis unless a `copy` modifier is used. Figure 3.12 shows examples of mirroring.

The `scale` operation requires a scale factor. Alternative names of this operation which are also used by CAD/CAM systems are zoom or magnify. The operation is usually used to adjust the size of displayed geometry within a view window. If the geometry is too small, the operation can enlarge it to fill the view window. If it is too large and some entities are clipped against the window boundary, the `scale` operation can be used to scale the size down. Figure 3.12 shows examples of scaling.

3.13 Geometric Measurements

A `measure` operation can be used to obtain basic geometric measurements such as:

- The minimum distance between two entities, two points, or an entity and a point
- The angle between two lines
- The angle specified by three points
- The length of a contour of a connected set of entities

If either two geometric entities, two points, or one of each are specified, the minimum distance between the two is given. The **minimum distance** is defined as the measurement along a line mutually perpendicular to both entities or between the two closest points of the entities. If two line entities are specified, the following results are obtained:

- For two lines which intersect, the minimum distance is zero and the measurement of the angle (usually acute) between the lines is displayed on the screen.

- For two parallel lines, the the distance between the two lines is displayed.
- For two lines which do not intersect in space, the measurement of the distance between the lines at the two points of closest proximity is given. The measurement operation has a modifier which can project the two lines temporarily into a given plane (construction plane) of the WCS before any measurements are obtained. The resulting projected lines are treated as planar lines.

If three points are specified, the measurement of the angle defined by the three points is displayed on the screen. The angle used is usually the counterclockwise angle from the second point to the third point, using the first point as the origin point of the angle.

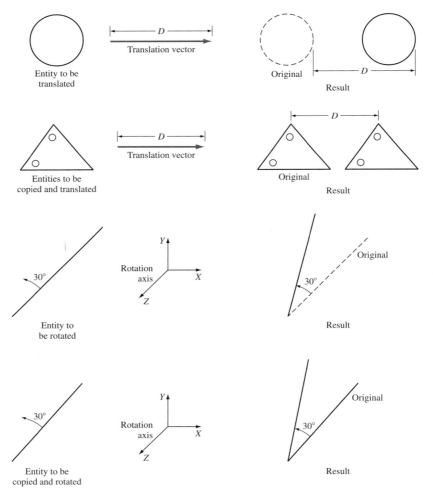

Figure 3.11   Translation and rotation of entities.

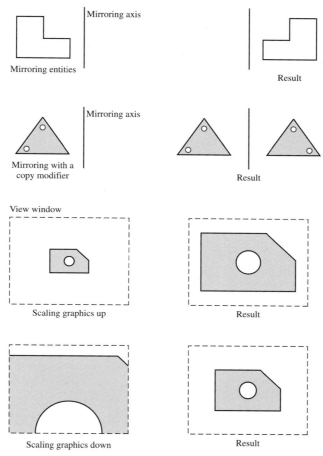

Figure 3.12   Mirroring and scaling of entities.

The measure operation allows the user to measure the length of a contour. The curves that make up the contour may be any curves. Successive curves in the sequence, except for the first and the last, either match at an end point or should intersect one another. The sequence of curves may form an open or a closed contour. Figure 3.13 shows some examples of the measure operation.

3.14 Offsetting

Entity offsetting allows the user to construct an offset of a planar entity. Uniform and tapered offsets can be obtained. In uniform offsetting, the offset entity is of the same type as the original entity. For example, offsetting a circle produces a circle. Tapered offsetting of an entity results in an entity that may have a different type than the type of the original entity. For example, tapered offsets of arcs and conics result in B-spline entities.

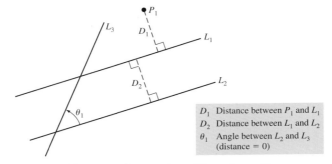

D_1	Distance between P_1 and L_1
D_2	Distance between L_1 and L_2
θ_1	Angle between L_2 and L_3 (distance = 0)

(*a*) Measurements between points and lines or lines and lines

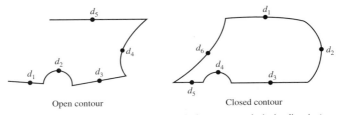

Open contour   Closed contour

Select contour entities in a given order (clockwise or counterclockwise direction)

(*b*) Contour measurements

Figure 3.13   Geometric measurements.

An offset operation requires the entities to be offset, an offsetting direction (the side for which the offset is to be constructed), and offsetting parameters. These parameters usually include the offset value (or values in tapered offset) and may include start and end points that may be different from those of the original entity.

An offset operation finds useful application in engineering design. It can be used to generate pipes and pressure vessels. Figure 3.14 shows some examples of entity offsetting. The following screenshot shows an offset operation on a CAD system.

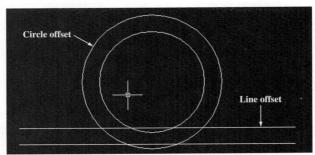

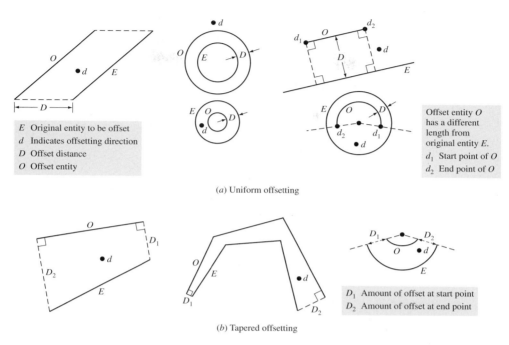

E   Original entity to be offset
d   Indicates offsetting direction
D   Offset distance
O   Offset entity

Offset entity *O* has a different length from original entity *E*.
d_1 Start point of *O*
d_2 End point of *O*

(*a*) Uniform offsetting

D_1 Amount of offset at start point
D_2 Amount of offset at end point

(*b*) Tapered offsetting

Figure 3.14   Entity offsetting.

3.15   Editing

This section covers the common editing operations available on CAD/CAM systems. Editing operations provide useful tools for design and engineering applications.

3.15.1   Entity Trimming

Trimming is used to stretch or contract end points of geometric entities such as lines within particular boundaries specified by the user. Trimming can be performed by using a `trim` operation.

A `trim` operation requires the entity or entities to be trimmed and trimming boundaries. These boundaries can be specified as a given length or location(s). Geometric modifiers are usually useful in defining trimming boundaries. Figure 3.15 shows examples of trimming open entities such as lines. It should be noticed that open splines can be trimmed within the original end points, but they should not be extended beyond those points because there is no definition beyond the spline ends. If extended, the spline is extended linearly along the directions of the end tangent vectors.

Trimming closed entities such as circles follows the counterclockwise direction rule. According to this rule, the part of the closed entity that defines the counterclockwise direction from the first trimming boundary to the second boundary remains after trimming. Figure 3.16 shows examples of trimming closed entities such as circles. Notice that if a trimming operation results in keeping the undesired portion of an entity, the operation can be repeated by using the same trimming boundaries used with the first operation but in a reverse order. If the original closed entity is to be obtained again, the two trimming boundaries can be chosen to be identical (same point).

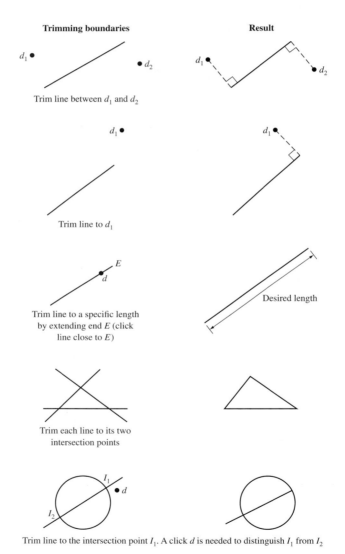

Figure 3.15   Trimming open entities.

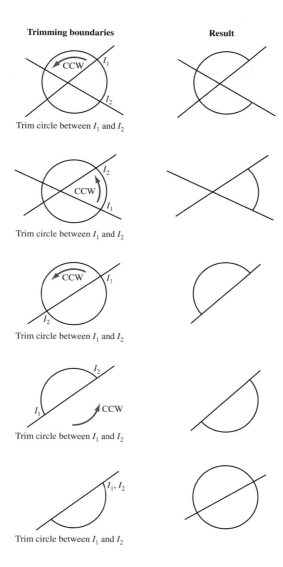

Figure 3.16   Trimming closed entities.

3.15.2 Entity Division

An existing entity can be divided into separate smaller entities of the same type. The division can be done by specifying the number of divisions or by specifying the division points. When a `divide entity` operation is applied to an entity, this original entity is deleted and is replaced by the newly formed entities. Figure 3.17 shows examples of entity division.

At the completion of a `divide entity` operation, the user cannot see a difference. To overcome this uncertainty, the user can proceed with the next operation to be   performed or an operation, such as `verify entity`, that results in temporary highlighting.

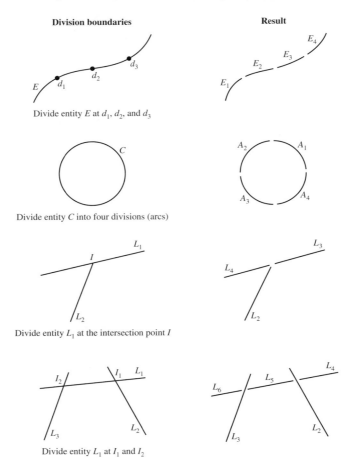

Division boundaries   **Result**

Divide entity E at d_1, d_2, and d_3

Divide entity C into four divisions (arcs)

Divide entity L_1 at the intersection point I

Divide entity L_1 at I_1 and I_2

Figure 3.17   Entity division.

3.15.3 Entity Stretching

Sometimes an entity is drawn with a wrong end point, or the end point may have to move to a new location. It is always inefficient to delete the entity and re-create it, especially if the entity is defined by many points. CAD/CAM systems provide an operation which can be used to stretch (modify) the undesired end point to the desired location.

3.15.4 Entity Editing

Some CAD/CAM systems provide their users with specialized operations which enable them to edit existing entities in special ways. By using these operations, the user does not have to delete entities and re-create them. While some of these commands can be replaced by the already introduced editing operations, some cannot. Some CAD/CAM systems, for example, provide their users with operations to edit existing fillets or change existing circles. A user can change the radius or diameter of an existing circle. These operations are useful, and you should investigate your own system for some of these useful commands.

3.16 Tutorials

3.16.1 Create a Spur Gear

This tutorial shows how to create a spur gear as an example of a circular array. The gear teeth form the circular array. We create one tooth using involute curves. We use the tooth to generate the array. The gear model is a 2½D uniform thickness model. After we create the front face with the teeth, we extrude it a thickness of 25 mm. All dimensions are in mm. The screenshot on the right shows the final model.

Modeling strategy:

1. **Create the tooth involute curve.** Using the Front sketch plane, create a spline curve through free points and enter the points shown in the table below, to generate the curve as shown.

x	y	z
0	−50	0
0.011068	−50.19	0
0.08834	−50.7558	0
0.297012	−51.6842	0
0.700275	−52.954	0
1.358341	−54.5355	0
2.327502	−56.3912	0
3.659237	−58.4765	0
5.399372	−60.7397	0
7.587302	−63.1233	0
10.2553	−65.5644	0

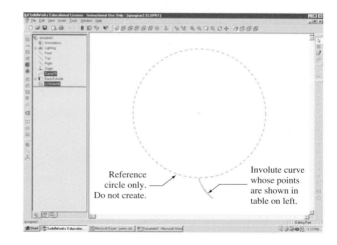

Reference circle only. Do not create.

Involute curve whose points are shown in table on left.

2. **Create the gear base and addendum circles.** Sketch the two circles in the Front sketch plane as shown below. Also trim the involute curve to the addendum circle.

3. **Create the tooths other curve and the gear root circle.** Mirror the trimmed curve from step 2 with respect to the center line shown in the screenshot below. Also create the root circle shown below.

Step 2

Step 3

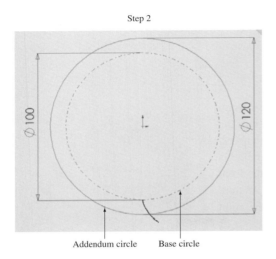

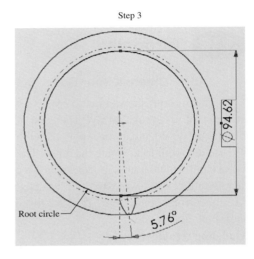

4. **Create the tooth shoulder and trim base circle.** Draw the two radial lines to join the end points of the involute curves with the origin. Trim the radial lines and the root circle to obtain the profile shown below.

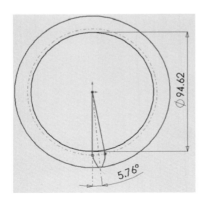

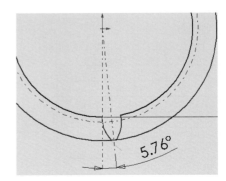

5. **Fillet the corners of the tooth shoulder and trim circles**. Zoom into the profile and fillet the edges shown. Use a fillet radius of 1 mm. Trim the outer circle to obtain the tooth profile.

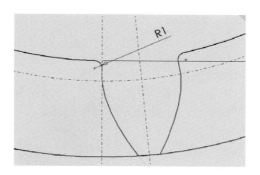
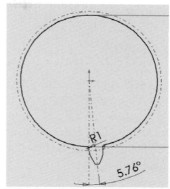

6. **Create the circular array.** Using the circular pattern option, array the tooth profile about the center of the gear. Using the number of teeth as 23, create the cross section of the gear. Trim the root circle at appropriate locations to get the final gear profile.

7. **Extrude the gear cross section.** Extrude the profile to get the solid model of the spur gear. Extrusion thickness is 25 mm.

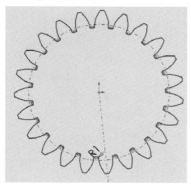
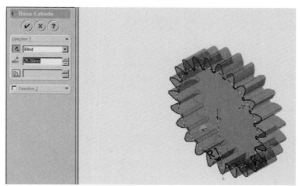

Tutorial 3.16.1 discussion:

The creation of the involute curve uses the involute equations. Any point on the involute has the following coordinates:

$$x = r(\sin\theta - \theta\cos\theta) \qquad \theta_0 \le \theta \le \theta_{max} \qquad (3.1)$$

$$y = -r(\cos\theta + \theta\sin\theta) \qquad \theta_0 \le \theta \le \theta_{max} \qquad (3.2)$$

We use $\theta = 0$ and $\theta_{max} = 50$. We use an increment of $\Delta\theta = 5$ degrees to generate points on the involute curve. The maximum value, 50, of θ is chosen to ensure that the involute curve clears the addendum circle. Note that we can write a C or Java program that uses Eqs. (3.1) and (3.2) to generate the table shown at the beginning of the tutorial. The program saves the points in a file that we can open to read from in a CAD system, such as SolidWorks. C trigonometric functions require the angles to be in radians and not degrees.

Tutorial 3.16.1 hands-on exercise:

Change the addendum, the base, and the root circles. This changes the number of teeth. Find out the new correct number by working out the geometry and using your CAD system.

3.16.2 Create a Telephone

This tutorial shows how to create a telephone as an example of a rectangular array. The phone keys on the keypad form the rectangular array. We create one key and use it to generate the array. The phone model is a 2½D composite model. All dimensions are in inches. The screenshot on the right shows the final model.

Modeling strategy:

1. **Extrude the base profile.** Select the `Right` sketch plane and sketch the following profile. Extrude the profile a distance of 4 inches.

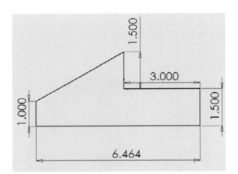
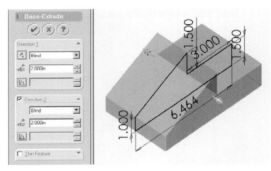

2. **Extrude the receiver holder.** Select the `Right` sketch plane and sketch the following profile. Extrude the profile a distance of 2.5 inches.

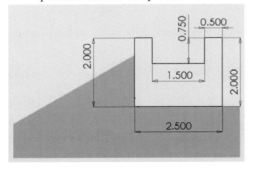
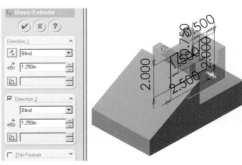

3. **Create the cut in the receiver holder.** Select the face of the receiver holder shown in the image on the left below as the sketch plane and draw a rectangle as in the image on the right below.

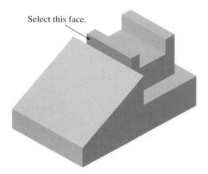

Select this face.

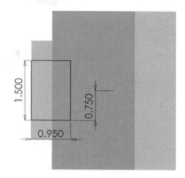

Now using the extruded-cut operation, cut the receiver holder using the parameters in the image on the left below to obtain the model as in the image on the right below.

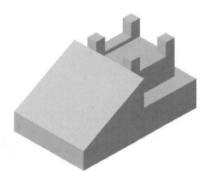

4. **Create the disconnectors.** Select the face shown in the image below as the sketch plane and sketch the two circles as shown below. Extrude these circles to a height of 0.7 inches.

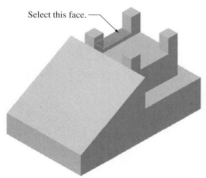

Select this face.

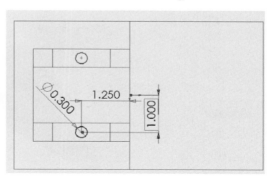

5. **Create the buttons.** Select the inclined plane of the model as the sketch plane and sketch the square as shown in the image below. Extrude this square profile of the button to a height of 0.2" to obtain the button as in the image below.

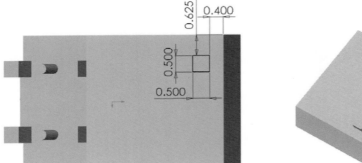
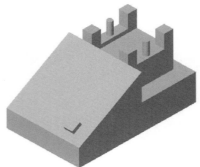

Now select the `Linear Pattern` command and set the parameters as in the image below. Finally, create the pattern. The phone model should appear as in the image below.

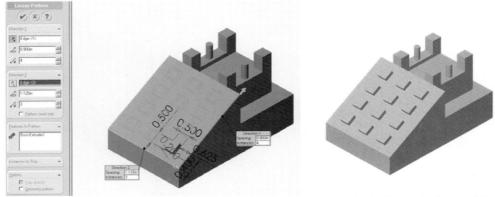

6. **Fillet the edges.** Use a 3D fillet operation and select the edges in the image below to fillet with appropriate radii also shown in the image. The final model is shown below.

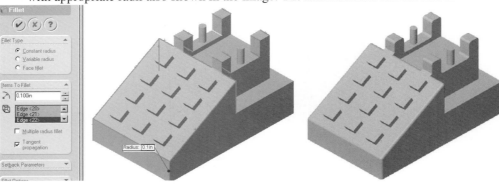

Tutorial 3.16.2 discussion:

Step 5 shows how we can create a rectangular array. First, we create the base unit, a square, and extrude it. Second, we pattern the extruded feature in a 3 × 4 array.

Tutorial 3.16.2 hands-on exercise:

Change the square cross section into a circle cross section, and regenerate the buttons. Also, add fillets to the other phone edges.

Problems

Part I: Theory

3.1   A user uses a line operation and clicks close to end points P_1 and P_2 to close the shown polygon. When the user utilized a later operation to calculate the area enclosed by the polygon, the system issued an error message that the polygon was not closed. How is this possible, and how can the user solve the problem?

3.2   Why does circle trimming follow the counterclockwise rule?

3.3   Give some examples where the layering concept is useful.

3.4   Give some examples where the grid concept is useful.

3.5   Give some examples where the offsetting concept is useful.

3.6   How can you use the geometric modifiers to create the following sketch without having to calculate any other dimensions explicitly:

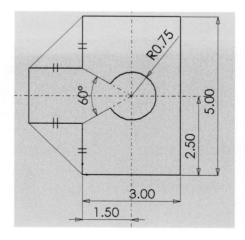

Part II: Laboratory

Use your in-house CAD/CAM system to answer the questions in this part.

3.7   Find the geometric modifiers available on your CAD/CAM system. What is their syntax? How do you use them? What is the default modifier?

3.8   Find all the layer-related commands on your system, specifically how to select/deselect layers, assign entities to layers, assign layers to entities, assign colors to layers, modify layer colors, and modify layers of existing entities.

3.9   For the model shown in problem 3.6, create the straight lines on layer 10, the circle on layer 20, and the dimensions on layer 30.

3.10 Find the commands related to rectangular and radial grids. How can you use them? Investigate their related modifiers.

3.11 With the aid of grids, sketch the following geometry:

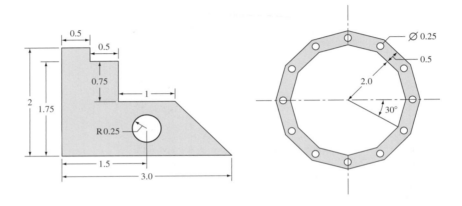

3.12 Create the following 4 × 4 rectangular array. Use a spacing of 1 in. in both directions.

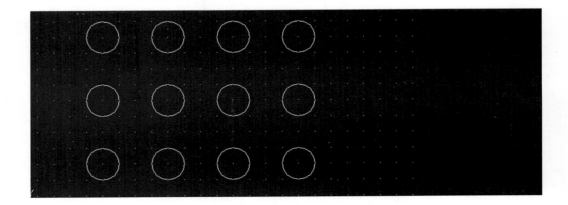

3.13 Create the following circular (polar) array that sweeps a 360-degree angle and has a center at $(x, y) = (4, 2)$.

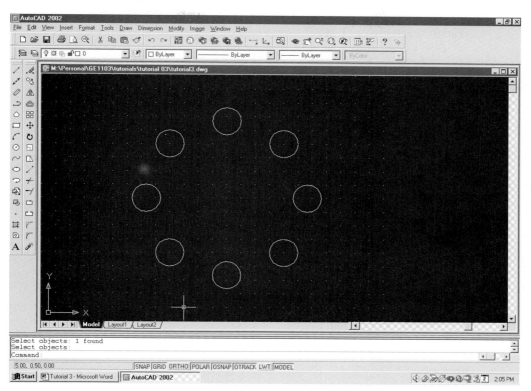

3.14 Modify Tutorial 3.16.1 to create a helical gear. The gear thickness is 25 mm. The gear helix angle is 15 degrees. Use Eqs. (3.1) and (3.2) to generate points on the involute curve. Use $\theta_0 = 0$, $\theta_{max} = 75$, and $\Delta\theta = 5$. Use a total of 16 points on the curve. Use 18 teeth for the gear. Offset the two cross sections of the gear by an angle of 4.08 degrees. The gear dimensions are shown in the screenshots below.

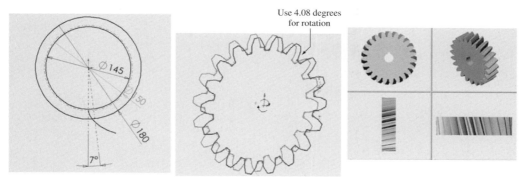

CHAPTER 4

Engineering Drawings

GOAL

Understand and master the creation of engineering drawings, their use in design documentation, their use in manufacturing activities, the information that a drawing stores, and how to avoid over- or under-dimensioning.

OBJECTIVES

After reading this chapter, you should understand the following concepts:

- Engineering drawings (blueprints)
- CAD drawings
- CAD views
- Title block
- Dimensioning
- Tolerances
- Notes and labels
- Manufacturing information

CHAPTER HEADLINES

4.1 Introduction

Engineering drawings are as old as engineering itself. They have been and still are used, by all engineering disciplines. An **engineering drawing** serves as a tool to document a design and communicate it to the entire engineering enterprise, from production, inspection, and assembly to sales and marketing. After designers finalize the design of a part or a product, they document their design ideas and specifications in a drawing accurately and thoroughly. The drawing becomes a reference that is used during the remainder of the product life cycle after design.

An engineering drawing comes in different sizes ranging from size A to size E. Each engineering discipline—mechanical, electrical, civil, industrial, or chemical—has specific formats and customs for engineering drawings. For mechanical and industrial disciplines, the parts and their models are 3D; thus their drawings contain different views of the part geometry. For electrical, the circuit and PCB (printed circuit board) design is usually 2D; thus their drawings contains wiring diagrams and schematics of components—no need for views. We only cover engineering drawings for mechanical and industrial designs in this book.

The generation of engineering drawings has changed over the years. Before the CAD age, draftsperson would use drafting tables, tools, and drafting papers to create a drawing. The drafting paper would then go through a chemical process that preserved its content. The drafting paper turned blue as a result of this process. This is why we refer to engineering drawings as blueprints.

With CAD, the generation of engineering drawings has changed a great deal. Instead of having a designer and a draftsperson, a CAD designer today may generate drawings as well. The CAD system makes the generation of an engineering drawing much less laborious than the manual approach. However, a designer must know the rules of drafting and how to present design content and information on a drawing—the subject of this chapter. We will use the term *drawing* to mean *engineering drawing* from now on in this book for simplicity.

How many drawings can we generate for a part? As many as we need. There is no limit. However, the fewer the better, to avoid confusion. Some CAD systems refer to a drawing as a sheet. Designers can create separate drawing sheets. CAD systems treat drawings as separate files from parts files. A drawing file typically has the *.drw* or *.drwg* extension.

CAD systems provide a drawing mode that designers must use to create drawings. The invocation of this mode depends on the CAD system. For example, SolidWorks and Pro/E let the user create a drawing file by selecting the drawing option—for example, `File => New => select drawing icon`. Legacy systems such as Unigraphics let users access the mode through a menu option while the part file is open—for example, `Application => Drafting`. In either case, the part whose drawing we create must be active, otherwise the CAD system would not know how to associate a drawing with a part file. In the case of SolidWorks and Pro/E, we open the part file first, and then a drawing. In the case of Unigraphics, the part file is already open.

CAD systems provide either drawing templates or they allow designers to create new ones or to customize the system template. For example, a drawing template may come with a generic title block that designers can customize for their own organizations. A typical file extension of a drawing template file is *.drwdot*.

4.2 Drawing Structure

A drawing has a universal structure. Figure 4.1 shows an example. The structure usually consists of views, a title block, a BOM (bill of material), and labels and notes. The views are laid out in an ordered fashion according to the rules of orthographic views. The front, top, and right views are located as shown in Figure 4.1. An isometric view may be added in the top right corner of the drawing. Each view displays dimensions and tolerances to fully define the part.

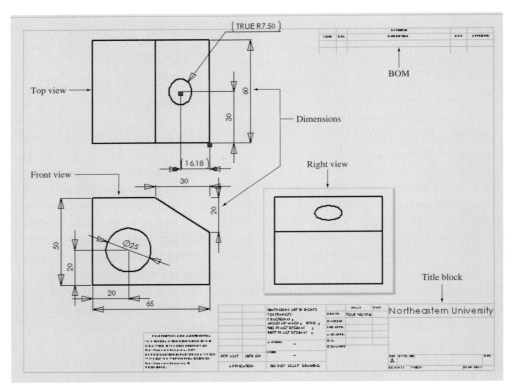

Figure 4.1   Structure of an engineering drawing.

The title block is usually located in the bottom right corner of the drawing. It contains information such as company name, part number, drawing number, revision number, sheet number, materials and finish, general tolerances, drawing scale, sheet size, revision block, and drawn by/checked by.

The BOM is a table that is usually located in the top right corner of the drawing. A BOM is usually used in assembly drawings to list item number, quantity, part number, and description.

The labels and notes provide additional information about the design that cannot be conveyed by dimensions. Examples include instructions on manufacturing, machining, assembly, surface finish, and so forth.

4.3 Model and Drawing Associativity

CAD systems provide a two-way associativity between model and drawing modes. If a designer makes a geometric change in the model mode, the change is reflected in the drawing mode. Alternatively, if the designer changes a model dimension in one of the views of a drawing, the model is automatically updated to reflect the change, after the designer uses a Rebuild or Update command.

Associativity between model and drawing modes makes it convenient and productive to produce drawings. Designers need to remember to update the model whenever they make changes in the drawing mode.

4.4 Drawing Content

While Figure 4.1 shows the structure of a drawing, it also shows the contents of a drawing. A drawing usually has model views, dimension, tolerances, annotations, BOM, assembly instructions, machining instructions, surface finish, and roughness symbols. We cover all of these topics in this chapter.

The generation of a drawing begins by defining and placing views in the drawing while its file is open on the computer screen. After placing the views, we can move them around to adjust their locations. We can scale them up or down. We can also enforce the orthographic projection rules — for example, a line in the front view becomes a point in the right view at the same height as the line, and it becomes a line in the top view with end points matching those of the line in the front view.

After being satisfied with the locations and sizes of the views, we begin adding dimensions and tolerances. CAD systems provide rich dimension menus that allow the control of dimension type (cartesian, ordinate, and so forth), height, font, units (mm or in), standard (ANSI or ISO), arrows, the number of decimal places, and orientation.

After adding dimensions and tolerances, we add labels and notes, fill in the title block and BOM, print the drawing, and then save it. We retrieve it later for updating or for other reasons.

In industry practice, drawings are checked by a lead person after they are created and finalized. They must be signed off by an authority before they can be released for production. Some CAD systems keep track of all the revisions and the changes that have been made in each revision. Sometimes, the lead person locks the drawing file, thus preventing others from opening it or making any changes to it. It is also possible that a drawing can be discussed in a conference call. Each person participating in the call can view the drawing.

4.5 Methods of Angle of Projection

When we create a view of a model such as the front, top, or right view, we project the model onto a plane parallel to the model face that we are viewing and normal to the line of sight of the observer. The lines that are visible from the viewing angle are rendered as solid lines, and the ones that are hidden are rendered as dashed.

The location of the projection plane relative to the observer and the model defines the method of angle of projection (also known as the type of projection). The projection plane can either be between the observer and the model, or it can be behind the model and away from the observer, as shown in Figure 4.2. A method of projection corresponds to each case.

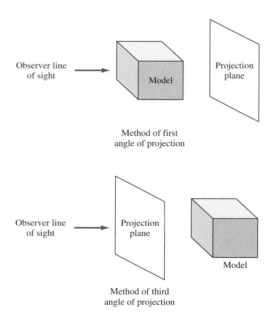

Figure 4.2   Methods of angle of projection.

We can determine the method of angle of projection used in a particular drawing by simply looking at the drawing views. In the first angle method, the projected view is the opposite view of what the view name indicates. For example, when we request a top view, we get a bottom (the opposite) view, as indicated by the hidden lines of the view. The same is true for other views.

In the third angle method, the projected view is the same as what the view name indicates. For example, when we request a top view, we get a top view, as indicated by the hidden lines of the view. The same is true for other views.

Observe that regardless of the method of angle of projection, the model views do not change; only their appearance changes. Moreover, the front view remains the same, unaffected by either method. Thus it serves as a reference view for placing other views. The screenshot below shows the effect of the method of angle of projection on model views. Some CAD systems such as SolidWorks allow users to set the type of projection. The following screenshots show examples of the setup options.

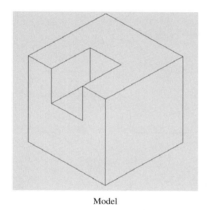

Model

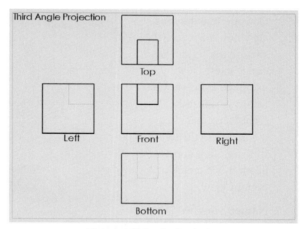

Method of third angle of projection
(we always use this method)

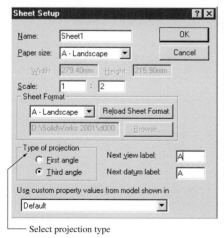

Select projection type

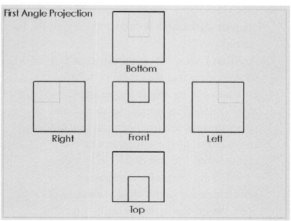

Method of first angle of projection

4.6 Types of Views

CAD systems provide us with various types of views that we can use in a drawing. Depending on the complexity of the model, we may use a few of them. The following list shows the available types of views that can be included in a drawing:

1. **Projected view.** This is a view that results from projecting an existing view in a given direction. Consider the drawing shown in the screen-shot on the right. We create a left view by using the front view as follows: Select the front view, and click the mouse to the left of it. If we click under-neath it, we create a bottom view.

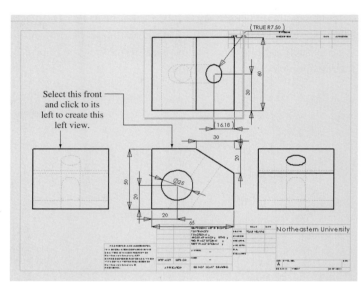

2. **Named view.** A named view is a standard predefined view. Examples of named views are shown on the screenshot on the right. These views include the six standard 2D views and the three standard 3D views. Named views allow us to generate the standard orthographic views. We can place these views anywhere on the drawing sheet. We can also create user-defined views and add them to the list of the standard named views.

Named views

3. **Auxiliary view.** An auxiliary view is a custom view that we create using a custom viewing angle. For example, we may want a view normal to an inclined plane to show the true location of a hole in the plane, as the screenshot on the right shows. We can define such a view by selecting the inclined edge as a reference edge in the front view, and click the mouse in the top right direction to place the auxiliary view.

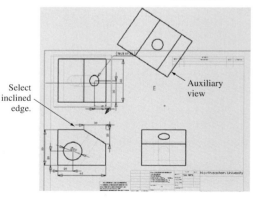

4. **Sectional view.** This is the view that we obtain after we cut open the model to reveal important hidden details of its geometry. We need to define a direction for the sectional view because we have two choices when we cut open the model. The screenshot on the right shows a sectional view B – B that we create by drawing a vertical line in the middle of the right view and clicking the mouse to the left of it to place the auxiliary view there. We can flip the direction of the section view if we need to.

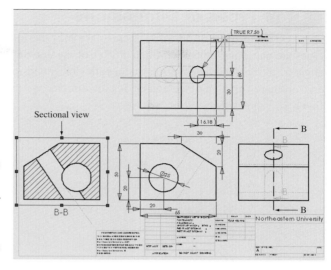

5. **Detailed view.** We use these views to magnify a small portion of a given view to show the details of the small portion only.We can define the portion with a circle. The screenshot on the right shows an example of a detailed view.

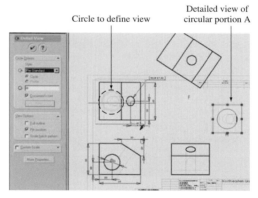

After we place views in a drawing, we can resize them and move them around to fit within the drawing sheet. We typically include the three standard views (front, top, and right) at a minimum in a drawing. We should use the same scale for these three views. The following screenshots show a drawing before and after scaling.

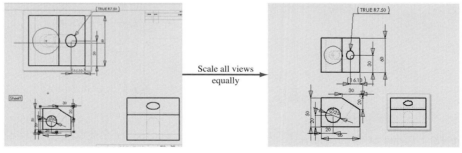

Scale all views equally

4.7 Types of Dimensions

Different types of dimensions exist. We may use all or some of them in a drawing. These types are cartesian (along the X and Y axes of the drawing), radial (a circle radius or diameter), angular (to dimension an angle), true length (for dimensions that are not along the X or Y axis), and ordinate (uses same reference for all dimensions that are in one direction) dimensions. Figure 4.3 shows these types.

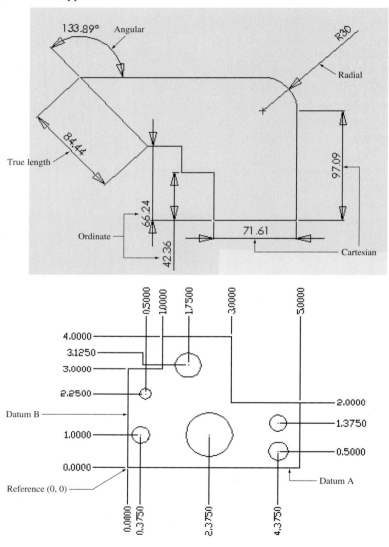

Figure 4.3   Types of dimensions.

Ordinate dimensions are important in engineering drawings. These dimensions are mainly used for dimensioning stepped profiles or an array of holes. They require a reference point and two datums from which the dimensions are measured. The two datums serve as reference surfaces for manufacturing the part. For example, the machinist finishes the two part surfaces A and B shown in Figure 4.3 and measures all the dimensions from it. This reduces the cost of machining the part.

4.8 Annotations

Sometimes we need to add information to a drawing above and beyond dimensions and tolerances. In this case, we "annotate" the drawing. Annotating a drawing means adding notes and labels to it, as shown in Figure 4.4. There is no clear distinction between notes and labels. Both have text that is placed on the drawing. In this book, we define notes as standalone text without leaders, and labels as notes with leaders.

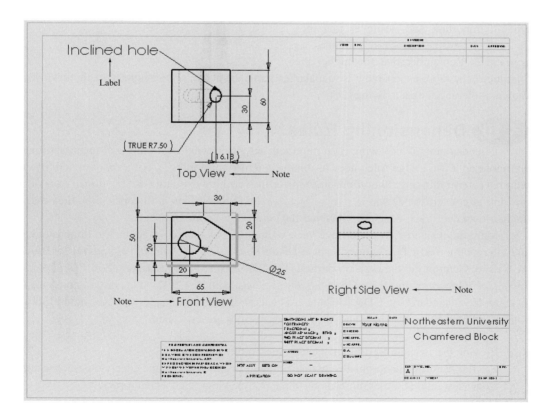

Figure 4.4   Notes and labels.

4.9 Tolerances

Tolerances are an important concept in engineering design and manufacturing. Tolerances allow for variability during manufacturing because there is no perfection in all manufacturing conditions. Tolerances produce normal distribution production lots. We cover tolerances and their types in more detail later in this book.

Designers are responsible for assigning tolerances to dimensions based on the functional requirements of the design. If tolerances are important, should we assign a tolerance to every dimension on a drawing? We only assign tolerances to important dimensions that directly affect the design functionality. For all other dimensions, we assign "general" tolerances. In essence, we tell production that the general tolerances are not important and whatever tolerance the manufacturing process produces is acceptable. Remember that the higher the tolerances, the more expensive it is to make the part. The above screenshot shows three ways of showing tolerances on a drawing.

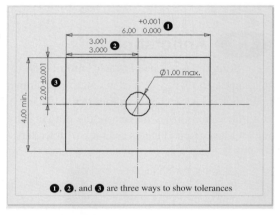

❶, ❷, and ❸ are three ways to show tolerances

4.10 Dimensioning Rules

Dimensioning of a drawing is an important activity if we think in terms of manufacturing, and more so if we assign tolerances to dimensions. Dimensioning is not a random activity. It requires careful thinking. Shop floor machinists and machine operators use the dimensions on a drawing to convert the design into a product. If they find conflicts in dimensioning, they make up their own decisions, which can increase the percentage of scrap.

Many books and references have been written on the subject of geometric dimensioning and tolerancing (GD&T). Samples include *Dimensioning and Tolerancing Handbook* by Bruce A. Wilson, Genium Publishing Corporation (ISBN 0-931690-80-3); Chapter 13 in *Principles of Technical Drawings* by F.E. Giesecke et al., MacMillian, 1992; and the famous *ANSI Y14.5M-1994* standards on GD&T. There are general dimensioning rules that we should follow when we dimension a drawing. The most important rules are listed here:

1. Provide the size and location of each feature in the drawing.
2. For flat features, such as plates, give the thickness dimension in the edge view, such as the front view, and all other dimensions in the outline view, such as the top view.
3. Dimension features in the view that shows their true size and shape.
4. Use diameter dimensions for circles and radial dimensions for arcs.
5. Omit unnecessary dimensions.

6. Ensure that dimensions are large enough to see, spaced out from each other, and are placed away from the profile.

7. Provide a gap between profile lines (features edges) and dimension extension lines. This eliminates any ambiguity about the profile entities.

8. Use a consistent size and style of leader lines, text, and arrows throughout the drawing.

9. Avoid over- or underdimensioning. The following shows a case of overdimensioning.

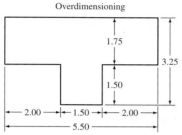

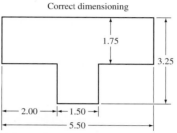

10. Use the maximum material condition (MMC) to display toler-anced dimensions on a drawing, as shown on the right. Show the larger dimension above the dimension line and the smaller dimension below the line to minimize the amount of production scrap. A machinist normally aims at producing the dimension above the dimension line. If the machinist cuts more material

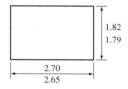

inadvertently, the par dimension is still within specification and should pass inspection.

4.11 Editing Drawings

After creating a drawing, we may need to edit it. Editing a drawing includes manipulating views, dimensions, and annotations. Manipulating views includes replacing, positioning, and scaling them as well as setting hidden line removal and tangent edge display. Manipulating dimensions includes positioning them, changing their text size, changing arrowheads from pointing inward to pointing outward, and changing their font.

4.12 Tutorials

4.12.1 Create a Drawing

In this tutorial, we create a drawing on a CAD system and show the required steps. The drawing has the three stan-dard views: front, top, and right. All dimensinsions are in mm.

Modeling strategy:

1. **Open a file in the drawing mode.** Click this command sequence: `File => New =>` select the drawing mode as shown in the screenshot on the right. This

sequence opens up a drawing sheet, with a template if your system is set up so.

2. **Open the part and add views to the drawing.** Open the part file whose drawing is being created. Once the part file is open, switch to the drawing sheet and insert the three standard views; use the drawing toolbars provided by your CAD system. The screenshots below show an example. Adjust the views in the drawing sheet by moving them around. Drag and move with the mouse. Also if needed, scale the views up or down to fit inside the drawing.

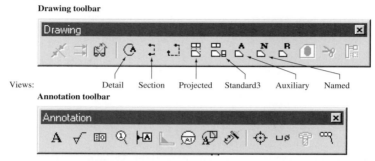

3. **Add dimensions to views.** Use the `Dimension` menu and add dimensions. To add a dimension, click the desired edge in a view, drag the mouse to the proper location, and release the mouse. The dimension should appear. If needed, move the dimension. Also if needed, modify the dimension properties. The screenshots below show the final drawing and an example of a `Dimensions properties` window.

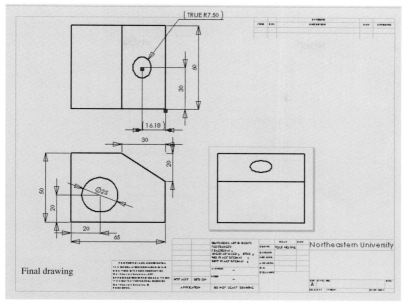

Final drawing

Tutorial 4.12.1 discussion:

Creating and editing drawings is an easy task to do on CAD systems. However, it is a time-consuming and mundane task due to the drafting details. It may take a lot longer to create a drawing of a model than to create the model itself. One task that takes time is dashing or hiding lines. If multiple lines from the model projection overlap, we must remove them manually. Some systems automate this task to save time.

Tutorial 4.12.1 hands-on exercise:

Add an isometric view in the top right corner of the drawing. Try to dimension the isometric view. Is it easy?

4.12.2   Use Drawing and Model Associativity

This tutorial shows the associativity between the `drawing` and `model` modes. We make a change in a drawing and show the change when we open the part (model) file. We use the drawing file of the chamfered block from Tutorial 4.12.1. All dimensions are in mm.

Modeling strategy:

1. **Open the drawing file.** Use `File => Open => to` select the file from Tutorial 4.12.1.
2. **Change the drawing.** Change the diameter of the horizontal hole to 40 mm.
3. **Rebuild the model.** Click a `Rebuild` or `Update` icon to update the model.
4. **Check model update.** Open the part file and note the change. The following screenshots show the modified drawing and model.

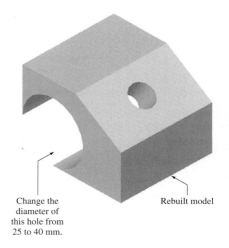

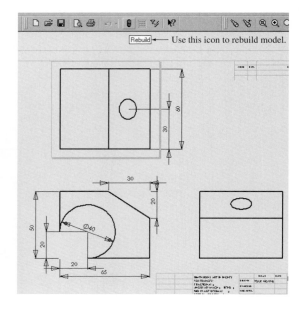

Use this icon to rebuild model.

Change the diameter of this hole from 25 to 40 mm.

Rebuilt model

Tutorial 4.12.2 discussion:

We see here that the model and the drawing files are associated, that is, when we change a dimension in a drawing file we may physically effect that change in the model file as well, and when we change a feature property in a model file we may effect that change in the drawing file as well. Note that changes due to associativity between drawing and model files will NOT be effected until you execute a `Rebuild` or `Update` command.

Tutorial 4.12.2 hands-on exercise:

Fillet some of the edges of the model in the front view of the drawing. Rebuild the model. Open the model file and check the isometric view. Do you see filleted edges?

PROBLEMS

Part I: Theory

4.1   What does a drawing contain?

4.2   List the different types of views. When do we use them?

4.3   What are the available types of dimensioning?

4.4   What is drawing annotation? What is the difference between notes and labels?

4.5   What are the best practices for dimensioning?

Part II: Laboratory

Use your in-house CAD/CAM system to answer the questions in this part.

4.6   Create the drawing shown below of a support bracket. Add an isometric view in the empty space in the top right corner of the drawing.

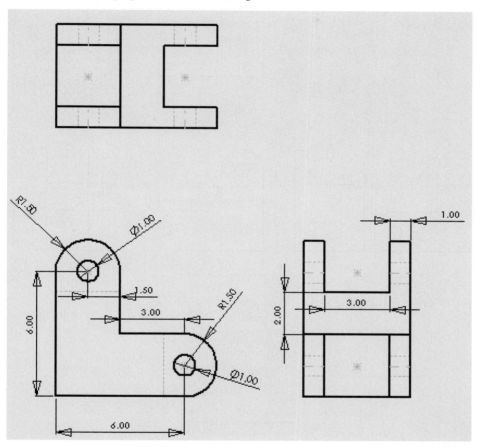

4.7   Create the following detailed view, with the rectangular area as shown on the right. You may create the part with approximate dimensions.

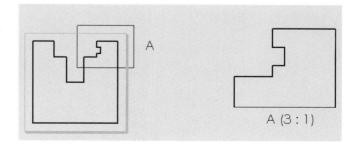

4.8   Open the drawing sheet of the support bracket that we created in Problem 4.6 and create the following drawing views: projected view (on the left-hand side), sectional view, and an isometric view, as shown in the following screenshot.

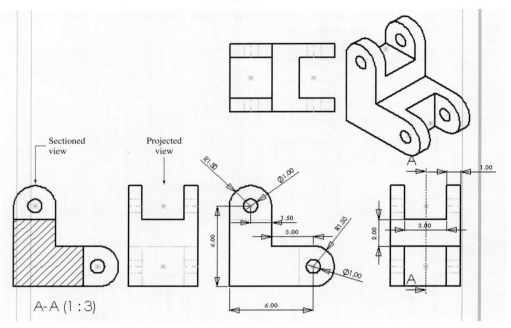

4.9   Create the following section view of a multiple-holed solid block.

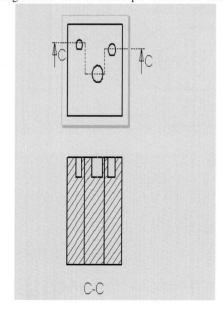

4.10 Create the following broken-out section in the drawing sheet. The part is a rectangular block with a countersink hole.

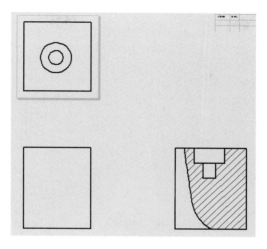

4.11 You can use a broken (or interrupted) view on the drawing of a long part that has a uniform cross section. This makes it possible to display the part with a larger scale in a smaller-size drawing sheet. Construct the following drawing with broken lines as shown below.

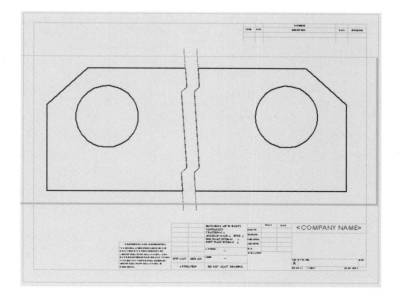

4.12 Create a drawing of the telephone model of Tutorial 3.16.2, in Chapter 3. Create the drawing with the following layout:

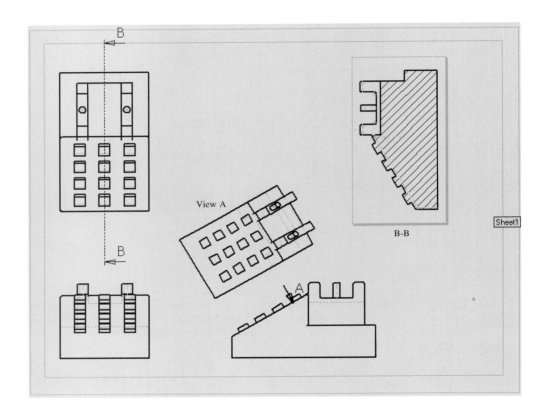

CHAPTER 5

CAD/CAM Programming

**OAL**

Understand and master the concepts of object-oriented programming and their use in parametric design, become familiar with customizing CAD/CAM systems, and know the various methods of CAD/CAM programming.

BJECTIVES

After reading this chapter, you should understand the following concepts:

- Object definition
- Object creation and use
- Object concepts
- Inheritance
- Macros
- CAD/CAM API functions
- C++ and OpenGL
- Java 3D

CHAPTER HEADLINES

5.1   Introduction

CAD/CAM systems are powerful software applications. They employ complex geometric modeling concepts, computer graphics algorithms, software techniques, design and analysis theories, manufacturing techniques, and database and management concepts. CAD/CAM software of any commercial system has been evolving over the years. The code itself is very long, complex, and difficult to follow.

Modifying and programming CAD/CAM software is a difficult, but not impossible, task. We can, and do, modify CAD/CAM software to meet our modeling needs. After all, it is engineers who work for CAD/CAM vendors and write the CAD/CAM software.

CAD/CAM software is written in a variety of languages. Legacy systems, such as Unigraphics and I-DEAS, started as Fortran code. Newer systems, such as Pro/E and SolidWorks, are written in C and C++. ACIS is written in C++ and also uses Scheme. Most of these systems use Java also in one form or another, especially for Web-based functionality.

How often do we need to program CAD/CAM systems? In comparison to using the systems as they are, the need for programming is not great. This is not to say that customizing these systems is hardly needed. There are different levels of customization and programming, ranging from writing a simple macro to automate repetitive tasks to using the system API (application programming interface) functions to extend the system functionality.

Users who are interested in programming their CAD/CAM systems must be familiar with system functionality, structure, hierarchy, terminology, and philosophy. Such familiarity is a precursor to being able to use the programming tools and functions offered by a system. For example, those who want to modify a spline curve by programming it would have an easier time doing so if they have used the system spline command already.

What does it take to program a CAD/CAM system? We need to know a programming language, geometry and graphics concepts, and database techniques. We may need more or less of each of these, depending on the programming or customization task at hand. For a programming language, C, C++, or Java is good. The effective use of C++ or Java requires a knowledge of object-oriented (OO) concepts and OO programming (OOP). Geometry and graphics concepts are covered in this book. The use of databases requires an understanding of relational databases and DBMSs (database management systems). We cover all these areas briefly in this chapter.

5.2   Relational and Object Databases

CAD/CAM systems utilize relational databases to store geometric, graphical, and other data that defines a CAD model. A model database is stored in the model or part file. The relational database model was developed by IBM researchers in 1968 in an attempt to apply mathematical rigor to the world of database management systems. The model became popular quickly both in industry and academia as its mathematical foundation made it easier to develop a query language.

A **relational database** is conceptually a set or collection of tables. A **table** is considered a relation. A database with four tables is considered a database with four relations. Each table stores data about a given entity. The data in a table is organized or stored as attributes in rows. Each row in a table is known as a record (or a tuple). For example, a database of a bank may have a table of customers (the entity). Each customer has a row (record) in the table. The record could hold the customer name and account number (the data attributes). Entities can be thought of as objects with attributes.

One can also view a table as a set of columns instead of rows. Each column holds the values of the attribute represented by such a column. For example, the column of account numbers in the bank database holds the valid values for the account number attribute. These values may range, for example, from 00000 to 99999. The integer values between these two limits inclusive represent the domain for the account number attribute.

A relational database is never one table. It typically consists of many tables. These tables are interrelated; that is, relationships exist among these tables (similar to relationships among objects). For example, for each customer in a table (CustomerName table) of a bank database, two other related tables (CustomerAddress and CustomerMonthlyActivities tables) could show the customer address and the account activities of a thirty-day period. In this case, the three tables are related.

Relationships between tables can be of three types: one-to-one, one-to-many, and many-to-one. In a one-to-one relationship, there is only one record in the second table for each record in the first table. In the one-to-many relationship, multiple records exist in the second table for each record in the first table. The one-to-many relationship is sometimes known as a parent-child relationship. The many-to-one relationship is opposite to the one-to-many relationship; for multiple records in the first table, only one record exists in the second table.

SQL (Structured Query Language) is the language commonly used to query and update relational databases. SQL can be used to retrieve, sort, and filter specific data of a database. SQL also allows users to define data in databases and later manipulate it. SQL was developed by IBM. SQL uses set logic to query databases. It is an easy language to learn; it has an English-like syntax. It requires few keywords and clauses to perform powerful data retrieval from the database. Report formatting is also simple using SQL. Sample statements include `CREATE DATABASE`, `CREATE TABLE`, `SELECT`, `ALTER`, `INSERT`, `DELETE`, and `UPDATE`.

The query `SELECT FirstName, LastName FROM CustomerName;` returns the two columns labeled `FirstName` and `LastName` in the table `CustomerName` as the query result. Note that the `SELECT` statement must end with a semicolon, as all SQL statements end with a semicolon.

We must create a database before we can use it. We need a database schema that defines the types of variables used in the database records in each table. We use this schema to create each table of the database. The following screenshot (of Microsoft Access) shows the definition of a schema for a bank database table.

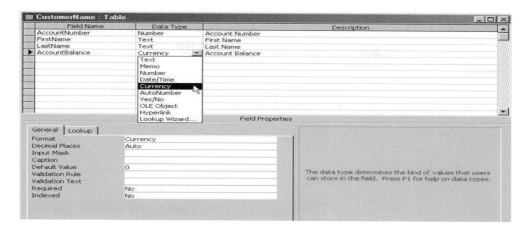

After we create the table schema, we populate the table with records and data as shown in the screenshot below. The type of data we enter in each field of a record must match the data type specified in the table schema.

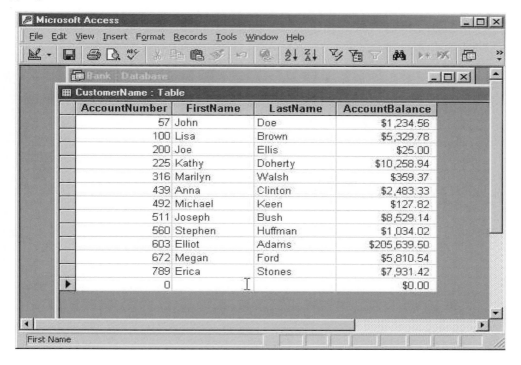

We can write SQL queries to retrieve data from the table. For example, the SQL query SELECT * FROM CustomerName; returns all the above records of the table as the result.

CAD/CAM systems utilize various types of data structures and databases to store CAD/CAM models and their related data. Many systems use relational databases to store model data. They store the boundary representation (B-rep) of the model, that is, the faces, edges, and vertices of the model. We cover B-rep in depth later. In such as a database, we use three relational tables: Faces, Edges, and Vertices as shown in Figure 5.1. The Faces table links (points) to the Edges table, and the Edges table links (points) to the Vertices table. Each of the three tables has one record per entity (face, edge, or vertex). Each record has the geometric definition of each entity and other information, such as color and layer, that we do not show in the tables.

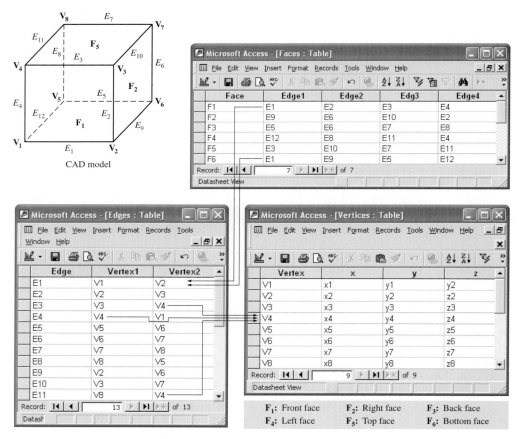

Figure 5.1   Relational database of a CAD model.

Think of the CAD user interface as an interface to the relational tables. When we create a box by extruding a rectangular sketch, the CAD system creates the three tables in Figure 5.1 as the extrusion database. When the user deletes entities, their records are deleted from the database tables.

While relational databases are pervasive in CAD systems, there exist object databases where the unit of data storage is an object, instead of a table as in traditional relational databases. We cover objects in the next few sections. An ODMBS (object database management system) is similar to the conventional DBMS in that it manages the database activities and transactions. Object databases extend SQL to OQL (object query language). OQL is a high-level interface that allows us to query object properties. In addition to OQL queries, object databases support relationships between objects.

Object databases have predictable performance and scalability. They provide an excellent use for complex applications. CAD/CAM applications are a natural fit for object databases. Actually, CAD/CAM is a complete misfit for relational databases.

There is a third type of database that provides object-relational mapping. This mapping converts in-memory objects to relational data. Each object is broken into its parts, and each part is stored individually as a field. For example, a Java class would map to an SQL table. An instance of the class would map to a record (row) in the table, while properties would map to individual fields (cells) in that row.

5.3 Object Definition

The basic unit of thinking, designing, building, and programming OO code is the object. Let us consider a car to define what an object is. If we think of what characterizes a car, we come up with make, model, year, color, drive, stop, maintain, and so forth. We can further divide these characteristics into two groups: attributes (properties) and behaviors. The attributes are make, model, year, and color. An **attribute** is just a value that we specify; it does not require any calculations — the year is 2005 and the color is red. Object attributes become its *variables* in an OOP program.

The behaviors are drive, stop, and maintain. A **behavior** is doing something; it takes input and produces output. Driving a car requires a sequence of steps. Think of a behavior as an algorithm. Object behaviors become its *methods* in OOP. A **method** is a function associated with an object.

An **object** is defined as an entity (construct) that has attributes and behaviors. The attributes are its variables, and the behaviors are its methods.

CAD/CAM programming utilizes OOP languages. We use these languages to create user-defined objects, or we use their built-in (predefined) objects provided by CAD/CAM systems. Understanding objects is a prerequisite to creating one's (custom) objects or using existing (predefined) ones.

Consider the ACIS solid modeling engine. It provides objects (classes) such as `FACE`, `EDGE`, `VERTEX`, `SPHERE`, `SPLINE`, `PLANE`, and `TORUS`. These objects make up the data structure of ACIS. We can build on these objects by using them or extending (inheriting from) them.

EXAMPLE 5.1   **Define objects.**
Write the definition of a spur gear and telephone objects.

SOLUTION   We have created two models in Tutorials 3.16.1 and 3.16.2: a spur gear and a telephone respectively. Each of these models is an object that has attributes and behaviors. Table 5.1 shows some of them. It also shows their mapping to OOP variables and methods.

Table 5.1   Object definition.

Spur gear object = {{root diameter, base diameter, addendum diameter, thickness, number of teeth}, {rotate, maintain}}
Telephone object = {{type, color, owner, number}, {dial a number, ring}}

Object	Definition		OOP mapping	
Spur gear	Attributes	root diameter	Variables	`rootDiameter`
		base diameter		`baseDiameter`
		addendum diameter		`addendumDiameter`
		thickness		`thickness`
		number of teeth		`numberTeeth`
	Behaviors	rotate	Methods	`rotate()`
		maintain		`maintain()`
Telephone	Attributes	type (cordless, wireless)	Variables	`type`
		color		`color`
		owner		`owner`
		number		`number`
	Behaviors	dial a number	Methods	`dialNumber()`
		ring		`ring()`

Observe that, for the most part, the variables and methods have the same names as those of the object attributes and behaviors. This is a common practice in OOP.

5.4 Object Concepts

We need to define, create, and use objects in our programs to realize the full benefits of OOP. Object definition as variables and methods sets the stage to implement objects. OOP languages use three concepts to implement objects: classes, instantiation, and membership access. Classes define objects, instantiation creates them, and membership access uses them.

5.4.1   Classes

OOP begins by creating a class for each object that a program uses. A **class** is a template that holds the object definition. A class is considered a generic object. It is one size fits all. It defines a family of objects. Inside the class we include all the variables and methods that define the object. Consider a class that defines a spur gear. A spur gear has a root diameter, a base diameter, an addendum diameter, a thickness, and a number of teeth. *Any* spur gear has these properties (attributes). As a result, the class is a generic object. A particular spur gear, `myCarClutchFirstGear`, still has these properties, but with *specific* values.

In class-based languages, such as Java and C++, we define an object using the `class` keyword. Inside the class, we define a constructor(s). A **constructor** is a special method that creates and initializes instances. It assigns initial values for the instance variables (properties). An **instance** is a specific copy of a class.

5.4.2   Instantiation

As the class is a generic template, we use it to create specific objects; that is, we customize it. These specific objects are the instances. We instantiate the class to create an instance. In C++, we use this syntax:

```
SpurGear mySpurGear;
```

which creates the `mySpurGear` instance. In Java, we use the `new` keyword (operator) to create an instance of an object — for example,

```
SpurGear mySpurGear = new SpurGear();
```

The class instantiation creates the instance and allocates memory space for it during runtime. Think of instantiation as the process that assigns specific values to the variables of the instance. In terms of procedural programming, instantiation is similar to writing `int x = 5`. This statement declares the type of the variable *x* as integer and initializes *x* to 5. Similarly, the C++ or the Java statement shown above declares the type of the instance variable `mySpurGear` as `SpurGear`, and the constructor `SpurGear()` initializes all the variables of `mySpurGear` (which come from the `SpurGear` class) to their corresponding values as assigned in the `SpurGear` class.

We can create as many instances of a class as we need. Each instance has its own specific values of the variables of its class. For example, we can repeat the above C++ or Java statement to create other gear instances such as `yourSpurGear`, `machineSpurGear`, and so forth. We can change the values of the variables of an instance after it is created by using the concept of membership access covered in the next section.

5.4.3   Membership Access

Membership access allows an instance to access its class members (variables and methods). Membership access is implemented in OOP languages via the member operator (also known as the dot notation). For example the instance `mySpurGear` can access its variable `rootDiameter` as follows: `mySpurGear.rootDiameter`—which reads as the rootDiameter of mySpurGear. The member operator is the dot. To the left of the dot must come an instance name — mySpurGear. To the right of the dot must come a variable name — rootDiameter. Once we access the instance variables, we can assign it a new value—for example, mySpurGear.rootDiameter = 100.

We can call (access) the methods of a class of an instance by using the member operator in a way similar to the way in which we accessed the variables—for example, mySpurGear.rotate(). Here we call the rotate() method of the SpurGear class of the mySpurGear instance.

OOP programming languages such as C++ and Java provide control over membership access by allowing programmers to declare variables and methods as public or private. Instances can only access the public members of their corresponding classes. Good practice of OOP recommends that we declare all the variables of a class private and provide a pair of public methods for each private variable: one that sets (known as the set method) the variable value to a new one, and one (known as the get method) that gets the current value of the variable.

5.5 Inheritance

Inheritance promotes code modularity and reuse. **Inheritance** allows one object to inherit variables and methods from another object. The class (object) that we inherit from is known as a **superclass** in Java or **base class** in C++, and the class (object) that inherits is a **subclass** in Java or **derived class** in C++. A subclass can define its own variables and methods, above and beyond what it inherits from its superclass(es). Inheritance creates an inheritance tree as shown in Figure 5.2. The classes at the bottom of the tree are more specific than those at the top of the tree.

Java or C++ implements inheritance by allowing one class to use another class in its declaration. Java uses the keyword extends to indicate inheritance, while C++ uses a ": ".

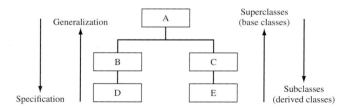

Figure 5.2   Inheritance tree.

5.6 Parametric Design and Objects

The philosophy of parametric modeling and design goes hand-in-hand with OOP. As shown in Figure 2.7, a parametric solid model is defined by its parameters P_1 to P_4, R, and an extrusion thickness T (not shown in the figure). The different values of these six parameters define a family of parts; each part in the family has specific values for the six parameters (variables). A CAD designer creates the solid sketch shown in Figure 2.7. When the designer sets the values of the parameters and regenerates the solid, a new part in the family is created.

We can explain the parametric design as described above in terms of the object concepts we cover in Sections 5.4 and 5.5 as follows: The model as defined by its parameters is the generic object (one size fits all) or the class — call it the Extrude object. The six parameters define the attributes (variables) of the class. In this case, there are no methods for the class. Each part in the family is an instance of the class. Each part has specific values for the six parameters.

The CAD software makes the object concepts transparent to the CAD designer. It hides all the OOP concepts and their implementation away from the designer. If we need to customize and/or program a CAD system, we must face and understand these object concepts to be able to program the system.

The following pseudocode illustrates the definition of the Extrude class:

```
class Extrude {
  //define class variables (object attributes)
  private P1, P2, P3, P4, R, T;
  //define set methods
  public void setP1 (P1Input) {
    P1 = P1Input;
  }
  //repeat the above set method for P2 to P4, R and T
  //define get methods
  public Point getP1() {
    return P1;
  }
  //repeat the above get method for P2 to P4, R and T
} //Extrude
```

Each time a designer assigns specific values to the six parameters of the Extrude class to create a part, the CAD system creates an instance — extrude1, extrude2, and so forth. When the designer deletes the part, the CAD system deletes the instance. However, the Extrude class is still there and can be instantiated to create a new part.

When the CAD designer clicks a parameter to assign it a new value (dimension), the designer is using the object concept of membership access. For example, selecting P_1 of the instance extrude1 is equivalent to calling the extrude1.getP1() method. Changing the current value of P_1 to, for example, 5, and hitting Enter or OK is equivalent to calling extrude1.setP1(5).

5.7 Programming Levels

We divide CAD/CAM programming into two groups. The first group is programming the three-dimensional geometric and graphics concepts we cover in this book. The second group is programming the CAD/CAM systems themselves. Programming the first group is a prerequisite to programming the second group because we need to learn the programming concepts without having to deal with the high level of sophistication of the programming interfaces of CAD/CAM systems and the complex data structure and large size.

We can use one of two approaches to program the first group: programming languages or symbolic equation solvers. The approach of programming languages requires a programming language such as C, C++, or Java, and a rendering system to display the modeling and graphics results in 3D space. Two major rendering systems exist: OpenGL and Java 3D. The OpenGL-based system is used with C or C++. We write a C or C++ program to code the geometric and graphics concepts, and we call functions from the OpenGL library to render the results.

The Java 3D-based system performs a similar task. It uses Java instead of C++ to write the code. It also uses Java 3D instead of OpenGL to render the results. Actually, Java 3D is built on top of OpenGL; thus it provides higher-level functionality to applications than OpenGL. (Another system that is built on top of OpenGL is Hoops3D, which is good for CAD/CAM and scientific visualization.) Java 3D comes in two versions. One version uses OpenGL. We must download and install OpenGL from http://www.opengl.org before we can use Java 3D. The other version uses Windows DirectX (which uses Direct3D) objects that we download from http://www.microsoft.com/directx.

The choice of OpenGL or Java 3D depends on the application at hand. The general practice suggests using C or C++ and OpenGL for non–Web-related developments and for high performance. We tend to use Java and Java 3D for Web-related applications. For example, SDRC I-DEAS uses Java and Java 3D in its Web-based design viewer. The viewer allows users to exchange product designs over the Internet. Java 3D should be a perfect fit for collaborative design and engineering activities that are Web-based inherently.

While Java 3D provides high-level APIs as compared with OpenGL and DirectX to shield programmers from low-level rendering details, it also permits low-level control similar to OpenGL and DirectX. Like VRML, Java 3D allows developers to focus on what to draw (scene composition) and not how to draw it. However, it offers low-level control that VRML does not offer. Figure 5.3 shows the relationship between Java 3D and OpenGL and DirectX.

The approach of symbolic equation solvers requires software packages that allow us to manipulate and solve the equations of geometric modeling and graphics concepts. These symbolic equation solvers include MATLAB, Maple, Mathcad, and Mathematica. The use of symbolic equation solvers is quite common in engineering practice. These solvers are used for teaching and research in various disciplines in academia and industry.

A symbolic equation solver solves differential and integral equations and performs linear algebra, including vector manipulations. It also solves matrix (system of) equations and performs matrix computations such as adding, multiplying, and inverting matrices.

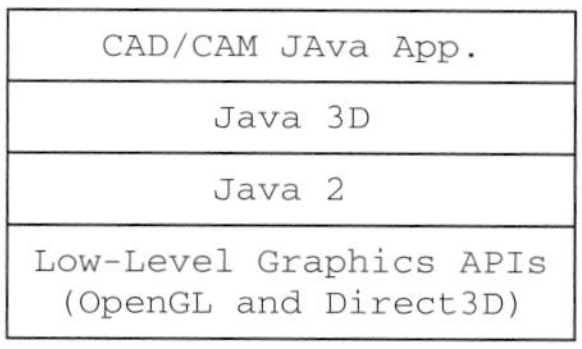

```
CAD/CAM JAva App.
Java 3D
Java 2
Low-Level Graphics APIs
(OpenGL and Direct3D)
```

Figure 5.3   Java 3D Layer.

The use of symbolic equation solvers appeals to those engineers who might not be keen on learning and using programming languages while learning the concepts of geometric modeling and computer graphics. These solvers might also be the right tools to use while teaching CAD/CAM courses (depending on the course philosophy and focus) or performing basic research in the subject. However, we must keep in mind that this approach of using symbolic equation solvers is just a start to the eventual learning and use of OOP concepts and languages that are required for the true programming of CAD/CAM systems, and that are part of today's computing.

The second set of CAD/CAM programming tasks involves programming CAD/CAM systems themselves. This group builds on what we learned about programming the geometry and graphics. But the focus here is on learning the syntax and the use of the system APIs. This adds one level of complexity to the programming environment. However, the knowledge we gained with the first group helps with this group.

CAD/CAM systems provide two types of programming tools: macros and APIs. We cover macros in this book. We cover Java 3D instead of CAD/CAM APIs to avoid becoming too involved with a particular CAD/CAM system syntax and programming environment.

EXAMPLE 5.2   **Use symbolic equation solvers.**
Create and display a line and a circle using an equation solver.

SOLUTION   The formulation of geometric modeling, as used in CAD/CAM systems, is based on the use of parametric equations for curves, surfaces, and solids as we show in Part II of the book. For example, the parametric equation of a line defined by two end points is given by:

$$\mathbf{P}(u) = \mathbf{P}_1 + u(\mathbf{P}_2 - \mathbf{P}_1) \qquad 0 \le u \le 1$$

where P [defined by the vector $\mathbf{P}(u)$] is any point on the line, P_1 and P_2 are the two end points of the line, and u is the parameter.

Similarly the parametric equation of a circle with center at point P_c and radius R, and a parameter u is give by:

$$x(u) = x_c + R\cos(u)$$
$$y(y) = y_c + R\sin(u) \qquad 0 \le u \le 2\pi$$
$$z(u) = z_c$$

We use Maple to enter these two equations in their symbolic forms and plot them within Maple. The screenshots below show the symbolic forms, in Maple format, and the two Maple plots.

```
> restart;
  with (linalg):
  with (plots):
Warning, the protected names norm and trace have been redefined and unprotected

Warning, the name changecoords has been redefined

> P(U):= P(1) + u*(P(2)-P(1));
```

$$P(U) := P(1) + u\,(P(2) - P(1))$$

```
> P(1) := Matrix([[1],[2],[0]]);
  P(2) := Matrix([[7],[8],[0]]);
```

$$P(1) := \begin{bmatrix} 1 \\ 2 \\ 0 \end{bmatrix}$$

$$P(2) := \begin{bmatrix} 7 \\ 8 \\ 0 \end{bmatrix}$$

```
> P(U) := simplify(eval(P(U)));
```

$$P(U) := \begin{bmatrix} 1 + 6\,u \\ 2 + 6\,u \\ 0 \end{bmatrix}$$

```
> x(U):=P(U)[1,1]; y(U):=P(U)[2,1]; z(U):=P(U)[3,1];
```

$$x(U) := 1 + 6\,u$$
$$y(U) := 2 + 6\,u$$
$$z(U) := 0$$

```
> plot3d([x(U),y(U),z(U)], u=0..1,v=0..1, axes=NORMAL, orientation=[-90,0]);
```

```
> restart;
  with (linalg):
  with (plots):
Warning, the protected names norm and trace have been redefined and unprotected

Warning, the name changecoords has been redefined

> x(U) := x(c) + R*cos(u);
  y(U) := y(c) + R*sin(u);
  z(U) := z(c);
  x(U) := x(c) + R cos(u)
```

$$x(U) := x(c) + R\cos(u)$$
$$y(U) := y(c) + R\sin(u)$$
$$z(U) := z(c)$$

```
> x(c):= 0;y(c):= 0;z(c):= 0;
  R:= 4;
>
>
```

$$x(c) := 0$$
$$y(c) := 0$$
$$z(c) := 0$$
$$R := 4$$

```
> x(U) := eval(x(U)); y(U):= eval(y(U)); z(U):= eval(z(U));
```

$$x(U) := 4\cos(u)$$
$$y(U) := 4\sin(u)$$
$$z(U) := 0$$

```
> plot3d([x(U),y(U),z(U)], u=0..2*Pi,v=0..1, axes=NORMAL, orientation=[-90,0]);
```

Example 5.2 discussion:

We use Maple to program the line and circle parametric equations in a symbolic form. The two end points of the line are $P_1(1, 2, 0)$ and $P_2(7, 8, 0)$. The center point of the circle is $P_c(0, 0, 0)$ and its radius R is 4. The units of these dimensions (inches or otherwise) are implied in the user input.

As shown in the screenshot, Maple takes the user input and displays it in a vector and matrix form, ready for symbolic manipulation. When the user requests plots of the symbolic equations, Maple evaluates the symbolic equation numerically by using an increment Δu of u. It uses Δu to generate points on the curve and then connects them with line segments. Maple computes Δu by using the u range of each parametric equation to ensure a smooth plot of the curve.

Example 5.2 hands-on exercise:

Use your own symbolic equation solver and redo this example. Also, use the ranges of $0 \le u \le \frac{\pi}{2}$, $0 \le u \le \pi$, and $0 \le u \le \frac{3\pi}{2}$ for the circle to generate arcs with different lengths.

5.8 Macros

A **macro** is a simple string of commands that records a sequence of actions that a designer performs from the time the designer turns on the macro command on a CAD system until it is turned off. The macro is saved in a file that can be played back to execute the sequence of commands. Macros are good to use to automate repetitive design tasks. Some macros can be edited to change parameters of a feature, thus enabling the designer to create a family of parts.

Different CAD systems use different languages to create macro files. For example, SolidWorks uses VB (Visual Basic) to record macros. We usually perform three activities: create, run, and edit. We first create a macro, and then run it. If necessary, we edit the macro file to make changes, save the file and run it again. Think of a macro as a simple computer program. Moreover, CAD systems allow their users to create a button to access the macro.

EXAMPLE 5.3   **Create, run, and edit macros.**
Show an example of creating, running, and editing a macro.

SOLUTION   In this example, we use a CAD system in a pseudo fashion to illustrate dealing with a macro. Follow these steps:

1. **Create macro.** Turn on the macro command to begin recording. For example, click this sequence, or its equivalent: `Tools` (menu on menu bar) `=> Macro => Record`.

2. **Perform activities.** Begin geometric modeling as if macro were not on. For example, select sketch planes, draw a sketch, and extrude it to create a feature.

3. **Stop macro and save it.** Stop the macro recording by turning the macro off. Click this sequence, or its equivalent: `Tools` (menu on menu bar) => `Macro` => `Stop`. A window opens up asking for a macro filename and a directory to save the macro in, as shown in the screenshot on the right.

4. **Exit the part without saving.** This way we can run the macro to test it.

5. **Run the macro.** Open a new part file. Then run the macro. The macro should execute its sequence of commands and create the extruded feature in the open part file. For example, click this sequence, or its equivalent: `Tools` (menu on menu bar) => `Macro` => `Run`. A dialog box opens to allow the selection of the macro, in case more than one exists, as shown in the screenshot on the right.

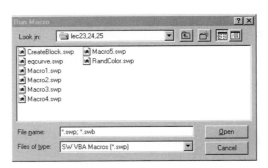

6. **Assign a toolbar button to a macro.** If needed, follow the system directions to add a macro button to the CAD system toolbar. For example, click this sequence, or its equivalent: `Tools` (menu on menu bar) => `Customize` => `Macros` (tab). Select the desired macro from the popup window shown in the screenshot at right. After the procedure is complete, the macro is added to the toolbar as shown on the right.

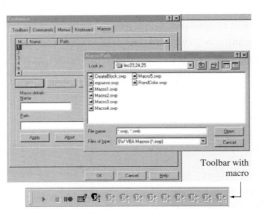

Toolbar with macro

7. **Edit macro.** If necessary, a macro can be edited if we know the language it is written in. We open its file, edit it, save it, and run it again. For example, click this sequence, or its equivalent: `Tools` (menu on menu bar) => `Macro` => `Edit`. After you select the macro to edit, its file opens up in an editor for editing. For example, SolidWorks uses Microsoft Visual Basic editor, as shown in the forthcoming screenshot. Recognize the VB statements and the SolidWorks API calls. Other systems use system editors. For example, Pro/E uses the `vi` editor on Unix systems. Users who know the APIs well can produce the macro files by writing the code manually.

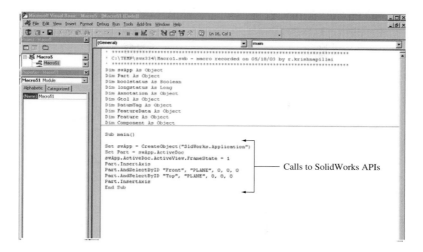

Calls to SolidWorks APIs

Example 5.3 discussion:

This example shows how to deal with macros in a system-independent fashion. It illustrates the macro concepts through some SolidWorks commands and screenshots. As the example shows, dealing with a macro on a CAD system is an easy task to accomplish.

Example 5.3 hands-on exercise:

Apply the preceding seven macro steps to your CAD system to record the creation of an extruded feature using a rectangular cross section. Change the macro to accept input from the user for the rectangle's width and height, as well as the extrusion depth. For example, SolidWorks lets users use VB forms to receive input from them.

5.9 Programming the CAD and CAM Interface

The interface between CAD and CAM includes how CAD communicates with CAM on one end, and how CAM communicates with CNC (computer numerical control) machine tools on the other end. Programming at the CAD and CAM interface offers interesting challenges. The interface usually requires user intervention. For example, the user of a CAD/CAM system may have to write a sequence to automate the setup procedure of a workpiece before machining.

Programming the CAD/CAM interface lends itself to using macros. Users can write macros to automate different machining setup procedures. In some cases, they can use their CAD/CAM systems themselves, using a macro function to generate the program. In other cases, they may have to generate files from the CAD/CAM system, read them into their own programs to process them, and then generate the setup procedure. In such cases, they can use a programming language such as VB (Visual Basic), C, C++, or Java.

One example of programming the CAD/CAM interface is CNC machining. A user can generate a toolpath file from a CAD/CAM system that may require conversion into the native language of the CNC controller for the machine tool, or the file may need editing to add some specific controller setups at the beginning.

5.10 Java 3D

Figure 5.3 shows that we need to download and install the following components before we can use and run Java 3D. The installation of these components is simple because all files are self-extracting.

1. **Download and install OpenGL or DirectX.** Download either OpenGL or DirectX, depending on the Java 3D version to be used, as we discuss in Section 5.7. We recommend using DirectX over OpenGL because DirectX drivers are usually faster than OpenGL drivers, and DirectX is supported on a wider range of hardware. Download DirectX from http://www.microsoft.com/directx, for Windows platforms.

2. **Download and install Java 2 platform, Standard Edition (J2SE).** Java 3D requires J2SE version 1.2 or higher. Download from http://java.sun.com.

3. **Download and install Java 3D.** Download from http://java.sun.com.

After installation is complete, run the demo by typing the following two commands in an OS window (DOS for Windows or a Terminal for Unix):

```
>cd \j2sdk1.4.2\deom\java3d\HelloUniverse
>java HelloUniverse
```

The first line sets the path to the demo file. The second line runs the `HelloUniverse` demo. The demo displays a window with a rotating colored cube in it.

To use Java 3D, we write a Java application (program) and run it. Visit http://www.j3d.org to jump-start learning Java 3D. However, we need to understand Java 3D graphics scenes before we can write applications. Figure 5.4 shows the tree structure of a scene graph. Java 3D uses scene graph structure. That means that a 3D display scene consists of a *virtual universe* that has objects that are related to each other. Objects are attached to the universe via a *Locale* which provides the coordinate system for attaching the objects to the universe. A unit of length in *Locale* is equivalent to 1 meter in the real world.

Subgraphs or scene graph branches exist under a *Locale* as shown in Figure 5.4. Each subgraph contains objects such as behaviors, transforms, view objects, and 3D shape objects. We add objects to the scene and then render them to see displays on the screen. Typically, each Java 3D program has two branch groups. One group (node) contains the 3D content that makes up the scene, such as shapes, textures, and lights. The other branch is used to render and view the scene (first) branch. Viewing include activities such as zooming, panning, and so forth.

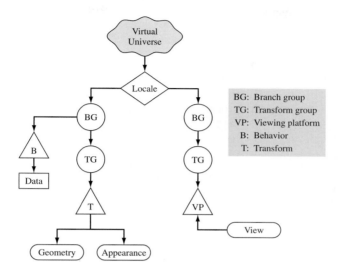

Figure 5.4   Java 3D scene graph.

Follow these steps to create and render geometry in Java 3D:

1. **Create a canvas.** Use this Java method:
```
void init {
  Canvas3D c = new Canvas3D(config); //Config is a graphics object
  Add ("North", c); //assuming a Border layout
   ...
}
```
2. **Create a simple universe.** Use this Java code:
```
private SimpleUniverse u = null;
 ...
u = new SimpleUniverse(c);
```
3. **Create a BG and connect it to the `SimpleUniverse`:**

   a. Create the BG root.
```
BranchGroup objRoot = new BranchGroup();
```
   b. Create the TG node, and enable write capability.
```
TransformGroup objTrans = new TransformGroup();
```

```
        ObjTrans.setCapability(TransformGroup.ALLOW_Transform
_Write);
```

c. Create a 3D object such as a cube and add it to the scene graph.

```
        ObjTrans.addChild(new ColorCube(0.5));
```

d. Set up the behavior.

```
        Transform3D y = new Transform3D();
```

e. Compile.

```
        objRoot.Compile();
```

f. Create behavior such as rotations; then add to `TransformGroup`.

```
        objTrans.addChild(behavior);
```

EXAMPLE 5.4	**Create lines and points using Java 3D.**

Write a Java application that displays lines and points on the screen.

SOLUTION	Using a text editor, type the code shown below, and save it as

J3DLinesPoints.java.

```
 1:   /*  J3DLinesPoints.java  */                           J3DLinesPoints.java
 2:
 3:   import java.applet.Applet;
 4:   import java.awt.*;
 5:   import com.sun.j3d.utils.applet.MainFrame;
 6:   import com.sun.j3d.utils.universe.*;
 7:   import com.sun.j3d.utils.geometry.*;
 8:   import javax.media.j3d.*;    //for BranchGroup
 9:   import javax.vecmath.*;      //for vertex data
10:
11:   public class J3DLinesPoints extends Applet  {
12:       private SimpleUniverse u = null;
13:
14:       public BranchGroup createScene() {
15:          BranchGroup objRoot = new BranchGroup();
16:           TransformGroup tg = new TransformGroup();
17:          Shape3D shapeLine = new Shape3D();
18:        Shape3D shapePoint = new Shape3D();
19:           Transform3D t3d2 = new Transform3D();
20:
21:           PointArray pa = new PointArray(2,
        PointArray.COORDINATES|PointArray.COLOR_3);
22:           LineArray la = new LineArray(2,
        LineArray.COORDINATES|LineArray.COLOR_3);
23:           Point3d[] ptsPoints = new Point3d[2];
24:        Point3d[] ptsLine = new Point3d[2];
25:           PointAttributes ptAtt = new PointAttributes();
26:           Color3f[] colors = new Color3f[2];
27:           Appearance appPoint = new Appearance();
28:
```

```
29:                  t3d2.setScale(0.2);
30:                  tg.setTransform(t3d2);
31:                  objRoot.addChild(tg);
32:
33:                  ptsPoints[0] = new Point3d(2, 3, -2);
34:                  ptsPoints[1] = new Point3d(2, 4, -2);
35:                  ptsLine[0] = new Point3d(1, 1, -2);
36:                  ptsLine[1] = new Point3d(-2, 2, -2);
37:
38:                  colors[0] = new Color3f(1.0f, 0.0f, 0.0f);   //red
39:                  colors[1] = new Color3f(0.0f, 0.0f, 1.0f);   //blue
40:
41:          pa.setCoordinates(0, ptsPoints);   //set the coordinates
         of the points into the PointArray
42:              pa.setColors(0, colors);
43:          ptAtt.setPointSize(5.0f);   //increase size of points to
         make them more visible
44:          appPoint.setPointAttributes(ptAtt);
45:          shapePoint.setGeometry(pa);
46:          shapePoint.setAppearance(appPoint);
47:
48:          la.setCoordinates(0, ptsLine);   //set the coordinates
         of the line points into the LineArray
49:              la.setColors(0, colors);
50:              shapeLine.setGeometry(la);
51:
52:              tg.addChild(shapePoint);
53:              tg.addChild(shapeLine);
54:              objRoot.compile();
55:              return objRoot;
56:          }
57:
58:      public void init() {
59:              setLayout(new BorderLayout());
60:              GraphicsConfiguration config =
         SimpleUniverse.getPreferredConfiguration();
61:              Canvas3D canvas = new Canvas3D(config);
62:      canvas.setBackground(Color.yellow);
63:              add("Center", canvas);
64:              BranchGroup scene = createScene();
65:              u = new SimpleUniverse(canvas);
66:              u.getViewingPlatform().setNominalViewingTransform();
67:              u.addBranchGraph(scene);
68:          }
69:
70:      public static void main(String[] args)  {
71:          new MainFrame(new J3DLinesPoints(), 500, 400);
72:          }
73:      }
```

Example 5.4 discussion:

This example uses a Java applet, code line 11, to create and display points and lines using Java 3D. Code lines 3–9 import the required Java classes. Code line 12 creates u as a `SimpleUniverse`. The `createScene()` method, code lines 14–56, follows the structure shown in Figure 5.4 to create the scene and render it. We declare two points, code line 21, two lines, code line 22, and two colors, code line 26. We create the two points in code lines 33 and 34. We create the points for the lines in code lines 35 and 36.

The `init()` method, code lines 58–72, is an applet method that initializes the applet. It creates a graphics configuration in code line 60. It also creates a canvas, to draw on, in code line 61. It attaches the canvas to the `SimpleUniverse` in code line 66. It finally attaches the scene to the `SimpleUniverse`, u, in code line 67; so we can see it on the screen. The Java `main()` method, code lines 70–72, creates an instance of the applet, in code line 71, and runs the application. The following screenshot shows the results.

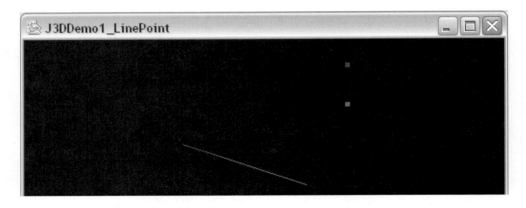

Example 5.4 hands-on exercise:

Starting with the foregoing code, add one more 3D point and one more line.

5.11 Tutorials

5.11.1 Create Primitives Using Java 3D

This tutorial shows how to use Java 3D to create geometric primitives such as box, cylinder, and sphere. Java 3D provides them as built-in primitives. We use them by assigning them sizes and orientation. This resembles CAD/CAM systems that support primitives such as Unigraphics and CATIA.

Using a text editor, type the following code, and save it as *J3DCube.java*.

```
1:    /* J3DCube.java */                                    J3DCube.java
2:
3:    import java.applet.Applet;
4:    import java.awt.*;
5:    import com.sun.j3d.utils.applet.MainFrame;
6:    import com.sun.j3d.utils.universe.*;
7:    import com.sun.j3d.utils.geometry.*;
8:    import javax.media.j3d.*;      //for BranchGroup
9:    import javax.vecmath.*;        //for vertex data
10:
11:   public class J3DCube extends Applet {
12:
13:       private SimpleUniverse u = null;
14:
15:       public BranchGroup createScene()   {
16:          BranchGroup objRoot = new BranchGroup();
17:          TransformGroup tg = new TransformGroup();
18:          Appearance app = new Appearance();
19:       //TransformGroup, used for box
20:           Transform3D t3d = new Transform3D();
21:       t3d.rotX(Math.toRadians(-45));
22:       t3d.setScale(0.1);
23:           tg.setTransform(t3d);
24:           objRoot.addChild(tg);
25:
26:          Box box = new Box(1.0f, 0.5f, 1.0f,
      Box.GENERATE_NORMALS, app);
27:
28:       Material mat = new Material();
29:
30:       mat.setLightingEnable(true);
31:           app.setMaterial(mat);
32:
33:           DirectionalLight lightD = new DirectionalLight();
34:           lightD.setInfluencingBounds(new BoundingSphere(new
      Point3d(-3.0,-1.0, 0), 35.1));
35:           lightD.setDirection(new Vector3f(1f,-1.0f,-5f));
36:           lightD.setColor(new Color3f(1.0f, 1.0f, 1.0f));
37:
38:           objRoot.addChild(lightD);
39:           tg.addChild(box);
40:
41:           objRoot.compile();
42:           return objRoot;
43:       }
44:
45:       public void init() {
46:           setLayout(new BorderLayout());
```

```
47:              GraphicsConfiguration config =
         SimpleUniverse.getPreferredConfiguration();
48:              Canvas3D canvas = new Canvas3D(config);
49:              add("Center", canvas);
50:              BranchGroup scene = createScene();
51:              u = new SimpleUniverse(canvas);
52:              u.getViewingPlatform().setNominalViewingTransform();
53:              u.addBranchGraph(scene);
54:          }
55:
56:          public void destroy() {
57:          u.cleanup();
58:          }
59:
60:          public static void main(String[] args) {
61:              new MainFrame(new J3DCube(), 500, 400);
62:          }
63:      }
```

Tutorial 5.11.1 discussion:

The setup of the Java 3D environment of this tutorial is the same as that of Example 5.4. This tutorial creates a box, in code line 26, that has a size of $1 \times 0.5 \times 1$. We assign a material to the box in code lines 28–31. We apply lighting to the scene in code lines 33–38.

Tutorial 5.11.1 hands-on exercise:

Starting with the preceding code, add more primitives: cylinder, sphere, and a cone.

Problems

Part I: Theory

5.1   Use Microsoft Access, or another database program, to extend the relational database shown in Figure 5.1 to include a square hole in the CAD model (block) shown in the figure.

5.2   Define a `HelicalGear` object that inherits from the `SpurGear` object of Example 5.1.

5.3   What are the different levels of CAD/CAM programming?

5.4   Describe the structure that Java 3D uses to render geometry and graphics.

5.5   Macros are used as recording tools, or as parametrization tools that create families of parts. Explain each option and provide examples.

Part II: Laboratory

Use your in-house CAD/CAM system to answer the questions in this part.

5.6   Write a macro that changes the color of a solid.

5.7   Write a macro that parametrizes the creation of a block.

5.8   Write a macro that creates the curve whose equation is:
$$y = 2x^3 + 5x^2 - 7x - 5 \qquad 0 \le x \le 2$$

Part III: Programming

5.9   Write a Java or C++ code to implement the `SpurGear` object definition of Example 5.1.

5.10   Write a Java or C++ code to create the `HelicalGear` object of Problem 5.2.

5.11   Write a Java application that uses Java 3D to create a cube that has a side of 1 unit of length.

Geometric Modeling

This part covers all the geometric modeling concepts that are behind commercial CAD/CAM systems. This part should help all users of these systems whether they are CAD designers (use CAD systems) or CAD programmers (program CAD systems). To achieve this goal, we must achieve the following objectives:

1. Understand curve theory **(Chapter 6)**.
2. Understand surface theory **(Chapter 7)**.
3. Understand NURBS theory **(Chapter 8)**.
4. Understand solids theory **(Chapter 9)**.
5. Understand features and parametrics **(Chapter 10)**.

Curves

GOAL

Understand and master the theory and practice of curves, their types, parametric formulation, vector analysis, their implementation by CAD/CAM systems, and their use in geometric modeling.

OBJECTIVES

After reading this chapter, you should understand the following concepts:

- Geometric modeling
- Modeling entities
- Curve implicit eqautions
- Curve parametric equations
- Curve properties
- Analytic curves
- Synthetic curves
- Curve manipulations

CHAPTER HEADLINES

6.1   Introduction

CAD tools have been defined in Chapter 1 as the melting pot of three disciplines: design, geometric modeling, and computer graphics. While the latter discipline is covered in Part III, this part discusses geometric modeling and its relevance to CAD/CAM. A geometric model should be an unambiguous representation of its corresponding object. That is to say that the model should be unique and complete to all engineering functions, from documentation to engineering analysis to manufacturing.

To convey the importance of geometric modeling to the CAD/CAM process, one may refer to other engineering disciplines and make the following analogy. Geometric modeling to CAD/CAM is as important as governing equilibrium equations to classical engineering fields such as mechanics and thermalfluids. From an engineering point of view, modeling of objects is by itself unimportant. Rather, it is a means (tool) to enable useful engineering analysis and judgment. Actually, the amount of time and effort a designer spends to create a geometric model cannot be justified unless the resulting database is utilized by engineering applications.

There are many reasons to study the mathematical basis of geometric modeling. From a strictly modeling point of view, it provides a good understanding of terminology encountered in the CAD/CAM field as well as in CAD/CAM systems. It also enables users to decide intelligently which type of entity to use in a particular model to meet geometric requirements such as slopes and/or curvatures. In addition, users become able to interpret any unexpected results they may encounter from using a particular CAD/CAM system. Moreover, those who are involved in the decision-making process and evaluations of CAD/CAM systems become equipped with better evaluation criteria.

While the ultimate goal of geometric modeling as we practice it on CAD/CAM systems is to create solid models, the construction of these models begins with points, lines, and curves regardless of the complexity of the models. Complex models, such as bottles and hair dryers, require surfaces. We extend curves to create surfaces, as we discuss in Chapter 7. We then use surfaces to create solids. Thus, CAD designers must master curve, surface, and solid concepts before they can truly create complex solid models and be productive in using CAD systems.

6.2   Curve Entities

All existing CAD/CAM systems provide users with curve entities, which can be divided into analytic and synthetic entities. Analytic entities are points, lines, arcs and circles, fillets and chambers, and conics (ellipses, parabolas, and hyperbolas). Synthetic entities include various types of spline (cubic spline and B-spline) and Bezier curves. The mathematical properties of each entity and how it is used in geometric modeling are covered in the remainder of the chapter.

Tables 6.1 to 6.5 show the most common methods utilized by CAD/CAM systems to create curve entities. You should compare and/or modify these methods to your system and its user interface. You should also also find commands that correspond to these tables on your own CAD/CAM system.

Table 6.1   Methods of defining points.

Explicit methods	Implicit methods
1. Absolute cartesian coordinates	**A click d (with or without active grid)**

Y ↑ $\bullet P\,(x, y, z)$

                                        $\bullet\ d$

X

Z

Coordinates of resulting point can be obtained by using a `Verify` command. Coordinates are measured relative to the MCS or WCS.

---

2. Absolute cylindrical coordinates   **End points of an existing entity**

$\bullet P\,(R, \theta, z)$

R

θ

(Cylindrical coordinates are seldom used in practice)

(Cylindrical coordinates are seldom used in practice)

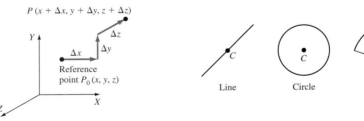

Line   Circle   Arc   Curve

---

3. Incremental cartesian coordinates   **Center point (origin) of an existing entity**

$P\,(x + \Delta x, y + \Delta y, z + \Delta z)$

Y ↑ Δz

        Δx Δy

Reference
point $P_0\,(x, y, z)$

X

Z

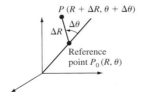

Line   Circle   Arc

---

4. Incremental cylindrical coordinates   **Intersection point of two existing entities**

$P\,(R + \Delta R, \theta + \Delta \theta)$

ΔR $\Delta \theta$

Reference
point $P_0\,(R, \theta)$

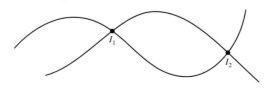

Table 6.2   Methods of defining lines.

Method	Illustration
1. Points	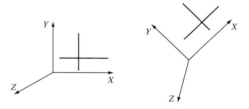
2. Horizontal (parallel to the _X_ axis of current WCS) or vertical (parallel to the _Y_ axis of current WCS)	
3. Parallel or perpendicular to an existing line	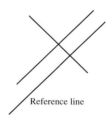

Reference line |
| **4. Tangent to existing entities** |

One of the four possibilities is obtained, depending on locations of user clicks

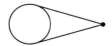

One of the two possibilities is obtained |

Table 6.3   Methods of defining circles.

Method	Illustration
1. Radius or diameter and center. In the case of an arc, beginning and ending angles are required.	
2. Three points	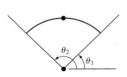
3. Center and a point on the circle	
4. Tangent to a line, pass through a given point, and with a given radius.	

Table 6.4   Methods of defining ellipses and parabolas.

Method	Illustration

1. Ellipses

a. Center and axes lengths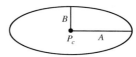

b. Four points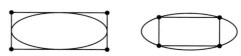

c. Two conjugate diameters

2. Parabolas

a. Vertex and focus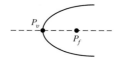

b. Three points

Table 6.5   Methods of defining synthetic curves.

Method	Illustration

1. A cubic spline
    A given set of data points

2. Bezier curve
    A given set of data points

3. B-spline curves: curve interpolates
(passes through) data points

4. B-spline curves: curve extrapolates
(does not pass through) data points

6.3 Curve Representation

Geometric description of curves defining an object can be tackled in several ways. A curve can be described by arrays of coordinate data or by an analytic equation. The coordinate array method is impractical for obvious reasons. The storage required can be excessively large, and the computation to transform the data from one form to another is cumbersome. In addition, the exact shape of the curve is not known, thus impairing exact computations such as the intersections of curves and the physical properties of objects—for example volume calculations. From a design point of view, it becomes difficult to redesign shapes of existing objects via the coordinate array method. Analytic equations of curves provide designers with information such as the effect of data points on curve behavior, control, continuity, and curvature.

The treatment of curves in computer graphics and CAD/CAM is different than that in analytic geometry or approximation theory. Curves describing engineering objects are generally smooth and well-behaved. In addition, not every available form of a curve equation is efficient to use in CAD/CAM software due to either computation or programming problems. For example, a curve equation that results in a division by zero while calculating the curve slope causes overflow and errors in calculations. Similarly, if the intersection of two curves is to be found by solving their two equations numerically, the forms of the two equations may be inadequate to program due to the known problems with numerical solutions. In addition, considering that most design data for objects are available in a discrete form, mainly key points, the curve equation should be able to accept points and/or tangent values as input from the designer.

Curves can be described mathematically by nonparametric or parametric equations. Nonparametric equations can be explicit or implicit. Explicit nonparametric representation of a general 3D curve takes the form:

$$\mathbf{P} = \begin{bmatrix} x & y & z \end{bmatrix}^T = \begin{bmatrix} x & f(x) & g(x) \end{bmatrix}^T \tag{6.1}$$

where $\mathbf{P}$ is the position vector of point P as shown in Figure 6.1. Equation (6.1) is a one-to-one relationship. Thus, this form cannot be used to represent closed curves, such as circles, or multivalued curves, such as parabolas. The implicit nonparametric representation can solve this problem and is given by the intersection of two surfaces as:

$$\begin{aligned} F(x, y, z) &= 0 \\ G(x, y, z) &= 0 \end{aligned} \tag{6.2}$$

However, Eq. (6.6.2) must be solved to find its roots (y and z values) if a certain value of x is given. This may be inconvenient and lengthy. Nonparametric representations of curves have other limitations such as:

1. If the slope of a curve at a point is vertical or near vertical, its value becomes infinity or very large, a difficult condition to deal with both computationally and programming-wise. Other ill-defined mathematical conditions may result.
2. Shapes of most engineering objects are intrinsically independent of any coordinate system. What determines the shape of an object is the relationship between its data points and not the relationship between these points and some arbitrary coordinate system.
3. If the curve is to be displayed as a series of points or straight-line segments, the computations involved could be extensive.

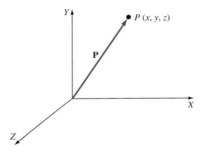

Figure 6.1   Position vector of point P.

Parametric representation of curves overcomes all of these difficulties. It allows closed and multiple-valued functions to be easily defined and replaces slopes with tangent vectors. And in the case of commonly used curves such as conics and cubics, these equations are polynomials rather than equations involving roots. Hence, the parametric form is not only more general, but it is also well suited for computations and display. In addition, this form has properties which are attractive to CAD and interactive environments.

In parametric form, each point on a curve is expressed as a function of a parameter u. The parameter acts as a local coordinate for points on the curve. The parametric equation for a 3D curve in space takes the following vector form:

$$\mathbf{P}(u) = \begin{bmatrix} x & y & z \end{bmatrix}^T = \begin{bmatrix} x(u) & y(u) & z(u) \end{bmatrix}^T, \qquad u_{min} \leq u \leq u_{max} \tag{6.3}$$

Equation (6.3) implies that the coordinates of a point on the curve are the components of its position vector. It is a one-to-one mapping from the parametric space (Euclidean space E^1 in u values) to the cartesian space (E^3 in x, y, z values) as shown in Figure 6.2. The parametric curve is bounded by two parametric values u_{min} and u_{max}. It is, however, convenient to normalize the parametric variable u to have the limits 0 and 1. The positive sense on the curve is the sense in which u increases (Figure 6.2).

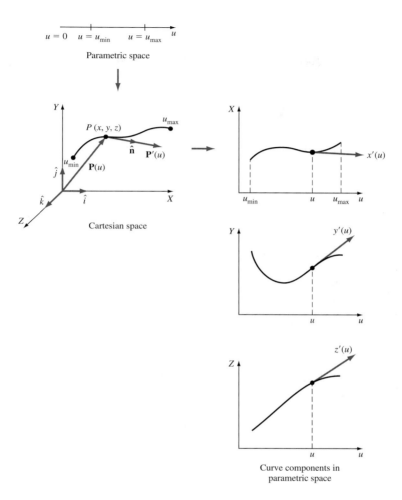

Figure 6.2 Parametric representation of a 3D curve.

The parametric form as given by Eq. (6.3) facilitates many of the useful related computations in geometric modeling. To check whether a given point lies on the curve or not reduces to finding the corresponding u value and checking whether that value lies in the stated u range. Points on the curve can be computed by substituting the proper parametric values into Eq. (6.3). Geometrical transformations can be performed directly on parametric equations. Curves defined by Eq. (6.3) are inherently bounded and, therefore, no additional geometric data is needed to define boundaries.

To evaluate the slope of a parametric curve at an arbitrary point on it, the concept of the tangent vector must be introduced. As shown in Figure 6.2, the tangent vector is defined as vector $\mathbf{P}'(u)$ in the cartesian space such that:

$$\mathbf{P}'(u) = \frac{d\mathbf{P}(u)}{du} \tag{6.4}$$

Substituting Eq. (6.3) into (6.4) yields the components of the tangent vector in the parametric space as

$$\mathbf{P}'(u) = \begin{bmatrix} x' & y' & z' \end{bmatrix}^T = \begin{bmatrix} x'(u) & y'(u) & z'(u) \end{bmatrix}^T, \qquad u_{min} \le u \le u_{max} \tag{6.5}$$

where $x'(u)$, $y'(u)$, and $z'(u)$ are the first parametric derivatives (with respect to u) of the position vector components $x(u)$, $y(u)$, and $z(u)$, respectively. The slopes of the curve are given by the ratios of the components of the tangent vector:

$$\frac{dy}{dx} = \frac{dy/du}{dx/du} = \frac{y'}{x'} \tag{6.6}$$

$$\frac{dz}{dy} = \frac{z'}{y'} \qquad \text{and} \qquad \frac{dx}{dz} = \frac{x'}{z'}$$

The tangent vector has the same direction as the tangent to the curve, hence the name tangent vector. The magnitude of the vector is given by:

$$|\mathbf{P}'(u)| = \sqrt{x'^2 + y'^2 + z'^2} \tag{6.7}$$

and the direction cosines of the vector are given by:

$$\hat{\mathbf{n}} = \frac{\mathbf{P}'(u)}{|\mathbf{P}'(u)|} = n_x \hat{\mathbf{i}} + n_y \hat{\mathbf{j}} + n_z \hat{\mathbf{k}} \tag{6.8}$$

where $\hat{\mathbf{n}}$ is the unit vector (Figure 6.2) with cartesian space components n_x, n_y, and n_z. The magnitudes of the tangent vector at the two ends of a curve affects its shape and can be used to control it as will be seen later.

There are two categories of curves that can be represented parametrically: analytic and synthetic. **Analytic curves** are defined as those that can described by analytic equations such as lines, circles, and conics. **Synthetic curves** are the ones that are described by sets of data points (control points) such as splines and Bezier curves. Parametric polynomials usually fit the control points. While analytic curves provide a very compact form in which to represent shapes and simplify the computation of related properties such as areas and volumes, they are not attractive to deal with interactively. Synthetic curves provide designers with greater flexibility and control of a curve shape by changing the positions of the control points.

EXAMPLE 6.1   **Convert an implicit equation.**

The nonparametric implicit equation of a circle with a center at the origin and radius R is given by $x^2 + y^2 = R^2$. Find the circle parametric equation. Using the resulting equation, find the slopes at the angles 0, 45, and 90 degrees.

SOLUTION   In general, the parametric equation of a curve is not unique and can take various forms. For a circle, let

$$x = R\cos 2\pi u , \qquad\qquad 0 \le u \le 1$$

Substituting into the circle equation gives $y = R\sin 2\pi u$, and the parametric equation of the circle becomes:

$$\mathbf{P}(u) = \begin{bmatrix} R\cos 2\pi u & R\sin 2\pi u \end{bmatrix}^T, \qquad\qquad 0 \le u \le 1$$

and the tangent vector is

$$\mathbf{P}'(u) = \begin{bmatrix} -2\pi R\sin 2\pi u & 2\pi R\cos 2\pi u \end{bmatrix}^T, \qquad\qquad 0 \le u \le 1$$

The parameter u takes the values 0, 0.125, and 0.25 at the angles 0, 45, and 90 degrees, respectively, and the corresponding tangent vectors are given by:

$$\mathbf{P}'(0) = \begin{bmatrix} 0 & 2\pi R \end{bmatrix}^T$$

$$\mathbf{P}'(0.125) = \begin{bmatrix} -2\pi R/\sqrt{2} & 2\pi R/(\sqrt{2}) \end{bmatrix}^T$$

$$\mathbf{P}'(0.25) = \begin{bmatrix} -2\pi R & 0 \end{bmatrix}^T$$

For a 2D curve, the slope is y'/x'. The slopes at the given angles are calculated as ∞, -1, 0.

Example 6.1 discussion:

Finding parametric equations from an implicit equation is a trial-and-error process. Select a parametric equation for x and find the equation for y. It also becomes harder to find equations of 3D curves, where we need to find the parametric equations for x, y, and z coordinates of a point on the curve.

Example 6.1 hands-on exercise:

Change the parametrization of the circle so that the limits on u become $0 \le u \le 2\pi$.

EXAMPLE 6.2   **Convert a parametric equation.**

The parametric equation of the helix shown in Figure 6.3 with a radius a and pitch b is given by

$$\mathbf{P}(u) = \begin{bmatrix} a\cos 2\pi u & a\sin 2\pi u & 2b\pi u \end{bmatrix}^T, \qquad\qquad 0 \le u \le 1$$

Find the implicit equation of the helix.

SOLUTION   The preceding helix equation gives

$$x = a\cos 2\pi u, \qquad y = a\sin 2\pi u, \qquad z = 2b\pi u$$

Solving the first equation for u and substituting in the other two, we obtain the following nonparametric equation:

$$P = \begin{bmatrix} x & a\sin(\cos^{-1}(x/a)) & b\cos^{-1}(x/a) \end{bmatrix}^T$$

Solving the z component for u and substituting in the x and y equations gives the following implicit equation:

$$x - a\cos(z/b) = 0$$
$$y - a\sin(z/b) = 0$$

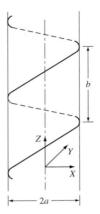

Figure 6.3   A helix curve.

6.4   Analytic Curves

The following sections cover the basics of the parametric equations of analytic curves that are most widely utilized in geometric modeling. These developments are related, whenever possible, to the common practice encountered on CAD/CAM systems. This should enable users to fully realize the input parameters they deal with. It should also help developers who may be interested in writing their own CAD software.

Parametric equations and their development are presented in vector forms. The benefits of vector forms include developing a unified approach and consistent notation to treat both 2D and 3D curves, in addition to yielding concise equations that are more convenient to program. Appendix B provides a review of the most relevant vector and linear algebra and analysis needed here. Additional equations can be found in standard linear algebra textbooks.

6.5   Lines

We derive the vector parametric equations of straight lines for various line definition cases. Consider the following two cases.

1. **A line connecting two end points.** Figure 6.4 shows P_1 and P_2 as the end points. Define a parameter u such that it has the values 0 and 1 at P_1 and P_2, respectively. Utilizing the triangle OPP_1, the following equation can be written:

$$\mathbf{P} = \mathbf{P}_1 + (\mathbf{P} - \mathbf{P}_1) \tag{6.9}$$

But the vector $(\mathbf{P} - \mathbf{P}_1)$ is proportional to the vector $(\mathbf{P}_2 - \mathbf{P}_1)$ such that

$$\mathbf{P} - \mathbf{P}_1 = u(\mathbf{P}_2 - \mathbf{P}_1) \tag{6.10}$$

Thus the equation of the line becomes

$$\mathbf{P} = \mathbf{P}_1 + u(\mathbf{P}_2 - \mathbf{P}_1) \qquad 0 \le u \le 1 \tag{6.11}$$

In scalar form, this equation can be written as

$$\left. \begin{array}{l} x = x_1 + u(x_2 - x_1) \\ y = y_1 + u(y_2 - y_1) \\ z = z_1 + u(z_2 - z_1) \end{array} \right\} \qquad 0 \le u \le 1 \tag{6.12}$$

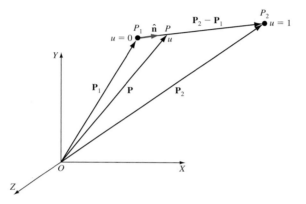

Figure 6.4   Line connecting two points: P_1 and P_2.

Equation (6.11) defines a line bounded by the end points P_1 and P_2 whose associated parametric values are 0 and 1, respectively. Any other point on the line or its extension has a certain value of u which is proportional to the point location, as Figure 6.5 shows. The coordinates of any point in the figure are obtained by substituting its corresponding u value in Eq. (6.11).

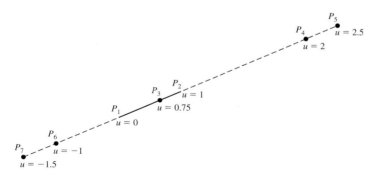

Figure 6.5   Locating points on an existing line.

The tangent vector of the line is given by

$$\mathbf{P}' = \mathbf{P}_2 - \mathbf{P}_1 \tag{6.13}$$

or, in scalar form,

$$
\begin{aligned}
x' &= x_2 - x_1 \\
y' &= y_2 - y_1 \\
z' &= z_2 - z_1
\end{aligned}
\tag{6.14}
$$

The independence of the tangent vector from u reflects the constant slope of the straight line. For a 2D line, the known infinite (vertical line) and zero (horizontal line) slope conditions can be generated from Eq. (6.14).

The unit vector in the direction of the line is given by

$$\hat{\mathbf{n}} = \frac{\mathbf{P}_2 - \mathbf{P}_1}{L} \tag{6.15}$$

where L is the length of the line,

$$L = |\mathbf{P}_2 - \mathbf{P}_1| = \sqrt{(x_2 - x_1)^2 + (y_2 - y_1)^2 + (z_2 - z_1)^2} \tag{6.16}$$

Regardless of the user input to create a line, a line database stores its two end points and additional information such as its font, width, color, and layer. Equations (6.11) and (6.13) show that the end points are enough to provide all geometric properties and characteristics of the line. They are also sufficient to construct and display the line. For reference purposes, CAD/CAM software usually identifies the first point input by the user during line construction as P_1 where $u = 0$.

2. **A line passing through a point P_1 in a direction defined by unit vector $\hat{\mathbf{n}}$.** Figure 6.6 shows this case. This case and others usually result in generating the end points from user input or given data.

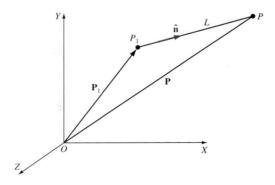

Figure 6.6   Line passing through P_1 in direction $\hat{\mathbf{n}}$.

To develop the line equation for this case, consider a general point P on the line at a distance L from P_1. The vector equation of the line becomes (see triangle OP_1P)

$$P = P_1 + L\hat{n}, \qquad 0 \le L \le L_{max} \tag{6.17}$$

and L is given by

$$L = |P - P_1| \tag{6.18}$$

L is the parameter in equation (6.17). Thus, the tangent vector is $\hat{n}$.

Once the user inputs P_1, $\hat{n}$, and L, the other end point P_2 is calculated using Eq. (6.17), and the line has the two end points P_1 and P_2 with u values of 0 and 1 as discussed in case 1.

Some of the following examples show how parametric equations of various line forms can be developed. The examples relate to the most common line commands offered by CAD/CAM software.

EXAMPLE 6.3   **Find the line equation.**

Given the two lines L_1 and L_2 and their end points shown on the right;

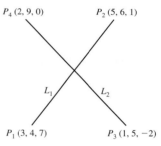

a. Find the equations of the lines. Show the parametrization directions.
  b. Are the two lines parallel or perpendicular?
   c. Find the coordinates of the intersection point.

SOLUTION   Using Eq. (6.11), we write the following equations:

For L_1,

$$P = P_1 + u(P_2 - P_1) = \begin{bmatrix} 3 \\ 4 \\ 7 \end{bmatrix} + u \begin{bmatrix} 2 \\ 2 \\ -6 \end{bmatrix}$$

For L_2,

$$P = P_3 + \upsilon(P_4 - P_3) = \begin{bmatrix} 1 \\ 5 \\ -2 \end{bmatrix} + \upsilon \begin{bmatrix} 1 \\ 4 \\ 2 \end{bmatrix}$$

The parametrization directions for L_1 and L_2 go from P_1 to P_2, and P_3 to P_4, respectively.

If lines L_1 and L_2 are perpendicular, the dot product of their tangent vectors should be zero, $P'_{L_1} \cdot P'_{L_2} = 0$. Using the two equations of the two lines, the two tangent vectors are $P'_{L_1} = P_2 - P_1 = 2\hat{i} + 2\hat{j} - 6\hat{k}$ and $P'_{L_2} = P_4 - P_3 = \hat{i} + 4\hat{j} + 2\hat{k}$. The dot product is equal to -2. Thus L_1 and L_2 are not perpendicular.

If L_1 and L_2 are parallel, the cross product of their tangent vectors should be zero, $P'_{L_1} \times P'_{L_2} = 0$. The cross product is equal to $28\hat{i} - 10\hat{j} + 26\hat{k}$. Thus L_1 and L_2 are not parallel.

To find the intersection point of the two lines, we equate the two line equations. This produces three equations (x, y, z) in two unknowns u and v. Let us use the x and y components to solve for u and v,

$$3 + 2u = 1 + v$$
$$4 + 2u = 5 + 4v$$

Solving gives $u = -1.5$ and $v = -1$. Substituting u in the L_1 equation, or v in the L_2 equation gives the intersection point as $(0, 1, 16)$.

Example 6.3 discussion:

Solving this problem using analytic geometry would be more difficult than the parametric approach presented here.

Example 6.3 hands-on exercise:

Use the x and z or the y and z components to solve for the intersection point. Do you get the same intersection point?

EXAMPLE 6.4   **Find the line equation.**

Find the equations and end points of two lines, one horizontal and the other vertical. Each line begins at and passes through a given point and is clipped by another given point.

SOLUTION   Horizontal and vertical lines are usually defined in reference to the current WCS axes. Horizontal lines are parallel to the X axis and vertical lines are parallel to the Y axis. Figure 6.7 shows a typical user working environment where the WCS has a different orientation from the MCS. In this case, the WCS is equivalent to the coordinate system used to develop the line equations. Once the end points are calculated from these equations with respect to the WCS, they are transformed to the MCS before the line display or storage.

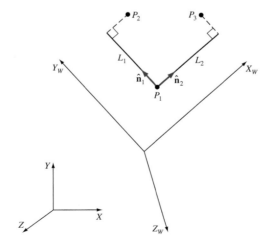

Figure 6.7   Horizontal and vertical lines.

Assume that three points P_1, P_2, and P_3 are given. The vertical line passes through P_1 and ends at P_2, while the horizontal line passes through P_1 and ends at P_3. In general, the two lines cannot pass through points P_2 or P_3. Therefore, the ends are determined by projecting the points to the lines as shown. Using Eq. (6.17), the line equations are

Vertical:
$$\mathbf{P} = \mathbf{P}_1 + L\hat{\mathbf{n}}_1, \qquad 0 \le L \le L_1$$

or
$$x_w = x_{1w}$$
$$y_w = y_{1w} + L$$
$$z_w = z_{1w}$$

where
$$L_1 = y_{2w} - y_{1w}$$

and the end points are (x_{1w}, y_{1w}, z_{1w}) and $(x_{1w}, y_{1w} + L_1, z_{1w})$.

Horizontal:
$$\mathbf{P} = \mathbf{P}_1 + L\hat{\mathbf{n}}_2, \qquad 0 \le L \le L_2$$

or
$$x_w = x_{1w} + L$$
$$y_w = y_{1w}$$
$$z_w = z_{1w}$$

where
$$L_2 = x_{3w} - x_{1w}$$

and the end points are (x_{1w}, y_{1w}, z_{1w}) and $(x_{1w} + L_2, y_w, z_w)$.

EXAMPLE 6.5 **Find line equations.**

Find the equation and end points of a line that passes through a point P_1, parallel to an existing line L_1, and is trimmed by point P_2 as shown in Figure 6.8.

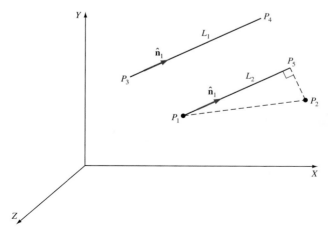

Figure 6.8   Line parallel to an existing line.

SOLUTION   To minimize confusion, the differentiation between the orientations of the WCS and the MCS is ignored and the position vectors of the various points are omitted from the figure. Assume that the existing line has the two end points P_3 and P_4, a length L_1, and a direction defined by the unit vector $\hat{\mathbf{n}}_1$. The new line has the same direction $\hat{\mathbf{n}}_1$, a length L_2, and end points P_1 and P_5. P_5 is the projection of P_2 on the line.

The equation of the new line is found by substituting the proper vectors into Eq. (6.17). The unit vector $\hat{\mathbf{n}}_1$ is given by

$$\hat{\mathbf{n}}_1 = \frac{\mathbf{P}_4 - \mathbf{P}_3}{L_1}$$

where L_1 is a known value [Eq. (6.16)]. L_2 can be found from the following equation

$$L_2 = \hat{\mathbf{n}}_1 \cdot (\mathbf{P}_2 - \mathbf{P}_1) = \frac{\mathbf{P}_4 - \mathbf{P}_3}{L_1} \cdot (\mathbf{P}_2 - \mathbf{P}_1)$$

$$= \frac{(x_4 - x_3)(x_2 - x_1) + (y_4 - y_3)(y_2 - y_1) + (z_4 - z_3)(z_2 - z_1)}{\sqrt{(x_4 - x_3)^2 + (y_4 - y_3)^2 + (z_4 - z_3)^2}}$$

The equation of the new line becomes

$$\mathbf{P} = \mathbf{P}_1 + L\hat{\mathbf{n}}_1, \qquad 0 \le L \le L_2$$

and its two end points are $P_1(x_1, y_1, z_1)$ and $P_5(x_1 + L_2 n_{1x}, \ y_1 + L_2 n_{1y}, \ z_1 + L_2 n_{1z})$. As numerical examples, consider the simple 2D case of constructing lines parallel to existing horizontal and vertical ones. Verify the equations presented here for the horizontal lines case $P_1(3, 2)$, $P_2(7, 4)$, $P_3(2, 3)$, and $P_4(5, 3)$. Repeat for the vertical lines case $P_1(8, 9)$, $P_2(9, 2)$, $P_3(5, 3)$, and $P_4(5, 9)$. Readers can also utilize the corresponding command to this example to construct the line on their CAD/CAM systems, verify the line to obtain its length and end points, and then compare with the results obtained here.

EXAMPLE 6.6   **Find line equations.**

Relate the following CAD commands to their mathematical foundations:

a. The command that measures the angle between two lines.
b. The command that measures the distance between a point and a line.

SOLUTION   (a) If the end points of the two lines are P_1, P_2 and P_3, P_4 (Figure 6.9), the angle measurement command uses the equation

$$\cos\theta = \frac{(\mathbf{P}_2 - \mathbf{P}_1) \cdot (\mathbf{P}_4 - \mathbf{P}_3)}{|\mathbf{P}_2 - \mathbf{P}_1||\mathbf{P}_4 - \mathbf{P}_3|}$$

$$= \frac{(x_2 - x_1)(x_4 - x_3) + (y_2 - y_1)(y_4 - y_3) + (z_2 - z_1)(z_4 - z_3)}{\sqrt{[(x_2 - x_1)^2 + (y_2 - y_1)^2 + (z_2 - z_1)^2][(x_4 - x_3)^2 + (y_4 - y_3)^2 + (z_4 - z_3)^2]}}$$

(b) Figure 6.9b shows the distance D from point P_3 to the line whose end points are P_1 and P_2 and direction is $\hat{\mathbf{n}}$. Its length is L. D is given by

$$D = \left|(P_3 - P_1) \times \hat{\mathbf{n}}\right| = \left|(P_3 - P_1) - \frac{(P_2 - P_1)}{(P_2 - P_1)}\right| = \begin{vmatrix} \hat{\mathbf{i}} & \hat{\mathbf{j}} & \hat{\mathbf{k}} \\ x_3 - x_1 & y_3 - y_1 & z_3 - z_1 \\ \dfrac{x_2 - x_1}{L} & \dfrac{y_2 - y_1}{L} & \dfrac{z_2 - z_1}{L} \end{vmatrix}$$

(where, on the furthest right-hand side, the inner vertical bars indicate the determinant and the outer vertical bars indicate the magnitude of the resulting vector) or

$$D = \frac{1}{L} \text{SQRT}\{[(y_3 - y_1)(z_2 - z_1) - (z_3 - z_1)(y_2 - y_1)]^2$$
$$+ [(z_3 - z_1)(x_2 - x_1) - (x_3 - x_1)(z_2 - z_1)]^2$$
$$+ [(x_3 - x_1)(y_2 - y_1) - (y_3 - y_1)(x_2 - x_1)]^2\}$$

where **SQRT** is the square root. For a two-dimensional case, this equation reduces to

$$D = \frac{1}{L} [(x_3 - x_1)(y_2 - y_1) - (y_3 - y_1)(x_2 - x_1)]$$

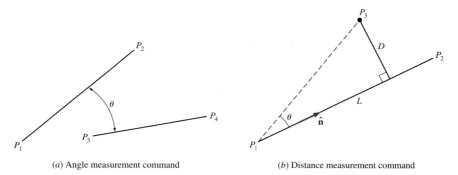

(a) Angle measurement command   (b) Distance measurement command

Figure 6.9   Data related to line commands.

EXAMPLE 6.7   **Find tangent vectors.**
Find the unit tangent vector in the direction of a line:

a. parallel to an existing line.
b. perpendicular to an existing line.

SOLUTION   The conditions of parallelism or perpendicularity of two lines given in the vector algebra review in Appendix B are useful if the vector equations of the two lines exist. If one equation is not available, which is mostly the case in practical problems, the conditions should be reduced to find the unit tangent vector of the missing line in terms of the existing one.

Figure 6.10 shows the existing line as L_1 with a known unit tangent vector $\hat{\mathbf{n}}_1$. The unit tangent vector $\hat{\mathbf{n}}_2$ is to be found in terms of $\hat{\mathbf{n}}_1$.

    (a) For L_1 and L_2 to be parallel (Figure 6.10a),

$$\hat{\mathbf{n}}_2 = \hat{\mathbf{n}}_1$$

or

$$\begin{bmatrix} n_{2x} & n_{2y} & n_{2z} \end{bmatrix}^T = \begin{bmatrix} n_{1x} & n_{1y} & n_{1z} \end{bmatrix}^T$$

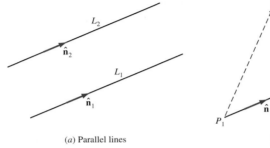

(a) Parallel lines   (b) Perpendicular lines

Figure 6.10   Unit tangent vectors of various lines.

    This equation defines an infinite number of lines in an infinite number of planes in space. Additional geometric conditions are required to define a specific line. This equation is equivalent to the condition of the equality of the two slopes of the two lines in the 2D case. Neglecting the Z component, the preceding equation gives the condition for this case as:

$$\begin{bmatrix} n_{2x} & n_{2y} \end{bmatrix}^T = \begin{bmatrix} n_{1x} & n_{1y} \end{bmatrix}^T$$

or

$$\frac{n_{2y}}{n_{2x}} = \frac{n_{1y}}{n_{1x}}$$

or

$$m_2 = m_1$$

where m_1 and m_2 are the slopes of the lines.

    (b) For L_1 and L_2 to be perpendicular (Figure 6.10b),

$$\hat{\mathbf{n}}_2 \cdot \hat{\mathbf{n}}_1 = 0$$

or $$n_{1x}n_{2x} + n_{1y}n_{2y} + n_{1z}n_{2z} = 0 \qquad (6.19)$$

    Additional equations are needed to solve for $\hat{\mathbf{n}}_2$. The following two equations can be written:

$$\left| \hat{\mathbf{n}}_1 \times \hat{\mathbf{n}}_2 \right| = 1$$

or

$$(n_{1y}n_{2z} - n_{1z}n_{2y})^2 + (n_{1z}n_{2x} - n_{1x}n_{2z})^2 + (n_{1x}n_{2y} - n_{1y}n_{2x})^2 = 1 \tag{6.20}$$

and

$$\hat{\mathbf{n}}_2 \cdot \hat{\mathbf{n}}_2 = |\hat{\mathbf{n}}_2|^2$$

or

$$n_{2x}^2 + n_{2y}^2 + n_{2z}^2 = 1 \tag{6.21}$$

However, only two equations out of the preceding three are independent. For example, squaring Eq. (6.19) and adding it to Eq. (6.20) results in Eq. (6.21) if the identity $n_{1x}^2 + n_{1y}^2 + n_{1z}^2 = 1$ is used. Therefore, only two equations are available to solve for n_{2x}, n_{2y}, and n_{2z} which implies that only two of these components can be obtained as functions of the third. This situation results from the fact that the perpendicular line L_2 to L_1 at point P shown in Figure 6.10 is not unique. A plane perpendicular to L_1 at P defines the locus of L_2.

As a more defined case, assume that point P_3 is known and L_2 is the perpendicular line from P_3 to L_1. Using Eqs. (6.19) to (6.21) to solve for $\hat{\mathbf{n}}_2$ is usually cumbersome. Instead, utilizing the triangle P_1PP_3, the following equation can be written:

$$\mathbf{P}_1\mathbf{P}_3 = \mathbf{P}_1\mathbf{P} + \mathbf{P}\mathbf{P}_3$$

or

$$\mathbf{P}_3 - \mathbf{P}_1 = [(\mathbf{P}_3 - \mathbf{P}_1) \cdot \hat{\mathbf{n}}_1]\hat{\mathbf{n}}_1 + D\hat{\mathbf{n}}_2 \tag{6.22}$$

where D is the perpendicular distance between P_3 and L_1 and is given in the previous example. Equation (6.22) gives $\hat{\mathbf{n}}_2$ as:

$$\hat{\mathbf{n}}_2 = \frac{1}{D}\{\mathbf{P}_3 - \mathbf{P}_1 - [(\mathbf{P}_3 - \mathbf{P}_1) \cdot \hat{\mathbf{n}}_1]\hat{\mathbf{n}}_1\}$$

In scalar form,

$$n_{2x} = \frac{1}{D}[(x_3 - x_1)(1 - n_{1x}^2) - (y_3 - y_1)n_{1x}n_{1y} - (z_3 - z_1)n_{1x}n_{1z}]$$

$$n_{2y} = \frac{1}{D}[(y_3 - y_1)(1 - n_{1y}^2) - (x_3 - x_1)n_{1x}n_{1y} - (z_3 - z_1)n_{1y}n_{1z}]$$

$$n_{2z} = \frac{1}{D}[(z_3 - z_1)(1 - n_{1z}^2) - (y_3 - y_1)n_{1y}n_{1z} - (x_3 - x_1)n_{1x}n_{1z}]$$

For the 2D case, Eq. (6.22) becomes:

$$x_3 - x_1 = [(x_3 - x_1)n_{1x} + (y_3 - y_1)n_{1y}]n_{1x} + Dn_{2x}$$
$$y_3 - y_1 = [(x_3 - x_1)n_{1x} + (y_3 - y_1)n_{1y}]n_{1y} + Dn_{2y}$$

Multiplying the first and second equations by n_{1x} and n_{1y}, respectively, and adding them utilizing the identity $n_{1x}^2 + n_{1y}^2 = 1$, we get:

$$D(n_{2x}n_{1x} + n_{2y}n_{1y}) = 0$$

or

$$\frac{n_{2y}}{n_{2x}} = -\frac{n_{1x}}{n_{1y}}$$

or

$$\frac{n_{2y}n_{1y}}{n_{2x}n_{1x}} = -1$$

or

$$m_1 m_2 = -1$$

which is a known result from 2D analytic geometry. In the 2D case, $\hat{n}_2$ can be chosen such that

$$n_{2x} = n_{1y} \qquad n_{2y} = -n_{1x}$$

or

$$n_{2x} = -n_{1y} \qquad n_{2y} = n_{1x}$$

Note: It is left to the reader to show that Eq. (6.22) can be reduced to the condition $\hat{n}_1 \cdot \hat{n}_2 = 0$.

6.6 Circles

Circles and circular arcs are among the most common entities used in geometric modeling. Circles, and circular arcs together with straight lines, are sufficient to construct many existing mechanical parts and components in practice. Besides other information, a circle database stores its radius and center as its essential geometric data. If the plane of the circle cannot be defined from the user input data, as in the case of specifying a center and a radius, it is typically assumed by the software to be the *XY* plane of the current WCS at the construction time. Regardless of the user input information to create a circle, such information is always converted into a radius and center by the software. This section presents some cases to show how such conversion is possible.

The basic parametric equation of a circle can be written as (refer to Figure 6.11):

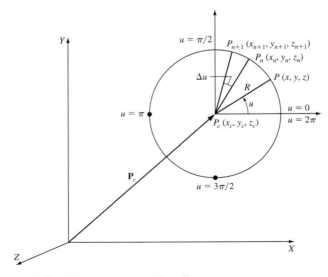

Figure 6.11   **A circle defined by a center and radius.**

$$
\left.\begin{array}{l}
x = x_c + R\cos u \\
y = y_c + R\sin u \\
z = z_c
\end{array}\right\} \quad 0 \le u \le 2\pi
\tag{6.23}
$$

assuming that the plane of the circle is the XY plane for simplicity. In this equation, the parameter u is the angle measured from the X axis to any point P on the circle. This parameter is used by commercial software to locate points at certain angles on the circle for construction purposes. Some certain values of u are shown in Figure 6.11.

Equation (6.23) can be used to generate points on the circle's circumference, for display purposes, by incrementing u from 0 to 360. These points are in turn connected with line segments to display the circle. However, this is an inefficient way due to the need to compute the trigonometric functions in the equation for each point. A less computational method is to write Eq. (6.23) in an incremental form. Assuming there is an increment Δu between two consecutive points $P(x_n, y_n, z_n)$ and $P(x_{n+1}, y_{n+1}, z_{n+1})$ on the circle's circumference, the following recursive relationship can be written.

$$
\begin{array}{rl}
x_n =& x_c + R\cos u \\
y_n =& y_c + R\sin u \\
x_{n+1} =& x_c + R\cos(u+1) \\
y_{n+1} =& y_c + R\sin(u+1) \\
z_{n+1} =& z_n
\end{array}
\tag{6.24}
$$

Expanding the x_{n+1} and y_{n+1} equations gives

$$
\begin{array}{rl}
x_{n+1} =& x_c + (x_n - x_c)\cos\Delta u - (y_n - y_c)\sin\Delta u \\
y_{n+1} =& y_c + (y_n - y_c)\cos\Delta u - (x_n - x_c)\sin\Delta u \\
z_{n+1} =& z_n
\end{array}
\tag{6.25}
$$

Thus, the circle can start from an arbitrary point and successive points with equal spacing can be calculated recursively. $\cos\Delta u$ and $\sin\Delta u$ have to be calculated only once, which eliminates the computation of trigonometric functions for each point. This algorithm is useful for hardware implementation to speed up the circle generation and display.

Circular arcs are considered a special case of circles. Therefore, all discussions covered here regarding circles can easily be extended to arcs. A circular arc equation can be written as:

$$x = x_c + R\cos u$$
$$y = y_c + R\sin u \quad\Bigg\} \quad u_s \le u \le u_e \tag{6.26}$$
$$z = z_c$$

where u_s and u_e are the starting and ending angles of the arc, respectively. An arc database includes its center and radius, as a circle, as well as its starting and ending angles. Most user inputs offered by software to create arcs are similar to these offered to create circles. As indicated from Eq. (6.26) and Figure 6.11, the arc always connects its beginning and ending points in a counterclockwise direction.

Following are some examples that show how various geometric data and constraints, which can be thought of as user inputs required by CAD systems, can be converted to a radius and center before their storage by the software in the corresponding circle database. They also show that three constraints (points and/or tangent vectors) are required to create a circle except for the obvious case of a center and radius. In all these examples, the reader can verify the derived equations by comparing their results with those of an accessible CAD/CAM system.

EXAMPLE 6.8   **Find circle parameters.**
Find the radius and the center of a circle whose diameter is given by two points.

SOLUTION   Assume the circle diameter is given by the two points P_1 and P_2. The circle radius and center are

$$R = \frac{1}{2}\sqrt{(x_2 - x_1)^2 + (y - y_1)^2 + (z_2 - z_1)^2}$$

$$\mathbf{P}_c = \frac{1}{2}(\mathbf{P}_1 + \mathbf{P}_2)$$

or $$[x_c \ \ y_c \ \ z_c]^T = \left[\frac{x_1 + x_2}{2} \ \ \frac{y_1 + y_2}{2} \ \ \frac{z_1 + z_2}{2}\right]^T$$

EXAMPLE 6.9   **Find circle parameters.**
Find the radius and the center of a circle passing through three points.

SOLUTION   Figure 6.12 shows the three given points P_1, P_2, and P_3 that the circle must pass through. The circle center and radius are shown as P_c and R, respectively. From analytic geometry, P_c is the intersection of the perpendicular lines to the chords P_1P_2, P_2P_3, and P_1P_3 from their midpoints P_4, P_5, and P_6, respectively. The unit vectors defining the directions of these chords in space are known and given by:

$$\hat{\mathbf{n}}_1 = \frac{\mathbf{P}_2 - \mathbf{P}_1}{|\mathbf{P}_2 - \mathbf{P}_1|} \qquad \hat{\mathbf{n}}_2 = \frac{\mathbf{P}_3 - \mathbf{P}_2}{|\mathbf{P}_3 - \mathbf{P}_2|} \qquad \hat{\mathbf{n}}_3 = \frac{\mathbf{P}_1 - \mathbf{P}_3}{|\mathbf{P}_1 - \mathbf{P}_3|}$$

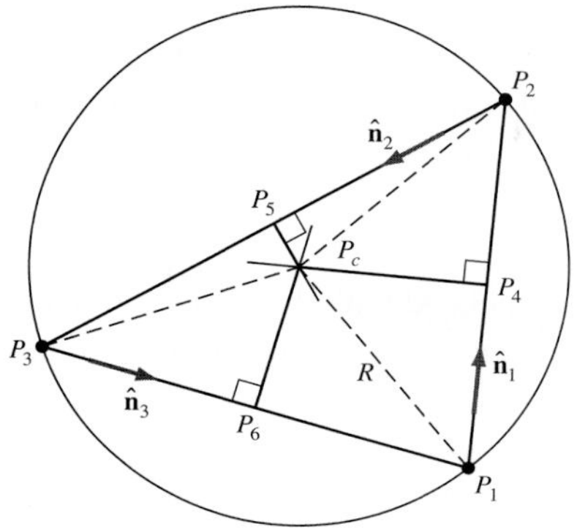

Figure 6.12   A circle passing through three points.

To find the center of the circle, P_c, the following three equations can be written:

$$(\mathbf{P}_c - \mathbf{P}_1) \cdot \hat{\mathbf{n}}_1 = \frac{|\mathbf{P}_2 - \mathbf{P}_1|}{2}$$

$$(\mathbf{P}_c - \mathbf{P}_2) \cdot \hat{\mathbf{n}}_2 = \frac{|\mathbf{P}_3 - \mathbf{P}_2|}{2} \tag{6.27}$$

$$(\mathbf{P}_c - \mathbf{P}_3) \cdot \hat{\mathbf{n}}_3 = \frac{|\mathbf{P}_1 - \mathbf{P}_3|}{2}$$

Each of these vector equations implies that the component of a vector radius in the direction of any of the three chords is equal to half the chord length. These equations can be used to solve for P_c (x_c, y_c, z_c). Expanding and rearranging the equations in a matrix form yield

$$\begin{bmatrix} n_{1x} & n_{1y} & n_{1z} \\ n_{2x} & n_{2y} & n_{2z} \\ n_{3x} & n_{3y} & n_{3z} \end{bmatrix} \begin{bmatrix} x_c \\ y_c \\ z_c \end{bmatrix} = \begin{bmatrix} b_1 \\ b_2 \\ b_3 \end{bmatrix} \tag{6.28}$$

where

$$b_1 = \frac{|\mathbf{P}_2 - \mathbf{P}_1|}{2} + (x_1 n_{1x} + y_1 n_{1y} + z_1 n_{1z})$$

$$b_2 = \frac{|\mathbf{P}_3 - \mathbf{P}_2|}{2} + (x_2 n_{2x} + y_2 n_{2y} + z_2 n_{2z})$$

$$b_3 = \frac{|\mathbf{P}_2 - \mathbf{P}_1|}{2} + (x_3 n_{3x} + y_3 n_{3y} + z_3 n_{3z})$$

The matrix Eq. (6.28) has the form $[A]\mathbf{P}_c = \mathbf{b}$. Therefore,

$$\mathbf{P}_c = [A]^{-1}\mathbf{b} = \frac{\text{Adj}\,([A])}{|A|}\mathbf{b}$$

where Adj ($[A]$) and $|A|$ are the adjoint matrix and determinant of $[A]$, respectively. Adj ($[A]$) is the matrix $[C]^T$ where $[C]$ is the matrix formed by the cofactors C_{ij} of the elements a_{ij} of $[A]$. The cofactor C_{ij} is given by

$$C_{ij} = (-1)^{i+j}M_{ij}$$

where M_{ij} is a unique scalar associated with the element a_{ij} and is defined as the determinant of the $(n-1) \times (n-1)$ matrix from the $n \times n$ matrix $[A]$ by crossing out the ith row and j^{th} column. Thus

$$\mathbf{P}_c = \frac{[C]^T}{|A|}\mathbf{b}$$

and $\quad |A| = n_{1x}(n_{2y}n_{3z} - n_{2z}n_{3y}) - n_{1y}(n_{2x}n_{3z} - n_{2z}n_{3x}) + n_{1z}(n_{2x}n_{3y} - n_{2y}n_{3x})$

$$[C] = \begin{bmatrix} C_{11} & C_{12} & C_{13} \\ C_{21} & C_{22} & C_{23} \\ C_{31} & C_{32} & C_{33} \end{bmatrix}$$

The elements of $[C]$ are given by

$$C_{11} = n_{2y}n_{3z} - n_{2z}n_{3y} \qquad C_{12} = n_{2z}n_{3x} - n_{2x}n_{3z} \qquad C_{13} = n_{2x}n_{3y} - n_{2y}n_{3x}$$
$$C_{21} = n_{1z}n_{3y} - n_{1y}n_{3z} \qquad C_{22} = n_{1x}n_{3z} - n_{1z}n_{3x} \qquad C_{23} = n_{1y}n_{3x} - n_{1x}n_{3y}$$
$$C_{31} = n_{1y}n_{2z} - n_{1z}n_{2y} \qquad C_{32} = n_{1z}n_{2x} - n_{1x}n_{2z} \qquad C_{33} = n_{1x}n_{2y} - n_{1y}n_{2x}$$

The coordinates of the center can now be written as:

$$x_c = \frac{1}{|A|}(C_{11}b_1 + C_{21}b_2 + C_{31}b_3)$$

$$y_c = \frac{1}{|A|}(C_{12}b_1 + C_{22}b_2 + C_{32}b_3)$$

$$z_c = \frac{1}{|A|}(C_{13}b_1 + C_{23}b_2 + C_{33}b_3)$$

The radius R is the distance between P_c and any of the three data points, that is,

$$R = |\mathbf{P}_c - \mathbf{P}_1| = |\mathbf{P}_c - \mathbf{P}_2| = |\mathbf{P}_c - \mathbf{P}_3|$$

or, for example,

$$R = \sqrt{(x_c - x_1)^2 + (y_c - y_1)^2 + (z_c - z_1)^2}$$

For the two-dimensional case, only two of the three relationships shown in Eq. (6.27) are sufficient to find the center $P_c(x_c, y_c)$. The third equation can be used to check the results. Using the first two relationships, the following matrix equations can be written:

$$\begin{bmatrix} n_{1x} & n_{1y} \\ n_{2x} & n_{2y} \end{bmatrix} \begin{bmatrix} x_c \\ y_c \end{bmatrix} = \begin{bmatrix} b_1 \\ b_2 \end{bmatrix}$$

where

$$b_1 = \frac{|P_2 - P_1|}{2} + (x_1 n_{1x} + y_1 n_{1y})$$

$$b_2 = \frac{|P_3 - P_2|}{2} + (x_2 n_{2x} + y_2 n_{2y})$$

Similar to the three-dimensional case, the center is given by

$$x_c = \frac{n_{2y} b_1 - n_{1y} b_2}{n_{1x} n_{2y} - n_{1y} n_{2x}}$$

$$y_c = \frac{n_{1x} b_2 - n_{2x} b_1}{n_{1x} n_{2y} - n_{1y} n_{2x}}$$

EXAMPLE 6.10   **Find circle parameters.**
Find the center of a circle tangent to two known lines with a given radius.

SOLUTION   This case is shown in Figure 6.13. The two existing lines are defined by the point pairs (P_1, P_2) and (P_3, P_4). The unit vectors $\hat{n}_1$ and $\hat{n}_2$ define the directions of the lines in

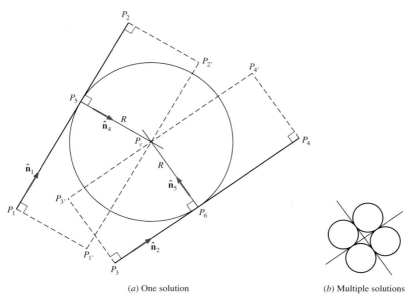

(a) One solution   (b) Multiple solutions

Figure 6.13   Circle tangent to two lines with given radius R.

space. The center of the circle P_c is the intersection of the two normals to the two lines at the tangency points. The unit vectors $\hat{\mathbf{n}}_4$ and $\hat{\mathbf{n}}_5$ define the directions of these two normals in space.

Conceptually, P_c can be found by finding the intersection of the two normals by solving their two vector equations. However, this route is cumbersome. Instead, P_c is found as the intersection of the two lines P_1P_2, and P_3P_4, that are parallel to P_1P_2 and P_3P_4, respectively, at distance R. The following development is based on the observation that the two lines P_1P_2 and P_3P_4 define the plane of the circle and all the other lines and points shown in Figure 6.13 lie in this plane.

The unit vectors can be defined as

$$\hat{\mathbf{n}}_1 = \frac{\mathbf{P}_2 - \mathbf{P}_1}{|\mathbf{P}_2 - \mathbf{P}_1|} \qquad \hat{\mathbf{n}}_2 = \frac{\mathbf{P}_4 - \mathbf{P}_3}{|\mathbf{P}_4 - \mathbf{P}_3|} \qquad \hat{\mathbf{n}}_3 = \frac{\hat{\mathbf{n}}_1 \times \hat{\mathbf{n}}_2}{|\hat{\mathbf{n}}_1 \times \hat{\mathbf{n}}_2|}$$

$$\hat{\mathbf{n}}_4 = \hat{\mathbf{n}}_3 \times \hat{\mathbf{n}}_1 \qquad \hat{\mathbf{n}}_5 = \hat{\mathbf{n}}_2 \times \hat{\mathbf{n}}_3$$

where $\hat{\mathbf{n}}_3$ is a vector perpendicular to the plane of the circle. The end points of the parallel lines are given by:

$$\mathbf{P}_{1'} = \mathbf{P}_1 + R\hat{\mathbf{n}}_4 \qquad\qquad \mathbf{P}_{2'} = \mathbf{P}_2 + R\hat{\mathbf{n}}_4$$

$$\mathbf{P}_{3'} = \mathbf{P}_3 + R\hat{\mathbf{n}}_5 \qquad\qquad \mathbf{P}_{4'} = \mathbf{P}_4 + R\hat{\mathbf{n}}_5$$

Thus, the parametric vector equations of these parallel lines become

$$\mathbf{P} = \mathbf{P}_1 + u(\mathbf{P}_2 - \mathbf{P}_1) + R\hat{\mathbf{n}}_4 \tag{6.29}$$

$$\mathbf{P} = \mathbf{P}_3 + v(\mathbf{P}_4 - \mathbf{P}_3) + R\hat{\mathbf{n}}_5 \tag{6.30}$$

Therefore, the intersection point of the two lines is defined by the equation

$$\mathbf{P}_1 + u(\mathbf{P}_2 - \mathbf{P}_1) + R\hat{\mathbf{n}}_4 = \mathbf{P}_3 + v(\mathbf{P}_4 - \mathbf{P}_3) + R\hat{\mathbf{n}}_5 \tag{6.31}$$

This vector equation can be solved for either u or v which, in turn, can be substituted into Eq. (6.29) or (6.30) to find P_c. If this equation is solved in scalar form, two components, say in X and Y directions, give u and v. The third component, in the Z direction, can be used to check the computations involved.

To find u, take the scalar product of Eq. (6.31) with $\hat{\mathbf{n}}_5$. This gives

$$u = \frac{(\mathbf{P}_3 - \mathbf{P}_1) \cdot \hat{\mathbf{n}}_5 + (1 - (\hat{\mathbf{n}}_4 \cdot \hat{\mathbf{n}}_5))R}{(\mathbf{P}_2 - \mathbf{P}_1) \cdot \hat{\mathbf{n}}_5} \tag{6.32}$$

Substituting this value into Eq. (6.29) gives

$$\mathbf{P}_c = \mathbf{P}_1 + \left[\frac{(\mathbf{P}_3 - \mathbf{P}_1) \cdot \hat{\mathbf{n}}_5 + (1 - (\hat{\mathbf{n}}_4 \cdot \hat{\mathbf{n}}_5))R}{(\mathbf{P}_2 - \mathbf{P}_1) \cdot \hat{\mathbf{n}}_5} \right](\mathbf{P}_2 - \mathbf{P}_1) + R\hat{\mathbf{n}}_4 \tag{6.33}$$

which can easily be written in scalar form to yield the coordinates x_c, y_c, and z_c of the center point P_c. The 2D case exhibits no special characteristics from the 3D case.

The development just described is not concerned with the existence of the four multiple solutions shown in Figure 6.13b if the two known lines, and not their extensions, intersect. In such a case, the software often follows a certain convention or asks the user to click the quadrant where the circle is to reside. One common convention is that the circle becomes tangent to the closest line segments chosen by the user while the user selects the two known lines to identify them. For all solutions, the development is valid.

Two special cases related to Eqs. (6.32) and (6.33) are discussed here. First, consider the case when the two known lines are parallel as shown in Figure 6.14a. In this case $\hat{\mathbf{n}}_1 = \hat{\mathbf{n}}_2$, $\hat{\mathbf{n}}_4 = -\hat{\mathbf{n}}_5$, $\hat{\mathbf{n}}_4 \cdot \hat{\mathbf{n}}_5 = -1$, and $(\mathbf{P}_2 - \mathbf{P}_1) \cdot \hat{\mathbf{n}}_5 = 0$. Therefore $u \to \infty$, and P_c is not defined. However, the locus of P_c is defined as the line parallel to the known lines at the middle distance between them as shown in Figure 6.14. In this case the software can override the user input of R and replace it by half the perpendicular distance between the two lines. Such distance can be computed as shown in Section 6.5. The two end points of the locus of P_c are given by

$$\mathbf{P}_{L1} = \frac{\mathbf{P}_1 + \mathbf{P}_3}{2} \qquad \mathbf{P}_{L2} = \frac{\mathbf{P}_2 + \mathbf{P}_4}{2}$$

An infinite number of circles exists with centers on the locus. The software can either display a warning message to the user or choose point P_{L1} or P_{L2} as a default center.

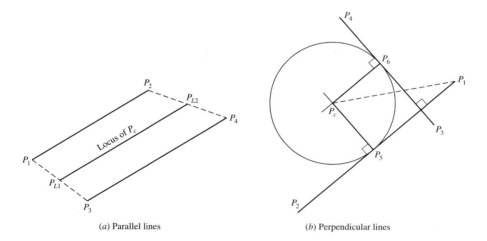

(a) Parallel lines   (b) Perpendicular lines

Figure 6.14   Circles tangent to parallel and perpendicular lines.

The second case is shown in Figure 6.14b, where the two known lines are perpendicular to each other. In this case $\hat{\mathbf{n}}_4 = \hat{\mathbf{n}}_2$, $\hat{\mathbf{n}}_5 = \hat{\mathbf{n}}_1$, and $\hat{\mathbf{n}}_4 \cdot \hat{\mathbf{n}}_5 = 0$. Equations (6.32) and (6.33) become

$$u = \frac{(\mathbf{P}_3 - \mathbf{P}_1) \cdot \hat{\mathbf{n}}_1 + R}{|\mathbf{P}_2 - \mathbf{P}_1|}$$

$$P_c = P_1 + [(P_3 - P_1) \cdot \hat{n}_1 + R]\hat{n}_1 + R\hat{n}_2$$

This equation for P_c is also obvious if we consider the triangle $P_1P_cP_5$ and write the vector equation $P_1P_c = P_1P_5 + P_5P_c$.

This development can be extended to fillets connecting lines as follows. Once P_c is known from Eq. (6.33), the two points P_5 and P_6 are given by:

$$P_5 = P_c - R\hat{n}_4$$
$$P_6 = P_c - R\hat{n}_5$$

These two points define the beginning and the end of the fillet and can be used to construct the proper part of the circle that forms the fillet.

EXAMPLE 6.11   **Find intersection point.**

Find the intersection point between a line and a circle.

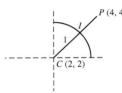

SOLUTION   We use points C and P to write the line equation that follows. The parametrization direction of the line is from C to P.

$$P_L(u) = \begin{bmatrix} x \\ y \end{bmatrix} = \begin{bmatrix} 2 + 2u \\ 2 + 2u \end{bmatrix}, \qquad 0 \le u \le 1$$

The equation of the arc, parametrized in a counterclockwise direction, is:

$$P_a(u) = \begin{bmatrix} x \\ y \end{bmatrix} = \begin{bmatrix} 2 + \cos\frac{\pi}{2}u \\ 2 + \sin\frac{\pi}{2}u \end{bmatrix}, \qquad 0 \le u \le 1$$

The intersection point, I, of the arc and the line is given by equating their equations:

$$P_L(u) = P_a(u)$$

The X and Y components of this equation give, respectively,

$$2 + 2u = 2 + \cos\frac{\pi}{2}u$$
$$2 + 2u = 2 + \sin\frac{\pi}{2}u$$

The two right-hand sides must be equal because the two left-hand sides are equal. Thus, we get:

$$\cos\frac{\pi}{2}u = \sin\frac{\pi}{2}u$$

The solution to this equation is $u = 0.5$. Substituting in the line or arc equation gives the intersection point I as $(3, 3)$, which agrees with visual inspection.

6.7 Ellipses

Mathematically the ellipse is a curve generated by a point moving in space such that at any position the sum of its distances from two fixed points (foci) is constant and equal to the major diameter. Each focus is located on the major axis of the ellipse at a distance from its center equal to $\sqrt{A^2 + B^2}$ (A and B are the major and minor radii). Circular holes and forms become ellipses when they are viewed obliquely relative to their planes.

The development of the parametric equation and other related characteristics of ellipses, elliptic arcs, and fillets is similar to those of circles, circular arcs, and fillets. However, four conditions (points and/or tangent vectors) are required to define the geometric shape of an ellipse as compared to three conditions to define a circle. The default plane of an ellipse, as in a circle, is the XY plane of the current WCS at the time of construction if the user input is not enough to define the ellipse plane as in the case of inputting center, half of the length of the major axis, and half of the length of the minor axis. The database of an ellipse usually stores user input as a center point, half the length of the major axis, half the length of the minor axis, and other information (orientation, starting and ending angles, layer, font, name, color, and so forth).

Figure 6.15 shows an ellipse with point P_c as the center and the lengths of half of the major and minor axes are A and B, respectively. The parametric equation of an ellipse can be written as:

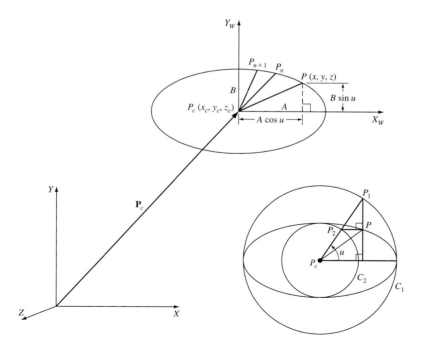

Figure 6.15   An ellipse defined by a center, major, and minor axes.

$$
\left. \begin{array}{l}
x = x_c + A\cos u \\
y = y_c + B\sin u \\
z = z_c
\end{array} \right\} \qquad 0 \le u \le 2\pi
\qquad\qquad (6.34)
$$

assuming the plane of the ellipse is the XY plane. The parameter u is the angle as in the case of a circle. However, for a point P shown in Figure 6.15, it is not the angle between the line PP_c and the major axis of the ellipse. Instead, it is defined as shown. To find point P on the ellipse that corresponds to an angle u, the two concentric circles C_1 and C_2 are constructed with centers at P_c and radii of A and B, respectively. A radial line is constructed at the angle u to intersect both circles at points P_1 and P_2, respectively. If a line parallel to the minor axis is drawn from P_1 and a line parallel to the major axis is drawn from P_2, the intersection of these two lines defines the point P.

Similar to circles, the following recursive relationships can be developed for an ellipse:

$$
\begin{array}{l}
x_{n+1} = x_c + (x_n - x_c)\cos\Delta u - \dfrac{A}{B}(y_n - y_c)\sin\Delta u \\[2mm]
y_{n+1} = y_c + (y_n - y_c)\cos\Delta u + \dfrac{A}{B}(x_n - x_c)\sin\Delta u \\[2mm]
z_{n+1} = z_n
\end{array}
\qquad\qquad (6.35)
$$

If the ellipse major axis is inclined with an angle α relative to the X axis as shown in Figure 6.16, the ellipse equation becomes

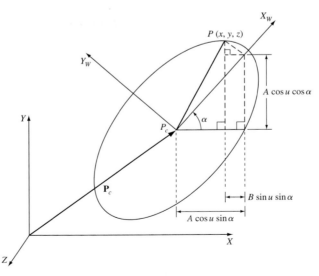

Figure 6.16   An inclined ellipse.

$$\left.\begin{array}{l} x = x_c + A\cos u \cos\alpha - B\sin u \sin\alpha \\ y = y_c + A\cos u \sin\alpha + B\sin u \cos\alpha \\ z = z_c \end{array}\right\} \quad 0 \le u \le 2\pi \qquad (6.36)$$

Equation (6.36) cannot be reduced to a recursive relationship similar to what is given in Eq. (6.35). Instead Eq. (6.36) can be written as

$$\begin{aligned} x_{n+1} &= x_c + A\cos(u_n + \Delta u)\cos\alpha - B\sin(u_n + \Delta u)\sin\alpha \\ y_{n+1} &= y_c + A\cos(u_n + \Delta u)\sin\alpha + B\sin(u_n + \Delta u)\cos\alpha \\ z_{n+1} &= z_n \end{aligned} \qquad (6.37)$$

where $u_n = (n-1)u$. The first point corresponds to $n = 0$, which lies at the end of the major axis. In addition, $\cos(u_n + \Delta u)$ and $\sin(u_n + \Delta u)$ are evaluated from the double angle formulas for cosine and sine. If the calculations from the previous point for $\sin(u_n)$ and $\cos(u_n)$ are stored temporarily, $\cos(u_n + \Delta u)$ and $\sin(u_n + \Delta u)$ can be evaluated without calculating trigonometric functions for each point. $\cos(u_n)$ and $\sin(u_n)$ need to be calculated only once. Therefore, computational savings similar to the circle case can be achieved.

EXAMPLE 6.12   **Find the parameters of an ellipse.**

Find the center, the lengths of half the axes, and the orientation in space of an ellipse defined by:

         a. its circumscribing rectangle.
         b. one of its internal rectangles.

SOLUTION   Both cases are equivalent to defining an ellipse by four points. The user interface can let the user input the three corner points of the rectangle or one corner point, the lengths of the rectangle sides, and an orientation.

         (a) From Figure 6.17a:

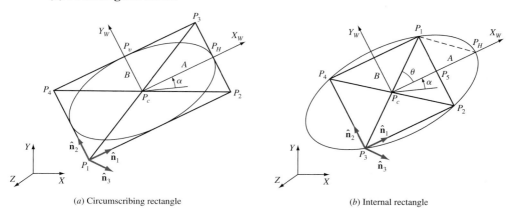

            (a) Circumscribing rectangle   (b) Internal rectangle

Figure 6.17   An ellipse defined by four points.

$$P_c = \tfrac{1}{2}(P_1 + P_3) = \tfrac{1}{2}(P_2 + P_4)$$

$$P_H = \tfrac{1}{2}(P_2 + P_3)$$

$$P_y = \tfrac{1}{2}(P_3 + P_4)$$

$$A = |P_c - P_4| = \sqrt{(x_c - x_4)^2 + (y_c - y_4)^2 + (z_c - z_4)^2}$$

$$B = |P_c - P_y| = \sqrt{(x_c - x_y)^2 + (y_c - y_y)^2 + (z_c - z_y)^2}$$

The orientation of the ellipse in space can be defined by the unit vectors $\hat{n}_1$, $\hat{n}_2$, and $\hat{n}_3$ instead of the angle α. These vectors are given by

$$\hat{n}_1 = \frac{P_2 - P_1}{|P_2 - P_1|} \qquad \hat{n}_2 = \frac{P_4 - P_1}{|P_4 - P_1|} \qquad \hat{n}_3 = \hat{n}_1 \times \hat{n}_2$$

(b) P_c, $\hat{n}_1$, $\hat{n}_2$, and $\hat{n}_3$ can be calculated as in case (a). To find A and B, one can write

$$P_5 = \tfrac{1}{2}(P_2 + P_3)$$

In reference to the WCS coordinate system, the x and y coordinates of point P_3 are $|P_5 - P_c|$ and $|P_3 - P_5|$, respectively. Substituting into the ellipse equation,

$$\frac{|P_5 - P_c|^2}{A^2} + \frac{|P_3 - P_5|^2}{B^2} = 1 \tag{6.38}$$

To write another equation in A and B, consider the triangle $P_c P_3 P_H$ and write the law of cosines as:

$$|P_3 - P_H|^2 = |P_3 - P_c|^2 + A^2 - 2A|P_3 - P_c|\cos\theta \tag{6.39}$$

The angle θ can be computed as the angle between the two vectors $(P_5 - P_c)$ and $(P_3 - P_c)$. To calculate the length $|P_3 - P_H|$, P_H, which has the coordinates $(A, 0, 0)$ in the WCS system must be transformed to the MCS. Consequently its MCS coordinates become $(x_c + n_{1x}A, y_c + n_{1y}A, z_c + n_{1z}A)$. Substituting these coordinates into Eq. (6.39), a second-order equation in A results which can be solved for A. Then Eq. (6.38) can be used to solve for B. The remainder of the solution (finding the ellipse orientation) is identical to case (a).

EXAMPLE 6.13   **Find parameters of an ellipse.**
Find the center, the lengths of half the axes, and the orientation of an ellipse given two of its conjugate diameters.

SOLUTION   This is a case of defining an ellipse by four points that form two of its conjugate diameters. Figure 6.18 shows the conjugate diameters and the relevant unit vectors.

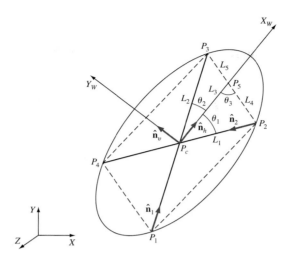

Figure 6.18   An ellipse defined by two conjugate diameters.

The center of the ellipse is given by

$$\mathbf{P}_c = \tfrac{1}{2}(\mathbf{P}_1 + \mathbf{P}_3) = \tfrac{1}{2}(\mathbf{P}_2 + \mathbf{P}_4)$$

The unit vectors $\hat{\mathbf{n}}_1$, $\hat{\mathbf{n}}_2$, and $\hat{\mathbf{n}}_3$ are given by

$$\hat{\mathbf{n}}_1 = \frac{\mathbf{P}_3 - \mathbf{P}_1}{|\mathbf{P}_3 - \mathbf{P}_1|} \qquad \hat{\mathbf{n}}_2 = \frac{\mathbf{P}_4 - \mathbf{P}_2}{|\mathbf{P}_4 - \mathbf{P}_2|} \qquad \hat{\mathbf{n}}_3 = \frac{\hat{\mathbf{n}}_1 \times \hat{\mathbf{n}}_2}{|\hat{\mathbf{n}}_1 \times \hat{\mathbf{n}}_2|}$$

Utilizing the ellipse equation for points P_2 and P_3, the following two equations can be written

$$\frac{(L_1 \cos\theta_1)^2}{A^2} + \frac{(L_1 \sin\theta_1)^2}{R^2} = 1 \tag{6.40}$$

$$\frac{(L_2 \cos\theta_2)^2}{A^2} + \frac{(L_2 \sin\theta_2)^2}{B^2} = 1 \tag{6.41}$$

where

$$L_1 = |\mathbf{P}_2 - \mathbf{P}_c| \qquad\qquad L_2 = |\mathbf{P}_3 - \mathbf{P}_c|$$

The angles θ_1 and θ_2 can be related together as follows:

$$\hat{\mathbf{n}}_1 \cdot \hat{\mathbf{n}}_2 = \cos(\theta_1 + \theta_2)$$

or

$$\theta_1 + \theta_2 = K \tag{6.42}$$

where K is a known value. In addition, applying the law of sines to the triangles $P_c P_3 P_5$ and $P_c P_2 P_5$, we can write

$$\frac{L_1}{\sin\theta_3} = \frac{L_4}{\sin\theta_1} \tag{6.43}$$

$$\frac{L_2}{\sin(180-\theta_3)} = \frac{L_5}{\sin\theta_2} \tag{6.44}$$

where

$$L_4 = |\mathbf{P}_2 - \mathbf{P}_5| \qquad L_5 = L_6 - L_4 \qquad L_6 = |\mathbf{P}_3 - \mathbf{P}_2|$$

Dividing Eq. (6.44) by (6.43) and using Eq. (6.42) gives

$$\frac{\sin\theta_1}{\sin(K-\theta_1)} = \frac{L_2 L_4}{L_1 (L_6 - L_4)} \tag{6.45}$$

Applying the Law of cosines to the same triangles gives

$$L_4^2 = L_1^2 + L_3^2 - 2L_1 L_3 \cos\theta_1 \tag{6.46}$$

$$(L_6 - L_4)^2 = L_2^2 + L_3^2 - 2L_2 L_3 \cos\theta_2 \tag{6.47}$$

Equations (6.40), (6.41), (6.42), (6.45), (6.46), and (6.47) form six equations to be solved for A, B, θ_1, θ_2, L_3, and L_4. Subtracting Eqs. (6.47) and (6.46) and using (6.42) gives

$$L_3 = \frac{L_2^2 - L_1^2 - L_6^2 + 2L_4 L_6}{2[L_2 \cos(K-\theta_1) - L_1 \cos\theta_1]} \tag{6.48}$$

If Eq. (6.48) is substituted into (6.46) and the resulting L_4 is substituted into Eq. (6.45), a nonlinear equation in θ_1 results which can be solved for θ_1. Consequently, θ_2 can be found from Eq. (6.42). Equations (6.40) and (6.41) can therefore be solved for A and B.

The orientation of the ellipse can be found by determining the unit vectors $\hat{\mathbf{n}}_h$ and $\hat{\mathbf{n}}_v$ that define the directions of the major and the minor axes, respectively. To find $\hat{\mathbf{n}}_h$, the following three equations can be written:

$$\hat{\mathbf{n}}_1 \cdot \hat{\mathbf{n}}_h = \cos\theta_2$$
$$\hat{\mathbf{n}}_2 \cdot \hat{\mathbf{n}}_h = \cos\theta_1$$
$$\hat{\mathbf{n}}_3 \cdot \hat{\mathbf{n}}_h = 0$$

These equations can be solved for the components n_{hx}, n_{hy}, and n_{hz} using the matrix approach utilized in Example 6.9. The unit vector $\hat{\mathbf{n}}_v$ can be found as

$$\hat{\mathbf{n}}_v = \hat{\mathbf{n}}_3 \times \hat{\mathbf{n}}_h$$

With $\hat{\mathbf{n}}_h$, $\hat{\mathbf{n}}_v$, and $\hat{\mathbf{n}}_3$ known, the orientation of the ellipse is completely defined in space.

Note: Case (b) of Example 6.12 is a special case of this example.

EXAMPLE 6.14   **Find the tangent to an ellipse.**

Find the tangent to an ellipse from a given point P_1 outside the ellipse.

SOLUTION   Two tangents can be drawn to the ellipse from P_1 as shown in Figure 6.19. Assume the tangency point is P_T. First, transform P_1 from the MCS to the ellipse local WCS system using the equation

$$\mathbf{P}_1 = [T]\mathbf{P}_{1W} \tag{6.49}$$

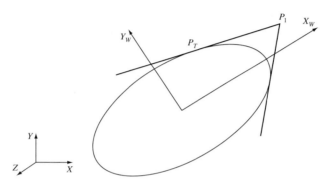

Figure 6.19   Tangents to an ellipse from an outside point.

The transformation matrix $[T]$ is known because the orientation of the ellipse is known. P_{1w} holds the local coordinates of P_1, which should be $[x_{1w} \; y_{1w} \; 0]$. Therefore:

$$\mathbf{P}_{1w} = [T]^{-1}\mathbf{P}_1$$

The tangent vector to the ellipse is given by

$$\mathbf{P}' = \begin{bmatrix} -A\sin u & B\cos u & 0 \end{bmatrix}^T$$

At point P_T, this vector becomes

$$\mathbf{P}' = \begin{bmatrix} -A\sin u_T & B\cos u_T & 0 \end{bmatrix}^T$$

and the slope of the tangent is given by

$$S = -\frac{B\cos u_T}{A\sin u_T}$$

This slope S can also be found using the two points P_1 and P_T as:

$$S = \frac{y_{1w} - B\sin u_T}{x_{1w} - A\cos u_T} \tag{6.50}$$

Therefore:

$$\frac{y_{1w} - B\sin u_T}{x_{1w} - A\cos u_T} = -\frac{B\cos u_T}{A\sin u_T}$$

which gives

$$x_1 B\cos u_T + y_1 A\sin u_T = AB$$

or

$$K_1 \cos u_T + K_2 \sin u_T = AB$$

where $K_1 = x_1 B$ and $K_2 = y_1 A$. Define an angle γ such that $\gamma = tan^{-1} (K_1/K_2)$. Thus, the preceding equation can be rewritten as

$$\sin(\gamma + u_T) = \frac{AB}{\sqrt{K_1^2 + K_2^2}}$$

or

$$u_T = \sin^{-1}\left(\frac{AB}{\sqrt{K_1^2 + K_2^2}}\right) - \gamma$$

The arcsine function gives two angles that result in two tangents.

Once u_T is known, the local coordinates of P_T become:

$$\mathbf{P}_{TW} = [x_{TW} \quad y_{TW} \quad 0]^T = [A\cos u_T \quad B\sin u_T \quad 0]^T$$

P_{Tw} can be substituted into Eq. (6.49) to obtain P_T. Therefore, the tangent is defined and stored in the database by the two end points P_1 and P_T. In practice, the CAD/CAM software can ask the user to click close to the desired tangent so that the other one is eliminated.

6.8 Parabolas

The parabola is defined mathematically as a curve generated by a point which moves such that its distance from a fixed point (the focus P_F) is always equal to its distance to a fixed line (the directrix) as shown in Figure 6.20. The vertex P_v is the intersection point of the parabola with its axis of symmetry. It is located midway between the directrix and the focus. The focus lies on the axis of symmetry. Useful applications of the parabola curve in engineering design include its use in parabolic sound and light reflectors, radar antennas, and bridge arches.

Three conditions are required to define a parabola, a parabolic curve, or a parabolic arc. The default plane of a parabola is the XY plane of the current WCS at the time of construction. The database of a parabola usually stores the coordinates of its vertex, distances y_{Hw} and y_{Lw} that define its end points as shown in Figure 6.20, the distance A between the focus and the vertex (the focal distance), and the orientation angle α. Unlike the ellipse, the parabola is not a closed curve. Thus, the two end points determine the amount of the parabola to be displayed.

Assuming the local coordinate system of the parabola as shown in Figure 6.20, its parametric equation can be written as:

$$\left.\begin{array}{l} x = x_v + Au^2 \\ y = y_v + 2Au \\ z = z_v \end{array}\right\} \quad 0 \le u \le \infty \qquad (6.51)$$

If the range of the y coordinate is limited to y_{Hw} and y_{Lw} for positive and negative values respectively, the corresponding u values become

$$u_H = \frac{y_{HW}}{2A}$$

(6.52)

$$u_L = \frac{y_{LW}}{2A}$$

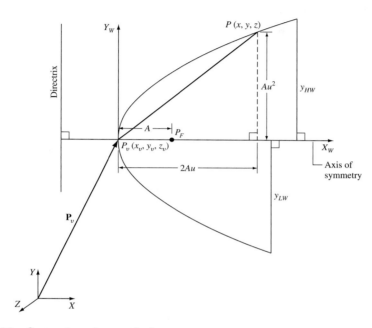

Figure 6.20   Geometry of a parabola.

The recursive relationships to generate points on the parabola are obtained by substituting $u_n + \Delta u$ for $n + 1$ points . This gives

$$x_{n+1} = x_n + (y_n - y_v)\,\Delta u + A(\Delta u)^2$$
$$y_{n+1} = y_n + 2A\,\Delta u$$ (6.53)
$$z_{n+1} = z_n$$

If the parabola axis of symmetry is included with an angle α as shown in Figure 6.21, its equation becomes:

$$x = x_v + Au^2\cos\alpha - 2Au\sin\alpha$$
$$y = y_v + Au^2\sin\alpha + 2Au\cos\alpha$$ (6.54)
$$z = z_v$$

and the recursive relationships reduce to

$$x_{n+1} = x_n\cos\alpha + (1 - \cos\alpha)x_v + (\Delta u\cos\alpha - \sin\alpha)(y_n - y_v)$$
$$+ A\Delta u(\Delta u\cos\alpha - 2\sin\alpha)$$

$$y_{n+1} = (\cos\alpha + \Delta u \sin\alpha)y_n + (1 - \cos\alpha - \Delta u \sin\alpha)y_v \qquad (6.55)$$
$$+ (x_n - x_v)\sin\alpha + A\Delta u(\Delta u \sin\alpha + 2\cos\alpha)$$
$$z_{n+1} = z_n$$

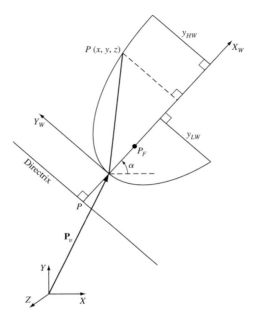

Figure 6.21   An inclined parabola.

EXAMPLE 6.15   **Find parameters of a parabola.**
Find the focal distance and the orientation in space of a parabola that passes through three points, one of which is the vertex.

SOLUTION   Figure 6.22 shows two cases where point P_1 is the vertex. In Figure 6.22a, the other two points P_2 and P_3 define a line perpendicular to the X_w axis of the parabola while Figure 6.22b shows the general case. The solution for the first case can be found as follows. The angles θ_1 and θ_2 in this case are equal. Thus:

$$P_4 = \frac{P_2 + P_3}{2}$$

The angle θ_1 can be found from

$$\tan\theta_1 = \frac{|P_2 - P_4|}{|P_4 - P_1|}$$

Using the parabolic equation $y_w^2 = 4Ax_w$, the focal distance A can be written as

$$A = \frac{|P_2 - P_4|^2}{4|P_4 - P_1|}$$

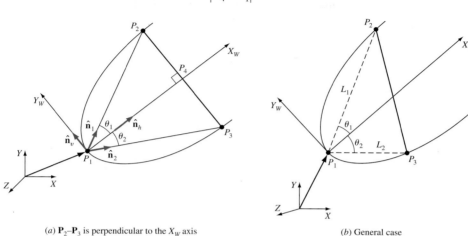

(a) P_2–P_3 is perpendicular to the X_W axis   (b) General case

Figure 6.22 A parabola passing through three points.

The orientation of the parabola is determined by the unit vectors $\hat{n}_h$ and $\hat{n}_v$ along the X_w and Y_w axes and the vector $\mathbf{n}_3$ which is perpendicular to its plane. These vectors are given as

$$\hat{n}_h = \frac{\hat{n}_1 + \hat{n}_2}{|\hat{n}_1 + \hat{n}_2|}$$

$$\hat{n}_3 = \frac{\hat{n}_2 \times \hat{n}_1}{|\hat{n}_2 \times \hat{n}_1|} = \frac{\hat{n}_2 \times \hat{n}_1}{\sin(\theta_1 + \theta_2)}$$

$$\hat{n}_v = \hat{n}_3 \times \hat{n}_h$$

where the unit vectors $\hat{n}_1$ and $\hat{n}_2$ can easily be computed from the given points.

Unlike the preceding case, the angles θ_1 and θ_2 are not equal for the second case (Figure 6.22b). Applying the parabolic equations to points P_2 and P_3, we can write

$$L_1 \sin^2\theta_1 = 4A\cos\theta_1 \tag{6.56}$$

$$L_2 \sin^2\theta_2 = 4A\cos\theta_2 \tag{6.57}$$

where $L_1 = |P_2 - P_1|$ and $L_2 = |P_3 - P_1|$. In addition,

$$\hat{n}_1 \cdot \hat{n}_2 = \cos(\theta_1 + \theta_2)$$

or

$$\theta_1 + \theta_2 = K \tag{6.58}$$

where

$$K = \cos^{-1}(\mathbf{n}_1 \cdot \hat{\mathbf{n}}_2)$$

The solution of Eqs. (6.56), (6.57), and (6.58) gives A, θ_1, and θ_2. If we divide Eq. (6.56) by (6.57) and use (6.58), we obtain

$$\frac{L_1}{L_2}\left[\frac{\sin\theta_1}{\sin(K-\theta_1)}\right]^2 = \frac{\cos\theta_1}{\cos(K-\theta_1)}$$

which is a nonlinear equation in θ_1. One way to solve it would be by plotting both the left- and right-hand sides as functions of θ_1 $(0 \le \theta_1 \le 90)$ and find the intersection point using the intersection modifier provided by the CAD software. Once θ_1 is found, Eq. (6.56) can be solved for A and (6.58) solved for θ_2. To find the orientation of the parabola, write

$$\hat{\mathbf{n}}_1 \cdot \hat{\mathbf{n}}_h = \cos\theta_1$$

$$\hat{\mathbf{n}}_2 \cdot \hat{\mathbf{n}}_h = \cos\theta_2$$

$$\hat{\mathbf{n}}_3 \cdot \hat{\mathbf{n}}_h = 0$$

which can be solved for the components of $\hat{\mathbf{n}}_h$. Then

$$\hat{\mathbf{n}}_v = \hat{\mathbf{n}}_3 \times \hat{\mathbf{n}}_h$$

6.9 Hyperbolas

A hyperbola is described mathematically as a curve generated by a point moving such, that at any position the difference of its distances from the fixed points (foci) F and F' is a constant and equal to the transverse axis of the hyperbola. Figure 6.23 shows a hyperbola.

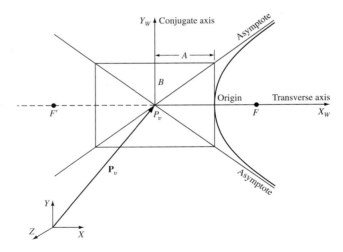

Figure 6.23   Hyperbola geometry.

The parametric equation of a hyperbola is given by:

$$x = x_v + A \cosh u$$
$$y = y_v + B \sinh u \tag{6.59}$$
$$z = z_v$$

Equation (6.59) is based on the nonparametric implicit equation of the hyperbola, which can be written as:

$$\frac{(x - x_v)^2}{A^2} - \frac{(y - y_v)^2}{B^2} = 1 \tag{6.60}$$

by utilizing the identity $\cosh^2 u - \sinh^2 u = 1$.

Similar to the ellipse developments, equations of an inclined hyperbola and recursive relationships can be derived. Also, the examples covered in the ellipse section can be extended to hyperbolas.

6.10 Conics

Conic curves or conic sections form the most general type of quadratic curves. Lines, circles, ellipses, parabolas, and hyperbolas covered in the previous sections are all special forms of conic curves. They all can be generated when a right circular cone of revolution is cut by planes at different angles relative to the cone axis; thus the derivation of the name conics. Straight lines result from intersecting a cone with a plane parallel to its axis and passing through its vertex. Circles result when the cone is sectioned by a plane perpendicular to its axis, while ellipses, parabolas, and hyperbolas are generated when the plane is inclined to the axis by various angles.

The general implicit nonparametric quadratic equation that describes a planar conic curve has five coefficients if the coefficient of the term x^2 is made equal to 1 — that is, we normalize the equation by dividing it by this coefficient if it is not. Thus, five conditions are required to completely define a conic curve. As presented in the previous sections, these reduce to two conditions to define lines, three for circles and parabolas, and four for ellipses and hyperbolas.

The conic parametric equation can be developed if five conditions are specified. Two cases are discussed here: specifying five points or three points and two tangent vectors. The development is based on the observation that a quadratic equation can be written as the product of two linear equations. Figure 6.24 shows a conic curve passing through points P_1 to P_5. Define the two pairs of lines (L_1, L_2) and (L_3, L_4) shown in the figure. Their equations are

$$L_1 = 0 \qquad L_2 = 0 \qquad L_3 = 0 \qquad L_4 = 0 \tag{6.61}$$

The four intersection points of these pairs are given by points P_1 to P_4. Let us define the two conics:

$$L_1 L_2 = 0 \qquad L_3 L_4 = 0 \tag{6.62}$$

Each one of these conics passes through points P_1 to P_4 but not necessarily through point P_5. However, any linear combination of these two conics represents another conic (since it is quadratic) which passes through their intersection points P_1 to P_4. For example, the equation

$$aL_1L_2 + bL_3L_4 = 0 \qquad (6.63)$$

represents such a conic. For the conic to pass through point P_5, its coordinates are substituted into Eq. (6.63) to find the ratio b/a. A more convenient form of Eq. (6.63) is

$$(1 - u)L_1L_2 + uL_3L_4 = 0, \qquad 0 \le u \le 1 \qquad (6.64)$$

Equation (6.64) gives the parametric equation of a conic curve with the parameter u. Changing the value of u results in a family (or pencil) of conics, two of which are $L_1L_2 = 0$ ($u = 0$) and $L_3L_4 = 0$ ($u = 1$). To use Eq. (6.64), four data points are used to find the equations for the lines L_1 to L_4 and the fifth point is used to find the u value.

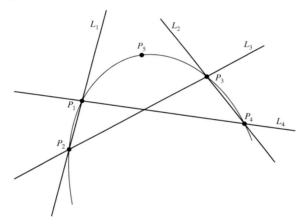

Figure 6.24 **A conic curve defined by five points.**

The case of a conic defined by three points and two tangent vectors is considered an adaptation of the preceding case. If we make lines L_1 and L_2 tangent to the conic curve, points P_1 and P_2 become one point (the tangent point) and P_3 and P_4 become another point. Consequently, the two lines L_3 and L_4 are merged into one line. Figure 6.25 shows the conic geometry in this case. The definition of the conic curve, in this case, is equivalent to a four-point definition: the two points of tangency P_1 and P_2, the intersection point P_4 of the two tangents L_1 and L_2, and a fourth point P_3, known as the shoulder point. P_3 must always be chosen inside the triangle $P_1P_2P_4$ to ensure the continuity of the conic curve segment that lies inside the triangle between P_1 and P_2. Equation (6.64) is then reduced for this case to:

$$(1 - u)L_1L_2 + uL_3^2 = 0, \qquad 0 \le u \le 1 \qquad (6.65)$$

Similar to the first case, points P_1, P_2, and P_4 determine the equations for the lines L_1 to L_3, and the point P_3 is used to find the u value.

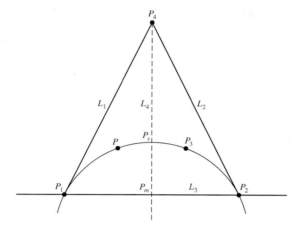

Figure 6.25   A conic curve defined by three points and two tangent curves.

The choice of the position of P_3 determines the type of the resulting conic curve. If P_3 is the midpoint of the line joining the midpoints of the tangents L_1 and L_2, then the conic is a parabola. P_3 becomes the vertex of the parabola and the line connecting P_3 and P_4 becomes its axis of symmetry. It is also obvious that the distance between P_3 and P_4 is equal to the distance between P_3 and the intersection point of the axis of symmetry and L_3 (a known characteristic of parabolas).

If P_3 lies inside the parabola and line L_3, the resulting conic is an ellipse. If it is outside the parabola, a hyperbolic curve results. To check which type of a conic curve results from a given input data, let P_m be the midpoint of the line L_3 and let the line L_4 intersect the conic curve at the point P_s. The parametric equation of line L_4 is $\mathbf{P} = \mathbf{P}_m + u\,(\mathbf{P}_4 - \mathbf{P}_m)$. Let the parameter u take the value S at P_s. Then $(\mathbf{P}_s - \mathbf{P}_m) = S\,(\mathbf{P}_4 - \mathbf{P}_m)$. Therefore, the conic curve is parabolic if $S = 1/2$, elliptic if $S < 1/2$, and hyperbolic if $S > 1/2$. A further test is necessary to determine whether an elliptic curve is circular. If $|\mathbf{P}_4 - \mathbf{P}_1| = |\mathbf{P}_2 - \mathbf{P}_4|$ and

$$\frac{s^2}{1 - s^2} = \frac{|\mathbf{P}_2 - \mathbf{P}_1|^2}{4|\mathbf{P}_4 - \mathbf{P}_1||\mathbf{P}_4 - \mathbf{P}_2|} \tag{6.66}$$

then the arc is circular.

EXAMPLE 6.16   **Find a conic equation.**
Find the equation of a conic curve defined by five points P_1 to P_5.

SOLUTION   Equation (6.64) must be reduced further to be able to utilize it to generate points on the conic curve for display purposes. The approach taken in this example is to create a local coordinate system (WCS) in the plane of the conic curve (Figure 6.26), write Eq. (6.64) in this system, and generate points. The input point P_2 is taken as the origin of the local system and the vector $(P_4 - P_2)$ as its X_w axis. To define the system, the unit vectors along the axes are calculated as

$$\hat{n}_1 = \frac{P_4 - P_2}{|P_4 - P_2|} \qquad \hat{n}_4 = \frac{P_1 - P_2}{|P_1 - P_2|}$$

$$\hat{n}_3 = \frac{\hat{n}_1 \times \hat{n}_4}{|\hat{n}_1 \times \hat{n}_4|} \qquad\qquad (6.67)$$

$$\hat{n}_2 = \hat{n}_3 \times \hat{n}_1$$

At this point, it might be useful to ensure that the five points the user has inputted lie in one plane (the conic curve plane). This can simply be achieved by checking whether the scalar (dot) products of the vector n_3 with the vectors $(P_3 - P_2)$ and $(P_3 - P_5)$ are zeros or not.

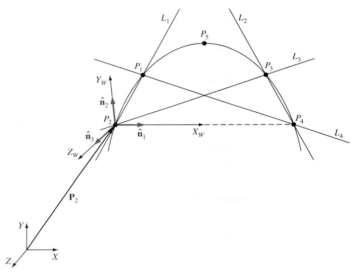

Figure 6.26   **Local coordinate system of a conic curve.**

6.11 Synthetic Curves

Analytic curves, described in Section 6.10, are usually not sufficient to meet the geometric design requirements of mechanical parts. Products such as car bodies, ship hulls, airplane fuselages and wings, propeller blades, shoe insoles, and bottles are a few examples that require free-form, or synthetic, curves and surfaces. The need for synthetic curves in design arises when a curve is represented by a collection of measured data points.

Mathematically, synthetic curves represent the problem of constructing a smooth curve that passes through given data points. Therefore the typical form of these curves is a polynomial.

Various continuity requirements can be specified at the data points to impose various degrees of smoothness upon the resulting curve. The order of continuity becomes important when a complex curve is modeled by several curve segments pieced together end-to-end. Zero-order continuity (C^0) yields a position-continuous curve. First- (C^1) and second- (C^2) order continuities imply slope- and curvature-continuous curves, respectively. Figure 6.27 shows a geometrical interpretation of these orders of continuity.

A cubic polynomial is the minimum-order polynomial that can guarantee the generation of C^0, C^1, or C^2 curves. In addition, the cubic polynomial is the lowest-degree polynomial that permits inflection within a curve segment and that allows representation of nonplanar (twisted) 3D curves in space. Higher-order polynomials are not commonly used in CAD because they tend to oscillate about control points, are computationally inconvenient, and are uneconomical of storing curve and surface representations in the computer.

Major CAD/CAM systems provide three types of synthetic curves: cubic spline, Bezier curve, and B-spline curve. The cubic spline curve passes through the data point and therefore is an interpolant. Bezier and B-spline curves can approximate (do not pass through) or interpolate (pass through) the data points.

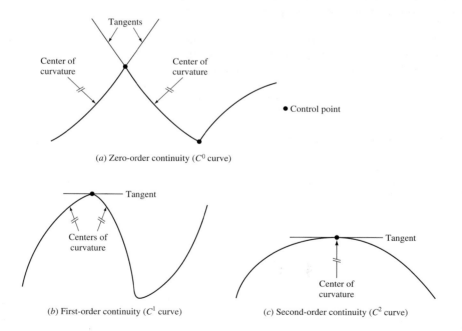

(a) Zero-order continuity (C^0 curve)

(b) First-order continuity (C^1 curve)

(c) Second-order continuity (C^2 curve)

Figure 6.27   Order of continuity of curves.

6.12 Hermite Cubic Spline

Splines draw their name from the traditional drafting tool called "French curves" or "splines." Cubic splines use cubic polynomials. A cubic polynomial has four coefficients and thus requires four conditions to evaluate. These conditions could be a combination of points and tangent vectors. A cubic spline uses four data points. The Hermite cubic spline uses two data points at its ends and two tangent vectors at these points.

The parametric equation of a cubic spline segment is given by:

$$\mathbf{P}(u) = \sum_{i=0}^{3} \mathbf{C}_i u^i, \qquad 0 \le u \le 1 \tag{6.68}$$

where u is the parameter and C_i are the polynomial (also called algebraic) coefficients. In scalar form this equation is written as:

$$\begin{aligned}
x(u) &= C_{3x}u^3 + C_{2x}u^2 + C_{1x}u + C_{0x} \\
y(u) &= C_{3y}u^3 + C_{2y}u^2 + C_{1y}u + C_{0y} \\
z(u) &= C_{3z}u^3 + C_{2z}u^2 + C_{1z}u + C_{0z}
\end{aligned} \tag{6.69}$$

In an expanded vector form, Eq. (6.68) can be written as:

$$\mathbf{P}(u) = \mathbf{C}_3 u^3 + \mathbf{C}_2 u^2 + \mathbf{C}_1 u + \mathbf{C}_0 \tag{6.70}$$

This equation (6.70) can also be written in a matrix form as

$$\mathbf{P}(u) = \mathbf{U}^T \mathbf{C} \tag{6.71}$$

where $\mathbf{U} = [u^3 \ u^2 \ u \ 1]^T$ and $\mathbf{C} = [\mathbf{C}_3 \ \mathbf{C}_2 \ \mathbf{C}_1 \ \mathbf{C}_0]^T$. $\mathbf{C}$ is called the coefficients vector.

The tangent vector to the curve at any point is given by differentiating Eq. (6.68) with respect to u to give

$$\mathbf{P}'(u) = \sum_{i=0}^{3} \mathbf{C}_i i u^{i-1}, \qquad 0 \le u \le 1 \tag{6.72}$$

In order to find the coefficients $\mathbf{C}_i$, consider the cubic spline segment with the two end points P_0 and P_1 shown in Figure 6.28. Applying the boundary conditions ($\mathbf{P}_0$, $\mathbf{P}_0'$ at $u = 0$ and P_1, $\mathbf{P}_1'$ at $u = 1$), Eqs. (6.68) and (6.72) give

$$\begin{aligned}
\mathbf{P}_0 &= \mathbf{C}_0 \\
\mathbf{P}_0' &= \mathbf{C}_1 \\
\mathbf{P}_1 &= \mathbf{C}_3 + \mathbf{C}_2 + \mathbf{C}_1 + \mathbf{C}_0 \\
\mathbf{P}_1' &= 3\mathbf{C}_3 + 2\mathbf{C}_2 + \mathbf{C}_1
\end{aligned} \tag{6.73}$$

Solving these four equations simultaneously for the coefficients gives

$$\mathbf{C}_0 = \mathbf{P}_0$$
$$\mathbf{C}_1 = \mathbf{P}_0'$$
$$\mathbf{C}_2 = 3(\mathbf{P}_1 - \mathbf{P}_0) - 2\,\mathbf{P}_0' - \mathbf{P}_1'$$
$$\mathbf{C}_3 = 2(\mathbf{P}_0 - \mathbf{P}_1) + \mathbf{P}_0' + \mathbf{P}_1'$$

(6.74)

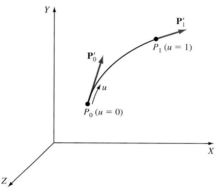

Figure 6.28   Hermite cubic spline curve.

Substituting Eq. (6.74) into Eq. (6.70) and rearranging gives:
$$\mathbf{P}(u) = (2u^3 - 3u^2 + 1)\mathbf{P}_0 + (-2u^3 + 3u^2)\mathbf{P}_1$$
$$+ (u^3 - 2u^2 + u)\,\mathbf{P}_0' + (u^3 - u^2)\mathbf{P}_1', \qquad 0 \le u \le 1$$

(6.75)

$\mathbf{P}_0$, $\mathbf{P}_1$, $\mathbf{P}_0'$, and $\mathbf{P}_1'$ are called geometric coefficients. The tangent vector becomes
$$\mathbf{P}'(u) = (6u^2 - 6u)\mathbf{P}_0 + (-6u^2 + 6u)\mathbf{P}_1$$
$$+ (3u^2 - 4u + 1)\,\mathbf{P}_0' + (3u^2 - 2u)\mathbf{P}_1', \qquad 0 \le u \le 1$$

(6.76)

The functions of u in Eqs. (6.75) and (6.76) are called blending functions. The first two functions blend $\mathbf{P}_0$ and $\mathbf{P}_1$, and the second two blend $\mathbf{P}_0'$ and $\mathbf{P}_1'$.

Equation (6.75) can be rewritten in a matrix form as

$$\mathbf{P}(u) = \mathbf{U}^T[M_H]\mathbf{V}, \qquad 0 \le u \le 1$$

(6.77)

where $[M_H]$ is the Hermite matrix and $\mathbf{V}$ is the geometry (or boundary conditions) matrix. Both are given by

$$[M_H] = \begin{bmatrix} 2 & -2 & 1 & 1 \\ -3 & 3 & -2 & -1 \\ 0 & 0 & 1 & 0 \\ 1 & 0 & 0 & 0 \end{bmatrix}$$

(6.78)

$$\mathbf{V} = [\mathbf{P}_0 \quad \mathbf{P}_1 \quad \mathbf{P}_0' \quad \mathbf{P}_1']^T$$

(6.79)

Comparing Eq. (6.71) and (6.77) shows that $\mathbf{C} = [M_H]\,\mathbf{V}$ or $\mathbf{V} = [M_H]^{-1}\,\mathbf{C}$ where

$$[M_H]^{-1} = \begin{bmatrix} 0 & 0 & 0 & 1 \\ 1 & 1 & 1 & 1 \\ 0 & 0 & 1 & 0 \\ 3 & 2 & 1 & 0 \end{bmatrix}$$

(6.80)

Similarly, Eq. (6.76) can be written as

$$\mathbf{P}'(u) = \mathbf{U}^T [M_H]^u \mathbf{V}$$

(6.81)

where $[M_H]^u$ is given by

$$[M_H]^u = \begin{bmatrix} 0 & 0 & 0 & 0 \\ 6 & -6 & 3 & 3 \\ -6 & 6 & -4 & -2 \\ 0 & 0 & 1 & 0 \end{bmatrix}$$

(6.82)

Equation (6.75) describes the Hermite cubic spline curve in terms of its two end points and their tangent vectors. The equation shows that the curve passes through the end points ($u = 0$ and 1). It also shows that the curve's shape can be controlled by changing its end points or its tangent vectors. If the two end points P_0 and P_1 are fixed in space, the designer can control the shape of the spline by changing either the magnitudes or the directions of the tangent vectors $\mathbf{P}'_0$ and $\mathbf{P}'_1$.

For planar splines, tangent vectors can be replaced by slopes. In this case, a default value, such as 1, for the lengths of the tangent vectors might be assumed by the software to enable using Eq. (6.75). For example, if the slope at P_0 is given as 30 degrees, then $\mathbf{P}'_0$ becomes [cos30 sin30 0]. It is obvious that the slope angle and the components of $\mathbf{P}'_0$ are given relative to the axes of the WCS that is active when the spline segment is created.

Equation (6.75) is for one cubic spline segment. It can be generalized for any two adjacent spline segments of a spline curve that is to fit a given number of data points. This introduces the problem of blending or joining cubic spline segments, which can be stated as follows: Given a set of n points, $P_0, P_1, ..., P_{n-1}$ and the two end tangent vectors $\mathbf{P}'_0$ and $\mathbf{P}'_1$, connect the points with a cubic spline curve.The spline curve is created as a blend of spline segments connecting the set of points starting from P_0 and ending at P_{n-1}. Interested readers can develop the blending equation.

The use of cubic splines in design applications is not very popular due to the need for tangent vectors or slopes to define the curve. Also, the control of the curve is not very obvious from the input data due to its global control characteristics. For example, changing the position of a data point or an end slope changes the entire shape of the spline, which does not provide the intuitive feel required for design. Figure 6.29 shows the control aspects of Hermite cubic spline curve.

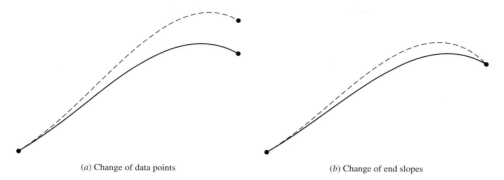

(a) Change of data points   (b) Change of end slopes

Figure 6.29   Control of Hermite cubic spline curve.

EXAMPLE 6.17   **Find a cubic spline equation.**
Find the equation of the Hermite cubic spline that connects P_0 and P_2 and that is tangent to the two line segments shown.

SOLUTION   The two end tangent vectors of the spline are the two lines shown. Thus, $\mathbf{P}_0' = \mathbf{P}_1 - \mathbf{P}_0$ and $\mathbf{P}_1' = \mathbf{P}_2 - \mathbf{P}_1$. The two end points are $\mathbf{P}_0$ and $\mathbf{P}_2$. Substituting into Eq. (6.75) gives

$$\mathbf{P}(u) = (u^3 - u^2 - u + 1)\mathbf{P}_0 - u(u-1)\mathbf{P}_1 + u^2(-u+2)\mathbf{P}_2$$

6.13 Bezier Curve

A Bezier curve is defined by a set of data points. The curve may interpolate or extrapolate the data points. Some CAD systems offer both options; others offer the interpolation version only. In both cases, the data points are used to control the shape of the resulting curves.

Bezier curves and surfaces are credited to P. Bezier of the French car firm Regie Renault, who developed (about 1962) and used them in his software system called UNISURF, which designers used to define the outer panels of several Renault cars. These curves, known as Bezier curves, were also independently developed by P. DeCasteljau of the French car company Citroen (about 1959), which used them as part of its CAD system. The Bezier UNISURF system was soon published in the literature; this is the reason that the curves now bear Bezier's name.

As its mathematics show, the major characteristics of the Bezier curve are:

1. The shape of the Bezier curve is controlled by its defining points. Tangent vectors are not used in the curve development as is the case with the cubic spline. This allows the designer a much better feel for the relationship between input (points) and output (curve).

2. The order or the degree of Bezier curve is variable and is related to the number of points defining it. $n + 1$ points define an nth degree curve, which permits higher-order continuity. This is not the case for cubic splines, where the degree is always cubic for a spline segment.

The data points of a Bezier curve are called control points. They form the vertices of what is called the control or characteristic polygon, which uniquely defines the curve shape as shown in Figure 6.30. Only the first and the last control points or vertices of the polygon actually lie on the curve. The other vertices define the order, derivatives, and shape of the curve. The curve is also always tangent to the first and last polygon segments. In addition, the curve shape tends to follow the polygon shape.

These three observations should enable the user to sketch or predict the curve shape once its control points are given as illustrated in Figure 6.31. The figure shows that the order of defining the control points changes the polygon definition, which changes the resulting curve shape. The arrow shown on each curve shows its parametrization direction.

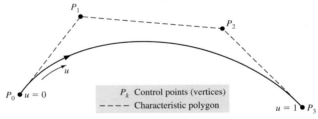

Figure 6.30   Cubic Bezier curve.

Mathematically, for $n + 1$ control points, the Bezier curve is defined by the following polynomial of degree n:

$$\mathbf{P}(u) = \sum_{i=0}^{n} \mathbf{P}_i B_{i,n}(u), \qquad 0 \le u \le 1 \tag{6.83}$$

where $\mathbf{P}(u)$ is a point on the curve and $\mathbf{P}_i$ is a control point. $B_{i,n}$ are the Bernstein polynomials. Thus, the Bezier curve has a Bernstein basis. The Bernstein polynomial serves as the blending or basis function for the Bezier curve and is given by

$$B_{i,n}(u) = C(n, i)u^i(1 - u)^{n-i} \tag{6.84}$$

where $C(n, i)$ is the binomial coefficient

$$C(n, i) = \frac{n!}{i!(n-i)!} \tag{6.85}$$

Utilizing Eqs. (6.84) and (6.85) and observing that $C(n, 0) = C(n, n) = 1$, Eq. (6.83) can be expanded to give:

$$\mathbf{P}(u) = \mathbf{P}_0(1-u)^n + \mathbf{P}_1 C(n, 1)u(1-u)^{n-1} + \mathbf{P}_2 C(n, 2)u^2(1-u)^{n-2}$$

$$+ \cdots + \mathbf{P}_{n-1} C(n, n-1)u^{n-1}(1-u) + \mathbf{P}_n u^n, \qquad 0 \le u \le 1 \tag{6.86}$$

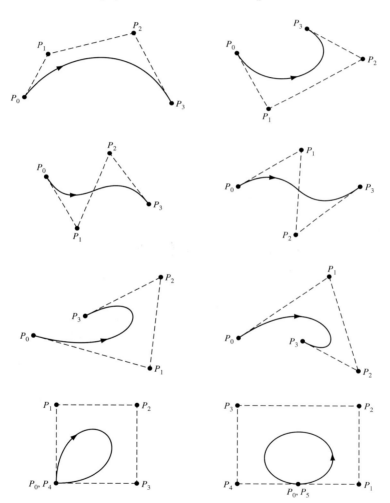

Figure 6.31   **Cubic Bezier curve for various control points.**

The characteristics of the Bezier curve are based on the properties of the Bernstein polynomials and can be summarized as follows:

1. The curve interpolates the first and last control points; that is it passes through P_0 and P_n if we substitute $u = 0$ and 1 in Eq. (6.86).

2. The curve is tangent to the first and last segments of the characteristic polygon. Using Eqs. (6.83) and (6.84), the rth derivatives at the starting and ending points are given by, respectively:

$$\mathbf{P}^r(0) = \frac{n!}{(n-r)!} \sum_{i=0}^{r} (-1)^{r-i} C(r,i)\mathbf{P}_i \tag{6.87}$$

$$\mathbf{P}^r(1) = \frac{n!}{(n-r)!} \sum_{i=0}^{r} (-1)^{i} C(r,i)\mathbf{P}_{n-i} \tag{6.88}$$

Therefore, the first derivatives at the end points are:

$$\mathbf{P}'(0) = n(\mathbf{P}_1 - \mathbf{P}_0) \tag{6.89}$$

$$\mathbf{P}'(1) = n(\mathbf{P}_n - \mathbf{P}_{n-1}) \tag{6.90}$$

where $(\mathbf{P}_1 - \mathbf{P}_0)$ and $(\mathbf{P}_n - \mathbf{P}_{n-1})$ define the first and last segments of the curve polygon. Similarly, it can be shown that the second derivative at P_0 is determined by $\mathbf{P}_0$, $\mathbf{P}_1$, and $\mathbf{P}_2$. Or, in general, the rth derivative at an end point is determined by its r neighboring vertices.

3. The curve is symmetric with respect to u and $(1 - u)$. This means that the sequence of control points defining the curve can be reversed without changing the curve shape; that is, reversing the direction of parametrization does not change the curve shape. This can be achieved by substituting $1 - u = v$ in Eq. (6.86) and noticing that $C(n, i) = C(n, n - i)$. This is a result of the fact that $B_{i,\,n}(u)$ and $B_{n-i,\,n}(u)$ are symmetric if they are plotted as functions of u.

4. The interpolation polynomial $B_{i,\,n}(u)$ has a maximum value of $C(n, i)\,(i/n)^i\,(1 - i/n)^{n-i}$ occurring at $u = i/n$, which can be obtained from the equation $d(B_{i,\,n})/du = 0$. This implies that each control point is most influential on the curve shape at $u = i/n$. For example, for a cubic Bezier curve, P_0, P_1, P_2, and P_3 are most influential when $u = 0$, 1/3, 2/3, and 1, respectively. Therefore, each control point is weighed by its blending function for each u value.

5. The curve shape can be modified by either changing one or more vertices of its polygon or by keeping the polygon fixed and specifying multiple coincident points at a vertex as shown in Figure 6.32. In Figure 6.32a, the vertex P_2 is pulled to the new position P_2^* and

in Figure 6.32b, P_2 is assigned a multiplicity K. The higher the multiplicity, the more the curve is pulled toward P_2.

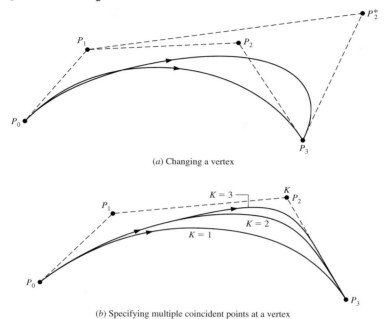

(a) Changing a vertex

(b) Specifying multiple coincident points at a vertex

Figure 6.32   **Modifications of a cubic Bezier curve.**

6. A closed Bezier curve can simply be generated by closing its characteristic polygon or choosing P_0 and P_n to be coincident. Figure 6.31 shows examples of closed curves.

7. For any valid value of u, the sum of the $B_{i,n}$ functions associated with the control points is always one for any degree Bezier curve. This fact can be used to check numerical computations and software developments.

A most desirable feature for any curve defined by a polygon such as a Bezier curve is the convex hull property. This property relates the curve to its characteristic polygon. This is what guarantees that incremental changes in control point positions produce intuitive geometric changes. A curve is said to have the convex hull property if it lies entirely within the convex hull defined by the polygon vertices. In a plane, the convex hull is a closed polygon, and in 3D it is a polyhedron. The shaded area shown in Figure 6.33 defines the convex hull of a Bezier curve. The hull is formed by connecting the vertices of the characteristic polygon.

Curves that possess the convex hull property enjoy some important consequences. If the polygon defining a curve segment degenerates to a straight line, the resulting segment must therefore be linear. Thus a Bezier curve may have locally linear segments embedded in it, which is a useful design feature.

Also, the size of the convex hull is an upper bound on the size of the curve itself; that is, the curve always lies inside its convex hull. This is a useful property for graphics functions such as displaying or clipping the curve. For example, instead of testing the curve itself for clipping, its convex hull is tested first, and only if it intersects the display window boundaries should the curve itself be examined.

A third consequence of the convex hull property is that the curve never oscillates wildly away from its defining control points because the curve is guaranteed to lie within its convex hull.

While the Bezier curve seems superior to the cubic spline curve, it still has some disadvantages. The curve lacks local control. It only has the global control nature. If one control point is changed, the whole curve changes. Therefore, the designer cannot selectively change part of the curve. Moreover, the curve degree depends on the number of data points that defines the curve. CAD systems limit the number of points used to define a Bezier curve.

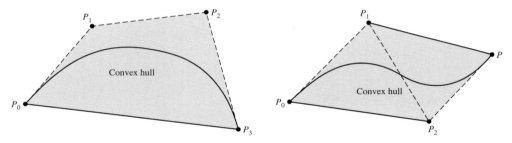

Figure 6.33   The convex hull of a Bezier curve.

EXAMPLE 6.18   **Find a Bezier curve equation.**
The coordinates of four points are given by $\mathbf{P}_0 = [2 \quad 2 \quad 0]^T$, $\mathbf{P}_1 = [2 \quad 3 \quad 0]^T$, $\mathbf{P}_2 = [3 \quad 3 \quad 0]^T$, and $\mathbf{P}_3 = [3 \quad 2 \quad 0]^T$. Find the equation of the Bezier curve. Also, find points on the curve for $u = 0, 0.25, 0.5, 0.75$ and 1.

SOLUTION   Equation (6.83) gives

$$\mathbf{P}(u) = \mathbf{P}_0 B_{0,3} + \mathbf{P}_1 B_{1,3} + \mathbf{P}_2 B_{2,3} + \mathbf{P}_3 B_{3,3}, \qquad 0 \le u \le 1$$

Using Eqs. (6.84) and (6.85), the preceding equation becomes

$$\mathbf{P}(u) = \mathbf{P}_0(1-u)^3 + 3\mathbf{P}_1 u(1-u)^2 + 3\mathbf{P}_2 u^2(1-u) + \mathbf{P}_3 u^3, \qquad 0 \le u \le 1$$

Substituting the u values into this equation gives

$$\mathbf{P}(0) = \mathbf{P}_0 = [2 \quad 2 \quad 0]^T$$

$$\mathbf{P}\left(\frac{1}{4}\right) = \frac{27}{64}\mathbf{P}_0 + \frac{27}{64}\mathbf{P}_1 + \frac{9}{64}\mathbf{P}_2 + \frac{1}{64}\mathbf{P}_3 = [2.156 \quad 2.563 \quad 0]^T$$

$$\mathbf{P}\left(\frac{1}{2}\right) = \frac{1}{8}\mathbf{P}_0 + \frac{3}{8}\mathbf{P}_1 + \frac{3}{8}\mathbf{P}_2 + \frac{1}{8}\mathbf{P}_3 = [2.5 \quad 2.75 \quad 0]^T$$

$$\mathbf{P}\left(\frac{3}{4}\right) = \frac{1}{64}\mathbf{P}_0 + \frac{9}{64}\mathbf{P}_1 + \frac{27}{64}\mathbf{P}_2 + \frac{27}{64}\mathbf{P}_3 = [2.844 \quad 2.563 \quad 0]^T$$

$$\mathbf{P}(1) = \mathbf{P}_3 = [3 \quad 2 \quad 0]^T$$

Observe that $\sum\limits_{i=0}^{3}$ is always equal to 1 for any u value. Figure 6.34 shows the curve and the points.

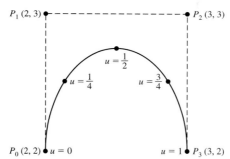

Figure 6.34   Cubic Bezier curve and generated points.

6.14 B-Spline Curve

B-spline curves provide another effective method, besides Bezier, of generating curves defined by data points. In fact, B-spline curves are the proper and powerful generalization of Bezier curves. In addition to sharing most of the characteristics of Bezier curves, they enjoy some other unique advantages. They provide local control of the curve shape as opposed to global control by using blending functions which provide local influence. They also provide the ability to separate the curve degree from the number of data points that defines it.

B-spline curves can interpolate or approximate a set of given data points. Interpolation is useful in displaying design or engineering results such as stress or displacement distribution in a part, while approximation is good to design free from curves. Interpolation is also useful if the designer has measured data points which must lie on the resulting curve. This section covers only B-spline curves as used for approximation.

In contrast to Bezier curves, the theory of B-spline curves separates the degree of the resulting curve from the number of the given control points. While four control points can always produce a cubic Bezier curve, they can generate a linear, quadratic, or cubic B-spline curve. This flexibility in the degree of the resulting curve is achieved by choosing the basis (blending) functions of B-spline curves with an additional degree of freedom that does not exist in Bernstein polynomials. These basis functions are the B-splines; thus the name B-spline curves.

Similar to Bezier curves, the B-spline curve defined by $n + 1$ control points P_i is given by

$$\mathbf{P}(u) = \sum_{i=0}^{n} \mathbf{P}_i N_{i,\,k}(u), \qquad 0 \le u \le u_{\max} \tag{6.91}$$

$N_{i,\,k}(u)$ are the B-spline functions. The control points (sometimes called deBoor points) form the vertices of the control or deBoor polygon. There are two major differences between Eqs. (6.91) and (6.83). First, the parameter k controls the degree $(k-1)$ of the resulting B-spline curve and is usually independent of the number of control points. Second, the maximum limit of the parameter u is no longer 1 as was arbitrarily chosen for Bezier curves. The B-spline functions have the following properties:

$$\text{Partition of unity:} \qquad \sum_{i=0}^{n} N_{i,k}(u) = 1$$

Positivity: $\qquad\qquad N_{i,k}(u) \ge 0$

Local support: $\qquad\; N_{i,k}(u) = 0 \qquad$ if $u \notin [u_i,\, u_{i+k+1}]$

Continuity: $\qquad\qquad N_{i,k}(u)$ is $(k-2)$ times continuously differentiable

The first property ensures that the relationship between the curve and its defining control points is invariant under affine transformations. The second property guarantees that the curve segment lies completely within the convex hull of P_i. The third property indicates that each segment of a B-spline curve is influenced by only k control points or each control point affects only k curve segments. It is useful to notice that the Bernstein polynomial, $B_{i,\,n}(u)$, has the same first two properties mentioned for the B-spline functions.

The B-spline functions also have the property of recursion, which defines them as:

$$N_{i,\,k}(u) = (u - u_i) \frac{N_{i,\,k-1}(u)}{u_{i+k-1} - u_i} + (u_{i+k} - u) \frac{N_{i+1,\,k-1}(u)}{u_{i+k} - u_{i+1}} \tag{6.92}$$

where

$$N_{i,\,1} = \begin{cases} 1, & u_i \le u \le u_{i+1} \\ 0, & \text{otherwise} \end{cases} \tag{6.93}$$

Choose $0/0 = 0$ if the denominators in Eq. (6.92) become zero. Equation (6.93) shows that $N_{i,\,1}$ is a unit step function.

Because $N_{i,\,1}$ is constant for $k = 1$, a general value of k produces a polynomial in u of degree $(k-1)$ [see Eq. (6.92)] and therefore a curve of order k and degree $(k-1)$. The u_i are called parametric knots or knot values. These values form a sequence of nondecreasing integers called the knot vector. The values of the u_i depend on whether the B-spline curve is an open (nonperiodic) or closed (periodic) curve. For an open curve, they are given by

$$u_j = \begin{cases} 0, & j < k \\ j - k + 1, & k \le j \le n \\ n - k + 2, & j > n \end{cases} \tag{6.94}$$

where

$$0 \le j \le n + k \tag{6.95}$$

and the range of u is

$$0 \le u \le n - k + 2 \tag{6.96}$$

Relation (6.95) shows that $(n + k + 1)$ knots are needed to create a $(k - 1)$ degree curve defined by $(n + 1)$ control points. These knots are evenly spaced over the range of u with unit separation ($\Delta u = 1$) between noncoincident knots. Multiple (coincident) knots for certain values of u may exist.

While the degree of the resulting B-spline curve is controlled by k, the range of the parameter u as given by Eq. (6.96) implies that there is a limit on k that is determined by the number of the given control points. This limit is found by requiring the upper bound in Eq. (6.96) to be greater than the lower bound for the u range to be valid, that is,

$$n - k + 2 > 0 \tag{6.97}$$

Equation (6.97) shows that a minimum of two, three, and four control points are required to define linear, quadratic, and cubic B-spline curves respectively.

The characteristics of B-spline curves that are useful in design can be summarized as follows:

1. The local control of the curve can be achieved by changing the position of a control point(s), using multiple control points by placing several points at the same location, or by choosing a different degree $(k - 1)$. Changing one control point affects only k segments. Figure 6.35 shows the local control for a cubic B-spline curve by moving P_3 into P_3^* and P_3^{**}. Only the four curve segments surrounding P_3 change.
2. A nonperiodic B-spline curve passes through the first and last control points P_0 and P_{n+1} and is tangent to the first $(\mathbf{P}_1 - \mathbf{P}_0)$ and last $(\mathbf{P}_{n+1} - \mathbf{P}_n)$ segments of the control polygon, similar to a Bezier curve, as shown in Figure 6.35.
3. Increasing the degree of the curve tightens it. In general, the less the degree, the closer the curve gets to the control points as shown in Figure 6.36. When $k = 1$, a zero-degree curve results. The curve then becomes the control points themselves. When $k = 2$, the curve becomes the polygon segments themselves.
4. A second-degree curve is always tangent to the midpoints of all the internal polygon segments (see Figure 6.36). This is not the case for other degrees.
5. If k equals the number of control points $(n + 1)$, then the resulting B-spline curve becomes a Bezier curve. In this case the range of u becomes 0 to 1 as expected.
6. Multiple control points induce regions of high curvature in a B-spline curve, that is the curve is pulled more toward a control point by increasing its multiplicity (see Figure 6.37). This multiplicity is useful for creating sharp corners in the curve.
7. Increasing the degree of the curve makes it more difficult to control and to calculate accurately. Therefore, a cubic B-spline is sufficient for a large number of applications.

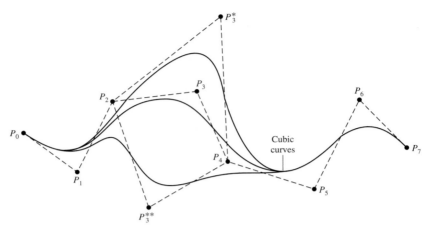

Figure 6.35   **Local control of B-spline curves.**

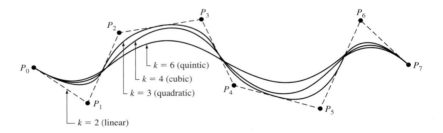

Figure 6.36   **Effect of the degree of B-spline curve on its shape.**

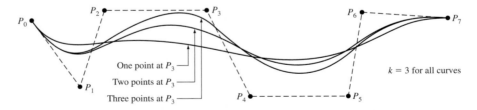

Figure 6.37   **Effect of point multiplicity of B-spline curve on its shape.**

Thus far, open or nonperiodic B-spline curves have been discussed. The same theory can be extended to cover closed or periodic B-spline curves. The only difference between open and closed curves is in the choice of the knots and the basic functions. Equations (6.94) to (6.96) determine the knots and the spacing between them for open curves.

Closed curves utilize periodic B-spline functions as their basis with knots at the integers. These basis functions are cyclic translates of a single canonical function with a period (interval) of k for support. For example, in a closed B-spline curve of order 2 ($k = 2$) or a degree 1 ($k - 1$), the basis function is linear, has nonzero value in the interval (0, 2) only, and has a maximum value of 1 at $u = 1$ as shown in Figure 6.38. The knot vector in this case is $[0 \ 1 \ 2]^T$. Quadratic and cubic closed curves have quadratic and cubic basis functions with intervals of (0, 3) and (0, 4) and knot vectors of $[0 \ 1 \ 2 \ 3]^T$ and $[0 \ 1 \ 2 \ 3 \ 4]^T$, respectively.

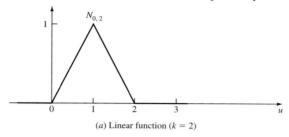

(a) Linear function ($k = 2$)

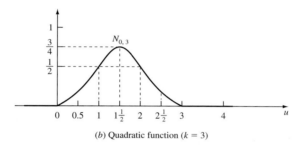

(b) Quadratic function ($k = 3$)

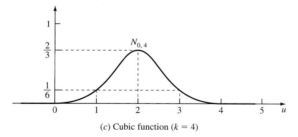

(c) Cubic function ($k = 4$)

Figure 6.38   Periodic B-spline basis functions.

The closed B-spline curve of degree ($k - 1$) or order K defined by ($n + 1$) control points is given by Eq. (6.91) as the open curve. However, for closed curves, Eqs. (6.92) to (6.96) become

$$N_{i,k}(u) = N_{0,k}((u - i + n + 1) \bmod (n + 1)) \tag{6.98}$$

$$u_j = j, \qquad 0 \le j \le n + 1 \tag{6.99}$$

$$0 < j \le n + 1 \tag{6.100}$$

and the range of u is

$$0 \le u \le n + 1 \qquad (6.101)$$

The mod $(n + 1)$ in Eq. (6.98) is the modulo function. It is defined

$$A \bmod n = \begin{cases} A, & A < n \\ 0, & A = n \\ \text{remainder of } A/n, & A > n \end{cases} \qquad (6.102)$$

For example, 3.5 mod 6 = 3.5, 6 mod 6 = 0, and 7 mod 6 = 1. The mod function enables the periodic (cyclic) translation [mod $(n + 1)$] of the canonical basis function $N_{0,k}$. $N_{0,k}$, which is the same as for open curves and can be calculated using Eqs. (6.92) and (6.93).

Like open curves, closed B-spline curves enjoy the properties of partition of unity, positivity, local support, and continuity. They also share the same characteristics of the open curves except that they do not pass through the first and last control points and therefore are not tangent to the first and last segments of the control polygon.

In representing closed curves, closed polygons are used where the first and last control points are connected by a polygon segment. It should be noted, however, that a closed B-spline curve cannot be generated by simply using an open curve with the first and last control points being the same (coincident). The resulting curve is only C^0 continuous; if the first and last segments of the polygon are collinear, a C^1 continuous curve results. In either case, the curve is still an open B-spline curve.

EXAMPLE 6.19   **Find the equation of an open B-spline curve.**

Find the equation of a cubic B-spline curve defined by the same control points as in Example 6.18. How does the curve compare with the Bezier curve?

SOLUTION   This cubic spline has $k = 4$ and $n = 3$. Eight knots are needed to calculate the B-spline functions. Equation (6.94) gives the knot vector $[u_0 \; u_1 \; u_2 \; u_3 \; u_4 \; u_5 \; u_6 \; u_7]^T$ as $[0 \; 0 \; 0 \; 0 \; 1 \; 1 \; 1 \; 1]^T$. The range of u [Eq. (6.96)] is $0 \le u \le 1$. Equation (6.91) gives

$$\mathbf{P}(u) = \mathbf{P}_0 N_{0,4} + \mathbf{P}_1 N_{1,4} + \mathbf{P}_2 N_{2,4} + \mathbf{P}_3 N_{3,4}, \qquad 0 \le u \le 1 \qquad (6.103)$$

To calculate the B-spline functions, use Eqs. (6.92) and (6.93) together with the knot vector as follows:

$$N_{0,1} = N_{1,1} = N_{2,1} = \begin{cases} 1, & u = 0 \\ 0, & \text{elsewhere} \end{cases}$$

$$N_{3,1} = \begin{cases} 1, & 0 \le u \le 1 \\ 0, & \text{elsewhere} \end{cases}$$

$$N_{4,1} = N_{5,1} = N_{6,1} = \begin{cases} 1, & u = 1 \\ 0, & \text{elsewhere} \end{cases}$$

$$N_{0,2} = (u - u_0)\frac{N_{0,1}}{u_1 - u_0} + (u_2 - u)\frac{N_{1,1}}{u_2 - u_1} = \frac{u N_{0,1}}{0} + \frac{(-u)N_{1,1}}{0} = 0$$

$$N_{1,2} = (u - u_1)\frac{N_{1,1}}{u_2 - u_1} + (u_3 - u)\frac{N_{2,1}}{u_3 - u_2} = \frac{u N_{1,1}}{0} + \frac{(-u)N_{2,1}}{0} = 0$$

$$N_{2,2} = (u-u_2)\frac{N_{2,1}}{u_3-u_2} + (u_4-u)\frac{N_{3,1}}{u_4-u_3} = \frac{uN_{2,1}}{0} + \frac{(1-u)N_{3,1}}{1} = (1-u)N_{3,1}$$

$$N_{3,2} = (u-u_3)\frac{N_{3,1}}{u_4-u_3} + (u_5-u)\frac{N_{4,1}}{u_5-u_4} = uN_{3,1} + \frac{(1-u)N_{4,1}}{0} = uN_{3,1}$$

$$N_{4,2} = (u-u_4)\frac{N_{4,1}}{u_5-u_4} + (u_6-u)\frac{N_{5,1}}{u_6-u_5} = (u-1)\frac{N_{4,1}}{0} + \frac{(1-u)N_{5,1}}{0} = 0$$

$$N_{5,2} = (u-u_5)\frac{N_{5,1}}{u_6-u_5} + (u_7-u)\frac{N_{6,1}}{u_7-u_6} = \frac{(u-1)N_{5,1}}{0} + \frac{(1-u)N_{6,1}}{0} = 0$$

$$N_{0,3} = (u-u_0)\frac{N_{0,2}}{u_2-u_0} + (u_3-u)\frac{N_{1,2}}{u_3-u_1} = u\frac{0}{0} + (-u)\frac{0}{0} = 0$$

$$N_{1,3} = (u-u_1)\frac{N_{1,2}}{u_3-u_1} + (u_4-u)\frac{N_{2,2}}{u_4-u_2} = u\frac{N_{1,2}}{0} + \frac{(1-u)N_{2,2}}{1} = (1-u)^2 N_{3,1}$$

$$N_{2,3} = (u-u_2)\frac{N_{2,2}}{u_4-u_2} + (u_5-u)\frac{N_{3,2}}{u_5-u_3} = uN_{2,2} + (1-u)N_{3,2} = 2u(1-u)N_{3,1}$$

$$N_{3,3} = (u-u_3)\frac{N_{3,2}}{u_5-u_3} + (u_6-u)\frac{N_{4,2}}{u_6-u_4} = u^2 N_{3,1} + (1-u)\frac{N_{4,2}}{0} = u^2 N_{3,1}$$

$$N_{4,3} = (u-u_4)\frac{N_{4,2}}{u_6-u_4} + (u_7-u)\frac{N_{5,2}}{u_7-u_5} = (u-1)\frac{N_{4,2}}{0} + (1-u)\frac{N_{5,2}}{0} = 0$$

$$N_{0,4} = (u-u_0)\frac{N_{0,3}}{u_3-u_0} + (u_4-u)\frac{N_{1,3}}{u_4-u_1} = (1-u)^3 N_{3,1}$$

$$N_{1,4} = (u-u_1)\frac{N_{1,3}}{u_4-u_1} + (u_5-u)\frac{N_{2,3}}{u_5-u_2} = 3u(1-u)^2 N_{3,1}$$

$$N_{2,4} = (u-u_2)\frac{N_{2,3}}{u_5-u_2} + (u_6-u)\frac{N_{3,3}}{u_6-u_3} = 3u^2(1-u)N_{3,1}$$

$$N_{3,4} = (u-u_3)\frac{N_{3,3}}{u_6-u_3} + (u_7-u)\frac{N_{4,3}}{u_7-u_4} = u^3 N_{3,1}$$

Substituting $N_{i,4}$ into Eq. (6.103) gives

$$\mathbf{P}(u) = [\mathbf{P}_0(1-u)^3 + 3\mathbf{P}_1 u(1-u)^2 + 3\mathbf{P}_2 u^2(1-u) + \mathbf{P}_3 u^3]N_{3,1}, \qquad 0 \le u \le 1$$

Substituting $N_{3,1}$ into this equation gives the curve equation as

$$\mathbf{P}(u) = \mathbf{P}_0(1-u)^3 + 3\mathbf{P}_1 u(1-u)^2 + 3\mathbf{P}_2 u^2(1-u) + \mathbf{P}_3 u^3, \qquad 0 \le u \le 1$$

This equation is the same as the one for the Bezier curve in Example 6.18. Thus the cubic B-spline curve defined by four control points is identical to the cubic Bezier curve defined by the same points. This fact can be generalized for $(k-1)$ degree curve as mentioned earlier.

There are two observations that are worth mentioning here. First, the sum of the two subscripts (i, k), of any B-spline function $N_{i,k}$ cannot exceed $(n + k)$. This gives a control on how far to go to calculate $N_{i,\,k}$. In this example, six functions of $N_{i,1}$, five of $N_{i,\,2}$, and four of $N_{i,\,3}$ were needed such that $(6 + 1)$ for the first, $(5 + 2)$ for the second, and $(4 + 3)$ for the last are always equal to 7 $(n + k)$. Second, whenever the limits of u for any $N_{i,\,1}$ are equal, the u range becomes one point.

EXAMPLE 6.20   **Find the equation of a closed B-spline curve.**
Find the equation of a closed (periodic) B-spline curve defined by four control points.

SOLUTION   This closed cubic spline has $k = 4$, $n = 3$. Using Eqs. (6.99) to (6.101), the knot vector $[u_0 \; u_1 \; u_2 \; u_3 \; u_4]^T$ is the integers $[0 \; 1 \; 2 \; 3 \; 4]^T$ and the range of u is $0 \le u \le 4$. Equation (6.91) gives the curve equation as:

$$P(u) = P_0 N_{0,\,4} + P_1 N_{1,\,4} + P_2 N_{2,\,4} + P_3 N_{3,\,4}, \qquad 0 \le u \le 4 \qquad (6.104)$$

To calculate the B-spline functions, use Eq. (6.98) to obtain:

$$N_{0,\,4}(u) = N_{0,\,4}((u + 4) \bmod 4)$$
$$N_{1,\,4}(u) = N_{0,\,4}((u + 3) \bmod 4)$$
$$N_{2,\,4}(u) = N_{0,\,4}((u + 2) \bmod 4)$$
$$N_{3,\,4}(u) = N_{0,\,4}((u + 1) \bmod 4)$$

In the preceding equations, $N_{0,\,4}$ on the right-hand side is the function for the open curve and the one on the left-hand side is the periodic function for the closed curve. Substituting these equations into Eq. (6.104) we get

$$P(u) = P_0 N_{0,\,4}((u + 4) \bmod 4) + P_1 N_{0,\,4}((u + 3) \bmod 4)$$
$$+ P_2 N_{0,\,4}((u + 2) \bmod 4) + P_3 N_{0,\,4}((u + 1) \bmod 4), \qquad 0 \le u \le 4 \qquad (6.105)$$

In Eq. (6.105), the function $N_{0,\,4}$ has various arguments which can be found if specific values of u are used. To find $N_{0,4}$, similar calculations to those in Example 6.19 are performed using the preceding knot vector as follows:

$$N_{0,\,1} = \begin{cases} 1, & 0 \le u \le 1 \\ 0, & \text{elsewhere} \end{cases}$$

$$N_{1,\,1} = \begin{cases} 1, & 1 \le u \le 2 \\ 0, & \text{elsewhere} \end{cases}$$

$$N_{2,\,1} = \begin{cases} 1, & 2 \le u \le 3 \\ 0, & \text{elsewhere} \end{cases}$$

$$N_{3,\,1} = \begin{cases} 1, & 3 \le u \le 4 \\ 0, & \text{elsewhere} \end{cases}$$

$$N_{0,\,2} = (u - u_0)\frac{N_{0,\,1}}{u_1 - u_0} + (u_2 - u)\frac{N_{1,\,1}}{u_2 - u_1} = uN_{0,\,1} + (2 - u)N_{1,\,1}$$

$$N_{1,\,2} = (u - u_1)\frac{N_{1,\,1}}{u_2 - u_1} + (u_3 - u)\frac{N_{2,\,1}}{u_3 - u_2} = (u - 1)N_{1,\,1} + (3 - u)N_{2,\,1}$$

$$N_{2,2} = (u - u_2)\frac{N_{2,1}}{u_3 - u_2} + (u_4 - u)\frac{N_{3,1}}{u_4 - u_3} = (u - 2)N_{2,1} + (4 - u)N_{3,1}$$

$$N_{0,3} = (u - u_0)\frac{N_{0,2}}{u_2 - u_0} + (u_3 - u)\frac{N_{1,2}}{u_3 - u_1} = \frac{1}{2}uN_{0,2} + \frac{1}{2}(3 - u)N_{1,2}$$

$$= \frac{1}{2}u^2 N_{0,1} + \frac{1}{2}[u(2 - u) + (3 - u)(u - 1)]N_{1,1} + \frac{1}{2}(3 - u)^2 N_{2,1}$$

$$N_{1,3} = (u - u_1)\frac{N_{1,2}}{u_3 - u_1} + (u_4 - u)\frac{N_{2,2}}{u_4 - u_2} = \frac{1}{2}(u - 1)N_{1,2} + \frac{1}{2}(4 - u)N_{2,2}$$

$$= \frac{1}{2}(u - 1)^2 N_{1,1} + \frac{1}{2}[(u - 1)(3 - u) + (u - 2)(4 - u)]N_{2,1} + \frac{1}{2}(4 - u)^2 N_{3,1}$$

$$N_{0,4} = (u - u_0)\frac{N_{0,3}}{u_3 - u_0} + (u_4 - u)\frac{N_{1,3}}{u_4 - u_1} = \frac{1}{3}uN_{0,3} + \frac{1}{3}(4 - u)N_{1,3}$$

$$= \frac{1}{6}\{u^3 N_{0,1} + [u^2(2 - u) + u(3 - u)(u - 1) + (4 - u)(u - 1)^2]N_{1,1}$$

$$+ [u(3 - u)^2 + (4 - u)(u - 1)(3 - u) + (4 - u)^2(u - 2)]N_{2,1} + (4 - u)^3 N_{3,1}\}$$

or

$$N_{0,4} = \frac{1}{6}[u^3 N_{0,1} + (-3u^3 + 12u^2 - 12u + 4)N_{1,1} + (3u^3 - 24u^2 + 60u - 44)N_{2,1}$$

$$+ (-u^3 + 12u^2 - 48u + 64)N_{3,1}]$$

Due to the nonzero values of the functions $N_{i,1}$ for various intervals of u, the preceding equation can be written as:

$$N_{0,4}(u) = \begin{cases} \frac{1}{6}u^3, & 0 \leq u \leq 1 \\ \frac{1}{6}(-3u^3 + 12u^2 - 12u + 4), & 1 \leq u \leq 2 \\ \frac{1}{6}(3u^3 - 24u^2 + 60u - 44), & 2 \leq u \leq 3 \\ \frac{1}{6}(-u^3 + 12u^2 - 48u + 64), & 3 \leq u \leq 4 \end{cases} \tag{6.106}$$

To check the correctness of Eq. (6.106), one would expect to obtain Figure 6.38c if this function is plotted. Indeed this figure is the plot of $N_{i,4}$. If $u = 0, 1, 2, 3$, and 4 are substituted into this function, the corresponding values of $N_{0,4}$ that are shown in Figure 6.38 are obtained.

Equations (6.105) and (6.106) together can be used to evaluate points on the closed B-spline curve for display or plotting purposes. As an illustration, consider the following points:

$$\mathbf{P}(0) = \mathbf{P}_0 N_{0,4}(4 \bmod 4) + \mathbf{P}_1 N_{0,4}(3 \bmod 4)$$

$$+ \mathbf{P}_2 N_{0,4}(2 \bmod 4) + \mathbf{P}_3 N_{0,4}(1 \bmod 4)$$

$$= \mathbf{P}_0 N_{0,4}(0) + \mathbf{P}_1 N_{0,4}(3) + \mathbf{P}_2 N_{0,4}(2) + \mathbf{P}_3 N_{0,4}(1)$$

$$= \frac{1}{6}\mathbf{P}_1 + \frac{2}{3}\mathbf{P}_2 + \frac{1}{6}\mathbf{P}_3$$

Similarly

$$\mathbf{P}(0.5) = \mathbf{P}_0 N_{0,4}(0.5) + \mathbf{P}_1 N_{0,4}(3.5) + \mathbf{P}_2 N_{0,4}(2.5) + \mathbf{P}_3 N_{0,4}(1.5)$$

$$= \frac{1}{48}\mathbf{P}_0 + \frac{1}{48}\mathbf{P}_1 + \frac{23}{48}\mathbf{P}_2 + \frac{23}{48}\mathbf{P}_3$$

$$\mathbf{P}(1) = \mathbf{P}_0 N_{0,4}(1) + \mathbf{P}_1 N_{0,4}(0) + \mathbf{P}_2 N_{0,4}(3) + \mathbf{P}_3 N_{0,4}(2)$$

$$= \tfrac{1}{6}\mathbf{P}_0 + \tfrac{1}{6}\mathbf{P}_2 + \tfrac{2}{3}\mathbf{P}_3$$

$$\mathbf{P}(2) = \mathbf{P}_0 N_{0,4}(2) + \mathbf{P}_1 N_{0,4}(1) + \mathbf{P}_2 N_{0,4}(0) + \mathbf{P}_3 N_{0,4}(3) = \tfrac{2}{3}\mathbf{P}_0 + \tfrac{1}{6}\mathbf{P}_1 + \tfrac{1}{6}\mathbf{P}_3$$

$$\mathbf{P}(3) = \mathbf{P}_0 N_{0,4}(3) + \mathbf{P}_1 N_{0,4}(2) + \mathbf{P}_2 N_{0,4}(1) + \mathbf{P}_3 N_{0,4}(0) = \tfrac{1}{6}\mathbf{P}_0 + \tfrac{2}{3}\mathbf{P}_1 + \tfrac{1}{6}\mathbf{P}_2$$

$$\mathbf{P}(4) = \mathbf{P}_0 N_{0,4}(0) + \mathbf{P}_1 N_{0,4}(3) + \mathbf{P}_2 N_{0,4}(2) + \mathbf{P}_3 N_{0,4}(1) = \tfrac{1}{6}\mathbf{P}_1 + \tfrac{2}{3}\mathbf{P}_2 + \tfrac{1}{6}\mathbf{P}_3$$

In these calculations, notice the cyclic rotation of the $N_{0,4}$ coefficients of the control points for the various values of u excluding $u = 0.5$. Notice also the effect of the canonical (symmetric) form of $N_{0,4}$ on the coefficients of the control points. If the u values are 0.5, 1.5, 2.5, and 3.5, or other values separated by unity, a similar cyclic rotation of the coefficients is expected. Finally, notice that $\mathbf{P}(0)$ and $\mathbf{P}(4)$ are equal, which ensures obtaining a closed B-spline curve.

6.15 Curve Manipulations

Analytic and synthetic curves essential to geometric modeling have been presented. The effective use of these curves in a design and manufacturing environment depends mainly on their manipulation to achieve goals in hand. A user might want to blend curves with certain continuity requirements. Or the intersection of two curves in space might provide the coordinates of an important point to engineering calculations or modeling of a part. This section covers some useful features of curve manipulations.

6.15.1 Evaluating Points on Curves

Points on curves are generated for different modeling needs. For example, finite element modeling requires generating nodal points (nodes) on the model to be used later in finite element analysis.

It is also mentioned in Section 6.14 that a curve parametric equation is used to evaluate points on it. Evaluation methods must be efficient and fast for interactive purposes as well as capable of producing enough points to display a smooth curve. The obvious method of calculating the coordinates of points on a curve by substituting successive values of its parameter u into its equation is inefficient. Incremental methods prove to be more efficient. Using the forward difference technique to evaluate a curve polynomial equation at equal intervals of its parameter is the most common incremental method.

Evaluating a point on a curve by using its parametric value, u, is sometimes called the direct point solution. It entails evaluating three polynomials in u, one for each coordinate of the point. The inverse problem is another form of evaluating a point on a curve. Given a point on or close to a curve in terms of its cartesian coordinates x, y, and z, find the corresponding u value. This problem arises, for example, if tangent vectors are to be evaluated at certain locations on the curve. The solution to this problem is called the inverse point solution and requires the solution of a nonlinear polynomial in u via numerical methods.

6.15.2 Blending

The construction of composite curves from the various types of parametric curves forms the core of the blending problem, which can be stated as follows: Given two curves $P_1(u_1)$, $0 < u_1 < a$, $P_2(u_2)$, $0 < u_2 < b$, find the conditions for the two curves to be continuous at the joint. Notice that the upper limits on u_1 and u_2 are taken as a and b, and not 1, for generality. Three classes of continuity at the joint can be considered. The first is C^0 continuity, where the ending point of the first curve and the starting point of the second curve are the same. This gives

$$P_1(a) = P_2(0) \tag{6.107}$$

If the two segments are to be C^1 continuous as well, they must have slope continuity,

$$P_1'(a) = \alpha_1 T$$
$$P_2'(0) = \alpha_2 T \tag{6.108}$$

where α_1 and α_2 are constants and T is the common unit tangent vector at the joint. Consider the blending of a Bezier curve and an open B-spline curve. For slope continuity at the joint, the last segment of the control polygon of the former and the first segment of the control polygon of the latter must be collinear.

The third useful class of continuity is if curvature (C^2 continuity) is to be continuous at the joint in addition to position and slope. To achieve curvature continuity is less straightforward and requires the binormal vector to a curve at a point. Figure 6.39 shows the tangent unit vector T, the normal unit vector N, the center of curvature O, and the radius of curvature ρ at point P on a curve.

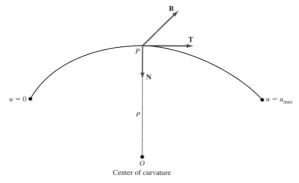

Figure 6.39   **Binormal vector to a curve.**

The curvature at P is defined as $\frac{1}{\rho}$. The binormal vector B is defined as

$$B = T \times N \tag{6.109}$$

The curvature is related to the curve derivatives through the vector B by the following equation:

$$\frac{1}{\rho}B = \frac{P' \times P''}{|P'|^3} \tag{6.110}$$

where $\mathbf{P}''$ is the second derivative with respect to the parameter u. The following condition can then be written for curvature continuity at the joint

$$\frac{\mathbf{P}_1'(a) \times \mathbf{P}_1''(a)}{|\mathbf{P}_1'(a)|^3} = \frac{\mathbf{P}_2'(0) \times \mathbf{P}_2''(0)}{|\mathbf{P}_2'(0)|^3} \tag{6.111}$$

Substituting Eq. (6.108) into Eq. (6.111) gives

$$\mathbf{T} \times \mathbf{P}_1''(a) = \left(\frac{\alpha_1}{\alpha_2}\right)^2 \mathbf{T} \times \mathbf{P}_2''(0) \tag{6.112}$$

This equation can be satisfied if

$$\mathbf{P}_2''(0) = \left(\frac{\alpha_2}{\alpha_1}\right)^2 \mathbf{P}_1''(a) \tag{6.113}$$

or, in general, if

$$\mathbf{P}_2''(0) = \left(\frac{\alpha_2}{\alpha_1}\right)^2 \mathbf{P}_1''(0) + \gamma \mathbf{P}_1'(a) \tag{6.114}$$

where γ is an arbitrary scaler which is chosen as zero for practical purposes.

6.15.3 Segmentation

Segmentation or curve splitting is defined as replacing one existing curve by one or more curve segments of the same curve type such that the shape of the composite curve is identical to that of the original curve. Segmentation is a very useful feature for CAD/CAM systems. It is implemented as a `Divide  Entity` command. Model cleanup for drafting purposes is an example where a curve might be divided into two at the line of sight of another. One of the resulting segments is then deleted. Another example is when a closed curve has to be split for modeling purposes.

Mathematically, curve segmentation is a reparametrization or parameter transformation of the curve. Splitting lines, circles, and conics is a simple problem. To split a line connecting two points P_0 and P_1 at a point P_2, all that is needed is to define two new lines connecting the point pairs (P_0, P_2) and (P_2, P_1). For circles and conics, the angle corresponding to the splitting point together with the starting and ending angles of the original curve defines the proper range of the parameter u for the resulting two segments.

Polynomial curves such as cubic splines, Bezier curves, and B-spline curves require a different parameter transformation. If the degree of the polynomial defining a curve is to be unchanged, which is the case in segmentation, the transformation must be linear. Let us assume that a polynomial curve is defined over the range $u = u_0, u_m$. To split the curve at a point defined by $u = u_1$ means that the first and the second segments are to be defined over the range $u = u_0$, u_1 and $u = u_1, u_m$, respectively. A new parameter v is introduced for each segment such that its range is $v = 0, 1$. (See Figure 6.40.) The parameter transformation takes the form:

$$u = u_0 + (u_1 - u_0)v \qquad \text{for the first segment}$$

and

$$u = u_1 + (u_m - u_1)v \qquad \text{for the second segment}$$

(6.115)

It is clear that $v = 0$, 1 corresponds to the proper u values as required. If Eq. (6.115) is substituted for the equation of a given curve, the proper equation of each segment can be obtained in terms of the parameter v.

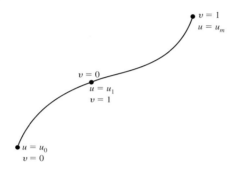

Figure 6.40   Reparametrization of a segmented curve.

6.15.4 Trimming

Trimming curves or entities is a very useful function provided by all CAD/CAM systems. Trimming can truncate or extend a curve. Trimming is mathematically identical to segmentation. The only difference between the two is that the result of trimming a curve is only one segment of the curve bounded by the trimming boundaries. Figure 6.41 shows the reparametrization of a trimmed curve. Extending general curves (Figure 6.41b) such as cubic splines, Bezier curves, and B-spline curves is not recommended because the curve behavior outside its original interval $u_0 < u < u_m$ may not be predictable.

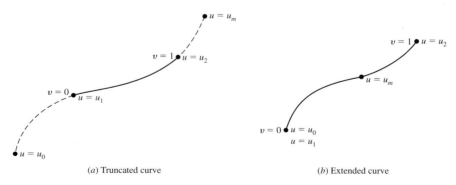

(a) Truncated curve   (b) Extended curve

Figure 6.41   Reparametrization of a trimmed curve.

6.15.5 Intersection

The intersection point (see Figure 6.42) of two parametric curves $P(u)$ and $Q(v)$ in 3D space requires the solution of the following equation in the parameters u and v:

$$P(u) - Q(v) = 0 \qquad (6.116)$$

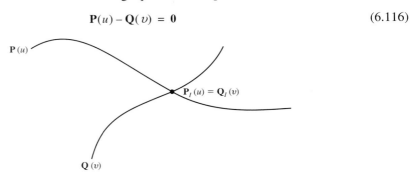

Figure 6.42   Intersection of two parametric curves in space.

Equation (6.116) represents three scalar equations that take the polynomial nonlinear form in the two parameter unknowns. One way to find u and v is to solve the x and y components of the equation simultaneously,

$$
\begin{aligned}
P_x(u) - Q_x(v) &= 0 \\
P_y(u) - Q_y(v) &= 0
\end{aligned}
\qquad (6.117)
$$

and then use the z component, $P_z(u) - Q_z(v) = 0$ to verify the solution.

The roots of Eqs. (6.116) or (6.117) can be found by numerical analysis methods such as Newton-Raphson method. This method requires an initial guess which can be determined interactively. The user is usually asked by the CAD/CAM software to click the two curves whose intersection is required. These clicks could be changed to an initial guess to start the solution. If more than one intersection point exists, the user must click close to the desired intersection point.

6.15.6 Transformation

Transformation allows the designer to translate, rotate, mirror, and scale various entities. These operations are known as rigid-body transformations. Homogeneous transformation, as discussed later in the book, offers concise matrix form to perform all rigid-body transformations as matrix multiplications, which is a desired feature from a software development point of view.

Equation (2.3) gives the general homogeneous transformation matrix $[T]$. The proper choice of the elements of this matrix produces the various rigid-body transformations. One of the main characteristics of rigid-body transformations is that geometric properties of curves, surfaces, and solids are invariant under these transformations. For example, parallel or perpendicular lines remain so after transformations. Intersection points of curves are transformed into the new intersection points, that is, one-to-one transformation.

6.16 Tutorials

6.16.1 Create a Guide Bracket

This tutorial illustrates the use of the analytic curves (lines and circles) in geometric modeling. It shows how to create the guide bracket shown below. All the sketches of the bracket use lines and circles of different sizes. The bracket is a 2½D uniform thickness model. All dimensions are in inches.

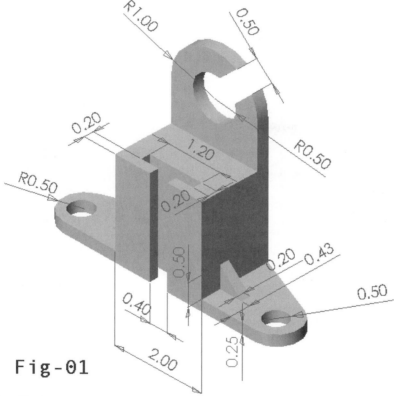

Fig-01

Modeling strategy:

1. Select the Top sketch plane.
2. Sketch the profile shown in the forthcoming screenshot.
3. Use extrusion with a thickness of 0.25 inch, to create the base shown in the forthcoming screenshot.
4. Select the top face of the base feature as the sketch plane and construct the sketch shown in the forthcoming screenshot.
5. Extrude the sketch in the upward direction a distance of 2.0 inches as shown in the forthcoming screenshot.

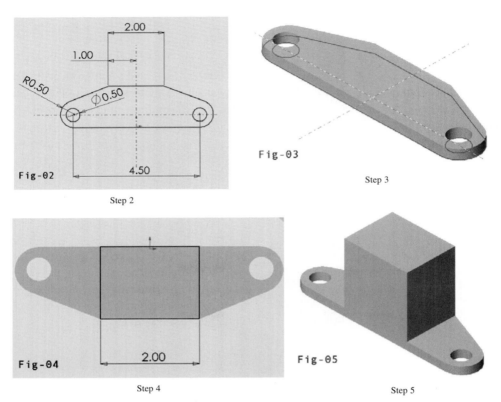

Fig-02   Step 2

Fig-03   Step 3

Fig-04   Step 4

Fig-05   Step 5

6. Select the top face of the block feature as the sketch plane. Sketch the profile shown the upcoming screenshot.

7. Cut extrude the profile (use `through all` option), to finish the feature as shown in the following screenshot.

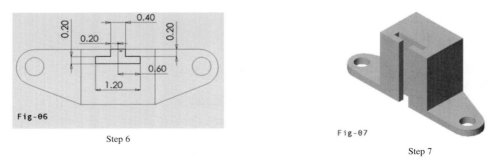

Fig-06   Step 6

Fig-07   Step 7

8. Select the back face of the block feature, shown in the forthcoming screenshot, as the sketch plane.

9. Construct the profile shown in the forthcoming screenshot.

10. Extrude the profile to the front of the existing block feature a distance of 0.25 inch.

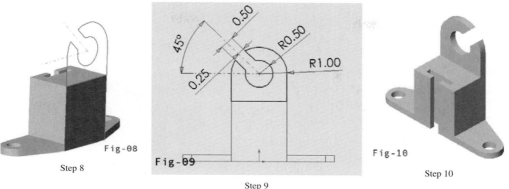

Fig-08

Step 8

Fig-09

Step 9

Fig-10

Step 10

11. To create the small rib features at the base, construct a new plane offset inside from the front face of the block feature up to distance of 0.425 inch as shown below.

12. Construct the profile shown below in the sketch plane of step 11.

13. Extrude the profile a distance of 0.2 inch toward the rear end of the part as shown below.

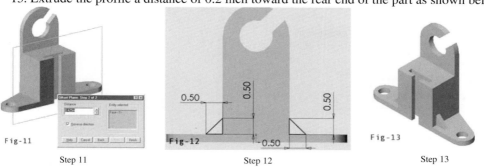

Fig-11

Step 11

Fig-12

Step 12

Fig-13

Step 13

The complete model is shown on the right.

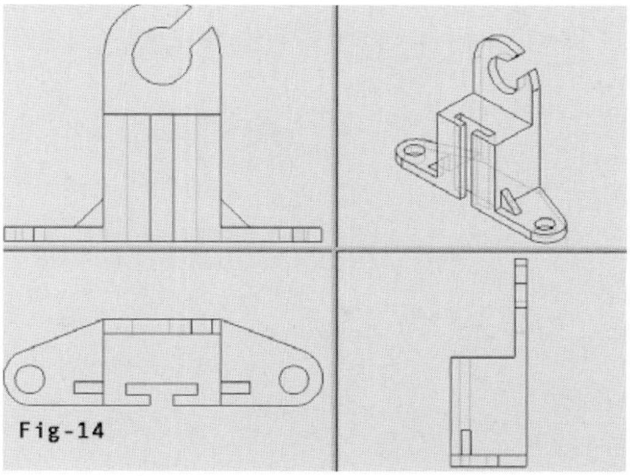

Tutorial 6.16.1 discussion:

The guide bracket model uses analytic curves such as lines and arcs.

As we construct the model, the CAD software shields us from the curve theory presented in this chapter. However, understanding the theory and being aware of it are important to productivity.

Fig-14

Tutorial 6.16.1 hands-on exercise:

Modify the model to use rectangular, instead of tapered, legs. Change the profile in step 2.

6.16.2　Create a Stop Block

This tutorial shows how to create the stop block shown on the right. The block is a 2½D uniform thickness model. All dimensions are in inches.

Modeling strategy:

1. Select the Front sketch plane.
2. Sketch the profile shown below.
3. Use extrusion with a thickness of 4.0 inches to create the bracket shown below.
4. Select the face of one of the legs as the sketch plane as shown below.
5. Create the profile shown below.
6. Cut extrude (use through all option) to create the feature shown below.

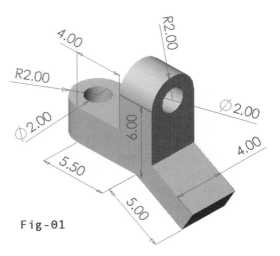

Fig-01

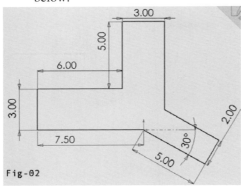

Fig-02

Step 2

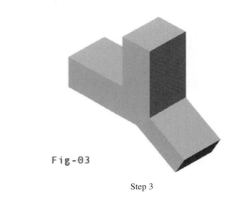

Fig-03

Step 3

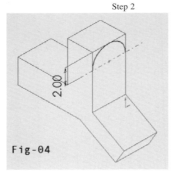

Fig-04

Steps 4 and 5

Fig-05

Step 6

7. Select the second-leg face, shown in the forthcoming screenshot, as the sketch plane.

8. Sketch the profile shown below.

9. Cut extrude (use `through all` option) the profile to complete the feature.

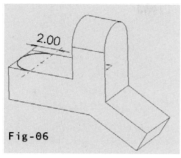

Fig-06

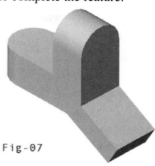

Fig-07

Steps 7 and 8   Step 9

10. Create holes on two of the legs as shown below. While creating the profile, apply concentric constraint between the circle sketch entity and the arc entity in order to make it concentric with the previously created feature.

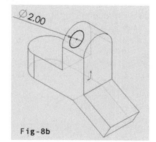

Fig-8b

Fig-09

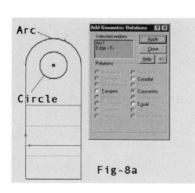

Fig-8a

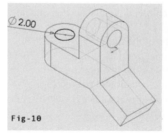

Fig-10

Fig-11

The complete model is shown on the right.

Tutorial 6.16.2 discussion:

The stop block model uses analytic curves, including lines, circles, and arcs.

Tutorial 6.16.2 hands-on exercise:

Round off the right leg of the model and create a hole in the round. Use the same dimensions of the round and the hole of the left leg.

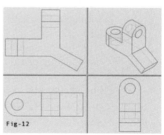

Fig-12

PROBLEMS

Part I: Theory

6.1 Find the equations of the three lines L_1, L_2, and L_3 shown. Are L_1 and L_2 perpendicular? Are L_1 and L_3 parallel? Prove your answers.

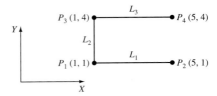

6.2 A line equation is given by:

$$\mathbf{P} = \begin{bmatrix} 3 \\ 4 \\ 0 \end{bmatrix} + u \begin{bmatrix} 2 \\ 2 \\ 0 \end{bmatrix}$$

Find the end points of the line. Sketch the line with the end points and the u direction. Find the intersection points between the line and a circle with a center at $(1, 2, 0)$ and radius of 2.

6.3 Find the angle θ between the two lines for both cases shown below.

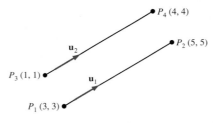 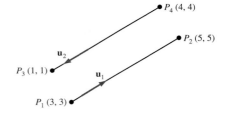

6.4 Two lines with their end points and an arc are shown on the right.
   a. Find the equations of the lines.
   b. Using the equations, find the intersection point I of the two lines.
   c. Find the parametric equation of the arc C_1 whose center is I and whose radius is 1 inch.
   d. Find the intersection point II between C_1 and L_1.

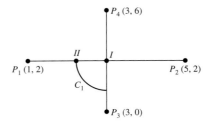

6.5 Two lines with their end points are shown on the right.
   a. Find the equations of lines L_1 and L_2. P_3 is the midpoint of L_1, that is, $u = 0.5$ at P_3.
   b. Find a point on each line where $u = 0.25$.
   c. Find the tangent vector for each line. Are they constant? What is your conclusion?

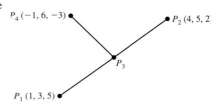

6.6 Find the equation of the circle shown. Using this equation, find the coordinates of point P_1 on the vertical centerline. Find the equation of the line that connects P_1 and P_2. What is the tangent vector of the line?

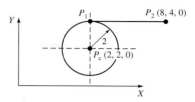

6.7 Find the equation of a Hermite cubic spline that passes through points $(1, 2)$ and $(3, 4)$ and whose tangent vectors are the two lines connecting these two points with point $(2, 7)$.

6.8 Find the equation and end points of each of the following lines:

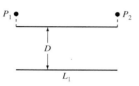

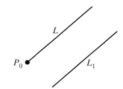

(a) Parallel to L_1 at distance D and bounded by points P_1 and P_2

(b) Parallel to L_1, passes through P_0, and has length L

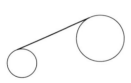

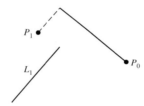

(c) Tangent to two given circles

(d) Perpendicular to L_1, passes through P_0, and bounded by P_1

(e) Tangent to a given circle from a given point P_0

(f) Tangent to a given ellipse from a given point P_0

6.9 Find the length of the common perpendicular to two skew lines.

6.10 Find the radius and the center of the following circles:

    a. Tangent to a given circle and a given line with a given radius.

    b. Tangent to two lines passing through a given point.

    c. Passing through two points and tangent to a line

    d. Tangent to a line, passing through a point, and with a given radius.

6.11 Find the center and the major and minor radii of an ellipse defined by two points and two slopes.

6.12 Find the intersection point of two tangent lines at two known points on an ellipse.

6.13 Find the tangent to an ellipse:

    a. at any given point on its circumference.

    b. from a point outside the ellipse.

6.14 Solve Problem 6.13 for a parabola.

6.15 The figure on the right shows a cubic spline curve consisting of two segments 1 and 2 with end conditions $\mathbf{P}_0$, $\mathbf{P'}_0$, $\mathbf{P}_1$, $\mathbf{P'}_{11}$, and $\mathbf{P}_1$, $\mathbf{P'}_{12}$, $\mathbf{P}_2$, $\mathbf{P'}_2$. The second subscripts in the tangent vectors at P_1 refers

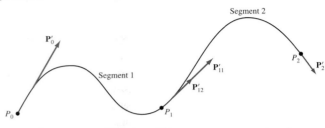

to the segment number. If $\mathbf{P'}_{11} = R$ and $\mathbf{P'}_{12} = KR$ where K is a constant, prove that the curve can only be C^1 continuous at P_1 and C^2 elsewhere.

6.16 Find the normal vector to a cubic spline curve at any of its points.

6.17 For a cubic Bezier curve, carry a similar matrix formulation to a cubic spline. Compare $[M_B]$ and V for the two curves.

6.18 Find the condition that degenerates a cubic Bezier curve to a line connecting P_0 and P_3.

6.19 Derive a method by which you can force a Bezier curve to pass through a given point in addition to the starting and end points of its polygon. Achieve that by changing the position of only one control point, say P_2.

6.20 Investigate the statement "each segment of a B-spline curve is influenced by only k control points or each control point affects only k curve segments." Use $n = 3$, $k = 2, 3, 4$.

6.21 Find the equation of an open quadratic B-spline curve defined by five control points.

6.22 For an open cubic B-spline curve defined by n control points, carry a similar matrix formulation to a cubic spline. Compare $[M_S]$ and V with those of the cubic spline and Bezier curves. Hint: Start with $n = 5$, that is six control points, and then generalize the resulting $[M_S]$.

6.23 Solve Problem 6.21 for a closed B-spline curve.

6.24 Given a point Q and a parametric curve in the cartesian space, find the closest point P on the curve to Q. Hint: Find P such that $(\mathbf{Q} - \mathbf{P})$ is perpendicular to the tangent vector.

6.25 A cubic Bezier curve is to be divided by a designer into two segments. Find the modified polygon points for each segment.

Part II: Laboratory

Use your in-house CAD/CAM system to answer the questions in this part.

6.26 Investigate the line, circle, ellipse, parabola, conics, cubic spline, Bezier curve, and B-spline curve commands and their related modifiers. Relate these modifiers to their theoretical background.

6.27 Choose a mechanical element such as a gear and generate its geometric model.

6.28 Create the following models. If dimensions are not shown, simply sketch in the sketcher so that the model looks similar to what is here, and then adjust the dimensions to finalize. Generate an engineering drawing for each model. Follow the drawing structure format shown in Figure 4.4.

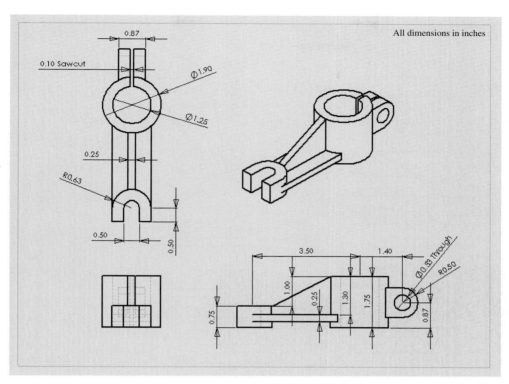

(a) Shifter

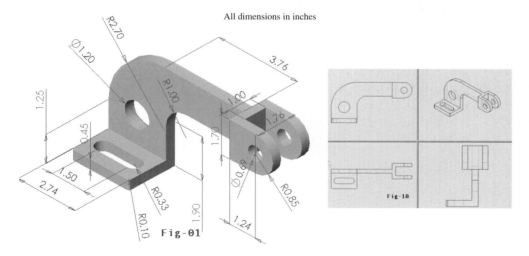

(b) Arm bracket

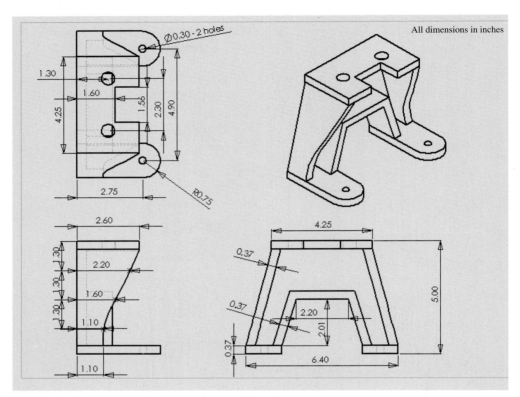

(*c*) Lathe leg

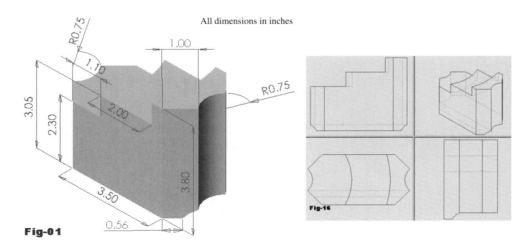

(*d*) Lathe jaw

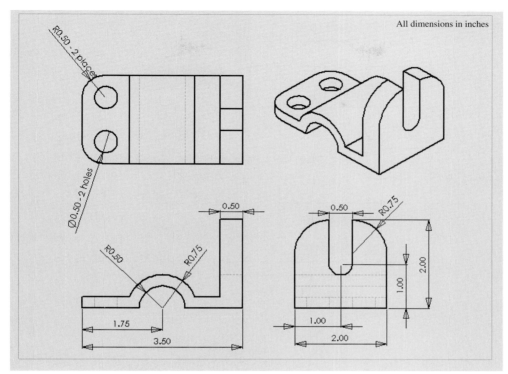

(e) Mounting bracket

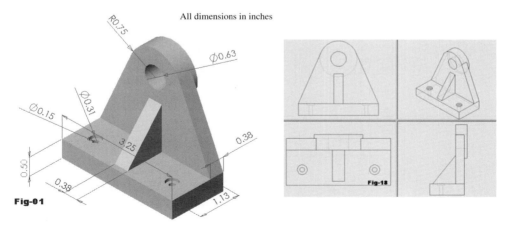

(f) Pipe bracket

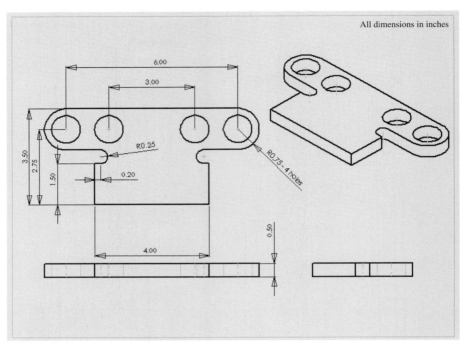

(*g*) Profile plate

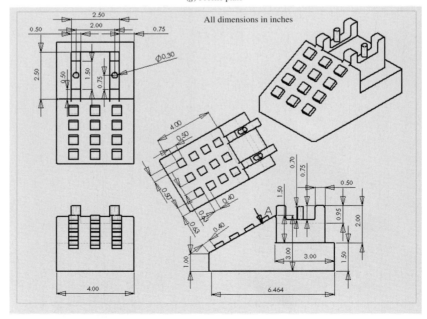

(*h*) Telephone

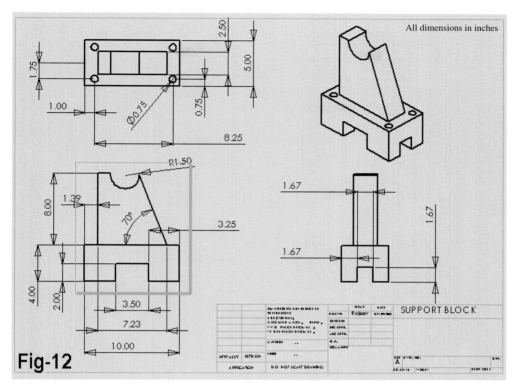

Fig-12

(*i*) Support block

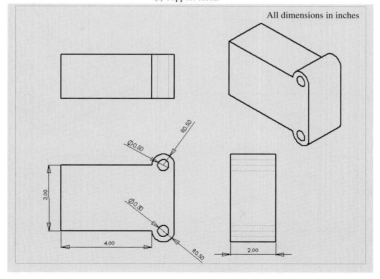

(*j*) Simple block

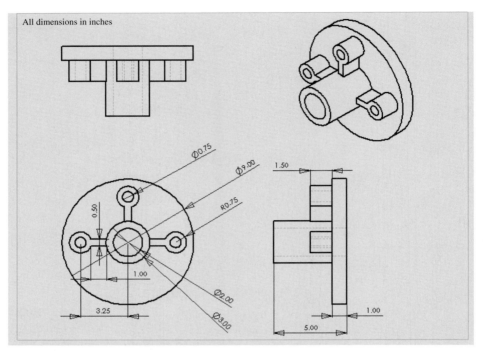

All dimensions in inches

(*k*) Circular plate

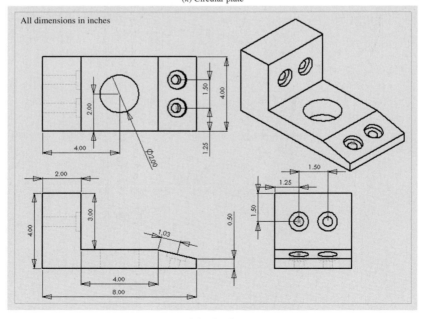

All dimensions in inches

(*l*) Leg bracket

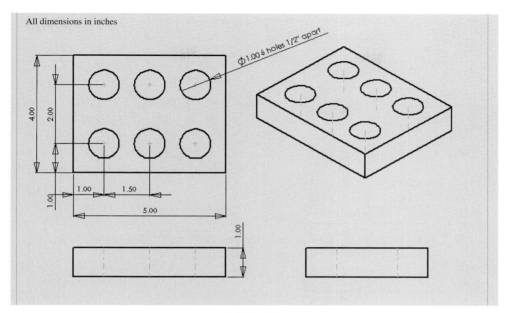

All dimensions in inches

(*m*) Base plate

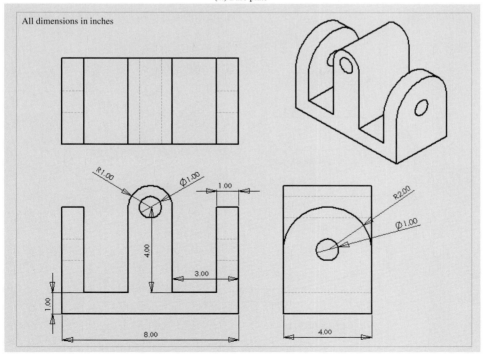

All dimensions in inches

(*n*) Side block

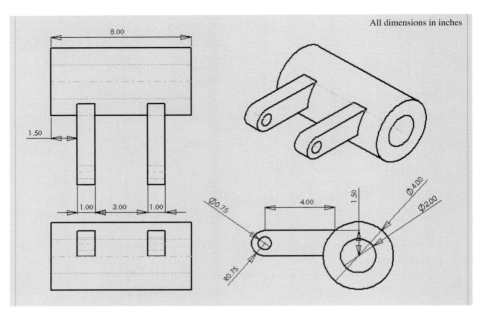

(*o*) Pipe support

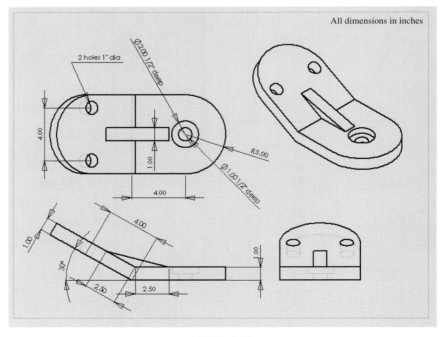

(*p*) Inclined plate

Part III: Programming

Use the parametric equations of curves in all the following questions. In addition, write the code so that the user can input values to control the curve shape and behavior as we have discussed in the chapter. The program should display the results. Use OpenGL with C or C++, or use Java 3D.

6.29 Write a Java or C++ code to create and display a line.

6.30 Write a Java or C++ code to create and display a circle and an arc. Write the code for an arc so that a circle results when the starting and ending angles become 0 and 360, respectively.

6.31 Write a Java or C++ code to create and display an ellipse.

6.32 Write a Java or C++ code to create and display a parabola.

6.33 Write a Java or C++ code to create and display a hyperbola.

6.34 Write a Java or C++ code to create and display a Hermite cubic spline.

6.35 Write a Java or C++ code to create and display a cubic spline defined by four points.

6.36 Write a Java or C++ code to create and display a Bezier curve.

6.37 Write a Java or C++ code to create and display an open B-spline.

6.38 Write a Java or C++ code to create and display a closed (periodic) B-spline.

CHAPTER 7

Surfaces

OAL

Understand and master the theory and practice of surfaces, their types, parametric formulation, their implementation by CAD/CAM systems, and their use in geometric modeling.

BJECTIVES

After reading this chapter, you should understand the following concepts:

● Surfaces and solid modeling

● Surface entities

● Extending curves to surfaces

● Surface parametric equations

● Surface properties

● Analytic surfaces

● Parametric surfaces

● Surface manipulations

CHAPTER HEADLINES

7.1   Introduction

Shape design and the representation of complex objects such as car, ship, and airplane bodies as well as castings cannot be achieved utilizing the curves covered in Chapter 6. In such cases, surfaces must be utilized to describe objects precisely and accurately. We create surfaces, and then we use them to cut and trim solid features and primitives to obtain the models of the complex objects. Surface creation usually begins with data points or curves.

Surface creation on CAD/CAM systems usually requires curves as a start. A surface might require two boundary curves, as in the case of a ruled surface that we cover in this chapter. All curves covered in Chapter 6 can be used to generate surfaces. In order to visualize surfaces on a computer screen, a mesh, say $m \times n$ in size, is usually displayed. The mesh size is controllable by the user.

This chapter covers both theoretical and practical aspects of surfaces. It shows their underlying theory and how to use them in geometric modeling. Throughout the chapter, issues related to surface creation on CAD/CAM systems are covered.

7.2   Surface Entities

During surface creation on a CAD/CAM system, you should follow the modeling guidelines and strategies discussed in Chapter 2. Moreover, you should be careful when selecting curves to create surfaces. Selecting the mismatching ends of curves results in twisted surfaces as shown in Figure 7.1. The figure shows how the wrong ruled surface is created if its defining curves are selected near the wrong ends. The +'s in the figure indicate the selection locations. In such a case, the user deletes the surface and re-creates it by selecting the matching ends. As a general rule, a CAD system uses the midpoint of a curve to interpret the user's click on a curve. If the click is on the right half of the curve, its right end point is selected, and vice versa.

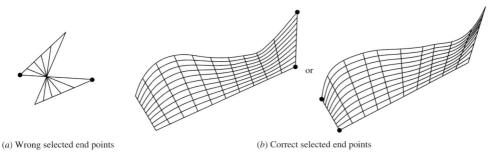

(*a*) Wrong selected end points   (*b*) Correct selected end points

Figure 7.1   Construction of improper and proper surfaces.

Visualization of a surface is aided by the addition of artificial fairing lines (called mesh), which crisscross the surface and so break it up into a network of interconnected patches. The default setting of a CAD system does not display a surface mesh — the surface is displayed with its four boundary curves only. In such a case, the mesh size is 2×2 . (All surfaces that we create define rectangular patches.) We can change the default mesh size. CAD systems provide users with a menu that allows them to specify the mesh size.

Figure 7.2 shows surfaces of revolutions with mesh sizes of 4×4 and 20×20. It should be mentioned that a finer mesh size for a surface does not improve its mathematical representation; it only improves its visualization. Finally, some CAD/CAM systems do not permit their users to delete curves used to create surfaces unless the latter are deleted first.

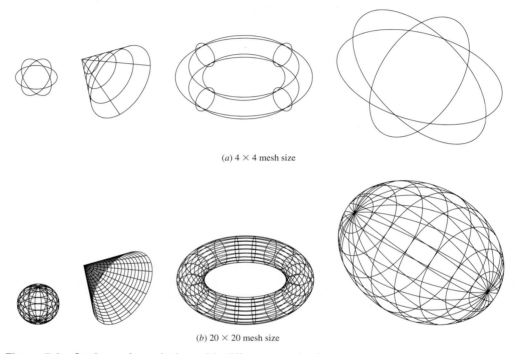

(*a*) 4 × 4 mesh size

(*b*) 20 × 20 mesh size

Figure 7.2   Surface of revolution with different mesh sizes.

As with curves, CAD/CAM systems provide designers with both analytic and synthetic surface entities. Analytic entities include plane surface, ruled surface, surface of revolution, and tabulated cylinder. Synthetic entities include bicubic Hermite spline surface, B-spline surface, rectangular and triangular Bezier patches, rectangular and triangular Coons patches, and NURBS (nonuniform B-splines). The mathematical properties of some of these entities are covered in this chapter. Following are descriptions of major surfaces:

1. **Plane surface:** It is the simplest surface. It requires three non-coincident points to define an infinite plane. The plane surface can be used to generate cross sections by intersecting a solid with it. Figure 7.3 shows planar surfaces.
2. **Ruled (lofted) surface:** It is a linear surface. It interpolates linearly between two boundary curves that define the surface (rails). Rails can be any curves. This surface is

ideal for representing surfaces that do not have any twists or kinks. Figure 7.4 gives some examples.

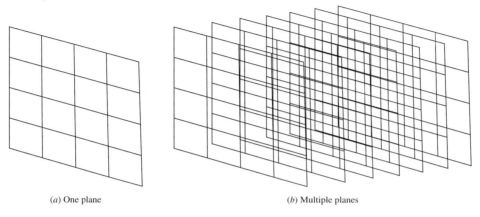

      (*a*) One plane   (*b*) Multiple planes

Figure 7.3   Plane surface.

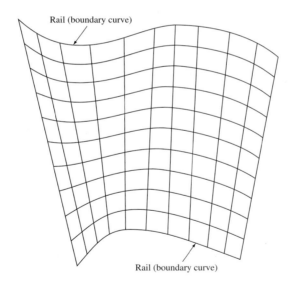

Figure 7.4   Ruled surface.

3. **Surface of revolution:** It is an axisymmetric surface that can model axisymmetric objects. It is generated by rotating a planar curve in space about the axis of symmetry a certain angle as shown in Figure 7.5.

4. **Tabulated cylinder:** It is a surface generated by translating a planar curve a certain distance along a specified direction (axis of the cylinder or directrix) as shown in

Figure 7.6. The plane of the curve is perpendicular to the directrix. This surface is not literally a cylinder. It is used to generate extruded surfaces that have identical cross sections.

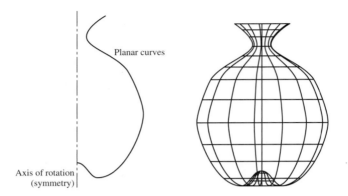

Figure 7.5   Surface of revolution.

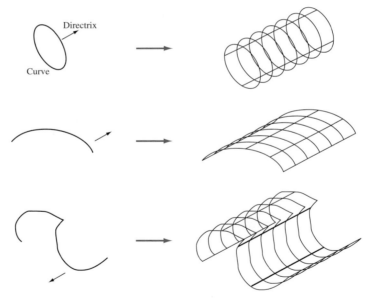

Figure 7.6   Tabulated surface.

5. **Bezier surface:** It is a surface that approximates or interpolates given input data. It is different from the previous surfaces in that it is a synthetic surface. It extends the Bezier curve to surfaces. It is a general surface that permits twists, and kinks. Bezier surface allows only global control of the surface. Figure 7.7 shows a Bezier surface.

6. **B-spline surface:** It is a surface that can approximate or interpolate given input data. Figure 7.8 shows an interpolating example. It is a synthetic surface. It is a general surface like a Bezier surface but with the advantage of permitting local control of the surface.

7. **Coons surface:** The previously described surfaces are used with either open boundaries or given data points. A Coons patch is used to create a surface using curves that form closed boundaries as shown in Figure 7.9.

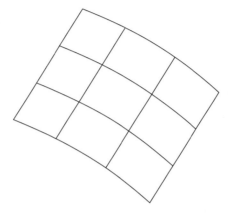

Figure 7.7   Bezier surface.

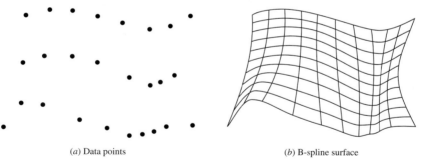

(*a*) Data points   (*b*) B-spline surface

Figure 7.8   B-spline surface.

8. **Fillet surface:** It is a B-spline surface that blends two surfaces together as shown in Figure 7.10. The two original surfaces may or may not be trimmed.

9. **Offset surface:** Existing surfaces can be offset to create new ones identical in shape but with different dimensions. It is a useful surface to use to speed up surface creation. For example, to create a hollow cylinder, the outer or inner cylinder can be created using a `cylinder` command and the other one can be created by an `offset` command. The `offset surface` command becomes very efficient to use if the original surface is a composite one. Figure 7.11 shows an offset surface.

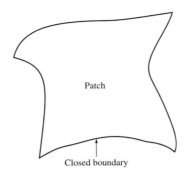

Figure 7.9   Coons surface.

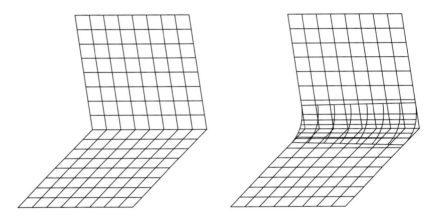

Figure 7.10   Fillet surface.

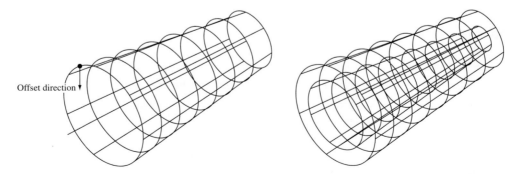

Figure 7.11   Offset surface.

7.3 Surface Representation

Surface representation is an extension of curve representation covered in Chapter 6. The nonparametric and parametric forms of curves can be extended to surfaces. Surfaces can be described mathematically in 3D space by nonparametric or parametric equations. In both cases, the equation of the surface or surface patch is given by:

$$\mathbf{P} = \begin{bmatrix} x & y & z \end{bmatrix}^T = \begin{bmatrix} x & y & f(x, y) \end{bmatrix}^T \tag{7.1}$$

where $\mathbf{P}$ is the position vector of a point on the surface as shown in Figure 7.12. The natural form of the function $f(x,y)$ for a surface to pass through all the given data points is a polynomial, that is,

$$z = f(x, y) = \sum_{m = 0}^{p} \sum_{n = 0}^{q} a_{mn} x^m y^n \tag{7.2}$$

where the surface is described by an XY grid of size $(p + 1) \times (q + 1)$ points.

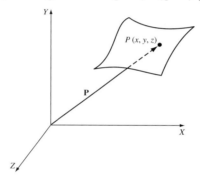

Figure 7.12   Position vector of point P on a surface.

As with curves, we use the parametric representation to develop surfaces. The surface parametric representation means a continuous, vector-valued function $\mathbf{P}(u, v)$ of two variables, or parameters, u and v. The variables are allowed to range over some connected region of the $u{-}v$ plane and as they do so, $\mathbf{P}(u, v)$ assumes every position on the surface. The function $\mathbf{P}(u, v)$ at certain u and v values is the point on the surface at these values. The most general way to describe the parametric equation of a 3D curved surface in space is

$$\mathbf{P}(u, v) = \begin{bmatrix} x & y & z \end{bmatrix}^T = \begin{bmatrix} x(u, v) & y(u, v) & z(u, v) \end{bmatrix}^T,$$

$$u_{min} \le u \le u_{max}, \, v_{min} \le v \le v_{max} \tag{7.3}$$

As with curves, Eq. (7.3) gives the coordinates of a point on the surface as the components of its position vector. It uniquely maps the parametric space (E^2 in u and v values) to the cartesian space (E^3 in x, y, and z) as shown in Figure 7.13. The parameters u and v are constrained to intervals bounded by minimum and maximum values. In most surfaces, these intervals are [0,1].

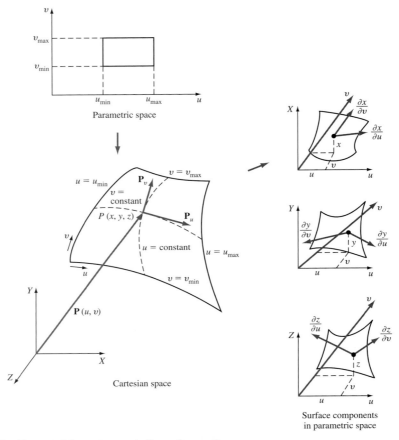

Figure 7.13 Parametric representation of a surface.

Equation (7.3) suggests that a general 3D surface can be modeled by dividing it into an assembly of topological patches. A **patch** is considered the basic mathematical element to create a composite surface. Some surfaces may consist of one patch only, while others may be a few patches connected together. Figure 7.14 shows a two-patch surface where the u and v values are [0,1]. The topology of a patch may be rectangular or triangular as shown in Figure 7.15. Triangular patches add more flexibility in surface modeling because they do not require ordered rectangular arrays of data points to create the surface as the rectangular patches do.

The tensor product method is widely used in surface formulation. Its widespread use is largely due to its simple separable nature involving only products of univariate basis functions, usually polynomials. It introduces no new conceptual complications due to the higher dimensionality of a surface over a curve. The properties of tensor product surfaces can easily be deduced from properties of the underlying curve schemes.

The tensor product formulation is a mapping of a rectangular domain described by the u and v values; for example, $0 \le u \le 1$ and $0 \le v \le 1$. Tensor product surfaces fit naturally onto rectangular patches. In addition, they have an explicit unique orientation (triangulation of a surface is not unique) and special parametric or coordinate directions associated with each independent parametric variable.

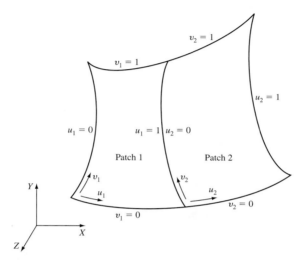

Figure 7.14   Two-patch parametric surface.

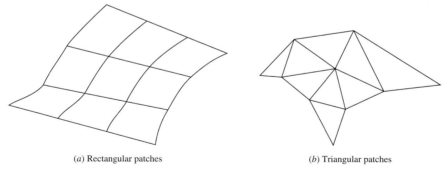

(a) Rectangular patches   (b) Triangular patches

Figure 7.15   Rectangular and triangular surface patches.

There is a set of boundary conditions associated with a rectangular patch. These are sixteen vectors and four boundary curves as shown in Figure 7.16. The vectors are four position vectors for the four corner points $P(0,0)$, $P(1,0)$, $P(1,1)$, and $P(0,1)$; eight tangent vectors (two at each corner); and four twist vectors at the corner points (twist vectors are discussed in Section 7.4.2). The four boundary curves are described by holding one parametric variable fixed at one of its limiting values and allowing the other to change freely. The boundary curves are then defined by the curve equations $u = 0$, $u = 1$, $v = 0$, and $v = 1$.

To generate curves on a surface patch, one can fix the value of one of the parametric variables, for example u, to obtain a curve in terms of the other variable, v. By continuing this process first for one variable and then for the other using a certain set of arbitrary values in the permissible domain, a network of two parametric families of curves is generated. Only one curve of each family passes through any point $P(u, v)$ on the surface. The positive sense of any of these curves is the sense in which its nonfixed parameter increases.

You can specify a mesh size, for example $m \times n$, to display a surface on a computer screen. That mesh size determines the number of curves in each family. The curves of each family are usually equally spaced in the permissible interval of the corresponding parametric variable.

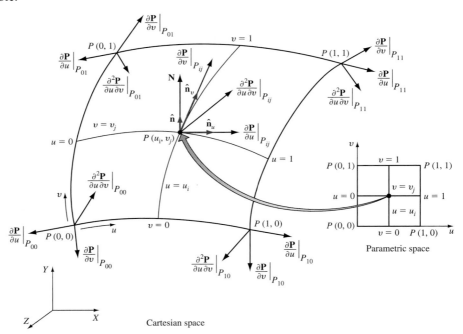

Figure 7.16   Boundary conditions of a surface.

7.4 Surface Analysis

Geometric surface analysis is an important part of using surfaces. For example, knowing the tangent vectors to a surface enables you to drive a cutting tool along the surface to machine it. And knowing the normal vectors to the surface provides the proper directions for the cutting tool to approach and retract from the surface.

Differential geometry plays a central role in the analysis of surfaces. The measurements of lengths and areas, the specification of directions and angles, and the definition of curvature on a surface are formulated in differential terms. The parametric surface $P(u, v)$ is directly amenable to differential analysis. There are intrinsic differential characteristics of a surface that are independent of parametrization, such as the unit normal, the principal curvatures, and directions.

7.4.1 Tangent Vectors

The tangent vector concept that is introduced in Chapter 6 can be extended to surfaces. The tangent vector at any point $P(u, v)$ on the surface is obtained by holding one parameter constant and differentiating the surface equation with respect to the other. Therefore, there are two tangent vectors, a tangent to each of the intersecting curves passing through the point as shown in Figure 7.16. These vectors are given by

$$\mathbf{P}_u(u, v) = \frac{\partial \mathbf{P}}{\partial u} = \frac{\partial x}{\partial u}\hat{\mathbf{i}} + \frac{\partial y}{\partial u}\hat{\mathbf{j}} + \frac{\partial z}{\partial u}\hat{\mathbf{k}}, \qquad u_{min} \le u \le u_{max},\ v_{min} \le v \le v_{max} \tag{7.4}$$

along the $v = $ constant curve, and

$$\mathbf{P}_v(u, v) = \frac{\partial \mathbf{P}}{\partial v} = \frac{\partial x}{\partial v}\hat{\mathbf{i}} + \frac{\partial y}{\partial v}\hat{\mathbf{j}} + \frac{\partial z}{\partial v}\hat{\mathbf{k}}, \qquad u_{min} \le u \le u_{max},\ v_{min} \le v \le v_{max} \tag{7.5}$$

along the $u = $ constant curve. These two equations can be combined to give

$$\begin{bmatrix} \mathbf{P}_u \\ \mathbf{P}_v \end{bmatrix} = \begin{bmatrix} \dfrac{\partial x}{\partial u} & \dfrac{\partial y}{\partial u} & \dfrac{\partial z}{\partial u} \\ \dfrac{\partial x}{\partial v} & \dfrac{\partial y}{\partial v} & \dfrac{\partial z}{\partial v} \end{bmatrix} \tag{7.6}$$

The components of each unit vector are shown in Figure 7.13. If the dot product of the two tangent vectors given by Eqs. (7.4) and (7.5) is equal to zero at a point on a surface, then the two vectors are perpendicular to one another at that point. Figure 7.16 shows the tangent vectors at the corner points of a rectangular patch and at point P_{ij}. The notation $\partial \mathbf{P}/\partial u|_{P_{ij}}$, for example, means that the derivative is calculated at the point P_{ij} defined by $u = u_i$ and $v = v_j$. The magnitudes and unit vectors of the tangent vectors are given by:

$$|\mathbf{P}_u| = \sqrt{\left(\frac{\partial x}{\partial u}\right)^2 + \left(\frac{\partial y}{\partial u}\right)^2 + \left(\frac{\partial z}{\partial u}\right)^2}$$
$$|\mathbf{P}_v| = \sqrt{\left(\frac{\partial x}{\partial v}\right)^2 + \left(\frac{\partial y}{\partial v}\right)^2 + \left(\frac{\partial z}{\partial v}\right)^2} \tag{7.7}$$

and

$$\hat{\mathbf{n}}_u = \frac{\mathbf{P}_u}{|\mathbf{P}_u|} \qquad \hat{\mathbf{n}}_v = \frac{\mathbf{P}_v}{|\mathbf{P}_v|} \tag{7.8}$$

The slopes to a given curve on a surface can be evaluated from the tangent vectors, although they are not significant and are seldom used in surface analysis.

7.4.2   Twist Vectors

The twist vector at a point on a surface is said to measure the twist in the surface at the point. It is the rate of change at the tangent vector $\mathbf{P}_u$ with respect to v or $\mathbf{P}_v$ with respect to u. Or it is the cross (mixed) derivative vector at the point. Figure 7.17 shows the geometric interpretation of the twist vector.

If we increment u and v by Δu and Δv, respectively, and draw the tangent vectors as shown in Figure 7.17, the incremental changes in $\mathbf{P}_u$ and $\mathbf{P}_v$ at point P, whose position vector is $\mathbf{P}(u,\ v)$, are obtained by translating $\mathbf{P}_u(u,\ v + \Delta v)$ and $\mathbf{P}_u(u,\ v + \Delta v)$ to P and forming the two triangles shown. The incremental rates of change of the two tangent vectors become $\Delta \mathbf{P}_u / \Delta v$ and $\Delta \mathbf{P}_v / \Delta u$. The infinitesimal rate of change is given by the following limits:

$$\underset{\Delta v \to 0}{\text{Limit}}\ \frac{\Delta \mathbf{P}_u}{\Delta v} = \frac{\partial \mathbf{P}_u}{\partial v} = \frac{\partial^2 \mathbf{P}}{\partial u\, \partial v} = \mathbf{P}_{uv}$$

$$\underset{\Delta u \to 0}{\text{Limit}}\ \frac{\Delta \mathbf{P}_v}{\Delta u} = \frac{\partial \mathbf{P}_v}{\partial u} = \frac{\partial^2 \mathbf{P}}{\partial u\, \partial v} = \mathbf{P}_{uv} \tag{7.9}$$

The twist vector can be written in terms of its cartesian components as:

$$\mathbf{P}_{uv} = \left[\frac{\partial^2 x}{\partial u\, \partial v}\ \frac{\partial^2 y}{\partial u\, \partial v}\ \frac{\partial^2 z}{\partial u\, \partial v} \right]^T = \frac{\partial^2 x}{\partial u\, \partial v}\hat{\mathbf{i}} + \frac{\partial^2 y}{\partial u\, \partial v}\hat{\mathbf{j}} + \frac{\partial^2 z}{\partial u\, \partial v}\hat{\mathbf{k}},$$

$$u_{min} \le u \le u_{max},\ v_{min} \le v \le v_{max} \tag{7.10}$$

The twist vector depends on both the surface geometric characteristics and its parametrization. Due to the latter dependency, interpreting the twist vector in geometrical terms may be misleading since $\mathbf{P}_{uv} \ne 0$ does not necessarily imply a twist in a surface. For example, a flat plane is not a twisted surface. However, depending on its parametric equation, $\mathbf{P}_{uv}$ may or may not be zero.

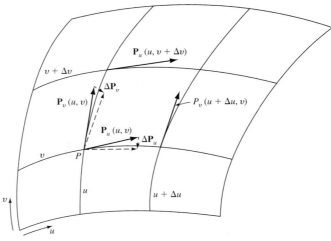

Figure 7.17   Geometric interpretation of twist vectors.

7.4.3   Normal Vectors

The normal to a surface is another important analytical property. It is used to calculate cutter offsets for 3D NC (numerical control) programming to machine surfaces, to make volume calculations, and to shade a surface model. The surface normal at a point is a vector which is perpendicular to both tangent vectors at the point (Figure 7.16); that is,

$$N(u, v) = \frac{\partial P}{\partial u} \times \frac{\partial P}{\partial v} = P_u \times P_v \tag{7.11}$$

and the unit normal vector is given by

$$\hat{n} = \frac{N}{|N|} = \frac{P_u \times P_v}{|P_u \times P_v|} \tag{7.12}$$

The order of the cross product in Eq. (7.11) can be reversed and still defines the normal vector. The sense of N, or $\hat{n}$, is chosen to suit the application. In machining, the sense of $\hat{n}$ is usually chosen so that $\hat{n}$ points out of the surface being machined. In volume calculations, the sense of $\hat{n}$ is chosen positive when pointing toward existing material and negative when pointing to holes in the part.

The surface normal is zero when $P_u \times P_v = 0$. This occurs at points lying on a cusp, ridge, or self-intersecting surface. It can also occur when the two derivatives P_u and P_v are parallel, or when one of them has a zero magnitude. The latter two cases correspond to a pathological parametrization which can be remedied.

7.4.4   Distance Calculations

The calculation of the distance between two points on a curved surface forms an important part of surface analysis. In general, two distinct points on a surface can be connected by many different paths, of different lengths, on the surface. The paths that have minimum lengths are analogous to straight lines, connecting two points in Euclidean space and are known as geodesics. Surface geodesics can, for example, provide optimized motion planning across a curved surface for numerical control machining, robot programming, and winding of coils around a rotor. The infinitesimal distance between two points $P(u, v)$ and $P(u + \Delta u,\ v + \Delta v)$ on a surface is given by

$$ds^2 = P_u \cdot P_u \, du^2 + 2 P_u \cdot P_v \, du \, dv + P_v \cdot P_v \, dv^2 \tag{7.13}$$

Equation (7.13) is often called the first fundamental quadratic form of a surface and is written as

$$ds^2 = E \, du^2 + 2F \, du \, dv + G \, dv^2 \tag{7.14}$$

where

$$E(u, v) = P_u \cdot P_u \qquad F(u, v) = P_u \cdot P_v \qquad G(u, v) = P_v \cdot P_v \tag{7.15}$$

E, F, and G are the first fundamental, or metric, coefficients of the surface. These coefficients provide the basis for the measurement of lengths and areas, and the specification of directions and angles on a surface.

The distance between two points $P(u_a, v_a)$ and $P(u_b, v_b)$ shown in Figure 7.18 is obtained by integrating Eq. (7.14) along a specified path $\{u = u(t), v = v(t)\}$ on the surface to give

$$S = \int_{t_a}^{t_b} \sqrt{Eu'^2 + 2Fu'v' + Gv'^2}\, dt \tag{7.16}$$

where $u' = du/dt$ and $v = dv/dt$. The minimum distance S is the geodesic between the two points.

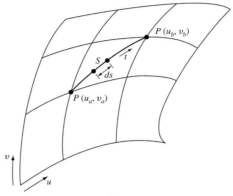

Figure 7.18   Surface geodesics.

The first fundamental form gives the distance element ds which lies in the tangent plane to the surface at $P(u, v)$ and, therefore, yields no information on how the surface curves away from the tangent plane at that point. To investigate surface curvature, another distance perpendicular to the tangent plane at $P(u, v)$ is introduced and given by

$$\tfrac{1}{2} dh^2 = \hat{\mathbf{n}} \cdot \mathbf{P}_{uu}\, du^2 + 2\hat{\mathbf{n}} \cdot \mathbf{P}_{uv}\, du\, dv + \hat{\mathbf{n}} \cdot \mathbf{P}_{vv}\, dv^2 \tag{7.17}$$

Similar to Eq. (7.13), Eq. (7.17) is often called the second fundamental quadratic form of a surface and is written as

$$\tfrac{1}{2} dh^2 = L\, du^2 + 2M\, du\, dv + N\, dv^2 \tag{7.18}$$

where

$$L(u, v) = \hat{\mathbf{n}} \cdot \mathbf{P}_{uu} \qquad M(u, v) = \hat{\mathbf{n}} \cdot \mathbf{P}_{uv} \qquad N(u, v) = \hat{\mathbf{n}} \cdot \mathbf{P}_{vv} \tag{7.19}$$

L, M, and N are the second fundamental coefficients of the surface and form the basis for defining and analyzing the curvature of a surface.

7.4.5   Curvatures

A **surface curvature** at a point $P(u, v)$ is defined as the curvature of the normal-section curve that lies on the surface and passes by the point. A normal-section curve is the intersection curve of a plane passing through the surface normal $\hat{\mathbf{n}}$ at the point, and the surface. The surface curvature at a point on a normal-section curve given by the form $\{u = u(t), \ v = v(t)\}$ can be written as:

$$\chi = \frac{Lu'^2 + 2Mu'v' + Nv'^2}{Eu'^2 + 2Fu'v' + Gv'^2} \tag{7.20}$$

and the radius of curvature at the point is $\rho = 1/\chi$. The sense of curvature must be chosen. One convention is to choose the sign in Eq. (7.20) to give positive curvature for convex surfaces and negative curvature for concave surfaces.

Few types of surface curvatures exist. The Gaussian curvature K and mean curvature H are defined by

$$K = \frac{LN - M^2}{EG - F^2} \tag{7.21}$$

$$H = \frac{EN + GL - 2FM}{2(EG - F^2)}$$

Equation (7.20) gives the surface curvature in any direction at point $P(u, v)$. However, it can be used to obtain the principal curvatures which are the upper (maximum) and lower (minimum) bounds on the curvature at the point:

$$\chi_{max} = H + \sqrt{H^2 - K} \tag{7.22}$$

$$\chi_{min} = H - \sqrt{H^2 - K}$$

The gaussian and mean curvatures can be written in terms of Eq. (7.22) as

$$K = \chi_{max}\,\chi_{min} \tag{7.23}$$

$$H = \tfrac{1}{2}(\chi_{max} + \chi_{min})$$

The gaussian curvature can be positive (as in a hill), negative (as in a saddle point), or zero (as in ruled or cylindrical surfaces) depending on the signs at χ_{max} and χ_{min}. Surfaces that have zero gaussian curvature everywhere are called developable, that is, they can be laid flat on a plane without stretching, tearing, or distorting them. CAD/CAM systems display gaussian curvature contour maps to convey information about surface shape.

From a practical point of view, the principal curvatures are of primary interest. For example, to machine a surface with a spherical cutter, it is important to ensure that the cutter radius is smaller than the smallest concave radius of curvature of the surface if gouging is to be avoided.

7.4.6 Tangent Planes

Analogous to a plane that is perpendicular to a surface and contains the normal $\hat{n}$, a tangent plane to a surface at one of its points can be defined. It is the common plane on which all the tangent vectors to the surface at the point lie. To develop the equation of a tangent plane, consider a point Q in a plane tangent to a given surface at a given point P as shown in Figure 7.19. The vectors $\mathbf{P}_u$, $\mathbf{P}_v$, and $\mathbf{Q} - \mathbf{P}$ lie on the plane. The normal $\hat{n}$ is normal to any vector in the plane. Thus, we can write

$$\hat{\mathbf{n}} \cdot (\mathbf{Q} - \mathbf{P}) = (\mathbf{P}_u \times \mathbf{P}_v) \cdot (\mathbf{Q} - \mathbf{P}) = 0 \tag{7.24}$$

which gives the equation of the tangent plane.

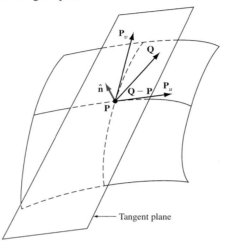

Figure 7.19 Tangent plane to a surface.

EXAMPLE 7.1   **Convert a parametric equation.**

The parametric equation of a sphere with a radius R and a center at point $P_0(x_0, y_0, z_0)$ is given by

$$\begin{aligned}
x &= x_0 + R\cos u \cos v \\
y &= y_0 + R\cos u \sin v \\
z &= z_0 + R\sin u
\end{aligned} \qquad , \qquad -\frac{\pi}{2} \le u \le \frac{\pi}{2}, \, 0 \le v \le 2\pi$$

Find the implicit equation of the sphere.

SOLUTION   Equation (7.1) suggests a useful and systematic way of developing the implicit equation of a surface from its parametric form. Implicit equations of surfaces are useful when solving surface intersection problems. Based on Eq. (7.1), the z coordinate of a point on the sphere is to be rewritten as a function of x and y. Squaring and adding the first two equations give

$$\frac{(x - x_0)^2}{R^2} + \frac{(y - y_0)^2}{R^2} = \cos^2 u = 1 - \sin^2 u$$

which gives:

$$\sin u = \sqrt{1 - \frac{(x - x_0)^2}{R^2} - \frac{(y - y_0)^2}{R^2}}$$

By substituting $\sin u$ into the z coordinate equation, $f(x, y)$ for a sphere becomes

$$z = z_0 + \sqrt{R^2 - (x - x_0)^2 - (y - y_0)^2}$$

The sphere implicit equation can now be written as

$$\mathbf{P} = [x \quad y \quad z_0 + \sqrt{R^2 - (x - x_0)^2 - (y - y_0)^2}]^T$$

It is obvious that the square of the z coordinate gives the classical sphere equation:

$$(x - x_0)^2 + (y - y_0)^2 + (z - z_0)^2 = R^2$$

7.5   Analytic Surfaces

This following sections cover the basics of the parametric equations of analytic surfaces. The distinction between the WCS and MCS has been ignored in the development of the parametric equations of the various surfaces to avoid confusion. The transformation between the two systems is obvious. In addition, the terms "surface" and "patch" are used interchangeably in this chapter. However, in a more general sense a surface is considered the superset since a surface can contain one or more patches.

7.6   Plane Surface

The parametric equation of a plane can take different forms depending on the given data. Consider first the case of a plane defined by three points P_0, P_1, and P_2 as shown in Figure 7.20. Assume that the point P_0 defines $u = 0$ and $v = 0$ and the vectors $(\mathbf{P}_1 - \mathbf{P}_0)$ and $(\mathbf{P}_2 - \mathbf{P}_0)$ define the u and v directions, respectively. Assume also that the domains for u and v are $[0,1]$. The position vector of any point P on the plane can be now written as:

$$\mathbf{P}(u, v) = \mathbf{P}_0 + u(\mathbf{P}_1 - \mathbf{P}_0) + v(\mathbf{P}_2 - \mathbf{P}_0), \qquad 0 \le u \le 1, 0 \le v \le 1 \qquad (7.25)$$

Equation (7.25) can be seen as the bilinear form of Eq. (6.11). Utilizing Eqs. (7.4) and (6.5), the tangent vectors at point P are:

$$\mathbf{P}_u(u, v) = \mathbf{P}_1 - \mathbf{P}_0 \qquad \mathbf{P}_v(u, v) = \mathbf{P}_2 - \mathbf{P}_0, \qquad 0 \le u \le 1, 0 \le v \le 1 \qquad (7.26)$$

and the surface normal is

$$\hat{\mathbf{n}}(u, v) = \frac{(\mathbf{P}_1 - \mathbf{P}_0) \times (\mathbf{P}_2 - \mathbf{P}_0)}{|(\mathbf{P}_1 - \mathbf{P}_0) \times (\mathbf{P}_2 - \mathbf{P}_0)|}, \qquad 0 \le u \le 1, 0 \le v \le 1 \qquad (7.27)$$

which is constant for any point on the plane. As for the curvature of the plane, it is equal to zero [see Eq. (7.20)] because all the second fundamental coefficients of the plane are zeros [see Eq. (7.19)].

Another case of constructing a plane surface occurs when the surface passes through a point P_0 and contains two directions defined by the unit vectors $\hat{\mathbf{r}}$ and $\hat{\mathbf{s}}$ as shown in Figure 7.21. Similar to the preceding case, the plane equation can be written as

$$\mathbf{P}(u, v) = \mathbf{P}_0 + uL_u\hat{\mathbf{r}} + vL_v\hat{\mathbf{s}}, \qquad 0 \le u \le 1, 0 \le v \le 1 \qquad (7.28)$$

Equation (7.28) is also considered the bilinear form of Eq. (6.17). The equation assumes a plane of dimensions L_u and L_v that may be set to unity.

The preceding two cases can be combined to provide the equation of a plane surface that passes through two points P_0 and P_1 and is parallel to the unit vector $\hat{\mathbf{r}}$. In this case, we can write

$$\mathbf{P}(u, v) = \mathbf{P}_0 + u(\mathbf{P}_1 - \mathbf{P}_0) + vL_v\hat{\mathbf{r}}, \qquad 0 \leq u \leq 1, 0 \leq v \leq 1 \tag{7.29}$$

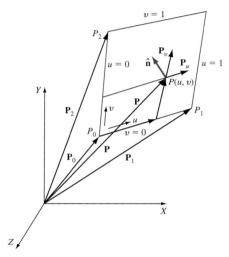

Figure 7.20   A plane surface defined by three points.

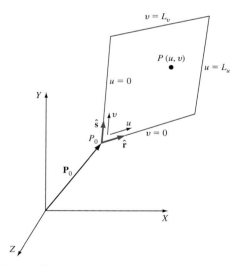

Figure 7.21   A plane surface defined by a point and two directions.

The last case to be considered is for a plane that passes through a point P_0 and is perpendicular to a given direction $\hat{n}$. Figure 7.22 shows this case. The vector $\hat{n}$ is normal to any vector in the plane. Thus,

$$(P - P_0) \cdot \hat{n} = 0 \tag{7.30}$$

which is a nonparametric equation of the plane surface. A parametric equation can be developed by using Eq. (7.30) to generate two points on the surface which can be used with P_0 in Eq. (7.25). Planes that are perpendicular to the axes of a current WCS are special cases of Eq. (7.30). For example, in the case of a plane perpendicular to the X axis, $\hat{n}$ is $(1,0,0)$ and the plane equation is $x = x_0$.

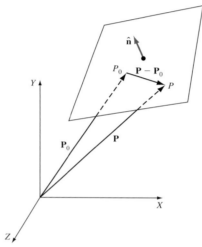

Figure 7.22   A plane surface defined by a point and a normal.

EXAMPLE 7.2   Find the minimum distance between a point in space and a plane.

SOLUTION   The minimum distance between a point and a plane is the perpendicular distance from the point to the plane. Let us assume the plane equation is given by:

$$P = P_0 + u\hat{r} + v\hat{s}, \qquad 0 \le u \le 1, 0 \le v \le 1 \tag{7.31}$$

From Figure 7.23, it is obvious that the perpendicular vector from point Q to the plane is parallel to $\hat{n}$. Thus, we can write:

$$P = Q - D\hat{n} \tag{7.32}$$

By using Eq. (7.31) in (7.32), we get

$$P_0 + u\hat{r} + v\hat{s} = Q - D\hat{n} \tag{7.33}$$

Equation (7.33) can be rewritten in a matrix form as:

$$\begin{bmatrix} r_x & s_x & n_x \\ r_y & s_y & n_y \\ r_z & s_z & n_z \end{bmatrix} \begin{bmatrix} u \\ v \\ D \end{bmatrix} = \begin{bmatrix} x_Q - x_0 \\ y_Q - y_0 \\ z_Q - z_0 \end{bmatrix}$$

(7.34)

where r_x, r_y, and r_z are the components of the unit vector $\hat{\mathbf{r}}$. Similarly, the components of the other vectors are given in Eq. (7.34). Solving Eq. (7.34) (see Chapter 6) gives the normal distance D and u and v, which can give the point P.

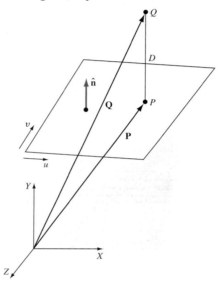

Figure 7.23   Minimum distance between a point and a plane.

7.7 Ruled Surface

A ruled surface is generated by joining corresponding points on two space curves (rails) $\mathbf{G}(u)$ and $\mathbf{Q}(u)$ by straight lines (also called rulings or generators) as shown in Figure 7.24. The main characteristic of a ruled surface is that it is linear in one direction, usually the v direction. In addition, every developable surface is a ruled surface. Cones and cylinders are examples of ruled surfaces, and the plane surface covered in Section 7.6 is considered the simplest of all ruled surfaces.

To develop the parametric equation of a ruled surface, consider the ruling $u = u_i$ joining points G_i and Q_i on the rails $\mathbf{G}(u)$ and $\mathbf{Q}(u)$, respectively, as shown in Figure 7.24. Using Eq. (6.11), the equation of the ruling becomes

$$\mathbf{P}(u_i, v) = \mathbf{G}_i + v(\mathbf{Q}_i - \mathbf{G}_i)$$

(7.35)

where v is the parameter along the ruling. Generalizing Eq. (7.35) for any ruling, the parametric equation of a ruled surface defined by two rails, $\mathbf{G}(u)$ and $\mathbf{Q}(u)$, is

$$P(u, v) = G(u) + v[Q(u) - G(u)] = (1 - v)G(u) + vQ(u), \qquad 0 \le u \le 1, 0 \le v \le 1 \qquad (7.36)$$

Holding the u value constant in Eq. (7.36) produces the rulings given by Eq. (7.35) in the v direction of the surface, while holding the v value constant yields curves in the u direction which are a linear blend of the rails. In fact, $G(u)$ and $Q(u)$ are $P(u, 0)$ and $P(u, 1)$, respectively. Therefore, the closer the value of v to 0, the greater the influence of $G(u)$ on the surface shape, and the less the influence of $Q(u)$ on the shape. Similarly, the influence of $Q(u)$ on the surface shape increases when the v value approaches 1.

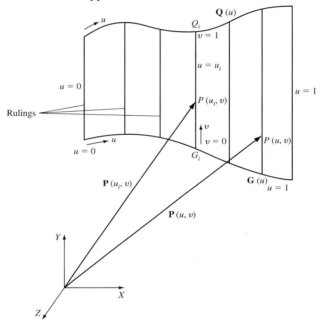

Figure 7.24   Ruled surface.

Based on Figure 7.24 and Eq. (7.35), it is now obvious why selecting the wrong ends of the rails produces the undesirable ruled surface shown in Figure 7.1a. In addition, Eq. (7.36), together with Eqs. (7.15), (7.19), and (7.20), shows that a ruled surface can only allow curvature in the u direction of the surface provided that the rails have curvatures. The surface curvature in the v direction (along the rulings) is zero, and thus a ruled surface cannot be used to model surface patches that have curvatures in two directions.

EXAMPLE 7.3 **Find the equation of a ruled surface.**
Find the equation of the ruled surface that covers the region R. Also, find the tangent and twist vectors of the surface.

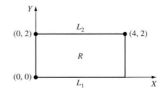

SOLUTION   We use L_1 and L_2 as the $G(u)$ and $Q(u)$ rules, respectively. The equations of the two lines and the ruled surface are:

$$P_{L1} = \begin{bmatrix} 0 \\ 0 \end{bmatrix} + u \begin{bmatrix} 4 \\ 0 \end{bmatrix} \qquad 0 \leq u \leq 1$$

$$P_{L2} = \begin{bmatrix} 0 \\ 2 \end{bmatrix} + u \begin{bmatrix} 4 \\ 0 \end{bmatrix}$$

$$\mathbf{P}(u,\, v) = (1 - v)\mathbf{P}_{l1} + v\mathbf{P}_{L2}$$

Substituting the equations of the lines and reducing gives the ruled surface equation as follows:

$$P(u,\, v) = \begin{bmatrix} 4u \\ 2v \end{bmatrix} \qquad 0 \leq u \leq 1,\, 0 \leq v \leq 1$$

The surface vectors are $\mathbf{P}_u = \begin{bmatrix} 4 & 0 \end{bmatrix}^T$, $\mathbf{P}_v = \begin{bmatrix} 0 & 2 \end{bmatrix}^T$, $\mathbf{P}_{uv} = \begin{bmatrix} 0 & 0 \end{bmatrix}^T$

Example 7.3 discussion:

The tangent vectors are constants and the twist vector is zero, as expected because the ruled surface is planar.

Example 7.3 hands-on exercise:

Derive the surface equation using the vertical sides of the region R. Is the equation the same as the above equation?

7.8 Surface of Revolution

The rotation of a planar curve an angle v about an axis of rotation creates a circle (if $v = 360$) for each point on the curve. The center of each circle lies on the axis of rotation and its radius $r_z(u)$ is variable as shown in Figure 7.25. The planar curve and the circles are called the profile and parallels, respectively, while the various positions of the profile around the axis are called meridians.

The planar curve and the axis of rotation form the plane of zero angle, that is, $v = 0$. To derive the parametric equation of a surface of revolution, a local coordinate system with a Z axis coincident with the axis of rotation is assumed, as shown in Figure 7.25. This local system shown by the subscript L can be created as follows. Choose the perpendicular direction from the point $u = 0$ on the profile as the X_L axis and the intersection point between X_L and Z_L as the origin of the local system. The Y_L axis is automatically determined by the right-hand rule.

Now, consider a point $G(u) = P(u, 0)$ on the profile that rotates an angle v about Z_L when the profile rotates the same angle. Considering the hatched triangle which is perpendicular to the Z_L axis, the parametric equation of the surface of revolution can be written as:

$$\mathbf{P}(u,\, v) = r_z(u) \cos v \hat{\mathbf{n}}_1 + r_z(u) \sin v \hat{\mathbf{n}}_2 + z_L(u)\hat{\mathbf{n}}_3, \qquad 0 \leq u \leq 1,\, 0 \leq v \leq 2\pi \qquad (7.37)$$

If we choose $z_L(u) = u$ for each point on the profile, Eq. (7.37) gives the local coordinates $(x_L,\, y_L,\, z_L)$ of a point $P(u,\, v)$ as $[r_z(u) \cos v,\, r_z(u) \sin v,\, u]^T$. The local coordinates are transformed to MCS coordinates before displaying the surface using Eq. (2.3), where the rotation matrix is formed from $\hat{\mathbf{n}}_1$, $\hat{\mathbf{n}}_2$, and $\hat{\mathbf{n}}_3$, and the position of the origin of the local system is given by $\mathbf{P}_L$ (see Figure 7.25).

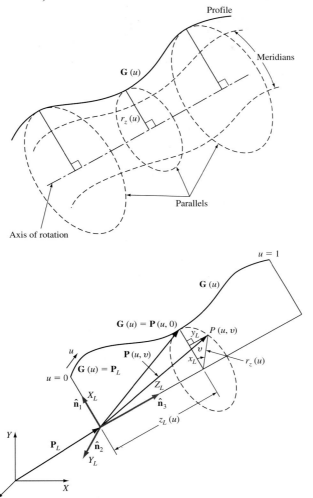

Figure 7.25 Surface of revolution.

7.9 Tabulated Cylinder

A **tabulated cylinder** is defined as a surface that results from translating a space planar curve along a given direction. It can also be defined as a surface that is generated by moving a straight line (called generatrix) along a given planar curve (called directrix). The straight line always stays parallel to a fixed given vector that defines the v direction of the cylinder as shown in Figure 7.26. The planar curve $\mathbf{G}(u)$ can be any curve. The position vector of any point $P(u, v)$ on the surface can be written as:

$$\mathbf{P}(u, v) = \mathbf{G}(u) + v\hat{\mathbf{n}}_v, \qquad 0 \le u \le u_{max}, 0 \le v \le v_{max} \tag{7.38}$$

From a user point of view, $\mathbf{G}(u)$ is an existing curve that the user selects to create the cylinder, v is the cylinder length, and $\hat{\mathbf{n}}_v$ is the cylinder axis. The cylinder length v is specified by the user.

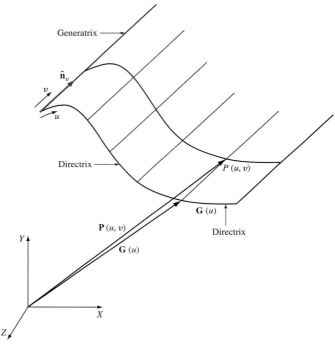

Figure 7.26   Tabulated cylinder.

EXAMPLE 7.4   **Find the equation of a cylinder.**

Find the equation of the cylinder that is defined by the half circle shown. The cylinder axis is in the $-Z$ direction and is 3 inches long.

SOLUTION   The equation of the half circle is:

$$\mathbf{G}(u) = \begin{bmatrix} 3 + 2\cos u \\ 4 + 2\sin u \end{bmatrix} \qquad 0 \le u \le \pi$$

Using Eq. (7.38), the equation of the cylinder is

$$\mathbf{P}(u, v) = \mathbf{G}(u) + 3v(-\hat{k}) = \begin{bmatrix} 3 + 2\cos u \\ 4 + 2\sin u \\ 0 \end{bmatrix} - v\begin{bmatrix} 0 \\ 0 \\ 3 \end{bmatrix} = \begin{bmatrix} 3 + 2\cos u \\ 4 + 2\sin u \\ -3v \end{bmatrix} \qquad 0 \le u \le \pi, 0 \le v \le 1$$

Example 7.4 discussion:

The equation of the half circle uses a 180° angle range.

Example 7.4 hands-on exercise:

Change the formulation of the cylinder so that the v range becomes $0 \le v \le 3$.

7.10 Synthetic Surfaces

Synthetic surfaces provide designers with better surface design tools than analytic surfaces. Consider the design of blade surfaces in jet aircraft engines. The design of these surfaces is usually based on aerodynamic and fluid-flow simulations often incorporating thermal and mechanical stress deformation. These simulations yield ordered sets of discrete streamline points which must then be connected accurately by surfaces. Any small deviation in these aerodynamic surfaces can degrade engine performance. Another example is the creation of blending surfaces typically encountered in designing dies for the injection molding of plastics.

For continuity purposes, the parametric representation of synthetic surfaces is presented in the following sections in a form similar to that of curves. Surfaces covered are bicubic, Bezier, B-spline, Coons, blending, offset, and triangular. All these surfaces are based on polynomial forms. Surfaces using other forms such as Fourier series are not considered here. Although Fourier series can approximate any curve given sufficient conditions, the computations involved with them are greater than with polynomials. Therefore, they are not suited to general use in CAD/CAM.

7.11 Hermite Bicubic Surface

The parametric bicubic surface patch connects four corner data points and utilizes a bicubic equation. Therefore, sixteen vector conditions (or 48 scalar conditions) are required to find the coefficients of the equation. When these coefficients are the four corner data points, the eight tangent vectors at the corner points (two at each point in the u and v directions), and the four twist vectors at the corner points, a Hermite bicubic surface patch results. The bicubic equation can be written as:

$$\mathbf{P}(u, v) = \sum_{i=0}^{3} \sum_{j=0}^{3} \mathbf{C}_{ij} u^i v^j, \qquad 0 \le u \le 1, 0 \le v \le 1 \tag{7.39}$$

Equation (7.39) can be expanded and written in the following matrix form:

$$\mathbf{P}(u, v) = \mathbf{U}^T[C]\mathbf{V}, \qquad 0 \le u \le 1, 0 \le v \le 1 \tag{7.40}$$

where $\mathbf{U} = [u^3 \ u^2 \ u \ 1]^T$, $\mathbf{V} = [v^3 \ v^2 \ v \ 1]^T$, and the coefficient matrix $[C]$ is given by

$$[C] = \begin{bmatrix} \mathbf{C}_{33} & \mathbf{C}_{32} & \mathbf{C}_{31} & \mathbf{C}_{30} \\ \mathbf{C}_{23} & \mathbf{C}_{22} & \mathbf{C}_{21} & \mathbf{C}_{20} \\ \mathbf{C}_{13} & \mathbf{C}_{12} & \mathbf{C}_{11} & \mathbf{C}_{10} \\ \mathbf{C}_{03} & \mathbf{C}_{02} & \mathbf{C}_{01} & \mathbf{C}_{00} \end{bmatrix} \tag{7.41}$$

In order to determine the coefficients $\mathbf{C}_i$, consider the patch shown in Figure 7.16. Applying the boundary conditions into Eq. (7.40), solving for the coefficients, and rearranging give the following final equation of a bicubic patch:

$$\mathbf{P}(u, v) = \mathbf{U}^T[M_H][B][M_H]^T\mathbf{V}, \qquad 0 \le u \le 1, 0 \le v \le 1 \tag{7.42}$$

where $[M_H]$ is given by Eq. (6.78) and $[B]$, the geometry or boundary condition matrix, is

$$[B] = \begin{bmatrix} \mathbf{P}_{00} & \mathbf{P}_{01} & \mathbf{P}_{v00} & \mathbf{P}_{v01} \\ \mathbf{P}_{10} & \mathbf{P}_{11} & \mathbf{P}_{v10} & \mathbf{P}_{v11} \\ \mathbf{P}_{u00} & \mathbf{P}_{u01} & \mathbf{P}_{uv00} & \mathbf{P}_{uv01} \\ \mathbf{P}_{u10} & \mathbf{P}_{u11} & \mathbf{P}_{uv10} & \mathbf{P}_{uv11} \end{bmatrix} \tag{7.43}$$

The matrix $[B]$ is partitioned as shown in Eq. (7.43) to indicate the grouping of the similar boundary conditions. It can also be written as:

$$[B] = \begin{bmatrix} [P] & [P_v] \\ [P_u] & [P_{uv}] \end{bmatrix} \tag{7.44}$$

where $[P]$, $[P_u]$, $[P_v]$, and $[P_{uv}]$ are the submatrices of the corner coordinates, corner u-tangent vectors, corner v-tangent vectors, and the corner twist vectors, respectively.

The tangent and twist vectors at any point on the surface are given by:

$$\mathbf{P}_u(u, v) = \mathbf{U}^T[M_H]^u[B][M_H]^T\mathbf{V} \tag{7.45}$$

$$\mathbf{P}_v(u, v) = \mathbf{U}^T[M_H][B][M_H]^{vT}\mathbf{V} \tag{7.46}$$

$$\mathbf{P}_{uv}(u, v) = \mathbf{U}^T[M_H]^u[B][M_H]^{vT}\mathbf{V} \tag{7.47}$$

where $[M_H]^u$ or $[M_H]^v$ is given by Eq. (6.82).

Similar to the cubic spline, the bicubic form permits C^1 continuity from one patch to the next. The necessary two conditions are to have the same curves (C^0 continuity) and the same direction of the tangent vectors (C^1 continuity) across the common edge between the two patches. The magnitudes of the tangent vectors do not have to be the same.

EXAMPLE 7.5   **Find the equation of a bicubic surface.**

Find the equation of the bicubic surface that is defined by the four points shown in the forthcoming drawing.

SOLUTION The boundary conditions given (only four points, $P_1 - P_4$) are not enough to define the bicubic surface. We need sixteen conditions in total. This suggests that we use the four lines connecting the four points to evaluate eight tangent vectors and four twist vectors. The twist vectors are all zeros because the four points define a plane. The four lines are shown dashed above. We assume the u and v directions as shown.

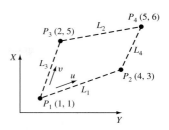

The four line equations are:

$$\mathbf{P}_{L1}(u) = \mathbf{P}_1 + u(\mathbf{P}_2 - \mathbf{P}_1) = \begin{bmatrix} 1 \\ 1 \end{bmatrix} + u\begin{bmatrix} 3 \\ 2 \end{bmatrix} \qquad 0 \le u \le 1$$

$$\mathbf{P}_{L2}(u) = \mathbf{P}_3 + u(\mathbf{P}_4 - \mathbf{P}_3) = \begin{bmatrix} 2 \\ 5 \end{bmatrix} + u\begin{bmatrix} 3 \\ 1 \end{bmatrix} \qquad 0 \le u \le 1$$

$$\mathbf{P}_{L3}(v) = \mathbf{P}_1 + v(\mathbf{P}_3 - \mathbf{P}_1) = \begin{bmatrix} 1 \\ 1 \end{bmatrix} + v\begin{bmatrix} 1 \\ 4 \end{bmatrix} \qquad 0 \le v \le 1$$

$$\mathbf{P}_{L4}(v) = \mathbf{P}_2 + v(\mathbf{P}_4 - \mathbf{P}_2) = \begin{bmatrix} 4 \\ 3 \end{bmatrix} + v\begin{bmatrix} 1 \\ 3 \end{bmatrix} \qquad 0 \le v \le 1$$

The tangent and twist vectors of the four lines are:

$$\mathbf{P}_{u00} = \mathbf{P}_{u10} = \frac{d\mathbf{P}_{L1}}{du} = \begin{bmatrix} 3 \\ 2 \end{bmatrix}, \frac{\partial^2 \mathbf{P}_{L1}}{\partial u \partial v} = \begin{bmatrix} 0 \\ 0 \end{bmatrix}$$

$$\mathbf{P}_{u01} = \mathbf{P}_{u11} = \frac{d\mathbf{P}_{L2}}{du} = \begin{bmatrix} 3 \\ 1 \end{bmatrix}, \frac{\partial^2 \mathbf{P}_{L2}}{\partial u \partial v} = \begin{bmatrix} 0 \\ 0 \end{bmatrix}$$

$$\mathbf{P}_{v00} = \mathbf{P}_{v01} = \frac{d\mathbf{P}_{L3}}{dv} = \begin{bmatrix} 1 \\ 4 \end{bmatrix}, \frac{\partial^2 \mathbf{P}_{L3}}{\partial u \partial v} = \begin{bmatrix} 0 \\ 0 \end{bmatrix}$$

$$\mathbf{P}_{v10} = \mathbf{P}_{v11} = \frac{d\mathbf{P}_{L4}}{dv} = \begin{bmatrix} 1 \\ 3 \end{bmatrix}, \frac{\partial^2 \mathbf{P}_{L4}}{\partial u \partial v} = \begin{bmatrix} 0 \\ 0 \end{bmatrix}$$

The four corner points are

$$\mathbf{P}_{00} = \mathbf{P}_1 = \begin{bmatrix} 1 \\ 1 \end{bmatrix}, \mathbf{P}_{10} = \mathbf{P}_2 = \begin{bmatrix} 4 \\ 3 \end{bmatrix}, \mathbf{P}_{01} = \mathbf{P}_3 = \begin{bmatrix} 2 \\ 5 \end{bmatrix}, \mathbf{P}_{11} = \mathbf{P}_4 = \begin{bmatrix} 5 \\ 6 \end{bmatrix}$$

Substituting all of the above values into Eq. (7.43), we get:

$$[B_x] = \begin{bmatrix} 1 & 2 & 1 & 1 \\ 4 & 5 & 1 & 1 \\ 3 & 3 & 0 & 0 \\ 3 & 3 & 0 & 0 \end{bmatrix}, [B_y] = \begin{bmatrix} 1 & 5 & 4 & 4 \\ 3 & 6 & 3 & 3 \\ 2 & 1 & 0 & 0 \\ 2 & 1 & 0 & 0 \end{bmatrix}$$

Substituting these values into Eq. (7.42) gives:

$$x(u, v) = U^T[M_H][B_x][M_H]^T V$$
$$= 3u - 4v^3 + 6v^2 - v + 1 \qquad 0 \le u \le 1, 0 \le v \le 1$$

and

$$y(u, v) = U^T[M_H][B_y][M_H]^T V$$
$$= u^3(4v^3 - 6v^2 + 2v) + u^2(-6v^3 + 9v^2 - 3v) \qquad 0 \le u \le 1, 0 \le v \le 1$$
$$+ u(2v^3 - 3v^2 + 2) + 4v + 1$$

Example 7.5 discussion:

We use the four corner points to generate 16 boundary conditions. The equations for $x(u, v)$ and $y(u, v)$ are cubic; thus they produce nonlinear boundary curve.

Example 7.5 hands-on exercise:

Using the above equations for $x(u, v)$ and $y(u, v)$, verify all the 16 boundary conditions of the bicubic surface by substituting the proper u and v values.

7.12 Bezier Surface

A tensor product Bezier surface is an extension of the Bezier curve in two parametric directions u and v. An orderly set of data or control points is used to build a topologically rectangular surface as shown in Figure 7.27. The surface equation can be written by extending Eq. (6.83); that is,

$$\mathbf{P}(u, v) = \sum_{i=0}^{n} \sum_{j=0}^{m} \mathbf{P}_{ij} B_{i,n}(u) B_{j,m}(v), \qquad 0 \le u \le 1, 0 \le v \le 1 \qquad (7.48)$$

where $\mathbf{P}(u, v)$ is any point on the surface and $\mathbf{P}_{ij}$ are the control points. These points form the vertices of the control or characteristic polyhedron (shown dashed in Figure 7.27) of the resulting Bezier surface. The points are arranged in a $(n + 1) \times (m + 1)$ rectangular array as seen from Eq. (7.48). We expand the double summation in Eq. (7.48) in two stages. First, we hold the outer sum over i and expand the inner one over j. We then expand the sum over i to get:

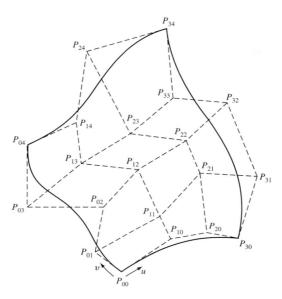

Figure 7.27   4 x 5 Bezier surface.

$$\mathbf{P}(u, v) = \sum_{i=0}^{n} B_{i,n}(u)[\mathbf{P}_{i0}B_{0,m}(v) + \mathbf{P}_{i1}B_{1,m}(v) + \ldots + \mathbf{P}_{im}B_{m,m}(v)]$$

$$= B_{0,n}(u)[\mathbf{P}_{00}B_{0,m}(v) + \mathbf{P}_{01}B_{1,m}(v) + \ldots + \mathbf{P}_{0m}B_{m,m}(v)] \tag{7.49}$$

$$+ B_{1,n}(u)[\mathbf{P}_{10}B_{0,m}(v) + \mathbf{P}_{11}B_{1,m}(v) + \ldots + \mathbf{P}_{1m}B_{m,m}(v)]$$

$$+ \ldots$$

$$+ B_{n,n}(u)[\mathbf{P}_{n0}B_{0,m}(v) + \mathbf{P}_{n1}B_{1,m}(v) + \ldots + \mathbf{P}_{nm}B_{m,m}(v)]$$

The characteristics of a Bezier surface are the same as those of a Bezier curve. The surface interpolates the four corner control points (see Figure 7.27) if we substitute the (u, v) values of $(0,0)$, $(1,0)$ $(0,1)$, and $(1,1)$ into Eq. (7.49). The surface is also tangent to the corner segments of the control polyhedron. The tangent vectors at the corners are:

$$\begin{array}{lll}
\mathbf{P}_{u00} = n(\mathbf{P}_{10} - \mathbf{P}_{00}) & \mathbf{P}_{un0} = n(\mathbf{P}_{n0} - \mathbf{P}_{(n-1)0}) & \text{along } v = 0 \text{ edge} \\
\mathbf{P}_{u0m} = n(\mathbf{P}_{1m} - \mathbf{P}_{0m}) & \mathbf{P}_{unm} = n(\mathbf{P}_{nm} - \mathbf{P}_{(n-1)m}) & \text{along } v = 1 \text{ edge} \\
\mathbf{P}_{v00} = m(\mathbf{P}_{01} - \mathbf{P}_{00}) & \mathbf{P}_{v0m} = m(\mathbf{P}_{0m} - \mathbf{P}_{0(m-1)}) & \text{along } u = 0 \text{ edge} \\
\mathbf{P}_{vn0} = m(\mathbf{P}_{n1} - \mathbf{P}_{n0}) & \mathbf{P}_{vnm} = m(\mathbf{P}_{nm} - \mathbf{P}_{n(m-1)}) & \text{along } u = 1 \text{ edge}
\end{array} \tag{7.50}$$

A Bezier surface possesses the convex hull property. The convex hull in this case is the polyhedron formed by connecting the furthest control points on the control polyhedron. The convex hull includes the control polyhedron of the surface as it includes the control polygon in the case of Bezier curve.

The shape of a Bezier surface can be modified by either changing some vertices of its polyhedron or by keeping the polyhedron fixed and specifying multiple coincident points (multiplicity) of some vertices. Moreover, a closed Bezier surface can be generated by closing its polyhedron or choosing coincident corner points.

The normal to a Bezier surface at any point can be calculated by substituting Eq. (7.48) into (7.11) to obtain:

$$
\mathbf{N}(u, v) = \sum_{i=0}^{n} \sum_{j=0}^{m} \mathbf{P}_{ij} \frac{\partial B_{i,n}(u)}{\partial u} B_{j,m}(v) \times \sum_{k=0}^{n} \sum_{l=0}^{m} \mathbf{P}_{kl} B_{k,n}(u) \frac{\partial B_{l,m}(v)}{\partial v}
$$

$$
= \sum_{i=0}^{n} \sum_{j=0}^{m} \sum_{k=0}^{n} \sum_{l=0}^{m} \frac{\partial B_{i,n}}{\partial u} B_{j,m}(v) B_{k,n}(u) \frac{\partial B_{l,m}(v)}{\partial v} \mathbf{P}_{ij} \times \mathbf{P}_{kl} \tag{7.51}
$$

When expanding this equation, it should be noted that $\mathbf{P}_{ij} \times \mathbf{P}_{kl} = \mathbf{0}$ if $i = k$ and $j = l$, and that $\mathbf{P}_{ij} \times \mathbf{P}_{kl} = -\mathbf{P}_{kl} \times \mathbf{P}_{ij}$.

As with a Bezier curve, the degree of Bezier surface is tied to the number of control points. Surfaces requiring great design flexibility need a large control point array and would, therefore, have a high polynomial degree. To achieve the required design flexibility while keeping the surface degree manageable, large surfaces are generally designed by piecing together smaller surface patches of lower degrees. This keeps the overall degree of the surface low but requires special attention to ensure that appropriate continuity is maintained across patch boundaries.

A composite Bezier surface can have C^0 (positional) and/or C^1 (tangent) continuity. A positional continuity between, say, two patches requires that the common boundary curve between the two patches must have a common boundary polygon between the two characteristic polyhedrons as shown in Figure 7.28a. For tangent continuity across the boundary, the segments, attached to the common boundary polygon, of one patch polyhedron must be collinear with the corresponding segments of the other patch polyhedron as shown in Figure 7.28b. This implies that the tangent planes of the patches at the common boundary curve are coincident.

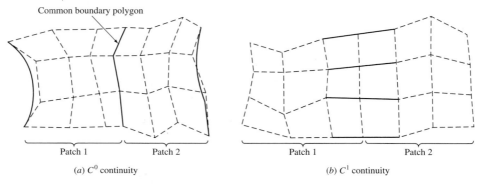

Common boundary polygon

Patch 1   Patch 2   Patch 1   Patch 2

(a) C^0 continuity   (b) C^1 continuity

Figure 7.28 **Continuity of composite Bezier surface.**

In a design environment, Bezier surface is superior to a bicubic surface in that it does not require tangent or twist vectors to define the surface. However, its main disadvantage is the lack of local control. Changing one or more control points affects the shape of the whole surface. Therefore, the user cannot selectively change the shape of part of the surface.

EXAMPLE 7.6   **Find the equation of a Bezier surface.**
Find the equation of the Bezier surface that covers the region R. Also, find the surface vectors and its midpoint.

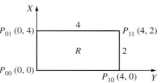

SOLUTION   The 4×2 region R generates the four corner points shown above. We name the points according to the convention used by a Bezier surface. We generate a 2×2 surface. Applying Eq. (7.48) gives:

$$\mathbf{P}(u, y) = \mathbf{P}_{00}B_{0,1}(u)B_{0,1}(v) + \mathbf{P}_{01}B_{0,1}(u)B_{1,1}(v)$$
$$+ \mathbf{P}_{10}B_{1,1}(u)B_{0,1}(v) + \mathbf{P}_{11}B_{1,1}(u)B_{1,1}(v)$$

where:

$$B_{0,1}(u) = 1 - u, B_{0,1}(v) = 1 - v, B_{1,1}(u) = u, B_{1,1}(v) = v$$

These two equations produce the following surface equation:

$$\mathbf{P}(u, v) = \begin{bmatrix} x(u, v) \\ y(u, v) \end{bmatrix} = u\begin{bmatrix} 4 \\ 0 \end{bmatrix} + v\begin{bmatrix} 0 \\ 2 \end{bmatrix} \qquad 0 \le u \le 1, 0 \le v \le 1$$

The surface vectors are:

$$\mathbf{P}_u = \begin{bmatrix} 4 \\ 0 \end{bmatrix}, \mathbf{P}_v = \begin{bmatrix} 0 \\ 2 \end{bmatrix}, \mathbf{P}_{uv} = \begin{bmatrix} 0 \\ 0 \end{bmatrix}, \mathbf{N} = \mathbf{P}_u \times \mathbf{P}_v = \begin{vmatrix} \hat{i} & \hat{j} & \hat{k} \\ 4 & 0 & 0 \\ 0 & 2 & 0 \end{vmatrix} = 8\hat{k} = \begin{bmatrix} 0 \\ 0 \\ 8 \end{bmatrix}$$

The midpoint of the surface is at $(u, v) = (0.5, 0.5)$. Substituting into the preceding Bezier surface equation gives:

$$\mathbf{P}_{midpoint} = \begin{bmatrix} 2 \\ 1 \end{bmatrix}$$

Example 7.6 discussion:
We use the four corner points to generate the Bezier surface. The equations for $x(u, v)$ and $y(u, v)$ are linear; thus the surface is planar as it should be. The tangent vectors are constants, and therefore there is no twist vector. The midpoint of the surface is the midpoint of the region R.

Example 7.6 hands-on exercise:
Using the equations for $x(u, v)$ and $y(u, v)$, verify all the four corner points of the Bezier surface by substituting the proper u and v values. Also verify the tangent vectors at these points.

7.13 B-Spline Surface

The same tensor product method used with Bezier curves can extend B-spline curves to describe B-spline surfaces. A rectangular set of data (control) points creates the surface. This set forms the vertices of the characteristic polyhedron that approximates and controls the shape of the resulting surface.

A B-spline surface can approximate or interpolate the vertices of the polyhedron as shown in Figure 7.29. The degree of the surface is independent of the number of control points, and continuity is automatically maintained throughout the surface by virtue of the form of blending functions. As a result, surface intersections can easily be managed.

A B-spline surface defined by $(n + 1) \times (m + 1)$ array of control points is given by extending Eq. (6.91) into two dimensions:

$$\mathbf{P}(u, v) = \sum_{i = 0}^{n} \sum_{j = 0}^{m} \mathbf{P}_{ij} N_{i, k}(u) N_{j, l}(v), \qquad 0 \le u \le u_{max}, \ 0 \le v \le v_{max} \qquad (7.52)$$

where $\mathbf{P}(u, v)$ is any point on the surface and $\mathbf{P}_{ij}$ are the control points. These points form the vertices of the control or characteristic polyhedron (shown dashed in Figure 7.29a) of the resulting B-spline surface. The points are arranged in a $(n + 1) \times (m + 1)$ rectangular array as seen from Eq. (7.52). In addition, the surface has an order of k (and a degree of $k - 1$) in the u direction, and an order of l (and a degree of $l - 1$) in the v direction.

We expand the double summation in Eq. (7.52) in two stages. First, we hold the outer sum over i and expand the inner one over j. We then expand the sum over i to get Eq. (7.53).

$$
\begin{aligned}
\mathbf{P}(u, v) = \ & \sum_{i = 0}^{n} N_{i, k}(u)[\mathbf{P}_{i0} N_{0, l}(v) + \mathbf{P}_{i1} N_{1, l}(v) + \ldots + \mathbf{P}_{im} N_{m, l}(v)] \\
= \ & N_{0, k}(u)[\mathbf{P}_{00} N_{0, l}(v) + \mathbf{P}_{01} N_{1, l}(v) + \ldots + \mathbf{P}_{0m} N_{m, l}(v)] \\
& + N_{1, k}(u)[\mathbf{P}_{10} N_{0, l}(v) + \mathbf{P}_{11} N_{1, l}(v) + \ldots + \mathbf{P}_{1m} N_{m, l}(v)] \\
& + \ldots \\
& + N_{n, k}(u)[\mathbf{P}_{n0} N_{0, l}(v) + \mathbf{P}_{n1} N_{1, l}(v) + \ldots + \mathbf{P}_{nm} N_{m, l}(v)]
\end{aligned}
\qquad (7.53)
$$

All the related discussions to Eqs. (6.92) and (6.93) apply to the Eq. (7.52). Equation (7.52) implies that knot vectors in both u and v directions are constant but not necessarily equal. Other formulations could allow various knot vectors in a given direction to increase the flexibility of local control.

B-spline surfaces have the same characteristics as B-spline curves. Their major advantage over Bezier surfaces is the local control. Composite B-spline surfaces can be generated with C^0 and/or C^1 continuity.

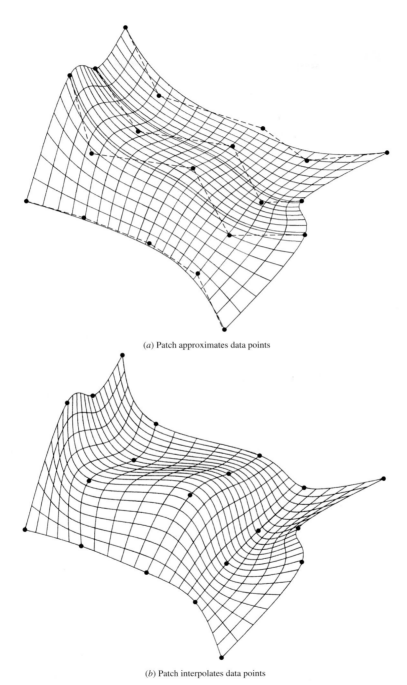

(*a*) Patch approximates data points

(*b*) Patch interpolates data points

Figure 7.29   4 × 5 **B-spline surface patches.**

EXAMPLE 7.7	**Find the equation of a B-spline surface.**

Find the equation of the B-spline surface that covers the region R. Also, find the surface vectors and its midpoint.

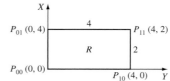

SOLUTION	This example uses the same data points as in Example 7.6. The 4×2 region

R generates the four corner points shown above. We name the points according to the convention used by a B-spline surface. We generate a 2×2 surface.

The four data points produce a linear spline surface in both directions. Thus, $k - 1 = 1$, which gives $k = 2$. Similarly, $l - 1 = 1$, which gives $l = 2$. Also, we have $n = 1$ and $m = 1$. Substituting these values into Eq. (7.52) gives:

$$
\begin{aligned}
\mathbf{P}(u, v) &= \sum_{i=0}^{1} \sum_{j=0}^{1} \mathbf{P}_{ij} N_{i,2}(u) N_{j,2}(v) \\
&= \mathbf{P}_{00} N_{0,2}(u) N_{0,2}(v) + \mathbf{P}_{01} N_{1,2}(u) N_{0,2}(v) \\
&\quad + \mathbf{P}_{01} N_{0,2}(u) N_{1,2}(v) + \mathbf{P}_{11} N_{1,2}(u) N_{1,2}(v)
\end{aligned}
\tag{7.54}
$$

Note that the maximum sum of the subscripts of any $N_{i,2}(u)$ is $k + n$, and of any $N_{j,2}(v)$ is $l + m$.

To find the $N_{i,2}(u)$ and $N_{j,2}(v)$, we need to calculate them recursively starting with $N_{i,1}$ given by Eq. (6.93). This requires the calculation of the u and v ranges and the knot vectors in the u and v directions. The calculations for both u and v are identical for this example because the number of points is the same (two) in each direction. We show the u direction here. Using the equations from the B-spline curve, we calculate the u range as $0 \le u \le 1$. The knot vector length is 4, and the knot vector itself is $\begin{bmatrix} u_0 & u_1 & u_2 & u_3 \end{bmatrix}^T = \begin{bmatrix} 0 & 0 & 1 & 1 \end{bmatrix}^T$.

We can now calculate the basis functions as:

$$
N_{0,1} = \begin{cases} 1 & u = 0 \\ 0 & \text{else} \end{cases}
\qquad
N_{1,1} = \begin{cases} 1 & 0 \le u \le 1 \\ 0 & \text{else} \end{cases}
\qquad
N_{2,1} = \begin{cases} 1 & u = 1 \\ 0 & \text{else} \end{cases}
$$

$$
N_{0,2} = \frac{u - u_0}{u_1 - u_0} N_{0,1} + \frac{u_2 - u}{u_2 - u_1} N_{1,1} = (1 - u) N_{1,1}
\tag{7.55}
$$

$$
N_{1,2} = \frac{u - u_1}{u_2 - u_1} N_{1,1} + \frac{u_3 - u}{u_3 - u_2} N_{2,1} = u N_{1,1}
\tag{7.56}
$$

These calculations hold for v; simply replace u with v to obtain the v functions. Substituting Eqs. (7.55) and (7.56) and the same equations in v (not shown) into Eq. (7.54), the surface equation becomes

$$\mathbf{P}(u, v) = \mathbf{P}_{00}(1-u)(1-v)N_{1,1}^2 + \mathbf{P}_{10}u(1-v)N_{1,1}^2 + \mathbf{P}_{01}(1-u)vN_{1,1}^2 + \mathbf{P}_{11}uvN_{1,1}^2$$

$$= \begin{bmatrix} x(u, v) \\ y(u, v) \end{bmatrix} = \begin{bmatrix} 4u(1-v) + 0 + 4uv \\ 0 + 2v(1-u) + 2uv \end{bmatrix} N_{1,1}^2 = \begin{bmatrix} 4u \\ 2v \end{bmatrix} \qquad 0 \le u \le 1, 0 \le v \le 1$$

The surface vectors are:

$$\mathbf{P}_u = \begin{bmatrix} 4 \\ 0 \end{bmatrix}, \mathbf{P}_v = \begin{bmatrix} 0 \\ 2 \end{bmatrix}, \mathbf{P}_{uv} = \begin{bmatrix} 0 \\ 0 \end{bmatrix}, \mathbf{N} = \mathbf{P}_u \times \mathbf{P}_v = \begin{vmatrix} \hat{i} & \hat{j} & \hat{k} \\ 4 & 0 & 0 \\ 0 & 2 & 0 \end{vmatrix} = 8\hat{k} = \begin{bmatrix} 0 \\ 0 \\ 8 \end{bmatrix}$$

The midpoint of the surface is at $(u, v) = (0.5, 0.5)$. Substituting into the preceding B-spline surface equation gives:

$$\mathbf{P}_{midpoint} = \begin{bmatrix} 2 \\ 1 \end{bmatrix}$$

Example 7.7 discussion:

We use the four corner points to generate the B-spline surface. The equations for $x(u, v)$ and $y(u, v)$ are linear; thus the surface is planar as it should be. The tangent vectors are constant, and therefore there is no twist vector. The midpoint of the surface is the midpoint of the region R. Moreover, the B-spline surface is identical to the Bezier surface of Example 7.6 as expected.

Example 7.7 hands-on exercise:

Starting with this example, change the surface degree along the u direction to become quadratic, by adding two midpoints at $(2, 0)$ and $(2, 2)$. How does the surface change?

7.14 Coons Surface

All the surface methods introduced thus far share one common philosophy; that is, they all require a finite number of data points to generate the respective surfaces. In contrast, a Coons surface patch is a form of "transfinite interpolation," which indicates that the Coons scheme interpolates to an infinite number of data points, that is, to all points of a curve segment, to generate the surface. Coons patch is particularly useful in blending four prescribed intersecting curves which form a closed boundary, as shown in Figure 7.30. In practice, Coons surface can be used to create windshield designs for cars and skyline windows for houses.

Figure 7.30 shows the given four boundary curves as $\mathbf{P}(u, 0)$, $\mathbf{P}(1, v)$, $\mathbf{P}(u, 1)$, and $\mathbf{P}(0, v)$. It is assumed that u and v range from 0 to 1 along these boundaries and that each pair of opposite boundary curves is identically parametrized. The development of Coons surface patch centers on answering the following question: What is a suitable, well-behaved function $\mathbf{P}(u, v)$ which blends the four given boundary curves and which satisfies the boundary conditions, that is, reduces to the correct boundary curves when $u = 0$, $u = 1$, $v = 0$, and $v = 1$?

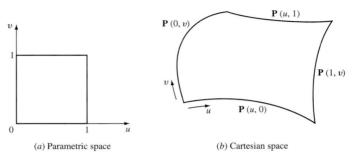

(a) Parametric space   (b) Cartesian space

Figure 7.30   Boundaries of a Coons surface patch.

Let us first consider the case of bilinearly blended Coons patch which interpolates to the four boundary curves shown in Figure 7.30. For this case, it is useful to recall that a ruled surface interpolates linearly between two given boundary curves in one direction, as shown by Eq. (7.36). Therefore, the superposition of two ruled surfaces connecting the two pairs of boundary curves might satisfy the boundary curve conditions and produce the Coons patch. Let us investigate this claim. Utilizing Eq. (7.36) in the v and u directions gives, respectively,

$$\mathbf{P}_1(u, v) = (1 - u)\mathbf{P}(0, v) + u\mathbf{P}(1, v) \tag{7.57}$$

and

$$\mathbf{P}_2(u, v) = (1 - v)\mathbf{P}(u, 0) + v\mathbf{P}(u, 1) \tag{7.58}$$

Adding these two equations gives the surface:

$$\mathbf{P}(u, v) = \mathbf{P}_1(u, v) + \mathbf{P}_2(u, v) \tag{7.59}$$

The resulting surface patch described by Eq. (7.59) does not satisfy the boundary conditions. For example, substituting $v = 0$ and 1 into this equation gives respectively

$$\mathbf{P}(u, 0) = \mathbf{P}(u, 0) + [(1 - u)\mathbf{P}(0, 0) + u\mathbf{P}(1, 0)] \tag{7.60}$$

$$\mathbf{P}(u, 1) = \mathbf{P}(u, 1) + [(1 - u)\mathbf{P}(0, 1) + u\mathbf{P}(1, 1)] \tag{7.61}$$

Equations (7.60) and (7.61) show that the terms in square brackets are extra and should be eliminated to recover the original boundary curves. These terms define the boundaries of an unwanted surface $\mathbf{P}_3(u, v)$ which is embedded in Eq. (7.59). This surface can be defined by linear interpolation in the v direction, that is,

$$\mathbf{P}_3(u, v) = (1 - v)[(1 - u)\mathbf{P}(0, 0) + u\mathbf{P}(1, 0)] + v[(1 - u)\mathbf{P}(0, 1) + u\mathbf{P}(1, 1)] \tag{7.62}$$

subtracting $\mathbf{P}_3(u, v)$ (called "correction surface") from Eq. (7.59) gives

$$\mathbf{P}(u, v) = \mathbf{P}_1(u, v) + \mathbf{P}_2(u, v) - \mathbf{P}_3(u, v) \tag{7.63}$$

or

$$\mathbf{P}(u, v) = \mathbf{P}_1(u, v) \oplus \mathbf{P}_2(u, v) \tag{7.64}$$

where $P_1 \oplus P_2$ defines the "Boolean Sum" which is $P_1 + P_2 - P_3$. The surface $P(u, v)$ given by Eq. (7.64) defines the bilinear Coons patch connecting the four boundary curves shown in Figure 7.30. Figure 7.31 shows the graphical representation of Eq. (7.64). And its matrix form is

$$\mathbf{P}(u, v) = -[-1 \quad (1-u) \quad u]\begin{bmatrix} 0 & \vdots & \mathbf{P}(u, 0) & \mathbf{P}(u, 1) \\ \mathbf{P}(0, v) & \vdots & \mathbf{P}(0, 0) & \mathbf{P}(0, 1) \\ \mathbf{P}(1, v) & \vdots & \mathbf{P}(1, 0) & \mathbf{P}(1, 1) \end{bmatrix}\begin{bmatrix} -1 \\ 1-v \\ v \end{bmatrix} \qquad (7.65)$$

The left column and the upper row of the matrix in Eq. (7.65) represent $\mathbf{P}_1(u, v)$ and $\mathbf{P}_2(u, v)$, respectively, while the lower right block represents the correction surface $\mathbf{P}_3(u, v)$. The functions -1, $1-u$, u, $1-v$, and v are called blending functions because they blend together four separate boundary curves to give one surface.

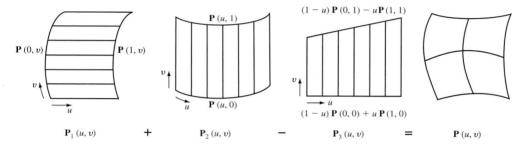

Figure 7.31   **Bilinearly blended Coons patch (Boolean sum).**

The main drawback of the bilinearly blended Coons patch is that it only provides C^0 continuity (position continuity) between adjacent patches even if their boundary curves form a C^1 continuity network. For example, if two patches are to be connected along the boundary curve $\mathbf{P}(1, v)$ of the first patch and $\mathbf{P}(0, v)$ of the second, it can be shown that the continuity of the cross-boundary derivatives (Figure 7.32) $\mathbf{P}_u(1, v)$ and $\mathbf{P}_u(0, v)$ of the two patches cannot be made equal (refer to Problem 7.17 at the end of the chapter).

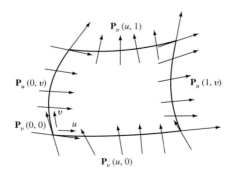

Figure 7.32   **Cross-boundary derivatives.**

Gradient continuity across boundaries of patches of a composite Coons surface is essential for practical applications. If, for example, a network of C^1 curves is given as shown in Figure 7.33, it becomes desirable to form a composite Coons surface which is smooth or C^1 continuous, that is, which provides continuity of cross-boundary derivatives between patches.

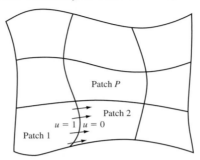

Figure 7.33   A composite Coons surface.

Investigating Eq. (7.65) shows that the choice of the blending functions controls the behavior of the resulting Coons patch. If we use two cubic polynomials $F_1(x)$ and $F_2(x)$ instead of the linear polynomials $(1 - u)$ and u, respectively, a bicubically blended Coons patch results. This patch guarantees C^1 continuity between patches. $F_1(x)$ and $F_2(x)$ can be written as:

$$F_1(x) = 2x^3 - 3x^2 + 1$$
$$F_2(x) = -2x^3 + 3x^2 \tag{7.66}$$

Substituting $F_1(x)$ and $F_2(x)$ into Eqs. (7.57) and (7.58) gives:

$$\mathbf{P}_1(u, v) = (2u^3 - 3u^2 + 1)\mathbf{P}(0, v) + (-2u^3 + 3u^2)\mathbf{P}(1, v) \tag{7.67}$$

$$\mathbf{P}_2(u, v) = (2v^3 - 3v^2 + 1)\mathbf{P}(u, 0) + (-2v^3 + 3v^2)\mathbf{P}(u, 1) \tag{7.68}$$

Similar to the bilinear patch, the Boolean sum $\mathbf{P}_1 \oplus \mathbf{P}_2$ can be formed, and the matrix equation of the bicubic Coons patch becomes:

$$\mathbf{P}(u, v) = -[-1 \quad F_1(u) \quad F_2(u)]\begin{bmatrix} 0 & \mathbf{P}(u, 0) & \mathbf{P}(u, 1) \\ \mathbf{P}(0, v) & \mathbf{P}(0, 0) & \mathbf{P}(0, 1) \\ \mathbf{P}(1, v) & \mathbf{P}(1, 0) & \mathbf{P}(1, 1) \end{bmatrix}\begin{bmatrix} -1 \\ F_1(v) \\ F_2(v) \end{bmatrix} \tag{7.69}$$

As can easily be seen from Eqs. (7.65) and (7.69), the correction surface is usually formed by applying the blending functions to the corner data alone.

To check continuity across patch boundaries, let us consider Patch 1 and Patch 2 in Figure 7.33. For C^0 and C^1 continuity we should have:

$$[\mathbf{P}(0, v)]_{\text{Patch 2}} = [\mathbf{P}(1, v)]_{\text{Patch 1}} \tag{7.70}$$

$$[\mathbf{P}_u(0, v)]_{\text{Patch 2}} = [\mathbf{P}_u(1, v)]_{\text{Patch 1}} \tag{7.71}$$

C^0 continuity is satisfied automatically between the two patches because they share the same boundary curve. For C^1 continuity, differentiating Eq. (7.69) with respect to u, and noticing that $dF_1/du|_{u=0} = dF_2/du|_{u=0} = dF_1/du|_{u=1} = dF_2/du|_{u=1} = 0$, we can write:

$$\mathbf{P}_u(u, v) = F_1(v)\mathbf{P}_u(u, 0) + F_2(v)\mathbf{P}_u(u, 1) \tag{7.72}$$

At the common boundary curve between the two patches, Eq. (7.72) becomes

$$[\mathbf{P}_u(0, v)]_{\text{Patch 2}} = F_1(v)\mathbf{P}_u(0, 0) + F_2(v)\mathbf{P}_u(0, 1) \tag{7.73}$$

and

$$[\mathbf{P}_u(1, v)]_{\text{Patch 1}} = F_1(v)\mathbf{P}_u(1, 0) + F_2(v)\mathbf{P}_u(1, 1) \tag{7.74}$$

Based on Eqs. (7.73) and (7.74), Eq. (7.71) is satisfied if the network of boundary curves is C^1 continuous because this makes $[\mathbf{P}_u(0, 0)]_{\text{Patch2}}$ and $[\mathbf{P}_u(0, 1)]_{\text{Patch2}}$ equal to $[\mathbf{P}_u(1, 0)]_{\text{Patch1}}$ and $[\mathbf{P}_u(1, 1)]_{\text{Patch1}}$, respectively. Therefore, continuity of cross-boundary derivatives is automatically satisfied for bicubic Coons patches if the boundary curves are C^1 continuous.

The bicubic Coons patch as defined by Eq. (7.69) is easy to use in a design environment because only the four boundary curves are needed. However, a more flexible composite C^1 bicubic Coons surface can be developed if, together with the boundary curves, the cross-boundary derivatives $\mathbf{P}_u(0, v)$, $\mathbf{P}_u(1, v)$, $\mathbf{P}_v(u, 0)$, and $\mathbf{P}_v(u, 1)$ are given (see Figure 7.32). Note that at the corners these derivatives must be compatible with the curve information. Similar to Eqs. (7.67) and (7.68), we can define the following cubic Hermite interpolants:

$$\mathbf{P}_1(u, v) = F_1(u)\mathbf{P}(0, v) + F_2(u)\mathbf{P}(1, v) + F_3(u)\mathbf{P}_u(0, v) + F_4(u)\mathbf{P}_u(1, v) \tag{7.75}$$

and

$$\mathbf{P}_2(u, v) = F_1(v)\mathbf{P}(u, 0) + F_2(v)\mathbf{P}(u, 1) + F_3(v)\mathbf{P}_v(u, 0) + F_4(v)\mathbf{P}_v(u, 1) \tag{7.76}$$

where the functions F_1 and F_2 are given by Eq. (7.66), and F_3 and F_4 are given by

$$F_3(x) = x^3 - 2x^2 + x \tag{7.77}$$
$$F_4(x) = x^3 - x^2$$

Forming the Boolean sum, $\mathbf{P}_1 \oplus \mathbf{P}_2$, of Eqs. (7.75) and (7.76) results in the bicubic Coons patch that incorporates cross-boundary derivatives. The introduction of the cross-boundary derivatives causes the twist vectors at the corners of the patch to appear (see Problem 7.18 at the end of the chapter). The matrix equation of this Coons patch can be written as

$$\mathbf{P}(u, v) = -[-1 \quad F_1(u) \quad F_2(u) \quad F_3(u) \quad F_4(u)]$$

$$\times \begin{bmatrix} 0 & \vdots & \mathbf{P}(u, 0) & \mathbf{P}(u, 1) & \vdots & \mathbf{P}_v(u, 0) & \mathbf{P}_v(u, 1) \\ \hline \mathbf{P}(0, v) & \vdots & \mathbf{P}(0, 0) & \mathbf{P}(0, 1) & \vdots & \mathbf{P}_v(0, 0) & \mathbf{P}_v(0, 1) \\ \mathbf{P}(1, v) & \vdots & \mathbf{P}(1, 0) & \mathbf{P}(1, 1) & \vdots & \mathbf{P}_v(1, 0) & \mathbf{P}_v(1, 1) \\ \hline \mathbf{P}_u(0, v) & \vdots & \mathbf{P}_u(0, 0) & \mathbf{P}_u(0, 1) & \vdots & \mathbf{P}_{uv}(0, 0) & \mathbf{P}_{uv}(0, 1) \\ \mathbf{P}_u(1, v) & \vdots & \mathbf{P}_u(1, 0) & \mathbf{P}_u(1, 1) & \vdots & \mathbf{P}_{uv}(1, 0) & \mathbf{P}_{uv}(1, 1) \end{bmatrix} \begin{bmatrix} -1 \\ F_1(v) \\ F_2(v) \\ F_3(v) \\ F_4(v) \end{bmatrix} \tag{7.78}$$

The upper 3×3 matrix of Eq. (7.78) determines the patch defined previously by Eq. (7.69). The left column and the upper row represent $\mathbf{P}_1(u, v)$ and $\mathbf{P}_2(u, v)$, respectively, while the lower right 4×4 matrix represents the bicubic tensor product surface discussed in Section 7.11 [see Eq. (7.43)]. Equation (7.78) shows that every bicubically blended Coons surface reproduces a bicubic surface. On the other hand, bicubically blended Coons patches cannot be described in general by bicubic tensor product surfaces. Thus, Coon formulation can describe a much richer variety of surfaces than do tensor product surfaces.

EXAMPLE 7.8   **Planar Coons patch.**
Show that if the boundary curves of a bilinear Coons patch are coplanar, the resulting patch is also planar.

SOLUTION   We need to show that the surface normal at any point on the surface is normal to the plane of the boundary curves. Based on Eq. (7.63), the surface tangent vectors are:

$$\mathbf{P}_u(u, v) = [\mathbf{P}_u(u, 0) + \mathbf{P}(1, v) - \mathbf{P}(0, v) + \mathbf{P}(0, 0) - \mathbf{P}(1, 0)]$$
$$+ v[\mathbf{P}_u(u, 1) - \mathbf{P}_u(u, 0) + \mathbf{P}(1, 0) - \mathbf{P}(0, 0) - \mathbf{P}(1, 1) + \mathbf{P}(0, 1)]$$
$$= \mathbf{A} + v\mathbf{B}$$

and
$$\mathbf{P}_v(u, v) = [\mathbf{P}_v(0, v) + \mathbf{P}(u, 1) - \mathbf{P}(u, 0) + \mathbf{P}(0, 0) - \mathbf{P}(0, 1)]$$
$$+ u[\mathbf{P}_v(1, v) - \mathbf{P}_v(0, v) - \mathbf{P}(0, 0) + \mathbf{P}(0, 1) + \mathbf{P}(1, 0) - \mathbf{P}(1, 1)]$$
$$= \mathbf{C} + u\mathbf{D}$$

It can easily be shown that $\mathbf{P}_u(u, v)$ and $\mathbf{P}_v(u, v)$ lie in the plane of the boundary curves. Considering $\mathbf{P}_u(u, v)$, the vectors $\mathbf{P}_u(u, 0)$, $\mathbf{P}(1, v) - \mathbf{P}(0, v)$, and $\mathbf{P}(0, 0) - \mathbf{P}(1, 0)$ lie in the given plane. Also, by investigating the coefficient of v, the vectors $\mathbf{P}_u(u, 1)$, $\mathbf{P}_u(u, 0)$, $\mathbf{P}(1, 0) - \mathbf{P}(0, 0)$, and $\mathbf{P}(0, 1) - \mathbf{P}(1, 1)$ lie in the given plane. Therefore, vectors $\mathbf{A}$ and $\mathbf{B}$ lie in the plane. Consequently the tangent vector $\mathbf{P}_u(u, v)$ to the surface at any point (u, v) lies in the plane of the boundary curves. The same argument can be extended to $\mathbf{P}_v(u, v)$.

The surface normal is given by:
$$\mathbf{N}(u, v) = \mathbf{P}_u \times \mathbf{P}_v = (\mathbf{A} + v\mathbf{B}) \times (\mathbf{C} + u\mathbf{D})$$

which is perpendicular to the plane of $\mathbf{P}_u$ and $\mathbf{P}_v$ and, therefore, the plane of the boundary curves. Thus for any point on the surface, the direction of the surface normal is constant (the magnitude depends on the point) or the unit normal is fixed in space. Knowing that the plane surface is the only surface that has a fixed unit normal, we conclude that a bilinear Coons patch degenerates to a plane if its boundary curves are coplanar. Thus, this patch can be used to create planes with curved boundaries.

EXAMPLE 7.9   **Find the equation of a Coons surface.**
Find the equation of the Coons surface that covers the region R. Also, find the surface vectors and its midpoint.

SOLUTION   The 4×2 region R is bounded by the four boundary curves, L_1 to L_4. We name the points according to the convention used by a Coons surface. We generate a bilinear Coons surface. We derive the surface equation by using the Boolean sum instead of using Eq. (7.65). The equations of the four boundary lines are:

$$P_{L1} = \begin{bmatrix} 0 \\ 0 \end{bmatrix} + u \begin{bmatrix} 4 \\ 0 \end{bmatrix}$$

$$P_{L2} = \begin{bmatrix} 0 \\ 2 \end{bmatrix} + u \begin{bmatrix} 4 \\ 0 \end{bmatrix}$$

$$0 \le u \le 1, \; 0 \le v \le 1$$

$$P_{L3} = \begin{bmatrix} 0 \\ 0 \end{bmatrix} + v \begin{bmatrix} 0 \\ 2 \end{bmatrix}$$

$$P_{L4} = \begin{bmatrix} 4 \\ 0 \end{bmatrix} + v \begin{bmatrix} 0 \\ 2 \end{bmatrix}$$

The ruled surface connecting $\mathbf{P}_{L1}$ and $\mathbf{P}_{L2}$ has this equation:

$$\mathbf{P}_1(u, v) = \begin{bmatrix} 4u \\ 2v \end{bmatrix}$$

The ruled surface connecting $\mathbf{P}_{L3}$ and $\mathbf{P}_{L4}$ has this equation:

$$\mathbf{P}_2(u, v) = \begin{bmatrix} 4u \\ 2v \end{bmatrix}$$

Using Eq. (7.62), the equation of the corrective surface can be written as:

$$\mathbf{P}_3(u, v) = \begin{bmatrix} x(u, v) \\ y(u, v) \end{bmatrix} = (1-u)(1-v)\begin{bmatrix} 0 \\ 0 \end{bmatrix} + u(1-v)\begin{bmatrix} 4 \\ 0 \end{bmatrix} + v(1-u)\begin{bmatrix} 0 \\ 2 \end{bmatrix} + uv\begin{bmatrix} 4 \\ 2 \end{bmatrix} = \begin{bmatrix} 4u \\ 2v \end{bmatrix}$$

The equation of the Coons surface can now be written as:

$$\mathbf{P}(u, v) = \mathbf{P}_1(u, v) + \mathbf{P}_2(u, v) - \mathbf{P}_3(u, v) = \begin{bmatrix} 4u \\ 2v \end{bmatrix}$$

The surface vectors are:

$$\mathbf{P}_u = \begin{bmatrix} 4 \\ 0 \end{bmatrix}, \mathbf{P}_v = \begin{bmatrix} 0 \\ 2 \end{bmatrix}, \mathbf{P}_{uv} = \begin{bmatrix} 0 \\ 0 \end{bmatrix}, \mathbf{N} = \mathbf{P}_u \times \mathbf{P}_v = \begin{vmatrix} \hat{i} & \hat{j} & \hat{k} \\ 4 & 0 & 0 \\ 0 & 2 & 0 \end{vmatrix} = 8\hat{k} = \begin{bmatrix} 0 \\ 0 \\ 8 \end{bmatrix}$$

The midpoint of the surface is at $(u, v) = (0.5, 0.5)$. Substituting into the above Coons surface equation gives:

$$\mathbf{P}_{midpoint} = \begin{bmatrix} 2 \\ 1 \end{bmatrix}$$

Example 7.9 discussion:

We use the four lines that make up a closed boundary to generate the Coons surface. The equations for $x(u, v)$ and $y(u, v)$ are linear; thus the surface is planar as it should be. The tangent vectors are constants, and therefore there is no twist vector. The midpoint of the surface is the midpoint of the region R.

Note that both $\mathbf{P}_1(u, v)$ and $\mathbf{P}_3(u, v)$ surfaces are identical because they are both ruled surfaces defined by the same rails.

Example 7.9 hands-on exercise:

Use Eq. (7.65) to derive the equation of the Coons surface. Compare with this example.

7.15 Blending Surface

This is a surface that connects two nonadjacent surfaces or patches. The blending surface is usually created to manifest C^0 and C^1 continuity with the two given patches. The fillet surface shown in Figure 7.10 is considered a special case of a blending surface. Figure 7.34 shows a general blending surface. A bicubic surface can be used to blend Patch 1 and Patch 2 with both C^0 and C^1 continuity. The corner points P_1, P_2, P_3, and P_4 of the blending surface and their related tangent and twist vectors are readily available from the two patches. Therefore, the $[B]$ matrix of the blending surface can be evaluated. A bicubic blending surface is suitable to blend cubic patches; that is, bicubic Bezier or B-spline patches.

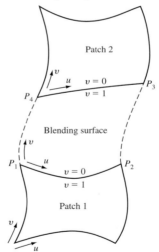

Figure 7.34   A blending surface.

For patches of other orders, a B-spline blending surface may be generated in the following scenario. A set of points and their related v-tangent vectors beginning with P_1 and ending with P_2 can be generated along the $v = 1$ edge of Patch 1. Similarly, a corresponding set can be generated along the $v = 0$ edge of Patch 2. Cubic spline curves can now be created between the two sets. These curves can be used to generate an ordered rectangular set of points that can be connected with B-spline surface, which becomes the blending surface.

Some CAD/CAM systems allow users to connect a given set of curves with a B-spline surface directly. In the case of the fillet surface shown in Figure 7.10, a fillet radius is used to generate the surface. Here, the rectangular set of points to create the B-spline surface can be generated by creating fillets between corresponding $u = constant$ curves on both patches. In turn, points can be generated on these fillets.

7.16 Offset Surface

If an original patch and an offset direction are given as shown in Figure 7.11, the equation of the resulting offset patch can be written as:

$$\mathbf{P}(u, v)_{\text{offset}} = \mathbf{P}(u, v) + \hat{\mathbf{n}}(u, v)\, d(u, v) \tag{7.79}$$

where $\mathbf{P}(u, v)$, $\hat{\mathbf{n}}(u, v)$ and $d(u, v)$ are, respectively, the original surface, the unit normal vector at point (u, v) on the original surface, and the offset distance at point (u, v) on the original surface. The unit normal $\hat{\mathbf{n}}(u, v)$ is the offset direction shown in Figure 7.11. The distance $d(u, v)$ permits generating uniform or tapered thickness surfaces depending on whether $d(u, v)$ is constant or varies linearly in u, v, or both.

7.17 Triangular Patches

Triangular patches are useful if the data points of a surface form a triangle or if a surface cannot be modeled by rectangular patches only and may require at least one triangular patch.

In tensor product surfaces, the parameters are u and v and the parametric domain is defined by the unit square of $0 \le u \le 1$ and $0 \le v \le 1$. In triangulation techniques, three parameters u, v, and w are used, and the parametric domain is defined by a symmetric unit triangle of $0 \le u \le 1$, $0 \le v \le 1$, and $0 \le w \le 1$ as shown in Figure 7.35.

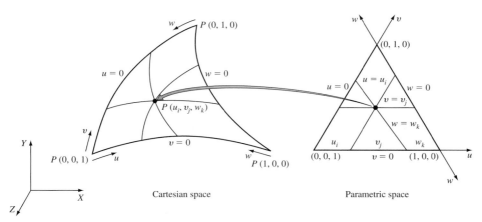

Figure 7.35 Representation of a triangular patch.

The coordinates u, v, and w are called "barycentric coordinates." While the coordinate w is not independent of u and v (note $u + v + w = 1$ for any point in the domain), it is introduced to emphasize the symmetry properties of the barycentric coordinates.

The formulation of triangular polynomial patches follows somewhat similar pattern to that of tensor product patches. For example, a triangular Bezier patch is defined by:

$$\mathbf{P}(u, v, w) = \sum_{i,j,k} \mathbf{P}_{ijk} B_{i,j,k,n}(u, v, w), \qquad 0 \le u \le 1, 0 \le v \le 1, 0 \le w \le 1 \tag{7.80}$$

where $i, j, k \ge 0$, $i + j + k = n$ and n is the degree of the patch. The $B_{i,j,k}$ are Bernstein polynomials of degree n:

$$B_{i,j,k,n} = \frac{n!}{i!j!k!} u^i v^j w^k \tag{7.81}$$

The coefficients $\mathbf{P}_{i,j,k}$ in Eq. (7.80) are the control or data points that form the vertices of the control polyhedron. The number of data points required to define a Bezier patch of degree n is given by $(n + 1)(n + 2)/2$.

Figure 7.36 shows cubic and quartic triangular Bezier patches with their related Bernstein polynomials. The order of inputting data points should follow the pyramid structure of Bernstein polynomial shown in Figure 7.36. For example, 15 points are required to create a quartic Bezier patch, and they must be input in five rows. The first row has five points $(n + 1)$, and each successive row has one point less than its predecessor until we reach the final row, which has only one point.

This pattern of input can be achieved symmetrically from any direction as shown. Note that the degree of the triangular Bezier patch is the same in all directions, contrary to the rectangular patch that can have n and m degree polynomials in u and v directions, respectively. However, all the characteristics of the rectangular patch hold true for the triangular patch.

A rectangular Coons patch can be modified in a similar fashion as described previously to develop a triangular Coons patch. It is left to the reader, as an exercise, to extend the preceding formulation to a triangular Coons patch.

7.18 Surface Manipulations

As with curves, surface manipulation enables the designer to use surfaces effectively in design applications. For example, if two surfaces are to be welded by a robot, their intersection curve provides the path the robot should follow. Therefore, the robot trajectory planning becomes very accurate. This section covers some useful features of surface manipulations.

7.18.1 Displaying

The simplest method for displaying surfaces is to generate a mesh of curves on the surface by holding one parameter constant at a time. Most CAD/CAM systems have a mesh size function. Using this size and the surface equation, the curves of the mesh are evaluated. For example, to display a Bezier surface with mesh size of 3×4, three curves (in the v direction) that have constant u values ($u = 0, 1/2, 1$) and four curves (in the u direction) that have constant v values ($v = 0, 1/3, 2/3, 1$) are evaluated using Eq. (7.48) and displayed. See Tutorial 7.19.5.

This method is not very efficient because fine details of the surface may be lost unless the mesh is very dense, in which case it becomes slow and expensive to generate, display, and/or update the surface. This method of displaying a surface by a mesh of curves is sometimes referred to as wireframe display of a surface.

To improve the visualization of a surface, surface normals can be displayed as straight lines in addition to the wireframe display. If the lengths of these straight lines are made long enough, any small vibration in surface shape becomes evident. The other best method to display surfaces is shading. Shading techniques are discussed later in the book. Various surfaces can be shaded with various colors. Surface curvatures and other related properties can also be displayed via shading. If surfaces are nested, inner shaded surfaces can be viewed by using X-ray and transparency techniques.

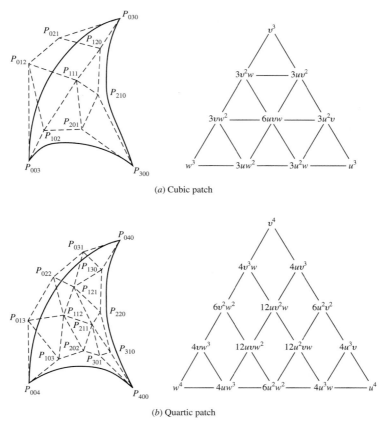

(a) Cubic patch

(b) Quartic patch

Figure 7.36   Triangular Bezier patches.

7.18.2 Evaluating Points and Curves on Surfaces

Generating points on surfaces is useful for applications such as finite element modeling and analysis. The obvious method of calculating points on a surface is by substituting their respective u and v values into the surface equation.

As with curves, we can evaluate a point on a surface for given u and v values with the direct point solution, which consists of evaluating three polynomials in u and v, one for each coordinate of the point. The inverse problem, which requires finding the corresponding u and v values if the x, y, and z coordinates of the point are given, arises if tangent vectors at the point are to be evaluated or if two coordinates of the point are given and we must find the third coordinate. The solution to this problem (the inverse point solution) requires solving two nonlinear polynomials in u and v simultaneously via numeric methods (see Problem 7.19 at the end of the chapter).

7.18.3 Segmentation

The segmentation problem of a given surface is identical to that of a curve. Surface segmentation is a reparametrization or parameter transformation of the surface while keeping the degree of its polynomial in u and v unchanged.

Let us assume that a surface patch is defined over the range $0 \le u \le u_m$ and $0 \le v \le v_m$ and we wish to split it at a point P_1 defined by (u_1, v_1). If the point is defined by its cartesian coordinates, the inverse point solution must be found first to obtain u and v. Let us introduce the new parameters u^1 and v^1 such that their ranges are $[0,1]$ for each resulting subpatch (see Figure 7.37).

The parameter transformation takes the form:

$$\left. \begin{array}{l} u^1 = u_0 + (u_1 - u_0)u \\ v^1 = v_0 + (v_1 - v_0)v \end{array} \right\} \quad \text{for subpatch 1} \tag{7.82}$$

Similar equations can be written for the other subpatches. Notice that $u^1 = 0, 1$ and $v^1 = 0, 1$ correspond to the proper values of u and v for the subpatch.

If the surface is to be divided into two segments or subpatches along the $u = u_1$ curve instead of four segments, Eq. (7.82) becomes

$$\left. \begin{array}{l} u^1 = u_0 + (u_1 - u_0)u \\ v^1 = v_0 + (v_m - v_0)v \end{array} \right\} \quad \text{for first segment} \tag{7.83}$$

If segmentation is done along the $v = v_1$ curve, Eq. (7.82) becomes

$$\left. \begin{array}{l} u^1 = u_0 + (u_m - u_0)u \\ v^1 = v_0 + (v_1 - v_0)v \end{array} \right\} \quad \text{for first segment} \tag{7.84}$$

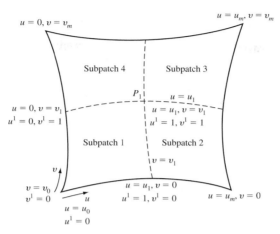

Figure 7.37   Representation of a segmented surface patch.

7.18.4 Trimming

Trimming surface entities is useful for engineering applications. It helps eliminate unnecessary calculations on the user's part. Surface trimming can be treated as a segmentation or intersection problem. If the surface is to be trimmed between two point trimming boundaries, we then have a segmentation problem at hand. If, on the other hand, a surface is to be trimmed to another surface, the intersection curve of the two surfaces must be found first and then the desired surface trimmed to it. Figure 7.38 shows an example. A Bezier surface (identified by click d_1) is trimmed between the two points identified by the two clicks d_2 and d_3.

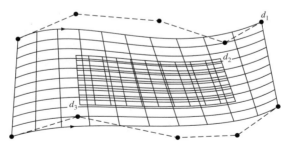

Figure 7.38   Surface trimming.

7.18.5 Intersection

The intersection problem involving surfaces is complex and nonlinear in nature. It depends on whether it is a surface/curve or surface/surface intersection. In a surface-to-curve intersection, the intersection problem is defined by the equation

$$P(u,\ v) - P(w) = 0 \tag{7.85}$$

where $P(u,\ v) = 0$ and $P(w) = 0$ are the parametric equations of the surface and the curve, respectively. Equation (7.85) is a system of three scalar equations, generally nonlinear, in three unknowns u, v, and w. There are efficient iterative methods that can solve Eq. (7.85), such as the Newton-Raphson method. The detailed solution of Eq. (7.85) is left to the reader.

The problem of surface-to-surface intersection is defined by the equation

$$P(u,\ v) - P(t,\ w) = 0 \tag{7.86}$$

where the $P(u,\ v) = 0$ and $P(t,\ w) = 0$ are the equations of the two surfaces.

The vector equation (7.86) corresponds to three scalar equations in four variables u, v, t, and w. To solve this overspecified problem, some methods hold one of the unknowns constant and therefore reduce the original surface/surface intersection problem into a surface/curve intersection. Other methods introduce a new constrained function of the four variables. In both approaches, the solutions are iterative and may yield any combination of curves (closed or open) and isolated points.

Solutions for sculptured (composed of multiple surface patches) surfaces usually employ curve-tracing, subdivision (divide-and-conquer method), or a combination of both techniques. Some of these solutions assume a given class of surface types, such as Bezier or B-spline representation. Figure 7.39 shows an example of surface/surface intersection.

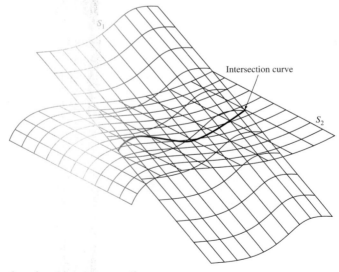

Figure 7.39   Surface/surface intersection.

7.18.6 Projection

Projecting an entity onto a plane or a surface is useful in applications such as determining shadows or finding the position of the entity relative to the plane or the surface. Entities that can be projected include points, lines, curves, or surfaces. Projecting a point onto a plane or a surface forms the basic problem which is shown in Figure 7.40a. Point P_0 is projected along the direction $\hat{\mathbf{r}}$ onto the given plane. We wish to calculate the coordinates of the projected point Q.

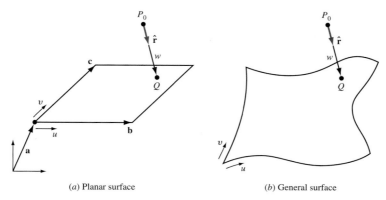

(a) Planar surface   (b) General surface

Figure 7.40   Projecting a point onto a surface.

The plane equation, based on Eq. (7.28), can be written as follows:

$$\mathbf{P}(u,\, v) = \mathbf{a} + u\mathbf{b} + v\mathbf{c} \tag{7.87}$$

where the vectors $\mathbf{a}$, $\mathbf{b}$, and $\mathbf{c}$ are shown in Figure 7.40a.

The equation of the projection line is given by:

$$\mathbf{P}(w) = \mathbf{P}_0 + w\hat{\mathbf{r}} \tag{7.88}$$

The projection point Q is the intersection point between the line and the plane; that is, the following equation must be solved for u, v, and w:

$$\mathbf{P}(u,\, v) - \mathbf{P}(w) = \mathbf{0} \tag{7.89}$$

or

$$\mathbf{a} + u\mathbf{b} + v\mathbf{c} = \mathbf{P}_0 + w\hat{\mathbf{r}} \tag{7.90}$$

To solve for w, dot-multiply both sides of Eq. (7.90) by $(\mathbf{b} \times \mathbf{c})$ to get:

$$(\mathbf{b} \times \mathbf{c}) \cdot \mathbf{a} = (\mathbf{b} \times \mathbf{c}) \cdot (\mathbf{P}_0 + w\hat{\mathbf{r}}) \tag{7.91}$$

since $(\mathbf{b} \times \mathbf{c})$ is perpendicular to both $\mathbf{b}$ and $\mathbf{c}$, Eq. (7.91) gives:

$$w = \frac{(\mathbf{b} \times \mathbf{c}) \cdot (\mathbf{a} - \mathbf{P}_0)}{(\mathbf{b} \times \mathbf{c}) \cdot \hat{\mathbf{r}}} \tag{7.92}$$

Similarly, we can write

$$u = \frac{(\mathbf{c} \times \hat{\mathbf{r}}) \cdot (\mathbf{P}_0 - \mathbf{a})}{(\mathbf{c} \times \hat{\mathbf{r}}) \cdot \mathbf{b}} \tag{7.93}$$

and

$$v = \frac{(\mathbf{b} \times \hat{\mathbf{r}}) \cdot (\mathbf{P}_0 - \mathbf{a})}{(\mathbf{b} \times \hat{\mathbf{r}}) \cdot \mathbf{c}} \tag{7.94}$$

The projection point Q results by substituting Eq. (7.92) into (7.88) or Eqs. (7.93) and (7.94) into Eq. (7.87).

If point P_0 is to be projected onto a general surface as shown in Figure 7.40b, Eq. (7.87) is replaced by the surface equation and Eq. (7.89) becomes a nonlinear equation similar to Eq. (7.85).

The preceding approach can be extended to projecting curves and surfaces onto a given surface as shown in Figures 7.41 and 7.42. The projection of a straight line passing by the two end points P_0 and P_1 (see Figure 7.41a) along the direction $\hat{\mathbf{r}}$ involves projecting the two points using Eqs. (7.92) to (7.94) and then connecting Q_0 and Q_1 by a straight line that, of course, must lie in the given plane.

The projection of a general curve $\mathbf{P}(s)$ onto a plane (Figure 7.41b) or a general surface (Figure 7.41c) requires repetitive solution of Eq. (7.85). One simple strategy is to generate a set of points on $\mathbf{P}(s)$, project them onto the given plane or surface, and then connect them by a B-spline curve to obtain the projection curve $\mathbf{Q}(t)$.

The projection of a surface $\mathbf{P}(u_1, v_1)$ onto a surface $\mathbf{P}(u, v)$ (Figure 7.42) can be seen as an extension of the preceding strategy. A set of points is first generated on the surface $\mathbf{P}(u_1, v_1)$. Utilizing Eq. (7.85), this set is projected into the surface $\mathbf{P}(u, v)$. The projection points are then connected by a B-spline surface to produce the projection surface $\mathbf{Q}(s, t)$. Note here that the more the number of the projection points, the closer the surface $\mathbf{Q}(s, t)$ to the surface $\mathbf{P}(u, v)$.

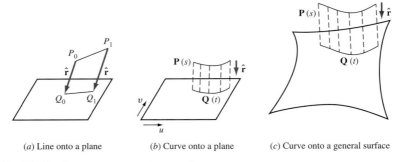

(a) Line onto a plane   (b) Curve onto a plane   (c) Curve onto a general surface

Figure 7.41   Projecting a curve onto a surface.

7.18.7   Transformation

Homogeneous transformations of surfaces are extensions of transformations of curves. They offer very useful tools to designers for creating surfaces. Functions such as translation, rotation, mirror, and scaling are offered by most CAD/CAM systems and are based on transformation concepts. More on transformation is covered in Chapter 12.

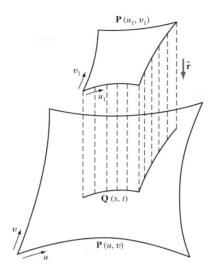

Figure 7.42   Projecting a surface onto a surface.

7.19 Tutorials

7.19.1 Create a B-Spline Surface and Solid

This tutorial shows how to create the closed surface shown on the right, using a set of B-spline curves. We also convert the closed surface into a solid. We use four B-spline curves to define the surface. Each curve is planar and is defined by four points. All dimensions are in mm.

Modeling strategy:

1. **Create the first B-spline**. Select the `Front` sketch plane. Use the four points shown below to create the B-spline curve.
2. **Create the second B-spline.** Define a sketch plane that is 20 mm offset from the `Front` sketch plane. Use the four points shown below to create the B-spline curve.
3. **Create the third B-spline.** Define a sketch plane that is 40 mm offset from the `Front` sketch plane. Use the four points shown below to create the B-spline curve.
4. **Create the fourth B-spline.** Define a sketch plane that is 60 mm offset from the `Front` sketch plane. Use the four points shown below to create the B-spline curve.
5. **Create the surface via loft.** Use a loft surface command on your CAD system and select the four B-spline curves to create the surface.
6. **Close the surface ends.** Use the two edges of the surface of step 5 to create two planar end caps.

Step 1

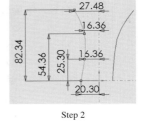

Step 2

Step 3

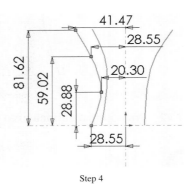

Step 4

7. **Convert surfaces to a solid.** Sew the three surfaces of step 6 (the B-spline surface and the two end caps) together. Then convert the resulting closed surface to the solid that is shown on the right.

Tutorial 7.19.1 discussion:

We use surfaces to create complex solids that do not fit any predefined feature on a CAD system, such as the solid of this tutorial. We create surfaces and then sew (stitch or knit) them to form a closed body. Sewing is a common operation offered by CAD systems. Find it on yours. If the surfaces have gaps across their common boundaries (edges), the sewing operation fails. Surfaces must be created properly for the operation to succeed. After sewing, an operation that converts the surface to a solid is used. This operation requires the user to select the sewed (knitted) surface.

Tutorial 7.19.1 hands-on exercise:

Change the location of some of the data points. Re-create the surface and the solid. Observe the local control property of the B-spline curves and the surface.

7.19.2   Create a Surface and Solid of Revolution

This tutorial shows how to create the surface of revolution shown below, using a B-spline curve. We also create a solid of revolution using the same curve. We create the spline curve using the seven circular cross sections shown below. All dimensions are in inches.

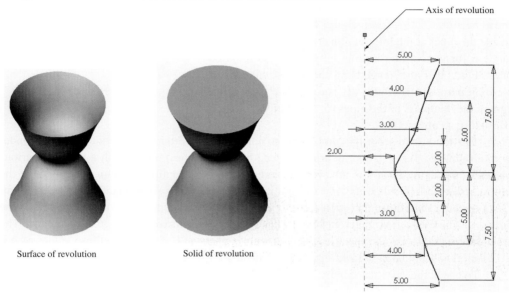

Surface of revolution   Solid of revolution

Modeling strategy:

1. **Create the B-spline curve.** Select the Front sketch plane. Construct the curve using the seven points shown on the foregoing sketch.

2. **Create the axis of revolution.** Use an axis command to create a vertical axis passing through the origin as shown on the foregoing sketch.

3. **Create the surface of revolution.** Use a surface of revolution operation to revolve the curve 360° around the axis of revolution for a full revoultion.

4. **Create the solid of revolution.** Repeat step 3, but use the revolution operation to create solids instead of surfaces. The solid is shown in the foregoing screenshot.

Tutorial 7.19.2 discussion:

The model in this tutorial is axisymmetric. We create it as a surface or solid model. Many CAD systems offer the same feature as a surface or a solid to increase their modeling domains. Make sure to check the surface menu of your CAD system.

Tutorial 7.19.2 hands-on exercise:

Convert the surface to a solid by capping (closing its two ends) and using a sewing operation. Also, investigate the local control of the curve and surface, by changing some of the data points that define the B-spline.

7.19.3 Design a Pipe Elbow

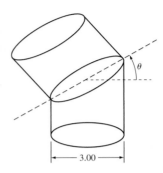

Two pipes whose interesection plane forms an angle θ are shown on the right. Each pipe has a radius of 1.5 inch. The two pipes are to be butt welded along the common ellipse of intersection. The area of the ellipse must be 10 in^2 ±1% for pressure drop considerations inside the pipes at the intersection cross section. Find the angle θ for which the resulting ellipse has the required area. Find also the lengths of the ellipse axes. For welding purposes, find also the perimeter of the ellipse. Ignore the thickness of the pipes for simplicity.

Modeling strategy:

The strategy to solve this problem is to construct the vertical pipe as a tabulated cylinder with a radius of 1.5 inch and an arbitrary length (use 6 inches), cut it with a plane at various values of θ ranging from 0 to 90°, verify the resulting ellipse entity to obtain the lengths of the ellipse axes, and calculate its area using a `calculate area` command provided by your CAD system. The ellipse that produces the given area is the desired one and the angle θ is the solution. The ellipse perimeter can be evaluated using a `measure length` command. The final design that meets the specified requirements happens at θ = 45° with ellipse values of $A = 2.121$ in, $B = 1.5$ in, area = 9.995 in^2, and perimeter = 11.46 in.

To check the system accuracy, compare the preceding results with those resulting from the following equations:

Ellipse area $a = \pi AB$

Ellipse perimeter $p = 2\pi\sqrt{\dfrac{A^2 + B^2}{2}} = 4aE$ (approximate)

where E is the complete elliptic integral at $\Re = \sqrt{A^2 - B^2}/A$. E can be obtained from the table of elliptic integrals available from CRC standard mathematical tables. For this problem E is found to be 1.3506.

7.19.4 Create a Magician Hat

This tutorial shows how to create the magician hat shown on the right, using a B-spline curve. All dimensions are in mm.

Modeling strategy:

1. **Create the B-spline curve.** Select the `Front` sketch plane. Construct the curve using the seven points shown in the forthcoming screenshot.

2. **Create the axis of revolution.** Use an `axis` command to create a vertical axis passing through the origin as shown on the right.

3. **Create the surface of revolution.** Use a surface of revolution operation to revolve the curve 360° around the axis of revolution.

4. **Close the hat top.** Use a `fill surface` command to close the hat top.

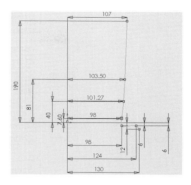

Tutorial 7.19.4 discussion:

The magician hat in this tutorial is axisymmetric. We create it as a surface model.

Tutorial 7.19.4 hands-on exercise:

Edit the hat model on your CAD system to have straight (linear) profile.

7.19.5 Edit and Control Surfaces

This tutorial creates the B-spline surface shown on the right, using a total of four B-spline curves and 32 points, eight per curve. It then shows the surface mesh, surface extension, the local control property of the curves and the surface, and how to convert the surface to a solid. All dimensions are in mm.

Modeling strategy:

1. **Create the first B-spline curves.** Select the `Front` sketch plane. Use the 32 points shown next page to create the four B-spline curves.

2. **Create the surface via loft.** Use a `loft surface` command on your CAD system and select the four B-spline curves to create the surface shown on the next page.

3. **Change the size of the surface mesh.** We change the surface default mesh size to provide better visualization of the surface. We use two sizes (see next page) : 20×8 and 40×16.

4. **Extend surface beyond its current boundary.** Find a CAD command that extends the surface across its boundary (along its four B-spline edges) by 5 mm. See screenshot on the next page.

5. **Convert the surface to a solid.** We extrude (thicken) the surface by 1.0 mm. See screenshot on the next page. There is a maximum thickness beyond which the operation fails, as the resulting faces of the solid begin to self-intersect.

6. **Change the surface shape.** Change the third point (third row in the spline data tables; see next page) for all the four splines as follows: Spline #1 => (25, 25, 0), Spline #2 => (30, 40, 20), Spline #3 => (35, 30, 40), and Spline #4 => (40, 40, 60). Regenerate all.

Step 1

Spline #1		
5mm	6mm	0mm
15mm	16mm	0mm
25mm	20mm	0mm
35mm	16mm	0mm
45mm	8mm	0mm
55mm	2mm	0mm
65mm	8mm	0mm
75mm	2mm	0mm

Spline #2		
5mm	6mm	20mm
20mm	16mm	20mm
30mm	20mm	20mm
40mm	16mm	20mm
45mm	8mm	20mm
55mm	2mm	20mm
65mm	8mm	20mm
75mm	2mm	20mm

Spline #3		
5mm	6mm	40mm
30mm	16mm	40mm
35mm	20mm	40mm
45mm	16mm	40mm
55mm	8mm	40mm
60mm	2mm	40mm
65mm	8mm	40mm
75mm	2mm	40mm

Spline #4		
5mm	6mm	60mm
20mm	16mm	60mm
40mm	20mm	60mm
50mm	16mm	60mm
60mm	8mm	60mm
70mm	2mm	60mm
85mm	8mm	60mm
95mm	2mm	60mm

Step 2

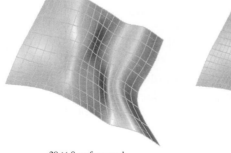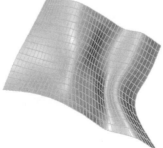

20 × 8 surface mesh   40 × 16 surface mesh

Step 3

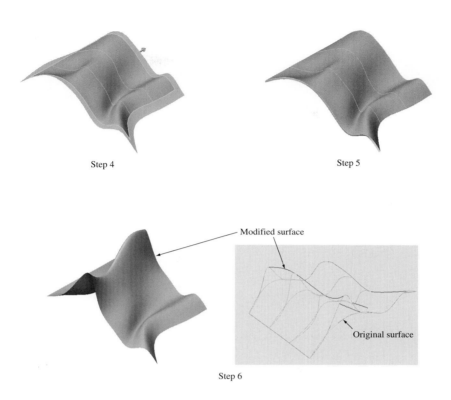

Step 4   Step 5

Step 6

Tutorial 7.19.5 discussion:

A close investigation of the results of step 4 shows that the surface extension is linear along the tangent vectors of the surface along its edges. We expect that from the theory of the surface. Step 6 shows clearly the local control of the B-spline surface. The spline did not change away from the region of the third point of each B-spline curve. Refer to Chapter 6.

Tutorial 7.19.5 hands-on exercise:

Using your CAD system and starting with the tutorial surface, change point 4 of each curve and observe the local control. Moreover, find the maximum thickness mentioned in step 5.

7.19.6 Create a Slipper

This tutorial creates the slipper shown on the right. It has a sole and a top. All dimensions are in mm. The sole has a uniform thickness of 20 mm. The top is a solid defined by a B-spline surface that is, in turn, defined by 14 B-spline curves. The top has a thickness of 3 mm and is symmetric with respect to the vertical midplane of the slipper. Thus we create seven curves and mirror copy them.

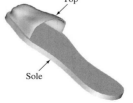

Modeling strategy to create the slipper sole:

1. **Create the B-spline curve of the sole.** Select the `Top` sketch plane. Use the 18 points shown below to create the B-spline curve.

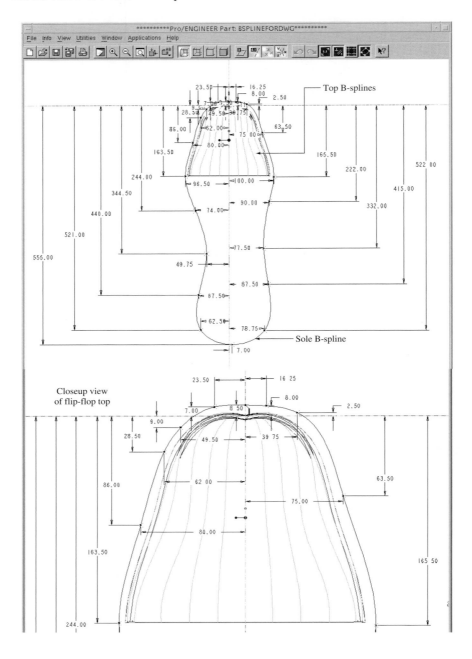

2. **Create the sole via extrusion.** Extrude the B-spline from step 1 a distance of 20 mm to create the sole shown in the screenshot at the beginning of the tutorial.

Modeling strategy to create the slipper top:

1. **Create the B-spline curves of the top**. We use 14 B-spline curves in total to define the top of the slipper; seven of them are defined while the other seven are copies. Each spline curve lies in an inclined plane perpendicular to the Front sketch plane as shown below.

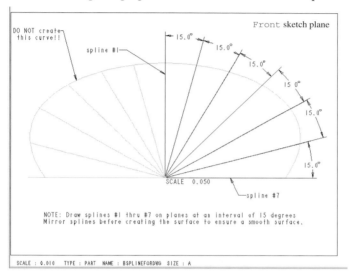

The coordinates of the points of each spline follow.

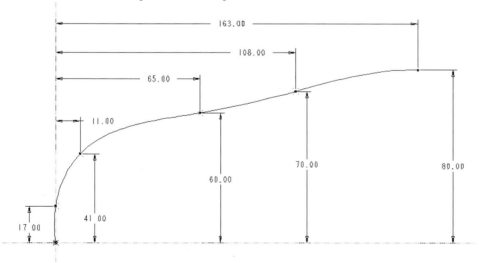

Spline #1

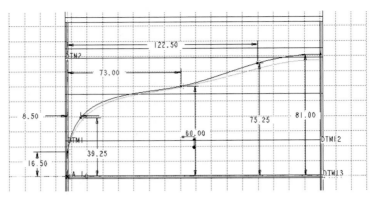

Spline #2

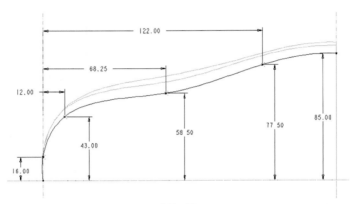

Spline #3

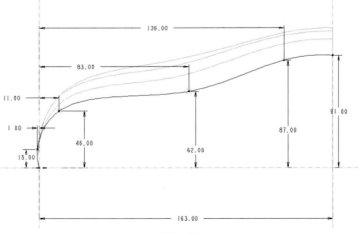

Spline #4

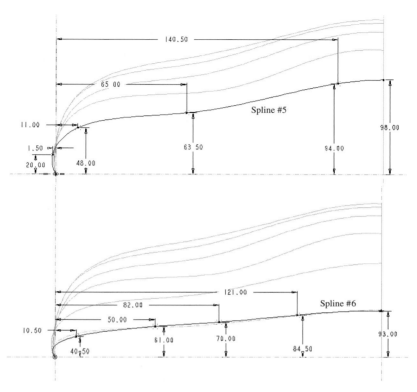

Spline #5

Spline #6

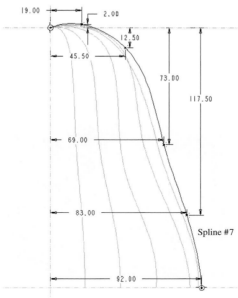

Spline #7

2. **Create the other seven B-spline curves.** Mirror copy the preceding seven B-spline curves with respect to the vertical plane passing through the WCS origin.
3. **Create the B-spline surface.** Connect the 14 B-spline curves with a B-spline surface.
4. **Create the solid of the slipper top.** Extrude (or offset) the surface of step 3 with a thickness of 3 mm. The final solid model of the slipper is shown below.

Tutorial 7.19.6 discussion:

While the sole of the slipper is a simple extrusion, its top is intricate and can crash the CAD system or result in a failure of geometric operations because of its design. Observe that all the 14 B-spline curves must meet at a point on its symmetric plane (refer to the screenshots of the seven splines). This results in "crumbling" the surface to meet at this point. If you use a thickness larger than 3 mm, the extrusion (or offset) operation may fail. Go ahead and try it.

Tutorial 7.19.6 hands-on exercise:

Modify the sole of the slipper sole so that it is tapered, with a thickness of 20 mm at the heel area (back), and 15 mm at the toe area (front).

7.19.7 Create an Air Conditioning Duct

This tutorial creates the AC duct shown below. It is a composite solid, consisting of three extensions. One extension is a pipe. The other two extensions are lofts. Loft 1 connects two circular cross sections. Loft 2 connects a circular and a rectangular cross section. The duct is closed on one end. The AC duct is 0.5 inch thick. All dimensions are in inches. The dimensions of the duct are shown in the following screenshot.

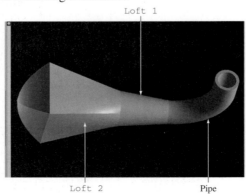

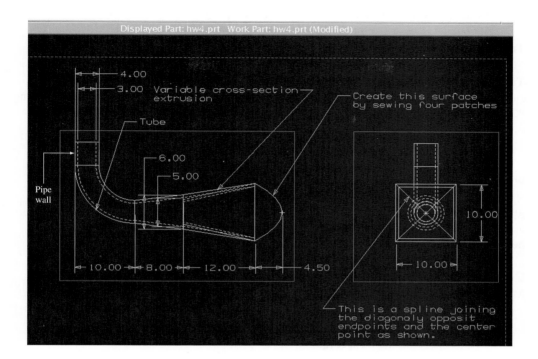

Modeling strategy:

1. **Create the pipe.** Select the `Front` sketch plane. Construct the cross section of the pipe wall that is shown above. Create an axis with a line segment and an arc. The axis is parallel to the pipe wall and passes through the WCS origin. Revolve the pipe wall around the axis to create the pipe.

2. **Create `Loft` 1.** Define the vertical end of the pipe as a sketch plane. Sketch two circles with diameters of 3 and 4 inches. Create a new sketch plane by offsetting the vertical plane by a distance of 8 inches to the right. Sketch two circles with diameters of 5 and 6 inches. Use a loft operation to create `Loft` 1.

3. **Create `Loft` 2.** Define the vertical end of `Loft` 1 as a sketch plane. Sketch two circles with diameters of 5 and 6 inches. Create a new sketch plane by offsetting the vertical plane by a distance of 12 inches to the right. Sketch two squares with sides of 9 and 10 inches. Use a loft operation to create `Loft` 2.

4. **Create the closed end.** Create a point in space that is in the center of the square end, but with an offset of 4.5 inches as shown in the screen above. Create two splines, each connecting two ends of the square diagonals and the point. Divide (split) each spline at the point to generate four splines. Create four surface patches; each patch uses two splines and

one side (line) of the square. Sew the four patches to the duct solid to close it. The final duct model is shown at the beginning of the tutorial.

Tutorial 7.19.7 discussion:

The four surface patches that close one end of the duct are just a sheet; it has no thickness.

Tutorial 7.19.7 hands-on exercise:

How can you add a thickness to the four surface patches? Can you find a better modeling way to create the closed end of the duct?

7.19.8 Create a Household Fan

This tutorial creates the household fan shown on the right. It consists of three blades and a hub. The creation of the blade requires surfaces. The blade profile is given at three sections of the blade (5.0 inch apart). We also have the front profile of a blade. The blade thickness is 0.1 inch.

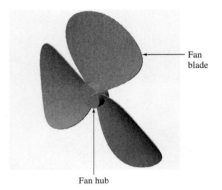

Fan blade

Fan hub

The hub is an extrusion. The hub diameter is 3.0 inches. Its thickness is 1.5 inches.

Modeling strategy:

1. **Create the fan profile.** Select the Front sketch plane. Construct the fan profile shown below.

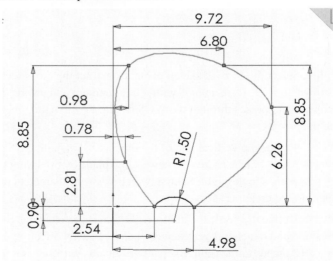

2. **Create the blade lofted surface.** Select the Top sketch plane. Construct the following three B-spline curves at planes parallel to the Top plane at offsets of 5 and 10 inches above it. The first plane is the Top plane itself.

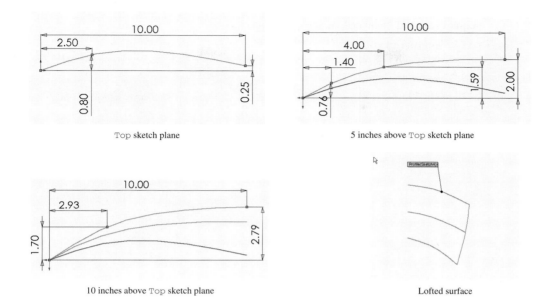

Top sketch plane

5 inches above Top sketch plane

10 inches above Top sketch plane

Lofted surface

3. **Create the blade final profile.** Using the blade profile, shown in step 1, as the cutting tool, trim the lofted surface created before to finish the fan blade as shown below.

4. **Create the fan hub.** Create the hub circle in the Front sketch plane and extrude it using a distance of 1.5 inches, as shown below.

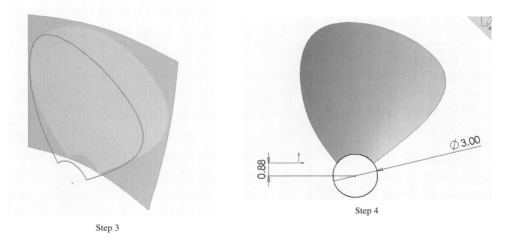

Step 3

Step 4

5. Create the blade solid model. Extrude (thicken) the trimmed surface of step 3, by 0.1 inch as shown in the forthcoming screenshot.

6. **Create the other fan blades.** Using the circular array concept, array the fan blade with three equally spaced instances, as shown below.

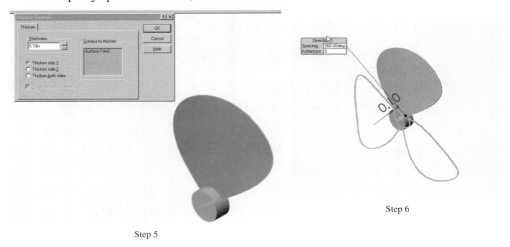

Step 5

Step 6

Tutorial 7.19.8 discussion:

This tutorial introduces two important modeling concepts. The first is trimming a surface by a profile (see step 3). It allows us to create the complex boundary shape of any surface. The second is the use of planar curves that are in different planes to define a surface (see step 2).

Tutorial 7.19.8 hands-on exercise:

Change the fan design so that the fan has five blades instead of three. Change dimensions as needed.

7.19.9 Create a Hair Dryer

This tutorial creates the hair dryer shown on the right. It consists of the handle and the air barrel. All dimensions are in mm.

This tutorial shows how to use surfaces to cut solids to create solids with complex faces. Here,

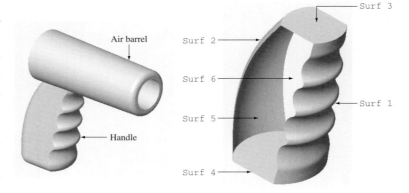

we cut a block with a knitted surface to create the handle of the hair dryer.

The knitted surface consists of six lofted surfaces (Surf 1 to Surf 6) as shown above, the four side ones, the top of the handle, and its bottom. After we create the six individual surfaces we knit (sew) them together to create a closed surface. Construction errors in the individual surfaces may prevent the creation of the knitted surface.

Modeling strategy:

1. **Create Surf 1.** In the Front sketch plane, create the surface profile, which consists of connected arcs as shown below. Then use the Top sketch plane to create two arcs. Loft the two arcs along the arced profile to create the lofted surface.

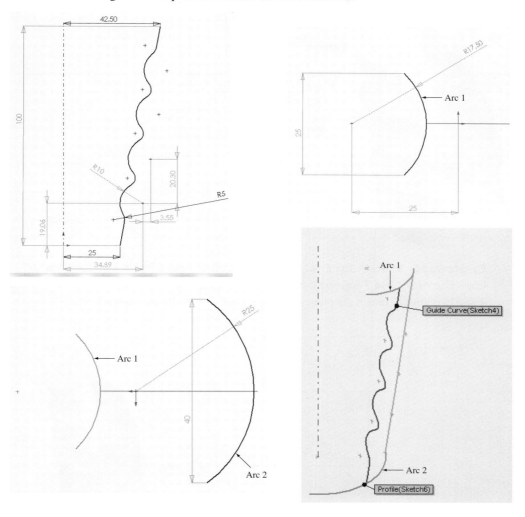

2. **Create Surf 2.** In the Front sketch plane, create the surface profile, which consists of an arc as shown in the upcoming screenshot. Then use the Top sketch plane to create two arcs. Loft the two arcs along the arced profile to create the lofted surface.

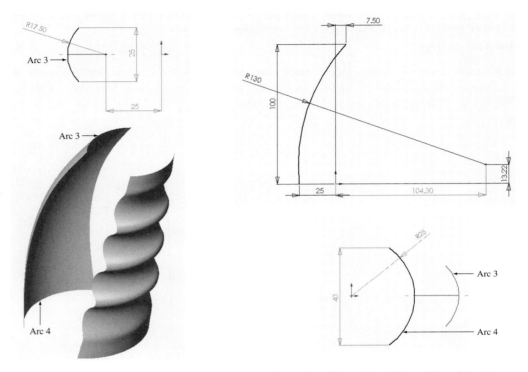

3. **Create Surf 3 and Surf 4.** Connect the ends of Arcs 1 and 3, and 2 and 4 with lines. Use Arcs 1 and 3 and two lines to create the planar surface, Surf 3. Use Arcs 2 and 4 and two lines to create the planar surface, Surf 4. Results are shown below.

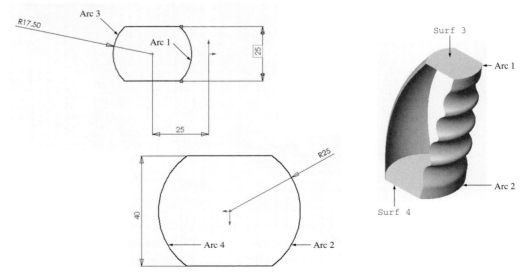

4. **Create** `Surf 5 and Surf 6`**.** Use a lofted surface operation to create `Surf 5` and `Surf 6`. All the needed boundaries are there. Results are shown below.

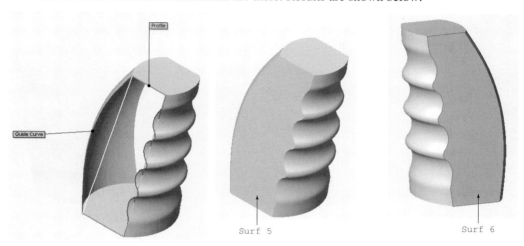

Surf 5   Surf 6

5. **Create the knitted surface of the handle.** Use a knit surface icon and select all `Surf 1` to `Surf 6`. The surfaces will be knitted but will not appear any different. The way in which you can verify the difference between the individual surfaces and the knitted surface is by clicking on each of the surface entities in the feature manager tree and observing what is selected. Note that for the knitted surface the entire skin is selected. After knitting, the model is still a surface model and not a solid model.

6. **Create a block.** Now select the `Top` sketch plane and sketch a rectangle with larger dimensions than the knitted surface. Extrude the rectangle to a height of 100 mm to create a rectangular block as shown below.

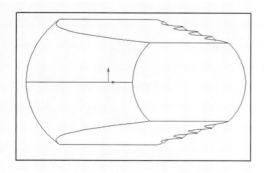
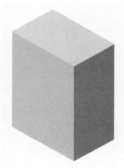

7. **Create the handle at last.** Now we cut the block of step 6 with the knitted surface of step 5. Make sure to select the appropriate direction to cut. Fillet all the edges, as shown in the following screenshots.

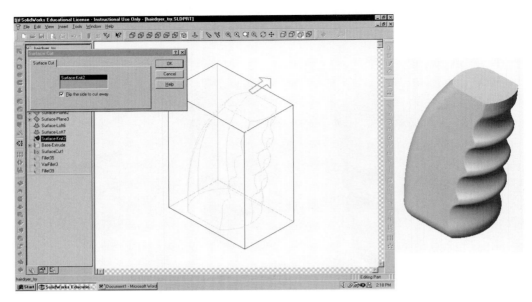

8. **Create the air barrel.** The barrel is created as a lofted feature. We create two circles at a plane that is offset from the `Right` sketch plane a distance of 125 mm, as shown below. We then loft them. We fillet the edges of the barrel. We cut extrude a circle from the barrel to create its hole as shown below.

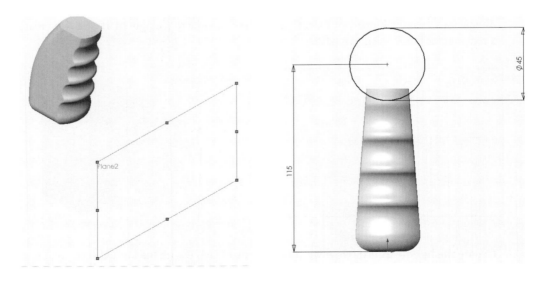

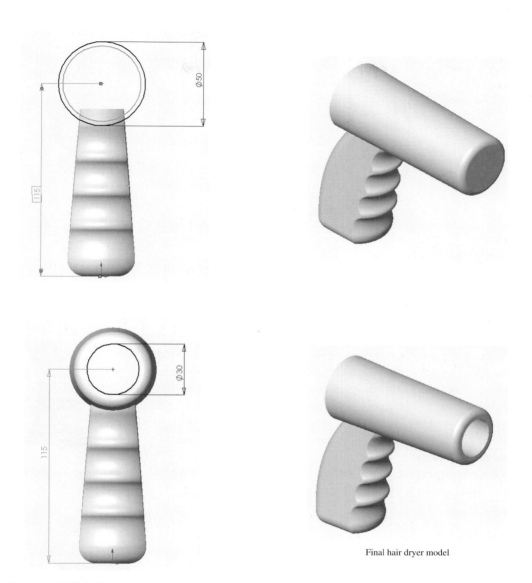

Final hair dryer model

Tutorial 7.19.9 discussion:

This tutorial shows how to cut a solid with a surface to add a complex boundary to the solid. We can use surfaces in another way to create solids: we extrude (offset or thicken) them.

Tutorial 7.19.9 hands-on exercise:

We need to add a ventilation grill to cool the blower motor of the hair dryer. The grill is added to the back of the hair dryer barrel. Refer to the following for steps and dimensions.

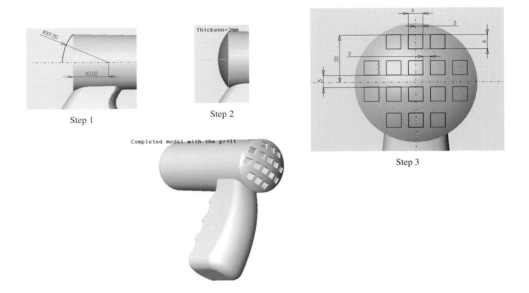

Step 1

Step 2

Step 3

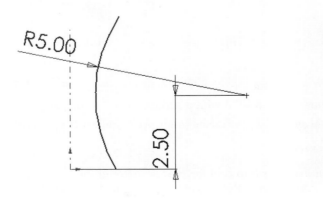

7.19.10 Create Free-Form Springs

This tutorial creates the free-form spring shown on the right. All dimensions are in inches. The main idea here is to create two surfaces and intersect them to generate the correct helix curve shown on the right. We then sweep a circle along the helix curve. One of the surfaces, Surf 1, creates the double-tapered helix shape shown on the right. The other surface, Surf 2, creates the helix curve. The detailed steps follow.

Modeling strategy:

1. **Create Surf 1.** In the Front sketch plane, construct the following sketch, and use a surface revolve operation to revolve the sketch.

2. **Create a helix curve.** Construct a circle of diameter 3.0 inches in the `Top` sketch plane. Using the circle as a base circle, create the helix with 5.0 inches height and a pitch of 0.4 inch. Use a starting angle of 180°. Results are shown below.

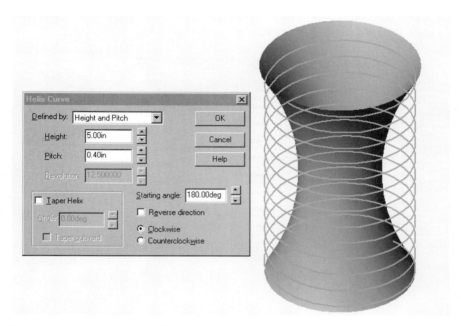

3. **Create `Surf 2`.** Construct a line, for sweeping, in the `Top` sketch plane, as shown below, with a length equivalent to the helix base radius. Now using a surface sweep operation, sweep the line profile with the helix path.

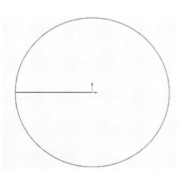

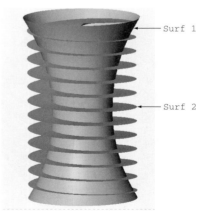

4. **Create the spring helix curve.** Select both the `Surf 1` and `Surf 2`, and use a curve intersection operation to generate the intersection 3D curve shown below.

5. **Create the free-form spring.** Construct the circular profile of the spring cross section in the `Front` sketch plane. Make sure the helix curve starts at the center of the cross section. Use a sweep operation, and sweep the circular cross section along the 3D helix curve as the path. The resulting solid body is as shown below.

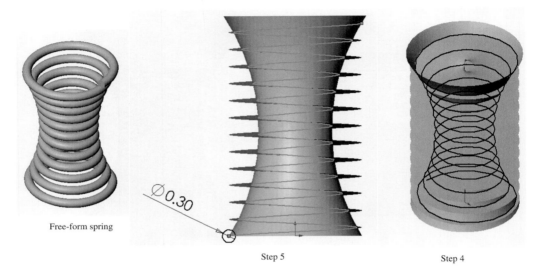

Free-form spring

∅0.30

Step 5   Step 4

Tutorial 7.19.10 discussion:

We can generate many other shapes for free-form springs using the idea of this tutorial.

Tutorial 7.19.10 hands-on exercise:

Modify the tutorial to create the following spring shape. Assume any required dimension.

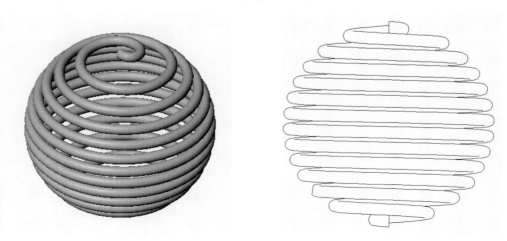

P ROBLEMS

Part I: Theory

7.1 CAD/CAM systems use parametric representation for surfaces as they do for curves. Explain how a surface is represented.

7.2 What are the types of surfaces that CAD/CAM systems use?

7.3 Sketch the geometric parameters required to create these surface operations: tabulated cylinder (extruded), revolve, sweep, loft, offset, and knit.

7.4 Explain how surfaces can aid in creating solid models, that is, when must you use surfaces in solid modeling?

7.5 Sketch the geometric parameters required to define a helix.

7.6 Find the equation of the plane that passes through points $P_0(1, 2, 3)$, $P_1(3, 4, -1)$ and $P_2(1, -2, 2)$. Find the coordinates of the center point of the plane. Find the tangent vectors and the normal vector at this point.

7.7 The figure on the right shows two half circles C_1 and C_2 with centers at P_1 and P_2, respectively.

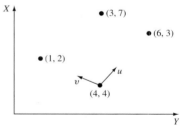

   a. Find the equation of the ruled surface that uses C_1 and C_2 as rails.
   b. Sketch the surface to show the u and v directions.
   c. Find the geometric center point of the rules surface using its equation.

7.8 The figure on the right shows four points.

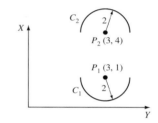

   a. Find the equation of the Bezier surface defined by the four data points. Use the u and v directions shown. Find the surface tangent and normal vectors. Are they constant? What is your conclusion?
   b. Using the same data points, find the equation of the ruled surface. Is it identical to Bezier surface?

7.9 The figure on the right shows six points.

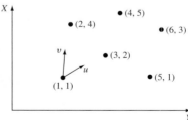

   a. Find the equation of the Bezier surface defined by the six data points. Use the u and v directions shown. Find the surface tangent and normal vectors at point $u = 0.5$ and $v = 0.5$.
   b. Find the two quadratic Bezier curves that bound the Bezier surface. Find the equation of the ruled surface that uses the two curves as its rails. Is it identical to Bezier surface?

7.10 Equation (7.36) gives the equation of a ruled surface joining two space curves. Find the equation of a ruled surface that passes through one space curve along a given set of direction vectors defining the surface rulings.

7.11 Derive the parametric equations of an ellipsoid, a torus, and a circular cone.

7.12 Find the tangent and normal vectors to a surface of revolution in terms of its profile equation.

7.13 Repeat Problem 7.12 but for a tabulated cylinder in terms of its directrix.

7.14 Prove that the curvature of a circular cylinder is zero. What is the radius of curvature at any point on its surface?

7.15 Develop the relationships between the position, tangent, and twist vectors at the corner points of a bicubic surface patch and C_{ij} coefficients used in Eq. (7.39).

7.16 Derive the conditions for C^0 and C^1 continuity of a cubic Bezier composite surface of two patches.

7.17 Show that a bilinear Coons patch cannot provide C^1 continuity between adjacent patches even if their boundary curves form a C^1 continuity network.

7.18 Derive Eq. (7.78) for a bicubic Coons patch whose boundary curves and cross-boundary derivatives are given.

7.19 Given a point Q and a parametric surface in the carte-sian space, find the closest point P on the surface to Q.
       Hint: find P such that $(Q - P)$ is perpendicular to the tangent vectors P_u and P_v at P.

7.20 Find the minimum distance between

   a. a point and a surface
   b. a curve and a surface
   c. two surfaces

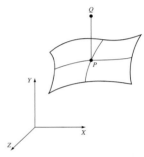

Part II: Laboratory

Use your in-house CAD/CAM system to answer the questions in this part.

7.21 Create the surface of the following magician hat.

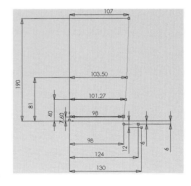

7.22 Create the surface of the following magician hat. Sweep the B-spline curve (profile) along the ellipse as the guide curve.

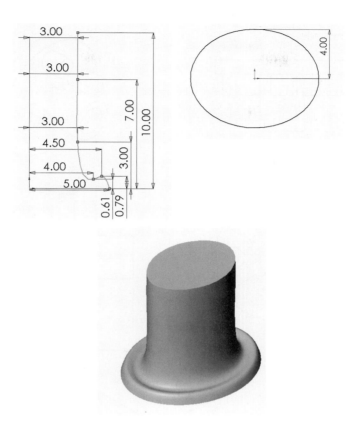

7.23 How will you extract a midsurface from a solid body? For example, the hollow cylinder on the right is part of a pressure vessel (assume dimension). Extract the midsurface of the cylinder for finite element analysis.

7.24 Create the following models. If dimensions are not shown, simply sketch in the sketcher so that the model looks similar to what is here, and then adjust the dimensions to finalize. Generate an engineering drawing for each model. Follow the drawing structure format shown in Figure 4.4. All dimensions are in inches.

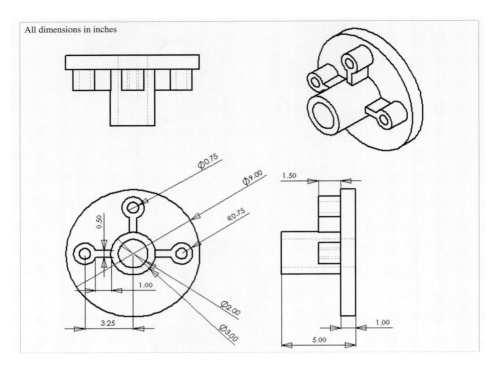

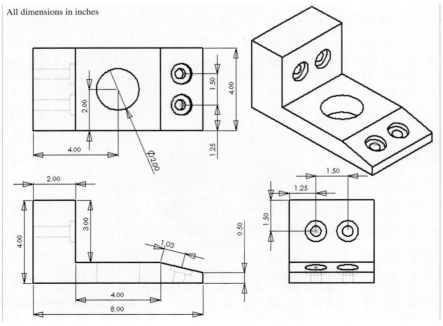

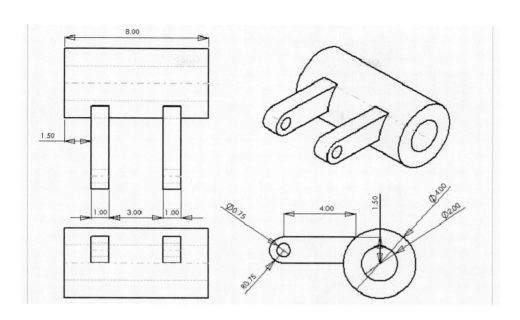

Model 1

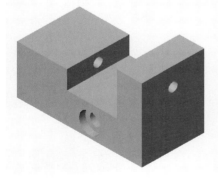

Model 3

Model 2

Model 3 (unshaded)

Model 4

Model 4 (unshaded)

Model 5

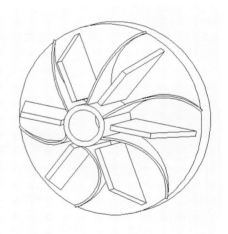

Model 5 (unshaded)

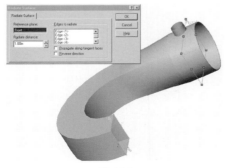

Model 6

Model 6 (with flange)

7.25   Create the following quadric (quadratic) surfaces on your CAD/CAM system.

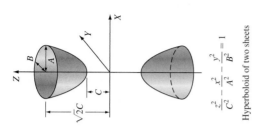

$$\frac{z^2}{C^2} - \frac{x^2}{A^2} - \frac{y^2}{B^2} = 1$$

Hyperboloid of two sheets

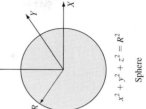

$$x^2 + y^2 + z^2 = R^2$$

Sphere

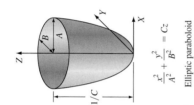

$$\frac{x^2}{A^2} + \frac{y^2}{B^2} = Cz$$

Elliptic paraboloid

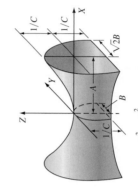
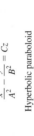

$$\frac{x^2}{A^2} - \frac{y^2}{B^2} = Cz$$

Hyperbolic paraboloid

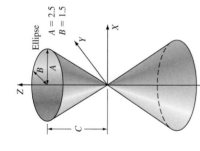

Ellipse $A = 2.5$
                $B = 1.5$

$A = 2.5, B = 1.5, C = 3$

$$\frac{x^2}{A^2} + \frac{y^2}{B^2} - \frac{z^2}{C^2} = 0$$

Elliptic cone

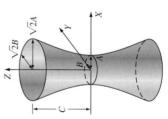
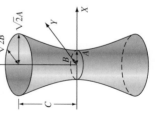

$$\frac{x^2}{A^2} + \frac{y^2}{B^2} - \frac{z^2}{C^2} = 1$$

Hyperboloid of one sheet

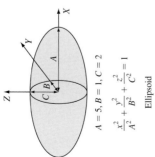

$A = 5, B = 1, C = 2$

$$\frac{x^2}{A^2} + \frac{y^2}{B^2} + \frac{z^2}{C^2} = 1$$

Ellipsoid

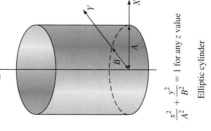

$$\frac{x^2}{A^2} + \frac{y^2}{B^2} = 1 \text{ for any z value}$$

Elliptic cylinder

Part III: Programming

Use the parametric equations of surfaces in all the following questions. In addition, write the code so that the user can input values to control the curve shape and behavior as we have discussed in this chapter. The program should display the results. Use OpenGL with C or C++, or use Java 3D. Write a computer program to generate a:

7.26 plane surface that passes through three points
7.27 ruled surface that connects two given rails
7.28 surface of revolution
7.29 tabulated cylinder
7.30 bicubic surface
7.31 cubic rectangular Bezier patch
7.32 open cubic B-spline surface patch
7.33 closed cubic B-spline surface patch
7.34 bilinear Coons patch
7.35 cubic Coons patch
7.36 quadratic triangular Bezier patch
7.37 cubic triangular Bezier patch

CHAPTER 8

NURBS

GOAL

Understand and master the theory and practice of NURBS curves and surfaces, their formulation, their implementation by CAD/CAM systems, and the advantages of their use in geometric modeling.

OBJECTIVES

After reading this chapter, you should understand the following concepts:

- NURBS modeling
- NURBS basics (knot vectors and weights)
- NURBS curves
- NURBS lines
- NURBS arcs and circles
- NURBS surfaces
- NURBS bilinear surface
- NURBS ruled surface

CHAPTER HEADLINES

8.1   Introduction

NURBS (nonuniform rational B-splines) are almost exclusively used by CAD/CAM systems as the internal representation of all geometric entities that these systems provide. NURBS provide a unified approach to formulate and represent curves and surfaces. NURBS provide a convenient design tool to create smooth curves and surfaces interactively. As we cover in this chapter, NURBS are generalizations of the curve and surface theories we have covered in Chapters 6 and 7.

The development of NURBS dates back to the late 1970s when Boeing developed the Tiger CAD system based on rational B-splines. These splines were developed by integrating the representations of B-splines with rational Bezier curves. In 1983, SDRC released the first NURBS-based commercial modeler, Geomod.

NURBS have many advantages. Any curve or surface can be formulated using NURBS. NURBS are considered a unified canonical representation that can define both synthetic (Bezier, B-splines) and analytic (circles and conics) curves and surfaces. The premise is that they can represent all curve, surface, and solid entities, allowing unification and conversion from one CAD system to another via exchange standards (STEP and IGES). Their related algorithms are stable and accurate. They enhance manufacturing and machining accuracy and speed. They are intuitive and flexible to use in design and geometric modeling. NURBS are also invariant under affine and perspective transformations such as translation, rotation, and projections.

NURBS have some problems. The definition of simple curves such as arcs, circles, and conics is verbose; they require more data to define as NURBS than the traditional way. For example, the traditional definition of a circle is a center (x, y, z), radius and the circle plane defined by a normal vector (n_x, n_y, n_z); that is a total of seven numbers. The NURBS definition requires 38 numbers. The loss of information on simple shapes is another problem. For example, if a circular cylinder (hole) is represented by a B-spline, some data on the specific curve type may be lost unless it is carried along. Data, including that the part feature was a cylinder, would be useful in manufacturing to identify it as a hole to be drilled or bored rather than a surface to be milled. The improper use of the extra flexibility that NURBS offer (such as weights, as we cover later) can produce ill-behaved NURBS. Moreover, some algorithms, such as surface/surface intersection and inverse-point solution, work better under the non-NURBS representation. Nonetheless, NURBS advantages far outweigh their disadvantages.

8.2   Basics

The key to understanding NURBS comes from the name itself: nonuniform (NU), rational (R), and B-splines (BS). While we understand B-splines from Chapters 6 and 7, we need to understand *rational* and *nonuniform*. A rational curve is defined by the algebraic ratio of two polynomials while a nonrational curve [Eq. (6.91) gives an example] is defined by one polynomial. Rational curves draw their theories from projective geometry. They are important because of their invariance under projective transformation, that is, the perspective shape of a rational curve is a rational curve. Rational Bezier curves, rational B-spline curves, rational conic sections, rational cubics, and rational surfaces have been formulated. The most widely used rational curves are NURBS.

The formulation of rational curves requires the introduction of homogeneous space and the homogeneous coordinates. We cover this subject in detail in Chapter 12. The homogeneous space is four-dimensional (4D) space. A point in E^3 with coordinates (x, y, z) is represented in the homogeneous space by the coordinates (x^*, y^*, z^*, h) where h is a scalar factor. When $h = 1$, we obtain the point in E^3. Thus, the homogeneous coordinates of a point in E^3 is $(x, y, z, 1)$. The relationship between the two types of coordinates is obtained by normalizing h to 1. Thus,

$$x = \frac{x^*}{h} \qquad y = \frac{y^*}{h} \qquad z = \frac{z^*}{h} \tag{8.1}$$

A rational B-spline curve defined by $n + 1$ control points $\mathbf{P}_i$ is given by:

$$\mathbf{P}(u) = \sum_{i=0}^{n} \mathbf{P}_i R_{i,k}(u) \qquad\qquad 0 \le u \le u_{max} \tag{8.2}$$

$R_{i,k}(u)$ are the rational B-spline basis functions and are defined by

$$R_{i,k}(u) = \frac{w_i N_{i,k}(u)}{\sum\limits_{i=0}^{n} w_i N_{i,k}(u)} \tag{8.3}$$

where $N_{i,k}(u)$ are the B-spline basis functions given by Eqs. (6.92) and (6.93). w_i are weights associated with the control points, $\mathbf{P}_i$, of the rational B-spline curve. Each control point has an associated weight, w_i, with it. These weights serve the same purpose as h in Eq. (8.1). The control point P_i has the homogeneous coordinates $(w_i x_i, w_i y_i, w_i z_i, w_i)$ in the 4D projective space, and the coordinates $(x_i, y_i, z_i, 1)$ in the 3D Euclidean space.

Equation (8.3) shows that $R_{i,k}(u)$ are a generalization of the nonrational basis functions $N_{i,k}(u)$. If we substitute $w_i = 1$ in the equation and use the partition-of-unity property of $N_{i,k}(u)$ covered in Chapter 6, $R_{i,k}(u) = N_{i,k}(u)$, and the rational curve becomes nonrational. The rational basis functions $R_{i,k}(u)$ have nearly all the analytic and geometric properties of their nonrational B-spline counterparts, $N_{i,k}(u)$. All the discussions covered in Section 6.14 apply here. We summarize these properties here:

Partition of unity: $\quad \sum\limits_{i=0}^{n} R_{i,k}(u) = 1$

Positivity: $\qquad\quad R_{i,k}(u) \ge 0 \qquad$ if all $w_i > 0$

Local support: $\qquad R_{i,k}(u) = 0 \qquad$ if $u \notin [u_i, u_{i+k+1}]$

Continuity: $\qquad\quad R_{i,k}(u)$ is $(k-2)$ times continuously differentiable

Generality: $\qquad\quad$ *if all $w_i = 1$, then*

$$\begin{cases} B_{i,k}(u) & \text{if all values of knot vector are 0s and 1s} \\ N_{i,k}(u) & \text{otherwise} \end{cases}$$

where $B_{i,k}(u)$ is the Bernstein basis functions of Bezier curve. The generality property shows that nonrational B-spline and Bezier curves are special cases of NURBS.

Substituting Eq. (8.3) into Eq. (8.2) gives:

$$\mathbf{P}(u) = \sum_{i=0}^{n} \mathbf{P}_i \frac{w_i N_{i,k}(u)}{\sum\limits_{i=0}^{n} w_i N_{i,k}(u)} = \frac{\mathbf{P}_0 w_0 N_{0,k}(u)}{\sum\limits_{i=0}^{n} w_i N_{i,k}(u)} + \frac{\mathbf{P}_1 w_1 N_{1,k}(u)}{\sum\limits_{i=0}^{n} w_i N_{i,k}(u)} + \dots + \frac{\mathbf{P}_n w_n N_{n,k}(u)}{\sum\limits_{i=0}^{n} w_i N_{i,k}(u)} \tag{8.4}$$

Equation (8.4) can be written in a compact form as:

$$\mathbf{P}(u) = \frac{\sum\limits_{i=0}^{n} \mathbf{P}_i w_i N_{i,k}(u)}{\sum\limits_{i=0}^{n} w_i N_{i,k}(u)} \qquad 0 \le u \le u_{max} \tag{8.5}$$

Equations (8.2) and (8.3) together are equivalent to Eq. (8.5).

The main difference between rational and nonrational B-spline curves is the ability to use w_i at each control point to control the behavior of the rational curves. Thus, similar to the knot vector, one can define a homogeneous coordinate or weight vector $\mathbf{W} = [w_0\ w_1\ w_2\ w_3\ \dots\ w_n]^T$ at the control points $P_0, P_1, P_2, P_3, \dots, P_n$ of the rational B-spline curve. Each control point, P_i, has a weight, w_i, associated with it. The weights are usually positive numbers. The choice of the weight vector controls the behavior of the curve. When a curve's control points all have the same weight (usually 1), the curve becomes nonrational.

8.2.1 The Knot Vector

The B-spline curve we covered in Chapter 6 is uniform. That means that the knots in the knot vector are evenly spaced over the range of u with unit separation ($\Delta u = 1$) between noncoincident knots. Multiple (coincident) knots for certain values of u may exist. Relation (6.95) shows that $(n + k + 1)$ knots are needed to create a $(k - 1)$ degree curve defined by $(n + 1)$ control points.

While the degree of the resulting B-spline curve is controlled by k, the range of the parameter u as given by Eq. (6.96) implies that there is a limit on k that is determined by the number of the given control points. This limit is found by requiring the upper bound in Eq. (6.96) to be greater than the lower bound for the u range to be valid, as shown in Eq. (6.97). Equation (6.97) shows that a minimum of two, three, and four control points are required to define linear, quadratic, and cubic B-spline curves, respectively.

The nonuniform B-spline curve uses knots and has a knot vector (exactly as we have covered in Chapter 6) expressed as:

$$\mathbf{KV} = \begin{bmatrix} u_0\ u_1\ \dots\ u_v \end{bmatrix}^T \tag{8.6}$$

where the number of knots (length of the knot vector) $m = n + k + 1$ as given by Eq. (6.95). Equation (8.6) applies to both uniform and nonuniform B-splines. For the uniform B-spline curve, the knot values of Eq. (8.6) depend on whether the curve is open (nonperiodic) or closed (periodic). Equations (6.94) and (6.99) are used to calculate the knot values for the open and closed curves, respectively. The knot values are equally spaced in the u space by a unit value, that is, $\Delta u = 1$. For example, the knot vector $[0\ 0\ 0\ 1\ 2\ 3\ 4\ 4\ 4]^T$ is uniform while the knot vector $[0\ 0\ 0\ 1\ 2\ 5\ 6\ 6\ 6]^T$ is nonuniform.

The knot values for a nonuniform B-spline curve are not uniform (not equally spaced in the u space), thus the name nonuniform B-splines. A common form for the knot vector of nonuniform B-splines is for the first k knots to be 0 and the last k to be 1, that is, Eq. (8.6) becomes:

$$\mathbf{KV} = \begin{bmatrix} u_0\ u_1\ \ldots\ u_\upsilon \end{bmatrix}^T = \begin{bmatrix} u_0\ \ldots\ u_0\ u_k\ \ldots\ u_{m-k}\ \ldots u_\upsilon \end{bmatrix}^T \qquad (8.7)$$

$$= \begin{bmatrix} \underbrace{0\ldots0}_{k}\ u_k\ \ldots\ u_{m-k}\ \underbrace{1\ldots1}_{k} \end{bmatrix}^T$$

These knots are known as the end knots of the B-spline, and their values as 0 or 1 are adequate for most practical applications. The multiplicity of these knots is known to be k because the value 0 or 1 is repeated k times. [Note that the knot vector index begins at 0 and not at 1 in Eq. (8.6) or (8.7).] The remaining values (interior knots) of the knot vector must form a nondecreasing sequence of real numbers.

Duplicate values (knots) in the middle of the knot vector make the NURBS curve less smooth because they create sharp kinks (discontinuities) in the curve. However, multiplicity of the end knots is desirable because it can be used to ensure that the resulting NURBS curve interpolates the first and the last data points (vertices). Examples of a uniform knot vector are $\mathbf{KV}=[0\ 0\ 0\ 1/3\ 2/3\ 1\ 1\ 1]^T$, $\mathbf{KV}=[0\ 0\ 0\ 1/4\ 1/2\ 3/4\ 1\ 1\ 1]^T$, $\mathbf{KV}=[0\ 0\ 0\ 0\ 1/5\ 2/5\ 3/5\ 4/5\ 1\ 1\ 1\ 1]^T$, $\mathbf{KV}=[0\ 0\ 0\ 1/8\ 2/8\ 3/8\ 4/8\ 5/8\ 6/8\ 7/8\ 1\ 1\ 1]^T$ and so forth. These knot vectors are also known as **open knot vectors** because the knot spacing is uniform along the curve except at the end points to ensure that the curve interpolates the first and the last control point. In contrast, nonuniform knot vectors have unequally spaced knot values.

Knot multiplicity determines the degree of the segment of the NURBS curve that passes through the first and the last data points. For example, the multiplicity of 3 ensures that a third-degree B-spline passes through the end points. Multiplicities of 2 and 4 produce second and fourth degrees, respectively.

8.2.2   The Weights

The weights in Eq. (8.3) or (8.5) provide CAD designers with an effective design tool to control NURBS curves locally. Weights are normally chosen to be positive. The effect of w_i on a NURBS curve is that a higher (lower) value of w_i at control point P_i pulls the curve closer to (farther away from) P_i as shown in Figure 8.1. Thus, increasing the value of w_i tightens the curve or increases the pull toward P_i. Each weight w_i affects the curve in the parametric interval defined by $u \in [u_i, u_{i+k})$; this notation means that the lower limit u_i is included in the interval while the upper limit u_{i+k} is excluded.

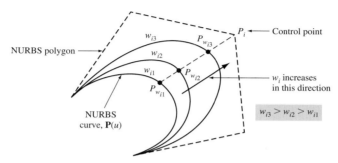

Figure 8.1   Effect of changing the weight w_i.

The movement of the NURBS curve $P(u)$ for a fixed value of u and a changing value of w_i is linear as shown in Figure 8.1. Let us assume that point P_{wi} has a fixed u value. P_{wi} moves along a straight line when w_i changes as shown in Figure 8.1. This linear motion is attributed to the fact that w_i is a scale factor in the projective 4D space. The line of motion connects the origin of the projective coordinate system and the P_{wi} points.

The choice of weight values at control points is up to the NURBS curve designer. Much research has been done trying to find the optimum values of these weights. One approach is to formulate a system of equations and solve it for the values

$$\mathbf{w} = [C]\mathbf{P} \qquad\qquad (8.8)$$

where $\mathbf{w}$ (weight vector) = $[w_1\ w_2 ... w_n]^T$, $[C]$ is a coefficient matrix, and $\mathbf{P}$ is some vector, for example the control points.

For our purposes here, we use any positive arbitrary number for weights, $0 \le w_i < \infty$ — for example, 0, 1/2, 5/8, 1, 1.5, 2, 2.3, and so forth. The 0 and 1 values have special meaning. When $w_i = 0$, the curve segment in the interval defined by the vertices of $P \in [P_{u-i}, P_i, P_{i+1}]$ becomes linear. If all the weights are set to zeros, the NURBS curve becomes piecewise linear. When all weights are set to 1 or any other equal value, we can factor w_i from Eq. (8.3) or (8.5), which gives $R_{i,k}(u) = N_{i,k}(u)$, that is, the NURBS curve degenerates to a nonrational B-spline curve.

All of these concepts and discussions for NURBS curves are applicable to NURBS surfaces in each of the surface parametric directions, u and v.

The effect of the weights on a NURBS curve is somewhat similar to that of the effect of point multiplicity for uniform nonrational B-splines as we covered in Chapter 6. Combining both point multiplicity and changing w_i can provide great control of NURBS curves. Actually we can combine all the control tools (vertex location, degree, and vertex multiplicity) of uniform nonrational B-splines with the control tools of NURBS (knot values and weights). While having these five control tools is great in theory, NURBS control could be frustrating at times.

In the context of Sections 8.2.1 and 8.2.2, we can define three classes of B-spline curves:

1. Uniform nonrational B-splines. These splines are covered in Chapter 6.
2. Nonuniform, nonrational B-splines. This class is similar to class 1, but it uses nonuniform knot values. Use Section 8.2.1 as a guide to find the knot vector for this class.
3. Uniform rational B-splines. This class is similar to NURBS but it uses uniform knot values. Use Chapter 6 to find the knot vector.
4. NURBS curves. We cover these curves in this chapter.

8.3 Curves

The sections that follow show how to formulate some of the analytic and synthetic curves of Chapter 6 using the NURBS theory. We follow a systematic and methodical approach in our formulation to ensure a good understanding of the NURBS basics covered in Section 8.2 and how to apply them to different problems. We emphasize the choice of the knot vectors and weights. As we show shortly, the formulation and development of even the simplest curve become complex and mundane. It is easy to make errors. To deal with this observation, we verify the results and compare them with the results of Chapter 6.

In all NURBS formulations in this chapter, we follow the same approach we used to develop B-spline curves in Section 6.14 of Chapter 6. Thus, we need to determine the curve order (k), curve degree $(k - 1)$, its number of control points $(n + 1)$, the length of the knot vector (m), the values of the knots, and the weights at the control points. The curve degree, $k - 1$ (or order k) and its control points $(n + 1)$ are given data. The length of the knot vector (m) is determined from Eq. (6.95) as

$$m = n + k + 1 \tag{8.9}$$

The values of the knots of the knot vector are determined from Eq. (8.7). The values of the end knots are already known. We must choose the values of the remaining $(m - 2k)$ knots $u_k, \ldots, u_m$. While there may be many ways to choose these values, we follow this consistent approach. We need to bridge the two end-knot values 0 and 1 using $(m - 2k)$ values. We simply define a uniform parametric increment as follows:

$$\Delta u = \frac{1}{m - 2k + 1} \tag{8.10}$$

For example, if $k = 3$ and $m = 12$, then $\Delta u = 1/7$ as Section 8.6 shows when we develop the NURBS circle. In some cases, the entire knot vector consists of end-knot values only; that is, values of 0s and 1s as Sections 8.4 and 8.5 show.

To evaluate the rational B-spline basis functions $R_{i,\,k}(u)$, we first calculate the B-spline basis functions $N_{i,\,k}(u)$ via using Eqs. (6.92) and (6.93), as we have done in Chapter 6, and then substitute them into Eq. (8.3). The calculations of $N_{i,\,k}(u)$ are lengthy and error-prone due to their recursive and algorithmic nature. Writing a computer program would be helpful. However, we recommend manual calculations first until a good understanding of NURBS theory is achieved.

The calculations of $N_{i,k}(u)$ follow a pyramid scheme. At the bottom of the pyramid are the $N_{i,1}(u)$ functions given by Eq. (6.93). These functions propagate up the pyramid as shown on the right for $n = 1$ and $k = 2$. At any level k, *the last function is always $N_{i,j}(u)$ such that $i + j = n + k$.*

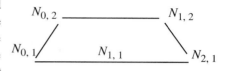

We finally substitute $R_{i,k}(u)$ and w_i into Eq. (8.2) to find the equation of the NURBS curve under development.

Summarizing, use the following steps to develop the NURBS equation of any curve:

1. **Find curve order, k.** The curve degree is $(k - 1)$ and is always given.
2. **Find the number of control points, $n + 1$.** The number of data points given in the problem is equal to $(n + 1)$.
3. **Find the length of the knot vector, m.** Use Eq. (8.9).
4. **Calculate the knot values (end and interior) of the knot vector.** Use Eqs. (8.7) and (8.10).
5. **Calculate the B-spline basis functions $N_{i,k}(u)$.** Use Eqs. (6.92) and (6.93).
6. **Calculate the rational B-spline basis functions $R_{i,k}(u)$.** Use Eq. (8.3).
7. **Find the equation of the NURBS curve.** Use Eq. (8.2).
8. **Test the NURBS equation.** The NURBS equation of step 7 is a vector equation. It is usually complex and is expressed in terms of the position vectors, $\mathbf{P}_i$, of the curve control points and the weights, w_i. In this step, write the equation in a scalar form, use equal weights, for example set all $w_i = 1$, and choose key values for the parameter u. For example, each NURBS curve passes through the first ($u = 0$) and the last ($u = 1$) data (control) points. If we substitute these values into the NURBS equation, we should get $\mathbf{P}_0$, and $\mathbf{P}_1$, respectively.

8.4 Lines

Section 6.5, Chapter 6, shows the uniform nonrational parametric equation [Eq. (6.11)] of a line defined by two end points, P_0 and P_1, as shown in Figure 6.4. In this section, we develop the NURBS formulation of the same line and show that Eq. (6.11) is a special case. Using Eq. (8.2), the line equation becomes:

$$\mathbf{P}(u) = \sum_{i=0}^{1} \mathbf{P}_i R_{i,2}(u) = \mathbf{P}_0 R_{0,2}(u) + \mathbf{P}_1 R_{1,2}(u) \tag{8.11}$$

The two key issues to NURBS development are the choice of the knots of the knot vector and the weights at the control points P_0 and P_1. Given the line data (linear and two end points), we have $k = 2$ and $n = 1$. Utilizing Eq. (8.9) gives $m = 4$. Utilizing Eq. (8.7) gives the knot vector as $KV = [u_0 \ u_1 \ u_2 \ u_3]^T = [0 \ 0 \ 1 \ 1]^T$. In this case, all the knot values are end values because $m = 4$ and $k = 2$. We use this knot vector and Eqs. (6.92) and (6.93) to calculate $N_{i,\,k}(u)$ as follows:

$$N_{0,1} = \begin{cases} 1 & u = 0 \\ 0 & \text{elsewhere} \end{cases}, \quad N_{1,1} = \begin{cases} 1 & 0 \leq u \leq 1 \\ 0 & \text{elsewhere} \end{cases}, \quad N_{2,1} = \begin{cases} 1 & u = 1 \\ 0 & \text{elsewhere} \end{cases}$$

$$N_{0,2} = \frac{(u - u_0)N_{0,1}}{u_1 - u_0} + \frac{(u_2 - u)N_{1,1}}{u_2 - u_1} = 0 + (1 - u)N_{1,1} = (1 - u) \qquad 0 \leq u \leq 1$$

$$N_{1,2} = \frac{(u - u_1)N_{1,1}}{u_2 - u_1} + \frac{(u_3 - u)N_{2,1}}{u_3 - u_2} = uN_{1,1} = u \qquad 0 \leq u \leq 1$$

Substituting $N_{i,\,2}$ into Eq. (8.3) gives

$$R_{0,2} = \frac{w_1 N_{0,2}}{w_1 N_{0,2} + w_2 N_{1,2}} = \frac{w_1(1 - u)}{w_1(1 - u) + uw_2}$$

$$R_{1,2} = \frac{w_2 N_{1,2}}{w_1 N_{0,2} + w_2 N_{1,2}} = \frac{w_2 u}{w_1(1 - u) + uw_2}$$

Substituting $R_{i,\,2}$ into Eq. (8.11) gives

$$\mathbf{P}(u) = \frac{w_1 \mathbf{P}_0(1 - u) + w_2 \mathbf{P}_1 u}{w_1(1 - u) + w_2 u} \qquad 0 \leq u \leq 1 \qquad (8.12)$$

Equation (8.12) is the rational form of a line. We now investigate the effect of the weights on the line. If we set $w_1 = w_2 = 1$ or any other same value, Eq. (8.12) degenerates to Eq. (6.11). If we use $w_1 = 0$ and $w_2 = 1$, Eq. (8.12) becomes $\mathbf{P}(u) = \mathbf{P}_1$ — suggesting that the line becomes the point P_1. Similarly, setting $w_1 = 1$ and $w_2 = 0$ gives $\mathbf{P}(u) = \mathbf{P}_2$. Using $w_1 = 1$ and $w_2 = 0.5$ gives

$$\mathbf{P}(u) = \frac{\mathbf{P}_0 + u(0.5\mathbf{P}_1 - \mathbf{P}_0)}{1 - 0.5u}$$

This equations shows that P_0 has more influence on the line than P_1.

8.5   Arcs

A circular arc is considered part of a circle. It has the same quadratic equation as a circle but with a limited angle. A NURBS circular arc has its own equation. In this section, we develop the NURBS equation for the right arc (quarter of a circle) shown in Figure 8.2. The arc is defined by three points P_0, P_1, and P_2 that have weights w_1, w_2, and w_3, respectively. This data suggests that $k = 3$, $n = 2$, and the length of the knot vector $m = 6$. Utilizing Eq. (8.7) gives the knot vector as $\mathbf{KV} = [u_0 \ u_1 \ u_2 \ u_3 \ u_4 \ u_5]^T = [0 \ 0 \ 0 \ 1 \ 1 \ 1]^T$ — all the knots are end knots.

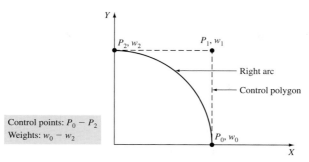

Figure 8.2   A right NURBS arc.

The equation of the NURBS arc is written as, based on Eq. (8.2),

$$\mathbf{P}(u) = \sum_{i=0}^{2} \mathbf{P}_i R_{i,3}(u) = \mathbf{P}_0 R_{0,3}(u) + \mathbf{P}_1 R_{1,3}(u) + \mathbf{P}_2 R_{2,3}(u) \tag{8.13}$$

The pyramid of the basis functions $N_{i,k}(u)$ is three stories high as shown on the right — we must recur from $N_{i,1}$ level to $N_{i,3}$ level to be able to evaluate Eq. (8.13). The sum of the two indexes of the last basis function in each story (level) must equal 5 $(n + k)$. We evaluate the $N_{i,k}(u)$ basis functions as follows:

$$N_{0,3}\ \ N_{1,3}\ \ N_{2,3}$$
$$N_{0,2}\ \ N_{1,2}\ \ N_{2,2}\ \ N_{3,2}$$
$$N_{0,1}\ \ N_{1,1}\ \ N_{2,1}\ \ N_{3,1}\ \ N_{4,1}$$

$$N_{0,1} = \begin{cases} 1 & u = 0 \\ 0 & \text{elsewhere} \end{cases}, \qquad N_{1,1} = \begin{cases} 1 & u = 0 \\ 0 & \text{elsewhere} \end{cases}$$

$$N_{2,1} = \begin{cases} 1 & 0 \le u \le 1 \\ 0 & \text{elsewhere} \end{cases}, \qquad N_{3,1} = \begin{cases} 1 & u = 1 \\ 0 & \text{elsewhere} \end{cases}, \qquad N_{4,1} = \begin{cases} 1 & u = 1 \\ 0 & \text{elsewhere} \end{cases}$$

$$N_{0,2} = \frac{(u - u_0)N_{0,1}}{u_1 - u_0} + \frac{(u_2 - u)N_{1,1}}{u_2 - u_1} = 0$$

$$N_{1,2} = \frac{(u - u_1)N_{1,1}}{u_2 - u_1} + \frac{(u_3 - u)N_{2,1}}{u_3 - u_2} = (1 - u)N_{2,1}$$

$$N_{2,2} = \frac{(u - u_2)N_{2,1}}{u_3 - u_2} + \frac{(u_4 - u)N_{3,1}}{u_4 - u_3} = uN_{2,1}$$

$$N_{3,2} = \frac{(u - u_3)N_{3,1}}{u_4 - u_3} + \frac{(u_5 - u)N_{4,1}}{u_5 - u_4} = 0$$

$$N_{0,3} = \frac{(u - u_0)N_{0,2}}{u_2 - u_0} + \frac{(u_3 - u)N_{1,2}}{u_3 - u_1} = (1 - u)^2 N_{2,1}$$

$$N_{1,3} = \frac{(u - u_1)N_{1,2}}{u_3 - u_1} + \frac{(u_4 - u)N_{2,2}}{u_4 - u_2} = u(1 - u)N_{2,1} + (1 - u)uN_{2,1} = 2u(1 - u)N_{2,1}$$

$$N_{2,3} = \frac{(u - u_2)N_{2,2}}{u_4 - u_2} + \frac{(u_5 - u)N_{3,2}}{u_5 - u_3} = u^2 N_{2,1}$$

Substituting $N_{i,3}$ into Eq. (8.3) gives

$$\mathbf{P}(u) = \frac{\mathbf{P}_0 w_0 N_{0,3} + \mathbf{P}_1 w_1 N_{1,3} + \mathbf{P}_2 w_2 N_{2,3}}{w_1 N_{0,3} + w_2 N_{1,3} + w_3 N_{2,3}}$$

or

$$\mathbf{P}(u) = \frac{\mathbf{P}_0 w_0 (1 - u)^2 + 2\mathbf{P}_1 w_1 u(1 - u) + \mathbf{P}_2 w_2 u^2}{w_0 (1 - u)^2 + 2w_1 u(1 - u) + w_2 u^2} \qquad 0 \le u \le 1 \qquad (8.14)$$

If we use $w_0 = w_1 = w_2 = 1$, Eq. (8.14) reduces to

$$\mathbf{P}(u) = \mathbf{P}_0 (1 - u)^2 + 2\mathbf{P}_1 u(1 - u) + \mathbf{P}_2 u^2 \qquad (8.15)$$

If we use $w_0 = w_1 = 1$ and $w_2 = 2$, Eq. (8.14) gives

$$\mathbf{P}(u) = \frac{\mathbf{P}_0 (1 - u)^2 + 2\mathbf{P}_1 u(1 - u) + 2\mathbf{P}_2 u^2}{(1 - u)^2 + 2u(1 - u) + 2u^2} \qquad (8.16)$$

or

$$\mathbf{P}(u) = \frac{\mathbf{P}_0 (1 - u)^2 + \mathbf{P}_1 2u(1 - u) + 2\mathbf{P}_2 u^2}{1 + u^2} \qquad (8.17)$$

Let us assume a unit arc in the *XY* coordinate system shown below. Substituting the (x, y) coordinates of the control points shown below into Eq. (8.17), the arc equation becomes:

$$\mathbf{P}(u) = \begin{bmatrix} x(u) \\ y(u) \end{bmatrix} = \begin{bmatrix} \dfrac{1 - u^2}{1 + u^2} \\ \dfrac{2u}{1 + u^2} \end{bmatrix} \qquad (8.18)$$

Equation (8.18) satisfies the boundary conditions; at $u = 0$, $\mathbf{P} = \mathbf{P}_0$, and $u = 1$, $\mathbf{P} = \mathbf{P}_2$.

8.6 Circles

Section 6.6, Chapter 6, shows the uniform nonrational parametric equation [Eq. (6.23)] of a circle defined by a center P_c and a radius R as shown in Figure 6.11. Here, we develop the NURBS formulation of a circle with a center at $(0, 0)$ and a radius $R = 1$. The formulation of a NURBS circle requires a set of control points that forms a closed polygon, as shown in Figures 8.3 and 8.4. As the forthcoming formulation shows, a NURBS circle consists of arc segments that are pieced together. The number of segments depends on the number of the control points used in the development. Figures 8.3 and 8.4 show two cases: eight and five control points. We develop the NURBS equation for eight points in this section.

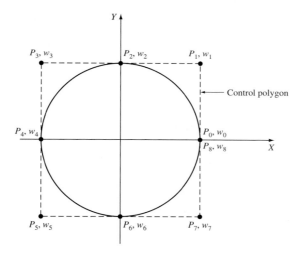

Figure 8.3   A NURBS circle defined by eight control points.

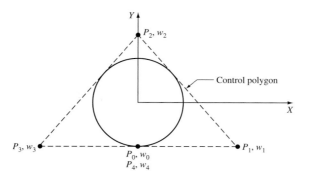

Figure 8.4   A NURBS circle defined by five control points.

Using Eq. (8.2), the circle equation becomes:

$$\mathbf{P}(u) = \sum_{i=0}^{8} \mathbf{P}_i R_{i,3}(u) = \mathbf{P}_0 R_{0,3}(u) + \mathbf{P}_1 R_{1,3}(u) + \mathbf{P}_2 R_{2,3}(u) + \mathbf{P}_3 R_{3,3}(u)$$

$$+ \mathbf{P}_4 R_{4,3}(u) + \mathbf{P}_5 R_{5,3}(u) + \mathbf{P}_6 R_{6,3}(u) + \mathbf{P}_7 R_{7,3}(u) + \mathbf{P}_8 R_{8,3}(u)$$

(8.19)

Based on Figure 8.3, $k = 3$ (circle is quadratic degree), $n = 8$, and $m = 12$. Utilizing Eq. (8.7) gives the knot vector as $KV = [u_0\ u_1\ u_2\ u_3\ u_4\ u_5\ u_6\ u_7\ u_8\ u_9\ u_{10}\ u_{11}]^T = [0\ 0\ 0\ 1/7\ 2/7\ 3/7\ 4/7\ 5/7\ 6/7\ 1\ 1\ 1]^T$. Here, we have six interior knots. Their values are calculated using Eq. (8.10).

The pyramid of the basis functions $N_{i,k}(u)$ is three stories high as shown in Figure 8.5. The sum of the two indexes of the last basis function in each story must equal 11 $(n + k)$.

$$N_{0,3} \quad N_{1,3} \quad N_{2,3} \quad N_{3,3} \quad N_{4,3} \quad N_{5,3} \quad N_{6,3} \quad N_{7,3} \quad N_{8,3}$$

$$N_{0,2} \quad N_{1,2} \quad N_{2,2} \quad N_{3,2} \quad N_{4,2} \quad N_{5,2} \quad N_{6,2} \quad N_{7,2} \quad N_{8,2} \quad N_{9,2}$$

$$N_{0,1} \quad N_{1,1} \quad N_{2,1} \quad N_{3,1} \quad N_{4,1} \quad N_{5,1} \quad N_{6,1} \quad N_{7,1} \quad N_{8,1} \quad N_{9,1} \quad N_{10,1}$$

Figure 8.5 **Basis functions, $N_{i,k}$, for NURBS circle of Figure 8.3.**

We evaluate the $N_{i,k}(u)$ basis functions shown in Figure 8.5 as follows:

$$N_{0,1} = \begin{cases} 1 & u = 0 \\ 0 & \text{elsewhere} \end{cases}, \quad N_{1,1} = \begin{cases} 1 & u = 0 \\ 0 & \text{elsewhere} \end{cases}, \quad N_{2,1} = \begin{cases} 1 & 0 \le u \le 1/7 \\ 0 & \text{elsewhere} \end{cases}$$

$$N_{3,1} = \begin{cases} 1 & 1/7 \le u \le 2/7 \\ 0 & \text{elsewhere} \end{cases}, \quad N_{4,1} = \begin{cases} 1 & 2/7 \le u \le 3/7 \\ 0 & \text{elsewhere} \end{cases}, \quad N_{5,1} = \begin{cases} 1 & 3/7 \le u \le 4/7 \\ 0 & \text{elsewhere} \end{cases}$$

$$N_{6,1} = \begin{cases} 1 & 4/7 \le u \le 5/7 \\ 0 & \text{elsewhere} \end{cases}, \quad N_{7,1} = \begin{cases} 1 & 5/7 \le u \le 6/7 \\ 0 & \text{elsewhere} \end{cases}, \quad N_{8,1} = \begin{cases} 1 & 6/7 \le u \le 1 \\ 0 & \text{elsewhere} \end{cases}$$

$$N_{9,1} = \begin{cases} 1 & u = 1 \\ 0 & \text{elsewhere} \end{cases}, \quad N_{10,1} = \begin{cases} 1 & u = 1 \\ 0 & \text{elsewhere} \end{cases}$$

$$N_{0,2} = \frac{(u - u_0)N_{0,1}}{u_1 - u_0} + \frac{(u_2 - u)N_{1,1}}{u_2 - u_1} = 0$$

$$N_{1,2} = \frac{(u - u_1)N_{1,1}}{u_2 - u_1} + \frac{(u_3 - u)N_{2,1}}{u_3 - u_2} = \frac{\left(\frac{1}{7} - u\right)N_{2,1}}{\frac{1}{7}} = (1 - 7u)N_{2,1}$$

$$N_{2,2} = \frac{(u - u_2)N_{2,1}}{u_3 - u_2} + \frac{(u_4 - u)N_{3,1}}{u_4 - u_3} = \frac{uN_{2,1}}{\frac{1}{7}} + \frac{\left(\frac{2}{7} - u\right)N_{3,1}}{\frac{1}{7}}$$

$$= 7u\,N_{2,1} + (2 - 7u)N_{3,1}$$

$$N_{3,2} = \frac{(u-u_3)N_{3,1}}{u_4-u_3} + \frac{(u_5-u)N_{4,1}}{u_5-u_4} = \frac{\left(u-\frac{1}{7}\right)N_{3,1}}{\frac{1}{7}} + \frac{\left(\frac{3}{7}-u\right)N_{4,1}}{\frac{1}{7}}$$

$$= (7u-1)N_{3,1} + (3-7u)N_{4,1}$$

$$N_{4,2} = \frac{(u-u_4)N_{4,1}}{u_5-u_4} + \frac{(u_6-u)N_{5,1}}{u_6-u_5} = \frac{\left(u-\frac{2}{7}\right)N_{4,1}}{\frac{1}{7}} + \frac{\left(\frac{4}{7}-u\right)N_{5,1}}{\frac{1}{7}}$$

$$= (7u-2)N_{4,1} + (4-7u)N_{5,1}$$

$$N_{5,2} = \frac{(u-u_5)N_{5,1}}{u_6-u_5} + \frac{(u_7-u)N_{6,1}}{u_7-u_6} = \frac{\left(u-\frac{3}{7}\right)N_{5,1}}{\frac{1}{7}} + \frac{\left(\frac{5}{7}-u\right)N_{6,1}}{\frac{1}{7}}$$

$$= (7u-3)N_{5,1} + (5-7u)N_{6,1}$$

$$N_{6,2} = \frac{(u-u_6)N_{6,1}}{u_7-u_6} + \frac{(u_8-u)N_{7,1}}{u_8-u_7} = \frac{\left(u-\frac{4}{7}\right)N_{6,1}}{\frac{1}{7}} + \frac{\left(\frac{6}{7}-u\right)N_{7,1}}{\frac{1}{7}}$$

$$= (7u-4)N_{6,1} + (6-7u)N_{7,1}$$

$$N_{7,2} = \frac{(u-u_7)N_{7,1}}{u_8-u_7} + \frac{(u_9-u)N_{8,1}}{u_9-u_8} = \frac{\left(u-\frac{5}{7}\right)N_{7,1}}{\frac{1}{7}} + \frac{(1-u)N_{8,1}}{\frac{1}{7}}$$

$$= (7u-5)N_{7,1} + 7(1-u)N_{8,1}$$

$$N_{8,2} = \frac{(u-u_8)N_{8,1}}{u_9-u_8} + \frac{(u_{10}-u)N_{9,1}}{u_{10}-u_9} = \frac{\left(u-\frac{6}{7}\right)N_{8,1}}{\frac{1}{7}}$$

$$= (7u-6)N_{8,1}$$

$$N_{9,2} = \frac{(u-u_9)N_{9,1}}{u_{10}-u_9} + \frac{(u_{11}-u)N_{10,1}}{u_{11}-u_{10}} = 0$$

$$N_{0,3} = \frac{(u-u_0)N_{0,2}}{u_2-u_0} + \frac{(u_3-u)N_{1,2}}{u_3-u_1} = \frac{\left(\frac{1}{7}-u\right)N_{1,2}}{\frac{1}{7}}$$

$$= (1-7u)N_{1,2} = (1-7u)^2 N_{2,1}$$

$$N_{1,3} = \frac{(u-u_1)N_{1,2}}{u_3-u_1} + \frac{(u_4-u)N_{2,2}}{u_4-u_2}$$

$$= \frac{u}{\frac{1}{7}}(1-7u)N_{2,1} + \frac{\left(\frac{2}{7}-u\right)}{\frac{2}{7}}[7uN_{2,1}+(2-7u)N_{3,1}]$$

$$= 7u(1-7u)N_{2,1} + \frac{7}{2}u(2-7u)N_{2,1} + \frac{1}{2}(2-7u)^2N_{3,1}$$

$$= \left(-\frac{147}{2}u^2+14u\right)N_{2,1} + \frac{1}{2}(2-7u)^2N_{3,1}$$

$$N_{2,3} = \frac{(u-u_2)N_{2,2}}{u_4-u_2} + \frac{(u_5-u)N_{3,2}}{u_5-u_3}$$

$$= \frac{u}{\frac{2}{7}}[7uN_{2,1}+(2-7u)N_{3,1}] + \frac{\left(\frac{3}{7}-u\right)}{\frac{2}{7}}[(7u-1)N_{3,1}+(3-7u)N_{4,1}]$$

$$= \frac{49}{2}u^2N_{2,1} + \frac{7}{2}u(2-7u)N_{3,1} + \frac{1}{2}(3-7u)(7u-1)N_{3,1} + \frac{1}{2}(3-7u)^2N_{4,1}$$

$$= \frac{49}{2}u^2N_{2,1} + \frac{1}{2}(-98u^2+42u-3)N_{3,1} + \frac{1}{2}(3-7u)^2N_{4,1}$$

$$N_{3,3} = \frac{(u-u_3)N_{3,2}}{u_5-u_3} + \frac{(u_6-u)N_{4,2}}{u_6-u_4}$$

$$= \frac{\left(u-\frac{1}{7}\right)N_{3,2}}{\frac{2}{7}} + \frac{\left(\frac{4}{7}-u\right)N_{4,2}}{\frac{2}{7}}$$

$$= \frac{1}{2}(7u-1)[(7u-1)N_{3,1}+(3-7u)N_{4,1}] + \frac{1}{2}(4-7u)[(7u-2)N_{4,1}+(4-7u)N_{5,1}]$$

$$= \frac{1}{2}(7u-1)^2N_{3,1} + \frac{1}{2}(7u-1)(3-7u)N_{4,1} + \frac{1}{2}(4-7u)(7u-2)N_{4,1} + \frac{1}{2}(4-7u)^2N_{5,1}$$

$$= \frac{1}{2}(7u-1)^2N_{3,1} + \frac{1}{2}(-98u^2+70u-11)N_{4,1} + \frac{1}{2}(4-7u)^2N_{5,1}$$

$$N_{4,3} = \frac{(u-u_4)N_{4,2}}{u_6-u_4} + \frac{(u_7-u)N_{5,2}}{u_7-u_5}$$

$$= \frac{\left(u-\frac{2}{7}\right)}{\frac{2}{7}}[(7u-2)N_{4,1}+(4-7u)N_{5,1}] + \frac{\left(\frac{5}{7}-u\right)}{\frac{2}{7}}[(7u-3)N_{5,1}+(5-7u)N_{6,1}]$$

$$= \frac{1}{2}(7u-2)^2N_{4,1} + \frac{1}{2}(7u-2)(4-7u)N_{5,1} + \frac{1}{2}(5-7u)(7u-3)N_{5,1} + \frac{1}{2}(5-7u)^2N_{6,1}$$

$$= \frac{1}{2}(7u-2)^2N_{4,1} + \frac{1}{2}(-98u^2+98u-23)N_{5,1} + \frac{1}{2}(5-7u)^2N_{6,1}$$

$$N_{5,3} = \frac{(u-u_5)N_{5,2}}{u_7-u_5} + \frac{(u_8-u)N_{6,2}}{u_8-u_6}$$

$$= \frac{\left(u-\frac{3}{7}\right)}{\frac{2}{7}}[(7u-3)N_{5,1}+(5-7u)N_{6,1}] + \frac{\left(\frac{6}{7}-u\right)}{\frac{2}{7}}[(7u-4)N_{6,1}+(6-7u)N_{7,1}]$$

$$= \frac{1}{2}(7u-3)^2N_{5,1}+\frac{1}{2}(7u-3)(5-7u)N_{6,1}+\frac{1}{2}(6-7u)(7u-4)N_{6,1}+\frac{1}{2}(6-7u)^2N_{7,1}$$

$$= \frac{1}{2}(7u-3)^2N_{5,1}+\frac{1}{2}(-98u^2+126u-39)N_{6,1}+\frac{1}{2}(6-7u)^2N_{7,1}$$

$$N_{6,3} = \frac{(u-u_6)N_{6,2}}{u_8-u_6} + \frac{(u_9-u)N_{7,2}}{u_9-u_7}$$

$$= \frac{\left(u-\frac{4}{7}\right)}{\frac{2}{7}}[(7u-4)N_{6,1}+(6-7u)N_{7,1}] + \frac{(1-u)}{\frac{2}{7}}[(7u-5)N_{7,1}+7(1-u)N_{8,1}]$$

$$= \frac{1}{2}(7u-4)^2N_{6,1}+\frac{1}{2}(7u-4)(6-7u)N_{7,1}+\frac{7}{2}(1-u)(7u-5)N_{7,1}+\frac{49}{2}(1-u)^2N_{8,1}$$

$$= \frac{1}{2}(7u-4)^2N_{6,1}+\frac{1}{2}(-98u^2+154u-59)N_{7,1}+\frac{49}{2}(1-u)^2N_{8,1}$$

$$N_{7,3} = \frac{(u-u_7)N_{7,2}}{u_9-u_7} + \frac{(u_{10}-u)N_{8,2}}{u_{10}-u_8}$$

$$= \frac{\left(u-\frac{5}{7}\right)}{\frac{2}{7}}[(7u-5)N_{7,1}+7(1-u)N_{8,1}] + \frac{(1-u)}{\frac{1}{7}}(7u-6)N_{8,1}$$

$$= \frac{1}{2}(7u-5)^2N_{7,1}+\frac{7}{2}(7u-5)(1-u)N_{8,1}+7(1-u)(7u-6)N_{8,1}$$

$$= \frac{1}{2}(7u-5)^2N_{7,1}+\frac{7}{2}(-21u^2+38u-17)N_{8,1}$$

$$N_{8,3} = \frac{(u-u_8)N_{8,2}}{u_{10}-u_8} + \frac{(u_{11}-u)N_{9,2}}{u_{11}-u_9}$$

$$= \frac{\left(u-\frac{6}{7}\right)}{\frac{1}{7}}(7u-6)N_{8,1}$$

$$= (7u-6)^2 N_{8,1}$$

Substituting $N_{i,3}$ into Eq. (8.3), and the resulting $R_{i,3}(u)$ into Eq. (8.19) gives

$$P(u) = \cfrac{1}{8 \displaystyle\sum_{i=0} w_i N_{i,3}} \times \begin{cases} \end{cases}$$

$$\mathbf{P}_0 w_0 (1-7u)^2 + \mathbf{P}_1 w_1\left(-\frac{147u^2}{2} + 14u\right) + \mathbf{P}_2 w_2 \frac{49u^2}{2} \qquad\qquad 0 \le u \le \frac{1}{7}$$

$$\frac{1}{2}\mathbf{P}_1 w_1 (2-7u)^2 + \frac{1}{2}\mathbf{P}_2 w_2(-98u^2 + 42u - 3) + \frac{1}{2}\mathbf{P}_3 w_3 (7u-1)^2 \qquad \frac{1}{7} \le u \le \frac{2}{7}$$

$$\frac{1}{2}\mathbf{P}_2 w_2 (3-7u)^2 + \frac{1}{2}\mathbf{P}_3 w_3(-98u^2 + 70u - 11) + \frac{1}{2}\mathbf{P}_4 w_4 (7u-2)^2 \qquad \frac{2}{7} \le u \le \frac{3}{7}$$

$$\frac{1}{2}\mathbf{P}_3 w_3 (4-7u)^2 + \frac{1}{2}\mathbf{P}_4 w_4(-98u^2 + 98u - 23) + \frac{1}{2}\mathbf{P}_5 w_5 (7u-3)^2 \qquad \frac{3}{7} \le u \le \frac{4}{7}$$

$$\frac{1}{2}\mathbf{P}_4 w_4 (5-7u)^2 + \frac{1}{2}\mathbf{P}_5 w_5(-98u^2 + 126u - 39) + \frac{1}{2}\mathbf{P}_6 w_6 (7u-4)^2 \qquad \frac{4}{7} \le u \le \frac{5}{7}$$

$$\frac{1}{2}\mathbf{P}_5 w_5 (6-7u)^2 + \frac{1}{2}\mathbf{P}_6 w_6(-98u^2 + 154u - 59) + \frac{1}{2}\mathbf{P}_7 w_7 (7u-5)^2 \qquad \frac{5}{7} \le u \le \frac{6}{7}$$

$$\frac{49}{2}\mathbf{P}_6 w_6 (1-u)^2 + \frac{7}{2}\mathbf{P}_7 w_7(-21u^2 + 38u - 17) + \mathbf{P}_8 w_8 (7u-6)^2 \qquad \frac{6}{7} \le u \le 1$$

$$(8.20)$$

We simplify Eq. (8.20) by using $w_i = 1$ for all the weights at the control points, $R = 1$ for the circle radius, and P_c (0, 0) for the circle center. The radius and the center produce the following coordinates for $P_0 - P_8$: $P_0(1, 0)$, $P_1(1, 1)$, $P_2(0, 1)$, $P_3(-1, 1)$, $P_4(-1, 0)$, $P_5(-1, -1)$, $P_6(0, -1)$, $P_7(1, -1)$, $P_8(1, 0)$. The circle, its polygon, and control points are shown below.

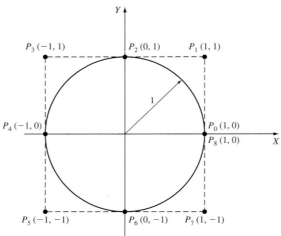

Substituting all these values into Eq. (8.20) gives the following two scalar equations for any point $P(x, y)$ on the circle:

$$
x(u) = \begin{cases}
(1-7u)^2 - \dfrac{147u^2}{2} + 14u & 0 \le u \le \dfrac{1}{7} \\[2mm]
\dfrac{1}{2}(2-7u)^2 - \dfrac{1}{2}(7u-1)^2 & \dfrac{1}{7} \le u \le \dfrac{2}{7} \\[2mm]
-\dfrac{1}{2}(-98u^2 + 70u - 11) - \dfrac{1}{2}(7u-1)^2 & \dfrac{2}{7} \le u \le \dfrac{3}{7} \\[2mm]
-\dfrac{1}{2}(4-7u)^2 - \dfrac{1}{2}(-98u^2 + 98u - 23) - \dfrac{1}{2}(7u-3)^2 & \dfrac{3}{7} \le u \le \dfrac{4}{7} \\[2mm]
-\dfrac{1}{2}(5-7u)^2 - \dfrac{1}{2}(-98u^2 + 126u - 39) & \dfrac{4}{7} \le u \le \dfrac{5}{7} \\[2mm]
-\dfrac{1}{2}(6-7u)^2 + \dfrac{1}{2}(7u-5)^2 & \dfrac{5}{7} \le u \le \dfrac{6}{7} \\[2mm]
\dfrac{7}{2}(-21u^2 + 38u - 17) + (7u-6)^2 & \dfrac{6}{7} \le u \le 1
\end{cases}
\tag{8.21}
$$

$$
y(u) = \begin{cases}
-\dfrac{98u^2}{2} + 14u & 0 \le u \le \dfrac{1}{7} \\[2mm]
\dfrac{1}{2}(2-7u)^2 + \dfrac{1}{2}(-98u^2 + 42u - 3) + \dfrac{1}{2}(7u-1)^2 & \dfrac{1}{7} \le u \le \dfrac{2}{7} \\[2mm]
\dfrac{1}{2}(3-7u)^2 + \dfrac{1}{2}(-98u^2 + 70u - 11) & \dfrac{2}{7} \le u \le \dfrac{3}{7} \\[2mm]
\dfrac{1}{2}(4-7u)^2 - \dfrac{1}{2}(7u-3)^2 & \dfrac{3}{7} \le u \le \dfrac{4}{7} \\[2mm]
-\dfrac{1}{2}(-98u^2 + 126u - 39) - \dfrac{1}{2}(7u-4)^2 & \dfrac{4}{7} \le u \le \dfrac{5}{7} \\[2mm]
-\dfrac{1}{2}(6-7u)^2 - \dfrac{1}{2}(-98u^2 + 154u - 59) - \dfrac{1}{2}(7u-5)^2 & \dfrac{5}{7} \le u \le \dfrac{6}{7} \\[2mm]
-\dfrac{49}{2}(1-u)^2 - \dfrac{7}{2}(-21u^2 + 38u - 17) & \dfrac{6}{7} \le u \le 1
\end{cases}
\tag{8.22}
$$

The vector equation, Eq. (8.20), or its scalar form, Eqs. (8.21) and (8.22), defines the NURBS circle as seven continuous curve segments. Each segment is defined over the range $\Delta u = 1/7$. This is a result of the existence range of the $N_{i,1}(u)$ basis functions. The parameter u has the range of $u \in [0,1]$. When we have a given value of u, we must use the right NURBS segment to find the coordinates of the corresponding point.

We verify the correctness of Eq. (8.20) or its scalar form, Eqs. (8.21) and (8.22), by substituting specific values of u. Let us use $u = 0$, 0.25 (1.75/7), 0.5 (3.5/7), 0.75 (5.25/7), and 1. The fractions in parentheses help us determine which curve segment in Eqs. (8.21) and (8.22) we should use to calculate the x and y coordinates. The table on the right shows

u	u range	x	y
0	$0 \le u \le 1/7$	1	0
1.75/7	$1/7 \le u \le 2/7$	−0.25	1
3.5/7	$3/7 \le u \le 4/7$	−1	0
5.25/7	$5/7 \le u \le 6/7$	−0.25	−1
1	$6/7 \le u \le 1$	1	0

the results. At $u = 0$, 0.5, and 1 we get respectively points P_0, P_4, and P_8. However, at $u = 0.25$ and 0.75, we do *not* get points P_2 and P_6, respectively. The two points seem to be offset to the left, in the X direction, by a distance of -0.25. We can find the u value at which $x = 0$ in the curve segment defined by $1/7 \le u \le 2/7$, by equating the x coordinate of this segment to zero and solving for u. This gives a u value of 1.5/7. Substituting this value into the y coordinate equation gives $y = 1$. Thus, the circle seems to be flat at its top and bottom.

8.7 Bilinear Surface

Section 7.6, Chapter 7, shows the analytic equation [Eq. (7.25)] of a plane defined by three points, P_0, P_1 and P_2, as shown in Figure 7.20. Here, we develop the NURBS formulation of the same plane. Figure 8.6 shows the points labeled as we need them in the tensor sum. A fourth point, P_{11}, is needed as shown in Eq. (8.23). It is known that three points (P_0, P_1, and P_2) define a plane. Thus, we define P_{11} as a linear combination of the three points, $\mathbf{P}_{11} = (\mathbf{P}_1 - \mathbf{P}_0) + \mathbf{P}_2$ or $\mathbf{P}_{11} = (\mathbf{P}_2 - \mathbf{P}_0) + \mathbf{P}_1$. Using Eq. (8.2), the equation of the bilinear surface becomes:

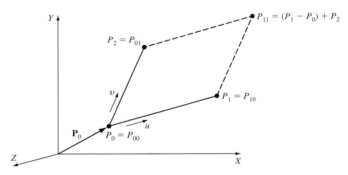

Figure 8.6   A NURBS bilinear surface.

$$\mathbf{P}(u, v) = \sum_{i=0}^{1}\sum_{j=0}^{1} \mathbf{P}_{ij}R_{i,k}(u)R_{i,l}(v) = \sum_{i=0}^{1}\sum_{j=0}^{1} \mathbf{P}_{ij}R_{i,2}(u)R_{i,2}(v) \quad 0 \le u \le 1, 0 \le v \le 1 \quad (8.23)$$

Given the surface data (linear and two end points in each direction), we have $k = 2, l = 2$, and $n = 1$. Utilizing Eq. (8.9) gives $m = 4$. Utilizing Eq. (8.7) gives the knot vectors as $\mathbf{KV}_u = [u_0\ u_1\ u_2\ u_3]^T = [0\ 0\ 1\ 1]^T$ and $\mathbf{KV}_v = [v_0\ v_1\ v_2\ v_3]^T = [0\ 0\ 1\ 1]^T$. In this case, all the knot values are end values because $m = 4$ and $k = 2$. We use these knot vectors and Eqs. (6.92) and (6.93) to calculate $N_{i,2}(u)$ and $N_{i,2}(v)$. $N_{i,2}(u)$ and $N_{i,2}(v)$ have the same form; $N_{i,2}(u)$ uses u and $N_{i,2}(v)$ uses v. Moreover, $N_{i,2}(u)$ and $N_{i,2}(v)$ are the same as those of Section 8.4, that is, $N_{0,2}(u) = u$, $N_{i,2}(v) = v$, $N_{1,2}(u) = (1 - u)$, and $N_{i,2}(v) = (1 - v)$.

Expanding Eq. (8.23) gives:

$$\mathbf{P}(u,\, v) = \mathbf{P}_{00}R_{0,\,2}(u)R_{0,\,2}(v) + \mathbf{P}_{01}R_{0,\,2}(u)R_{1,\,2}(v)$$
$$+ \mathbf{P}_{10}R_{1,\,2}(u)R_{0,\,2}(v) + \mathbf{P}_{11}R_{1,\,2}(u)R_{1,\,2}(v) \tag{8.24}$$

Substituting the $N_{i,\,2}(u)$ and $N_{i,\,2}(v)$ basis functions into Eq. (8.24) gives:

$$\mathbf{P}(u,\, v) = \cfrac{1}{\displaystyle\sum_{i=0}^{1}\sum_{j=0}^{1} w_{ij}w_{ij}N_{i,\,2}(u)N_{j,\,2}(v)} [w_{00}w_{00}\mathbf{P}_{00}N_{0,\,2}(u)N_{0,\,2}(v)$$
$$+ w_{01}w_{01}\mathbf{P}_{01}N_{0,\,2}(u)N_{1,\,2}(v) + w_{10}w_{10}\mathbf{P}_{10}N_{1,\,2}(u)N_{0,\,2}(v)$$
$$+ w_{11}w_{11}\mathbf{P}_{11}N_{1,\,2}(u)N_{1,\,2}(v)] \tag{8.25}$$

where w_{ij} indicates the weight at point P_{ij}. In relation to the boundary points, shown in Figure 8.6, these weights are $w_{00} = w_0$, $w_{10} = w_1$, $w_{01} = w_2$, and $w_{11} = w_1 - w_0 + w_2$. The weight, w_{11}, at point P_{11} is based on the relation, $\mathbf{P}_{11} = (\mathbf{P}_1 - \mathbf{P}_0) + \mathbf{P}_2$.

If all $w_i = 1$, Eq. (8.25) reduces to

$$\mathbf{P}(u,\, v) = \mathbf{P}_0 N_{0,\,2}N_{0,\,2} + \mathbf{P}_2 N_{0,\,2}N_{1,\,2} + \mathbf{P}_1 N_{1,\,2}N_{0,\,2} + [(\mathbf{P}_1 - \mathbf{P}_0) + \mathbf{P}_2]N_{1,\,2}N_{1,\,2}$$
$$= \mathbf{P}_0(1-u)(1-v) + \mathbf{P}_2(1-u)v + \mathbf{P}_1(1-v)u + [(\mathbf{P}_1 - \mathbf{P}_0) + \mathbf{P}_2]uv \tag{8.26}$$

This equation gives the following final form:

$$\mathbf{P}(u,\, v) = \mathbf{P}_0 + u(\mathbf{P}_1 - \mathbf{P}_0) + v(\mathbf{P}_2 - \mathbf{P}_0) \qquad 0 \le u \le 1, 0 \le v \le 1 \tag{8.27}$$

Equation (8.27) is the same as Eq. (7.25).

8.8   Ruled Surface

The development of the NURBS ruled surface is the same as the ruled surface we have developed in Section 7.7. The only difference is that the surface rails should be NURBS curves that have the same u range and identical knot vectors. If this is not the case, we must reparametrize the rails and obtain the new values for n, w_{ij} and P_{ij}.

PROBLEMS

Part I: Theory

8.1   Find the rational line that connects points P_0 (3, 5, 7) and P_1(0, 1, 4). Using $w_1 = 1$ and $w_2 = 2$, find the line midpoint. What is the midpoint of the line if $w_1 = w_2 = 1$? What is your conclusion?

8.2   Find the coordinates of the midpoint of the right arc developed in Section 8.5. Use both Eqs. (8.15) and (8.17). What is the effect of the weights?

8.3   Repeat Problem 8.2, but use P_0(2, 0), P_1(2, 2), and P_2(0, 2). Compare the results with those of Problem 8.2.

8.4   Find the coordinates of the point at $u = 1/8$ on the circle developed in Section 8.6.

8.5 Use the point calculated in problem 8.4 and find the radius of the circle.

8.6 Find the equation of the rational bilinear surface that covers the region R shown on the right.

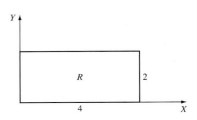

Part II: Laboratory

Use your in-house CAD/CAM system to answer the questions in this part.

8.7 Does your CAD system provide NURBS curves? If it does, can you find menus or commands that allow you to access to the knot vector and the weights?

Part III: Programming

Use the NURBS formulations in all of the following questions. In addition, write the code so that the user can input knot and weight values to control the curve shape and behavior as we have discussed in the chapter. The program should display the results. Use OpenGL with C or C++, or Use Java 3D. Write a computer program to generate a:

8.8 NURBS line.

8.9 NURBS arc.

8.10 NURBS circle.

8.11 NURBS bilinear surface.

8.12 NURBS ruled surface.

CHAPTER 9

Solids

OAL

Understand and master the concepts of solid modeling, the use of primitives in geometric modeling, the basics of Boolean operations, and the representation schemes of solids.

BJECTIVES

After reading this chapter, you should understand the following concepts:

- Geometry and topology
- Primitives
- Primitives' manipulations
- Regularized sets
- Set membership classification
- Constructive solid geometry (CSG)
- Boundary representation (B-rep)
- Sweeps

CHAPTER HEADLINES

9.1   Introduction

Solid models are known to be complete, valid, and unambiguous representations of objects. Simply stated, a **complete solid** is one which enables a point in space to be classified relative to the object, if it is inside, outside, or on the object. This classification is sometimes called spatial addressability. A **valid solid** is one that does not have dangling edges or faces. An **unambiguous solid** has one and only one interpretation. Solid modeling achieves completeness, validity, and unambiguity of geometric models. Therefore, the automation of tasks such as interference analysis, mass property calculations, finite element modeling and analysis, Computer Aided Process Planning (CAPP), machine vision, and NC machining is possible.

CAD systems offer two approaches to creating solid models: primitives and features. The former approach allows designers to use predefined shapes (primitives) as building blocks (like Lego pieces) to create complex solids. Designers must use Boolean operations to combine the primitives. This approach is limited by the restricted shapes of the primitives. The features are more flexible as they allow the construction of more complex and elaborate solids than what the primitives offer. Some CAD systems (such as Unigraphics, CATIA, and I-DEAS) offer both approaches, while others (such as SolidWorks and Pro/E) offer only the features approach.

Consider the object shown in Figure 9.1 to illustrate the two approaches. We can create a block and subtract six cylinders from it using the primitives approach. Or, we can create a rectangle with six circles inside it in the `Top` sketch plane and extrude it using the features approach. The resulting solid is the feature in this case.

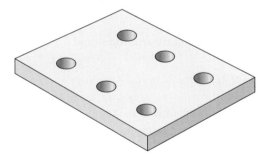

Figure 9.1 A typical solid model.

The features approach is considered by some to be a generalization of the primitives approach. Systems that offer the features hide away the Boolean operations from their users. These operations must be used to create the final solid model. A `boss` command is a form of Boolean union operation, and a `cut` command is a form of a Boolean subtract operation.

This chapter covers the theoretical and practical aspects of solids. Throughout the chapter, issues related to constructing solids on CAD/CAM systems are covered.

9.2 Geometry and Topology

A solid model of an object consists of both the topological and geometrical data of the object. The completeness and unambiguity of a solid model are attributed to the fact that its database stores both its geometry and its topology. The difference between geometry and topology is illustrated in Figure 9.2. **Geometry** (sometimes called metric information) is the actual dimensions that define the entities of the object. The geometry that defines the object shown in Figure 9.2 is the lengths of lines L_1, L_2, and L_3, the angles between the lines, and the radius R and the center P_1 of the half circle.

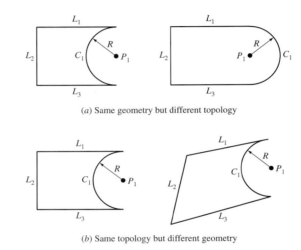

(a) Same geometry but different topology

(b) Same topology but different geometry

Figure 9.2 Difference between geometry and topology.

Topology (sometimes called combinatorial structure) is the connectivity and associativity of the object entities. It has to do with the notion of neighborhood; that is, it determines the relational information between object entities. The topology of the object shown in Figure 9.2b can be stated as follows: L_1 shares a vertex (point) with L_2 and C_1, L_2 shares a vertex with L_1 and L_3, L_3 shares a vertex with L_2 and C_1, L_1 and L_3 do not overlap, and P_1 lies outside the object. Based on these definitions, neither geometry nor topology alone can completely define objects.

While solid models are complete and unambiguous, they are not unique. An object may be constructed in various ways. Consider the object shown in Figure 9.3. Using the primitives approach, one can construct the solid model of the object by dividing it into two blocks and a cylinder. We can add the two blocks first and then subtract the cylinder (Figure 9.3b), or we can subtract the cylinder from a block and add the other block to the resulting subsolid (Figure 9.3c). Figure 9.4 shows two alternatives (create different cross sections and extrude them) if we use the features approach. Regardless of the order and method of construction, the resulting solid model of the object is always complete and unambiguous. However, there will always be one way that is more efficient than others to construct solid models, as is the case with curves and surfaces.

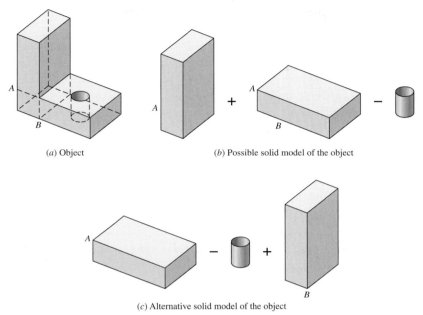

(a) Object   (b) Possible solid model of the object

(c) Alternative solid model of the object

Figure 9.3   Nonuniqueness of solid models: the primitives approach.

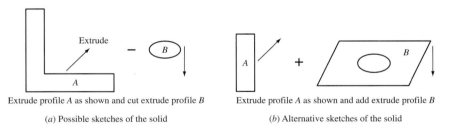

Extrude profile A as shown and cut extrude profile B   Extrude profile A as shown and add extrude profile B

(a) Possible sketches of the solid   (b) Alternative sketches of the solid

Figure 9.4   Nonuniqueness of solid models: the features approach.

9.3   Solid Entities

The entities we use to create solid models depend on the approach we use. The primitives approach uses primitives and the features approach uses sketches. Many CAD systems provide both approaches to increase their modeling domain. The entities required for sketches are curves and surfaces, which we covered in Chapters 6 and 7. This section covers the basics of primitives.

Primitives are considered building blocks. **Primitives** are simple, basic shapes which can be combined by a mathematical set of Boolean operations to create the solid. Primitives themselves are considered valid off-the-shelf solids. The user usually positions primitives as required before applying Boolean operations to construct the final solid.

There is a wide variety of primitives available commercially to users. However, the four most commonly used ones are the block, cylinder, cone, and sphere. These are based on the four natural quadrics: planes, cylinders, cones, and spheres. For example, the block is formed by intersecting six planes. These quadrics are considered natural because they represent the most commonly occurring surfaces in mechanical design which can be produced by rolling, turning, milling, cutting, drilling, and other machining operations used in industry.

Planar surfaces result from rolling, chamfering, and milling; cylindrical surfaces from turning or filleting; spherical surfaces from cutting with a ball-end cutting tool; conical surfaces from turning as well as from drill tips and countersinks. Natural quadrics are distinguished by the fact that they are combinations of linear motion and rotation. Other surfaces, except the torus, require at least dual axis control.

From a user-input point of view and regardless of a specific system syntax, a primitive requires a set of location data, a set of geometric data, and a set of orientation data to define it completely. Location data entails a primitive local coordinate system and an input point defining its origin. Geometrical data differs from one primitive to another and is user input. Orientation data is typically used to orient primitives properly relative to the MCS or WCS of the solid model under construction. Primitives are usually translated and/or rotated to position and oriented properly before applying Boolean operations. Following are descriptions of the most commonly used primitives (refer to Figure 9.5):

1. **Block.** This is a box or cube whose geometrical data is its width, height, and depth. Its local coordinate system $X_L Y_L Z_L$ is shown in Figure 9.5. Point P defines the origin of the $X_L Y_L Z_L$ system. The signs of W, H, and D determine the position of the block relative to its coordinate system. For example, a block with a negative value of W is displayed as if the block shown in Figure 9.5 is mirrored about the $Y_L Z_L$ plane.

2. **Cylinder.** This primitive is a right circular cylinder whose geometry is defined by its radius R (or diameter D) and length H. The length H is usually taken along the direction of the Z_L axis. H can be positive or negative.

3. **Cone.** This is a right circular cone or a frustum of a right circular cone whose base diameter R, top diameter (for truncated cone), and height H are user-defined.

4. **Sphere.** This is defined by its radius R or diameter D and is centered about the origin of its local coordinate system.

5. **Wedge.** This is a right angled wedge whose height H, width W, and base depth D form its geometric data.

6. **Torus.** This primitive is generated by the revolution of a circle about an axis lying in its plane Z_L axis in Figure 9.5. The torus geometry can be defined by the radius (or diameter)

of its body R_1 and the radius (or diameter) of the centerline of the torus body R_2, or the geometry can be defined by the inner radius (or diameter) R_I and outer radius R_O.

All these primitives can be created using the features approach. They are all 2½D objects. The block, cylinder, and wedge are uniform thickness. The cone, sphere, and torus are axisymmetric. This explains why some CAD systems such as Pro/E and SolidWorks do not offer them — the user can generate them via sketching. This simplifies software development as there is no need to write separate primitives' functions.

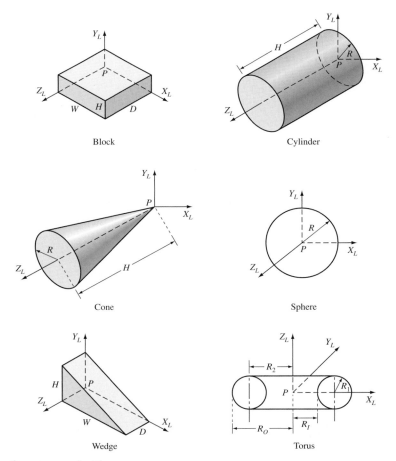

Figure 9.5   Common primitives.

For all the primitives, there are default values for the data defining their geometries. Most CDA systems use default values of 1. In addition, the local coordinate systems for the various primitives shown in Figure 9.5 may change from one system to another. Some systems assume that the origin, P, of the local coordinate system is coincident with that of the MCS or WCS and require the user to translate the primitive to the desired location, thus eliminating the input of point P by the user.

Two or more primitives can be combined to form a solid. To ensure the validity of the resulting solid, the allowed combinatorial relationships between primitives are achieved via Boolean (or set) operations. The available Boolean operators are union ($\cup$ or +), intersection ($\cap$ or I), and difference ($-$). The union operator is used to combine or add together two objects or primitives. Intersecting two primitives gives a shape equal to their common value. The difference operator is used to subtract one object from the other and results in a shape equal to the difference in their volumes. Figure 9.6 shows Boolean operations of a block A and a cylinder B.

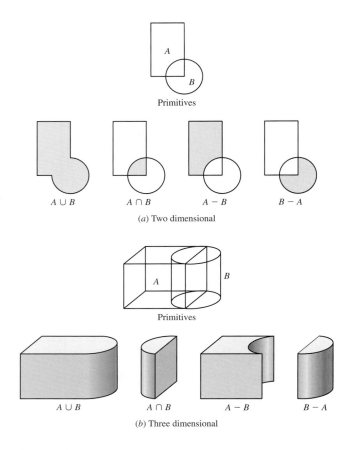

Figure 9.6   Boolean operations.

9.4   Solid Representation

Solid representation of an object can support reliably and automatically, at least in theory, related design and manufacturing applications due to its informational completeness. Such representation is based fundamentally on the fact that a physical object divides an n-dimensional space, E^n, into two regions, interior and exterior, separated by the object boundaries. A **region** is defined as a portion of space E^n, and the boundary of a region is a closed surface, as in the case of a sphere, or a collection of open surfaces connected at proper edges as in the case of a box.

In terms of the above notion, a solid model of an object is defined mathematically as a point set S in 3D Euclidean space (E^3). If we denote the interior and boundary of the set by iS and bS, respectively, we can write.

$$S = iS \cup bS \tag{9.1}$$

And if we let the exterior be defined by cS (complement of S), then

$$W = iS \cup bS \cup cS \tag{9.2}$$

where W is the universal set, which in the case of E^3 is all possible 3D points.

The solid definition given by Eq. (9.1) introduces the concept of geometric closure, which implies that the interior of the solid is geometrically closed by its boundary. That is, the boundary acts a tight "skin" that is entirely in contact with the solid interior. Thus, Eq. (9.1) can be rewritten as:

$$S = kS \tag{9.3}$$

where kS is the closure of the solid (point set S). Comparing Eqs. (9.1) and (9.3) gives kS as

$$kS = iS \cup bS \tag{9.4}$$

Figure 9.7 shows the geometric explanation of equations (9.1) to (9.4).

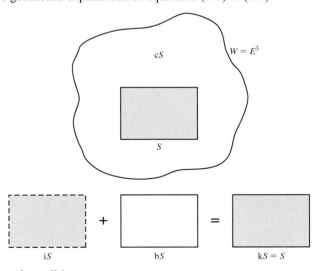

Figure 9.7   **Closure of a solid.**

The properties that a solid model should capture mathematically are:

a. **Rigidity.** This implies that the shape of a solid model is invariant and does not depend on the model's location or its orientation in space.

b. **Homogeneous three-dimensionality.** Solid boundaries must be in contact with the interior. No isolated or dangling boundaries (see Figure 9.8) should be permitted.

c. **Finiteness and finite describability.** The former property means that the size of the solid is not infinite, while the latter ensures that a limited amount of information can describe the solid. This latter property is needed to be able to store solid models in computers whose storage space is always limited. It should be noted that the former property does not include the latter and vice versa. For example, a cylinder which may have a finite radius and length may be described by an infinite number of planar faces.

d. **Closure under rigid motion and regularized Boolean operations.** This property ensures that manipulating solids by moving them in space or changing them via Boolean operations must produce other valid solids.

e. **Boundary determinism.** The boundary of a solid must contain the solid and hence must determine distinctively the interior of the solid.

The mathematical implication of these properties is that valid solid models are bounded, closed, regular, and semianalytic subsets of E^3. These subsets are called r-sets (regularized sets). Intuitively, r-sets are "curved polyhedra" with "well-behaved" boundaries. The point set S that defines a solid model and is given by Eq. (9.1) is always an r-set. Intuitively, a "closed regular set" means that the set is closed and has no dangling portions as shown in Figure 9.8, and a "semianalytic set" means that the set does not oscillate infinitely fast anywhere within the set. The concept of "semianalytic set" is important in choosing equations to describe surfaces or primitives of solid models. For example, the point set that satisfies $\sin(x) < 0$ is a semianalytic set, while the set that satisfies $\sin(1/x) < 0$ is not because the function $\sin(1/x)$ oscillates fast when x approaches zero.

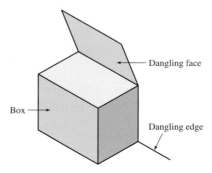

Figure 9.8   Isolated boundaries of a solid.

Conversion of a solid model to an edge representation is well understood and is used to generate orthographic views for display and drafting purposes. While the views are generated automatically, they are not dimensioned automatically, and a manual or semiautomatic dimensioning is required. We should mention that CAD systems can dimension views automatically; however, the resulting dimensions need manual cleanup such as repositioning and adjusting.

9.5 Fundamentals of Solid Modeling

Before covering the details of the various representation schemes, it is appropriate to discuss the details of some of the underlying fundamentals of solid modeling theory. These are geometry, topology, geometric closure, set theory, regularization of set operations, set membership classification, and neighborhood. Geometry and topology have been covered in Section 9.2 and geometric closure is introduced in Section 9.4. This section covers set theory, regularization, classification, and neighborhood. The significance of these topics to solid modeling stems from the definition of a solid model as a point set in E^3 as given in Eq. (9.1). They provide good, rigorous mathematical foundations for developing and analyzing solids.

9.5.1   Set Theory

We begin the review of set theory by introducing some definitions followed by set algebra (operations or sets) and laws (properties) of the algebra of sets. At the end, the concept of ordered pairs and cartesian product is introduced. A **set** is defined as a collection or aggregate of objects. The objects that belong to the set are called the elements or members of the set. For example, the digits 0,1,...,9 form a set (set of digits) D whose elements are 0,1, ..., 9. While the concept is relatively simple, the elements of a set must satisfy certain requirements. First, the elements must be well defined to determine unequivocally whether or not any object belongs to the set. Fuzzy sets are excluded. Second, the elements of a set must be distinct and no element may appear twice. Third, the order of the elements within the set must be immaterial. To realize the importance of these requirements in geometric modeling, the reader can apply them to a point set of eight elements which are the corner points of a block.

The elements of a set can be designated by one of two methods: the roster method or the descriptive method. The former involves listing within braces all the elements of the set, and the latter involves describing the condition(s) that every element in the set must meet. The set of digits D can be written using the roster and the descriptive methods, respectively, as:

$$D = \{1, 2, 3, 4, 5, 6, 7, 8, 9\} \tag{9.5}$$

and

$$D = \{x{:}x = 0, 1, 2, 3, 4, 5, 6, 7, 8, 9\} \tag{9.6}$$

Equation (9.5) reads as "D is equal to the set of elements 0, 1, 2, 3, 4, 5, 6, 7, 8, 9." Equation (9.6) reads as "D is equal to the set of elements x such that x equals to 0, 1, 2, 3, 4, 5, 6, 7, 8, 9." The colon in Eq. (9.6) may be replaced by a vertical bar, that is, $D = \{x| x = 0, 1, ..., 9\}$.

Regardless of set designation, set membership and nonmembership are customarily indicated by $\in$ and $\notin$, respectively. If we write $9 \in D$, we mean 9 is an element (or member) of the set of digits D or 9 belongs to D. Similarly, $-2 \notin D$ means that -2 is not an element of D, or -2 does not belong to D.

Two sets P and Q are equal, written $P = Q$, if the two sets contain exactly the same elements. For example, the two sets $P = \{1, 3, 5, 7\}$ and $Q = \{1, 5, 7, 3\}$ are equal, since every element in P is in Q and every element in Q is in P. The inequality is denoted by $\neq$ ($P \neq Q$ reads "P is not equal to Q").

A set R is a subset of another set S if every element in R is in S. The notation for subset is $\subseteq$, and $R \subseteq S$ reads "R is a subset of S." Analogous to $\in$ and $\notin$, the notation for not subset is $\not\subset$. If it happens that all elements in R are in S but all elements in S are not in R, then R is called a proper subset of S and is written $R \subset S$. This means that for R to be a proper subset of S, S must have all elements of R plus at least one element that is not in R. For example, given $S = \{1, 3, 5, 7\}$, then $R = \{1, 3, 5, 7\}$ is a subset of S and $R = \{5, 7\}$ is a proper subset of S. Formally, $R \subset S \Leftrightarrow R \cap S = R$ and $R \neq S$ ($\Leftrightarrow$ reads "if and only if") or $R \subset S \Leftrightarrow R \cup S = S$ and $R \neq S$.

There are two sets that usually come to mind when discussing sets and subsets. The universal set W is a set that contains all the elements that the analyst wishes to consider. It is a problem dependent. In solid modeling, W contains E^3 and all points in E^3 are the elements of W. In contrast, the null (sometimes referred to as the empty set) set is defined as a set that has no elements or members. It is designated by the Greek letter $\varnothing$. The null set is analogous to zero in ordinary algebra.

Having introduced the required definitions, we now discuss set algebra. Set algebra consists of certain operations that can be performed on sets to produce other sets. These operations are simple in themselves but are powerful when combined with the laws of set algebra to solve geometric modeling problems. The operations are most easily illustrated through the use of the Venn diagram, named after the English logician John Venn. It consists of a rectangle that conceptually represents the universal set. Subsets of the universal set are represented by circles drawn within the rectangle or the universal set.

The three essential set operations are complement, union, and intersection. The complement of P, denoted by cP (reads "P complement"), is the subset of elements of W that are not members of P, that is,

$$cP = \{x : x \notin P\} \tag{9.7}$$

The shaded portion of the Venn diagram in Figure 9.9a shows the complement of P.

The union of two sets $P \cup Q$ (reads "P union Q") is the subset of elements of W that are members of either P or Q, that is.

$$P \cup Q = \{x : x \in P \text{ or } x \in Q\} \tag{9.8}$$

The union is shown in Figure 9.9b as the shaded area.

The intersection of two sets $P \cap Q$ (reads "P intersect Q") is the subset of elements of W that are simultaneously elements of both P and Q, that is,

$$P \cap Q = \{x: x \in P \text{ and } x \in Q\} \tag{9.9}$$

The shaded portion in Figure 9.9c shows the intersection of P and Q. It is easy to realize that $P \cap W = P$ and $P \cap cP = \varnothing$. Sets that have no common elements are termed disjoint or mutually exclusive.

Two additional set operators that can be derived from the set operations just discussed are difference and exclusive union. The difference of two sets $P - Q$ (reads "P minus Q") is the subset of elements of W that belong to P and not Q, that is.

$$P - Q = \{x: x \in P \text{ and } x \notin Q\} \tag{9.10}$$

or

$$Q - P = \{x: x \in Q \text{ and } x \notin P\} \tag{9.11}$$

In solid modeling, Q and P in Eq. (9.11) are known as the *target* and the *tool* solids, respectively. Figures 9.9d and 9.9e show the difference operator. The difference can also be expressed as:

$$Q - P = P \cap cQ \tag{9.12}$$

The exclusive union (also known as symmetric difference) of two sets $P \cup Q$ (also written as $P \Delta Q$) is the subset of elements of W that are members of P or Q but not both, that is,

$$P \cup Q = \{x: x \notin P \cap Q\} \tag{9.13}$$

Figure 9.9f shows the exclusive union. Using a Venn diagram, it can be shown that $P \cup Q$ can also be expressed as $c(P \cap Q) \cap (P \cup Q)$, $(P \cap cQ) \cup (cP \cap Q)$, $(P - Q) \cup (Q - P)$, or $(P \cup Q) - (P \cap Q)$.

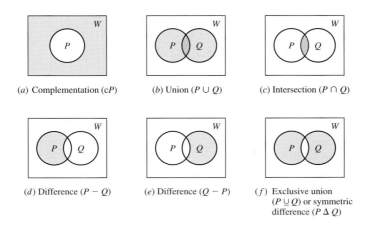

(a) Complementation (cP)   (b) Union ($P \cup Q$)   (c) Intersection ($P \cap Q$)

(d) Difference ($P - Q$)   (e) Difference ($Q - P$)   (f) Exclusive union ($P \cup Q$) or symmetric difference ($P \Delta Q$)

Figure 9.9   Nonuniqueness of solid models: the primitives approach.

The laws of set algebra are in some cases similar to the laws of ordinary algebra. Just as the latter can be used to simplify algebraic equations and expressions, the former can be used to simplify sets. The laws of set algebra are stated here without any mathematical proofs. Interested readers can prove most of them using a Venn diagram. These laws are:

the commutative law (similar to ordinary algebra $p + q = q + p$ and $pq = qp$):

$$P \cup Q = Q \cup P \tag{9.14}$$

$$P \cap Q = Q \cap P \tag{9.15}$$

the associative law [similar to ordinary algebra $p + (q + r) = (p + q) + r$ and $p(qr) = (pq)r$]:

$$P \cup (Q \cup R) = (P \cup Q) \cup R \tag{9.16}$$

$$P \cap (Q \cap R) = (P \cap Q) \cap R \tag{9.17}$$

the distributive law [similar to $p(q + r) = pq + pr$]:

$$P \cup (Q \cap R) = (P \cup Q) \cap (P \cup R) \tag{9.18}$$

$$P \cap (Q \cup R) = (P \cap Q) \cup (P \cap R) \tag{9.19}$$

the idemoptence law:

$$P \cap P = P \tag{9.20}$$

$$P \cup P = P \tag{9.21}$$

the involution law:

$$c(cP) = P \tag{9.22}$$

and

$$P \cup \varnothing = P \tag{9.23}$$

$$P \cap W = P \tag{9.24}$$

$$P \cup cP = W \tag{9.25}$$

$$P \cap cP = \varnothing \tag{9.26}$$

$$c(P \cup Q) = cP \cap cQ \tag{9.27}$$

$$c(P \cap Q) = cP \cup cQ \tag{9.28}$$

where Eqs. (7.26) and (7.27) are DeMorgan's Laws and Eqs. (9.14) to (9.28) provide the tools necessary to manipulate and simplify sets. For example, using Eqs. (9.14), (9.18), and (9.20) one can prove that the set $(P \cup Q) \cup (P \cap Q)$ is equal to the set $P \cup Q$. The Venn diagram can also be used as an informal method to reach the same conclusion. From a geometric modeling point of view, these equations, or the set theory in general, can operate on point sets that represent solids in E^3 or they can be used to classify other point sets in space against solids to determine which points in space are inside, on, or outside a given solid.

The concept of the cartesian product of two sets is useful to geometric modeling because it can be related to coordinates of points in space. The concept of an ordered pair must be introduced first. Let us assume that a and b are two elements. An ordered pair of a and b is denoted by (a, b); a is the first coordinate of the pair and b is the second coordinate. This guarantees that $(a, b) \neq (b, a)$. The **ordered pair** of a and b is a set and can be defined as:

$$(a, b) = \{\{a\}, \{a, b\}\} \tag{9.29}$$

Equation (9.29) implies that the first coordinate of the ordered pair is the first element $\{a\}$ and the second coordinate is the second element $\{a, b\}$; both elements form the set of the ordered pair (a, b). If $a = b$, then $(a, a) = \{\{a\}, \{a, a\}\} = \{\{a\},\{a\}\}=\{\{a\}\}$. Based on this definition, there is a theorem which states that two ordered pairs are equal if and only if their corresponding coordinates are equal, that is, $(a, b) = (c, d) \Leftrightarrow a = c$ and $b = d$.

The cartesian product is the concept that can be used to form ordered pairs. If A and B are two sets, the cartesian product of the sets, designated by $A \times B$, is the set containing all possible ordered pairs (a, b) such that $a \in A$ and $b \in B$, that is,

$$A \times B = \{(a, b): a \in A \text{ and } b \in B\} \tag{9.30}$$

If, for example, $A = \{1, 2, 3\}$ and $B = \{1, 4\}$, then $A \times B = \{(1, 1), (1, 4), (2, 1), (2, 4), (3, 1), (3, 4)\}$. Note that $A \times B \neq B \times A$. We denote $A \times A$ by A^2. The cartesian product of three sets can now be introduced as:

$$A \times B \times C = (A \times B) \times C = \{(a, b, c): a \in A, b \in B, c \in C\} \tag{9.31}$$

where (a, b, c) is an ordered triplet defined by $(a, b, c) = ((a, b), c)$. $A \times A \times A$ is usually denoted by A^3. Usually, an n-tuple can be defined as the cartesian product of n sets and takes the form $(a_1, a_2, ..., a_n)$. Ordered pairs and triples are considered 2-tuples and 3-tuples, respectively.

Equations (9.30) and (9.31) can be used to define points and their coordinates in the context of set theory. If we consider a set of points (set of real numbers) R in 1D Euclidean space E^1, then R^2 defines a set of points in E^2; each is defined by two numbers or an ordered pair. Similarly, R^3 defines a set of points in E^3; each is defined by three numbers or an ordered triple.

How does a CAD system implement the set theory? The system features, such as a boss or a hole, are the point sets, and the system operations, such as extrusion, are the Boolean (set) operators that manipulate the features to create the solid models of objects. All the software algorithms and code implement the set concepts covered here. Thus, system developers utilize all these concepts directly while system users need to understand them fully to utilize CAD systems effectively in design tasks .

EXAMPLE 9.1	**Manipulate two solids.**

Two solids S_1 and S_2 are defined by the following two point sets:
$S_1 = \{(1, 2), (3, 5), (6, 7), (2, -2)\}$ and $S_2 = \{(0, 0), (3, 5), (6, 7)\}$. Find $S_1 \cup S_2$, $S_1 \cap S_2$, $S_1 - S_2$, and $S_2 - S_1$.

SOLUTION	Equations (9.8) - (9.11) give $S_1 \cup S_2 = \{(1, 2), (3, 5), (6, 7), (2, -2), (0, 0)\}$,

$S_1 \cap S_2 = \{(3, 5), (6, 7)\}$, $S_1 - S_2 = \{(1, 2), (2, -2)\}$, and $S_2 - S_1 = \{(0, 0)\}$.

Example 9.1 discussion:

Each solid is defined by the set of its vertices (corner points). S_1 and S_2 have four and three corner points, respectively. Each vertex is defined by an ordered pair, (x, y), that are its coordinates. The two sets, as given, do not include the intersection points of the edges of S_1 and S_2.

Example 9.1 hands-on exercise:

Extend this example to 3D space. Add the following Z coordinates (one per point) to the points: 1, 2, 3, 4, 5, 6, and 7. Solve the example again.

EXAMPLE 9.2   **Find a solid via cartesian product.**
A point set S that defines a solid in E^3 is a set of ordered triples. Find the three sets whose cartesian product produces S.

SOLUTION   The point set can be written as

$$S = \{P_1, P_2, ..., P_n\} \tag{9.32}$$

where P_1, P_2, ..., P_n are points inside or on the solid. This set can also be written as

$$S = \{(x_1, y_1, z_1), (x_2, y_2, z_2), ..., (x_n, y_n, z_n)\} = \{(x_i, y_i, z_i): 1 \le i \le n\} \tag{9.33}$$

We can define three sets A, B, and C such that

$$A = \{x_1, x_2, ..., x_n\} \tag{9.34}$$

$$B = \{y_1, y_2, ..., y_n\} \tag{9.35}$$

$$C = \{z_1, z_2, ..., z_n\} \tag{9.36}$$

Let us define the set P as the cartesian product $A \times B \times C$, that is,

$$P = A \times B \times C = \{(x_i, y_j, z_k): 1 \le i \le n, 1 \le j \le n, i \le k \le n\} \tag{9.37}$$

The point set S of the solid given by Eq. (9.33) is clearly a (proper) subset of the set P, that is, $S \subset P$. The elements of S are equal to the elements of P only when $i = j = k$.

Let us introduce a new notion called *ordered* cartesian product. It is a more restricted special case of the cartesian product concept. It is applied only to sets that have the same number of elements. We denote it by "$\otimes$" to differentiate it from "$\times$" which is used for the unordered cartesian product. If we have two sets defined as $A = \{a_1, a_2, ..., a_n\}$ and $B = \{b_1, b_2, ..., b_n\}$, then

$$A \otimes B = \{(a_i, b_i): a_i \in A, b_i \in B, \text{ and } 1 \le i \le n\} \tag{9.38}$$

The ordered cartesian product of three sets is similarly given by:

$$A \otimes B \otimes C = (A \otimes B) \otimes C = \{(a_i, b_i, c_i): a_i \in A, b_i \in B, c_i \in C, \text{ and } 1 \le i \le n)\} \tag{9.39}$$

Comparing Eqs. (9.33) and (9.39) shows that the ordered cartesian product of the three sets A, B, and C given by Eqs. (9.34) to (9.36) gives the point set S of a solid. This observation that S can be related to A, B, and C might be useful in classification problems.

9.5.2 Regularized Set Operations

The set operations (c, $\cup$, $\cap$, and $-$) covered in Section 9.5.1 are known as the set-theoretic operations. When we use these operations in geometric modeling to build complex objects from primitive ones, the complement operation is usually dropped because it might create unacceptable geometry. Moreover, if we use the other operations without regularization in solid modeling, they may cause user inconvenience—for example a user may have overlapping faces in a solid. In addition, models resulting from these operations may lack geometric closure, may be difficult to validate, or may be inadequate for applications such as interference analysis.

To avoid these problems, the point sets that represent solids and the set operations that operate on them must be regularized. Regular sets and regularized set operations (Boolean operations) are considered Boolean algebra.

A **regular set** (r-set) is defined as a set which is geometrically closed as defined by Eq. (9.3). The notion of a regular set is introduced in geometric modeling to ensure the validity of the models they describe and therefore eliminate "nonsense" models. Under geometric closure, a regular set has interior and boundary subsets. More importantly, the boundary contains the interior, and any point on the boundary is in contact with a point in the interior. In other words, the boundary acts as a skin wrapped around the interior. The set S shown in Figure 9.7 is an example of a regular set, while Figure 9.8 shows a nonregular set because the dangling edge and face are not in contact with the interior of the set (in this case, the box).

Mathematically, a set S is regular if and only if:

$$S = \text{ki}S \tag{9.40}$$

Equation (9.40) states that if the closure of the interior of a given set yields that same given set, then the set is regular. Figure 9.10b shows that set S is not regular because $S' = \text{ki}S$ is not equal to S. Some modeling systems use regular sets that are open or do not have boundaries. A set S is regular open if and only if:

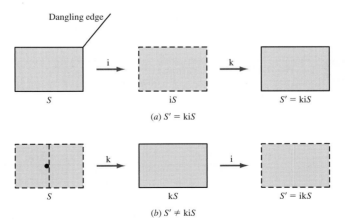

(a) $S' = \text{ki}S$

(b) $S' \neq \text{ki}S$

Figure 9.10 Set regularity.

$$S = \text{ik}S \tag{9.41}$$

Equation (9.41) states that a set is regular open if the interior of its closure is equal to the original set. Figure 9.10*b* shows that S is not regular open because $S' \neq S$.

Set (Boolean) operations must be regularized to ensure that their outcomes are always regular sets. For geometric modeling, this means that solid models built from well-defined primitives are always valid and represent valid (no-nonsense) objects. Regularized set operators preserve homogeneity and spatial dimensionality. The former means that no disconnected or dangling parts should result from using these operators, and the latter means that if two 3D models are combined by one of the operators, the resulting model should not be of lower dimensionality (2D or 1D). Regularization of set operators is particularly useful when users deal with overlapping faces of different models, that is, when dealing with tangent models.

Based on the preceding description, regularized set operators can be defined as follows:

$$P \cup^* Q = \text{ki}(P \cup Q) \tag{9.42}$$

$$P \cap^* Q = \text{ki}(P \cap Q) \tag{9.43}$$

$$P -^* Q = \text{ki}(P - Q) \tag{9.44}$$

$$\text{c}^* P = \text{ki}(\text{c}P) \tag{9.45}$$

where the superscript * to the right of each operator denotes regularization. The sets P and Q used in Eqs. (9.42) to (9.45) are assumed to be any arbitrary sets. However, if two sets X and Y are r-sets, which is always the case in geometric modeling, then Eqs. (9.42) to (9.45) become:

$$X \cup^* Y = X \cup Y \tag{9.46}$$

$$X \cap^* Y = X \cap Y \Leftrightarrow \text{b}X \text{ and b}Y \text{ do not overlap} \tag{9.47}$$

$$X -^* Y = \text{k}(X - Y) \tag{9.48}$$

$$\text{c}^* X = \text{k}(\text{c}X) \tag{9.49}$$

If bX and bY overlap in Eq. (9.47), Eq. (9.43) is used and the result is a null object. Figure 9.11 illustrates Eqs. (9.42) to (9.49) geometrically, excluding the complement operation.

EXAMPLE 9.3   **Use regularized set operators.**

Three 2D solids (two plates A and B and disk C) are shown below. Find and sketch $A \cup B \cup C, A \cap B \cap C, A - B - C, A \cup^* B \cup^* C, A \cap^* B \cap^* C$, and $A -^* B -^* C$.

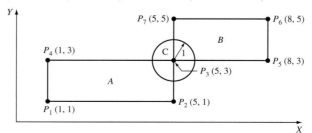

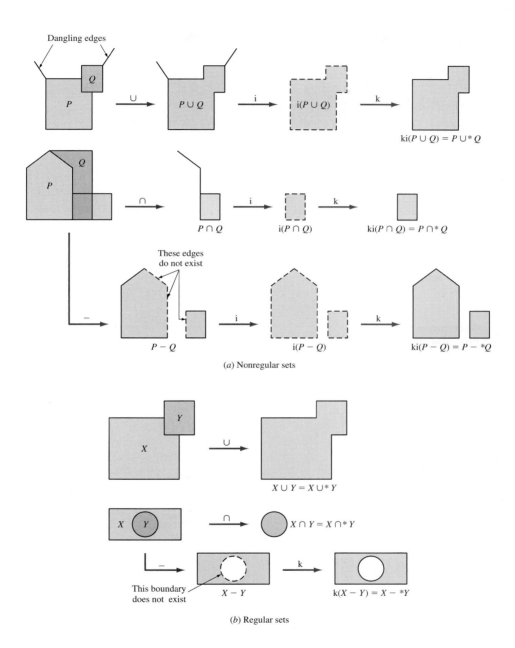

Figure 9.11 Regularized set operators.

SOLUTION   Using Eqs. (9.8) to (9.10) and Eqs. (9.46) to (9.48), we produce the follow-ing results and sketches.

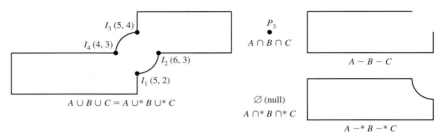

Example 9.3 discussion:

Each of the two solids A and B is defined by its four vertices (corner points). Solid C is defined by its center point, P_3, and radius. We calculate the four intersection points, I_1 to I_4, by inspection. We write some of the results in terms of sets as follows: $A \cup^* B \cup^* C = \{(1, 1),$ $(5, 1), (5, 2), (6, 3), (8, 3), (8, 5), (5, 5), (5, 4), (4, 3), (1, 3)\}$, $A \cap B \cap C = \{(5, 2)$, and $A -^* B -^* C = \{(1, 1), (5, 1), (5, 2), (6, 3), (4, 3), (1, 3)\}$.

Example 9.3 hands-on exercise:

Find $A \cap B, A \cap^* B, B \cap C, B \cap^* C, C -^* B -^* A$, and $A -^* C$.

EXAMPLE 9.4   **Verify if a solid is an r-set.**

For the two solids A and B shown on the right, find $A - B$, $A -^* B$, $i(A -^* B)$, and $k(A -^* B)$. Is $A -^* B$ regular? Why?

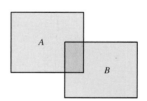

SOLUTION   Following the steps shown in Figure 9.11, we draw the following sketches:

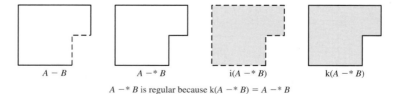

Example 9.4 discussion:

The dashed boundaries in the above sketches imply that the boundary does not exist. Solid boundaries do exist. Thus the difference between $A - B$ and $A -^* B$ is that the former is an open solid, that is, an invalid solid. The interior of a solid, iS, does not include its boundary, bS. The closure of a solid, kS, includes both the interior, iS, and the boundary, bS.

Example 9.4 hands-on exercise:

Redo the example for $A \cup^* B$ and $A \cap^* B$.

EXAMPLE 9.5   **Combine tangent solids.**

What are the results of applying the regularized set operations to solids A and B shown in Figure 9.12?

SOLUTION   The positions of solids A and B shown in Figure 9.12 are chosen to illustrate some tangency cases of objects. A and B are r-sets. The results applying Eqs. (9.38) to (9.40) are shown in Table 9.1 for each case. For all the cases, the results of the regularized union operations are obvious. However, the results of the intersection operations may be less obvious.

For case 1, $A \cap B$ is the common face which is eliminated by the regularization process. For case 2, the intersection does not exist; therefore the result is an empty set or a null solid. For case 3, $A \cap B$ is the common edge which is eliminated by the regularization process. For case 4, $A \cap B$ is the common block and the common face. The common face is eliminated after regularization. The results of the regularized difference operations are obvious. In cases 1, 2, and 3, $A -^* B$ is the solid A itself. For case 4, the difference is a disjoint solid. Such a solid should not be viewed as two separate solids. Any further set operation or rigid body motion treats it as one solid.

The reader is advised to carry out the details of these results following the steps illustrated in Figure 9.11. The reader should also try to use these cases to test any available CAD system.

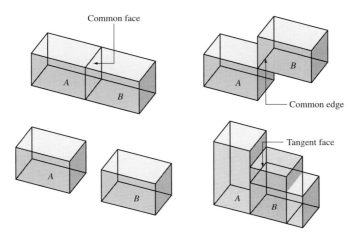

Figure 9.12   Tangent objects.

Table 9.1 Results of Example 9.5.

Case	Objects	Set operation	Results
1		$A \cup^* B$	
		$A \cap^* B$	$\varnothing$ (null object)
		$A -^* B$	
2		$A \cup^* B$	
		$A \cap^* B$	$\varnothing$ (null object)
		$A -^* B$	
3		$A \cup^* B$	
		$A \cap^* B$	$\varnothing$ (null object)
		$A -^* B$	
4		$A \cup^* B$	
		$A \cap^* B$	
		$A -^* B$	

9.5.3 Set Membership Classification

In various geometric problems involving solid models, we are often faced with the following question: Given a particular solid, which point, line segment, or portion of another solid intersects such a solid? These are all geometric intersection problems. For a point/solid, curve/solid, or solid/solid intersection, we need to know respectively which points, curve segments, or solid portions are inside, outside, or on the boundary of a given solid. These geometric intersection problems have useful practical engineering applications. For example, curve/solid intersection can be used to shade or calculate mass properties of given solids via ray tracing algorithms, while solid/solid intersection can be used for interference checking between two solids.

In each of the preceding problems, we are given two point sets: a reference set S and a candidate set X. The reference set is usually the given solid whose inside (interior) and boundary are iS and bS, respectively. The outside of S is its complement cS. The candidate set is the geometric entity that must be classified against S. **Set membership classification** is the process by which various parts of X (points, curve segments, or solid portions) are assigned to iS, bS, and/or cS.

The set membership classification function is introduced to provide a unifying approach to study the behavior of the candidate set X relative to the reference set S. The function is denoted by $M[.]$ and is defined as:

$$M[X, S] = (X \text{ in } S, X \text{ on } S, X \text{ out } S) \tag{9.50}$$

Equation (9.50) implies that the input to $M[.]$ is the two sets X and S, and the output is the classification of X relative to S as in, on, or out S. Figure 9.13 shows an example of classifying a line L against a polygon R.

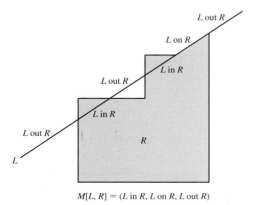

$$M[L, R] = (L \text{ in } R, L \text{ on } R, L \text{ out } R)$$

Figure 9.13   Line/polygon classification.

The implementation of the classification function given by equation (9.50) depends to a great extent on the representations of both X and S and their data structures. Let us consider the line/polygon classification problem when the polygon (reference solid) is stored as a B-rep (a B-rep solid is a set of faces, edges, and vertices) or a CSG (a CSG solid is a combination of primitives). Figure 9.14 shows the B-rep case. The line L is chosen such that no on segments result for simplicity. The algorithm for this case can be described as follows:

a. Utilizing a line/edge intersection routine, find the boundary crossing P_1 and P_2.
b. Sort the boundary crossings according to any agreed direction for L: Let the sorted boundary crossing list be given, by (P_0, P_1, P_2, P_3).
c. Classify L with respect to R. For this simple case, we know that the odd boundary crossings (such as P_1) flag in segments and the even boundary crossings (such as P_2) flag out segments. Thus, the classification of L with respect to R becomes:
$$[P_0, P_1] \subset L \text{ out } R$$
$$[P_1, P_2] \subset L \text{ in } R$$
$$[P_2, P_3] \subset L \text{ out } R$$

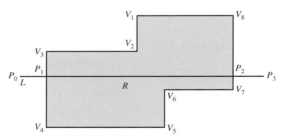

Figure 9.14   Line/polygon classification for B-rep.

If the line L contains an edge of the polygon, the preceding classification criterion of odd and even crossings would not work; another criterion should be found. In this case, a direction (clockwise or counterclockwise) to traverse the polygon boundaries is needed. Let us apply this idea to the problem at our hand to see how it would work. If we choose counterclockwise direction, polygon vertices would be numbered as shown in Figure 9.14. Now we know that iR is always to the left of any edge.

The new classification criterion can be stated as follows: Let us assume that an edge is defined by the two vertices V_i and V_{i+1}. Whenever there is a boundary crossing on an edge whose V_i is above L and V_{i+1} is below L, this crossing is flagged as in, and whenever V_i is below L and V_{i+1} is above, it is flagged out. This criterion obviously gives the same result as the previous criterion for this example.

Let us consider the same line/polygon classification problem when the polygon is stored as a CSG representation. The classification for this case is done at the primitive level and the algorithm becomes as follows:

a. Utilize a line/primitive intersection routine to find the intersection points of the line with each primitive of R.

b. Use these intersection points to classify the line against each primitive of R.

c. Combine the `in` and `on` line segments obtained in step b using the same Boolean operators that combine the primitives. For example, if two primitives A and B are unioned, then the `in` and `on` line segments are combined.

d. Find the `out` segments by taking the difference between the line (candidate set) and the `in` and `on` segments. Figure 9.15 shows the `classify` and `combine` strategy for the three Boolean operations of two blocks A and B. Notice that the polygon that results from the union operation is the same as the polygon R used in the classification of the B-rep case.

The classification of L relative to A and B is straightforward. To combine these classifications, we first combine L in A and L in B to obtain L in R using the proper Boolean operator. The L on R can result from combining three possibilities: L in A and L on B, L on A and L in B, and L on A and L on B. All these possibilities are obtained and then combined to give L on R. The remaining classification L out R is obtained by adding L in R and L on R and subtract the result from L itself.

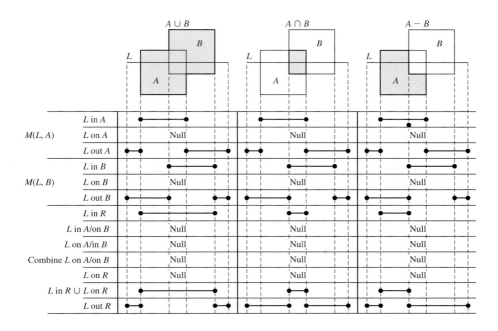

Figure 9.15   Line/polygon classification for CSG rep.

The preceding example considers the polygon case. The example purposely does not include on segments because they are ambiguous and need more information (neighborhoods) to resolve their ambiguities for both B-rep and CSG. Algorithms to classify candidate sets against 3D solids can follow similar steps to those described in the preceding example.

9.6 Half-Spaces

This section and the next three cover the common four representation schemes of solid modeling (half-space, B-rep, CSG, and sweep). Each scheme views a solid model differently. Thus, each scheme has a set of elements that act as building blocks and a set of operators that go along with them to combine (assemble) the elements to build (create) the solid model.

Half-spaces form a basic representation scheme for bounded solids. By combining half-spaces (using set operations) in a building block fashion, various solids can be constructed. Half-spaces are usually unbounded geometric entities; each one of them divides the representation space into two infinite halves; one is "filled with material" and the other is "empty." Surfaces are viewed as half-space boundaries, and half-spaces are viewed as directed surfaces.

A half-space is defined as a regular point set in E^3 as follows:

$$H = \{P : P \in E^3 \text{ and } f(P) < 0\} \tag{9.51}$$

where P is a point in E^3 and $f(P) = 0$ defines the surface equation of the half-space boundaries. Half-spaces can be combined using set operations to create complex objects.

9.6.1 Basic Elements

Various half-spaces can be described and created using Eq. (9.51). However, to make them useful for design and manufacturing applications, supporting algorithms and utility routines must be provided. For example, if one would add a cylindrical half-space to a CAD system, intersecting routines that enable this half-space to intersect itself as well as other existing half-spaces must be developed and added as well.

The most widely used half-spaces (unbounded) are planar, cylindrical, spherical, conical, and toroidal. They form the natural quadrics discussed earlier in Section 9.3 (with the exception of the torus that can be formed from the other half-spaces). The regular point set of each half-space is a set of ordered triplets (x, y, z) given by:

Planar half-space: $\qquad H = \{(x, y, z): z < 0\}$ $\qquad\qquad$ (9.52)

Cylindrical half-space: $\qquad H = \{(x, y, z): x^2 + y^2 < R^2\}$ $\qquad\qquad$ (9.53)

Spherical half-space: $\qquad H = \{(x, y, z): x^2 + y^2 + z^2 < R^2\}$ $\qquad\qquad$ (9.54)

Conical half-space: $\qquad H = \{(x, y, z): x^2 + y^2 < [(\tan\ \alpha/2)z]^2\}$ $\qquad\qquad$ (9.55)

Torodial half-space: $\qquad H = \{(x, y, z): (x^2 + y^2 + z^2 - R_2^2 - R_1^2)^2$

$$< 4R_2^2\ (R_1^2 - z^2)^2\} \tag{9.56}$$

Equations (9.52) to (9.56) are implicit equations and are expressed in terms of each half-space local coordinate system whose axes are X_H, Y_H, and Z_H. The implicit form is efficient for finding surface intersections. The corresponding surface of each half-space is given by its equation when the right and left sides are equal. For the planar half-space, Eq. (9.52) is based on the vertical plane $z = 0$. Other definitions can be written easily. Figure 9.16 shows the various half-spaces with their local coordinate systems and the limits on their configuration parameters.

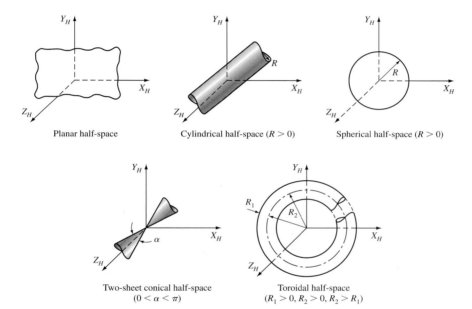

Planar half-space   Cylindrical half-space ($R > 0$)   Spherical half-space ($R > 0$)

Two-sheet conical half-space
$(0 < \alpha < \pi)$

Toroidal half-space
$(R_1 > 0, R_2 > 0, R_2 > R_1)$

Figure 9.16   Half-spaces.

9.6.2   Building Operations

Complex objects can be modeled as half-spaces combined by the set operations. As a matter of fact, half-spaces are treated as lower-level primitives. Regularized set operations are used to combine half-spaces to create solid models — for example $H_1 \cap H_2 \cap H_3$ combines the three half-spaces to create a solid. Most often, half-spaces may have to undergo rigid motion via homogenous transformations to be positioned properly before intersection.

Let us represent the solid S shown in Figure 9.17a using half-spaces. Parameters of the solid are shown. The hole is centered in the top face. The MCS of the solid model is chosen as shown in the figure. Figure 9.17b shows that nine half-spaces (eight planes and one cylinder) H_1 to H_9 are needed to represent S. Half-spaces H_7 and H_8 that model the front and back faces of the model are not shown in the figure. Utilizing the local coordinate systems shown in Figure 9.16, some half-spaces have to be positioned first. For example, rotate H given by

Eq. (9.52) an angle $-90°$ degrees about the X axis and translate it up in the Y direction a distance b to obtain H_1. In a similar fashion, the other half-spaces can be positioned using the proper rigid motion. Only H_7 needs no positioning. The positioned half-spaces can be intersected and then Boolean operations are used to combine them. H_1 to H_8 are unioned and H_9 is subtracted from the result.

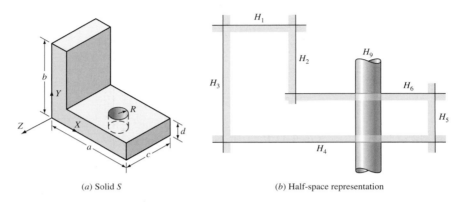

(a) Solid S (b) Half-space representation

Figure 9.17   Half-space representation of solid S.

EXAMPLE 9.6   **Create a fillet using half-spaces.**
How can we create a fillet using half-spaces?

SOLUTION   A surface fillet has been defined in Chapter 7 as a B-spline surface blending two given surfaces. Similarly, a solid fillet can be used to blend sharp edges of a solid as shown in Figure 9.18a. The solid fillet is defined by its radius r and length d. Six half-spaces H_1 to H_6 (see Figure 9.18b) are needed to construct the fillet. H_1, H_2, H_3, and H_4 represent the front, left, back, and bottom faces, respectively. H_6 is the cylindrical face. H_5 is an auxiliary half-space positioned at distance r from the origin of the $X_L Y_L Z_L$ local coordinate system of the fillet and oriented at $45°$ as shown in view A-A in Figure 9.18b. H_5 is used to intersect H_2, H_4, and H_6 so that the boundaries of the fillet can be evaluated. This is because the cylindrical half-space is tangent to H_2 and H_4.

Except for H_1, the other half-spaces must be positioned before intersection, and other set operations are performed to create the fillet. Let us look at positioning H_5 and H_6 as an example. Using Eq. (9.52), H has to rotate an angle of $90°$ about the Y axis, followed by a $45°$ rotation about the Z axis, and finally translated a distance r in the positive X direction. H_6 is obtained by translating the cylindrical half-space H given by Eq. (9.53) an equal distance r in both the positive X and Y directions. At this position, the complement of the cylindrical half-space, cH, is taken to obtain H_6.

Theoretically, cH is equal to E^3 minus the cylindrical half-space. For practical and implementation purposes, E^3 can be limited to a bounded volume, such as a box, enclosing the cylindrical half-space. Or the complement process can be replaced by choosing a surface normal to be positive in one side of the half-space and negative on the other side.

The intersections of H_1 to H_6 with each other can now be performed, and the results can be unioned to obtain the solid fillet. Notice that the solid fillet could have been created without the complement operation by subtracting the cylindrical half-space itself, after its positioning, from the intersection results of H_1 to H_5. However, the complement of a half-space is generally used to minimize the number of half-spaces used in modeling objects.

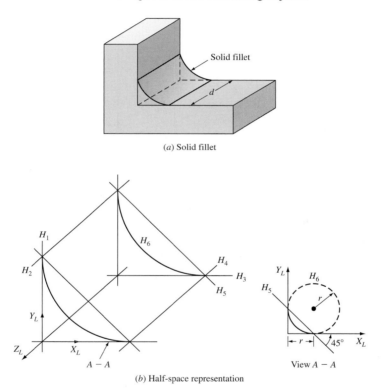

(a) Solid fillet

(b) Half-space representation

Figure 9.18   **Half-space representation of a fillet.**

It should be noted from this example that by using half-spaces and/or their complements or directed surface normals, any complex object can be modeled as the union of the intersection of half-spaces, that is,

$$S = \cup \left(\bigcap_{i=1}^{n} H_i \right)$$

(9.57)

where S is the solid and n is the number of half-spaces and/or their complements. As an example, a box is the union of six intersected half-spaces.

9.6.3   Remarks

The half-space representation scheme is the lowest level available to represent a complex object as a solid model. The main advantage of half-spaces is its conciseness in representing objects compared to other schemes such as CSG. However, it has few disadvantages. This representation can lead to unbounded solid models if the user is not careful. Such unboundedness can result in missing faces and abnormal shaded images. It can also lead to system crashes or wrong results if application algorithms attempt to access databases of unbounded models. Another major disadvantage is that modeling with half-spaces is cumbersome for casual users and designers to use and may be difficult to understand. Therefore, half-space representation is probably useful only for research purposes. CAD/CAM systems that use half-spaces shield their users from dealing directly with them.

9.7 Boundary Representation (B-rep)

Boundary representation is one of the two most popular and widely used schemes (the other is CSG) to create solid models of physical objects. A B-rep model or boundary model is based on the topological notion that an object is bounded by a set of faces. These faces are regions or subsets of closed and orientable surfaces. A **closed surface** is one that is continuous without breaks. An **orientable surface** is one where it is possible to distinguish two sides by using the direction of the surface normal to point inside or outside the solid model. Each face is bounded by edges, and each edge is bounded by vertices.

A **B-rep (boundary) model** of an object consists of faces, edges, vertices, loops, and handles. A **face** is a closed, orientable, and bounded (by edges) surface. An **edge** is a bounded (by two vertices) curve. A **vertex** is a point in E^3. A **loop** is a hole in a face. A **handle** is a through hole in a solid (body). Think of a loop as a 2D hole and of a handle as a 3D through hole. Figure 9.19 shows the B-rep model of a box with a hole.

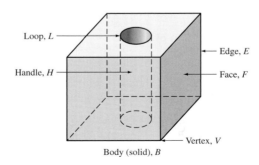

Figure 9.19   B-rep model of a solid S.

The database of a boundary model contains both its topology and its geometry. Topology is created by performing Euler operations, and geometry is created by performing Euclidean calculations. Euler operations are used to create, manipulate, and edit the faces, edges, and vertices of a boundary model. Euler operators, as Boolean operators, ensure the validity (closeness, no dangling faces or edges, and so forth) of boundary models. Geometry includes coordinates of vertices and rigid motion and transformation.

Topology and geometry are interrelated and cannot be separated entirely from each other. Figure 9.20 shows a square which, after dividing its top edges by introducing a new vertex, is still valid topologically but produces a nonsense object depending on the geometry of the new vertex.

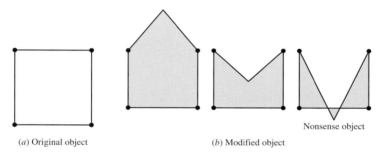

(*a*) Original object   (*b*) Modified object

Figure 9.20   Effect of topology and geometry on boundary models.

9.7.1   Basic Elements

Objects that are often encountered in engineering applications can be classified as either polyhedral or curved objects. A polyhedral object (plane-faced polyhedron) consists of planar faces (or sides) connected at straight (linear) edges which, in turn, are connected at vertices. A cube or a tetrahedron is an obvious example.

A curved object (curved polyhedron) is similar to a polyhedral object but with curved faces and edges instead. The identification of faces, edges, and vertices for curved closed objects such as a sphere or a cylinder needs careful attention as will be seen later in this section. Polyhedral objects are simpler to deal with and are covered first.

The basic elements of a B-rep are faces, edges, vertices, loops, handles, and bodies (solids themselves). B-rep models can be simple or complex, as Figure 9.21 shows. Polyhedral objects can be classified into four classes. The first class, Figure 9.21*a*, is the simple polyhedra. These do not have holes, and each face is bounded by a single set of connected edges.

The second class, Figure 9.21*b*, is similar to the first with the exception that a face may have loops. The third class, Figure 9.21*c*, includes objects with holes that do not go through the entire object. For this class, a hole may have a face coincident with the object boundary; in this case we call it a boundary hole. Or if it is an interior hole (as a void or crack inside the object), it has no faces on the boundary.

The fourth and the last class, Figure 9.21*d*, includes objects that have holes that go through the entire object. Topologically, these through holes are called handles. The topological name for the number of handles in an object is genus.

There exist topological constraints on the B-rep elements to ensure model validity. A vertex is always a unique point (an ordered triplet) in space. An edge is a nonself-intersecting, directed space curve bounded by two vertices that are not necessarily distinct (as in the case of a closed edge). A face is a finite connected, nonself-intersecting region of a closed oriented surface bounded by edges. A loop defines a nonself-intersecting, piecewise, closed space curve. The body (sometimes called a shell) is the solid itself. A minimum body is a point.

The object on the right of Figure 9.21*c* has two bodies (the exterior and interior cubes), and any other object in Figure 9.21 has only one body. A solid that has more than one body is known as a **disjoint solid**. Some CAD/CAM systems may not not allow the creation of disjoint solids. An example is cutting a block with another block to produce two disconnected blocks.

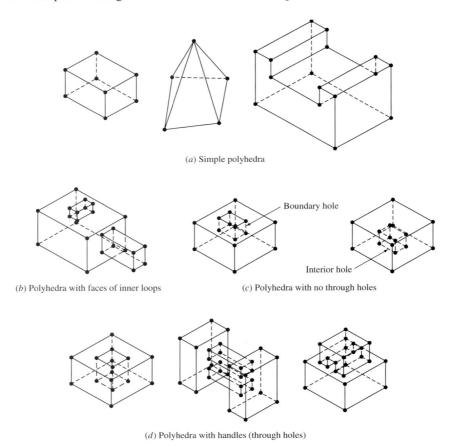

(*a*) Simple polyhedra

(*b*) Polyhedra with faces of inner loops

Boundary hole

Interior hole

(*c*) Polyhedra with no through holes

(*d*) Polyhedra with handles (through holes)

Figure 9.21   Types of polyhedral objects.

Faces of boundary models possess certain essential properties and characteristics that ensure the regularity of the model, that is, the model has an interior and a boundary. The face of a solid is a subset of the solid boundary, and the union of all faces of a solid defines such boundary. Faces are 2D homogeneous regions, so they have areas and no dangling edges. In addition, a face is a subset of some underlying closed oriented surface.

Figure 9.22 shows the relationship between a face and its surface. At each point on the face, there is a surface normal **N** that has a sign associated with it to indicate if it points into or away from the solid interior. One convention is to assume that **N** is positive if it points away from the solid. It is desirable, but not required, that a face has a constant surface normal.

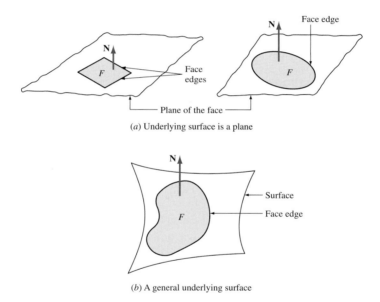

(a) Underlying surface is a plane

(b) A general underlying surface

Figure 9.22   Definition of a face.

The representation of a face must ensure that both the face and the solid interiors can be deduced from the representation. The direction of the face's surface normal is used to indicate the inside or outside of the model. The face interior can be determined by traversing the face loops in a certain direction or assigning flags to them. In traversing loops, the boundary edges of the face (known as the face outer loop) are traversed, for example, in a counterclockwise direction, and the edges of the inner loops are traversed in the opposite direction, clockwise. Figure 9.23 shows some traversal examples.

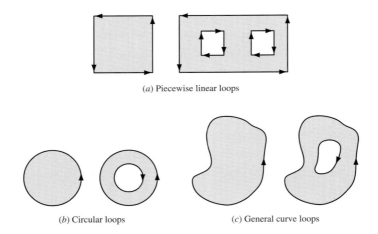

(a) Piecewise linear loops

(b) Circular loops   (c) General curve loops

Figure 9.23   Traversal of face's loops.

9.7.2   Euler Equations

Having defined the boundary model elements, how can we validate B-rep models? Euler (in 1752) proved that polyhedra that are homomorphic to a sphere, that is, their faces are nonself-intersecting and belong to closed orientable surfaces, are topologically valid if they satisfy the following equation:

$$F - E + V - L = 2(B - G) \tag{9.58}$$

where F, E, V, L, B, and G are the number of faces, edges, vertices, faces' inner loops, bodies, and genus. Eq. (9.58) is known as the Euler or Euler-Poincare Law. The simplest version of this equation is $F - E + V = 2$, which applies to polyhedra shown in Figure 9.21a. Table 9.2 shows the counts of the various variables of Eq. (9.58) for polyhedra shown in Figure 9.21. The numbering of these polyhedra in the table is taken from left to right, top to bottom with the top left cube being polyhedron number 1 and the bottom right object being number 9.

Euler's law given by Eq. (9.58) applies to closed polyhedral objects only. These are the valid 3D solid models that we deal with. However, open polyhedral objects do not satisfy Eq. (9.58). This class of objects include open polyhedra that may result from the construction of boundary models of closed objects as well as all 2D polygonal objects. Open objects satisfy the following Euler's law:

$$F - E + V - L = B - G \tag{9.59}$$

Figure 9.24 shows some examples of open objects. The reader can easily verify that they satisfy Eq. (9.59). In the equation, B refers to an open body which can be a wire, an area, or a volume. All the objects in Figure 9.24 have one body, and only the two bodies of Figure 9.24c have one genus each.

Table 9.2   Verifying the validity of the solids shown in Figure 9.21.

Apply Euler Eq. (9.58) to validate the values listed in this table						
Solid number	F	E	V	L	B	G
1	6	12	8	0	1	0
2		8	5	0	1	0
3	10	24	16	0	1	0
4	16	36	24	2	1	0
5	11	24	16	1	1	0
6	12	24	16	0	2	0
7	10	24	16	2	1	1
8	20	48	32	4	1	1
9	14	36	24	2	1	1

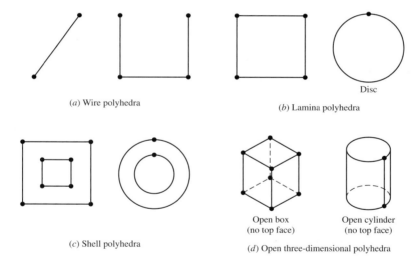

(a) Wire polyhedra

(b) Lamina polyhedra

Disc

(c) Shell polyhedra

Open box
(no top face)

Open cylinder
(no top face)

(d) Open three-dimensional polyhedra

Figure 9.24   Open polyhedral objects.

EXAMPLE 9.7   **Check the validity of 3D solids.**
Verify the Euler equation for the two forthcoming solids.

SOLUTION   Solid S_1 is a dome consisting of a cylinder and half a sphere on top of it. For this model, Euler variables are $F = 3$, $E = 3$, $V = 2$, $L = 0$, $B = 1$, and $G = 0$. Substituting these values into Eq. (9.58) satisfies the equation, which means that our numbers are correct.

Solid S_2 is a block with a blind hole and a boss. For this model, Euler variables are $F = 10$, $E = 18$, $V = 12$, $L = 2$, $B = 1$, and $G = 0$. Substituting these values into Eq. (9.58) satisfies the equation.

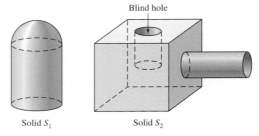

Blind hole

Solid S_1 Solid S_2

Example 9.7 discussion:

When calculating the number of faces, make sure *not* to include the internal faces that connect the various parts of the solid together. For S_1, the half sphere gives $F = 1$, $E = 0$, $V = 0$, and the cylinder gives $F = 2$, $E = 3$, $V = 2$. For S_2, the block gives $F = 6$, $E = 12$, $V = 8$, $L = 2$, the boss gives $F = 2$, $E = 3$, $V = 2$, and the blind hole gives $F = 2$, $E = 3$, $V = 2$.

Example 9.7 hands-on exercise:

Redo the example after adding a horizontal through hole in both S_1 and S_2.

EXAMPLE 9.8   **Check the validity of 2D solids.**
Verify the Euler equation for the 2D solid shown below.

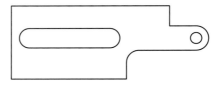

SOLUTION   This solid has two holes inside it. For this model, Euler variables are $F = 1$, $E = 14$, $V = 14$, $L = 2$, $B = 1$, and $G = 2$. Substituting these values into Eq. (9.59) satisfies the equation, which means that our numbers are correct.

Example 9.8 discussion:

Follow the following rule to obtain the right count of Euler variables. Imagine that you are sketching the model. Each time you start sketching a new curve, place a vertex there. Once the vertices are placed, the other counts are easy to calculate.

Example 9.8 hands-on exercise:

Redo the example after adding a circular hole in the model.

9.7.3   Curved and Faceted B-rep Models

We now turn from polyhedral objects to curved objects such as cylinders and spheres. The same rules and guidelines for boundary modeling discussed thus far apply. The major difference between the two types of objects results if closed curved edges or faces exist.

Consider, for example, the closed cylinder and sphere shown in Figure 9.25. A closed cylindrical face has one edge and two vertices and a spherical face has one vertex and no edges. The boundary model of a closed cylinder has three faces (top, bottom, and cylindrical face itself), two vertices, and one edge connecting the two vertices. The other "edges" are for visualization purposes. They are called limbs, virtual edges, or silhouette edges. The boundary model of a sphere consists of one face, one vertex, and no edges.

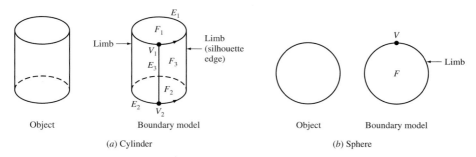

Figure 9.25   B-rep of a cylinder and a sphere.

If the curved objects are represented by storing the equations of curves and surfaces of edges and faces, respectively, the resulting boundary scheme is known as an exact B-rep scheme. Another alternative is the approximate or faceted B-rep (sometimes called tessellation rep). In this scheme, any curved face is divided into planar facets.

Figure 9.26 shows a faceted B-rep of a cylinder and sphere. The faceted cylinder is generated by rotating a line incrementally about the cylinder axis the desired total number of facets. A faceted sphere is formed in a similar way by rotating m connected-line segments about the sphere axis for a total of n sides. Faceted objects, although continuous, are no longer as smooth as the original curved objects they represent. Increasing the number of facets results in more accurate representation.

A data structure for a boundary model stores both topology and geometry. The structure shown in Figure 9.27 is based on Eq. (9.58). A relational database model is very effective for implementing such a data structure. Lists for bodies, faces, loops, edges, and vertices are generated and stored in tables. Each line in Figure 9.27 represents a pointer in the database.

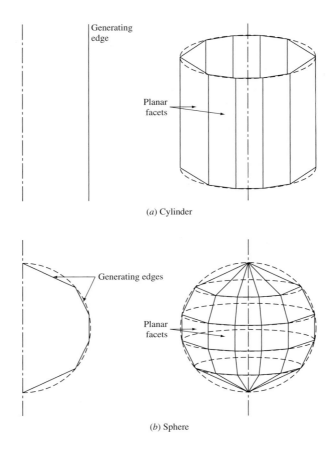

Figure 9.26   Faceted B-rep of a cylinder and a sphere.

9.7.4   Building Operations

Equation (9.58) forms the basis upon which to develop building operations to create B-rep models of complex objects. Euler operators, which are based on this equation, are not used in practice. CAD systems use B-rep only as an internal representation scheme for solid models. CAD designers create models using either the primitive or the features approach. The CAD software converts users' input to B-rep format before storing it in the model database.

9.7.5   Remarks

The B-rep scheme is popular and has a strong history in computer graphics because it is closely related to traditional drafting. The boundary model requires a large amount of storage because it stores the explicit definition of the model boundaries. The model is defined by its faces, edges, vertices, loops, and genus, which tend to grow fairly fast for complex models. Faceted B-rep is not suitable for many applications such as tool path generation.

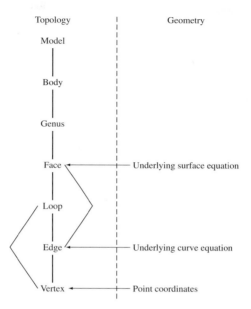

Figure 9.27   Data structure for B-rep models.

9.8 Constructive Solid Geometry (CSG)

A CSG model is based on the topological notion that a physical object can be divided into a set of primitives (basic elements or shapes) that can be combined in a certain order following a set of rules (Boolean operations) to form the object. Primitives themselves are considered valid CSG models. Each primitive is bounded by a set of faces. Faces are bounded by edges, and edges are bounded by vertices.

The database of a CSG model, similar to B-rep, stores its topology and geometry. Topology is created via the r-set (Boolean) operations that combine primitives. Therefore, the validity of the resulting model is reduced to the validity checks of the used primitives. These checks are usually simple and are in the form of *greater than zero*. For example, if the dimensions of a block are greater than zero and the radius and length of a cylinder are also greater than zero, combining them produces a valid solid. The geometry stored in the database of a CSG model includes the geometry of its primitives and rigid motion and transformation.

9.8.1   CSG Trees

Data structures of the CSG representation are based on the concept of graphs and trees. That means that the data of the solid model is stored in its database in a tree; we call it the CSG tree. A **graph** is defined as a set of nodes connected by a set of branches (lines). A graph may have closed loops in it; that is, some branches that connect nodes form closed regions.

A **CSG tree** is defined as an inverted, ordered binary tree whose leaf nodes are primitives (or other solids) and interior nodes are regularized set operations. *Inverted* means that the tree is upside down; its root is on the top. *Ordered* means that each tree node has a left branch and a right branch. *Binary* means that each tree node has only two branches going into it. Unlike graphs, trees do not have closed loops. We focus on trees because they are widely used in CAD systems.

Figure 9.28 shows a generic CSG tree with eight primitives (P_1 to P_8) and seven Boolean operators (OP_1 to OP_7). The CSG tree is shown with its full details, including arrows. In practice, the arrows are usually not shown, the leaf nodes are just shown as primitives" names without circles surrounding them, and a line extends from the tree root up to indicate the result of the final solid. Other styles of showing a CSG tree may replace primitive names by their sketches as well as showing each intermediate subsolid that results from an operator in the stream of the tree branches.

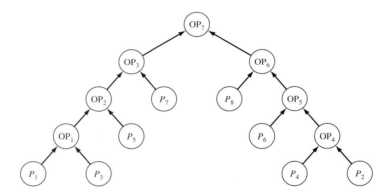

Figure 9.28 CSG tree of a solid.

The total number of nodes in a CSG tree of a given solid is directly related to the number of primitives the solid is decomposed to. The number of primitives decides automatically the number of Boolean operations required to construct the solid. If a solid has n primitives, then there are $(n - 1)$ Boolean operations for a total of $(2n - 1)$ nodes in its CSG tree. The balanced distribution of these nodes in the tree is a desired characteristic. A **balanced tree** is defined as a tree whose left and right subtrees have almost an equal number of nodes, that is, the absolute value of the difference $(n_L - n_R)$ is as minimal as possible where

$$n_L + n_R = 2n - 2 \tag{9.60}$$

The root node is not included in Eq. (9.60). n_L and n_R are the number of nodes of the left and right subtrees, respectively.

A perfect tree is one whose $n_L - n_R$ is equal to zero. A perfect tree results only if the number of primitives is even. The following equation applies to a perfect tree:

$$n_L = n_R = n - 1 \tag{9.61}$$

Each subtree has $n/2$ leaf nodes (primitives) and $(n - 2)/2$ interior nodes (Boolean operations). Figure 9.28 shows a perfect tree.

The creation of a balanced, unbalanced, or perfect CSG tree depends solely on the users and how they decompose a solid into its primitives. The general rule to create balanced trees is to start to build the model from an almost central position and branch out in two opposite directions or vice versa; that is, start from two opposite positions and meet in a central one.

Application algorithms must traverse a CSG tree, that is, they pass through the tree and visit each of its nodes. The order in which the nodes are visited in a traversal is clearly from the first node to the last one. However, there is no such natural linear order for the nodes of a tree. Thus, different orderings are possible for different cases.

There exist three main traversal methods: preorder, in order, and postorder. Sometimes, these methods are referred to as prefix, infix, and postfix traversals. The methods are all defined recursively so that traversing a binary tree involves visiting the root and traversing its left and right subtrees. The only difference among the methods is the order in which these three operations are performed.

The postfix traversal is the most efficient to use. To traverse a CSG tree in postorder:

1. Traverse the left subtree in postorder
2. Traverse the right subtree in postorder, and
3. Visit the root

Many CAD systems create and display a CSG tree (also known as manager tree or tree manager) for each model that a CAD designer creates. The tree saves the history of the model; that is, the steps of its creation. Some systems, such as SolidWorks and Pro/E, display the tree when the designer opens a part file. Other systems, such as CATIA, require the user to open the tree window via a command.

In either case, the designer can use the tree in various ways:

a. **Trace the creation steps.** The designer can walk up and down the tree by clicking any of its leaves or subsolids. Each time the designer clicks a leaf or a subsolid, the CAD system highlights the corresponding geometry in its display window.
b. **Edit a feature or a primitive.** When the designer selects a feature from the tree followed by an editing operation, the CAD system activates the feature sketch plane and parameters, allowing the designer to change/modify any values and re-create the feature.
c. **Prune the tree.** The designer can move an entire tree branch to a new location up or down the tree, or can delete the entire branch. However, the designer should be

careful because moving or deleting branches may result in creating invalid solids which cause the CAD system to crash. Pruning the tree by reordering its nodes is sometimes useful because changing the order of solid operations can overcome the failures. This is the case when the solid model is intricate and has very large and very small dimensions.

For example, filleting an edge between two complex faces may fail if we do it in one order, but it may work in another order. In such a case, we simply swap the leaves of the two faces, and regenerate the solid, hoping that the filleting process succeeds.

Tutorial 9.11.1 illustrates how to use CSG trees that CAD systems provide.

EXAMPLE 9.9   **Sketch a CSG tree.**
Sketch the CSG tree for each of the two solids shown below.

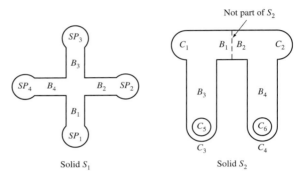

Solid S_1 Solid S_2

SOLUTION   The two CSG trees of S_1 and S_2 are shown below.

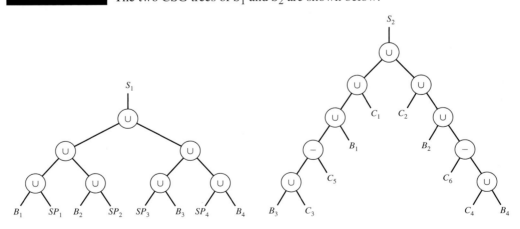

Example 9.9 discussion:

This example illustrates how we can create perfectly balanced CSG trees. If the solids are symmetric, as they are in this example, we can create a perfect CSG tree if the model is divided symmetrically, as we have done here. The solid S_1 consists of four blocks (B_1 to B_4), that make up a cross, and four spheres (SP_1 to SP_4) at the four ends of the cross. We create the four blocks and the four spheres in the proper positions and orientations. We union them in the order shown by the tree on the left.

Similarly, the solid S_2 is a U shape with cylinders at the ends and two holes at the bottom. We choose to divide S_2 into four blocks (B_1 to B_4) and six cylinders (C_1 to C_6). The dashed line shown inside S_2 in the example figure is not part of S_2. It is a hypothetical line that we use to divide the top part of S_2 into two blocks, B_1 and B_2. We do it this way to ensure that all the primitives follow the symmetry of S_2 which, in turn, ensures the symmetry of the S_2 CSG tree. We create the four blocks and the six spheres in the proper positions and orientations. We apply Boolean operations to them in the order shown by the tree on the right.

What is the value of balanced CSG trees? And is it practical to think that hard before embarking on the creation of a solid model of an object? The answer to the first question lies in the belief that applications that use solid models, such as volume calculations, traverse balanced CSG trees faster than nonbalanced trees. Regarding the second question, it is not practical, but keeping this fact in mind is helpful.

Can we apply all of the preceding CSG discussion to the features approach of creating solids? Yes—simply replace each primitive in Figure 9.28 by a feature. Use the notation F_n for the names of the tree leaves, instead of P_n.

Example 9.9 hands-on exercise:

For solid S_1, replace B_1 and B_3, and B_2 and B_4 by two blocks B_1 and B_2, respectively. Sketch the tree. Is it balanced? Repeat for S_2 by replacing B_1 and B_2 by one block B_1.

9.8.2 Basic Elements

Primitives are the basic elements or building blocks that a CSG scheme uses to build a solid model. Primitives can be viewed as parametric solids which are defined by two sets of geometric data. The first set is called configuration parameters, and it defines the primitive size. The second is the rigid motion parameters, and it defines the primitive orientation. The most common primitives are shown in Figure 9.5. For example, the configuration parameters of a block primitive are its W, H, and D, and its rigid motion is given by the location of its origin P relative to a coordinate system (WCS) or by explicit rigid motion values (translation and/or rotation).

Each primitive, viewed as a parametric object, corresponds to a family of elements. Each element of the family is called a primitive instance and corresponds to one and only one value set of the primitive configuration parameters. Each primitive has a valid configuration domain which is maintained by its solid modeler. User-input values of any primitive parameters are usually checked against its valid domain. For example, a block primitive instance of the triplet $(0, 0, 0)$ is not a valid instance because its parameters are not within the valid domain of a block.

Mathematically, each primitive is defined as a regular point set of ordered triplets (x, y, z). For the primitives shown in Figure 9.5, these point sets are given by:

Block: $\{(x, y, z): 0 < x < W, 0 < y < H, \text{ and } 0 < z < D\}$ (9.62)

Cylinder: $\{(x, y, z): x^2 + y^2 < R^2, \text{ and } 0 < z < H\}$ (9.63)

Cone: $\{(x, y, z): x^2 + y^2 < [(R/H)z]2, \text{ and } 0 < z < H\}$ (9.64)

Sphere: $\{(x, y, z): x^2 + y^2 + z^2 < R^2\}$ (9.65)

Wedge: $\{(x, y, z): 0 < x < W, 0 < y < H, 0 < z < D, \text{ and } yW + xH < HW\}$ (9.66)

Torus: $\{(x, y, z): (x^2 + y^2 + z^2 - R_2^2 - R_1^2) < 4R_2^2 \ (R_1^2 - z^2)\}$ (9.67)

Equations (9.62) to (9.67) give the interior, iS, of each primitive. The boundary, bS, of each primitive is derived from the underlying surface of the half-space that describes the primitive. We must bound the surface because the primitive has finite faces. The equations of these surfaces (point sets) are given by:

Planar surface: $P = \{(x, y, z): z = 0\}$ (9.68)

Cylindrical surface: $P = \{(x, y, z): x^2 + y^2 + z^2 = R^2\}$ (9.69)

Spherical surface: $P = \{(x, y, z): x^2 + y^2 + z^2 = R^2\}$ (9.70)

Conical surface: $P = \{(x, y, z): x^2 + y^2 = [(R/H)z]^2\}$ (9.71)

Toroidal surface: $P = \{(x, y, z): (x^2 + y^2 + z^2 - R_2^2 - R_1^2)^2 = 4R_2^2 \ (R_1^2 - z^2)\}$ (9.72)

Equation (9.68) describes a plane that is the XY plane. Replacing $z = 0$ by $x = 0$ or $y = 0$ gives the YZ or XZ plane respectively.

Equations (9.68) to (9.72) are infinite surfaces. Primitive faces (Pfaces) are subsets of these surfaces; that is, they are finite bounded regions. The boundary of any primitive may be represented as the union of a finite number of these Pfaces after being positioned properly in space. The set of primitive faces that is sufficient to represent the boundary of any of the primitives shown in Figure 9.5 consists of plate, tri-plate, disk, cylindrical, spherical, conical, and toroidal Pfaces. The equations of these Pfaces are given by:

Plate Pface: $F = \{(x, y, z): 0 < x < W, 0 < y < H, \text{ and } z = 0\}$ (9.73)

Tri-plate Pface: $F = \{(x, y, z): 0 < x < W, 0 < y < H, \text{ and } yW + xH < HW\}$ (9.74)

Disk Pface: $F = \{(x, y, z): x^2 + y^2 < R^2, \text{ and } z = 0\}$ (9.75)

Cylindrical Pface: $F = \{(x, y, z): x^2 + y^2 = R^2, \text{ and } 0 < z < H\}$ (9.76)

Spherical Pface: $F = \{(x, y, z): x^2 + y^2 + z^2 = R^2$ (9.77)

Conical Pface: $F = \{(x, y, z): x^2 + y^2 = [(R/H) z]^2, \text{ and } 0 < z < H\}$ (9.78)

Toroidal Pface: $F = \{(x, y, z): (x^2 + y^2 + z^2 - R_2^2 - R_1^2)^2 = 4R_2^2 \ (R_1^2 - z^2)\}$ (9.79)

Any of the preceding primitive faces is a subset of its underlying surface; $F \subset P$. For example, the plate primitive face, Eq. (9.73), is a subset of the planar surface given by Eq. (9.68). The subset is bounded in the x and y directions. Similarly, a cylindrical primitive face is a finite (but not bounded) region (between $z = 0$ and $z = H$) of the cylindrical surface given in Eq. (9.69).

In addition, the boundary of any primitive is a combination of these primitive faces. The boundary of a block primitive consists of six plate primitive faces positioned properly, while that of a sphere or a torus consists of one spherical or toroidal primitive face, respectively. The boundary of a cylinder consists of one cylindrical primitive face closed from each end by a disk primitive face. And lastly, a cone has a conical face closed by a disk, and a wedge has three plates and two tri-plates.

Similar to primitive faces, primitive edges (sometimes called face-bounding edges) are edges selected such that the boundary of any primitive face may be represented as the union of a finite number of these edges after being positioned properly in space. Each edge is a finite or bounded region of a corresponding underlying curve that may be unbounded and disjointed (in which case curve segments make up the total curve). The underlying curves are usually obtained by finding all possible intersections of the underlying surfaces.

Curves, and therefore edges, can be true curves or virtual (profile or silhouette) curves as in the case of a cylinder, spline, cone, and torus. For quadric surfaces, the virtual edges can be generated by intersecting the surface with a plane. This offers the advantage that true and virtual curves can be obtained via surface/surface intersection. It is useful to realize that edges and curves, like surfaces and half-spaces, have a single representation with a dual purpose. This implies that a curve equation is also an edge equation after imposing the proper parameter limits.

9.8.3 Surface/Surface Intersection

Surface/surface intersection is fundamental to geometric modeling in general and to implementing Boolean operations in particular. Because surface/surface intersections are computationally intensive, surface/surface intersection problems are solved algebraically beforehand and stored in the corresponding solid modeler.

A solid modeler that supports the primitives shown in Figure 9.5 must contain the intersections of the surfaces given by Eqs. (9.68) to (9.72). For example, a cylinder/plane intersection may give a circle, an ellipse, two infinite parallel lines, or no intersection at all. What usually complicates the surface/surface intersection is the position and orientation of the two intersecting surfaces in space.

One solution to the orientation problem is to intersect the two surfaces in a given standard position and orientation and then transform the result to the actual position and orientation. Intersection curves are usually represented in parametric form because quadric surfaces can be parametrized conveniently as shown in Chapter 7. Equations of edges and curves are not given here. The reader is referred to Chapter 6 for some of these equations.

Equations (9.68) to (9.72) are infinite surfaces whose intersections yield infinite curves. These curves are usually classified against given primitives using set membership classification to determine which curve segments lie within these primitives and consequently within the solid.

While the full details of surface/surface intersections of quadric surfaces are beyond the scope of this book, the essence of the problem can be described as follows. Any of the surfaces described by Eqs. (9.68) to (9.72) can be rewritten in the following polynomial form:

$$Ax^2 + By^2 + Cz^2 + 2Dxy + 2Eyz + 2Fxz + 2Gx + 2Hy + 2Jz + K = 0 \tag{9.80}$$

where $A, B, ..., K$ are arbitrary real constants. This equation can be expressed in a quadratic form as:

$$F(x, y, z) = \mathbf{V}^T[Q]\mathbf{V} = 0 \tag{9.81}$$

where $\mathbf{V}$ is a vector of homogeneous coordinates of a point on the surface and is given by $[x \quad y \quad z \quad 1]^T$. $[Q]$ is the coefficient matrix. It is symmetric and is given by:

$$[Q] = \begin{bmatrix} A & D & F & G \\ D & B & E & H \\ F & E & C & J \\ G & H & J & K \end{bmatrix} \tag{9.82}$$

The coefficient matrix $[Q]$ can be formed for any of the quadric surfaces [Eqs. (9.68) to (9.71)] by comparing the surface equation with Eq. (9.81). This gives:

Planar surface: $$[Q] = \begin{bmatrix} 0 & 0 & 0 & 0 \\ 0 & 0 & 0 & 0 \\ 0 & 0 & 0 & \frac{1}{2} \\ 0 & 0 & \frac{1}{2} & 0 \end{bmatrix} \tag{9.83}$$

Cylindrical surface: $$[Q] = \begin{bmatrix} 1 & 0 & 0 & 0 \\ 0 & 1 & 0 & 0 \\ 0 & 0 & 0 & 0 \\ 0 & 0 & 0 & -R^2 \end{bmatrix} \tag{9.84}$$

Spherical surface: $$[Q] = \begin{bmatrix} 1 & 0 & 0 & 0 \\ 0 & 1 & 0 & 0 \\ 0 & 0 & 1 & 0 \\ 0 & 0 & 0 & -R^2 \end{bmatrix} \tag{9.85}$$

Conical surface: $$[Q] = \begin{bmatrix} 1 & 0 & 0 & 0 \\ 0 & 1 & 0 & 0 \\ 0 & 0 & -(R/H) & 0 \\ 0 & 0 & 0 & 0 \end{bmatrix} \tag{9.86}$$

The coefficient matrix $[Q]$ depends directly on the surface orientation and position. The preceding matrices are valid only if the local coordinate system of the primitive is identical to the MCS of the solid model to whom the primitive belongs. Otherwise, each matrix has to be transformed by the transformation matrix that results from the rigid motion (rotation and/or translation) of the primitive. This gives:

$$[Q'] = [T]^T[Q][T] \tag{9.87}$$

where $[Q']$ is the transformed coefficient matrix and $[T]$ is the transformation matrix given by equation (3.3). For example, if the origin of a sphere is located at point (a, b, c) measured in the MCS, Eq. (7.76) is transformed by a translation vector to give:

$$[Q'] = \begin{bmatrix} 1 & 0 & 0 & a \\ 0 & 1 & 0 & b \\ 0 & 0 & 1 & c \\ a & b & c & a^2 + b^2 + c^2 - R^2 \end{bmatrix} \tag{9.88}$$

This idea of transformation suggests that one can solve the intersection problem of two primitives in any convenient coordinate system and then transform the results as needed. Actually, there exists a set of algebraic constructs derived from $[Q]$ that remain invariant under rigid motions and completely specify the shape of the quadric surface.

The intersection of two quadric surfaces can be solved as follows. If one surface has a coefficient matrix $[Q_1]$ and the other has $[Q_2]$, then the equation of the intersection curve is:

$$\mathbf{V}^T([Q_1] - [Q_2])\mathbf{V} = 0 \tag{9.89}$$

Equation (9.89) describes an infinite intersection curve. In order to determine the finite segments of this curve (edges) which belong to the intersecting primitives, we must find the appropriate bounding points. One way of finding these points is if one of the intersecting surfaces can be parametrized in terms of two parameters u and v, Eq. (9.89) can be solved for one of the parameters in terms of the other. In this case, the equation of the intersecting curve can be written in a parametric form in terms of one parameter.

To understand this approach, consider the cylinder/quadric intersection case. Assuming the cylinder is in a standard position, its parametric equation is known. In this case the vector $\mathbf{V}$ in Eq. (9.89) is $[R\cos u \ R\sin u \ v]^T$. Substituting Eqs. (9.82) and (9.84) for $[Q_1]$ and $[Q_2]$, respectively, and simplifying the result, we get:

$$a(u)v^2 + b(u)v + c(u) = 0 \tag{9.90}$$

Equation (9.90) could also be obtained by substituting the parametric equation of a cylinder directly into Eq. (9.80). Equation (9.90) is a quadratic equation which can be solved for υ in terms of u. Moreover, the proper range of u can be obtained by investigating the characteristics of the discriminant $b^2(u) - 4a(u) \times c(u)$. The same analysis can be done for the cone/quadric intersection. In this case the vector $\mathbf{V}$ (the cone parametric equation) is given by $[(R/H)\upsilon\cos u \;\; (R/H)\upsilon\sin u \;\; \upsilon]^T$.

What made it possible to reduce Eq. (9.89) to Eq. (9.90) in the case of a cylinder or a cone is the fact that either surface is a ruled quadric with υ being the parameter along the surface rulings. A sphere/quadric or torus/quadric intersection problem cannot be solved directly by following the preceding approach. Special transformation must be done first and is not discussed here.

9.8.4 Building Operations

The main building operations in CSG schemes are achieved via the set operators or, more specifically, the regularized operators: union ($\cup^*$), intersection ($\cap^*$), and difference ($-^*$). Set operators are also known as Boolean operators due to the close correspondence of the two. Union, intersection, and difference are equivalent to OR, AND, and NOT AND, respectively. Due to the deep roots of CSG schemes in set theory or Boolean algebra, CSG models are usually referred to as set-theoretic, Boolean, or combinatorial models. Set operations algorithms are among the most fundamental and delicate software components of solid modelers.

Regularized set operators are not based on a given law or equation. But they derive their properties from the set theory and the concept of closure. The validity checks for set operators are usually simple in the case of bounded primitives. They take the form of checking the user input of each primitive parameter. This is due to the fact that if primitives are valid and set operators are regularized, then the topology of the resulting solid is always valid.

How do we implement the regularized Boolean operators? We need to write an algorithm that must evaluate the boundary of the resulting solid from a desired operation. The general form of a set operation can be written as $A<OP>B$ where A and B are the primitives (operands) that we want to combine, and $<OP>$ is any regularized set operator. The algorithm's main task is to evaluate the boundary of the resulting solid. Figure 9.29 shows two primitives A and B. We assume that A and B do not have tangent (overlapping) boundaries (faces, edges, or vertices). Here are the steps of the Boolean algorithm:

1. **Intersect A and B.** We intersect the boundaries of A and B to find the intersection points I_1 and I_2 shown in Figure 9.29. In the 3D case, the primitives intersect at curves. We use two nested loops to find the intersections. For each face in A, we intersect it with all the faces of B. And, for each face in B, we intersect it with all the faces of A.

2. **Create new edges.** Divide the boundaries of A and B at intersection points and create new edges.

3. **Classify A with respect to B.** Use the membership classification function to classify A relative to B, that is, $M[A, B] = \{A \text{ in } B, A \text{ out } B\}$. See Figure 9.29.

4. **Classify B with respect to A.** Use the membership classification function to classify B relative to A, that is, $M[B, A] = \{B \text{ in } A, B \text{ out } A\}$. See Figure 9.29.

5. **Combine classifications of A and B.** In this last step, we create the result of the Boolean operation by combining the classification of A and B according to the Boolean operation as follows:

$$A \cup^* B = \{A \text{ out } B, B \text{ out } A\}$$
$$A \cap^* B = \{A \text{ in } B, B \text{ in } A\}$$
$$A -^* B = \{A \text{ out } B, B \text{ in } A\}$$
$$B -^* A = \{B \text{ out } A, A \text{ in } B\}$$

Note that the "B in A" and "A in B" in the last two operators must be used to close the resulting solid.

We must mention that the Boolean algorithms are much more complex than this coverage indicates because they must solve the surface/surface intersection problem, and they must change the geometry and topology of the two primitives (operands) at hand.

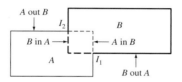

Figure 9.29   Boolean algorithm.

9.8.5   The Neighborhood Concept

In the Boolean algorithm discussed in Section 9.8.4, we made the assumption that A and B are not tangent, or none of their faces overlap. We have been avoiding this overlapping issue thus far. For example, Figure 9.15 does not show any on segments, and it also shows that combining in segments usually results in in segments. If overlapping occurs, an ambiguity arises in the classification function that must be resolved. The concept of neighborhoods is introduced to resolve the on/on ambiguities that result when combining on segments in classifications.

Figure 9.30 shows a case where a solid $S = A \cup^* B$. After classifying the edges of A and B against each other, combining the on segments of each primitive may result in in or on segments of S. The on/on ambiguities usually result when the two subsolids or primitives to be combined are tangent to each other along one or more faces. These ambiguities can be resolved using neighborhood information of any point on the on segments.

The neighborhood of a point P with respect to a solid S, denoted by $N(P, S)$ is the regularized intersection of a sphere with radius R centered at P with the solid. The value of the radius R is arbitrary and should be chosen to be sufficiently small. We can generalize the set membership classification function $M[X, S]$ given by Eq. (9.50) to include neighborhood information by assuming that the candidate set X has such information and therefore the resulting segment X on S has it.

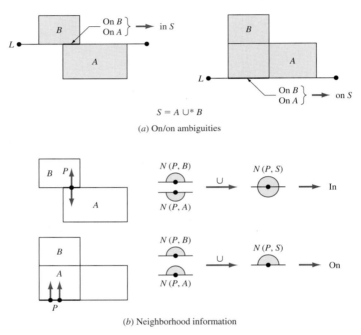

(b) Neighborhood information

Figure 9.30   On/on classification.

The representation of the neighborhood of a point is related to its position relative to the solid under investigation. Neighborhoods for points that are in the interior or outside of the solid are full or empty and can be represented easily. One alternative is to assign a variable, for example N, to the neighborhood of the point such that it can take the values -1, 0, and 1 if the point is outside, on, or inside the solid, respectively. If $N = -1$ or 1, no further information is needed.

For points on the boundaries of the solid ($N = 0$), three cases can arise. A point may be in the interior of a solid's face. This becomes a case of face neighborhoods, which can be represented using the face and surface normal signs.

The second case is edge neighborhoods, which result if the point lies on a solid's edge. Assuming that the edge is shared by two faces, the normal and tangent signs of the faces and their underlying surfaces serve to represent the neighborhood.

The third case arises when the point is a vertex, thus resulting in a vertex neighborhood. A vertex is typically shared by three faces, and its neighborhood is complex and difficult to manipulate and is not needed in most algorithms. Figure 9.31 shows face and edge neighborhoods.

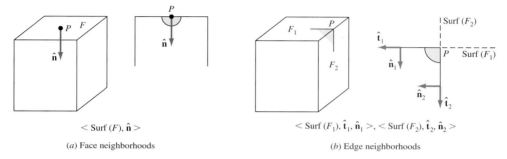

(a) Face neighborhoods   (b) Edge neighborhoods

Figure 9.31   Face and edge neighborhood representation.

9.8.6   Remarks

The CSG scheme is a powerful representation scheme. It has many advantages. It is easy to construct out of primitives and Boolean operations. It is concise and requires minimum storage to store solid definitions (CSG tree). Application algorithms based on CSG schemes are very reliable.

9.9 Sweeps

Sweep operations are popular in the features approach. We sketch a cross section and sweep it to create a solid. These operations are useful in creating solid models of 2½D objects. The class of 2½D objects includes both solids of uniform thickness in a given direction and axisymmetric solids. The former are known as extruded solids and are created via linear or translational sweep; the latter are solids of revolution which can be created via rotational sweep.

Sweep is based on the notion of moving a point, curve, or surface along a given path. There are three types of sweep: linear, nonlinear, and hybrid. In linear sweep, the sweeping path is a linear (for extrusions) or circular (for axisymmetric solids) vector described by a parametric linear equation. In nonlinear sweep, the path is a curve described by a higher order (quadratic, cubic, or higher) equation. Hybrid sweep combines linear and nonlinear sweeps.

Linear sweep can be divided into translational and rotational sweep. In translational sweep, a planar 2D point set described by its boundary (or contour) can be moved a given distance in space in a perpendicular direction (called directrix) to the plane of the set as shown in Figure 9.32a. The boundary of the point set must be closed otherwise invalid solids (open sets) result. In rotational sweep, the planar 2D point set is rotated about an axis of rotation (axis of symmetry of the object to created) a given angle as shown in Figure 9.32a.

Nonlinear sweep is similar to linear sweep but with the directrix being a curve instead of a vector as shown in Figure 9.32b. Hybrid sweep tends to utilize some form of set operations. Figure 9.32c shows the same object shown in Figure 9.32a but with a hole. In this case, two point sets are swept in two different directions and combined (added or subtracted) to create the final solid. Invalid solids or nonregular sets may result if the sweeping direction is not chosen properly as shown in Figure 9.32d.

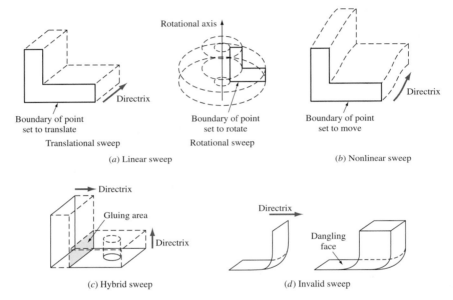

Figure 9.32   Types of sweep.

9.9.1   Basic Elements

Curves, both analytic and synthetic, covered in Chapter 6 are valid elements to create 2D contours for sweep operations. Users sketch the contours on various sketch planes. The contours (2D point set) used in sweep operations can consist of nested contours up to one level only (one inner contour) within the outer contour. This is allowed in order to create holes in the resulting solid. There may also exist a maximum number of entities that is allowed to create any one contour. The number is usually adequate for practical uses. It is usually set for purposes of implementing the sweep algorithm.

9.9.2   Building Operations

The building operations of linear and nonlinear sweeps are simple; generate the sketch and sweep it. How are sweep operations converted to B-rep? Consider the case of a translational sweep. Each entity in the swept boundary represents an edge, and each corner point represents a vertex in the corresponding boundary model. For a one-contour boundary, each entity in the boundary also indicates a face. The number of faces of the model is equal to $(N + 2)$, where N is the number of entities of the sketch (contour) and the 2 accounts for the front and back faces (see Figure 9.32a). The number of edges is equal to $(2N + M)$, where M is the number of corner points of the boundary. The number of vertices is equal to $2M$.

9.9.3   Remarks

Sweep representation is useful. Its modeling domain can be extended beyond 2½D objects via nonlinear (sometimes called general) sweep. Complex mechanical parts such as screws, springs, and other components that require helical and special loci can be represented by sweeps.

9.10 Solid Manipulations

Solid manipulations are useful during the creation of solid models. It is beneficial if the manipulation concepts utilized by curves and surfaces can be extended to solids. Most of these concepts can be viewed as solving intersection problems and/or using set membership classification. In addition, any solid manipulation involves manipulating both its geometry (as in curves and surfaces) and its topology to ensure that the resulting solids are valid.

9.10.1 Displaying

Displaying a solid can take two forms: wire (edge) display and shaded images. The wire display requires the B-rep of the solid, in which case the metric information of edges and vertices is used to generate the wireframe of the solid. This wireframe can be displayed, edited, or used to produce line drawings. If the underlying surfaces of the solid faces are utilized, a mesh can be added to the display to help visualization.

Displaying solids as shaded images provides realistic visual feedbacks to users. Shading can be performed directly from CSG by means of ray-casting (also called ray-tracing) algorithms, or depth-buffer (also called z-buffer) algorithms. Many improvements and alterations have been introduced to ray casting to speed up the algorithms.

9.10.2 Evaluating Points, Curves, and Surfaces on Solids

Evaluating points and curves on solids can be viewed as an intersection problem. Solutions of curve/solid and surface/solid intersection problems generate points and curves on solids, respectively. A plane/solid intersection generatse desired cross sections of a solid.

Evaluating surfaces on a solid can be regarded as extracting the underlying surfaces of the solid faces. These surfaces can be bounded by the solid edges. To keep the solid topology and geometry intact, the geometry and other related information of these surfaces must be copied. The parametric equations of the surfaces might have to be stored in the given solid modeler to facilitate editing the extracted surfaces.

The inverse problem involves checking whether given points, curves, and surfaces lie `in`, `on`, or `outside` a given solid. The solution of this problem is achieved by the set membership classification and neighborhoods. To classify a point against a solid, one passes a line through the point, intersects it with the solid, and classifies it with respect to the solid. A similar approach can be followed for curve/solid and surface/solid classification.

9.10.3 Segmentation

The segmentation concept introduced for curves and surfaces is applicable to solids. Segmenting a solid is equivalent to splitting it into two or four valid subsolids, depending on whether it is to be split by a plane or a point, respectively. Each resulting subsolid should have its own topology and geometry. In a B-rep model, new vertices, edges, and faces are created.

9.10.4 Trimming and Intersection

Trimming a solid entails intersecting the solid with the trimming boundaries, such as surfaces, followed by the removal of the solid portions outside these boundaries. In trimming a solid, it is split into two subsolids, one of which is removed.

The intersection problem uses Boolean operations. All that needs to be done is to use the intersection operator. No additional development or programming is required.

9.10.5 Transformation

Homogeneous transformations, or rigid motion, of solids involve translating, rotating, or scaling them. These transformations can be used on two different occasions. When constructing a solid, its primitives are positioned and oriented properly before applying Boolean operations. The same transformations can be applied to the solid later after its complete construction.

9.10.6 Editing

Editing a solid involves changing its existing topological and geometrical information. An efficient means of solids editing is to use its CSG tree. CAD systems must provide users with fast visual feedbacks when solids are edited. This implies that boundary representations must be updated and displayed rapidly.

9.11 Tutorials

9.11.1 Use Solid Trees

This tutorial shows how to use CSG trees displayed by CAD systems. All major CAD systems display a tree for each part that a CAD designer creates. The tree serves as a history of how the designer has created the part. It captures the creation steps sequentially, in the same sequence that the designer follows in creating the part. A solid tree serves three purposes:

a. It helps the designer to visualize the results of modeling steps. For example, when you click on a step in the tree, the corresponding geometry in the model window (display area) is highlighted.

b. More importantly, the designer may use the tree to change the current solid, by editing the tree. There are two distinct methods of using the tree to change the current solid. First, change the dimensions of a sketch and update the solid. In this case, the designer selects the sketch from the tree. The sketch geometry is displayed. The designer makes the changes — change the diameter of a circle. Once the designer accepts the sketch, the solid modeler updates the entire solid automatically, by traversing the tree and reconstructing every branch of it. If the changes are drastic, the resulting solid may be invalid because the topology becomes impossible to create. The following steps show an example of editing the solid tree of a spur gear to change its extrusion depth from 20 mm to 30 mm.

1. Select the feature to edit. Click the feature to edit (Base-Extrude here) from the tree manager => Edit definition (in the window that pops up as shown in the following screenshot).

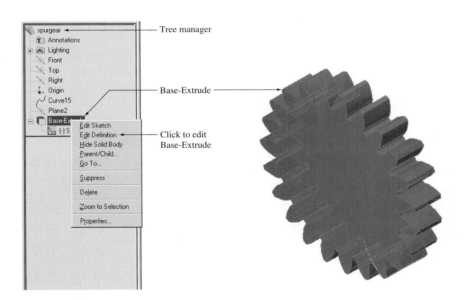

2. Change the depth of the spur gear. Type 30 for the depth and click the check mark (✓) as shown in the following screenshot.

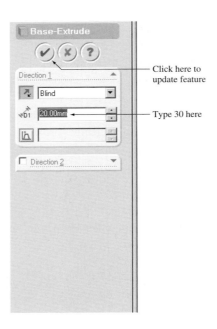

Click here to update feature

Type 30 here

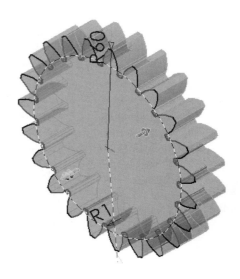

c. The designer can also prune the tree by deleting a branch or cutying part of it. Again, this may result in an invalid solid if the designer is not careful. One best way of using a solid tree is to back up the part by saving under a different name, and then cutting a part of the tree. The popup menu in the screenshot of step 1 has a `Delete` item that the designer can click to delete the `Base-Extrude`.

Tutorial 9.11.1 discussion:

More options to edit the solid tree are available as shown in the popup menu shown in step 1 and we show it here again on the right. The `Edit Sketch` edits the feature sketch. `Edit Definition` edits the feature depth or other definitions. `Hide Solid` Body is obvious. `Parent/Child` displays the children of any feature in the tree that you click. `Suppress` is similar to `Hide Solid` Body; however, a suppressed feature is not updated or regenerated. `Go To` searches the tree (if it is too long) for a feature. `Delete` deletes a selected feature. `Zoom to Selection` is obvious. `Properties` shows the properties of the feature and its sketch entities.

Tutorial 9.11.1 hands-on exercise:

Invoke the tree manager on your CAD system and do this tutorial.

9.11.2 Create a Candy Dish

This tutorial illustrates the use of the features approach to create solids as opposed to the primitives approach which is more limited. We create the candy dish shown on the right by using `extrude` and `revolve` features. Here are the detailed creation steps (modeling strategy); all dimensions in mm:

1. **Create the dish base.** Select the `Top` sketch plane and sketch the profile shown below. Extrude the profile with a draft of 20°, a distance of 15 mm.

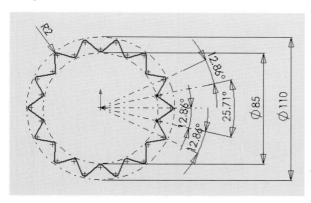

2. **Cut out the bottom of the dish base.** Create the profile shown below in the `Front` sketch plane and `revolve cut` it from the dish base of step 1.

3. **Cut out the top of the dish base.** Similarly `revolve cut` the profile shown below for the cut in the top of the base. To close the profile, use approximate dimensions.

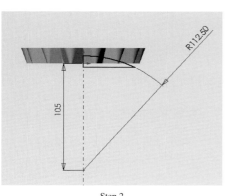

Step 2

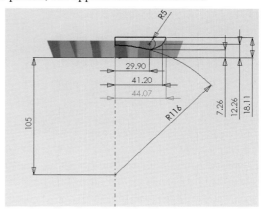

Step 3

4. **Create the profile of the dish wall (body).** Create the profile of the wall in the `Front` sketch plane, as shown below, and revolve it.
5. **Fillet the edge between the dish base and its wall.** Use a fillet radius of 15 mm as shown in the forthcoming screenshot.
6. **Create the cutout in the top edge of the dish wall.** Sketch the profile, shown in the forthcoming screenshot, in the `Front` sketch plane, and extrude cut it from the wall.
7. **Pattern the cutout using a circular array.** Cut out the pattern from the wall.

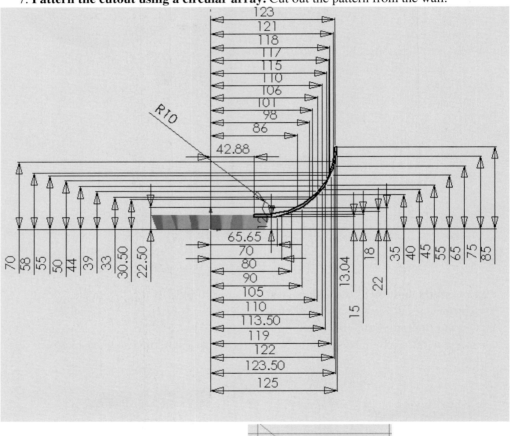

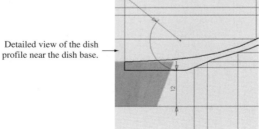

Detailed view of the dish profile near the dish base.

Step 4

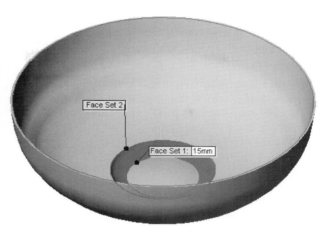

Step 5

The ISO view of the final model is shown below from different viewing angles for ease of understanding.

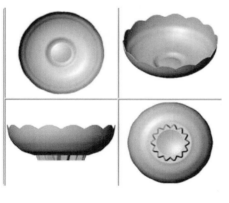

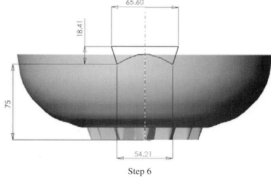

Step 6

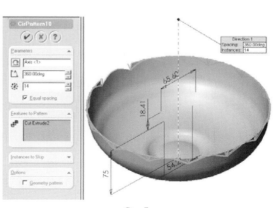

Step 7

Problems

Part I: Theory

9.1   A valid solid is defined as a point set that has an interior and a boundary as given by Eq. (9.1). A valid boundary must be in contact with the interior. Sketch a few 2D and 3D solids and identify iS and bS for each one. Is iS always joint for any S? Can bS be disjoint? What is your conclusion?

9.2   Three point sets in E^2 define three valid polygonal solids S_1, S_2, and S_3. The three solids are bounded by three boundary sets bS_1, bS_2, and bS_3 given by their corner points as: $bS_1 = \{(2, 2), (5, 2), (5, 5), (2, 5)\}$, $bS_2 = \{(3, 3), (7, 3), (7,6), (3, 6)\}$, and $bS_3 = \{(4,1), (6,1), (6,4), (4,4)\}$. Find $S_1 \cup S_2 \cup S_3$, $S_1 \cap S_2 \cap S_3$, and $S_1 - S_2 - S_3$.

9.3   Given the four solids S_1 to S_4 shown below, find $S_1 \cap S_2$, $S_1 \cap^* S_2$, $S_2 \cap S_3$, $S_2 \cap^* S_3$, $S_3 \cap S_4$, and $S_3 \cap^* S_4$.

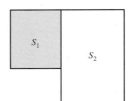

9.4   Given the two solids S_1 and S_2 shown below, find $S_1 \cup S_2$, $S_1 \cup^* S_2$, $S_1 \cap S_2$, $S_1 \cap^* S_2$, $S_1 - S_2$, and $S_1 -^* S_2$.

S_1: Block
S_2: Cylinder

S_1: L bracket overlapping with S_2
S_2: Block

9.5   Reduce the following set expressions:
   a. $c(P \cap Q) \cup P$
   b. $c(P \cup Q) \cup (P \cap cQ)$
   c. $(P \cap Q) \cap c(P \cup Q)$
   d. $P - (P - Q)$

Use the set laws given by Eqs. (9.14) to (9.28) as well as the Venn diagram.

9.6   Using the set membership classification, classify the line L with respect to the solid shown in the following figure, if the solid is given as a B-rep and a CSG.

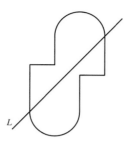

9.7　A solid S is defined as the union of two intersecting spheres, that is, $S = SP_1 \cup SP_2$. Sketch the solid S. What is $SP_1 \cap SP_2$ if the spheres have the same radius, and if they have different radii? Verify the Euler equation for the solid S.

9.8　Verify Euler equation for the following solids.

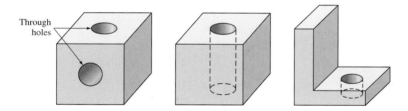

9.9　Implementing set operations given by Eqs. (9.42) to (9.49), or $S = A <OP> B$ in general, involves finding intersections between bA and bB, classifying the boundaries with respect to each operand, and combining classifications according to the rules:

$$<OP> = \cup^*: bS = (A \text{ out } B \text{ } UN^* B \text{ out } A \text{ } UN^* A \text{ on } B+)$$

$$<OP> = \cap^*: bS = (A \text{ in } B \text{ } UN^* B \text{ in } A \text{ } UN^* A \text{ on } B+)$$

$$<OP> = -^*: bS = (A \text{ out } B \text{ } UN^* B \text{ in } A^{-1} \text{ } UN^* A \text{ on } B-)$$

where UN^* is a regularized union operator in E^1. A on $B+$ consists of those parts of bA that lie on bB so that the face normals of the respective faces are equal, whereas A on $B-$ consists of the overlapping parts where the normals are opposite. B in $A-$ denotes the complement of B in A, that is, B in A with the orientation of all faces reversed. Notice that the + and −, and the exponent −1 are a form of neighborhoods. Using the above equations:

　　a. Explain the concept of closure used in Eqs. (9.42) to (9.48) and shown in Figure 9.11. What should be done to implement the concept of interior to eliminate nonregular sets?

　　b. Find $S = A <OP> B$ for A and B shown in the following figure:

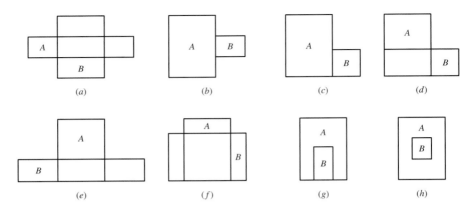

9.10 Solids in E^2 (that is, 2D) and their solid modelers are valuable in understanding many of the concepts needed to handle solids in E^3 (that is, 3D solids). A 2D solid is a point set of ordered pairs (x, y). A 2D solid modeler based on half-spaces is to be developed utilizing linear and disk half-spaces. Find the equation of the two half-spaces

Hint: see Eqs (9.52) and (9.53). Develop a parametric equation for their intersection.

9.11 We wish to develop bounded primitives for a 2D solid modeler based on the CSG scheme. A plate (rectangular plate and triplate) and disk primitives are to be developed. Find the mathematical definitions of these primitives

Hint: see Eqs. (9.62), (9.63), and (9.66).

Develop intersection equations (refer to Problem 9.10).

9.12 How can you use a cylinder primitive to generate a sphere?

9.13 How can you generate a torus using other natural quadrics?

9.14 Problems 9.10 and 9.11 introduce the mathematical foundations of a 2D solid modeler. The further development of such a modeler requires representations of orientable surfaces that represent surface normals, neighborhoods, classifications, and combining the classifications. Discuss the following for half-space, B-rep, and CSG schemes:

 a. how to represent surface normals and neighborhoods

 b. how to develop a classification algorithm

 c. how to combine classifications

Part II: Laboratory

Use your in-house CAD/CAM system to answer the questions in this part.

9.15 Check your CAD system. Which solid modeling approach does it support: the primitives approach, the features approach, or both?

Part III: Programming

Write the code so that the user can input values. The program should display the results. Use OpenGL with C or C++, or use Java 3D.

9.16 Classify lines with respect to 2D solids for B-rep and CSG (see Figures 9.1 4 and 9.15).

9.17 Combine the classifications that result from Problem 9.16.

9.18 Implement the results of Problem 9.10.

9.19 Implement the results of Problem 9.11.

9.20 Implement results of Problem 9.14.

9.21 Implement results of Problem 9.12 into an existing solid modeler.

9.22 Implement results of Problem 9.13 into an existing solid modeler.

Features

OAL

Understand and master the concept of features and their use in geometric modeling; the basics of parametrics, relations, and constraints; and the use of parametrics and relations in "what-if" design questions.

OBJECTIVES

After reading this chapter, you should understand the following concepts:

- Features and primitives

- Feature entities

- 3D sketching

- Feature representation

- Creating features

- Parametrics

- Relations and constraints

- Feature manipulations

CHAPTER   HEADLINES

10.1 Introduction

Features is one of the approaches to creating solid models, the other being the primitives approach (see Chapter 9). The features approach is more general than the primitives approach, that is, it provides a larger domain of geometric modeling. It allows designers to create complex parts.

A **feature** is defined as a shape and an operation to build parts. The **shape** is a two-dimensional sketch. Sample shapes are bosses, cuts, and holes. The sketch orientation is controlled in the three-dimensional modeling space (E^3) via the sketch planes. The sketch entities are curves and/or surfaces, depending on the complexity of the part under construction. The **operation** is an activity that converts the sketch into a three-dimensional shape. Sample operations include `extrude`, `revolve`, `fillet`, `shell`, `chamfer`, and `sweep`.

The Boolean operations are built into the features operations. Thus, feature-based systems such as Pro/E and SolidWorks do not provide explicit Boolean operations. (Early CAD systems such as Unigraphics, CATIA, and SDRC I-DEAS support both primitives and features.) Features are usually added to each other unless the feature operation specifies otherwise. For example, an `extrude cut` operation uses the `subtract` Boolean operation (–*), while a `boss` feature uses the union Boolean operation (∪*). Feature-based systems provide the intersection Boolean operation (∩*) as a feature—for example, Pro/E has an `Intersect` feature under its Features menu. The first feature we create in a model is sometimes known as the **base feature.**

Features have been a topic of much discussion and research. Efforts have focused on capturing design intent and manufacturing intent for products. Manufacturing features, for example, include manufacturing attributes of the feature. For example, a hole feature accompanies a drilling process. A fillet feature comes with a milling operation. The features that we cover here are known as geometric features. These features describe the product's geometric definition. Geometric features include form (shape), tolerance, assembly features, and so forth. We cover form features in this chapter.

Much, if not all, of the coverage in Chapter 9 applies to features. As a matter of fact, Chapter 9 is a prerequisite to understanding features. For example, CSG trees apply to features. In such an application, the tree leaves are the sketches, and its nodes are the features operations.

The use of CSG (history) trees in feature-based modeling is the same as in the primitives-based modeling. A CAD designer can reorder features in the tree to change the creation history of the solid. For example, the designer can move a shell operation down the tree to include a new hole feature added to the part. Or the designer may reorder the tree to fix failed geometry as discussed in Chapter 9. For example, reordering a failed fillet or chamfer in the model tree could solve the problem because ordering and reordering generate different numerical errors (round off and truncation) in the numerical algorithms that the CAD system uses to execute the required Boolean operations.

Using features to create parts involves three steps (shown in Figure 10.1):

1. **Create sketches.** A **sketch** is a two-dimensional profile or a cross section. We control the sketch orientation via the sketch planes.

2. **Create features.** Apply a feature operation to the sketch to create the feature. Sketches can be extruded, revolved, lofted, or swept along a path to create features. Feature operations convert two-dimensional sketches to three-dimensional shapes. By two-dimensional, we mean planar geometry in the sketch plane because this geometry is still three-dimensional in the MCS of the part under construction.

3. **Use features to build parts.** Combine features to build parts. Add or subtract features from the base feature during the building process.

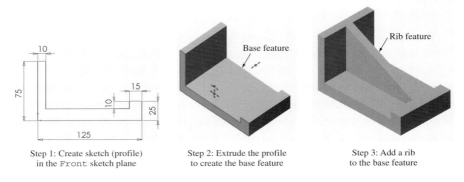

Step 1: Create sketch (profile)   Step 2: Extrude the profile   Step 3: Add a rib
in the Front sketch plane   to create the base feature   to the base feature

Figure 10.1   Three steps to create features.

10.2 Feature Entities

CAD/CAM systems provide designers with many features that extend the domain of geometric modeling far beyond primitives. While all features begin as a sketch, it is the feature operation that determines the features a CAD system supports. Most systems provide a features menu (with features as menu items), menu bar (with features as visual icons), command sequence, or both to access features. The command sequence may be Insert => Feature => select a feature from popup menu, or Feature => Create => select a feature from popup menu.

All CAD systems offer almost the same set of features. Each system may have a few unique features. Remember that if a CAD system does not have a feature, we can simply create it via the three steps shown in Figure 10.1. If we can sketch the feature profile, we can create it. Following are descriptions of major common features that CAD systems support:

1. **Extruded (protruded) feature.** It is the most basic feature. We use it to create solid models of 2½D objects with uniform thickness. It requires a profile (cross section), and an extrusion vector (length and direction). The extrusion direction is always perpendicular to

the sketch plane of the profile. Figure 10.2 shows an extrusion (protrusion) with its parameters.

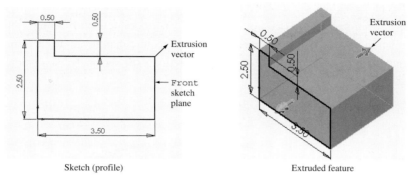

Sketch (profile)   Extruded feature

Figure 10.2   Extruded (protruded) feature.

2. **Revolved feature.** We use it to create solid models of axisymmetric 2½D objects. It requires a profile (cross section) and a revolution vector (axis and angle of revolution). The axis of revolution is always in the sketch plane of the profile. Figure 10.3 shows a revolution with its parameters.

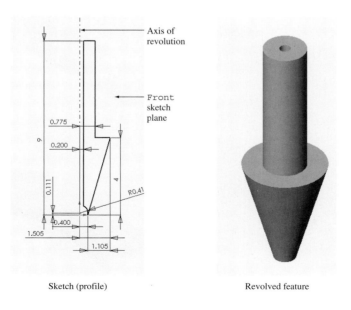

Sketch (profile)   Revolved feature

Figure 10.3   Revolved feature.

3. **Sweep.** Sweep is a generalization of extrusion. Linear and nonlinear sweeps exist. Linear sweep sweeps a cross section along a straight line. Linear sweep and extrusion produce the same feature. A nonlinear sweep has two types. One type sweeps one cross section (profile) along a guide curve (not a line) as shown in Figure 10.4a. The other type blends two cross sections linearly (as shown in Figure 10.4b) or nonlinearly along a guide curve (as shown in Figure 10.4c). All sweep operations require that the guide curve must be perpendicular to the sketch plane of the cross section to be swept. To blend more than two cross sections, we must use a loft feature.

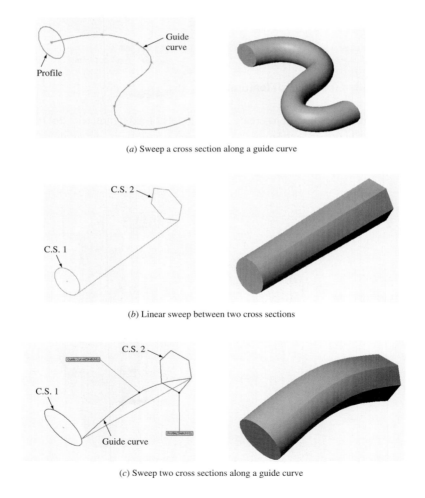

(a) Sweep a cross section along a guide curve

(b) Linear sweep between two cross sections

(c) Sweep two cross sections along a guide curve

Figure 10.4   Sweep feature.

4. **Loft.** Loft is a generalization of sweep. We use loft to blend multiple cross sections to create a solid. We classify loft features into linear and nonlinear. A linear loft blends two cross sections linearly along the lines connecting the corner points of the cross sections as shown in Figure 10.5a. A nonlinear loft has two types. One type blends two cross sections along a guide curve (not a line) as shown in Figure 10.5b. The other type blends a set of cross sections along a guide curve as shown in Figure 10.5c.

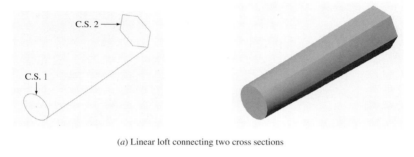

(a) Linear loft connecting two cross sections

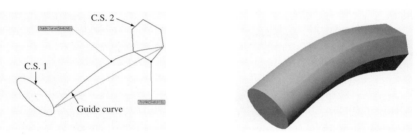

(b) Nonlinear loft connection two cross sections along a guide curve

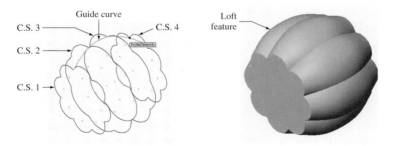

(c) Nonlinear loft connecting multiple cross sections along a guide curve

Figure 10.5   Loft feature.

5. **Boss.** A boss is an extruded or revolved feature that we add to a base feature or an existing solid.

6. **Shaft.** A shaft is an extruded feature with a circular cross section.

7. **Pipe.** A pipe is sweeping a circular cross section (solid or hollow) along a guide curve (the pipe axis). The pipe parameters are its axis and diameter (or wall thickness) as shown in Figure 10.6.

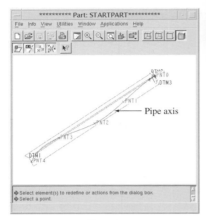

Pipe axis

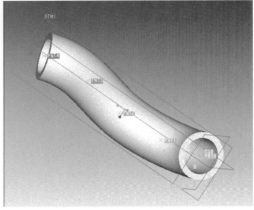

Pipe feature

Figure 10.6   Pipe feature.

8. **Cut.** This feature is the opposite of the boss feature. A cut is an extruded or revolved feature that we subtract from a base feature or an existing solid.

9. **Hole.** This feature is equivalent to subtracting a cylinder from a solid. Five types of hole features exist (through, blind, tapped, countersink, and counterbore) as shown in Figure 10.7. A through hole is the simplest hole. A blind hole is not a through hole. Holes are drilled with a drill bit that has a tapered end. If the end of the blind hole is flat, the hole must be milled. Milling a hole is more expensive than drilling it. A tapped hole has threads. After drilling the hole, a tapping machining process is used to create the threads. The difference between the countersink (C-sink) and counterbore (C-bore) holes comes in the shape of the hole end(s). The C-sink has tapered end(s) while the C-bore has vertical end(s) as shown in Figure 10.7.

10. **Slot.** Slots are like holes. They remove "material" from a solid. Five types of slots exist as shown in Figure 10.8. A slot can be a through (cutting through opposite sides of an object) or blind (contained within opposite sides of an object) slot. The rectangular slot has sharp edges at the bottom. The rectangular slot is created with a flat-end mill (cutting tool). The ball-end slot is created with a ball-end mill (cutting tool). The ball-end slot has a full radius bottom and corners. The U-slot has a "U" shape (rounded corners and floor radii). The T-slot is an inverted T shape. The dovetail slot has a dovetail shape (sharp corners and angled walls).

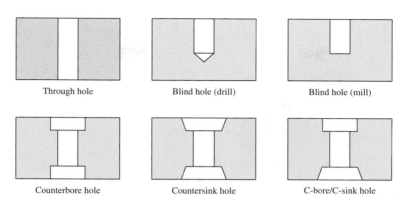

Figure 10.7   Hole feature.

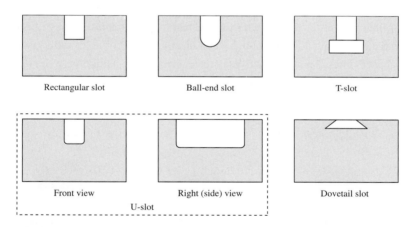

Figure 10.8   Slot feature.

11. **Pocket.** A pocket is a cavity in an existing body. Think of a pocket as a blind slot with any shape. For example, a cylindrical pocket has a circular cross section and depth, with or without blended floor. It may have straight or tapered sides. A rectangular pocket has a rectangular cross section and depth and has radii at the corners and on the floor, having straight or tapered sides. A general pocket can have any cross section. It can be generated via sketching and feature operations.

12. **Shell.** A shelling operation is used to create hollow (thin-walled) solids by carving (removing) "material" out. Essentially, shelling uses an offset operation to shell a solid. The input to a shelling operation is the faces to be removed and a wall (shelling) thickness. The wall thickness can be constant or variable. For multithickness shelling, select the faces and specify the thickness required for each face. Figure 10.9 shows a shell feature.

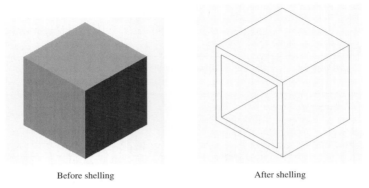

Before shelling   After shelling

Figure 10.9   Shell feature.

13. **Fillet.** Fillets are used to smoothen (round) sharp edges of solids. Many different types of fillets exist, including constant radius (default), multiple (variable) radius, setback, and round corner fillets. The setback fillets create a smooth transition between the blended faces, along the edge of the part, into the fillet corner. Figure 10.10 shows sample fillets. The input to create a fillet feature is to select the edges to be filleted and the fillet radius.

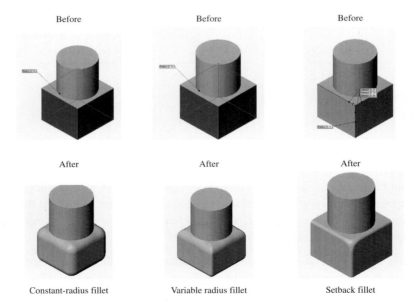

Before   Before   Before

After   After   After

Constant-radius fillet   Variable radius fillet   Setback fillet

Figure 10.10   Fillet feature.

14. **Chamfer.** We use chamfers to remove sharp edges (or corners) from parts by creating beveled edges on the selected edges, faces, or both. The input to create a chamfer is angle-distance, distance-distance, or a vertex chamfer. Figure 10.11 shows a distance-distance chamfer.

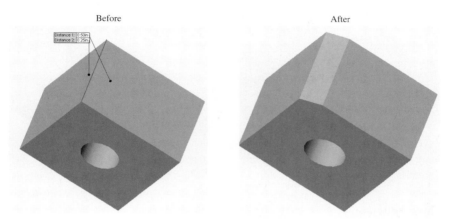

Figure 10.11   Chamfer feature.

15. **Draft.** A draft is most commonly used in injection molding of plastic parts. **Draft** is the angle between the direction of ejection of a part from the mold and the surface of the part. Its function is to facilitate the removal of the part from the mold. We need to select the face to be drafted, the draft angle, and a neutral plane to measure the draft angle from, to create a draft feature. We can create split-line draft as shown in Figure 10. 12.

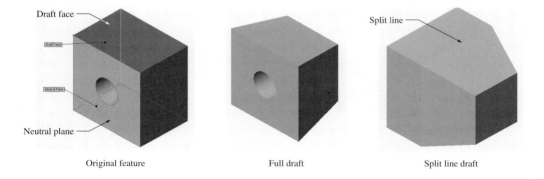

Figure 10.12   Draft feature.

16. **Rib.** A rib is a stiffener. A **rib** is a special type of extruded feature. It adds material to an existing feature. It can be created from an open or closed sketch (contour). The input to create a rib is a contour and a thickness. Figure 10. 13 shows an example.

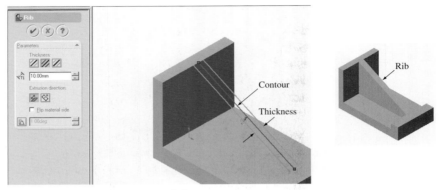

Input to create rib feature

Figure 10.13   Rib feature.

17. **Flange.** It is a revolved feature that adds material to ends of pipes or other revolved features. We usually drill holes in flanges to connect pipes (or other parts) together. Figure 10.14 shows an example.

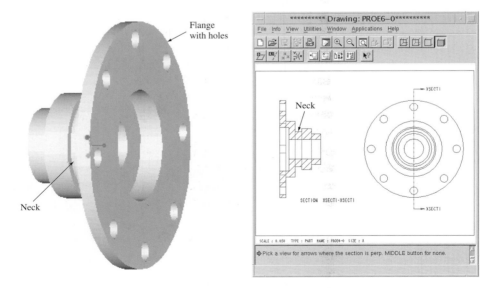

Figure 10.14   Flange and neck features.

18. **Neck.** A neck is the opposite of a flange. It removes material from an existing revolved feature. It is a revolved feature. It acts as a revolved cut. Necks are widely used in mechanical parts. Figure 10.14 shows a neck feature.

19. **Pattern.** A pattern features follows the concepts of geometric arrays we cover in Chapter 3. A pattern feature may be a rectangular (linear) or circular array. A pattern feature may be uniform or nonuniform; it may also have varying dimension. Figure 10.15 shows examples.

Uniform circular pattern   Nonuniform circular pattern   Varying-dimension linear pattern

Figure 10.15   Pattern feature.

20. **Spiral.** A spiral feature is a special case of the sweep feature. It creates spiral springs. It requires a profile to sweep and a sweeping (guide) curve. The curve is a helix. That is what makes the sweep a spiral. As in sweep, the guide curve must be perpendicular to the spiral profile (cross section) at their intersection point. Figure 10.16 shows a spiral feature.

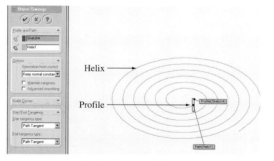

Input to create spiral feature   Spiral feature

Figure 10.16   Spiral feature.

21. **Spring.** The spring feature is identical to the spiral feature except that the profile (cross section) is circular. Figure 10.17 shows a helical spring feature.

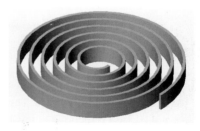

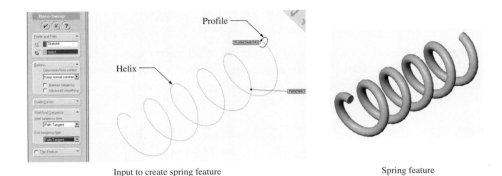

Input to create spring feature   Spring feature

Figure 10.17   Spring feature.

22. **Dimple.** A **dimple** is a small depression in a face of a solid. We usually create dimples in sheet metals or plates with small thicknesses. To create a dimple in a thin solid, cut it with an extrusion, create the profile of the depression, and boss revolve it as shown in Figure 10. 18.

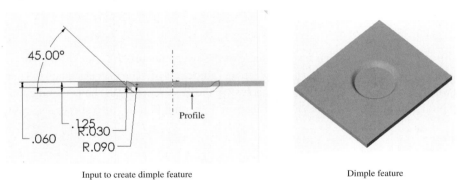

Input to create dimple feature   Dimple feature

Figure 10.18   Dimple feature.

23. **Thread.** A thread feature creates threads on cylindrical faces of holes, bosses, cylinders, or sweeps of circular curves. Threads can be right-handed (winds or tightens by rotating it in a clockwise direction) or left-handed (winds or tightens by rotating it in a counterclockwise direction). Thread parameters include length, thread angle, major diameter, minor diameter, and pitch.

24. **Tweak.** A **tweak** is a feature that modifies surfaces.

25. **Intersect.** This feature implements the Boolean intersection operation ($\cap^*$). Feature-based CAD systems such as Pro/E provide this operator to allow designers to intersect features together.

10.3 Feature Representation

The major goal of features when they originally appeared was to store a product definition—that is, design and manufacturing data in addition to shape information. Features representation should provide for this data. However, the majority, if not all, of features provided by CAD systems are shape (form) features. They only store shape information. We limit our coverage here to shape (form) features.

The representation of shape features usually contains the following properties and data:

1. **Shape.** The geometry and topology of a feature are the most essential data. Both geometry and topology are required because features are simply solid models. The shape of a feature is not limited, as in the case of primitives.

2. **Configuration parameters.** All the parameters, and their values, that fully define a feature are part of its representation.

3. **Default values.** These values become part of the GUI that CAD designers use to create features. Designers can accept or change the default values. Default values are useful when a designer tests a design concept and does not want to invest too much time working out the dimensions of a feature.

4. **Location and orientation parameters.** Similar to the primitives of Chapter 9, the location and orientation parameters of a feature locate and orient its local coordinate system with respect to the WCS during construction.

5. **Relations.** Features parameters can be related to each other via relations and equations. Section 10.6 covers relations in detail.

6. **Constraints.** Different types of constraints, such as geometric and manufacturing, exist. Section 10.7 covers geometric constraints in detail.

7. **Composite features.** Individual features are combined to form composite features. The feature we add to is known as a base feature.

8. **Symbolic or skeletal representation.** While not offered by CAD systems, symbolic representation of features is useful in the early stages of design and analysis. For example, we work with centerlines of rigid bodies and joints to analyze mechanisms and structures. We also use envelopes (convex hulls) of objects to perform a quick analysis to make design decisions.

9. **Feature validation.** When we delete (add) features from (to) existing features, the resulting composite feature (solid) must be valid. Feature validation includes attachment validation, verifying dimension and location limits, and interaction (with other features) validation.

10.4 Three-Dimensional Sketching

All features require a two-dimensional sketch. A designer may need to invoke a sketcher to sketch, depending on the CAD system. The designer selects a sketch plane and begins sketching the feature profile. We refer to this process as two-dimensional sketching.

Three-dimensional (3D) sketching is also possible and useful at times. All CAD systems, including feature-based systems such as SolidWorks and Pro/E, allow designers to create 3D sketches. 3D sketches can be used as sweep paths, as guide curves for lofts or sweeps, or as centerlines for lofts and pipes.

CAD systems may offer a 3D sketch icon or the user may simply input points by their (*x*, *y*, *z*) coordinates in a WCS. 3D sketching entities include points, lines, B-splines, fillets, centerlines, chamfers, and others. Figure 10.19 shows an example of a 3D curve. We combine the four planar edges of a plate to create a 3D composite single curve. We sweep a circle along the curve as a guide curve, to create a model that looks like a tray as shown in Figure 10.19.

Examples of 3D sketching include the creation of an oven rack frame. The outer frame of the rack is a 3D curve that requires 3D sketching. Another example is the pipe centerline shown in Figure 10.6.

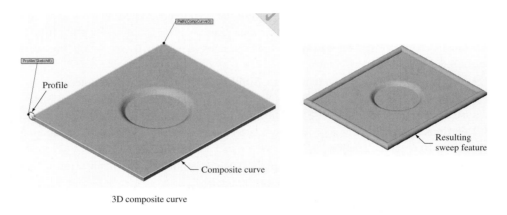

Figure 10.19   3D sketching.

10.5 Parametrics

Parametrics is a powerful characteristic of features. Parametric representation provides unlimited scaling capabilities. It creates families of parts. Parametrics supports the notion of changing and editing solid features as design changes occur in the product life cycle. Parametrics also supports the notion of model reuse. It allows designers to change dimensions in an existing model to use it again. Many design activities in practice involve modifying an existing design. Moreover, if the design is complex and requires a considerable time investment to create, parametrics can save time and money by making a few changes to modify the existing model instead of building it again entirely.

The creation of parametric features requires some steps that need to be carried out. Some CAD systems make some of the steps more transparent to the users than others. Figure 10.20

shows the steps of creating parametric features. The main idea is to parametrize the features first and then assign dimensions (values), relations, and constraints to the parameters. Some CAD systems may create feature parameters and assign values to them as the designer creates the features. Others wait until the designer defines the parameters explicitly.

Three separate, but related, sets of data are utilized in Figure 10.20: dimensions, relations, and constraints. These sets are bound to conflict with each other in one way or another at some time during the modeling process. Designers must change the data to resolve these conflicts.

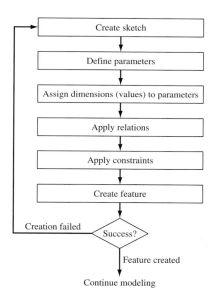

Figure 10.20   Creating parametric features.

Section 2.6 and Figure 2.7 discuss parameters in more detail. Depending on the number of parameters we define for a sketch, the sketch assumes one of three states as discussed in Section 2.5: underdefined, fully defined, or overdefined. Successful sketches must be fully defined. Moreover, the parametrization of a feature profile (sketch) is not always unique, as discussed in Section 2.6.

10.6 Relations

Primitives-based modeling (also known as Boolean modeling) has its limitations. A designer cannot incorporate intelligence (design intent) such as relations and constraints into a model. Changing one primitive of the model has no effect on other primitives. With parametrics, the designer can, for example, create a part with a hole positioned two inches away from two edges. If changes to the model move the edges or if the diameter of the hole changes, the parametric relationships relocate the hole to satisfy the original design intent. With a Boolean model, the designer must modify the hole manually, which could have adverse effects on the entire model, especially if it is complex.

Parts with algebraic or other relations between several characteristic dimensions are suited for feature-based modeling because the new values can be used to control and manipulate the model at any time. A **relation** is a mathematical way of describing how sketch parameters are related. Relations use variables, dimension names, expressions, and algebraic equations. As in programming language, a **variable** is a name that can assume multiple values, one at a time. A **dimension name** is a parameter name in features modeling. The difference between a variable and a dimension name is that the variable is not a sketch parameter while the dimension name is.

An **expression** is a collection of variables and dimension names that are combined via mathematical operators. An **equation** is a statement that has an equal sign; to the left of the equal sign is a variable, and to the right of it is an expression. The expression must evaluate to a number that is stored in the variable. The expression itself may have variables. Here are some examples:

> variables: *input1, input2, ...*
> dimension names: *length, diameter, ...*
> expression: $3*length - 1.5*diameter/3$
> equation: *height* $= 3*length - 1.5*diameter/3$

Relations are equations. We need to define variables and dimension names followed by using them to define expressions. We assign expressions to sketch parameters to create relations.

A CAD system creates relations implicitly any time a user enters information into the system to dimension a feature or to specify its location. Relations are also created when the user dimensions a sketch or changes its position.

Relations have their own grammar that usually mimics programming languages. For example, names of variables and dimension lines cannot contain spaces, and they are case sensitive. Relations may reference other relations; for example, $length = 2 + 3* p_1$, where p_1 is a relation itself.

CAD systems provide built-in mathematical functions to allow users to define relations. These functions make up system libraries. These functions are similar to what we find in most programming languages. Sample functions include absolute value (abs), the value of π, all trigonometric functions (for example, sin) and their inverses (for example, $\sin^{-1}$), the square root ($\sqrt{}$), logarithms (log), and so forth.

CAD users need a tool to type relations. Some CAD systems use the text editor of the operating system running the CAD software. For example, Pro/E uses the vi editor of the Unix system. Users may also configure their systems to use any editor of choice. There are config files that users can customize to fit their needs.

Other CAD systems, such as SolidWorks, use spreadsheets such as Microsoft Excel. These spreadsheets serve as relations tables. Users can use the columns of the spreadsheet to define variables and relations between them. Users can specify new values in the spreadsheet and then update the model to use the new values.

We identify a generic procedure to create and use relations and their tables on CAD systems. The steps of the procedure are:

1. **Rename features.** CAD systems assign generic names to features. Examples of generic names are `Base-Extrude`, `Fillet1`, `Revolve1`, and so forth. It is a good practice to replace these names with meaningful names such as `EngineMount`, `CornerFillet`, `Flange`, and so forth. Meaningful names prevent confusion about complex parts and make relations spreadsheets (tables) easy to follow and understand. Users may right-click on the features in the feature manager tree to rename them. It is also possible to assign a name to a feature when creating it.

2. **Rename feature parameters.** This step is similar to step 1. Feature parameters are also known as dimensions. To facilitate renaming parameters, CAD systems allow their users to hide or display all dimensions of a feature, or partially display them in an effort to make it manageable to handle them. Users may right-click a feature and select a `Hide All Dimensions` or `Show All Dimensions` from the popup menu. A user may also right click a given dimension to change its name. Dimension (parameter) names are displayed below the values of the dimensions.

3. **Link features' non-sketch values.** Steps 1 and 2 are applicable to the sketch parameters. These are the parameters that are part of the sketch. However, feature parameters that are not part of the sketch, such as the depth of an extrusion, must be linked to some variables to control their values. We define a variable, for example `depth`, in the spreadsheet and link the extrusion depth to it. We can link more than one feature parameter (dimension) to the same variable. We can also unlink any dimension without affecting the others that we want to remain linked.

4. **Change values and rebuild part.** At any time during step 1 or 2, change values of dimensions and rebuild the part by clicking a `Rebuild` icon. Observe the family of parts created for the different values of the dimensions.

5. **Create relations.** Define any required geometric relations to maintain the integrity of the part design and to convey the design intent of the part accurately. CAD systems permit modifying or deleting relations after they are created. Also, we can assign new values to the variables of the relations and rebuild the part to see their effects.

6. **Create a relations table.** Instead of manually changing values and rebuilding the part as in step 5, we can create a relations table that can store as many values as needed, and rebuild the model once. In this case, the CAD system executes the relations table and generates all the part instances in it, allowing the CAD designer to review all the instances at once for effective decision making. The relations table is embedded and saved in the part's file. The designer can always edit the relations table and change the values. We can think of the relations table as a database of the dimensions that is executed whenever we rebuild the part.

Users need to be aware that using the wrong set of values for dimensions or using incompatible relations could lead to a failure to rebuild the part or to a CAD system crash. Sometimes the wrong combination of values results in pathological topology or geometry. For example, an entire hole could be located completely outside a part, or self-intersecting faces or edges may result.

10.7 Constraints

Constraints add intelligence to a sketch, which helps to maintain design intent as the designer modifies the model. For example, constraining sketch geometry to stay symmetric lets the designer modify the length and width of the sketch on one side, then have the opposite side automatically stay identical. Designers can modify these intelligent constraints at any time. Constraints are applied to sketch entities, and therefore they are geometric in nature. Constraints include coincident, concentric, parallel, perpendicular, tangent, equal radius, equal distance, midpoint, symmetric, horizontal, and vertical. If we cannot find a built-in system constraint to meet our design needs, we write a relation. Relation that we create in a model, which we covered in Section 10.6, can be viewed as custom constraints. Sometimes it is hard to distinguish between relations and constraints. Some users and CAD systems use the terms interchangeably. We define a relation in Section 10.6. A **constraint** is defined as a geometric condition that relates two or more sketch entities.

Dragging sketch entities changes the sketch profile dramatically. However, entities react to dragging based on the constraints that control them. For example, a circle with a center at the origin remains at the origin, while its radius can change.

CAD systems allow their users to enable an auto constraints mode. Once enabled, the CAD system looks for geometric relationships between entities such as concentric, parallel, and coincident. Auto constraints is a useful tool for imported sketch geometry of existing parts because it quickly assigns logical constraints to sketch entities. If needed, the user can delete (or edit) the undesired constraints from the automatically generated set.

Unlike primitives, sketching and sketches lend themselves very well to relations and constraints. Sketches can capture the design intent because a designer can assign relations and constraints to sketch entities. A sketch is considered the best tool to create solids because of its ability to edit and change the shape of the solid in the future by using its relations and constraints. In addition, a sketch has the flexibility to create any complex shape.

Constraints are almost always geometric. Table 10.1 shows a list of the most common constraints offered by CAD systems, along with their geometric meaning. Most of them require the selection of two sketch entities; one acts as a reference. For example, the Concentric constraint requires two circles. Most of these constraints are self-explanatory. We explain the less obvious ones as follows.

●**Coincident.** Makes two distinct points have the same (x, y, z) coordinates. This constraint could be useful in designing mechanisms where two ends of two rods must be coincident all the time.

●**Concentric.** Makes two distinct circles have the same center all the time. The radius can be different or the same.

●**Coradial.** Makes a circle and an arc, whose centers are the same, tangent to each other. The circle radius changes to become equal to that of the arc.

●**Equal.** Makes two sketch entities (lines or angles) equal. Equal lines means they have the same length. Equal angles means they have the same value.

●**Fix.** Fixes the location of a sketch entity in space relative to other entities.

●**Merge points.** Makes two points become one. This constraint is useful for closing boundaries. For example, if we sketch two separate lines touching at two of their ends, the ends have a gap between them that is not visible to the naked eye. When we merge them, the two lines meet perfectly at the end.

●**Mirror.** Updates the mirrored part of a sketch when its parent (the one we mirrored) changes.

●**Pierce.** Makes one entity penetrate the other. This constraint is useful when features operations fail. For example, sweeping a profile along a guide curve fails, in some cases, if we do not use the Pierce constraint. The screenshot on the right shows an example of sweeping a circle along a guide curve that requires the use of a Pierce constraint.

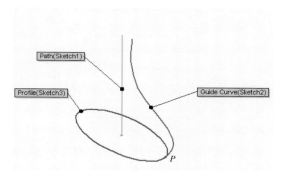

Must apply Pierce constraint at point *P* between profile and guide curve for sweep operation to work

●**Symmetric.** Makes an entity symmetric relative to another entity. For example, we can make a circle symmetric relative to an axis.

Table 10.1   Geometric constraints.

Constraints are coincident, collinear, concentric, constant angle, constant length, coradial, equal, fix, horizontal, intersection, merge points, midpoint, mirror, parallel, perpendicular, pierce, symmetric, tangent, and vertical.

Constraint	Before applying constraint	After applying constraint
Coincident	• P_2 • P_1 Different locations	• P_1, P_2 Same location
Collinear	L_1 ———— L_2 Different locations	L_1 ———————— L_2 Same location

Table 10.1   Geometric constraints. (Continued)

Constraints are coincident, collinear, concentric, constant angle, constant length, coradial, equal, fix, horizontal, intersection, merge points, midpoint, mirror, parallel, perpendicular, pierce, symmetric, tangent, and vertical.

Constraint	Before applying constraint	After applying constraint
Concentric	Different centers	Same center
Constant angle	θ can change	θ cannot change
Constant length	Length L can change	Length L cannot change
Coradial	Edges are not tangent	Edges must be tangent
Equal	$L_1 \neq L_2$ $\theta_1 \neq \theta_2$	$L_1 = L_2$ $\theta_1 = \theta_2$
Fix	P_1 can move	P_1 cannot move
Horizontal	Line is not horizontal	Line must be horizontal
Intersection	L_1 and L_2 may or may not intersect	L_1 and L_2 must intersect
Merge points	P_1 and P_2 are two different points	P_1 and P_2 are now one point P
Midpoint	P is any point	P must be midpoint M

Table 10.1   Geometric constraints. (Continued)

Constraints are coincident, collinear, concentric, constant angle, constant length, coradial, equal, fix, horizontal, intersection, merge points, midpoint, mirror, parallel, perpendicular, pierce, symmetric, tangent, and vertical.

Constraint	Before applying constraint	After applying constraint
Mirror	R_1 and R_2 are two different regions	R_2 must be a mirror of R_1
Parallel	L_1 and L_2 are not parallel	L_1 and L_2 must be parallel
Perpendicular	L_1 and L_2 are not perpendicular	L_1 and L_2 must be perpendicular
Pierce	Curve C_1 and circle C_2 are just touching at P	Curve C_1 and circle C_2 are overlapping slightly at P
Symmetric	Left and right halves may be symm.	Left and right halves must be symm.
Tangent	Entities are not tangent	Entities must be tangent
Vertical	Line is not vertical	Line must be vertical

10.8   Feature Manipulations

The solid manipulations covered in Chapter 9 apply to features; after all, features are solids. In addition, the sketches of features can be manipulated. This offers us more manipulation power, thus adding more flexibility to using features. Manipulating sketches allows us to use all curve and surface manipulation concepts covered in Chapters 6 and 7, respectively.

10.9 Tutorials

10.9.1 Create a Sweep Feature

This tutorial shows how to create a sweep feature. To create a sweep feature, we need a sweep path and a sweep section that is moved along the sweep path. Here, we sweep a circle along a B-spline curve to create the solid shown on the right. Here are the detailed creation steps (modeling strategy); all dimensions are in mm:

1. **Sketch the sweep path.** Select `Top` as the sketch plane. Orient the plane so that you can see it clearly. Create points by clicking in the sketch plane to generate the points, approximately in the locations shown below. Connect the points with B-spline as shown below.

2. **Sketch the sweep profile (cross section).** Create a circle, shown below, with center at the origin and diameter 25 mm in the `Front` sketch plane.
3. **Create the sweep feature.** Use a sweep feature operation to create the sweep. Select the profile (sweep section) first and then the path. The resulting feature is shown below.

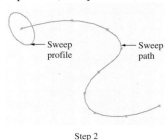

Step 2

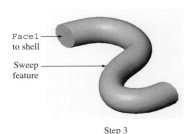

Step 3

Tutorial 10.9.1 discussion:

We ensure the perpendicularity of the sweep profile and the sweep path by sketching on two perpendicular sketch planes (`Top` for the profile and `Front` for the path). Otherwise, the sweep operation fails.

Tutorial 10.9.1 hands-on exercise:

Shell the sweep. Use `Face1` as the face to shell and use a shelling thickness of 5 mm.

10.9.2 Create a Loft Feature

We create a loft feature that blends four cross sections (profiles) along a B-spline guide curve. The feature is shown on the right. Here is the modeling strategy to create the loft; all dimensions are in mm:

1. **Sketch the first profile.** Select the `Front` sketch plane and sketch the geometry shown below. The geometry consists of two concentric circles, two axes at 45° each, and horizontal and vertical axes. Use the intersection of the horizontal axis and the inner circle as the center to draw the circle shown below. Trim the circle just created and use a circular array and copy it eight times. Trim all circles to get the final profile shown.

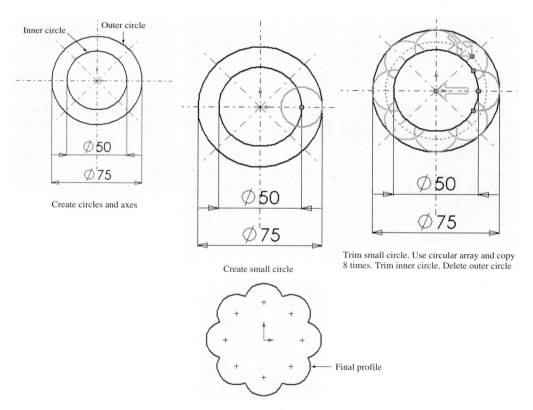

Inner circle Outer circle

Ø50
Ø75

Create circles and axes

Ø50
Ø75

Create small circle

Ø50
Ø75

Trim small circle. Use circular array and copy 8 times. Trim inner circle. Delete outer circle

Final profile

2. **Create the other three profiles.** The three profiles are derived from the first one. We first create three sketch planes by offsetting the `Front` plane a distance of 25 mm three times to create the forthcoming `Plane1`, `Plane2`, and `Plane3`. We copy the first profile onto `Plane1`. We offset the copied profile outward a distance of 20 mm to enlarge it. We copy the resulting offset profile onto `Plane2`. We also copy the first profile onto `Plane3`.

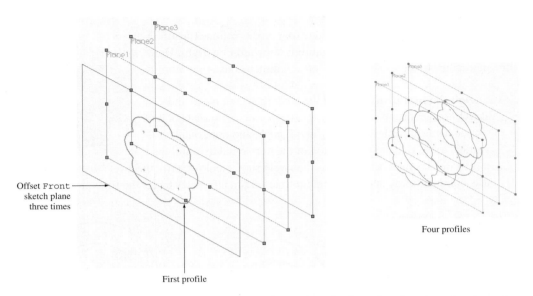

Offset Front
sketch plane
three times

First profile

Four profiles

3. **Create the loft.** Using a loft feature operation, select the four sketches, in order, in such a way that the point of selection is curvilinear as shown below. The loft feature is created.

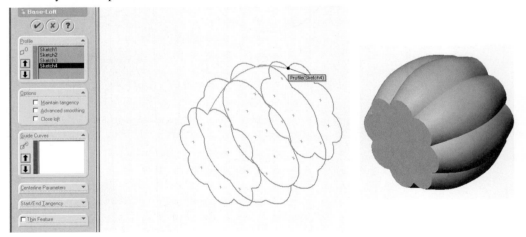

Tutorial 10.9.2 discussion:

If we select the profiles in noncurvilinear fashion, a twisted loft results, as shown in the forthcoming screenshot, because the CAD system interpolates the profiles starting from the user's selection clicks.

Tutorial 10.9.2 hands-on exercise:

Add a 130×150 mm rectangular profile in the middle of the four profiles. Create a new offset plane at a 25 mm distance. Re-create the loft using the five profiles. How different is it from the loft of this tutorial? Hint: Depending on your CAD system, you may have to divide (split) each rectangle side into two (equal) parts to create eight entities, in total, that match the eight arcs of each of the other profiles. The CAD system may need this to interpolate the five profiles correctly.

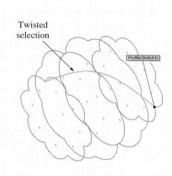

Twisted selection

Profile (Sketch4)

Twisted loft

10.9.3 Create a Spiral Feature

We create the spiral spring shown on the right. The spiral spring with a uniform cross-section is essentially a sweep feature. The profile is the cross section itself, and the sweep path is the spiral, which can be created using curves in a CAD system.

Here is the modeling strategy to create the spiral spring; all dimensions are in mm:

1. **Sketch the spiral curve.** Select Top as the sketch plane.
   Using a helix/spiral command on your CAD system, create the helix with the following parameters: the helix requires a pitch (4 mm), a number of revolutions (6 mm), a starting angle (90 °), and a direction (clockwise). Each revolution sweeps 360°. The **pitch** is the distance that separates the revolutions apart as shown in the forthcoming screenshot. The **starting angle** is the angle between the helix and the profile (to be swept along it) at their contact point.
2. **Sketch the spring profile (cross section).** Select Front as the sketch plane. Sketch the profile shown in the forthcoming screenshot.
3. **Create the spiral spring.** Use a sweep operation on your CAD system to create the spring. Select the rectangle as the profile and the spiral as the path. Set other parameters as shown the upcoming screenshot.

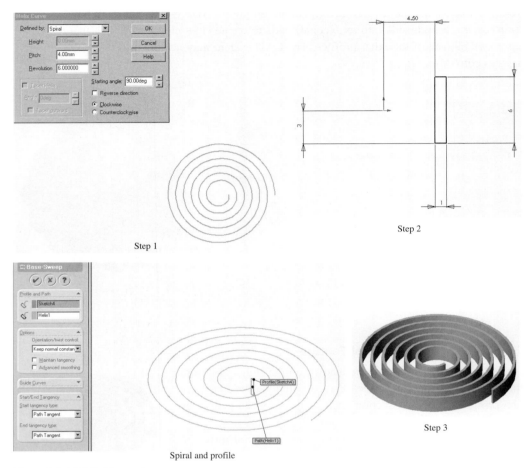

Step 1

Step 2

Step 3

Spiral and profile

Tutorial 10.9.3 discussion:

We use a spiral starting angle of 90° so that the starting point of the spiral coincides with the `Front` plane, or else we have to create a new datum plane such that it is coincident with the starting point of the helix.

Tutorial 10.9.3 hands-on exercise:

What is the maximum width of the rectangular profile before the spiral begins to intersect itself? Prove your answer using your CAD system. Also change the starting angle of the spiral to 30° and re-create the spiral.

10.9.4 Create a Helical Spring

The creation of a helical spring is the same as the spiral. We use a circular profile in this tutorial. We use the same three steps of Tutorial 10.9.3. We use five revolutions, a pitch of 10 mm, starting angle of 90°, and a clockwise direction. The following screenshots show them.

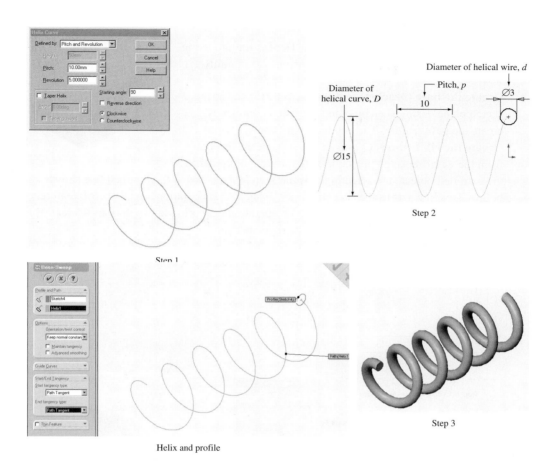

Diameter of helical curve, D

Diameter of helical wire, d

Pitch, p

Ø15

10

Ø3

Step 1

Step 2

Helix and profile

Step 3

Tutorial 10.9.4 discussion:

The spring requires one more parameter than the spiral: the diameter (D) of the helical curve as shown above. The profile diameter (d) is known as the wire diameter. Think of how springs are made. We use a wire of diameter d and coil it with a coil diameter D. Each coil is a revolution. The distance between two identical points on two consecutive coils is the *pitch, p*.

Tutorial 10.9.4 hands-on exercise:

Edit the spring feature definition to convert it into a taper spring. Use the feature-manager tree of your CAD system. Check the system for a `Taper Helix` command, or its equivalent. Select the `Taper outward` (or `Taper inward`) option. The screenshot on the right shows the result.

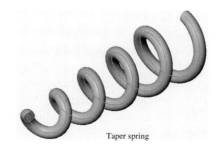

Taper spring

10.9.5 Create a Rib Feature

We create the rib shown on the right. Here is the modeling strategy to create the spiral spring; all dimensions are in mm:

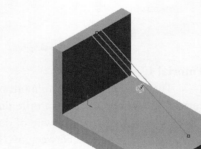

1. **Create the base (bracket) feature.** Select `Front` as the sketch plane. Using sketching and extrusion commands, create the base feature. The depth of extrusion is 75 mm.
2. **Create the rib feature.** Create an offset plane at 37.5 mm from the `Front` sketch plane, passing through the center of the extruded object, and selecting it as the sketch plane sketch the inclined line shown below. This line acts as the rib centerline. Using a `Rib` operation, create the rib feature. Use 10 mm for the rib thickness. The `Rib` operation may provide two options to create the rib: one side or both sides relative to the rib centerline.

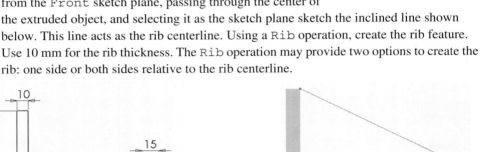

Step 1: Create extrusion profile

Step 2: Create rib centerline

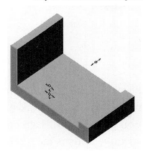

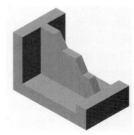

Step 1: Create base feature

Step 2: Create rib feature

Tutorial 10.9.5 discussion:

The rib feature is essentially an extrusion. We can create the rib in this tutorial by sketching the triangular cross section as the profile, and extrude it 10 mm. However, the `Rib` operation makes it less laborious and faster to create the feature.

Tutorial 10.9.5 hands-on exercise:

Edit the rib feature definition to convert it into a stepped rib as shown on the right. Assume the dimensions.

PROBLEMS

Part I: Theory

10.1 List the benefits of sketching and using sketches to create solid models.

10.2 List all the solids features that you know. What does each one create?

10.3 What is the difference between linear and nonlinear sweeps?

10.4 What are the two types of nonlinear lofts?

10.5 List the differences and the similarities between nonlinear sweeps and lofts.

10.6 What are the parameters that define a helix? Draw a sketch to illustrate your answer.

10.7 The concept of projected curves can be used to create nonplanar curves and embossing effects on a model. What are the geometric parameters needed to define each? How do you create each?

10.8 Describe how the variable-radius fillet works. When must you use one?

Part II: Laboratory

Use your in-house CAD/CAM system to answer the questions in this part.

10.9 Construct the following nonlinear sweep, with an elliptical profile, a centerline path, and a guide curve. Approximate dimensions. **Hint:** Each of the entities has to be a separate sketch. Use a `Pierce` constraint at the end of the guide curve (intersection with profile).

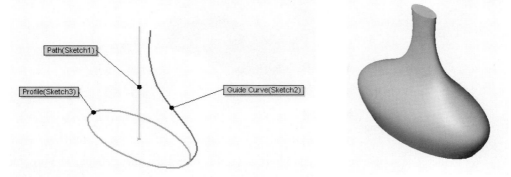

10.10 Construct the following sweep, with multiple profiles. Assume appropriate dimensions.

10.11 Create the following part with a 10° draft as shown below. Assume dimensions to create the part. Hint: use a `Parting Line` command with `Step Draft` option.

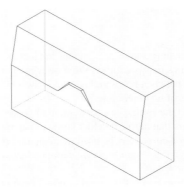

10.12 Create the following face fillet. Notice the elliptical shape of the fillet. Assume dimensions to create the part. **Hint:** Use a `Face Blend Fillet` with `Split Line`.

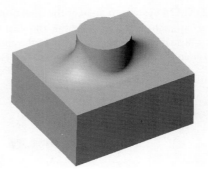

Part III: Programming

Write the code so that the user can input values to control the curve shape. The program should display the results. Use OpenGL with C or C++, or Use Java 3D.

10.13 Write a Java or C++ program to implement the parametric equation of the helix curve of Example 6.2.

Computer Graphics

This part covers the graphics concepts that are used in computer graphics in general. This part should help readers to understand much of the jargon utilized in the field. To achieve this goal, we accomplish the following objectives:

1. Understand graphics display **(Chapter 11)**
2. Understand geometric transformations **(Chapter 12)**
3. Understand rendering and visualization **(Chapter 13)**
4. Understand computer animation **(Chapter 14)**

Graphics Display

GOAL

Understand and master the concept of raster displays, the types of display monitors (including the digital flat panel displays), and the parameters of these displays.

OBJECTIVES

After reading this chapter, you should understand the following concepts:

- Impact of displays on CAD models
- CRTs
- Flat screen CRTs
- Flat panel displays
- LCDs
- Digital flat panel displays
- Comparing displays
- Specifications

CHAPTER HEADLINES

11.1   Introduction

CAD/CAM systems are considered off-the-shelf systems; that is, both hardware and software are standard open architecture. They are very much plug-and-play systems. Hardware is standard computer equipment with standard input (mouse and keyboard) and output devices (monitors, printers, and plotters). Software is written in standard programming languages such as C and C++, and it runs under standard operating systems.

The graphics display (monitor) of CAD hardware is considered an important component because the quality of the displayed image influences the perception of generated designs on the CAD/CAM system. In addition to viewing images, the graphics display enables the user to communicate with the displayed image by adding, deleting, blanking, and moving graphics entities on the display screen. As a matter of fact, this communication process is what gives interactive graphics its name to differentiate it from passive graphics, as in the case of a home TV set, that the user cannot change.

Knowledge of the basic hardware principles upon which graphics displays are designed is important to understanding how to develop and implement computer graphics algorithms to render and visualize CAD models. Visualizing these models is crucial to decision making in specialized applications that require high-resolution and high-quality images. These applications include medical imaging systems, analyzing high-terrain topology, recognizing and interpreting satellite images, and performing scientific visualization.

Outside the realm of science and engineering, understanding the principles of graphics display is very helpful in dealing with Web-based applications, including image generation and editing, and writing graphics applications. For example, Java 3D and OpenGL graphics systems are based on these principles. All the graphics software such as Adobe Photoshop and Microsoft Paint use the principles we cover in this chapter; the Java programming language uses them as well. Moreover, scanning software utilizes them. Finally, the color schemes and mixing color signals in Web applications and XHTML (Extensible HyperText Markup Language) are based on them.

Various display technologies are available to the user to choose from. They are all based on the concept of converting the computer's electrical signals, controlled by the corresponding digital information (CAD model data), into visible images at high speeds. Among the available technologies, the CRT (cathode ray tube) is still a formidable technology and has produced a wide range of effective graphics displays. Other technologies utilize laser, flat panel displays, or plasma panel displays. In the first, a laser beam, instead of an electron beam, is used to trace an image in a film. In the second, a liquid crystal display (LCD) and light-emitting diodes (LEDs) are used to generate images. The plasma display uses small neon bulbs arranged in a panel, which provides a medium-resolution display.

The LCDs are becoming more popular as their prices continue to drop. They compete well with CRTs because their quality is superior and their images are sharper. What is holding their dominance back is the price. They may very well replace CRTs if their prices drop down to the range of those of CRTs.

11.2 CRT

Figure 11.1 shows a schematic diagram of a typical CRT. The CRT uses a vacuum tube that provides high vacuum and voltage. Its operation is based on the concept of using an electron gun to fire a beam of electrons onto a color phosphor coating, in the back of the display surface, at very high speed. The energy transfer from the electron to the phosphor due to the impact, causes it to illuminate and glow. The electrons are generated by the electron gun, which contains the cathode, and they are focused into a beam by the focusing unit shown in Figure 11.1. The control of the beam's direction and intensity in a way that is related to the graphics information (CAD data) generated in the computer allows us to display meaningful and desired graphics on the screen.

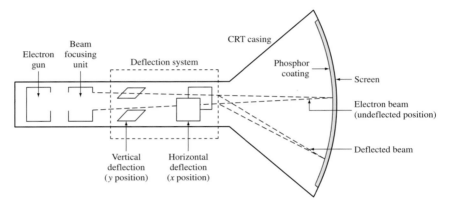

Figure 11.1   Schematic diagram of a CRT.

Different technologies exist to prevent the beam electrons from interfering with each other as they hit the phosphor. The old technology uses a shadow mask to break up light beams, so that each bit of light hits a phosphor dot and produces illumination. The **dot pitch** of a monitor is the distance between two dots (pixels) in the mask. The smaller the pitch, the sharper the image. A better technology has been developed by Sony. Called **aperture grill**, it features light that is painted in vertical stripes. The result is a crisper picture, equivalent to a finer dot pitch. Less interference with the light coming to the screen explains the noticeably brighter image associated with Sony's Trinitron tubes. Also, some CRTs use a single light gun instead of three separate guns, one for each color (red, green, and blue).

The deflection system of the CRT controls the x and y, or the horizontal and vertical, positions of the beam, which in turn are related to the graphics information through the display controller which typically sits between the computer and the CRT. The controller receives the information from the computer and converts it into signals acceptable to the CRT. Other names for the display controller are the display processor, the display logical processor, or the display processing unit.

The major tasks that the display processor performs are the voltage-level convergence between the computer and the CRT, the compensation for the difference in speed between the computer and the CRT (by acting as a buffer), and the generation of graphics and texts. More often, display processors use extra hardware to embed standard graphics software functions, such as shading and geometric transformations, into hardware to improve execution speed.

The graphics display can be divided into two types based on the scan technology used to control the electron beam to generate graphics on the screen. These are random and raster scans. In random scan (also referred to as stroke writing, vector writing, or calligraphic scan), graphics are generated by drawing vectors or line segments on the screen in a random order that is controlled by the user input and the software. The word "random" indicates that the screen is not scanned in a particular order. In the raster scan system, the screen is scanned from left to right, top to bottom all the time to display graphics, similar to the home TV scan system. Figure 11.2 shows the two types of scans. CRT displays are based on the raster scan technique.

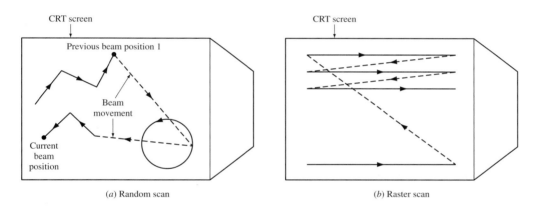

(a) Random scan   (b) Raster scan

Figure 11.2   Types of CRT screen scans.

The CRTs have some shortcomings. The most obvious one is the size. A CRT is big because of the space required by the electron gun and its focusing units. It is heavy and suffers from flicker. The actual viewable screen size is smaller than the measured glass face due to the casing overlap with it. A CRT consumes too much electricity because of the high voltage that is required by its tube. Its single electron beam is prone to misfocus. The displayed colors exhibit variations across the screen. Its clunky high-voltage electric circuits and strong magnetic fields create harmful electromagnetic variations. A CRT requires an analog signal; it uses a VGA (video graphics adapter) or SVGA (super VGA). Finally, the quality of image display suffers from the curved screen due to reflections and glare.

It is inevitable that one of the flat panel display technologies will win out in the long run and replace CRTs. The key factor that is preventing the demise of the CRT technology is the price.

11.3 Flat Screen CRT

Flat screen CRTs have flat screens. The flat display screen makes the monitors inherently comfortable to look at from wide viewing angles with the same bright, clean screen images associated with typical flat panel displays. A flat screen CRT catches fewer reflections. It also offers a larger viewable area than an equivalently sized curved CRT. Additionally, the flat screen CRT is well suited for touchscreen applications because it is more cost effective than the traditional curved CRT.

Flat screen CRTs still suffer from most of the problems of CRTs except for the screen curvature problem. The flat screen, however, improves the quality of the image display; it makes it crisper and reduces the amount of screen glare.

11.4 Analog Flat Panel Displays

Flat panel displays overcome all the shortcomings of CRTs. They eliminate the distortions caused by the curved CRTs. They provide a larger viewable screen area because the edges of the casing frame do not overlap with the glass of the screen. They use very little desk space. They provide clearer, brighter, and crisper images than CRTs. They are flicker free. They emit less radiation and use less energy than CRTs. They can display digital signals, thus eliminating the need to convert analog signals to digital ones as required by CRTs. Flat panel monitors can be adjusted to different heights, tilted, or swivelled 90 degrees from portrait to landscape. They can be detached from their base and mounted on a wall. This flexibility follows the trend in PCs toward smaller and more innovative form factors.

Some of the disadvantages of flat panel displays include the ease damaging of the screen's soft surface, no built-in speakers, poorer display (motion blur) at high speed action, and fixed native resolution. Resetting the resolution of a flat panel (as we do with CRTs through, for example, Windows Desktop) degrades its image quality. A key difference between CRT and LCD display technologies is that the LCD is a "fixed-pixel" display. If you must have a variable resolution monitor, then you should use a CRT and not a flat panel display.

Flat panel displays have other names that are derived from their look or the technologies they use. Common names are flat screens, LCD (liquid crystal display) displays, TFT (thin film transistor) panels, and thin screens.

Flat panel displays are used in many other devices than desktop display monitors. They are used in laptop computers, Palm Pilots, pocket PCs, PDAs (personal digital assistants), cell phones, and wall-mounted screens.

Flat panel displays do not use high-voltage or high-vacuum tubes. The most commonly available LCD displays use TFT technology to produce high quality images. A liquid crystal consists of long, rod-like molecules which, in their natural state, arrange themselves with their long axes roughly parallel. The molecules are almost transparent substances, exhibiting the properties of both solid and liquid matter. The property of solid matter makes light passing through the molecules follow the alignment of the solid molecules. The property of liquid matter make molecules change their molecular alignment when they are charged with electricity, and consequently this changes the way light passes through them.

The working and theory of an LCD display is based on two concepts: twisted molecules and polarized light. Molecules can be twisted (aligned) to pass light through them or to block it; molecules essentially redirect light. We can twist molecules from their natural vertical alignment to a horizontal alignment, or to any other angle (state) in between, by forcing them to pass through finely grooved surfaces. Vertical grooves orient the crystals vertically, while horizontal grooves orient them horizontally.

Polarizing light means controlling its orientation in space. We control light orientation via the use of polarizing filters. Natural light waves are oriented at random angles. A **polarizing filter** is a fine set of parallel lines that acts as a net, blocking all light waves from passing though the grid except those waves that are parallel to the lines. A polarizing filter works like a shutter, letting light in or blocking it out, depending on the angle of the light.

Figure 11.3 shows a schematic diagram of an LCD display. The figure shows a side view (perpendicular to its image-display surface) of the display. The diagram illustrates the two basic concepts behind LCD technology. A backlight panel is used to generate light. The display uses two grooved glass panels and two polarizing filters, forming two compatible pairs. The vertical pair has a vertical polarizing filter and a vertically grooved glass panel next to it. Similarly, the horizontal pair has a horizontally grooved glass panel and a horizontal polarizing filter next to it. Voltage is applied between the two glass panels to charge the molecules between the glass panels to change their alignment from vertical to horizontal.

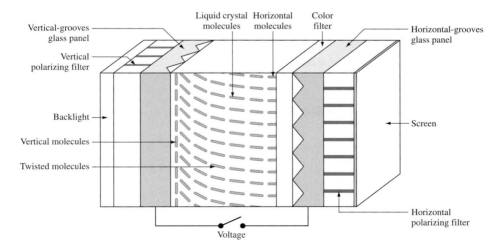

Figure 11.3   Schematic diagram (side view) of a flat panel display.

Based on Figure 11.3, light is polarized vertically via the vertical filter, therefore passing through the glass panel with vertical grooves and the molecules next to it. As the light passes through the molecules, it gets twisted. When the light reaches the second glass panel, it would have been oriented perfectly horizontal. It thus passes through the horizontal polarizing filter and causes the display to show the image.

The backlight is an assembly that can be side-lit, backlit, or direct-lit. The light of the backlight is provided by fluorescent lamps that have a finite life time. The aging process of the lamps (bulbs) reduces the brightness of the screen. When a backlight is replaced, it should be replaced as an entire assembly. This way the lamps age in unison. To keep constant brightness across the screen, the lamps need to be of the same age. A typical life of the bulbs is three to four years of leaving the monitor all the times. We can extend the life of the backlight by turning the monitor off when it is not in use, or by using a screen saver.

The color filter shown in Figure 11.3 is used to generate colors. Three cells are used in the filter, one for each of the red, green, and blue signals. This is similar to using three electron guns in a CRT. The color effect is created by controlling the level of brightness between all light and no light passing through.

The implementation of grooves in the glass panels is achieved by a special surface treatment given to the glass so that the grooves at one glass panel is perpendicular to the grooves at the other. This configuration sets up a 90-degree twist into the bulk of the liquid crystal. The most common LCD that is used for everyday items like watches and calculators is called the twisted nematic (TN) display. This device consists of a nematic liquid crystal sandwiched between two plates (panels) of glass.

The amount of twisting of LCD molecules controls the display sharpness. The more the molecules are twisted, the better the display contrast. TN twists the molecules 90 percent to improve contrast. STN (super TN) twists the molecules by 140 percent to make the contrast even better.

11.5 Digital Flat Panel Displays

Computers are inherently digital. They produce digital signals that display monitors use as input. If a monitor is analog, as in the case of CRTs and analog flat panel displays, the signals must be converted (via using a digital-to-analog converter, DAC) into analog signals before the monitor can display them. The DAC is part of the monitor interface, known as the video interface. The two types of hardware interfaces are analog video interface (AVI) and digital video interface (DVI).

Analog flat panel displays covered in Section 11.4 use AVIs. After a display receives the analog signal from its AVI, it must convert it back to digital before it can display it. The two conversions (digital to analog by the AVI, and analog back to digital by the display) result in what is known as pixel-color error. Pixel-color error is inherent in analog data, but analog displays can conceal errors by blurring or blending the phosphor in CRTs or the neighboring pixels in flat panel displays.

Digital flat panel displays do not suffer from the pixel-color error because they use DVIs. Maintaining the digital integrity of the signal from start to finish by the DVIs ensures that the digital-to-analog and analog-to-digital conversions do not introduce pixel-color error or signal degradation. Therefore, the digital flat-panel displays consistently provide crisp and sharp line edges and color values that are true to the original data, especially at higher resolutions.

All the concepts and hardware operations covered in Section 11.4 apply to digital flat-panel displays. The difference between analog and digital flat panels comes in the video interface each type uses.

11.6   Raster Displays

All types of displays covered so far, whether they are CRTs or flat panel displays, are raster displays. In raster displays, the display screen area is divided horizontally and vertically into a matrix of small elements called picture elements or pixels, as shown in Figure 11.4. A pixel is the smallest addressable area on a screen. An $N \times M$ resolution defines a screen with N rows and M columns. Each row defines a scan line. A rasterization process is needed in order to display either a shaded area or graphics entities. In this process, the area or entities are converted into their corresponding pixels whose intensity and color are controlled by the image display system.

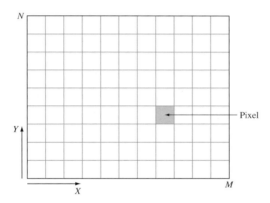

Figure 11.4   Pixel matrix of raster display.

Figure 11.5 shows a schematic of a typical CRT color raster display. Images (shaded areas or graphics entities) are displayed by converting geometric information into pixel values, which are then converted into electron beam deflections through the display processor and the deflection system shown in the figure. If the display is monochrome, a pixel value is used to control the intensity level or the gray level on the screen. For color displays, the value is used to control the color by mapping it into a color map.

In flat panel displays, we create colors and gray scale shades by changing the strength of the voltage applied to the molecules (Figure 11.3). In fact, the liquid molecules untwist at a speed directly proportional to the strength of the voltage, thereby allowing the amount of light passing through to be controlled. Similar to the amounts of beam deflections in CRTs, the voltage values become the pixel values, and they are related to the geometric information to be displayed.

The creation of raster-format data from geometric information is known as scan conversion or rasterization. A rasterizer that forms the image creation system as shown in Figure 11.5 for a CRT is mainly a set of scan-conversion algorithms. Due to the universal need for these algorithms, the scan conversion or rasterization process is hardware implemented and is done locally in the display. As an example, there are standard algorithms such as DDA (Digital Differential Analyzer) and Bresenham's method which are used to draw a line by generating pixels to approximate the line. Similar algorithms exist to draw arcs, text, and surfaces.

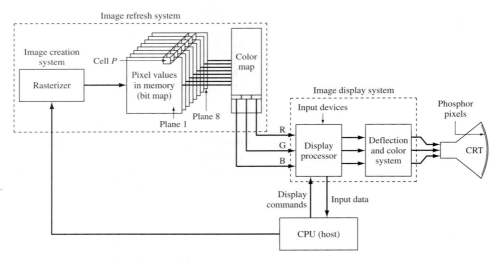

Figure 11.5 Color raster display with eight planes.

11.7 Specifications

CRTs and flat panel displays have many parameters each. Many of these parameters are shared by both types. Unless we specify otherwise, a parameter applies to both types. Here is a list of the parameters:

1. **Screen size.** The size of a monitor is specified by its diagonal length. Typical sizes are 15-, 17-, 19-, and 21-inch diagonal. In CRTs, the size of the viewable area is less than the monitor size due to the overlapping of the edges of the monitor casing with the monitor

screen. This problem does not exist for flat panel displays. Practice has it that a 15-inch flat panel is equivalent to a 19-inch CRT.

2. **Resolution.** The resolution of a monitor is determined by the number of pixels in both the horizontal and the vertical directions, as shown in Figure 11.4. The higher the resolution, the better the image quality on the display screen. Sample resolution are 640×512, 1280×1024, 1200×1600, and 4096×4096.

3. **dpi.** The resolution by itself is not enough to determine the quality of a monitor. The dpi (dots per inch) should also be considered; a dot is a pixel. The two together provide an excellent indication of the monitor's quality. A **dpi** figure is the number of pixels that a monitor has in a horizontal or vertical distance of one inch. It measures the density of pixels. The dpi is the inverse of the dot pitch; that is, $dpi = 1/$(dot pitch). Sample dpi are 78 and 105.

4. **Screen type.** This parameter describes the material used to make the display surface of the monitor. One popular type for flat panel displays is AMLCD (active-matrix liquid crystal display). AMLCD is popular due to its light weight, excellent image quality, wide color gamut (palette), and fast response time (refresh rate).

5. **Brightness and contrast.** These parameters indicate the quality of the monitor. The higher the values, the better.

6. **Viewing angle.** The **viewing angle** is the angle between a line of sight from an eye looking at the screen of a monitor and the monitor surface. The perfect viewing angle is 90°. We sometimes experience glare on laptop computer screens or on a PDA or cell phone if we look at it from the wrong angle. Flat panel displays have more problems with viewing angles than CRTs because the LCD technology is transmissive (passes light through the screen), while CRTs are emissive (phosphor emits light to screen). The emitted light in CRTs enables the screen to be viewed easily from greater angles. In a flat panel display, a light that passes through the intended pixel is obliquely emitted through adjacent pixels, causing color distortion and glare. A typical viewing angle for flat panel displays is 170° H/V (horizontal/vertical).

7. **Video interface.** CRTs use AVI, while flat panel displays can use AVI or DVI. DVI results in better-quality display for flat panels.

8. **VGA controller.** Both CRTs and flat panel displays should provide full VGA support. It is even better if they provide full SVGA (super VGA) support.

9. **Bitmap (frame buffer).** The values of the pixels are stored in a memory called a frame buffer or bitmap as shown in Figure 11.5. Each pixel value determines its color (or gray level) on the screen. There is one-to-one correspondence between every cell in the bitmap memory and every pixel on the screen. The display processor maps every cell into its corresponding screen pixel color.

10. **Display depth (number of bit planes).** The pixel cell mentioned in item 9 is a collection of bits. The values of these bits are the color values of the pixel. A bit can hold only two values (2^1): 0 or 1. Thus, one bit/pixel produces only a two level image (black or white),

which is unsatisfactory for any application. Practice suggests that 8 bits/pixel are needed to produce satisfactory continuous shades of gray for monochrome displays. For color displays, 24 bits/pixel would be needed: 8 bits for each primary color red, blue, and green. This would provide 2^{24} different colors, which are far more than needed in real applications. Typically, 4 to 8 bits/pixel are adequate for both monochrome and color displays used in most engineering applications. Some image processing applications may require more than that.

The bitmap memory is arranged conceptually as series of planes, one for each bit in the pixel value. The layout of each plane is identical to the pixel layout of the screen (Figure 11.4). Think of planes as layers of the monitor screen laid on top of each other. An eight-plane memory provides 8 bits/pixel as shown in Figures 11.5, which produces 2^8 different gray levels or different colors that can be displayed simultaneously in one image. The number of bits (planes) per pixel directly affects the quality of a display and consequently its price. This number is also known as the **display depth**.

11. **Lookup table.** The value of a pixel in the bitmap memory is translated to a gray level or a color through a lookup table (also called color table or color map for a color display). The pixel value is used as an index for this lookup table to find the corresponding table entry value, which is then used by the display system to control the gray level or color. Figure 11.6 shows how the pixel value is related to the lookup table in an eight-plane display. If cell P in the bitmap corresponds to pixel P at the location (x, y) on the screen, then the gray level of this pixel is 50 (00110010) or its corresponding color is 50.

12. **Color palette.** Figure 11.6 shows raster displays with what is called direct-definition systems in which the lookup table always has as many entries as there are pixel values in the bitmap. For the eight-plane display shown in the figure, the lookup table has 256 (2^8) entries, which correspond to all possible values a pixel may have. In practice, we use the indirect-definition system to increase the number of colors a display supports without increasing the size of its bitmap (the number of bits per pixel), thus keeping the monitor price down.

The indirect-definition system uses a lookup table with a number of entries far greater than the number of colors a monitor can display on its screen simultaneously. The total number of entries in a lookup table is known as the color palette of the monitor. The number of simultaneous colors a monitor can support is equal to 2^n, where n is the number of planes in its bitmap. An application can choose a maximum of 2^n colors from the color palette.

A common size of color palette is 2^{24}, or 16.7 million colors. As an example, a 4 bits/pixel monitor provides 16 (2^4) simultaneous colors in an image, which can be chosen from the color palette. This scheme, in this example, provides 16 simultaneous colors from a palette of 16.7 million. Sometimes we use the notation of 4 IN/24 OUT to indicate 4 bits/pixel (planes) and 2^{24} palette size. Also, for a monitor with a palette size of 2^{24} is known to support 24-bit true color, each red, green, and blue signal has 8 bits, or 256 values (2^8).

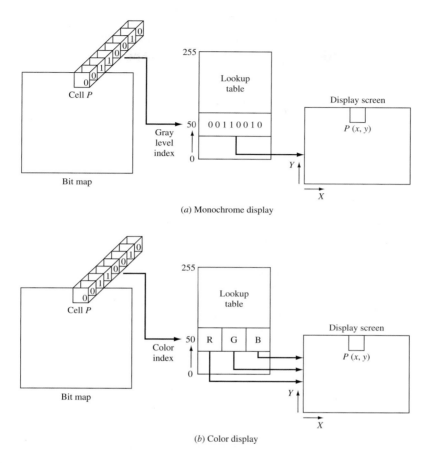

(a) Monochrome display

(b) Color display

Figure 11.6   Lookup table of a monitor.

13. **Refresh rate.** In order to maintain a flicker-free image on the screen, the screen must be refreshed frequently. For CRTs, the phosphor loses its brightness with time and must be energized again. For LCDs, the molecules (pixels) must change colors and display them. For CRTs, the refreshing process is achieved by the electron beam scanning the back of the monitor screen using the raster scan pattern shown in Figure 11.2. The screen is divided into horizontal scan lines; each scan line is one pixel high. Scan lines are numbered top to bottom. For LCDs, there are no electrons. The refreshing process is achieved via re-twisting of the molecules. There is a threshold refresh rate (how many times a monitor refreshes its screen per second) to avoid flicker problems. Below this threshold, our eyes notice the flicker on the screen.

For a CRT, the rate should be 70 Hz [70 hertz or 70 cycles (times) per second; a cycle is a full screen scan]. Other scan rates are 85 Hz to 100 Hz. There are two types of refresh rates:

noninterlaced and interlaced. The noninterlaced rate scans every scan line in one cycle. The interlaced divides the scan cycle into two subcycles. It first scans the odd scan lines in the first half of the cycle time, and it then scans the even scan lines in the second half. The net effect is to double the refresh rate. The interlaced type is based on the idea that the naked eye cannot recognize the missing scan lines if the image is continuous.

For flat panel displays, there is no flicker problem and therefore there is no threshold on the LCD refresh rate. But we have the problem of the image quality that depends on how fast the molecules can change colors. We measure this change-of-color speed by the *response time*. Some typical response times are 20 to 50 milliseconds (ms). However, LCDs use the refresh rate variable as CRTs. Some LCDs have a refresh rate that is programmable to 200 Hz.

14. **Antialiasing.** The aliasing problem is directly related to the resolution of the display, which determines how good or bad the raster approximation of geometric information is. The jaggedness of lines at angles other than multiples of 45° is called aliasing. Various methods of antialiasing exist which use various intensity levels to soften the edges of the lines or shades. Of course, the aliasing problem diminishes as the resolution increases, and it is only related to the screen image and not to the geometric representation in the computer or to drawings printed on paper.

EXAMPLE 11.1　　**Calculate monitor parameters.**
An eight-plane raster display has a resolution of 1280×1024 and a refresh rate of 60 Hz noninterlaced. Find:

　　a. the RAM size of the bitmap (refresh buffer).
　　b. the time required to display a scan line and a pixel.
　　c. the active display area of the screen if the resolution is 78 dpi.
　　d. the optimal design if the bitmap size is to be reduced by half.

SOLUTION　　The display resolution means 1280 and 1024 pixels in the horizontal and vertical directions, respectively.

　　a. The RAM size of the bitmap $= 8 \times 1280 \times 1024$ bits $= \dfrac{8 \times 1280 \times 1024}{8}$ bytes

$$= \dfrac{8 \times 1280 \times 1024}{8 \times 1024 \times 1024} \text{ megabytes} = 1.25 \text{ MB}$$

　　b. The time to display a scan line, t_s, and a pixel, t_p, are given by
$$t_s = [(1/60)/1280] \times 10^6 = 13 \text{ μsec (microseconds)}$$
$$t_p = (13/1024) \times 1000 = 13 \text{ nanoseconds}$$

　　c. The active display area $= 1280/78 \text{ horizontal} \times 1024/78 \text{ vertical} = 16.4 \times 13.1 \text{ inch.}$

　　d. The two solutions are to reduce the number of planes by half and keep the resolution as it is or vice versa. Thus the two choices are four-plane 1280×1024 display or eight-plane 640×512 display. The first choice is preferred, especially if 16 simultaneous colors are adequate for most applications that utilize the display.

| EXAMPLE 11.2 | **Calculate monitor resolution.** |

What is a reasonable resolution for an eight-plane display refreshed from a bitmap of 256 kilobytes of RAM?

| SOLUTION | Bitmap size per plane $= \dfrac{256 \times 1024 \times 8}{8} = 256{,}000$ bits |

This could support a display with a resolution of $505 \times 505, 640 \times 400$, or some other combination. If four planes are used instead, the resolution can be $715 \times 715, 640 \times 800$, or other combination yielding 512,000.

| EXAMPLE 11.3 | **Drawing onto a screen.** |

How can you draw a 500-pixel-wide square on a 1280×1024 screen whose aspect ratio is 4:3?

| SOLUTION | There are two ways to draw the square. The first is to assume a horizontal |

value of 500. The corresponding vertical value is given by $500 \times 1024/1280 \times 4/3 = 533$. In pixel value, the square corners become (assuming the bottom left corner is located at pixel (h, v)):

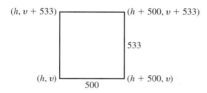

The second possibility is to assume a vertical value of 500. The horizontal value becomes $500 \times 1280/1024 \times 3/4 = 469$. The square corners become:

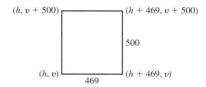

PROBLEMS

Part I: Theory

11.1 What are the differences between regular CRTs and flat screen CRTs?

11.2 What are the differences between analog and digital flat panel displays?

11.3 What does a refresh rate of 100 Hz noninterlaced and 100 Hz interlaced mean?

11.4 Why does flicker happen in CRTs and not in flat panel displays?

11.5 What does 8 IN/24 OUT mean for a monitor?

11.6 Why do monitors use color lookup tables and palettes?

11.7 Why do flat panel displays have problems with viewing angle while CRTs do not?

11.8 If a monitor has a dpi of 105, what is its dot pitch?

11.9 What does a display depth of 8 mean?

11.10 Three raster display monitors have the same bitmap size. They have resolutions of 512×512, 1024×1024, and 4096×4096. The 4096×4096 monitor has a depth of 3:

    a. Calculate the size of the bitmap (RAM)

    b. Calculate the display depth of the other two monitors.

    c. How many different colors can each monitor support? Assume no color lookups are available.

11.11 An eight-plane monitor has a resolution of $1024H \times 1280V$ and a refresh rate of 60 Hz non-interlaced. Find:

    a. bitmap size.

    b. the time required to display a scan line and a pixel.

    c. the number of simultaneous colors the monitor can support.

11.12 An eight-plane raster display has a resolution of $1024H \times 864V$ and a refresh rate of 60 Hz noninterlaced. Calculate:

    a. the size of the bitmap.

    b. the active display area of the screen if the display has 78 pixels per inch.

    c. the resolution of the display if both the number of planes and the size of the bitmap are reduced by half.

11.13 A raster display has four planes, 1024×1280, and 60 Hz noninterlaced refresh rate.

    a. Find the bitmap size and the time needed to scan a line and a pixel. How many simultaneous colors can the display show on the screen?

    b. If the display resolution is reduced by half in both directions, how many planes are needed to keep the same size of the bitmap? How many colors can the display show in this case?

11.14 Two raster displays have the following specifications:

    Display 1: $500H \times 500V$, eight planes, 60 Hz noninterlaced refresh rate.

    Display 2: $1000H \times 1000V$, four planes, 30 Hz noninterlaced refresh rate.

    a. Find the bitmap size of each display.

    b. Find the required time to scan a line and a pixel for each display.

    c. A line connecting two points of pixel coordinates of (100, 100) and (200, 300) is drawn on display 1. The line is to be drawn on display 2 with the same length. Assuming that both displays have the same active display area, find the new pixel coordinates of the two end points of the line.

11.15 A raster display has a graphics card with a size of 0.81 MB of RAM and a refresh rate of 60 Hz noninterlaced.

a. Find the resolution of the display if it has eight planes. What is the time needed to scan a line and a pixel? How many simultaneous colors can the display support?

b. If the number of planes is reduced to four, what is the resolution of the display to the nearest integer? How many simultaneous colors can you display now?

11.16 A black and white monitor has a resolution of 1000×1000 with a refresh rate of 60 Hz noninterlaced.

a. Find the bitmap size.
b. Find the time to scan a line and a pixel.
c. Find the number of simultaneous gray levels that can be displayed.
d. If the display is upgraded to display 256 gray level simultaneously, find the bitmap size.

11.17 Two raster displays have the following specifications:
>   Display A: 1000×1000 , eight planes.
>   Display B: 700×700 , 12 planes.

a. If the cost of one MB of RAM is $.05, which display will be less expensive to buy based on the above information?

b. Assuming that the display you choose in (a) has a refresh rate of 60 Hz, find t_s, t_p, and the number of simultaneous colors the display can support.

Part II: Laboratory

Use your in-house CAD/CAM system to answer the following question.

11.18 Find a regular CRT, flat screen CRT, analog flat panel display, and digital flat panel display. Write the specifications of each and compare them with those covered in Section 11.7 of this chapter.

Part III: Programming

Write the code so that the user can input values to control the curve shape The program should display the results. Use OpenGL with C or C++, or use Java 3D.

11.19 Write a Java or C++ code that draws shapes onto a CRT and a flat panel display.

Transformations

OAL

Understand and master the concepts of geometric transformations, their types, their use in geometric modeling, how they speed up the creation of CAD models, and how CAD systems use them to create views and engineering drawings.

BJECTIVES

After reading this chapter, you should understand the following concepts:

- Need for geometric transformations
- Translation, rotation, mirroring, scaling, and reflection
- Homogeneous transformations
- Concatenated transformations
- Inverse transformations
- Projections of geometric models
- Orthographic projections
- Perspective projections

CHAPTER   HEADLINES

12.1 Introduction

Geometric transformations play a central role in geometric modeling and viewing. They are used in modeling to express the locations of entities relative to others and to move them around in the modeling space. They are used in viewing to generate different views of a model for visualization and drafting purposes. Typical CAD operations to translate, rotate, zoom, and mirror entities are all based on geometric transformation covered in this chapter. Some of these operations have been utilized in Chapters 6 to 10 to construct solid models.

Geometric transformations can also be used to create animated files of geometric models to study their motion. For example, the motion of a spatial mechanism can be animated by first calculating its motion (displacements and/or rotations) using the proper kinematic and dynamic equations. The geometric model of the mechanism at the initial position is then constructed and transformed incrementally using the calculation results.

12.2 Formulation

By definition, geometric transformations move points in space from one location to another, or they change the description of a point from one coordinate system, such as a WCS, to another system, such as MCS. The simplest motion is the rigid-body motion in which the relative distances between modeling points remain constant; that is, the geometric model does not deform during the motion. Geometric transformations that describe this motion are often referred to as rigid-body transformations and typically include translation, rotation, scaling, reflection, and any combination of them. These transformations can be applied directly to the parametric representations of models including points, curves, surfaces, and solids. Matrix notation provides a very expedient way to develop and implement geometric transformations in CAD systems.

Transformation of a point represents the core problem in geometric transformations because it is the basic element in geometric modeling. For example, a line is represented by its two end points, and a general curve, surface or a solid is represented by a collection of points.

The problem of transforming a point can be stated as follows: Given a point P that belongs to a geometric model that undergoes a rigid-body motion, find the corresponding point $P*$ in the new position such that:

$$\mathbf{P*} = f(\mathbf{P}, \text{transformation parameters}) \tag{12.1}$$

The new position vector $\mathbf{P*}$ should be expressed in terms of the old position vector $\mathbf{P}$ and the motion parameters. One of the characteristics of Eq. (12.1) that should be emphasized here is that geometric transformations should be unique. A given set of transformation parameters must yield one and only one new point for each old point. Another characteristic is the concatenation, or combination, of transformations. Intuitively, two transformations can be concatenated to yield a single transformation, which should have the same effect as the sequential applications of the original two.

In order to implement Eq. (12.1) into CAD software, it is desirable to express it in terms of matrix notation as:

$$\mathbf{P*} = [T]\,\mathbf{P} \tag{12.2}$$

where $[T]$ is the transformation matrix. Its elements are functions of the transformation parameters. The matrix $[T]$ should have some important properties. It must apply to all rigid-body transformations (translation, rotation, scaling, and reflection). It should also be applicable to both 2D and 3D transformations.

Applying Eq. (12.2) repeatedly to key points of an entity enables us to transform of the entity. For example, to transform a straight line, its two end points are transformed and then connected to produce the transformed line. Similarly , to transform a curve, its data points are transformed and then connected to give the transformed curve.

12.3 Translation

Translating a model means that all of its points move an equal distance in a given direction. Translation can be specified by a vector, a unit vector and a distance, or two points that denote the initial and final positions of the model to be translated. Figure 12.1 shows the translation of a curve by a vector **d**.

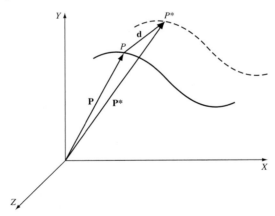

Figure 12.1 Translation.

To relate the final position vector **P*** of a point P on the curve to its initial position vector **P**, consider the triangle shown in Figure 12.1. In this case, Eq. (12.1) takes the form:

$$\mathbf{P*} = \mathbf{P} + \mathbf{d} \tag{12.3}$$

This equation is applicable to both two- and 3D points and can be written in a scalar form for the 3D case as:

$$\begin{aligned} x^* &= x + x_d \\ y^* &= y + y_d \\ z^* &= z + z_d \end{aligned} \tag{12.4}$$

If Eq. (12.3) is applied to each point on a curve, that is, pointwise transformation, the curve is then translated by the vector **d**. However, it is more efficient and useful to translate the entity data and then re-create the entity in the new position using the transformed data. For example, translating a circle or an ellipse requires translating its center only. And translating a parabola or hyperbola requires translating its vertex.

As expected intuitively, translating a curve does not change its tangent vector at any point. This can be seen by differentiating Eq. (12.3) with respect to the parameter u to obtain **P*'** = **P'** because the translation vector **d** is constant.

12.4 Rotation

Rotation enables users to view geometric models from different angles and also helps with many geometric operations such as creating circular patterns and revolved features. A rotation operation requires an entity (point, curve, or a solid) to rotate, a center of rotation, and an axis of rotation. Rotating a curve is equivalent to rotating its data and then reconstructing the curve using the rotated data.

12.4.1 Rotation About Coordinate System Axes

Rotating a point a given angle θ about the X, Y, or Z axis is sometimes referred to as rotation about the origin. A convention for choosing signs of angles of rotations must be established. In this book, we use the right-hand rule. Therefore, a rotation angle θ about a given axis is positive in a counterclockwise sense when viewed from a point on the positive portion of the axis toward the origin.

To develop the rotational transformation of a point (or a vector) about one of the principal axes, let us consider the rotation of point P a positive angle θ about the Z axis as shown in Figure 12.2. This case is equivalent to 2D rotation of a point in the XY plane about the origin. The final position of P after rotation is shown as point P^*. Equation (12.1) or (12.2) can be written here by relating the coordinates of P^* to those of P as follows:

$$x^* = r\cos(\theta + \alpha) = r\cos\alpha\,\cos\theta - r\sin\alpha\,\sin\theta$$
$$y^* = r\sin(\theta + \alpha) = r\sin\alpha\,\cos\theta + r\cos\alpha\,\sin\theta \tag{12.5}$$
$$z^* = z$$

where $r = |\mathbf{P}| = |\mathbf{P^*}|$. To eliminate the angle θ from Eq. (12.5), we can write (refer to the trigonometry in Figure 12.2):

$$x = r\cos\alpha \qquad y = r\sin\alpha \tag{12.6}$$

Substituting Eq. (12.6) into (12.5) gives:

$$x^* = x\cos\theta - y\sin\theta$$
$$y^* = x\sin\theta + y\cos\theta \tag{12.7}$$
$$z^* = z$$

Rewriting Eq. (12.7) in a matrix form gives:

$$\begin{bmatrix} x^* \\ y^* \\ z^* \end{bmatrix} = \begin{bmatrix} \cos\theta & -\sin\theta & 0 \\ \sin\theta & \cos\theta & 0 \\ 0 & 0 & 1 \end{bmatrix} \begin{bmatrix} x \\ y \\ z \end{bmatrix} \tag{12.8}$$

or,

$$\mathbf{P}^* = [R_Z]\mathbf{P} \tag{12.9}$$

where

$$[R_Z] = \begin{bmatrix} \cos\theta & -\sin\theta & 0 \\ \sin\theta & \cos\theta & 0 \\ 0 & 0 & 1 \end{bmatrix} \tag{12.10}$$

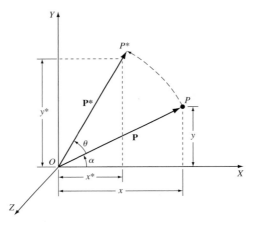

Figure 12.2 Rotation of a point about the Z axis.

Similarly, we can prove that matrices for rotations about X and Y axes are given by:

$$[R_X] = \begin{bmatrix} 1 & 0 & 0 \\ 0 & \cos\theta & -\sin\theta \\ 0 & \sin\theta & \cos\theta \end{bmatrix} \tag{12.11}$$

$$[R_Y] = \begin{bmatrix} \cos\theta & 0 & \sin\theta \\ 0 & 1 & 0 \\ -\sin\theta & 0 & \cos\theta \end{bmatrix} \tag{12.12}$$

Thus, in general, we can write

$$\mathbf{P}^* = [R]\mathbf{P} \tag{12.13}$$

where $[R]$ is the appropriate rotational matrix. Equations (12.10) to (12.13) show some popular forms of $[R]$. Other forms are covered in the next two sections.

The columns of the rotation matrix $[R]$ have some useful characteristics. If we substitute the unit vector $[1\ 0\ 0]^T$ in the X direction into Eq. (12.8), we obtain the first column of $[R_Z]$ as the components of the transformed unit vector. This implies that if we rotate the unit vector in the X direction an angle θ about the Z axis, the first column of $[R_Z]$ gives the coordinates of the transformed unit vector. Similarly, the second and third columns of $[R_Z]$ are the new coordinates of the unit vectors $[0\ 1\ 0]^T$ and $[0\ 0\ 1]^T$ in the Y and Z directions, respectively, after rotating them the same angle θ. Therefore, the columns of a rotation matrix $[R]$ represent the unit vectors that are mutually orthogonal in a right-hand system, that is, $\mathbf{C}_1 \times \mathbf{C}_2 = \mathbf{C}_3$, $\mathbf{C}_2 \times \mathbf{C}_3 = \mathbf{C}_1$, and $\mathbf{C}_3 \times \mathbf{C}_1 = \mathbf{C}_2$, where $\mathbf{C}_1$, $\mathbf{C}_2$, and $\mathbf{C}_3$ are the first, second, and third columns of $[R]$, respectively. From linear algebra, a matrix with orthonormal columns is an orthogonal matrix, and its inverse is equal to its transpose.

The effect of rotation on tangent vectors of a curve can be obtained from Eq. (12.13) as:

$$\mathbf{P}^{*\prime} = [R]\mathbf{P}^\prime \qquad (12.14)$$

Thus, for a Hermite cubic spline, it is easily seen that $\mathbf{V}^* = [R]\mathbf{V}$. Therefore, the rotation of the spline about a given axis is equivalent to rotating its end conditions about the same axis.

12.4.2 Two-Dimensional Rotation About an Arbitrary Axis

The rotation of a point, or an entity in general, about an axis passing through an arbitrary point that is not the origin occurs when one point rotates about another one. In fact, the rotation of a point about the origin covered in Section 12.4.1 is considered a special case of this problem we are about to solve. Rotation of a point or an entity about a point is useful in simulations of mechanisms, linkages, and robotics where links or members must rotate about their respective joints.

Figure 12.3 shows the rotation of point P about point P_1 in the XY plane. Figure 12.3a shows P and its rotation, to its final position P^*, an angle θ about an axis parallel to the Z axis and passing through P_1. In order to develop the rotation matrix correctly for this case, we can use Eq. (12.9) to rotate the vector $(\mathbf{P} - \mathbf{P}_1)$ (not $\mathbf{P}$) about P_1 to obtain $(\mathbf{P}^* - \mathbf{P}_1)$ (not $\mathbf{P}^*$). Thus, we can write:

$$\mathbf{P}^* - \mathbf{P}_1 = [R_Z](\mathbf{P} - \mathbf{P}_1) \qquad (12.15)$$

Rearranging Eq. (12.15) gives:

$$\mathbf{P}^* = [R_Z](\mathbf{P} - \mathbf{P}_1) + \mathbf{P}_1 \qquad (12.16)$$

Equation (12.16) can also be obtained by considering the rotation of point P about P_1 instead of considering the rotation of the vector $(\mathbf{P} - \mathbf{P}_1)$ about P_1 as we just did. From this point of view, the rotation of P and P_1 can be achieved in three steps as shown in Figures 12.3b to d. In the first step, translate P_1 to the origin O. In this position, we refer to point P_1 as P_{1t}. Also, translate point P to P_t by the translation vector $-\mathbf{P}_1$ as shown in Figure 12.3b. Therefore,

$$\mathbf{P}_t = \mathbf{P} - \mathbf{P}_1 \qquad (12.17)$$

In the second step, rotate P_t, in the XY plane, the angle θ about the origin as shown in Figure 12.3c. Consequently, Eq. (12.9) gives:

$$\mathbf{P}^*_t = [R_Z]\mathbf{P}_t = [R_Z](\mathbf{P} - \mathbf{P}_1) \tag{12.18}$$

In the last step, translate points P_{1t} and P_t back to their original positions P_1 and P, respectively, by the translation vector $\mathbf{P}_1$. This would require translating point P^*_t by the same vector to the position P^*, as shown in Figure 12.3d. Thus:

$$\mathbf{P}^* = \mathbf{P}^*_t + \mathbf{P}_1 = [R_Z](\mathbf{P} - \mathbf{P}_1) + \mathbf{P}_1 \tag{12.19}$$

Equation (12.19) is the same as Eq. (12.16). Equation (12.19) applies to 2D rotations in the XZ or YZ plane by replacing $[R_Z]$ by $[R_Y]$ or $[R_X]$, respectively.

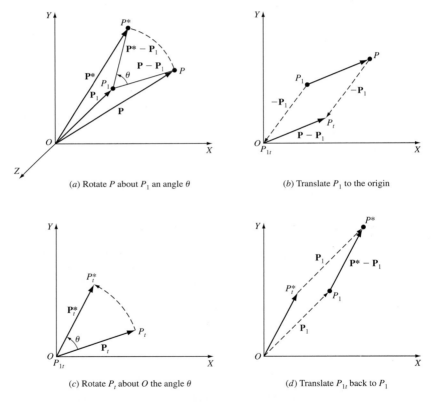

(a) Rotate P about P_1 an angle θ (b) Translate P_1 to the origin

(c) Rotate P_t about O the angle θ (d) Translate P_{1t} back to P_1

Figure 12.3   Two-dimensional rotation of a point about an arbitrary axis.

12.4.3 Three-Dimensional Rotation About an Arbitrary Axis

Points, undergoing rigid-body rotation, describe arcs in a plane perpendicular to a fixed line, the axis of rotation. In planar 2D rotation the axis is always perpendicular to the XY plane;

that is, parallel to the Z axis. Consequently, the axis is completely defined by its intersection with the XY plane (the origin O in Figure 12.2 and point P_1 in Figure 12.3). The orientation of the axis is implicitly defined (along the Z axis) and does not appear as a parameter in the rotation matrix $[R]$. As a result, the angle of rotation θ is the only transformation parameter required to completely define a 2D rotation and, therefore, the corresponding $[R]$. It will be shown later in this section that 2D rotation is a special case of 3D rotation.

In the general spatial (3D) case, rotation is not constrained to the XY plane, and the axis of rotation may be oriented in any direction. Therefore, the orientation of the axis must be incorporated into the rotation matrix in addition to the angle of rotation. If we define the orientation by the unit vector $\hat{\mathbf{n}}$ (Figure 12.4), Eq. (12.1) can be written as:

$$\mathbf{P}^* = f(\mathbf{P}, \hat{\mathbf{n}}, \theta) \tag{12.20}$$

In Eq. (12.20), it is assumed that the axis of rotation passes through the origin. If it does not, then a similar development to that presented in Section 12.4.2 should be followed. Equation (12.20) is derived below and recast in a matrix form for 3D rotation about an arbitrary axis. Two cases are considered: the axis passes through the origin and the axis is in an arbitrary location.

Figure 12.4 shows the 3D rotation of a point P an angle θ about an arbitrary axis that passes through the origin. The positions of the point before and after rotation are P and P^*, respectively. The orientation of the axis of rotation is defined by the unit vector $\hat{\mathbf{n}}$ such that:

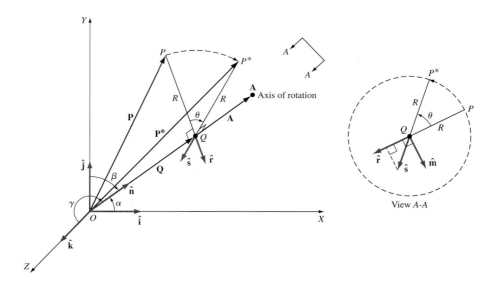

Figure 12.4   **Three-dimensional rotation of a point about an arbitrary axis.**

$$\hat{\mathbf{n}} = n_x\hat{\mathbf{i}} + n_y\hat{\mathbf{j}} + n_z\hat{\mathbf{k}} = \cos\alpha\,\hat{\mathbf{i}} + \cos\beta\,\hat{\mathbf{j}} + \cos\gamma\,\hat{\mathbf{k}} \tag{12.21}$$

where $n_x = \cos\alpha$, $n_y = \cos\beta$, and $n_z = \cos\gamma$ are the direction cosines of $\hat{\mathbf{n}}$. If the axis of rotation is defined as a line connecting the origin O and any point, for example A, on it (see Figure 12.4), then $n_x = x_A/|\mathbf{A}|$, $n_y = y_A/|\mathbf{A}|$, and $n_z = z_A/|\mathbf{A}|$, where x_A, y_A, and z_A are the coordinates of point A and $|\mathbf{A}| = \sqrt{x_A^2 + y_A^2 + z_A^2}$.

The rotation of P about the axis OA defines a circle whose plane is perpendicular to OA. Its center is point Q, which is the intersection between the axis and the plane. Its radius is R, which is the perpendicular distance between P and OA in any position, that is, $R = PQ = P*Q$. The angle of rotation θ is chosen in Figure 12.4 to be positive according to the agreed-upon convention adopted for 2D rotation. View A-A shows θ counterclockwise, that is, positive, if the observer is placed at A-A, that is, on the positive portion of the axis.

In order to facilitate the development, let us define the directions of the lines PQ and $P*Q$ by the unit vectors $\hat{\mathbf{r}}$ and $\hat{\mathbf{s}}$, respectively, as shown in Figure 12.4. From the figure, it is obvious that the final position vector $\mathbf{P}*$ of point P is the resultant of three vectors, that is,

$$\mathbf{P}* = \mathbf{P} + \mathbf{PQ} + \mathbf{QP}* \tag{12.22}$$

where the notation $\mathbf{PQ}$ indicates a vector going from point P to point Q. Utilizing $\hat{\mathbf{r}}$, $\hat{\mathbf{s}}$, and R, Eq. (12.22) can be written as:

$$\mathbf{P}* = \mathbf{P} + R\hat{\mathbf{r}} - R\hat{\mathbf{s}} \tag{12.23}$$

The remainder of the development that follows centers around expressing $\hat{\mathbf{r}}$ and $\hat{\mathbf{s}}$ in terms of $\mathbf{P}$, $\hat{\mathbf{n}}$, and θ, that is, in terms of the desired rotation parameters. Utilizing the right triangle OPQ, we can write:

$$\mathbf{PQ} = \mathbf{Q} - \mathbf{P} \tag{12.24}$$

Observing that $\mathbf{Q}$ is the component of $\mathbf{P}$ along the axis of rotation, we can write

$$\mathbf{Q} = (\mathbf{P} \cdot \hat{\mathbf{n}})\hat{\mathbf{n}} \tag{12.25}$$

Substituting Eq. (12.25) into (12.24) and dividing the result by R (the magnitude of $\mathbf{PQ}$) gives:

$$\hat{\mathbf{r}} = \frac{(\mathbf{P} \cdot \hat{\mathbf{n}})\hat{\mathbf{n}} - \mathbf{P}}{R} \tag{12.26}$$

In order to express $\hat{\mathbf{s}}$ in terms of $\mathbf{P}$ and $\hat{\mathbf{n}}$, we need to introduce the intermediate unit vector $\hat{\mathbf{m}}$ shown in view A-A in Figure 12.4. The vector is chosen to be perpendicular to $\hat{\mathbf{r}}$ and lies in the plane of the circle; thus it is also perpendicular to $\hat{\mathbf{n}}$. Utilizing the cross product definition of two vectors, we can write:

$$\hat{\mathbf{m}} = \hat{\mathbf{n}} \times \hat{\mathbf{r}} \tag{12.27}$$

The unit vector $\hat{\mathbf{s}}$ can now be written in terms of its components in the $\hat{\mathbf{r}}$ and $\hat{\mathbf{m}}$ directions as:

$$\hat{\mathbf{s}} = \cos\theta\,\hat{\mathbf{r}} + \sin\theta\,\hat{\mathbf{m}} \tag{12.28}$$

Substituting Eq. (12.27) into (12.28) and substituting the result together with Eq. (12.26) into (12.23), we obtain:

$$\mathbf{P}^* = (\mathbf{P} \cdot \hat{\mathbf{n}})\hat{\mathbf{n}} + [\mathbf{P} - (\mathbf{P} \cdot \hat{\mathbf{n}})\hat{\mathbf{n}}]\cos\theta + (\hat{\mathbf{n}} \times \mathbf{P}) \sin\theta - \hat{\mathbf{n}} \times (\mathbf{P} \cdot \hat{\mathbf{n}})\hat{\mathbf{n}} \sin\theta \qquad (12.29)$$

The last term in Eq. (12.29) is equal to zero because it represents the cross product of two collinear vectors $\hat{\mathbf{n}}$ and $(\mathbf{P}.\mathbf{n})\hat{\mathbf{n}}$ (also $\mathbf{Q}$). Thus we have:

$$\mathbf{P}^* = (\mathbf{P} \cdot \hat{\mathbf{n}})\hat{\mathbf{n}} + [\mathbf{P} - (\mathbf{P} \cdot \hat{\mathbf{n}})\hat{\mathbf{n}}]\cos\theta + (\hat{\mathbf{n}} \times \mathbf{P})\sin\theta \qquad (12.30)$$

To write Eq. (12.30) in a matrix form, we can write the following:

$$\mathbf{P} \cdot \hat{\mathbf{n}} = xn_x + yn_y + zn_z = [n_x \quad n_y \quad n_z]\begin{bmatrix} x \\ y \\ z \end{bmatrix} \qquad (12.31)$$

$$\hat{\mathbf{n}} \times \mathbf{P} = \begin{bmatrix} \hat{\mathbf{i}} & \hat{\mathbf{j}} & \hat{\mathbf{k}} \\ n_x & n_y & n_z \\ x & y & z \end{bmatrix} = (n_y z - n_z y)\hat{\mathbf{i}} + (n_z x - n_x z)\hat{\mathbf{j}} + (n_x y - n_y x)\hat{\mathbf{k}}$$

$$= \begin{bmatrix} 0 & -n_z & n_y \\ n_z & 0 & -n_x \\ -n_y & n_x & 0 \end{bmatrix}\begin{bmatrix} x \\ y \\ z \end{bmatrix} \qquad (12.32)$$

Substituting Eqs. (12.31) and (12.32) into Eq. (12.30) and rearranging, we get:

$$\mathbf{P}^* = \left\{(1 - \cos\theta)\begin{bmatrix} n_x \\ n_y \\ n_z \end{bmatrix}[n_x \quad n_y \quad n_z] + \cos\theta\begin{bmatrix} 1 & 0 & 0 \\ 0 & 1 & 0 \\ 0 & 0 & 1 \end{bmatrix}\right.$$

$$\left. + \sin\theta\begin{bmatrix} 0 & -n_z & n_y \\ n_z & 0 & -n_x \\ -n_y & n_x & 0 \end{bmatrix}\right\}\begin{bmatrix} x \\ y \\ z \end{bmatrix} \qquad (12.33)$$

or

$$\mathbf{P}^* = [R]\mathbf{P} \qquad (12.34)$$

After reducing Eq. (12.33) further, $[R]$ becomes:

$$[R] = \begin{bmatrix} n_x^2 \, v\theta + c\theta & n_x n_y v\theta - n_z s\theta & n_x n_z v\theta + n_y s\theta \\ n_x n_y v\theta + n_z s\theta & n_y^2 v\theta + c\theta & n_y n_z v\theta - n_x s\theta \\ n_x n_z v\theta - n_y s\theta & n_y n_z v\theta + n_x s\theta & n_z^2 v\theta + c\theta \end{bmatrix} \qquad (12.35)$$

where $c\theta = \cos\theta$, $s\theta = \sin\theta$, and $v\theta = $ versine $\theta = 1 - \cos\theta$.

The general rotation matrix $[R]$ given by Eq. (12.35) has two important characteristics. First, it is skew symmetric because the third term in Eq. (12.33) is skew symmetric. Second, its determinant $|R|$ is equal to 1. In general, $|R| = 1$ for any rotation matrix that describes rotation about the origin. The reader can verify this fact for Eqs. (12.10) to (12.12) and (12.35).

The rotation matrices given by Eqs. (12.10) to (12.12) can now be seen as special cases of Eq. (12.35) as follows: If the axis of rotation is the Z axis, then its orientation is given by $\mathbf{n} = \hat{\mathbf{k}}$, that is, $n_x = n_y = 0$ and $n_z = 1$. Substituting these values into Eq. (12.35) gives Eq. (12.10). Similarly, the X and Y axes of rotation are given by $\mathbf{n} = \hat{\mathbf{i}}$ ($n_x = 1$, $n_y = n_z = 0$) and $\mathbf{n} = \hat{\mathbf{j}}$ ($n_x = n_z = 0$, $n_y = 1$), respectively, and Eqs. (12.11) and (12.12) can be easily obtained from Eq. (12.35).

We now return to the case of 3D rotation about an arbitrary axis that does not pass through the origin. This case is conceptually similar to the 2D case covered in Section 12.4.2, and its development follows exactly the same steps. Therefore, Eq. (12.19) is applicable for the 3D case after replacing $[R_z]$ by $[R]$ given by Eq. (12.35), that is,

$$\mathbf{P}^* = [R](\mathbf{P} - \mathbf{P}_1) + \mathbf{P}_1 \tag{12.36}$$

Here the point P_1 can be any point in space and is not restricted to the XY plane as in the 2D case.

EXAMPLE 12.1   **Point invariance with rotation.**

Prove that if a point to be rotated about a given axis of rotation lies on the axis, the point does not change position in space and, therefore, its coordinates do not change.

SOLUTION   Figure 12.5 shows this problem. It is essential that P and P^* are identical and, therefore, $\mathbf{P} = \mathbf{P}^*$ regardless of the angle of rotation θ. Substituting Eq. (12.35) into Eq. (12.34) and expanding the result, we obtain for the x coordinate

$$x^* = (n_x^2 \, \upsilon\theta + c\theta)x + (n_x n_y \upsilon\theta - n_z s\theta)y + (n_x n_z \upsilon\theta + n_y s\theta)z \tag{12.37}$$

If point P lies on the axis of rotation, then we can write (see Figure 12.6):

$$x = |\mathbf{P}|n_x \qquad y = |\mathbf{P}|n_y \qquad z = |\mathbf{P}|n_z \tag{12.38}$$

where $|\mathbf{P}|$ is the magnitude of $\mathbf{P}$. Substituting Eq. (12.38) into (12.37) and reducing the result, we obtain:

$$x^* = |\mathbf{P}|[n_x \upsilon\theta(n_x^2 + n_y^2 + n_z^2) + n_x c\theta] \tag{12.39}$$

Using the identity $n_x^2 + n_y^2 + n_z^2 = 1$, Eq. (12.39) becomes:

$$x^* = |\mathbf{p}|[n_x(1 - \cos\theta) + n_x \cos\theta] = |\mathbf{P}|n_x = x \tag{12.40}$$

Similarly we can prove $y^* = y$ and $z^* = z$.

This example has useful practical implications. Typically, CAD/CAM systems let users define axes of rotation by inputting end points. If the axis of rotation happens to be an entity of a geometric model to be rotated, the coordinates of the end points of that entity should stay the same before and after rotation. The user can utilize the `verify entity` command available on the system before and after the rotation to display the coordinates and compare. If there are small differences, they usually result from the round-off errors. Coordinates should be the same within the given significant digits of the computer system used.

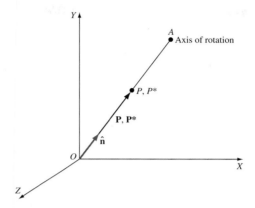

Figure 12.5   Point invariance with rotation.

12.5 Scaling

Scaling is used to change, increase or decrease, the size of an entity or a model, as shown in Figure 12.6. Pointwise scaling can be performed if the matrix $[T]$ in Eq. (12.2) is diagonal, that is,

$$\mathbf{P^*} = [S]\mathbf{P} \tag{12.41}$$

where $[S]$ is a diagonal matrix. In three dimensions, it is given by:

$$[S] = \begin{bmatrix} s_x & 0 & 0 \\ 0 & s_y & 0 \\ 0 & 0 & s_z \end{bmatrix} \tag{12.42}$$

Thus, Eq. (12.41) can be expanded to give:

$$x^* = s_x x \qquad y^* = s_y y \qquad z^* = s_z z \tag{12.43}$$

The elements S_x, S_y, and S_z of the scaling matrix $[S]$ are the scaling factors in the X, Y, Z directions, respectively. Scaling factors are always positive (negative factors produce reflection). If the scaling factors are smaller than 1, the geometric model or entity to which scaling is applied is compressed; if the factors are greater than 1, the model is stretched (see Figure 12.6). If the scale factors are equal, that is, $S_x = S_y = S_z = S$, the model changes in size only and not in shape; this is the case of uniform scaling. For this case, Eq. (12.41) becomes:

$$\mathbf{P^*} = S\,\mathbf{P} \tag{12.44}$$

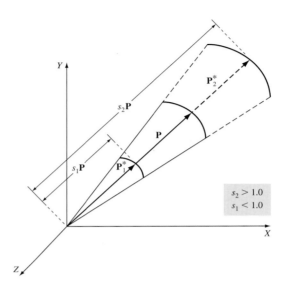

Figure 12.6   Scaling a curve relative to the origin.

Unlike translation, scaling uniformly changes the tangent vectors of a curve by the factor S as differentiating Eq. (12.44) with respect to u gives $\mathbf{P^{*\prime}} = S\mathbf{P^\prime}$. However, uniform scaling does not change the slope, or direction cosines, at any point.

Differential scaling occurs when $S_x \neq S_y \neq S_z$, that is, different scaling factors are applied in different directions. Differential scaling changes both the size and the shape of a geometric model or a curve. It also changes the direction cosines at any point. Differential scaling is seldom used in practical application.

The scaling we have discussed is said to be about the origin, that is, a model or a curve changes size and location with respect to the origin of the coordinate system as shown in Figure 12.6. The model or the curve gets closer to or further from the origin depending on whether the scaling factor is smaller or greater than 1, respectively. Scaling about any other point than the origin is possible, and its development is assigned as Problem 12.3 at the end of the chapter.

Uniform scaling is available on CAD/CAM systems in the form of a `zoom` command. The command requires users to input the scale factor S and clicks the entity or the view to be zoomed. Scaling is useful if a user needs to magnify a dense graphics area on the screen in order to visually identify the geometry in the area for picking and selection purposes. If a view is zoomed, a `set view` or `reset view` command is usually required to make the view scaling permanent or to return the view to its original size, respectively.

12.6 Reflection

Reflection (or mirror) transformation is useful in constructing symmetric models. If, for example, a model is symmetric with respect to a plane, then only half of its geometry is created, which can be copied by reflection to generate the full model.

A geometric entity can be reflected through a plane, a line, or a point in space as illustrated in Figure 12.7. Reflecting an entity through a principal plane ($x = 0$, $y = 0$, $z = 0$ plane) is equivalent to negating the corresponding coordinate of each point on the entity. Reflection through the $x = 0$, $y = 0$, or $z = 0$ planes can be achieved by negating the x, y, or z coordinate respectively.

Reflection through an axis is equivalent to reflection through two principal planes intersecting at the given axis. As shown in Figure 12.7b, an entity is reflected through the Y axis by reflecting through the $z = 0$ plane followed by a reflection through the $x = 0$ plane. In this case, reflection is accomplished by negating the x and z coordinates of each point on the entity.

Similarly, reflection through the X and Z axes require negating the y and z and, the x and y coordinates, respectively. Reflection through the origin is equivalent to reflection through the three principal planes that intersect at the origin. Figure 12.7c shows the reflection of an entity through the origin, which is accomplished by negating the three coordinates of any point on the entity.

Equation (12.41) can be used to describe reflection if the diagonal elements of $[S]$ are chosen to be 1. Thus, the reflection transformation can be expressed as follows:

$$\mathbf{P}^* = [M]\mathbf{P} \tag{12.45}$$

where $[M]$ (mirror matrix) is a diagonal matrix with elements of ±1, that is,

$$[M] = \begin{bmatrix} m_{11} & 0 & 0 \\ 0 & m_{22} & 0 \\ 0 & 0 & m_{33} \end{bmatrix} = \begin{bmatrix} \pm1 & 0 & 0 \\ 0 & \pm1 & 0 \\ 0 & 0 & \pm1 \end{bmatrix} \tag{12.46}$$

The reflection matrix $[M]$ given by Eq. (12.46) applies only to reflections relative to planes, axes, or the origin of a coordinate system. For reflection through the $x = 0$ plane, set $m_{11} = -1$, and $m_{22} = m_{33} = 1$. Similarly, setting $m_{11} = m_{33} = 1$ and $m_{22} = 1$, or $m_{11} = m_{22} = 1$ and $m_{33} = -1$ produces reflection through the $y = 0$ or $z = 0$ plane, respectively.

Reflection through the X axis requires $m_{11} = 1$ and $m_{22} = m_{33} = -1$, through the Y axis requires $m_{11} = m_{33} = -1$ and $m_{22} = -1$, and through the Z axis requires $m_{11} = m_{22} = -1$ and $m_{33} = 1$. Selecting all the diagonal elements to be negative, that is, $m_{11} = m_{22} = m_{33} = -1$, produces reflection through the origin. For the latter case, Eq. (12.45) becomes $\mathbf{P}^* = -\mathbf{P}$ and, therefore,

$\mathbf{P'}^{*} = -\mathbf{P'}$, that is, the magnitudes of tangent vectors remain constant but their directions are reversed.

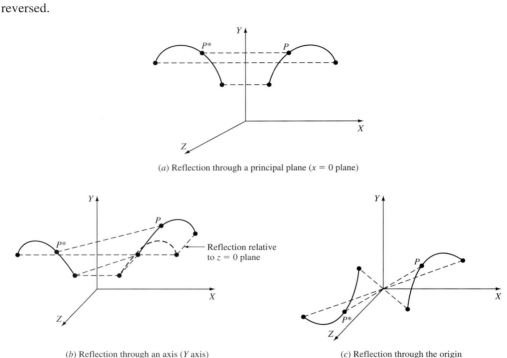

(a) Reflection through a principal plane ($x = 0$ plane)

(b) Reflection through an axis (Y axis)   (c) Reflection through the origin

Figure 12.7   Curve reflection relative to a coordinate system.

While reflections relative to a coordinate system have been discussed previously, reflections through general planes, lines, and points are possible and useful in practice. Figure 12.8 illustrates this general reflection of an entity. As seen from the figure, the common characteristic of the general reflection is that the distance from any point P to be reflected to the reflection mirror (plane, line, or point) is equal to that from the mirror to the image (reflected) point P^{*}. In the three cases, the triangle OPP^{*} can be identified and can be used to relate the coordinates of P^{*} to those of P as follows:

$$\mathbf{P}^{*} = \mathbf{P} + \mathbf{Q} \tag{12.47}$$

where $\mathbf{Q}$ is the vector connecting P and P^{*}.

In order to obtain $\mathbf{P}^{*}$ from Eq. (12.47), the vector $\mathbf{Q}$ must be evaluated first from the given geometry. This is when the difference between the three cases comes into play. Consider first the case shown in Figure 12.8a. The plane is defined by a point P_0 and two unit vectors $\hat{\mathbf{r}}$ and $\hat{\mathbf{s}}$. The vector $\mathbf{Q}$ is perpendicular to the plane, and its magnitude is double the normal distance between P and the plane. Utilizing Eq. (7.34), the normal distance D can be obtained (compare Figs. 12.8a and 7.23), and we can write:

$$\mathbf{Q} = 2D\hat{\mathbf{n}} \tag{12.48}$$

where $\hat{\mathbf{n}}$ is the surface normal to the plane. This normal can be calculated from Eq. (7.27) by using $\hat{\mathbf{r}}$ and $\hat{\mathbf{s}}$ in place of $(\mathbf{P}_1 - \mathbf{P}_0)$ and $(\mathbf{P}_2 - \mathbf{P}_0)$, respectively, in Eq. (7.27).

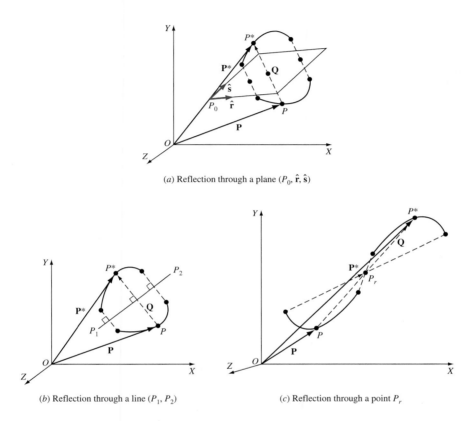

(a) Reflection through a plane $(P_0, \hat{\mathbf{r}}, \hat{\mathbf{s}})$

(b) Reflection through a line (P_1, P_2) (c) Reflection through a point P_r

Figure 12.8   General reflection of a curve.

The vector $\mathbf{Q}$ in the case of reflection through a general line shown in Figure 12.8*b* can be evaluated by using the results of Example 6.7, case *b*. Comparing Figure 6.10*b* and Figure 12.8*b*, we can rewrite Eq. (6.22) to give

$$\mathbf{Q} = -2\{(\mathbf{P} - \mathbf{P}_1) - [(\mathbf{P} - \mathbf{P}_1) \cdot \hat{\mathbf{n}}_1]\hat{\mathbf{n}}_1\} \tag{12.49}$$

The factor -2 is used in Eq. (12.49) because the magnitude of $\mathbf{Q}$ is twice the normal distance between P and the line, and its direction is opposite to $\hat{\mathbf{n}}_2$, shown in Figure 6.10*b*.

In the case of reflection through a general point, the vector $\mathbf{Q}$ can easily be written as (refer to Figure 12.8*c*)

$$\mathbf{Q} = 2(\mathbf{P}_r - \mathbf{P}) \tag{12.50}$$

The effect of reflection on tangent vectors of a curve depends on each individual given case. Only for reflection through the origin or a general point can we write:

$$\mathbf{P}^{*\prime} = -\mathbf{P}^\prime \tag{12.51}$$

that is, tangent vectors reverse directions and their magnitudes remain unchanged. For other cases, only appropriate component(s) reverse directions.

12.7 Homogeneous Representation

The various rigid-body geometric transformations have been developed in previous sections. Equations (12.3), (12.34), (12.41), and (12.45) represent translation, rotation, scaling, and mirroring, respectively. Equations (12.34), (12.41), and (12.45) are in the form of matrix multiplication, but the translation Eq. (12.3) takes the form of vector addition. This makes it inconvenient to concatenate (combine) transformation involving translation. Equation (12.19) shows an example. It is desirable to express all geometric transformations in the form of matrix multiplications only. Representing points by their homogeneous coordinates provides an effective way to unify the description of geometric transformations as matrix multiplications.

Homogeneous coordinates have been used in computer graphics and geometric modeling for a long time. With their aid, geometric transformations are customarily embedded into graphics hardware to speed their execution. Homogeneous coordinates are useful for other applications. They are useful for obtaining perspective views of geometric models. They are often used in projective geometry, mechanism analysis and design, and robotics for development and formulation. In addition, homogeneous coordinates remove many anomalous situations encountered in cartesian geometry, such as representing points of infinity and the non-intersection of parallel lines. Also, they greatly simplify expressing rational parametric curves and surfaces, as covered in Chapter 8.

In homogeneous coordinates, an n-dimensional space is mapped into $(n + 1)$-dimensional space, that is, a point (or a position vector) in n-dimensional space is represented by $(n + 1)$ coordinates (or components). In 3D space, a point P with cartesian coordinates (x, y, z) has the homogeneous coordinates (x^*, y^*, z^*, h) where h is any scalar factor $\neq 0$. The two types of coordinates are related to each other by the following equation:

$$x = \frac{x^*}{h} \qquad y = \frac{y^*}{h} \qquad z = \frac{z^*}{h} \tag{12.52}$$

Equation (12.52) is based on the fact that if the cartesian coordinates of a given point P are multiplied by a scalar factor h, P is scaled to a new point P^* and the coordinates of P and P^* are related by Eq. (12.52). Figure 12.9 shows point P scaled by two factors h_1 and h_2 to produce the two new points P_1^* and P_2^*, respectively. These two points could be interpreted in two different ways. From a cartesian-coordinates point of view, Eq. (12.44) can be used with $S = h_1$ and h_2. Once the cartesian coordinates of P_1^* and P_2^* are calculated, their relationships to P do not exist anymore. Moreover, the three points still belong to the cartesian space.

From a homogeneous-coordinates point of view, the original point P is represented by $(x,\ y,\ z, 1)$, and P_1^* and P_2^* are represented by $(x_1^*,\ y_1^*,\ z_1^*)$ and $(x_2^*,\ y_2^*,\ z_2^*)$, respectively, according to Eq. (12.52). More importantly, the three points belong to the homogeneous space, with the cartesian coordinates obtained when $h = 1$, and the relationship between P and P_1^* or $P*_2$ is maintained through the proper value of h. As a matter of fact, any two homogeneous-coordinates points P_1^* and P_2^* represent the same cartesian point if and only if $h_2 = ch_1$, for any nonzero constant c. Therefore, there is no unique homogeneous representation of a point. For the purpose of geometric transformations, the scale factor h used in Eq. (12.52) is taken to be 1 to avoid unnecessary division.

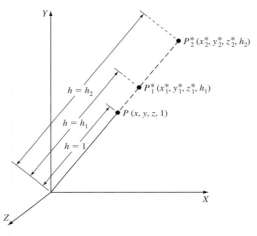

Figure 12.9 Homogeneous coordinates of point P.

The translation transformation given by Eq. (12.3) can now be written as a matrix multiplication by adding the component of 1 to each vector, and using a 4×4 matrix as follows:

$$[x^* \quad y^* \quad z^* \quad 1]^T = \begin{bmatrix} 1 & 0 & 0 & x_d \\ 0 & 1 & 0 & y_d \\ 0 & 0 & 1 & z_d \\ 0 & 0 & 0 & 1 \end{bmatrix} \begin{bmatrix} x \\ y \\ z \\ 1 \end{bmatrix} \tag{12.53}$$

or

$$\mathbf{P^*} = [D]\mathbf{P} \tag{12.54}$$

where $[D]$ is the translation matrix shown in Eq. (12.53).

While rotation, scaling, and mirroring are already expressed in terms of matrix multiplication, their corresponding matrices are changed from 3×3 to 4×4 by adding a column and a row of zero elements except the fourth, which is 1. Thus, the rotation matrix [Eq. (12.35)] becomes

$$[R] = \begin{bmatrix} r_{11} & r_{12} & r_{13} & 0 \\ r_{21} & r_{22} & r_{23} & 0 \\ r_{31} & r_{32} & r_{33} & 0 \\ 0 & 0 & 0 & 1 \end{bmatrix} \qquad (12.55)$$

Similarly, the scaling matrix [(Eq. (12.42)] becomes

$$[S] = \begin{bmatrix} s_x & 0 & 0 & 0 \\ 0 & s_y & 0 & 0 \\ 0 & 0 & s_z & 0 \\ 0 & 0 & 0 & 1 \end{bmatrix} \qquad (12.56)$$

and the reflection matrix [Eq.(12.46)] becomes

$$[M] = \begin{bmatrix} \pm 1 & 0 & 0 & 0 \\ 0 & \pm 1 & 0 & 0 \\ 0 & 0 & \pm 1 & 0 \\ 0 & 0 & 0 & 1 \end{bmatrix} \qquad (12.57)$$

Equations (12.10) to (12.12) can be rewritten in a similar fashion.

To illustrate the convenience gained from the homogeneous representation, Eq. (12.19) can be written as:

$$\mathbf{P}^* = [D_1][R_Z][D_2]\mathbf{P} = [T]\mathbf{P} \qquad (12.58)$$

where

$$[D_1] = \begin{bmatrix} 1 & 0 & 0 & -x_1 \\ 0 & 1 & 0 & -y_1 \\ 0 & 0 & 1 & -z_1 \\ 0 & 0 & 0 & 1 \end{bmatrix} \qquad (12.59)$$

$$[R_Z] = \begin{bmatrix} \cos\theta & -\sin\theta & 0 & 0 \\ \sin\theta & \cos\theta & 0 & 0 \\ 0 & 0 & 1 & 0 \\ 0 & 0 & 0 & 1 \end{bmatrix} \qquad (12.60)$$

$$[D_2] = \begin{bmatrix} 1 & 0 & 0 & x_1 \\ 0 & 1 & 0 & y_1 \\ 0 & 0 & 1 & z_1 \\ 0 & 0 & 0 & 1 \end{bmatrix} \qquad (12.61)$$

and

$$[T] = [D_1][R_Z][D_2] = \begin{bmatrix} \cos\theta & -\sin\theta & x_1(\cos\theta - 1) - y_1\sin\theta & 0 \\ \sin\theta & \cos\theta & x_1\sin\theta + y_1(\cos\theta - 1) & 0 \\ 0 & 0 & 1 & 0 \\ 0 & 0 & 0 & 1 \end{bmatrix} \qquad (12.62)$$

A closer look at the transformation matrices given in Eqs. (12.54) to (12.57) shows that they all can be embedded into a 4×4 matrix. This matrix takes the form:

$$[T] = \begin{bmatrix} t_{11} & t_{12} & t_{13} & t_{14} \\ t_{21} & t_{22} & t_{23} & t_{24} \\ t_{31} & t_{32} & t_{33} & t_{34} \\ t_{41} & t_{42} & t_{43} & t_{44} \end{bmatrix} = \begin{bmatrix} T_1 & T_2 \\ T_3 & 1 \end{bmatrix} \qquad (12.63)$$

The 3×3 submatrix $[T_1]$ produces rotation or scaling. The 3×1 column matrix $[T_2]$ generates translation. The 1×3 row matrix $[T_3]$ produces perspective projection covered in Section 12.10. The fourth diagonal element is the homogeneous-coordinates scale factor h used in Eq. (12.52) and is chosen to be 1 as mentioned earlier.

Equation (12.63) gives the explicit form of the transformation matrix $[T]$ used in Eq. (12.2). It is usually written for one geometric transformation at a time by using any of Eqs. (12.54) to (12.57). If more than one transformation is desired, the resulting matrices are multiplied to produce the total transformation as discussed in Section 12.8.

12.8 Concatenated Transformations

So far we have concentrated on one-step transformations of points such as rotating or translating a point. However, in practice a series of transformations may be applied to geometric models. Thus, combining or concatenating transformations are quite useful. Concatenated transformations are simply obtained by multiplying the $[T]$ matrices [Eq. (12.63)] of the corresponding individual transformations. However, because matrix multiplication may not be commutative in all cases, attention must be paid to the order in which transformations are applied to a given geometric model. In general, if we apply n transformations to a point starting with transformation 1, with $[T_1]$, and ending with transformation n, with $[T_n]$, the concatenated transformation of the point is given by:

$$\mathbf{P}^* = [T_n][T_{n-1}] \cdots [T_2][T_1]\mathbf{P} \qquad (12.64)$$

As an example, consider rotating a point by the following rotations and order: α about the Z axis, β about the Y axis, and γ about the X axis. Substituting α, β, and γ in Eqs. (12.10), (12.12) and (12.11), respectively, and multiplying, we obtain the concatenated transformation matrix as:

$$[T] = [T_X][T_Y][T_Z] \qquad (12.65)$$

or

$$\left[\begin{array}{c|c} [R] & 0 \\ \hline 0 & 1 \end{array}\right] = \left[\begin{array}{c|c} [R_X] & 0 \\ \hline 0 & 1 \end{array}\right]\left[\begin{array}{c|c} [R_Y] & 0 \\ \hline 0 & 1 \end{array}\right]\left[\begin{array}{c|c} [R_Z] & 0 \\ \hline 0 & 1 \end{array}\right] \tag{12.66}$$

or

$$[R] = [R_X][R_Y][R_Z] \tag{12.67}$$

Expanding Eq. (12.67) gives:

$$[R] = \begin{bmatrix} c\alpha\ c\beta & -s\alpha\ c\beta & s\beta \\ s\alpha\ c\gamma + c\alpha\ s\beta\ s\gamma & c\alpha\ c\gamma - s\alpha\ s\beta\ s\gamma & -c\beta\ s\gamma \\ s\alpha\ s\gamma - c\alpha\ s\beta\ c\gamma & c\alpha\ s\gamma + s\alpha\ s\beta\ c\gamma & c\beta\ c\gamma \end{bmatrix} \tag{12.68}$$

EXAMPLE 12.2   **Concatenate transformations.**

Given a point $P(1, 3, -5)$ find:

a. the transformed point P^* if P is translated by $\mathbf{d} = 2\hat{i} + 3\hat{j} - 4\hat{k}$ and then rotated by 30° about the Z axis.

b. Same as in (a) but point P is rotated first, then translated.

c. Is the final point P^* the same in both (a) and (b)? Explain your answer.

SOLUTION   (a)

$$\mathbf{P}_T = \begin{bmatrix} 1 & 0 & 0 & 2 \\ 0 & 1 & 0 & 3 \\ 0 & 0 & 1 & -4 \\ 0 & 0 & 0 & 1 \end{bmatrix}\begin{bmatrix} 1 \\ 3 \\ -5 \\ 1 \end{bmatrix} = \begin{bmatrix} 3 \\ 6 \\ -9 \\ 1 \end{bmatrix} \qquad \mathbf{P}_R = \begin{bmatrix} c30 & -s30 & 0 & 0 \\ s30 & c30 & 0 & 0 \\ 0 & 0 & 1 & 0 \\ 0 & 0 & 0 & 1 \end{bmatrix}\begin{bmatrix} 3 \\ 6 \\ -9 \\ 1 \end{bmatrix} = \begin{bmatrix} -0.402 \\ 6.696 \\ -9 \\ 1 \end{bmatrix}$$

(b)

$$\mathbf{P}_R = \begin{bmatrix} c30 & -s30 & 0 & 0 \\ s30 & c30 & 0 & 0 \\ 0 & 0 & 1 & 0 \\ 0 & 0 & 0 & 1 \end{bmatrix}\begin{bmatrix} 1 \\ 3 \\ -5 \\ 1 \end{bmatrix} = \begin{bmatrix} c30 \\ 3s30 \\ -5 \\ 1 \end{bmatrix} \qquad \mathbf{P}_T = \begin{bmatrix} 1 & 0 & 0 & 2 \\ 0 & 1 & 0 & 3 \\ 0 & 0 & 1 & -4 \\ 0 & 0 & 0 & 1 \end{bmatrix}\begin{bmatrix} c30 \\ 3s30 \\ -5 \\ 1 \end{bmatrix} = \begin{bmatrix} 2.866 \\ 5.598 \\ -9 \\ 1 \end{bmatrix}$$

(c) The final points shown in (a) and (b) are different because rotation and translation are not commutative, because matrix multiplication is not commutative.

Example 12.2 discussion:

In case (a), the concatenated matrix is $[T_R][T_d]$ while it is $[T_d][T_R]$ in case (b). This is why the final result is different in both cases.

Example 12.2 hands-on exercise:

Change the problem to make it 2D. Use $P(1, 3, 0)$ and $\mathbf{d} = 2\hat{i} + 3\hat{j}$. Solve the problem again. Is the final result different in both cases? What is your conclusion?

EXAMPLE 12.3   **Concatenate rotations.**

Using concatenated rotations about the axes of the coordinate systems shown in Figure 12.4, rederive Eq. (12.35).

SOLUTION   The basic idea to solve this example is to rotate the axis of rotation OA shown in Figure 12.4 to coincide with one of the axes, rotate the point P the angle θ about this coincident axis, and finally rotate OA in the opposite direction back to its original position.

The rotation of OA is achieved in two steps. In effect, this is equivalent to decomposing the rotation about the general axis into three rotations about the principal axes X, Y, and Z. Figure 12.10 shows one possible decomposition. In this decomposition, the following sequence of rotations is followed:

1. Rotate OA and point P about the Y axis an angle $-\Phi$ so that OA is collinear with the Z axis, where

$$\tan\phi = \frac{x_A}{z_A} = \frac{x_A/|A|}{z_A/|A|} = \frac{n_x}{n_z} \tag{12.69}$$

This then gives:

$$\cos\phi = \frac{n_z}{\sqrt{n_x^2 + n_z^2}} \qquad \sin\phi = \frac{n_x}{\sqrt{n_x^2 + n_z^2}} \tag{12.70}$$

Thus,

$$\mathbf{P}^* = [R_Y(-\phi)]\mathbf{P} \tag{12.71}$$

2. Following the preceding rotation, rotate OA and P about the X axis an angle ψ so that OA is collinear with the Z axis, where

$$\sin\psi = \frac{y_A}{|A|} = n_y \qquad \cos\psi = \sqrt{1 - n_y^2} \tag{12.72}$$

Equation (12.71) becomes:

$$\mathbf{P}^* = [R_X(\psi)][R_Y(-\phi)]\mathbf{P} \tag{12.73}$$

3. Rotate point P about the Z axis an angle θ. Note that P is now given by Eq. (12.73). After rotating it by the angle θ, it then becomes:

$$\mathbf{P}^* = [R_Z(\theta)][R_X(\psi)][R_Y(-\phi)]\mathbf{P} \tag{12.74}$$

4. Reverse step 2, that is, rotate about the X axis an angle $-\psi$. This modifies Eq. (12.74) to:

$$\mathbf{P}^* = [R_X(-\psi)][R_Z(\theta)][R_X(\psi)][R_Y(-\phi)]\mathbf{P} \tag{12.75}$$

5. Reverse step 1, that is, rotate about the Y axis an angle Φ. This modifies Eq. (12.75) to:

$$\mathbf{P}^* = [R_Y(\phi)][R_X(-\psi)][R_Z(\theta)][R_X(\psi)][R_Y(-\phi)]\mathbf{P} \tag{12.76}$$

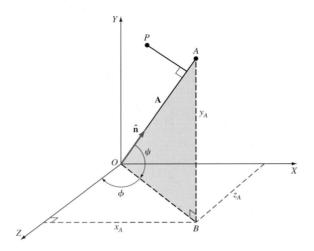

Figure 12.10   Rotation decomposition.

Equation (12.76) should be the same as Eq. (12.33) but in a different form, and comparing it with Eq. (12.34), we can write

$$[R] = [R_Y(\phi)][R_X(-\psi)][R_Z(\theta)][R_X(\psi)][R_Y(-\phi)] \tag{12.77}$$

If the matrix multiplications in Eq. (12.77) are performed and the result is reduced, Eq. (12.35) will be obtained. The reader can carry out the details by using Eqs. (12.10) to (12.12) and using the identity $n_x^2 + n_y^2 + n_z^2 = 1$. If other sequences of decompositions of the rotation are used, the right-hand side of Eq. (12.77) changes but the final result, that is, $[R]$, stays the same. The reader is encouraged to try these sequences.

Example 12.3 discussion:

Rotation has a unique characteristic that is not shared by translation, scaling, or reflection. That is non-cummutativeness. The final position and orientation of an entity after going through two subsequent translations, scalings, or reflections are independent of the order of operations, that is, they are commutative. On the other hand, two subsequent rotations of the entity about two different axes produce two different configurations of the entity depending on the order of the rotations. The reader can verify this by simply marking an edge of a box and then rotating it about two of its other edges and observing the final configuration of the marked edge. The same experiment can be performed on a CAD/CAM system.

Example 12.3 hands-on exercise:

Investigate the non-commutative property of rotations using an empty box of cereal or any other box. Use your CAD/CAM system to verify. What happens if you rotate an entity two angles about the same axis? Is rotation commutative in this case? Prove your answer.

12.9 Mapping of Geometric Models

In the previous sections, we have concerned ourselves with rigid-body transformations of geometric models. Thus, we have discussed transforming a point P to another location P^*. Both points P and P^* are described in the same coordinate system. Thus, the model position and orientation change with respect to the origin of the coordinate system, which stays unaltered in space. Transformations are what we have covered in Chapters 6, 7, 9, and 10 as manipulation operations (translation, rotation, scaling, and mirroring).

Mappings of geometric models are as useful in geometric modeling as transformations. We have covered mapping briefly in Section 2.4. While transformation involves one point and one coordinate system, mapping involves one point and two coordinate systems. Transforming the point results in moving it from one location to another within the same coordinate system. Mapping the point changes its description from one coordinate system to another, but it does not change its location in the modeling space.

We use mapping implicitly during CAD construction. Each time we define a sketch plane and sketch the profile of a feature, we use mapping. We input coordinates relative to the coordinate system (WCS) of the sketch plane, while CAD software maps the input to the model MCS coordinates before storing it in the model database. Mapping is also useful in assembly modeling, as we cover later in the book.

Mapping a geometric model, a collection of points, from one coordinate system to another does not change its position and orientation with respect to the origins of both systems. It only changes the description of such position and orientation. This is equivalent to transforming one coordinate system to another.

The same mathematical forms which we have developed for geometric transformations can be used to map points between coordinate systems. However, the interpretations of these forms are different. The problem of mapping a point from one coordinate system to another can be stated as follows: Given the coordinates of a point P measured in a given XYZ coordinate system, find the coordinates of the point measured in another coordinate system, $X^*Y^*Z^*$, such that:

$$\mathbf{P}^* = f(\mathbf{P}, \text{mapping parameters}) \tag{12.78}$$

where $\mathbf{P}$ and $\mathbf{P}^*$ are the position vectors of point P in the XYZ and $X^*Y^*Z^*$ systems respectively. The mapping parameters describe the relationship between the two systems and consist of the position of the origin and the orientation of the $X^*Y^*Z^*$ system relative to the XYZ system. Equations (12.1) and (12.78) are the same. Equation (12.78) can be expressed in the matrix form given by Eq. (12.2), where $[T]$ is referred to as the mapping matrix. We now consider the three possible cases of mapping and develop their corresponding matrices.

12.9.1 Translational Mapping

When the axes of the two coordinate systems are parallel, the mapping is defined to be translational. In Figure 12.11, the origins of the XYZ and the $X^*Y^*Z^*$ systems are different but their orientations in space are the same. The point P is described by the vectors $\mathbf{P}$ and $\mathbf{P}^*$ in the

XYZ and *X*Y*Z**, respectively. The vector **d** describes the position of the origin of the former system relative to the latter. Equation (12.78) can be written exactly as Eq. (12.3) or (12.53).

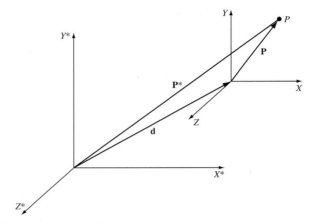

Figure 12.11 Translational mapping of a point.

12.9.2 Rotational Mapping

Figure 12.12 shows rotational mapping between two coordinate systems. The two systems share the same origin, and their orientations are different by the angle θ. In this figure, we assume that the *XY* and *X*Y** planes are coincident. Utilizing the trigonometric relationships shown in the figure, Eq. (12.7) can be derived. Therefore, the rotation matrix given by Eq. (12.10) or (12.60) is applicable to rotational mapping. Similarly, Eqs. (12.11), (12.12), (12.35), and (12.68) are applicable for their corresponding rotational mapping cases.

In rotational mapping, it is important to realize that the columns of a rotational matrix $[R]$ can be interpreted to describe the orientations of a given coordinate system in space. If we take the unit vectors $\hat{i}$, $\hat{j}$, and $\hat{k}$ in the directions of the axes of the *XYZ* system as shown in Figure 12.12, these vectors can be expressed in terms of the *X*Y*Z** system as follows:

$$\hat{i} = \cos\theta\hat{i}^* + \sin\theta\hat{j}^* + 0\hat{k}^*$$

$$\hat{j} = -\sin\theta\hat{i}^* + \cos\theta\hat{j}^* + 0\hat{k}^* \tag{12.79}$$

$$\hat{k} = \hat{k}^*$$

Rewriting Eqs. (12.79) in a matrix form, we obtain:

$$[\hat{i} \quad \hat{j} \quad \hat{k}]^T = \begin{bmatrix} \cos\theta & -\sin\theta & 0 \\ \sin\theta & \cos\theta & 0 \\ 0 & 0 & 1 \end{bmatrix}^T \begin{bmatrix} \hat{i}^* \\ \hat{j}^* \\ \hat{k}^* \end{bmatrix} \tag{12.80}$$

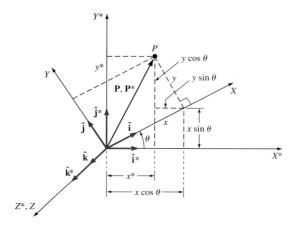

Figure 12.12   Rotational mapping of a point.

The matrix in Eq. (12.80) is $[R_Z]^T$, and each of its columns represents the components of a unit vector. Comparing the columns of this matrix with Eq. (12.79) shows that the first column represents the direction cosines (components) of the unit vector $\hat{i}$. The second and the third columns represent the direction cosines of the unit vectors $\hat{j}$ and $\hat{k}$, respectively. Therefore, the columns of any rotational matrix $[R]$ represent orthogonal unit vectors. This observation is useful in building $[R]$ from user input as explained in Example 12.4. The reader can show that the columns of any rotation matrix [Eqs. (12.10) to (12.12), (12.35), or (12.68)] all have unit magnitude and that they are orthogonal.

12.9.3   General Mapping

The general mapping combines both translational and rotational mappings as shown in Figure 12.13. The origins and the orientations of the XYZ and $X^*Y^*Z^*$ systems are different. In this case, the general mapping matrix $[T]$ is given by Eq. (12.63) with the submatrix $[T_3]$ set to zero, that is,

$$[T] = \begin{bmatrix} r_{11} & r_{12} & r_{13} & x_d \\ r_{21} & r_{22} & r_{23} & y_d \\ r_{31} & r_{32} & r_{33} & z_d \\ \hline 0 & 0 & 0 & 1 \end{bmatrix} = \begin{bmatrix} [R] & \mathbf{d} \\ \hline 0 & 1 \end{bmatrix} \tag{12.81}$$

where $[R]$ and $\mathbf{d}$ are the rotational and translational mapping parts of $[T]$, respectively.

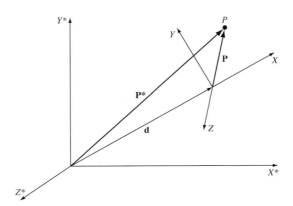

Figure 12.13   General mapping of a point.

EXAMPLE 12.4   **Mapping sketch planes.**

For the geometric model shown in Figure 12.14, a user defines the XYZ coordinate system shown as the MCS of the model database. The user later defines the two WCSs (sketch planes) W and W_1 shown to be able to construct two holes defined by the two circles whose centers are C and C_1. Both C and C_1 are the center points of the shown faces of the model. Find

a. the mapping matrices $[T_W]$ and $[T_{W1}]$ that map points from any one of the two WCSs to the MCS.

b. the coordinates of the centers C and C_1 as they are stored in the model database.

SOLUTION   (*a*) Using a typical CAD/CAM system, the user defines two sketch planes to construct the two circles and cut extrudes them to create the holes. In order to calculate $[T_W]$, we need to calculate $[R]$ and **d** given in Eq. (12.81). The coordinates of points P_1, P_2, and P_3 are (5, 0 0), (5, 0, −3), and (3, $2\sqrt{3}$, 0), respectively. The unit vectors of the WCS $\{W\}$ are calculated as follows:

$$\hat{i}_w = -\hat{k} \qquad (12.82)$$

Notice that Eq. (12.82) is obtained via inspection; there is no need to use $\dfrac{P_2 - P_1}{|P_2 - P_1|}$.

$$\hat{j}_W = \frac{P_3 - P_1}{|P_3 - P_1|} = \frac{1}{4}(-2\hat{i} + 2\sqrt{3}\hat{j}) = -0.5\hat{i} + 0.866\hat{j} \qquad (12.83)$$

$$\hat{k}_W = \hat{i}_W \times \hat{j}_W = 0.866\hat{i} + 0.5\hat{j} \qquad (12.84)$$

Writing Eqs. (12.82) to (12.84) in a matrix form, we obtain:

$$[R_W] = \begin{bmatrix} 0 & -0.5 & 0.866 \\ 0 & 0.866 & 0.5 \\ -1 & 0 & 0 \end{bmatrix} \qquad (12.85)$$

Substituting this equation into equation (12.81) and knowing that $\mathbf{d} = \mathbf{P_1}$, we obtain:

$$[T_W] = \left[\begin{array}{ccc|c} 0 & -0.5 & 0.866 & 5.0 \\ 0 & 0.866 & 0.5 & 0 \\ -1 & 0 & 0 & 0 \\ \hline 0 & 0 & 0 & 1 \end{array} \right] \qquad (12.86)$$

A similar way can be followed to find $[T_{W1}]$ by using the points P_3 (3, $2\sqrt{3}$, 0), P_4 (0, $2\sqrt{3}$, 0), and P_5 (0, 0, −3). However, by inspection we can see that $\hat{\mathbf{i}}_{w1} = \hat{\mathbf{i}}$, $\hat{\mathbf{j}}_{w1} = -\hat{\mathbf{k}}$, and $\hat{\mathbf{k}}_{w1} = \hat{\mathbf{j}}$. The vector $\mathbf{d}$ is equal to $\mathbf{P_4}$. Therefore,

$$[T_{W1}] = \left[\begin{array}{ccc|c} 1 & 0 & 0 & 0 \\ 0 & 0 & 1 & 3.464 \\ 0 & -1 & 0 & 0 \\ \hline 0 & 0 & 0 & 1 \end{array} \right] \qquad (12.87)$$

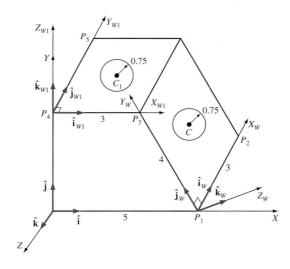

Figure 12.14   Mapping of sketch planes.

(b) The coordinates of the centers C and C_1 are expressed in terms of the MCS before they are stored in the model database. The coordinates of C relative to the WCS $\{W\}$ are (1.5, 2, 0) and those of C_1 relative to the WCS $\{W_1\}$ are (1.5, 1.5, 0). Their MCS coordinates are given, by utilizing Eqs. (12.86) and (12.87), as follows:

$$C = [T_W] \begin{bmatrix} 1.5 \\ 2.0 \\ 0 \\ 1 \end{bmatrix} = \begin{bmatrix} 4.0 \\ 1.732 \\ -1.5 \\ 1 \end{bmatrix}, \qquad C_1 = [T_{W1}] \begin{bmatrix} 1.5 \\ 1.5 \\ 0 \\ 1 \end{bmatrix} = \begin{bmatrix} 1.5 \\ 3.464 \\ -1.5 \\ 1 \end{bmatrix}$$

The reader can verify these results by constructing the model on a CAD/CAM system and using the `Verify Entity` command or its equivalent.

12.10 Inverse Transformations

Calculating the inverse of transformations and mappings is useful in both theoretical and practical aspects of geometric modeling. For example, using inverse mappings, a CAD system enables users to display coordinates of a point relative to a given WCS.

The transformation and mapping matrices developed thus far have inverses. The matrices have been collected into one general matrix given by Eq. (12.63). Thus, it is appropriate to find the inverse $[T]^{-1}$ of this matrix and then try to relate the result to the other matrices. Because $[T]$ is partitioned into four submatrices, then $[T]^{-1} = [A]$ also has four submatrices such that

$$[T][A] = [I] \tag{12.88}$$

where $[I]$ is the identity matrix. This equation can be rewritten as:

$$\begin{bmatrix} T_1 & T_2 \\ T_3 & T_4 \end{bmatrix} \begin{bmatrix} A_1 & A_2 \\ A_3 & A_4 \end{bmatrix} = \begin{bmatrix} I & 0 \\ 0 & I \end{bmatrix} \tag{12.89}$$

Here we replaced the element $t_{44} = 1$ of $[T]$ given by Eq. (12.63) with the submatrix T_4 of size 1×1. The partitioned form implies four separate matrix equations, two of which are $T_1 A_1 + T_2 A_3 = I$ and $T_3 A_1 + T_4 A_3 = 0$. These can be solved simultaneously for A_1 and A_3. The remaining two equations give A_2 and A_4 and lead to the following result:

$$[T]^{-1} = \begin{bmatrix} (T_1 - T_2 T_4^{-1} T_3)^{-1} & -T_1^{-1} T_2 (T_4 - T_3 T_1^{-1} T_2)^{-1} \\ -T_4^{-1} T_3 (T_1 - T_2 T_4^{-1} T_3)^{-1} & (T_4 - T_3 T_1^{-1} T_2)^{-1} \end{bmatrix} \tag{12.90}$$

The inverse $[T]^{-1}$ given by Eq. (12.90) does not take full advantage of the inherent structure of $[T]$ itself. First, T_4 is one element equal to 1. Therefore, $T_4^{-1} = 1$ also. Second, $[T_3]$ has all zero elements, $[T_3] = [0\ 0\ 0]$. Moreover $[T_1]$ and $[T_2]$ are the rotation matrix $[R]$ and the translational vector $\mathbf{d}$. Substituting all these properties into Eq. (12.90) gives

$$[T]^{-1} = \begin{bmatrix} [R]^{-1} & -[R]^{-1}\mathbf{d} \\ 0 \quad 0 \quad 0 & 1 \end{bmatrix} \tag{12.91}$$

The inverse of the rotational matrix $[R]$ is equal to its transpose because it is orthogonal,

$$[R]^{-1} = [R]^T \tag{12.92}$$

Substituting Eq. (12.92) into Eq. (12.91), we obtain the final form of $[T]^{-1}$ as:

$$[T]^{-1} = \begin{bmatrix} [R]^T & | & -[R]^T\mathbf{d} \\ 0 \quad 0 \quad 0 & | & 1 \end{bmatrix} \tag{12.93}$$

Equation (12.93) is general and useful for computing the inverse of a homogeneous transformation or mapping. The derivation we just followed to obtain Eq. (12.93) is quite general; other less general derivations may exist.

We now may ask the following questions. Does Eq. (12.93) agree with one's intuition for simple cases? And if it does, how? Let us consider translation. If a translational transformation or mapping is given by Eq. (12.53), the corresponding inverse is obtained by reversing the translational vector $\mathbf{d}$, that is, by negating its components x_d, y_d, and z_d. Here, no rotation is involved, and therefore $[R] = [R]^T = [I]$ and Eq. (12.93) yields exactly the same result.

Similarly, for scaling $\mathbf{d} = \mathbf{0}$ and $[R]$ is $[S]$ given by Eq. (12.57). In this case, Eq. (12.91) must be used, because $[S]$ is not orthogonal. The diagonal elements of $[S]^{-1}$ are the reciprocals of those of Eq. (12.57). For rotation, $\mathbf{d} = \mathbf{0}$ and $[R]^{-1} = [R]^T$. $[R]^{-1}$ is equivalent to negating the angle of rotation as one would expect. The reader can check this observation for Eqs. (12.10) to (12.12), (12.35), and (12.68).

One last useful inverse transformation problem is that of determining the direction of the axis of rotation $\hat{\mathbf{n}}$ and the angle of rotation θ from a given rotation matrix $[R]$. In general, the elements of $[R]$ are as shown by the top left submatrix of Eq. (12.81). They are also as shown in Eq. (12.35) in terms of $\hat{\mathbf{n}}$ and θ. To solve for $\hat{\mathbf{n}}$ and θ, we equate both forms, that is,

$$[R] = \begin{bmatrix} r_{11} & r_{12} & r_{13} \\ r_{21} & r_{22} & r_{23} \\ r_{31} & r_{32} & r_{33} \end{bmatrix} = \begin{bmatrix} n_x^2\, v\theta + c\theta & n_x n_y\, v\theta - n_z\, s\theta & n_x n_z\, v\theta + n_y\, s\theta \\ n_x n_y\, v\theta + n_z\, s\theta & n_y^2\, v\theta + c\theta & n_y n_z\, v\theta - n_x\, s\theta \\ n_x n_z\, v\theta - n_y\, s\theta & n_y n_z\, v\theta + n_x\, s\theta & n_z^2\, v\theta + c\theta \end{bmatrix} \tag{12.94}$$

Adding matrix elements (1, 1), (2, 2), and (3, 3) on both sides of Eq. (12.94) to each other and simplifying the result, we obtain:

$$\theta = \cos^{-1}\left(\frac{r_{11} + r_{22} + r_{33} - 1}{2}\right) \tag{12.95}$$

Subtracting element (2, 3) from (3, 2) yields

$$n_x = \frac{r_{32} - r_{23}}{2\sin\theta} \tag{12.96}$$

Similarly, we can find n_y and n_z, and write:

$$\hat{\mathbf{n}} = \left(\frac{1}{2\sin\theta}\right) \begin{bmatrix} r_{32} - r_{23} \\ r_{13} - r_{31} \\ r_{21} - r_{12} \end{bmatrix} \tag{12.97}$$

Equation (12.95) always computes a value of θ between 1 and 180°. Thus, for any axis-angle pair ($\hat{\mathbf{n}}$, θ), there is another pair ($-\hat{\mathbf{n}}$, $-\theta$) which results in the same orientation in space. Therefore, a choice always has to be made when converting from a rotation matrix into axis-angle representation. It is also obvious from Eq. (12.94) that the smaller the angle of rotation θ, the closer to zero the off-diagonal elements are, and consequently the more ill-defined the axis of rotation becomes as seen from Eq. (12.97). When $\theta = 0$ or 180°, the axis becomes completely undefined.

EXAMPLE 12.5 | **Derive an axis-angle pair ($\hat{\mathbf{n}}$, θ).**

An entity is rotated about the three principal axes of its MCS equal angles of 45° each. Find the equivalent axis and angle of rotation.

SOLUTION | Substituting $\alpha = \beta = \gamma = 445°$ into Eq. (12.68), we obtain:

$$[T] = \begin{bmatrix} 0.5 & -0.5 & 0.707 \\ 0.854 & 0.146 & -0.5 \\ 0.146 & 0.854 & 0.54 \end{bmatrix}$$

Substituting these values into Eqs. (12.95) and (12.97), we obtain:

$$\theta = 85.81° \text{ and } \hat{\mathbf{n}} = [0.679 \quad 0.281 \quad 0.679]^T.$$

12.11 Projections

Viewing a 3D models is a rather complex due to the fact that display devices are only 2D. This mismatch between 3D models and 2D screens can be resolved by utilizing projections, which transform 3D models onto a 2D projection plane. Various views of a model can be generated by using various projection planes.

To define a projection, a center of projection and a projection plane must be defined as shown in Figure 12.15. To obtain the projection of an entity (line connecting points P_1 and P_2 in the figure), projection rays (called projectors) are constructed by connecting the center of projection with each point of the entity. The intersections of these projectors with the projection plane define the projected points which are connected to produce the projected entity.

There are two types of projections based on the location of the center of projection relative to the projection plane. If the center is at a finite distance from the plane, perspective projection results (Figure 12.15a), and all the projectors meet at the center. If, on the other hand, the center is at infinite distance, all the projectors become parallel (meet at infinity) and parallel projection results (Figure 12.15b). Perspective projection is usually a part of perspective or projective geometry. Such geometry does not preserve parallelism, that is, no two lines are parallel. Parallel projection is a part of affine geometry, which is identical to Euclidean geometry. In affine geometry, parallelism is an important concept and therefore is preserved.

Perspective projection creates an artistic effect that adds some realism to perspective views. As can be seen from Figure 12.15a, the size of an entity is inversely proportional to its distance from the center of projection. That is, the closer the entity is to the center, the larger its size is.

Unlike perspective projection, parallel projection preserves actual dimensions and shapes of objects. It also preserves parallelism. Angles are preserved only on object faces that are parallel to the projection plane. There are two types of parallel projections based on the relation between the direction of projection and the projection plane. If this direction is normal to the projection plane, orthographic projection and views result. If the direction is not normal to the plane, oblique projection occurs. In this book, we cover orthographic projection.

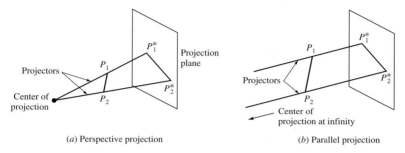

            (*a*) Perspective projection   (*b*) Parallel projection

Figure 12.15   Projections.

There are two types of orthographic projections. The most common type is the one that uses projection planes which are perpendicular to the principal axes of the MCS of the model, that is, the direction of projection coincides with one of these axes. The front, top, and right views that are used customarily in engineering drawings belong to this type. The other type of orthographic projections uses projection planes which are not normal to the principal axes and therefore show several faces of a model at once. This type is called axonometric projections. These projections preserve parallelism of lines but not angles. Thus, measurements can be made along each principal axis.

Axonometric projections are further divided into trimetric, dimetric, and isometric projections. The isometric projection is the most common axonometric projection. The isometric projection has the useful property that all three principal axes are equally foreshortened as will be seen in Section 12.11.3. Therefore, measurements along the axes can be made with the same scale; thus the name: *iso* for equal, *metric* for measure. In addition, the normal to the projection plane makes equal angles with each principal axis, and the principal axes make equal angles (120° each) with one another when projected onto the projection plane.

12.11.1 View Definition

A view has a viewing coordinate system (VCS). It is a 2D system with the X_v axis horizontal, pointing to the right, and the Y_v axis vertical, pointing upward as shown in Figure 12.16. The Z_v axis defines the viewing direction. To obtain views of a model, the viewing plane, the X_vY_v plane, is made coincident with the XY plane of the MCS such that the VCS origin is the same as that of the MCS. Model views can be obtained in two steps. First, we rotate the model (and its MCS) with respect to its MCS axes until the desired model plane coincides with the viewing plane. Second, we project the model in its final orientation onto the viewing plane. Figure 12.17 shows the relationship between the MCS and VCS for typical views of a geometric model. We can apply the two-step procedure just described to the figure. For the front view, the XY and X_vY_v planes are identical. To obtain this view, we simply project the geometry onto the viewing plane. For the top view, we must rotate the model about the X axis of the MCS $90°$ so that the XZ plane coincides with the X_vY_v plane.

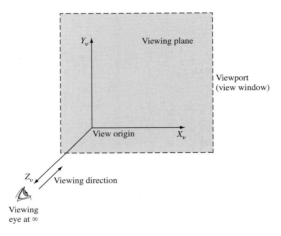

Figure 12.16   View definition.

12.11.2 Orthographic Projections

An orthographic projection of a view is obtained by setting to zero the coordinate value that coincides with the direction of projection (or viewing) after the model rotation. To obtain the front view (Figure 12.17b), we set $z = 0$ for all the key points of the model. Thus, Eq. (12.63) becomes:

$$[T] = \begin{bmatrix} 1 & 0 & 0 & 0 \\ 0 & 1 & 0 & 0 \\ 0 & 0 & 0 & 0 \\ 0 & 0 & 0 & 1 \end{bmatrix} \tag{12.98}$$

and Eq. (12.2) gives

$$\mathbf{P}_v = [T]\,\mathbf{P} \tag{12.99}$$

where P_v is the point expressed in the VCS. For the front view, Eq. (12.99) gives $x_v = x$, and $y_v = y$.

For the top view, the model is rotated $90°$ about the X axis followed by setting the y coordinates of the resulting points to zero. The y coordinate is the one to set to zero because the Y axis of the MCS coincides with the projection direction. In this case, $[T]$ becomes

$$[T] = \begin{bmatrix} 1 & 0 & 0 & 0 \\ 0 & 0 & -1 & 0 \\ 0 & 0 & 0 & 0 \\ 0 & 0 & 0 & 1 \end{bmatrix} \qquad (12.100)$$

Equation (12.99) gives $x_v = x$ and $y_v = -z$. If we use Eq. (12.100) to transform the MCS itself, the X axis ($y = z = 0$) transforms to $x_v = x$ and the Y axis ($x = z = 0$) transforms to $y_v = -z$. This result agrees with Figure 12.17b.

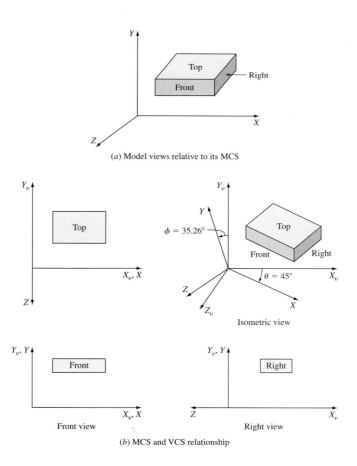

(a) Model views relative to its MCS

(b) MCS and VCS relationship

Figure 12.17   View of a geometric model.

The right view shown in Figure 12.17b can be obtained by rotating the model about the Y axis $-90°$ and setting the x coordinate to zero. Thus,

$$[T] = \begin{bmatrix} 0 & 0 & -1 & 0 \\ 0 & 1 & 0 & 0 \\ 0 & 0 & 0 & 0 \\ 0 & 0 & 0 & 1 \end{bmatrix}$$ (12.101)

which gives $x_v = -z$ and $y_v = y$.

Examining Eqs. (12.98), (12.100), and (12.101) shows that $[T]$ is singular with a column of zeros which corresponds to the projection or viewing direction. Once the viewpoints P_v are generated using these equations, they are clipped against the viewport boundaries and then mapped into the physical device coordinate system (SCS, Chapter 2) to display the view.

EXAMPLE 12.6   **Projecting points.**

The square shown lies in the XY plane. It is rotated 30° about the Y axis, followed by 45° rotation about the X axis, followed by translation of $\mathbf{d} = 2\mathbf{i} + 4\mathbf{j} - 3\mathbf{k}$. Find the coordinates of corner A in the final position of the square. Also, find the coordinates of A if the square is projected (in its final position) onto the XY plane (front view), XZ (top view), and YZ (right view).

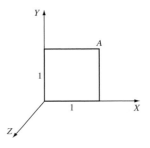

SOLUTION   Point A goes through three transformations as follows:

$\mathbf{P}^* = [T_Y]\mathbf{P}, \quad \mathbf{P}^{**} = [T_X]\mathbf{P}^* = [T_X][T_Y]\mathbf{P}, \quad \mathbf{P}^{***} = [T_d]\mathbf{P}^{**} = [T_d][T_X][T_Y]\mathbf{P}.$ This gives:

$$\mathbf{P}^{***} = \begin{bmatrix} 1 & 0 & 0 & 2 \\ 0 & 1 & 0 & 4 \\ 0 & 0 & 1 & -3 \\ 0 & 0 & 0 & 1 \end{bmatrix} \begin{bmatrix} 1 & 0 & 0 & 0 \\ 0 & 0.707 & -0.707 & 0 \\ 0 & 0.707 & 0.707 & 0 \\ 0 & 0 & 0 & 1 \end{bmatrix} \begin{bmatrix} 0.866 & 0 & 0.5 & 0 \\ 0 & 1 & 0 & 0 \\ -0.5 & 0 & 0.866 & 0 \\ 0 & 0 & 0 & 1 \end{bmatrix} = \begin{bmatrix} 0.866 & 0 & 0.5 & 2 \\ 0.354 & 0.707 & -0.612 & 4 \\ -0.354 & 0.707 & 0.612 & -3 \\ 0 & 0 & 0 & 1 \end{bmatrix} \begin{bmatrix} 1 \\ 1 \\ 0 \\ 1 \end{bmatrix} = \begin{bmatrix} 2.87 \\ 5.06 \\ -2.65 \\ 1 \end{bmatrix}$$

Front view: $x_v = x = 2.87, y_v = y = 5.06$
Top view: $x_v = x = 2.87, y_v = -z = 2.65$
Right view: $x_v = -z = 2.65, y_v = y = 5.06$

12.11.3 Isometric Projections

To obtain an isometric projection or view, the model and its MCS are customarily rotated an angle $\theta = \pm 45°$ about the Y_v axis followed by a rotation $\phi = \pm 35.26°$ about the X_v axis. These angles have been used for years in conventional manual drafting. In practice, the angle ϕ is taken as ± 30 to enable the use of drafting (plastic) triangles in the manual construction of isometric views. The values of these angles are based on the fact that the three axes are foreshortened equally in the isometric view. This can be explained as follows. The two rotations give:

$$P_v = [T_x][T_y]P = \begin{bmatrix} 1 & 0 & 0 & 0 \\ 0 & \cos\phi & -\sin\phi & 0 \\ 0 & \sin\phi & \cos\phi & 0 \\ 0 & 0 & 0 & 1 \end{bmatrix} \begin{bmatrix} \cos\theta & 0 & \sin\theta & 0 \\ 0 & 1 & 0 & 0 \\ -\sin\theta & 0 & \cos\theta & 0 \\ 0 & 0 & 0 & 1 \end{bmatrix} \begin{bmatrix} x \\ y \\ z \\ 1 \end{bmatrix} \qquad (12.102)$$

Applying Eq. (12.102) to transform the unit vectors in the X direction $[1\ \ 0\ \ 0\ \ 1]^T$, in the Y direction $[0\ \ 1\ \ 0\ \ 1]^T$, and in the Z direction $[0\ \ 0\ \ 1\ \ 1]^T$, and ignoring the z component because we are projecting onto the $z_v = 0$ plane, we obtain respectively:

$$x_v = \cos\phi \qquad\qquad y_v = \sin\phi\ \sin\theta$$
$$x_v = 0 \qquad\qquad\qquad y_v = \cos\theta \qquad\qquad\qquad (12.103)$$
$$x_v = \sin\phi \qquad\qquad y_v = -\cos\phi\ \sin\theta$$

If the three axes are to be foreshortened equally, the magnitudes of the unit vectors given by Eq. (12.103) must be equal. The first two equations give:

$$\cos^2\phi + \sin^2\phi\ \sin^2\theta = \cos^2\theta \qquad\qquad (12.104)$$

and the last two equations give:

$$\sin^2\phi + \cos^2\phi\ \sin^2\theta = \cos^2\theta \qquad\qquad (12.105)$$

Solving Eqs. (12.104) and (12.105) gives $\theta = \pm45°$ and $\phi = \pm35.26°$. The signs of the angles θ and ϕ result in four possible orientations of isometric views. Figure 12.17b shows the most common orientation where $\theta = -45°$ and $\phi = 35.26°$.

12.11.4 Perspective Projections

One common way to obtain a perspective view is to place the center of projection along the Z_v axis of the VCS and project onto the $z_v = 0$ or the X_vY_v plane. Figure 12.18 shows this case. The center of projection C is placed at distance d (measured along the Z_v axis) from the projection plane. We derive the matrix $[T]$ from the trigonometry shown in Figure 12.18. The viewing eye is located at the center C.

Figure 12.18 shows the perspective projection of point P as point P_v. To find y_v of P_v, the two similar triangles COP_2 and CP_3P_1 give:

$$\frac{y_v}{y} = \frac{d}{d-z} = \frac{1}{1-z/d} \qquad\qquad (12.106)$$

The two similar triangles CP_vP_2 and CPP_1 give x_v of P_v as:

$$\frac{x_v}{x} = \frac{r_2}{r_1} = \frac{d}{d-z} = \frac{1}{1-z/d} \qquad\qquad (12.107)$$

Rearranging Eqs. (12.106) and (12.107) to give y_v and x_v, respectively, and knowing $z_v = 0$, we can put the result in a homogeneous form as:

$$\mathbf{P}_v = \begin{bmatrix} 1 & 0 & 0 & 0 \\ 0 & 1 & 0 & 0 \\ 0 & 0 & 0 & 0 \\ 0 & 0 & -1/d & 1 \end{bmatrix} \begin{bmatrix} x \\ y \\ z \\ 1 \end{bmatrix} \qquad\qquad (12.108)$$

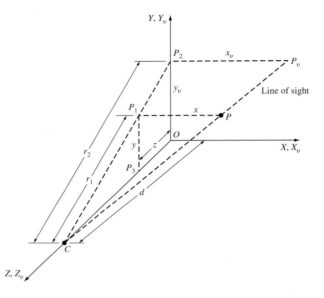

Figure 12.18   View of a geometric model.

If Eq. (12.108) is expanded, it gives $\mathbf{P}_v = [x \quad y \quad 0 \quad (1 - z/d)]^t$. This would require the division of x and y by $(1 - z/d)$ to obtain the corresponding cartesian coordinates from these homogeneous coordinates. Consequently, Eqs. (12.106) and (12.107) result. Thus, Eq. (12.108) gives the perspective projection onto the $z_v = 0$ plane when the center of projection is placed on the Z_v axis at a distance d from the origin. This result agrees with what we mentioned earlier that the matrix $[T_3]$ [Eq. (12.63)] produces perspective projection. If the center of projection is placed at a general point in space, the other elements of $[T_3]$ will be nonzero.

PROBLEMS

Part I: Theory

12.1 A general curve such as Bezier or B-spline is to be translated. Does translating the control points and then generating the curve give the same result as translating the original curve or not? Prove your answer.

12.2 Develop the translational transformation equation for a Hermite bicubic spline surface, a bicubic Bezier surface, and a bicubic B-spline surface. How can you extend the results to a cubic hyperpatch?

12.3 Derive the relationship between a point and its scaled counterpart $P*$ if P is scaled uniformly about a given point Q which is not the origin.

12.4 How can a Bezier curve, B-spline curve, Hermite bicubic surface, a Bezier surface, and a B-spline surface be scaled uniformly?

12.5 Show that Eqs. (12.47) to (12.50) can reduce to Eqs. (12.45) and (12.46); that is, show that reflection relative to a coordinate system is a special case of general reflection.

12.6 Develop the reflection transformation equations for Bezier and B-spline curves and surfaces as well as a Hermite bicubic surface. Carry the developments for the case of reflection through a general point.

12.7 The figure on the right shows a cube of length 2 inches. The cube is rotated an angle $\theta = 30°$ about the cube diagonal OD. If point B is the midpoint of side AD, find the coordinates of points A, B, and C before and after rotation. Verify your answer by solving the problem on your CAD/CAM system.

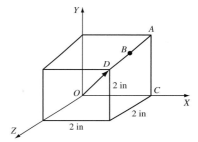

12.8 Show how the homogeneous representation can help represent points at infinity and can also be used to force parallel lines to intersect.

12.9 Show that parallel and perpendicular lines transform to parallel and perpendicular lines.

12.10 Show that the midpoint of a line transforms to the midpoint of the transformed lines.

12.11 A point is rotated about the Z axis two successive angles θ_1 and θ_2. Show that this is equivalent to rotating the point about the same axis once with any angle $\theta = \theta_1 + \theta_2$.

12.12 Show that:

a. translation is commutative.

b. mirror and 2D rotation about the Z axis and not commutative.

c. scaling and 2D rotation about the Z axis are commutative.

d. 3D rotations are not commutative.

12.13 Given a point $P = (2, 4, 8)$ and using the homogeneous representation:

a. Calculate the coordinates of the transformed point $P*$ if P is rotated about the X, Y, and Z axes at angles of 30, 60, and 90°, respectively.

b. If the point $P*$ obtained in part (a) is to be rotated back to its original position, find the corresponding rotation matrix. Verify your answer.

c. Calculate $P*$ if P is translated by $\mathbf{d} = 3\mathbf{i} - 4\mathbf{j} - 5\mathbf{k}$ and then scaled uniformly by $S = 1.5$.

d. Calculate the orthographic projection P_v of P.

e. Calculate the perspective projection P_v of P if the center of projection is at distance $d = 10$ inches from the origin along the Z axis.

12.14 Given three points P_1, P_2, and P_3 that belong to a geometric model and given three other points Q_1, Q_2, and Q_3, find the transformation matrix $[T]$ that:

a. transforms P_1 to Q_1.

b. transforms the direction of the vector $(\mathbf{P}_2 - \mathbf{P}_1)$ into the direction of the vector $(\mathbf{Q}_2 - \mathbf{Q}_1)$.

c. transforms the plane of the three points P_1, P_2, and P_3 into the plane of Q_1, Q_2, and Q_3.

This problem is sometimes called "three-point" transformation. It is useful for moving two geometric models, mainly solids, to coincide with one another or for positioning entities in a geometric model.

12.15 The following figure shows the rotation of a point P about an arbitrary axis of rotation that passes through the origin and lies in the XZ plane. Derive the rotation matrix $[R]$ for this case. Verify your answer by substituting the proper values in the general matrix $[R]$ given by Eq. (12.35).

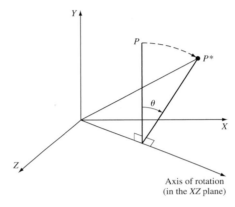

Axis of rotation
(in the XZ plane)

Part II: Laboratory

Use your in-house CAD/CAM system to answer the questions in this part.

12.16 Find the required user input to create 2D and 3D views on your CAD/CAM system. Can you define perspective views?

12.17 Show that 3D rotations are not commutative. Take a sequence of 90° rotations about the three principal axes and permutate them.

12.18 Redo Example 12.3 on your CAD/CAM system. Use $\hat{\mathbf{n}} = 0.5\hat{\mathbf{i}} + 0.707\hat{\mathbf{j}} + 0.5\hat{\mathbf{k}}$ and $P(1, 3, 5)$.

12.19 A line is connecting the origin of the MCS of a model and a point $P(1, 2, 3)$. Find three different ways to rotate the line so that it coincides with the Z axis of the MCS.

Part III: Programming

Write the code so that the user can input values to control the output. The program should display the results. Use OpenGL with C or C++, or use Java 3D.

12.20 Write a program to implement Eq. (12.63). Use the form $\mathbf{P}^* = [T]\mathbf{P}$ where $\mathbf{P}$ is a vector of points, tangent vectors, or any other vectors of interest. For example, P could be four points, that is, $[\mathbf{P}_1 \quad \mathbf{P}_2 \quad \mathbf{P}_3 \quad \mathbf{P}_4]^T$, or it could be $[\mathbf{P}_0 \quad \mathbf{P}_1 \quad \mathbf{P}_0' \quad \mathbf{P}_1']^T$ as for the Hermite cubic spline. Write the program for rotation about the Z axis, translation, and scaling uniformly.

12.21 Using the program developed in Problem 12.20, write a program to translate a cubic spline curve, a Bezier curve, and a B-spline curve.

Visualization

GOAL

Understand and master the concepts of rendering and visualization and their importance to geometric modeling, and understand the related algorithms.

OBJECTIVES

After reading this chapter, you should understand the following concepts:

- Need for rendering
- Model cleanup
- Hidden line removal
- Hidden surface removal
- Hidden solid removal
- Visualization algorithms
- Shading
- Colors

501

CHAPTER HEADLINES

13.1 Introduction

One of the most recognized benefits of CAD systems is their ability to provide their users with visual displays of the models and scenes they create. Visualization has always been recognized as the most effective wat to communicate new ideas among engineers and designers. Virtually all CAD-related application programs present their results to their users in a graphical form. CAD systems utilize visualization both in geometric modeling (projection, hidden line/surface/solid removal, shading, and colors) and in scientific computing (displaying analysis and manufacturing results).

The basic problem in visualization is how to display 3D data on 2D screens. How can the screen show depth? How can the visual complexities of the real environment such as lighting, color, shadows, and texture be represented as image attributes?

What complicates the display of 3D models even further is the centralized nature of their databases. If we project a complex 3D model onto a screen, we get a complex maze of lines and curves. To interpret this maze, curves and surfaces that cannot be seen from the given viewpoint must be removed. Visualization algorithms eliminate the ambiguities of rendering 3D models and are considered the first step toward visual realism.

The two main concepts utilized in visualization algorithms are coherence and sorting. Geometric models are more than a set of random discontinuities. They have some consistency both over the area of the image and over the time from one frame to the next. Thus, using this coherence between various frames can improve the efficiency of visualization algorithms. Similarly, all these algorithms sort or search through collections of surfaces, edges, or objects according to various criteria, finally discovering the visible items and displaying them.

13.2 Model Cleanup

Model cleanup is perhaps considered the oldest, most elementary way to achieve visual realism. The cleanup process begins by adding the proper orthographic views to a drawing. Visual realism is added to each view separately, in the drawing mode, by handling hidden lines and adding centerlines, dimensions, notes, and labels as discussed in Chapter 4. Handling hidden lines includes eliminating them in the view, changing their fonts from solid to dashed, or blanking (hiding) them. A major advantage of manual model cleanup is the control it gives the user over which entities should be removed and which should be dashed. The automatic elimination of hidden lines may result in a loss of depth information.

CAD systems provide users with useful options to deal with hidden lines (edges) in an automatic way upon view creation. A user can create a view with all hidden lines removed, all hidden lines visible (known as wireframe model), or showing hidden lines in a specific font (dashed or gray). The user can also create centerlines for holes and other entities automatically by using centerlines options provided by the system—for example, to show centerlines for circles. CAD systems also handle the partially visible entities automatically. They hide the invisible part of an entity and keep the visible part. Figure 13.1 shows examples of model cleanup.

CAD users still have the option for manual model cleanup, which is useful for complex models. A hybrid (a mix of manual and automatic activities) approach to model cleanup is the best and fastest way to prepare views in an engineering drawing. For example, a user may display all hidden lines of a view automatically in a dashed font, and then remove some manually from cluttered areas of the view.

The reason CAD systems can support all these automatic model cleanup options is that models are solids, that is, they are well defined. A hidden line algorithm should sort out precisely the edges and faces of a solid into hidden, visible, and partially hidden/visible. The algorithm uses a combination of membership classification function and depth analysis relative to the viewing point.

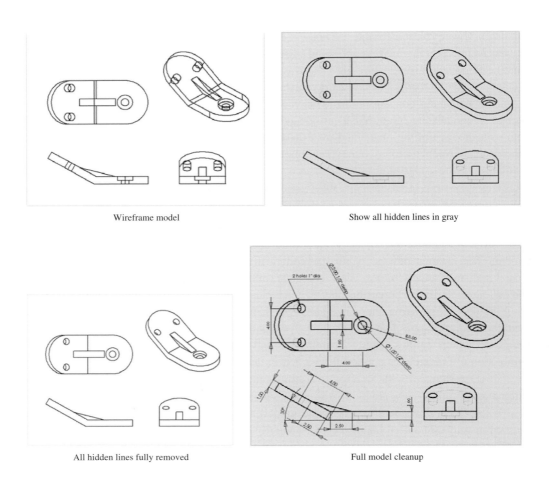

Wireframe model

Show all hidden lines in gray

All hidden lines fully removed

Full model cleanup

Figure 13.1   Model cleanup.

13.3 Hidden-Line Removal

Edges and surfaces of a given model may be hidden (invisible) or visible in a given image, depending on the viewing direction. The determination of hidden edges and surfaces is a challenging problem in computer graphics. The solution to this visibility problem typically demands large amounts of computer time and memory space. Removal algorithms are usually built into the local hardware processors of a graphics display to speed up their execution.

Removal algorithms are usually classified into hidden-line and hidden-surface algorithms. The former supports vector (line-drawing) devices such as printers and plotters, while the latter supports raster displays. Hidden-line algorithms can, of course, be used with raster displays. However, hidden-surface algorithms are not applicable to vector displays. From a geometric modeling point of view, this classification has a different meaning. Hidden-line removal deals with the visibility problem of edges of polyhedral models; these are models with orientable flat polygons as faces. Hidden-surface removal deals with the visibility problem of edges of models with orientable curved surfaces; these are models with orientable nonplanar faces. There is yet a third group, known as hidden-solid removal, which is similar to hidden-surface removal. The classification to hidden-line, hidden-surface, and hidden-solid removal reflects the historical order of the development of the related algorithms.

Hidden-line and hidden-surface algorithms have been classified as object-space methods, image-space methods, or a combination of both. Image-space algorithms can be further divided into raster and vector algorithms. The raster algorithms use a pixel matrix representation of the image, and the vector algorithms use line segment end point coordinates in representing the image. An object-space algorithm utilizes the spatial and geometrical relationships among the objects in the scene to determine which are the hidden and visible parts of these objects. An image-space algorithm, on the other hand, concentrates on the final image to determine what is visible within, for example, each raster pixel in the case of raster displays. Most hidden-surface algorithms use raster image-space methods, while most hidden-line algorithms use object-space methods.

The two approaches to achieving visual realism exhibit different characteristics. Object-space algorithms are more accurate than image-space algorithms. The former perform geometric calculations (such as intersections) using the floating-point precision of the computer hardware, while the latter perform calculations with accuracy equal to the resolution of the display screen used to render the image. Therefore, the enlargement of an object-space image does not degrade its quality of display as does the enlargement of an image-space image. As the complexity of the scene increases (large number of objects in the scene), the computation time grows more quickly for object-space algorithms than for image-space algorithms.

13.3.1 Visibility of Object Views

The visibility of parts of objects of a scene depends on the location of the viewing eye, the viewing direction, the type of projection (orthogonal or perspective), and the depth or the distance from various faces of various objects in the scene to the viewing eye. The hidden-line removal of perspective views is a fairly complex problem to solve. Many lines of sight (rays) from the viewing eye must be considered, and their points of intersection with objects' faces have to be calculated. The complexity of the problem is considerably reduced if orthographic views are utilized because no intersections would be necessary. It is the common practice to apply the perspective transformation given by Eq. (12.108) to the set of points in the scene, and then to apply orthographic hidden-line visibility algorithms to the resulting (transformed) set of points. This is equivalent to saying that the orthographic viewing of the transformed (perspective) objects is identical to the perspective viewing of the original (untransformed) objects. Hence, only orthographic hidden-line algorithms are discussed in this book.

The depth comparison is the central criterion utilized by hidden-line algorithms to determine visibility. Depth comparisons are typically done after the proper view transformation given by Eqs. (12.98) and (12.108) for orthographic and perspective projections, respectively. While these two equations destroy the depth information (the z_v coordinate of projected points) to generate views, such information can be saved for depth comparisons by hidden-line algorithms as follows: Replace the element t_{33} of $[T]$ in both Eqs. (12.98) and (12.108) by 1 instead of the current 0 to restore depth information. Similarly, set t_{32} and t_{31} of $[T]$ in Eqs. (12.100) and (12.101), respectively, to 1.

The depth comparison determines if a projected point P_{1v} (x_{1v}, y_{1v}) in a given view obscures another point P_{2v} (x_{2v}, y_{2v}). This is equivalent to determining if the two original corresponding points P_1 and P_2 lie on the same projector as shown in Figure 13.2 (the MCS and VCS are shown as the XYZ and the $X_vY_vZ_v$ systems, respectively). For orthographic projections, projectors are parallel. Therefore, two points P_1 and P_2 are on the same projector if $x_{1v} = x_{2v}$ and $y_{1v} = y_{2v}$. If they are, a comparison of z_{1v} and z_{2v} decides which point is closer to the viewing eye. Utilizing the VCS shown in Figures 12.16 and 13.2, the point with larger z_v coordinate lies closer to the viewer. Applying this depth comparison to points P_1, P_2, and P_3 of Figure 13.2 shows that point P_1 obscures P_2 (that is, P_1 is visible and P_2 is hidden), and P_3 is visible.

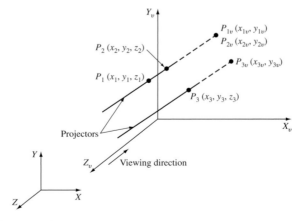

Figure 13.2   Depth comparison.

13.3.2 Visibility Techniques

If the depth comparison criterion is used solely with no other enhancements, the number of comparison grows rapidly (for n points, $\binom{n}{2}$ tests are required) which leads to difficulties storing and managing the results by the corresponding hidden-line algorithm. As a result, the algorithm might be slow in calculating the final image. Various visibility techniques exist to alleviate these problems. In general, these techniques attempt to establish relationships among polygons and edges in the viewing plane. The techniques normally check for overlapping of pairs of polygons (sometimes referred to as lateral comparisons) in the viewing plane (the screen). If overlapping occurs, depth comparisons are used to determine if part or all of one polygon is hidden by another. Both the lateral and depth comparisons are performed in the VCS.

13.3.2.1   Minimax Test

This test (also called overlap or bounding box test) checks if two polygons overlap. The test provides a quick method to determine if two polygons do not overlap. It surrounds each polygon with a box by finding its extents (minimum and maximum x and y coordinates) and then checks for the intersection of any two boxes in both the X and Y directions. If two boxes do not intersect, their corresponding polygons do not overlap as shown in Figure 13.3. In such a case, no further testing of the edges of the polygons is required.

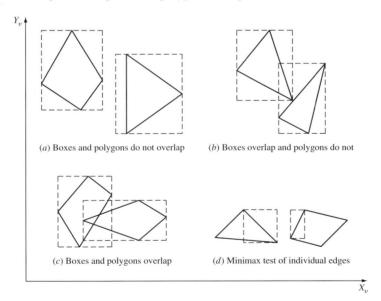

(a) Boxes and polygons do not overlap   (b) Boxes overlap and polygons do not

(c) Boxes and polygons overlap   (d) Minimax test of individual edges

Figure 13.3   Minimax test for polygons and edges.

If the minimax test fails (two boxes intersect), the two polygons may or may not overlap as shown in Figure 13.3. Each edge of one polygon is compared against all the edges of the other polygon to detect intersections. The minimax test can be applied first to any two edges to speed up this process.

The minimax test can be applied in the Z direction to check overlap in this direction. In all tests, finding the extents themselves is the most critical part of the test. Typically, this can be achieved by iterating through the list of vertex coordinates of each polygon and recording the largest and the smallest values for each coordinate.

13.3.2.2   Containment Test

Some hidden-line algorithms depend on whether a polygon surrounds a point or another polygon. The containment test determines whether a given point lies inside a given polygon or polyhedron. There are three methods to compute containment or surroundedness. For a convex polygon, one can substitute the x_v and y_v coordinates of the point into the line equation of each edge. If all substitutions result in the same sign, the point is on the same side of each edge and is therefore surrounded. This test requires that the signs of the coefficients of the line equations be chosen correctly.

For nonconvex polygons, two other methods can be used. In the first method, we draw a line from the point under testing to infinity as shown in Figure 13.4a. The semi-infinite line is intersected with the polygon edges. If the intersection count is even, the point is outside the polygon (P_2 in Figure 13.4a). If it is odd, the point is inside (P_1 in the figure). If one of the polygon edges lies on the semi-infinite line, a singular case arises which needs special treatment to guarantee the consistency of the results.

The second method for nonconvex polygons (Figure 13.4b) computes the sum of the angles subtended by each of the oriented edges as seen from the test point (P_1 or P_2). If the sum is zero, the point is outside the polygon. If the sum is 2π or -2π, the point is inside. The minus sign reflects whether the vertices of the polygon are ordered in a clockwise direction instead of counterclockwise.

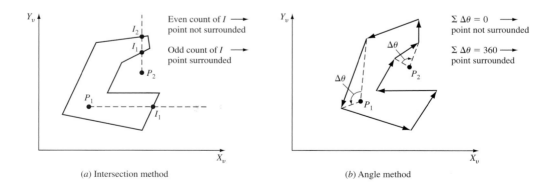

(a) Intersection method   (b) Angle method

Figure 13.4   Containment test.

13.3.2.3   Surface Test

This test (also called back face or depth test) provides an efficient method for implementing the depth comparison. Figure 13.5a shows face B obscures part of face A. In this case the equation of a plane is used to perform the test. The plane equation is given by:

$$ax_v + by_v + cz_v + d = 0 \qquad (13.1)$$

If a given point $P(x_v, y_v, z_v)$ is not on the plane, the sign of the left-hand side of Eq. (13.1) is positive if the point lies on one side of the plane, and negative if it lies on the other side. The equation coefficients a, b, c, and d can be arranged so that a positive value indicates a point outside the plane. The plane equation can also be used to compute the depth z_v of a face at a given point $P(x_v, y_v, z_v)$. The depths of two faces can, therefore, be computed at given points to decide which one is closer to the viewing eye.

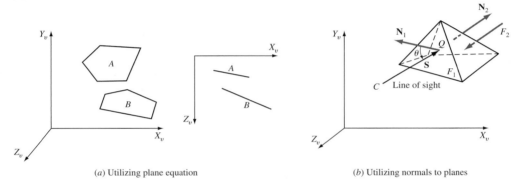

(a) Utilizing plane equation   (b) Utilizing normals to planes

Figure 13.5   Surface test.

Another important use of the plane equation in hidden-line removal is achieved by using the normal vector to the plane. The first three coefficients a, b, and c of Eq. (13.1) represent the normal to the plane, and the vector $[a, b, c, d]$ represents the homogeneous coordinates of this normal. The coefficient d is found by knowing a point on the plane. In Chapter 7, we discussed the various ways of finding the plane equation. Figure 13.5b shows how the normal to a face can be used to decide its visibility. The basic idea of the test is that faces whose outward normal vector points toward the viewing eye are visible (face F_1) while others are invisible (face F_2). This test is implemented by calculating the dot product of the normal vector $\mathbf{N}$ and the line-of-sight vector S (Figure 13.5b) as:

$$\mathbf{N} \cdot \mathbf{S} = |\mathbf{N}||\mathbf{S}| \cos\theta \qquad (13.2)$$

If, in Eq. (13.2), we assume that $\mathbf{N}$ points away from the solid and θ is measured from $\mathbf{N}$ to $\mathbf{S}$, the dot product is positive when $\mathbf{N}$ points toward the viewing eye or when the face and its edges are visible. The right-hand side of Eq. (13.2) gives the component of $\mathbf{N}$ along the direction of $\mathbf{S}$. For orthographic projection, this direction coincides with the Z_v axis. Thus the surface test can be stated as follows: Faces whose normals have positive components in the Z_v direction are visible, and those whose normals have negative Z_v components are invisible. The surface test by itself cannot solve the hidden-line problem except for single convex polyhedra. Moreover, the test may fail for perspective projection if more than one polyhedron exists in the scene.

13.3.2.4   Computing Silhouettes

A set of edges which separates visible faces from invisible faces of an object with respect to a given viewing direction is called silhouette edges (or silhouettes). The signs of the Z_v components of normal vectors of the object faces can be utilized to determine the silhouette. An edge that is part of the silhouette is characterized as the intersection of one visible face and one invisible face. An edge that is the intersection of two visible faces is visible, but it does not contribute to the silhouette. The intersection of two invisible faces produces an invisible edge. Figure 13.6a shows the silhouette of a cube.

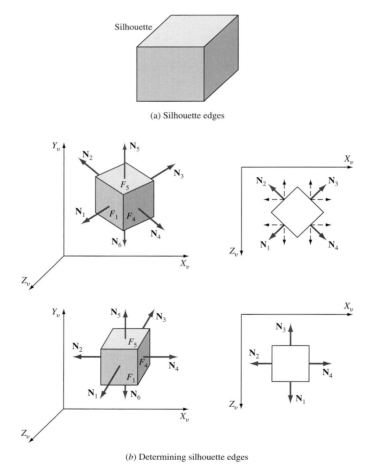

(a) Silhouette edges

(b) Determining silhouette edges

Figure 13.6   Silhouette edges of a polyhedron.

Figure 13.6*b* shows how to compute the silhouette edges. One cube is oriented at an angle with the axes of the VCS and the other is parallel to the axes. In the first case, the Z_v component of each normal is calculated (shown dashed in Figure 13.6). The edges between faces F_1 and F_2, F_1 and F_6, F_2 and F_5, F_3 and F_5, F_3 and F_4, and F_4 and F_6 are silhouette edges.

If a normal does not have a Z_v component, as in the second case of Figure 13.6*b*, additional information is needed to compute the silhouette. This case implies that the corresponding face is parallel to the Z_v axis and either the X_v or Y_v axis, and perpendicular to the remaining axis. For example, face F_4 is parallel to both the Z_v and Y_v axes and perpendicular to the X_v axis. Therefore, the face normal is parallel to one of the VCS axes. If this normal points in the positive direction of the axis, the face is visible. Face F_4 is visible and F_2 is not. Similarly, face F_5 is visible while F_6 is not.

Determining silhouette curves for curved surfaces follows a similar approach but is more involved. The silhouette curve of a surface is a curve on the surface along which the Z_v components of the surface normals are zeros, as shown in Figure 13.7. To obtain this curve, the equation of the Z_v component of the surface normal [Eq. (7.11)] is set to zero and solved for u and v. This approach is usually inconvenient because the resulting equation is difficult to solve. For a bicubic surface, the equation is a quintic polynomial in u and v. Other more efficient methods are available in the literature and are not discussed here.

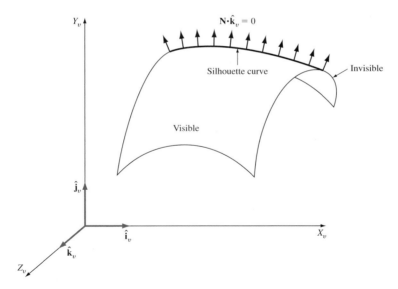

Figure 13.7   Silhouette curve of a surface.

13.3.2.5   Edge Intersections

All the visibility techniques discussed thus far cannot determine which parts are hidden and which are visible for partially hidden edges. To accomplish this, hidden-line algorithms first calculate edge intersections in two dimensions, that is, in the $X_v Y_v$ plane of the VCS. These intersections are used to determine edge visibility.

Consider the edge AB and the face F shown in Figure 13.8. The edges of the face F are directed in a counterclockwise direction. Let us consider the intersection I between AB and edge CD of face F. The visibility of AB with respect to F can fall into one of three cases: fully visible, I indicates the disappearance of AB, or I indicates the appearance of AB. In the first case, the depths z_v at point I are computed and compared.

If I is considered a point on F, its x_v and y_v coordinates are substituted into the plane equation to find z_v. If it is considered a point on AB, the line equation is used instead to find the other depth. If the depth of the line is larger (we are dealing with a right-handed VCS) than the depth of the face, the line AB is fully visible (Figure 13.8a). If the directed edge CD subtends a clockwise angle θ about A (Figure 13.8b), the edge disappears, otherwise it appears (Figure 13.8c). Notice that if the face edges are directed clockwise, the angle criterion reverses. The angle criterion is sometimes referred to as the vorticity of edge CD with respect to point A.

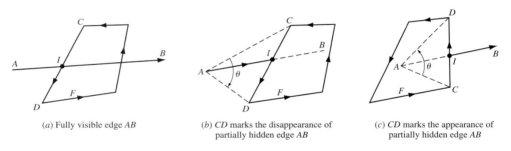

(a) Fully visible edge AB (b) CD marks the disappearance of partially hidden edge AB (c) CD marks the appearance of partially hidden edge AB

Figure 13.8 Edge visibility.

13.3.2.6   Segment (Scanline) Comparisons

This class of techniques is used to solve the hidden surface problem in the image (raster) space. It is covered in this section as another visibility technique. The techniques covered here are applicable to hidden-surface and hidden-solid algorithms as well. As discussed in Chapter 11, scan lines are traversed from top to bottom, left to right. Therefore, instead of computing the whole correct image at once, it can be computed scan line by scan line, that is in segments, and displayed in the same order of the scan lines. Computationally, the plane of the scan line defines segments where it intersects faces in the image as shown in Figure 13.9. Computing the correct image for one scan line is much simpler and uses less memory than computing all the scan lines at once.

The segment comparisons are performed in the $X_v Z_v$ plane as shown in Figure 13.9. The scan line is divided into spans (dashed line). The visibility is determined within each span by comparing the depths of the edge segments that lie within the span. Plane equations are used to compute these depths. Segments with maximum depth are visible throughout the span.

The strategy to divide a scan line into spans is a distinctive feature of any hidden-surface algorithm. One obvious strategy is to divide the scan line at each end point of each edge segment (lines *A, B, C,* and *D* in Figure 13.9). A better strategy is to choose fewer spans. In Figure 13.9, it is optimum to divide the scan line via line *C* into two spans only.

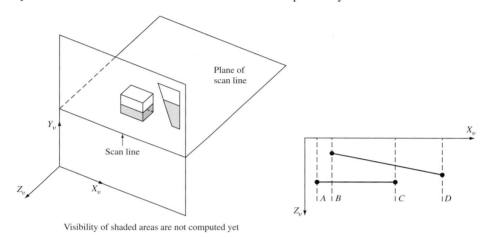

Visibility of shaded areas are not computed yet

Figure 13.9   Computing visibility via scan lines.

13.3.2.7   Homogeneity Test

The depth test described in Section 13.3.2.3 is concerned with comparing the depths of point sets (single points) to determine visibility. Computing the homogeneity of point sets is another test to determine visibility. The notion of neighborhood (discussed in Chapter 9) must be used to determine homogeneity. The neighborhood of a point P, denoted by $N(p)$, in a data set is all points in the set lying inside a sphere (or a circle for 2D point sets) around it.

Three types of points can be identified based on computing homogeneity: homogeneously visible, homogeneously invisible, and inhomogeneously visible. If a neighborhood of a point P can be bijectively projected onto a neighborhood of the projection of the point, then the neighborhood of P is visible or invisible, and P is called homogeneously visible or invisible respectively. Otherwise, P is inhomogeneously visible (or invisible).

If we denote the projection of P by $pr(p)$, P is homogeneously visible or invisible if $pr(N(P)) = N(pr(P))$ and inhomogeneously visible if $pr(N(P)) \neq N(pr(P))$. Using this test, inner points of scenes are homogeneously visible (covering) or invisible (hidden), and contour (edge) points are inhomogeneously visible, as shown in Figure 13.10. $N(P, F)$ is the neighborhood of a point that belongs to face F. The contour points (P_2) are inhomogeneously visible (covering) and inner points are homogeneously visible (P_1 on face F_2) or invisible (P_1 on face F_1).

Homogeneity is important for both covering and hiding. No point needs to be tested against any homogeneously visible (covering) point and no homogeneously hidden point needs to be tested against any other point, since these points are homogeneously invisible in both cases. Moreover, homogeneously invisible point sets are of no interest to the visibility problem. Homogeneously visible point sets can be completely displayed without further tests against other points. Therefore, the determination of homogeneous point sets reduces further computation. Some area-oriented algorithms (see Section 10.3.6.2) use this test to improve their efficiency.

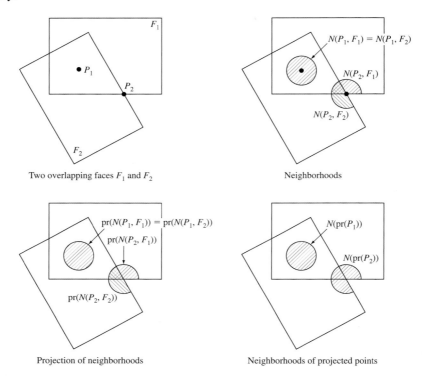

Two overlapping faces F_1 and F_2 Neighborhoods

Projection of neighborhoods   Neighborhoods of projected points

Figure 13.10   **Homogeneity test.**

13.3.3   Sorting

Many visibility algorithms (hidden-line, hidden-surface, and hidden-solid algorithms) make extensive use of sorting operations. Sorting and searching operate on the records of the scene database. These records typically contain geometrical, topological, and viewing information about the polygons and faces that make the scene. Sorting is an operation that orders a given set of records according to a selected criterion. The time required to perform the sort depends on the number of records to be processed, the algorithm that performs the sort, and the properties of the initial ordering of the records (whether it is random or semiordered). Various sorting techniques are available. They are not covered here, and the reader is referred to the literature.

13.3.4 Coherence

Naturally, the elements of a scene or its image have some interrelationships, known as coherence. Hidden-line algorithms that utilize coherence in their sorting techniques are more effective than other algorithms that do not. Coherence is a measure of how rapidly a scene or its image changes. It describes the extent to which a scene or its image is locally constant.

The coherence of a set of data can improve the speed of its sorting significantly. If, for example, an initially sorted deck of cards is slightly shuffled, the coherence remaining in the deck can be of great use in resorting it. Similarly, the gradual changes in the appearance of a scene or its image from one place to another can reduce the number of sorting operations greatly.

Several types of coherence can be identified in both the object space and the image space:

1. **Edge coherence.** The visibility of an edge changes only when it crosses another edge.
2. **Face coherence.** If a part of a face is visible, the entire face is probably visible. Moreover, the penetration of faces is a relatively rare occurrence, and therefore it is not usually checked by hidden-removal algorithms.
3. **Geometric coherence.** Edges that share the same vertex or faces that share the same edges have similar visibilities in most cases. For example, if three edges (the case of a box) share the same vertex, they may be all visible, all invisible, or two visible and one invisible. The proper combination depends on the angle between any two edges (less or greater than 180°) and on the location of any of the edges relative to the plane defined by the other two edges.
4. **Frame coherence.** A picture does not change very much from frame to frame.
5. **Scanline coherence.** Segments of a scene visible on one scan line are most probably visible on the next line.
6. **Area coherence.** A particular element (area) of an image and its neighbors are all likely to have the same visibility and to be influenced by the same face.
7. **Depth coherence.** The different surfaces at a given screen location are generally well separated in depth relative to the depth range of each.

The first three types of coherence are object space based, while the last four are image space based. If an image exhibits a particular predominant coherence, the coherence would form the basis of the related hidden-line removal algorithm.

13.3.5 Formulation and Implementation

The hidden-line removal problem can be stated as follows: For a given 3D scene, a given viewing point, and a given direction, eliminate from a 2D projection of the scene all parts of edges and faces which the observer cannot see. For orthographic projections, the location of the viewing point is not needed.

A set of generic steps that can implement a solution to the hidden-line removal problem is shown in Figure 13.11. The 3D scene is a set of 3D objects. Each object is defined by its geometry and topology. A solid model is an ideal representation. The root of the tree represents the 3D scene database.

The second step is to apply the proper geometric transformations (Chapter 12) based on the viewing direction of the 3D scene data to obtain the 2D image "raw" data. These transformations are modified as discussed earlier to also produce the depth information, which is stored in the image database for depth-comparison purposes later. At this stage, the image is a maze of all edges (visible and invisible).

The next two steps (sorting of image data and applying visibility techniques) are interrelated and may be difficult to identify as such. Nevertheless, we apply one or more of the visibility techniques covered earlier with the aid of a sorting technique. The surface test to eliminate the back faces is usually sufficient to solve the hidden-line problem if the scene is only one convex polyhedron without holes. Otherwise a combination of techniques is usually required. It is this combination and sorting techniques that differentiate existing algorithms. In order to apply the visibility techniques to the image data, the sorting of this data by either polygons or edges is required.

With the completion of the sorting according to the visibility criteria set by the visibility techniques, the hidden edges (or parts of edges) are identified and removed from the image data. The last step in the algorithm is to display and/or plot the final images.

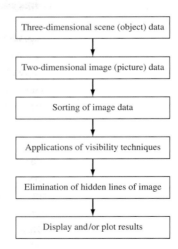

Figure 13.11   A generic hidden-line algorithm.

13.3.6 Sample Hidden-Line Algorithms

The purpose of a visibility algorithm is to test whether a picture element is hidden by another or not. A visibility algorithm can be described as va(H, C) where H and C are the sets of the hidden and the covering candidate edges of the scene under consideration, respectively. Since every edge in the H set has to be tested against all the edges in the C set, the algorithm requires $H \times C$ visibility tests.

The efficiency of a visibility algorithm depends on the choice of the H and C sets. Existing hidden-line removal algorithms can take one of three approaches: edge-oriented approach, silhouette (contour) -oriented approach, or area-oriented approach. To examine the efficiency of each approach, let E and S be the number of edges and silhouette edges in a scene, respectively.

The edge-oriented approach provides the basis for most existing hidden-line algorithms. The underlying strategy is the explicit computation of the visible segments of all individual edges. The visibility calculation consists of the test of all edges against all surfaces. Therefore, $H = E$ and $C = E$. Thus $E \times E$ visibility tests are required. This approach is inefficient because it tests all edges against each other whether they intersect or not. It does not recognize the structural information of a picture. Sorting of all edges improves the efficiency of the approach.

In the silhouette- (contour-) oriented approach, the silhouette edges are calculated first. The visibility calculation consists of testing all the edges against all the silhouette edges only. Thus, all the edges are hidden candidates but only the silhouette edges are covering candidates; that is, $H = E$, $C = S$. The calculation rate is estimated at $E \times S$. This approach is more efficient than the edge-oriented approach. However, some edges (after sorting) will still be tested against some non-intersecting silhouettes. Moreover, intersections of homogeneously (fully) hidden edges against hidden silhouettes are unnecessarily computed. Also, the tests of homogeneously covering (visible) edges against each other are avoided completely.

The area-oriented approach aims at the recognition of the visible areas of a picture. This approach calculates the silhouette edges and connects them to form closed polygons (areas). The connection can be done in $S \times S$ steps. The visibility calculations begin with the test of all silhouette edges against each other. Thus $H = S$ and $C = S$, and the calculation rate is $S \times S$. This approach works better than the silhouette-oriented approach. First, sorting is reduced to the silhouette edges only. Only silhouettes (contours) are tested against non-intersecting silhouettes. The computation of intersection points of both homogeneously hidden and homogeneously covering edges is avoided.

There exists a wealth of hidden-line algorithms that utilize one or more of the visibility techniques discussed in Section 13.3.2 and that follow one of the three approaches just described. These include the priority algorithm, the plane-sweep paradigm, algorithms for scenes of high complexities, algorithms for finite element models of planar elements, area-oriented algorithms, the overlay algorithm for surfaces defined by $u-v$ grids (Chapter 7), and algorithms for projected grid surfaces. In the remainder of this section, we discuss the priority algorithm and an area-oriented algorithm. Readers who are interested in details of other algorithms should consult the literature.

13.3.6.1   The Priority Algorithm

This algorithm is also known as the depth or Z algorithm. The algorithm is based on sorting all the faces (polygons) in the scene according to the largest z coordinate value of each.

This step is sometimes known as assignment of priorities. If a face intersects more than one face, other visibility tests besides the z depth are needed to resolve any ambiguities. This step constitutes determination of coverings.

To illustrate how the priority algorithm can be implemented, let us consider a scene of two boxes as shown in Figure 13.12. Figure 13.12 shows the scene in the standard VCS where the viewing eye is located at ∞—on the positive Z_v direction. The following steps provide guidance for implementing the algorithm:

1. Utilize the proper orthographic projection to obtain the desired view (whose hidden lines are to be removed) of the scene. This results in set a of vertices with coordinates of (x_v, y_v, z_v). To enable one to perform the depth test, the plane equation of any face (polygon) in the image can be obtained using Eq. (13.1). Given three points that lie in one face, Eq. (13.1) can be rewritten as:

$$z_v = Ax_v + By_v + C \qquad (13.3)$$

where

$$A = \frac{(z_{v1} - z_{v3})(y_{v2} - y_{v3}) - (z_{v2} - z_{v3})(y_{v1} - y_{v3})}{D} \qquad (13.4)$$

$$B = -\frac{(z_{v1} - z_{v3})(x_{v2} - x_{v3}) - (z_{v2} - z_{v3})(x_{v1} - x_{v3})}{D} \qquad (13.5)$$

$$C = z_{v1} - Ax_{v1} - By_{v1} \qquad (13.6)$$

and

$$D = (x_{v1} - x_{v3})(y_{v2} - y_{v3}) - (x_{v2} - x_{v3})(y_{v1} - y_{v3}) \qquad (13.7)$$

2. Utilize the surface test to remove back faces to improve the efficiency of the priority algorithm. Equation (13.2) can be used in this test. Any two edges of a given face can be used to calculate the face normal (refer to Chapter 7). Steps 1 and 2 result in a face list which will be sorted to assign priorities. For the scene shown in Figure 13.12a, six faces $F_1 - F_6$ form such a list. The order of the faces in the list is immaterial.

3. Assign priorities to the faces in the face list. The priority assignment is determined by comparing two faces at any one time. The priority list is continuously changed, and the final list is obtained after few interactions. Here is how priorities can be assigned. The first face in the face list (F_1 in Figure 13.12b) is assigned the highest priority 1. F_1 is intersected with the other faces in the list, that is, $F_2 - F_6$. The intersection between F_1 and another face may be an area, A, as in the case of F_1 and F_4 shown in Figure 13.12a, an edge as for faces F_1 and F_2, or an empty set (no intersection) as for faces F_1 and F_6. In the case of an area of intersection, the (x_v, y_v) coordinates of a point c inside A can be computed (notice the corner points of A are known). Utilizing Eq. (13.3) for both faces F_1 and F_4, the two corresponding z_v values of point c can be calculated and compared. The

face with the highest z_v values is assigned the highest priority. In the case of an edge of intersection, both faces are assigned the same priority. They obviously do not obscure each other, especially after the removal of the back faces. In the case of no face intersection, no priority is assigned.

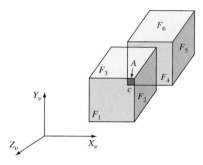

(*a*) Scene of two boxes

Face list	Priority list
F_1	1
F_2	1
F_3	1
F_4	2
F_5	
F_6	

Iteration 1

Face list	Priority list
F_2	1
F_3	1
F_4	2
F_5	
F_6	
F_1	

Iteration 2

Face list	Priority list
F_3	1
F_4	2
F_5	
F_6	
F_1	
F_2	

Iteration 3

Face list	Priority list
F_4	✗ 2
F_5	✗ 2
F_6	✗ 2
F_1	1
F_2	1
F_3	1

Iteration 4

(*b*) Assignment of priorities

(*c*) Image with hidden lines removed

Figure 13.12   The priority algorithm.

Let us apply the preceding strategy to Figure 13.12. F_1 intersects F_2 and F_3 in edges. Therefore both faces are assigned priority 1. F_1 and F_4 intersect in an area. Using the depth test, and assuming the depth of F_4 is less than that of F_1, F_4 is assigned priority 2. When we intersect faces F_1 and F_5, we obtain an empty set, that is, no priority assignment is possible. In this case, the face F_1 is moved to the end of the face list, and the sorting process to determine priority starts all over again. In each iteration, the first face in the face list is assigned priority 1. The end of each iteration is detected by no intersection. Figure 13.12*b* shows four iterations that yield the final priority list. In iteration 4, faces F_4 to F_6 are assigned the priority 1 first. When F_4 is intersected with F_1, the depth test shows that F_1 has higher priority. Thus, F_1 is assigned priority 1 and the priority of F_4 to F_6 is dropped to 2.

    4. Reorder the face and priority lists so that the highest priority is on top of the list. In this case, the face and priority lists are $[F_1, F_2, F_3, F_4, F_5, F_6]$ and $[1, 1, 1, 2, 2, 2]$, respectively.

5. In the case of a raster display, hidden-line removal is done by the hardware (frame buffer of the display). We simply display the faces in the reverse order of their priority. Any faces that would have to be hidden by others would thus first be displayed, but would be covered later either partially or entirely by faces of higher priority.

6. In the case of a vector display, the hidden-line removal must be done by software by determining coverings. For this purpose, the edges of a face are compared with all other edges of higher priority. An edge list can be created which maintains a list of all line segments that will have to be drawn as visible. Visibility techniques such as the containment test (Section 13.3.2.2) and edge intersection (Section 13.3.2.5) are useful in this case. Figure 13.12c shows the scene with hidden lines removed.

In some scenes, ambiguities may result after applying the priority test. Figure 13.13 shows an example. The figure shows a case in which the order of faces is cyclic. Face F_1 covers F_2, F_2 covers F_3, and F_3 covers F_1. The reader is encouraged to find the priority list that produces this cyclic ordering and coverage. To rectify this ambiguity, additional criteria to determine coverage must be added to the priority algorithm.

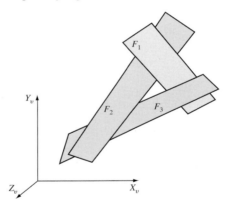

Figure 13.13   Cyclic faces.

13.3.6.2   Area-Oriented Algorithm

The area-oriented algorithm described here subdivides the data set of a given scene in a stepwise fashion until all visible areas in the scene are determined and displayed. In this algorithm as well as in the priority algorithm, no penetration of faces is allowed. Considering the same scene shown in Figure 13.12a, the area-oriented algorithm can be described as follows:

1. **Identify silhouette polygons.** Silhouette polygons are polygons whose edges are silhouette edges. First, silhouette edges in the scene are recognized as described in Section 13.3.2.4. Second, the connection of silhouette edges to form closed silhouette

polygons can be achieved by sorting all the edges for equal end points. For the scene shown in Figure 13.14, two closed silhouette polygons S_1 and S_2 are identified.

2. **Assign quantitative hiding (QH) values to edges of silhouette polygons.** This is achieved by intersecting the polygons (the containment test can be utilized first as a quick test). The intersection points define the points where the value of QH may change. Applying the depth test to the points of intersection (P_1 and P_2 in Figure 13.14), we determine the segments of the silhouette edges that are hidden. For example, if the depth test at P_1 shows that z_v at P_1 of S_1 is smaller than that of S_2, edge C_1C_2 is partially visible. Similarly, the depth test at P_2 shows that edge C_2C_3 is also partially visible. To determine which segment of an edge is visible, the visibility test of Section 13.3.2.5 can be used. Determination of the values of QH at the various edges or edge segments of silhouette polygons is based on the depth test. Step 2 in Figure 13.14 shows the values of QH. A value of 0 indicates that the edge or segment is visible, and a value of 1 indicates that the edge or segment is invisible.

3. **Determine the visible silhouette segments.** From the values of QH, the visibility of silhouette segments can be determined with the following rules in mind. If a closed silhouette polygon is completely invisible, it need not be considered any further. Otherwise, its segments with the lowest QH values are visible (step 3 in Figure 13.14).

4. **Intersect the visible silhouette segments with partially visible faces.** This step is used to determine if the silhouette segments hide or partially hide nonsilhouette edges in partially visible faces. In Figure 13.14, edges E_1 to E_6 of S_2 are intersected with the internal edges (edges of the square in the face) of F_1, and the visible segments of the internal edges are determined. By accessing only the silhouette edges of the covering silhouette polygon only and the partially visible face only, the algorithm avoids any unnecessary calculations.

5. **Display the interior of the visible or partially visible polygons.** This step can be achieved using a stack and simply enumerates all faces lying inside a silhouette polygon. The stack is initialized with a visible face which has a silhouette edge. We know this face belongs to a visible area. A loop begins with popping a face (F_2) from the stack. We examine all the edges of the face. If an edge (E_7) is not fully invisible, the neighboring face (F_3) also has visible edges and, therefore, is pushed into the stack if it has not already been pushed. The edge itself or its visible segments are displayed. The loop is repeated and the algorithm stops when the stack is empty.

The two hidden-line algorithms discussed in this section are sample algorithms to help you to the understand of the basic nature of the hidden-line removal problem. The area-oriented algorithm is more efficient than the priority algorithm because it hardly involves any unnecessary edge/face intersection.

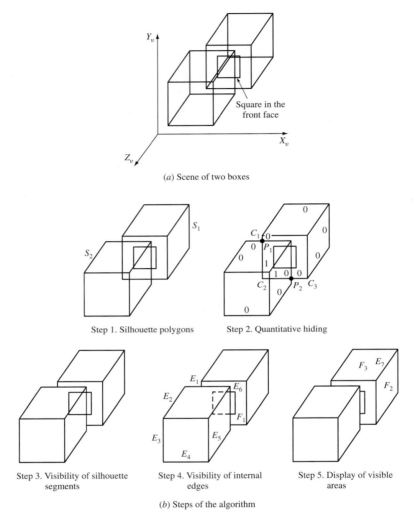

(a) Scene of two boxes

Step 1. Silhouette polygons   Step 2. Quantitative hiding

Step 3. Visibility of silhouette   Step 4. Visibility of internal   Step 5. Display of visible
segments   edges   areas

(b) Steps of the algorithm

Figure 13.14   Area-oriented algorithm.

13.3.7   Hidden-Line Removal for Curved Surfaces

The hidden-line algorithms described thus far are applicable to polyhedral objects with flat faces (planar polygons or surfaces). Fortunately, these algorithms are extendable to curved polyhedra by approximating them by planar polygons. The u–v grid offered by parametric surface representation (Chapter 7) offers such an approximation. This grid can be utilized to create a "grid surface" consisting of straight-edged regions, as shown in Figure 13.15, by approximating the u–v grid curves by line segments.

The overlay hidden-line algorithm mentioned in Section 13.3.6 is suitable for curved surfaces. The algorithm begins by calculating the u–v grid using the surface equation. It then creates the grid surface with linear edges. Various criteria discussed previously can be utilized to determine the visibility of the grid surface.

There is no best hidden-line algorithm. Many algorithms exist and some are more efficient and fast in rendering images than others for certain applications. Firmware and parallel processing computations of hidden-line algorithms is making it possible to render images in real time. This adds to the difficulty of deciding on a best algorithm.

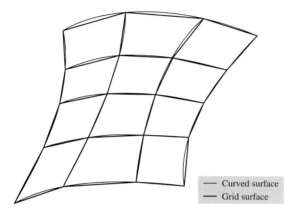

Curved surface
Grid surface

Figure 13.15   Approximation of a curved surface.

13.4   Hidden-Surface Removal

Hidden-surface removal and hidden-line removal are one problem. Most of the concepts and algorithms described in Section 13.3 are applicable here and vice versa. While we limited ourselves to object-space algorithms for hidden-line removal, we discuss image-space algorithms only for hidden-surface removal. A wide variety of these algorithms exist. They include the z-buffer algorithm, Watkin's algorithm, Warnock's algorithm, and Painter's algorithm. The Watkin's algorithm is based on scanline coherence, while the Warnock's algorithm is an area-coherence algorithm. The Painter's algorithm is a priority algorithm as described in Section 13.3 for raster displays. Two sample algorithms are covered in this section.

13.4.1   The z-Buffer Algorithm

This is also known as the depth-buffer algorithm. In addition to the frame buffer (see Chapter 11), this algorithm requires a z-buffer in which z values can be sorted for each pixel. The z-buffer is initialized to the smallest z-value, while the frame buffer is initialized to the background pixel value. Both the frame– and z-buffers are indexed by pixel coordinates (x, y). These coordinates are actually screen coordinates.

The z-buffer algorithm works as follows. For each polygon in the scene, find all the pixels (x, y) that lie inside or on the boundaries of the polygon when projected onto the screen. For each of these pixels, calculate the depth z of the polygon at (x, y). If $z >$ depth (x, y), the polygon is closer to the viewing eye than others already stored in the pixel. In this case, the z-buffer is updated by setting the depth at (x, y) to z. Similarly, the intensity of the frame-buffer location corresponding to the pixel is updated to the intensity of the polygon at (x, y). After all the polygons have been processed, the frame buffer contains the solution.

13.4.2 Warnock's Algorithm

This is one of the first area-coherence algorithms. Essentially, this algorithm solves the hidden-surface problem by recursively subdividing the image into subimages. It first attempts to solve the problem for a window that covers the entire image. Simple cases as one polygon in the window or none at all are easily solved. If polygons overlap, the algorithm tries to analyze the relationship between the polygons and generates the display for the window.

If the algorithm cannot decide easily, it subdivides the window into four smaller windows and applies the same solution technique to every window. If one of the four windows is still complex, it is further subdivided into four smaller windows. The recursion terminates if the hidden-surface problem can be solved for all the windows or if the window becomes as small as a single pixel on the screen. In this case, the intensity of the pixel is chosen equal to the polygon visible in the pixel. The subdivision process results in a window tree.

Figure 13.16 shows the application of Warnock's algorithm to the scene shown in Figure 10.13a. One would devise a rule that any window is recursively subdivided unless it contains two polygons. In such a case, comparing the z depth of the polygons determines which one hides the other.

While the subdivision of the original window is governed by the complexity of the scene in Figure 13.16, the subdivision of any window into four equal windows makes the algorithm inefficient. A better way would be to subdivide a window according to the complexity of the scene in the window. This is equivalent to subdividing a window into four unequal subwindows.

13.5 Hidden-Solid Removal

The hidden-solid removal problem involves the display of solid models with hidden lines or surfaces removed. Due to the completeness and unambiguity of solid models as discussed in Chapter 9, the hidden-solid removal is fully automatic. CAD systems provide users with menu choices to display models including shaded, hidden lines removed, or wireframe (no hidden lines removed),

For displaying CSG models, both the visibility problem and the problem of combining the primitive solids into one composite model have to be solved. There are three approaches to displaying CSG models. The first approach converts the CSG model into a boundary model that can be rendered with the standard hidden-surface algorithms. The second approach utilizes a spatial subdivision strategy. To simplify the combinational problems, the CSG tree is pruned simultaneously with the subdivision. This subdivision reduces the CSG evaluation to a simple preprocessing before the half-spaces are processed with standard rendering techniques. These half-spaces are the internal representation of the primitives as covered in Chapter 9.

The third approach uses a CSG hidden-surface algorithm, which combines the CSG evaluation with the hidden-surface removal on the basis of ray classification. The CSG ray-tracing and scanline algorithms utilize this approach. The attractiveness of the approach lies in the conversion of the complex 3D solid/solid intersection problem into 1D ray/solid intersection calculation. Due to its popularity and generality, the remainder of this section covers in more detail the ray-tracing (also called ray-casting) algorithm.

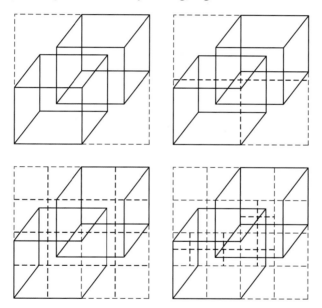

Figure 13.16   Warnock's algorithm.

13.5.1 Ray-Tracing Algorithm

The virtue of ray tracing is its simplicity and reliability. The most complicated numerical problem of the algorithm is finding the points at which lines (rays) intersect surfaces. Therefore a wide variety of surfaces and primitives can be covered. Ray tracing has been used to enhance the visual realism of solids by generating line drawings with hidden solids removed, animating solids, and shading pictures. It has also been utilized in solid analysis, mainly calculating mass properties.

The idea of using ray tracing to generate shaded images of solids is to emulate the photographic process in reverse. For each pixel in the screen, a light ray is cast through it into the scene to identify the visible surface. The first surface intersected by the ray, found by "tracing" along it, is the visible one. At the ray/surface intersection point, the surface normal is computed, and knowing the position of the light source, the brightness of the pixel can be calculated.

Ray tracing is considered a brute force method for solving problems. The basic ray-tracing algorithm is simple, but slow. The CPU usage of the algorithm increases with the complexity of the scene under consideration. Various alterations and refinements have been added to the algorithm to improve its efficiency. Moreover, the algorithm has been implemented into hardware (ray-tracing firmware) to speed its execution. In this book, we present the basic algorithm and some obvious refinements. More detailed work can be found in the literature.

13.5.1.1   Basics

The basics of tray tracing stems from light rays and camera models. The geometry of a simple camera model is analogous to that of projection of geometric models. Referring to Figure 12.15, the center of projection, projectors, and the projection plane represent the focal point, light rays, and the screen of the camera model, respectively. We assume that the camera model uses the VCS as described in Chapter 12 and shown in Figures 12.16 and 12.18. For each pixel of the screen, a straight light ray passes through it and connects the focal point with the scene.

When the focal length, the distance between focal point and screen, is infinite, parallel views result (Figure 12.16), and all light rays become parallel to the Z_v axis and perpendicular to the screen (the X_vY_v plane). Figure 13.17 shows the geometry of a camera model as described here. The XYZ coordinate system shown is the same as the VCS. We dropped the subscript "v" for simplicity. The origin of the XYZ system is taken to be the center of the screen.

A ray is a straight line which is best defined in a parametric form as a point (x_0, y_0, z_0) and a direction vector $(\Delta x, \Delta y, \Delta z)$. Thus, a ray is defined as $[(x_0, y_0, z_0)\ (\Delta x, \Delta y, \Delta z)]$. For a parameter t, any point (x, y, z) on the ray is given by:

$$
\begin{aligned}
x &= x_0 + t\Delta x \\
y &= y_0 + t\Delta y \\
z &= z_0 + t\Delta z
\end{aligned}
\tag{13.8}
$$

Equation (13.8) allows points on a ray to be ordered and accessed via a single parameter t. Thus, a ray in a parallel view that passes through the pixel (x, y) is defined as $[(x, y, 0)\ (0, 0, 1)]$. In a perspective view, the ray is defined by $[(0, 0, z_e)\ (x, y, -z_e)]$ given the screen center $(0, 0, 0)$ and the focal point $(0, 0, z_e)$. In the parallel view, t is taken to be zero at the pixel location while it is zero at the focal point (and 1 at the pixel location) in the perspective view.

A ray-tracing algorithm takes the ray definition given by Eq. (13.8) as an input and output information about how the ray intersects the scene. Knowing the camera model and the solid in the scene, the algorithm can find where the given ray enters and exits the solid as shown in Figure 13.18 for a parallel view. The output information is an ordered list of ray parameters, t_i, which denotes the enter/exit points, and a list of pointers, S_i, to the surfaces (faces) through which the ray passes. The ray enters the solid at point t_1, exits at t_2, enters at t_3, and finally exits at t_4. Point t_1 is closest to the screen and point t_4 is farthest. The lists of ray parameters and surface pointers suffice for various applications.

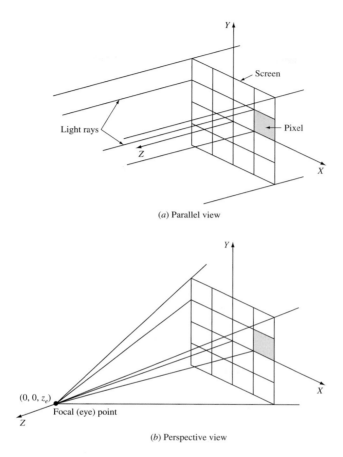

(a) Parallel view

(b) Perspective view

Figure 13.17   Camera model for ray tracing.

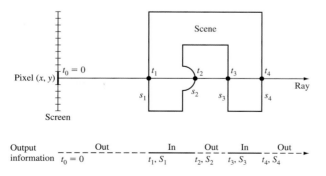

Figure 13.18   Cyclic faces.

13.5.1.2 Basic Ray-Tracing Algorithm

While the basics of ray tracing are simple, their implementation into a solid modeler is more involved and depends largely on the representation scheme of the modeler. When boundary representation is used in the object definition, the ray-tracing algorithm is simple. For a given pixel, the first face of the object intersected by the ray is the visible face at that pixel.

When the object is defined as a CSG model, the algorithm is more complicated because CSG models are compositions of primitive solids. Intersecting the primitive solids with a ray yields a number of intersection points, which requires additional calculations to determine which of these points are intersection points of the ray with the composite solid.

A ray-tracing algorithm for CSG models consists of three main modules: ray/primitive intersection, ray/primitive classification, and ray/solid classification.

Ray/Primitive Intersection

Utilizing the CSG tree structure, the general ray/solid intersection problem reduces to a ray/primitive intersection problem. The ray enters and exits the solid via the faces and the surfaces of the primitives.

For convex primitives (such as a block, cylinder, cone, or sphere), the ray/primitive intersection test has four possible outcomes: no intersection (the ray misses the primitive), ray is tangent (touches) to the primitive at one point, ray lies on a face of the primitive, or the ray intersects the primitive at two different points. In the case of a torus, the ray may be tangent to it at one or two points and may intersect it at as many as four points.

The ray/primitive intersection is evaluated in the local coordinate system of the primitive (see Figure 9.5) because it is very easy to find. The equation of a primitive [Eqs. (9.62) to (9.67)] expressed in its local coordinate system is simple and independent of any rigid-body motion the primitive may undergo to be oriented properly in a scene. Arbitrary elliptic cylinders in the scene are all the same primitive in its local coordinate system: a right circular cylinder at the origin. Similarly, ellipsoids are the same sphere, elliptic cones are the same right circular cone, and parallelepipeds are all the same block.

Given a ray originating in the screen coordinate system (SCS), it must be transformed into the primitive (local) coordinate system (PCS) via the model (scene) coordinate system (MCS) in order to find the ray/primitive intersection points. Each primitive has its local-to-scene transform and inverse, but there is only one scene-to-screen transform and inverse. The local-to-scene $[T_{LS}]$ and scene-to-screen $[T_{SS}]$ transformation matrices are determined from user input to orient primitives or define views of the scene respectively. The transformation matrix $[T]$ that transforms a ray from local to screen coordinate system is given by:

$$[T] = [T_{SS}][T_{LS}] \tag{13.9}$$

Therefore, a ray can be transformed to the PCS of a primitive by transforming its fixed point and direction vector:

$$\begin{bmatrix} x_0 & \Delta x \\ y_0 & \Delta y \\ z_0 & \Delta z \\ 1 & 1 \end{bmatrix}_{local} = [T]^{-1} \begin{bmatrix} x_0 & \Delta x \\ y_0 & \Delta y \\ z_0 & \Delta z \\ 1 & 1 \end{bmatrix}_{screen} \tag{13.10}$$

As discussed in Chapter 12, geometric transformation is one-to-one correspondence. Thus, the parameters t that designate the ray/primitive intersection points need not be transformed once the intersection problem is solved in the PCS. So, only rays need to be transformed between coordinate systems, not parameters.

The ray/plane intersection calculation is simple. For instance, to intersect the parameterized ray $[(x_0, y_0, z_0) (\Delta x, \Delta y, \Delta z)]$ with the XY plane, we simultaneously solve $z = 0$ and $z = z_0 + t\Delta z$ for t to get:

$$t = -\frac{z_0}{\Delta z} \tag{13.11}$$

Having found t, the point of intersection is $[x_0 + (-z_0/\Delta z)\Delta x, y_0 + (-z_0/\Delta z)\Delta y, 0]$. If Δz is zero, the ray is parallel to the plane, so they do not intersect. If the point of intersection lies within the bounds of the primitive, then it is a good ray/primitive intersection point. The bounds test for this point on the XY plane of a block given by Eq. (9.62) is:

$$0 \leq (x_0 + t\Delta x) \leq W \quad \text{and} \quad 0 \leq (y_0 + t\Delta y) \leq H \tag{13.12}$$

Finding ray/quadric intersection points is slightly more difficult. Consider a cylindrical surface given by Eq. (9.69). Substituting the x and y coordinates of the ray into Eq. (9.69) yields

$$(x_0 + t\Delta x)^2 + (y_0 + t\Delta y)^2 = R^2 \tag{13.13}$$

Rearranging Eq. (13.13) gives

$$t^2[(\Delta x)^2 + (\Delta y)^2] + 2t(x_0\Delta x + y_0\Delta y) + x_0^2 + y_0^2 - R^2 = 0 \tag{13.14}$$

Using the quadratic formula, we solve Eq. (13.14) to find t as

$$t = \frac{-B \pm \sqrt{B^2 - 4AC}}{2A} \tag{13.15}$$

where

$$\begin{aligned} A &= (\Delta x)^2 + (\Delta y)^2 \\ B &= 2(x_0\Delta x + y_0\Delta y) \\ C &= x_0^2 + y_0^2 - R^2 \end{aligned} \tag{13.16}$$

The ray intersects the cylinder only if $A \neq 0$ and $B^2 - 4AC > 0$. Having found the one or two value of t, the bounds test for the cylindrical surface is:

$$0 \leq (z_0 - t\Delta z) \leq H \tag{13.17}$$

Intersecting rays with a torus is more complicated because it is a quartic surface. It is left as an exercise (see problems at the end of the chapter) to the reader.

Ray/Primitive Classification

The classification of a ray with respect to a primitive is simple. Utilizing the set membership classification function introduced in Section 9.5.3 and the ray/primitive intersection points, the in, out, and on segments of the ray can be found. As shown in Figure 13.18, the odd intersection points signify the beginning of in segments and the end of out segments.

If a ray misses or touches a convex primitive at one point, it is classified as completely out. If the ray intersects the primitive in two different points, it is divided into three segments: out-in-out. If the ray lies on a face of the primitive, it is classified as out-on-out. For a torus, a ray classifies as out, out-in-out, or out-in-out-in-out.

Ray/Solid Classification

Combining ray/primitive classifications produces the ray/solid classification as in, on, and/or out segments of the ray with respect to the solid. It also reorders ray/primitive intersection points and gives the closest point and surface of the solid to the camera. To combine ray/primitive classifications, a ray-tracing algorithm starts at the top of the CSG tree, recursively descends to the bottom, classifies the ray with respect to the primitives, and then returns up the tree, combining the classifications of the left and right subtrees. Combining the on segments requires the use of neighborhood information as discussed in Chapter 9. Figures 13.19 and 13.20 illustrate ray/object classification. Solid lines in Figure 13.20 are in segments.

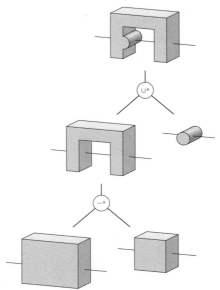

Figure 13.19   Ray/primitive intersection.

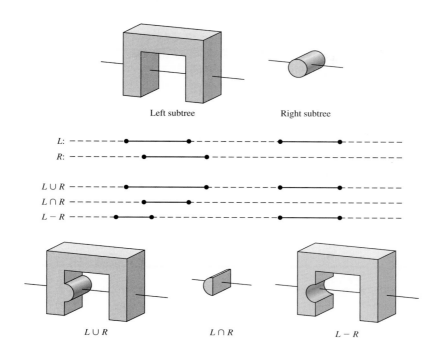

Figure 13.20   **Ray/solid classification.**

The combined operation is a three-step process. First, the ray/primitive intersection points from the left and right subtrees are merged in sorted order, forming a segmented composite ray. Second, the segments of the composite ray are classified depending on the Boolean operator and the classifications of the left and right rays along these segments. Third, the composite ray is simplified by merging contiguous segments with the same classification. Figure 13.21 illustrates these three steps for the union operator. Combining classifications is Boolean algebra. Table 13.1 defines the combine rules.

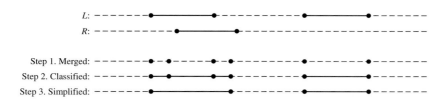

Figure 13.21   **Ray tracing for the union operator.**

Table 13.1   The combine values from Boolean algebra.

Operator	Left subtree	Right subtree	Combine
Union	IN	IN	IN
	IN	OUT	IN
	OUT	IN	IN
	OUT	OUT	OUT
Intersection	IN	IN	IN
	IN	OUT	OUT
	OUT	IN	OUT
	OUT	OUT	OUT
Difference	IN	IN	OUT
	IN	OUT	IN
	OUT	IN	OUT
	OUT	OUT	OUT

The Algorithm

To draw the visible edges of a solid, a ray per pixel is generated moving top-down, left-right in the screen. Each ray is intersected with the solid and the visible surface in the pixel is identified. If the visible surface at pixel (x, y) is different than the visible surface at pixel $(x - 1, y)$, then display a vertical line one pixel long centered at $(x - 0.5, y)$. Similarly, if the visible surface at pixel (x, y) is different than the visible surface at pixel $(x, y - 1)$, then display a horizontal line one pixel long centered at $(x, y - 0.5)$. The resulting line drawing with hidden solids removed will consist of horizontal and vertical edges only. Figure 13.22 shows a magnification of a drawing of a box with a hole. Figure 13.22a shows the pixel grid superimposed on the box and Figure 13.22b shows the drawing only. As shown in Figure 13.22a, the pixel-long horizontal and vertical lines may not coincide with the solid edges. However, the hidden solid still looks acceptable to the user's eyes because of the small size of each pixel.

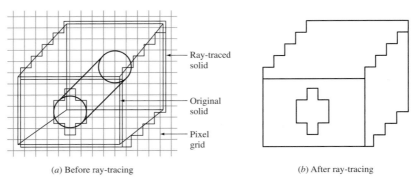

Ray-traced solid

Original solid

Pixel grid

(a) Before ray-tracing   (b) After ray-tracing

Figure 13.22   Results of ray tracing.

In pseudocode, a ray-tracing algorithm may be expressed as follows:

```
 1:  Procedure RAY TRACE
 2:  //classify ray against primitive
 3:  for each pixel (x, y) do
 4:    generate a ray through pixel (x, y)
 5:    if solid to be hidden is not a primitive
 6:    then
 7:      do //begin combine
 8:        classify ray against left subtree {L classify}
 9:        classify ray against right subtree {R classify)
10:        combine (L-classify and R-classify)
11:      end //end combine
12:    else
13:      do over primitives
14:        transform the ray equation from SCS to PCS
15:        branch to the proper primitive case:
16:          Block:
17:            do 6 ray-plane intersection tests;
18:          Sphere:
19:            do 1 ray-quadric intersection test;
20:          Cylinder:
21:            do 2 ray-plane & 1 ray-quadric intersection tests;
22:          Cone:
23:            do 1 ray-plane & 1 ray-quadric intersection tests;
24:          Torus:
25:            do 1 ray-quartic intersection test;
26:          end //branch
27:      classify ray against primitive
28:        end //do over primitive
29:    end //ray classification
30:  //find visible surface in pixel (x, y)
31:  find the first visible surface S₁ in pixel (x, y)
32:  if S₁ in pixel (x, y) is different than S₁ in pixel (x-1,y)
33:  then
34:    display a pixel-long vertical line centered at (x-0.5,y)
35:  else
36:    if S₁ in pixel (x, y) is different than S₁ in pixel (x, y-1)
37:    then
38:      display a pixel-long horizontal line centered at(x, y-0.5)
39:    end //if
40:  end //if
41: end //pixel loop
```

13.5.1.3   Improvements to the Basic Algorithm

The basic ray-tracing algorithm as described in Section 13.5.1.2 is slow and its memory and CPU usage are directly proportional to the scenes complexity measured by the number of primitives in the solid. In practice, the use of memory is of less concern than how fast the algorithm is.

To appreciate the cost of using ray tracing, consider the scenario of a scene of a solid composed of 300 primitives drawn on a raster display of 500×500 pixels. Since the solid is composed of 300 primitives, its CSG tree has 300 (actually 299) composite solids, making a total of 600 solids which a ray must visit via 600 calls of the algorithm. Thus $600 \times 500 \times 500$ calls are needed. At each composite solid in the tree, the left and right classifications must be combined, requiring $300 \times 500 \times 500$ classification combines. In addition, $300 \times 500 \times 500$ ray transformations from SCS to PCS are required.

Finally, assuming an average of four surfaces per primitive, a total of $4 \times 300 \times 500 \times 500$ ray-intersection tests are performed. So, the total cost for generating the hidden-solid line drawing (or the shaded image) is the sum of these four costs.

The high cost of algorithm is primarily due to the multipliers 300 and 500×500. Many applications may require casting a ray from every other pixel (or more), thus reducing the latter multiplier to 250×250. This is equivalent to using a raster display of 250×250 pixels. The former multiplier can be reduced significantly for a large class of solids by using box enclosures.

By using minimum bounding boxes around the solids in the CSG tree, the extensive search for ray/solid intersection becomes an efficient binary search. Figure 13.23 shows the tree of box enclosures for the solid shown in Figure 13.19. These boxes enable the algorithm to detect the "clear miss" cases between the ray and the solid. The CSG tree can be viewed as a hierarchical representation of the space that the solid occupies. The tree nodes would have enclosure boxes that are positioned in space. Quick ray/box intersection tests guide the search in the hierarchy. If the test fails at an intermediate node, the ray is guaranteed to be classified as out of the composite; thus recursing down the solid's subtrees to investigate further is unnecessary.

The ray/box intersection test is 2D because rays usually start at the screen and extend infinitely into the scene. When rays are bounded in depth, a ray/depth test can be added. Unlike the union and intersection operators, the subtraction operators do not obey the usual rules of algebra. The enclosure of $A - B$ is equal to the enclosure of A, regardless of B.

13.5.1.4   Remarks

The ray tracing algorithm to generate line drawings of hidden solids has advantages. It eliminates finding, parameterizing, classifying, and storing the curved edges formed by the intersection of surfaces. The silhouettes of curved surfaces are by-products, and they can be found whenever the view changes.

The main drawbacks of the algorithm are speed and aliasing. Aliasing causes edges to be jagged and surface "slivers" may be overlooked. Speed is particularly important to display hidden-solid line drawings in an interactive environment. If the user creates a balanced tree of the solid in the scene, the efficiency of ray tracing improves. The coherence of visible surfaces (surfaces visible at two neighboring pixels are more likely to be the same than different) can also speed up the algorithm.

13.6 Shading

Line drawings, still the most common means of communicating the geometry of parts, are limited in their ability to portray intricate shapes. Shaded images can convey complex shape information. They also can convey features other than shape such as surface finish or material type (plastic or metallic look). Shaded-image rendering algorithms filter information by displaying only the visible surface. Many spatial relationships that are unresolved in simple wireframe displays become clear with shaded displays. Shaded images are easier to interpret because they resemble the real objects.

Shaded images have viewing problems not present in wireframe displays. Solids of interest may be hidden or partially obstructed from view, in which case various shaded images may be obtained from various viewing points. Critical geometry such as lines, arcs, and vertices are not explicitly shown. Well-known techniques such as shaded-image/wireframe overlay, transparency, and sectioning can be used to resolve these problems.

In shading a scene (rendering an image), a pinhole camera model is almost universally used. Rendering begins by solving the hidden-surface removal problem to determine which objects and/or portions of objects are visible in the scene. As the visible surfaces are found, they must be broken down into pixels and shaded correctly. This process must take into account the position and color of the light sources and the position, orientation, and surface properties of the visible objects.

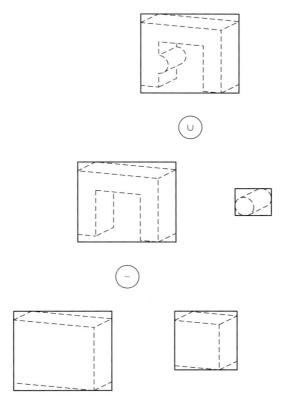

Figure 13.23   Tree of box enclosures.

13.6.1 Shading Models

Shading models simulates the way visible surfaces of objects reflect light. They determine the shade of a point of an object in terms of light sources, surface characteristics, and the positions and orientations of the surfaces and sources. Two types of light sources can be identified: point light source and ambient light. Objects illuminated with only point light sources look harsh because they are illuminated from one direction only. This produces a flashlight-like effect in a black room. Ambient light is a light of uniform brightness and is caused by the multiple reflections of light from the many surfaces present in real environments.

Shading models is simple. The inputs to a shading model include intensity and color of light source (S), surface characteristics at the point to be shaded, and the positions and orientations of surfaces and light sources. The output from a shading model is an intensity value at the point. Shading models are applicable to points only. To shade an object, a shading model is applied many times to many points on the object. These points are the pixels of the display. To compute a shade for each point on a 1024×1024 raster display, the shading model must be calculated over one million times. These calculations can be reduced by taking advantage of shading coherence, that is, the intensity of adjacent pixels is either identical or very close.

Let us examine the interaction of light with matter to gain insight into how to develop shading models. Particularly, we consider point light sources shining on the surfaces of objects. (Ambient light adds a constant intensity value to the shade at every point.) The light reflected off a surface can be divided into two components: diffuse and specular. When light hits an ideal diffuse surface, it is reradiated equally in all directions, so that the surface appears to have the same brightness from all viewing angles. Dull surfaces exhibit diffuse reflection. Examples of real surfaces that radiate mostly diffuse light are chalk, paper, and flat paints.

Ideal specular surfaces reradiate light in one direction only, the reflected light direction. Examples of specular surfaces are mirrors and shiny surfaces. Physically, the difference between these two components is that diffuse light penetrates the surface of an object and is scattered internally before emerging again, while specular light bounces off the surface. The absence of diffuse light makes a surface look shiny.

The light reflected from real objects contains both diffuse and specular components, and both must be modeled to create realistic images. A basic shading model that incorporates both a point light source and ambient light can be described as follows:

$$\mathbf{I}_p = \mathbf{I}_d + \mathbf{I}_s + \mathbf{I}_b \tag{13.18}$$

where $\mathbf{I}_p$, $\mathbf{I}_d$, $\mathbf{I}_s$, and $\mathbf{I}_b$ are respectively the resulting intensity (the amount of shade) at point p, the intensity due to the diffuse reflection component of the point light source, the intensity due to the specular reflection component, and the intensity due to ambient light. Equation (13.18) is written in a vector form to permit the modeling of colored surfaces. For the common red, green, and blue color system, Eq. (13.18) represents three scalar equations, one for each color. For simplicity of presentation, we develop Eq. (13.18) for one color and therefore refer to it as $I_p = I_d + I_s + I_b$ from now on (drop the vector notation).

To develop the intensity components in Eq. (13.18), consider the shading model shown in Figure 13.24. The figure shows the geometry of shading a point P on a surface S due to a point light source. An incident ray falls from the source to P at an angle θ (incidence angle) measured from the surface unit normal $\hat{n}$ at P. The unit vector $\hat{l}$ points from the light source to P. The reflected ray leaves P with an angle of reflection θ (equal to the angle of incidence) in the direction defined by the unit vector $\hat{r}$. The unit vector $\hat{v}$ defines the direction from P to the viewing eye.

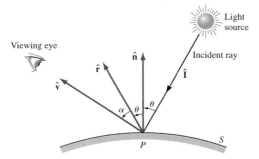

Figure 13.24   A shading model.

13.6.1.1   Diffuse Reflection

Lambert's cosine law governs the diffuse reflection. It relates the amount of reflected light to the cosine of the angle θ between $\hat{l}$ and $\hat{n}$. Lambert's law implies that the amount of reflected light seen by the viewer is independent of the viewer's position. The diffuse illumination is given by:

$$I_d = I_L K_d \cos\theta \qquad (13.19)$$

where I_L and K_d are the intensity of the point light source and the diffuse-reflection coefficient, respectively. K_d is a constant between 0 and 1 and varies from one material to another. Replacing $\cos\theta$ by the dot product of $\hat{l}$ and $\hat{n}$, we can rewrite Eq. (13.19) as:

$$I_d = I_L K_d (\hat{n} \cdot \hat{l}) \qquad (13.20)$$

Note that since diffuse light is radiated equally in all directions, the position of the viewing eye is not required by the computations, and the maximum intensity occurs when the surface is perpendicular to the light source. On the other hand, if the angle of incidence θ exceeds 90°, the surface is hidden from the light source and I_d must be set to zero. A sphere shaded with this model (diffuse reflection only) will be brightest at the point on the surface between the center of the sphere and the light source and will be completely dark on the far half of the sphere from the light.

Some shading models assume that the point source of light is coincident with the viewing eye, so no shadows can be cast. For parallel projection, this means that light rays striking a surface are all parallel. This means that $\hat{\mathbf{n}} \cdot \hat{\mathbf{I}}$ is constant for the entire surface, that is, the intensity I_d is constant for the surface as shown by Eq. (13.20).

13.6.1.2   Specular Reflection

Specular reflection is a characteristic of shiny surfaces. Highlights visible on shiny surfaces are due to specular reflection, while other light reflected from these surfaces is caused by diffuse reflection. The location of a highlight on a shiny surface depends on the directions of the light source and the viewing eye. If you illuminate an apple with a bright light, you can observe the effects of specular reflection. Note that at the highlight the apple appears to be white (not red), which is the color of the incident light.

The specular component is not as easy to compute as the diffuse component. Real objects are nonideal specular reflectors, and some light is also reflected slightly off axis from the ideal light direction (defined by vector $\hat{\mathbf{r}}$ in Figure 13.24). This is because the surface is never perfectly flat but contains microscopic deformations.

For ideal (perfect) shiny surfaces (such as mirrors), the angles of reflection and incidence are equal. This means that the viewer can only see specular reflected light when the angle α (Figure 13.24) is zero. For nonideal (nonperfect) reflectors, such as an apple, the intensity of the reflected light drops sharply as α increases. One of the reasonable approximations to the specular component is an empirical approximation and takes the form

$$I_s = I_L W(\theta) \cos^n \alpha \tag{13.21}$$

For real objects, as the angle of incidence θ changes, the ratio of incident light to reflected light also changes, and $W(\theta)$ is intended to model the change. In practice, however, $W(\theta)$ has been ignored by most implementors or very often is set to a constant K_s, which is selected experimentally to produce aesthetically pleasing results.

The value of n is the shininess factor and typically varies from 1 to 200, depending on the surface. For a perfect reflector, n would be infinite. $\cos^n \alpha$ reaches a maximum when the viewing eye is in the direction of $\hat{\mathbf{r}}$ ($\alpha = 0$). As n increases, the function dies off more quickly in the off-axis direction. Thus, a shiny surface with a concentrated highlight would have a large value of n, while a dull surface with the highlight covering a large area on the surface would have a low value of n as shown in Figure 13.25. Replacing $\cos \alpha$ by the dot product of $\hat{\mathbf{r}}$ and $\hat{\mathbf{v}}$, we can rewrite Eq. (13.21)as:

$$I_s = I_L W(\theta)(\hat{\mathbf{r}} \cdot \hat{\mathbf{v}})^n \tag{13.22}$$

If both the viewing eye and the point source of light are coincident at infinity, $\hat{\mathbf{r}} \cdot \hat{\mathbf{v}}$ becomes constant for the entire surface. This is because $\hat{\mathbf{n}} \cdot \hat{\mathbf{v}}$, that is, $\cos(\theta + \alpha)$ and $\hat{\mathbf{n}} \cdot \hat{\mathbf{I}}$, that is, $\cos(\theta)$, become constant.

Other, and more accurate, shading models for specular reflection have been developed and are not discussed in this book. Among these realistic models are Blinn and Cook and Torrance models.

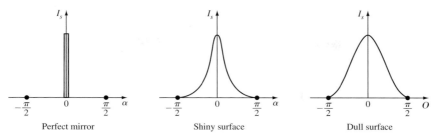

Figure 13.25   Surface reflectance as a function of α.

13.6.1.3   Ambient Light

Ambient light is a light with uniform brightness. It therefore has a uniform or constant intensity I_a. The intensity at point P due to ambient light can be written as:

$$I_b = I_a K_a \tag{13.23}$$

where K_a is a constant that ranges from 0 to 1. It indicates how much of the ambient light is reflected from the surface to which point P belongs.

Substituting Eqs. (13.20), (13.22), and (13.23) into Eq. (13.18), we obtain

$$I_P = I_a K_a + I_L[K_d(\hat{\mathbf{n}} \cdot \hat{\mathbf{l}}) + W(\theta)(\hat{\mathbf{r}} \cdot \hat{\mathbf{v}})^n] \tag{13.24}$$

If $W(\theta)$ is set to the constant K_s, this equation becomes:

$$I_P = I_a K_a + I_L[K_d(\hat{\mathbf{n}} \cdot \hat{\mathbf{l}}) + K_s(\hat{\mathbf{r}} \cdot \hat{\mathbf{v}})^n] \tag{13.25}$$

All the unit vectors can be calculated from the geometry of the shading model, while constants and intensities on the right-hand side of Eq. (13.25) are assumed by the model. Additional intensity terms can be added to the equation if more shading effects such as shadowing, transparency, and texture are needed. Some of these effects are discussed later in this section.

13.6.2 Shading Surfaces

Once we know how to shade a point [Eq. (13.25)], we can consider how to shade a surface. To calculate shading precisely, Eq. (13.25) can be applied to each point on the surface. Relevant points on the surface have the same locations in screen coordinates as the pixels of the raster display. Determining these points is an outcome of hidden-surface removal. The normal unit vector $\mathbf{n}$ used in Eq. (13.25) depends on the surface geometry and can be computed anew for each pixel of the display. This would require a large number of calculations. Sometimes they are evaluated incrementally.

Most surfaces, including those that are curved, are described by polygonal meshes when the visible-surface calculations are to be performed by the majority of rendering algorithms. The majority of shading techniques are therefore applicable to objects modeled as polyhedra. Among the many existing shading algorithms, we discuss three of them: constant shading, Gourand or first-derivative shading, and Phong or second-derivative shading.

13.6.2.1   Constant Shading

This is the simplest and least realistic shading algorithm. Since the unit normal vector of a polygon never changes, polygons will have just one shade. An entire polygon has a single intensity value calculated from Eq. (13.25). Constant shading makes the polygonal representation obvious and produces unsmooth shaded images (intensity discontinuities). Actually, if the viewing eye or the light source is very close to the surface, the shade of the pixels within the polygon will differ significantly.

The choice of point P (Figure 13.24) within the polygon becomes necessary if the light source and the viewing eye are not placed at infinity. Such choice affects the calculations of the vectors $\hat{\mathbf{l}}$ and $\hat{\mathbf{v}}$. P can be chosen to be the center of the polygon. On the other hand, $\hat{\mathbf{l}}$ and $\hat{\mathbf{v}}$ can be calculated at the polygon corners, and the average of these values is used in Eq. (13.25).

13.6.2.2   Gourand Shading

Gourand shading is a popular form of intensity-interpolation or first-derivative shading. Gourand proposed a technique to eliminate (although not completely) intensity discontinuities caused by constant shading. The first step in the Gourand algorithm is to calculate surface normals. We break down the surface into polygons and calculate one normal per polygon.

The second step is to calculate vertex normals. If more than one polygon shares the same vertex as shown in Figure 13.26a, the surface normals are averaged to give the vertex normal. If smooth shading between the four polygons shown is required, then

$$\mathbf{N}_\nu = \tfrac{1}{4}(\mathbf{N}_A + \mathbf{N}_B + \mathbf{N}_C + \mathbf{N}_D) \tag{13.26}$$

If shading discontinuities are to be introduced deliberately across an edge to show a crease or a sharp edge in the object, the proper surface normals can be dropped from Eq. (13.26). For example, shading discontinuities occur along the AD and BC boundaries shown in Figure 13.26a if we average only two face normals. $\mathbf{N}_\nu = \tfrac{1}{2}(\mathbf{N}_A + \mathbf{N}_B)$ and $\mathbf{N}_\nu = \tfrac{1}{2}(\mathbf{N}_C + \mathbf{N}_D)$ are used to interpolate shades between polygons A and B and C and D, respectively. Thus, smooth shading occurs along the AB and CD boundaries, while discontinuous shading occurs along the AD and BC boundaries.

The third step in the Gourand algorithm is to compute vertex intensities using the vertex normals and the desired shading model [Eq. (13.25)]. The fourth and the last step is to compute the shade of each polygon by linear interpolation of vertex intensities.

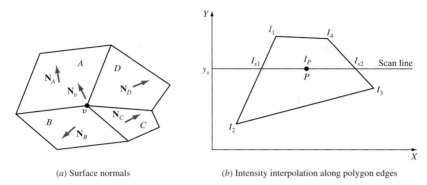

(a) Surface normals   (b) Intensity interpolation along polygon edges

Figure 13.26   Gourand shading.

If the Gourand algorithm is utilized with a scanline hidden-surface algorithm, the intensity at any point P inside any polygon (Figure 13.26b) is obtained by interpolating along each edge and then between edges along each scan line. This gives

$$I_P = \frac{x_{s2} - x_P}{x_{s2} - x_{s1}}I_{s1} + \frac{x_P - x_{s1}}{x_{s2} - x_{s1}}I_{s2} \qquad (13.27)$$

$$I_{s1} = \frac{y_s - y_2}{y_1 - y_2}I_1 + \frac{y_1 - y_s}{y_1 - y_2}I_2 \qquad (13.28)$$

$$I_{s2} = \frac{y_4 - y_s}{y_4 - y_3}I_3 + \frac{y_s - y_3}{y_4 - y_3}I_4 \qquad (13.29)$$

Gourand shading takes longer than constant shading and requires more bit planes of memory to get the smooth shading for each color. How the bits of each pixel are divided between the shade color and the shading grade? For example, one may use the red color to obtain a light red shade, dark red shade, or any variation in between.

Let us consider a display with 12 bits of color output, that is, 2^{12} or 4096 simultaneous colors. If we decide that 64 shading grades per color are required to obtain fairly smooth shades, then $4096/64 = 64$ gross different colors are possible. Actually only 63 colors are possible because the remaining one is reserved for the background. Within each color, 64 different shading grades are possible. This means that six bits of each pixel are reserved for colors and the other six for shades. The lookup table of the display would reflect this subdivision as shown in Figure 13.27. Usually, the six least significant bits (LSB) correspond to the shade and the six most significant bits (MSB) correspond to the color, so that when interpolation is performed to obtain shading grades the six most significant bits (i.e., the color) remain the same.

13.6.2.3   Phong Shading

While Gourand shading produces smooth shades, it has some disadvantages. If it is used to produce shaded animation (motion sequence), shading changes in a strange way because interpolation is based on intensities and not surface normals that actually change with motion. In addition, Mach bands (a phenomenon related to how the human visual system perceives and processes visual information) are sometimes produced, and highlights are distorted due to the linear interpolation of vertex intensities.

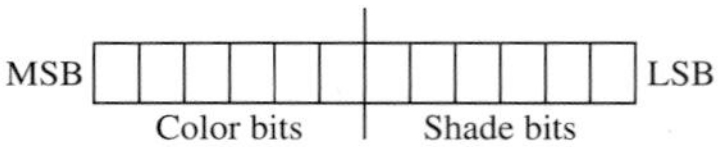

Address	Color	Shade
0 ⋮ 63	**Background**	–
64 ⋮ 127	1	Light shade ⋮ Dark shade
128 ⋮ 191	2	Light ⋮ Dark
⋮	⋮	⋮
4032 ⋮ 4095	63	Light ⋮ Dark

MSB ☐☐☐☐☐☐☐☐☐☐ LSB
Color bits   Shade bits

Figure 13.27   Splitting pixel bits between colors and shades.

Phong shading avoids all the problems associated with Gourand shading, although it requires more computational time. The basic idea behind Phong shading is to interpolate normal vectors at the vertices instead of the shade intensities and to apply the shading model [Eq. (13.25)] at each point (pixel) using the interpolated normal. To perform the interpolation, Eq. (13.26) can be used to obtain an average normal vector at each vertex. Phong shading is usually implemented with scanline algorithms. In this case, Figure 13.26b is applicable if we replace the intensities by the average normal vectors, N_v, at the vertices. Similarly, Eqs. (13.27) to (13.29) are applicable if the intensity variables are replaced by the normal vectors.

13.6.3 Shading Enhancements

The basic shading model described in Section 13.6.1 is usually enhanced to produce special effects for both artistic values and realism purposes. These effects include transparency, shadows, surface details, and texture.

1. **Transparency.** Transparency can be used to shade translucent material such as glass and plastics or to allow the user to see through the opaque material. Two shading techniques can be identified: opaque and translucent. In the opaque technique, hidden surfaces in every pixel are completely removed. In the translucent method, hidden surfaces are not

completely removed. This allows some of the back pixels to show through, producing a screen-door effect.

Consider the box shown in Figure 13.28. If the front face F_1 is made translucent, the back face F_2 can be seen through F_1. The intensity at a pixel coincident with the locations of points P_1 and P_2 can be calculated as a weighted sum of the intensities at these two points, that is,

$$I = KI_1 + (1 - K)I_2 \qquad (13.30)$$

where I_1 and I_2 are the intensities of front and back faces, respectively. They are calculated using, for example, Eq. (13.25). K is a constant that measures the transparency of the front face: when $K = 0$, the face is perfectly transparent and does not change the intensity of the pixel. When $K = 1$, the front face is opaque and transmits no light. Sometimes transparency is referred to as X-ray due to the similarity in effect.

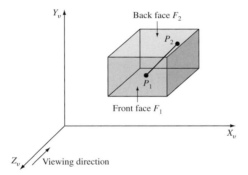

Figure 13.28   Transparency.

2. **Shadows.** Shadows are important in conveying realism to computer images. More importantly, they facilitate the comprehension of spatial relationships between objects of one image. The complexity of a shadow algorithm is related to the model of the light source. If it is a point source outside the field of view at infinity, the problem is simplified Finding which objects are in shadow is equivalent to solving the hidden surface problem as viewed from the light source. If several light sources exist in the scene, the hidden surface problem is solved several times; every time, one of the light sources is considered the viewing point. The surfaces that are visible to both the viewer and the light source are not shaded. Those that are visible to the viewer but not to the light source are shaded.

Surface details that are usually needed to add realism to the surface image are better treated as shading data than as geometrical data. Consider, for example, adding a logo of an object to its image. The logo can be modeled using "surface-detail" polygons. Polygons of the object geometric model point to these surface-detail polygons. If one of the geometric-model polygons is to be split for visibility reasons, its corresponding surface-detail polygon is split in the same fashion. Surface-detail polygons obviously cover the surface polygons when both overlap. When the shaded image is generated, the surface details are guaranteed to be visible with their desired color attributes. Separating the polygons of geometric models and surface details speeds up the rendering of images significantly and reduces the possibility of generating erroneous images.

3. **Texture.** Texture is important to provide the illusion of reality. For example, modeling of a rough casting should include the rough texture nature of its surfaces. These objects, rich in high frequencies, could be modeled by many individual polygons, but as the number of polygons increases, they can easily overflow the modeling and display programs. Texture mapping, shown in Figure 13.29, was introduced to solve this problem and to provide the illusion of complexity at a reasonable cost. It is a method of "wallpapering" the existing polygons. As each pixel is shaded, its corresponding texture coordinates are obtained from the texture map, and a lookup is performed in a 2D array of colors containing the texture. The value in this array is used as the color of the polygon at this pixel, thus providing the "wallpaper."

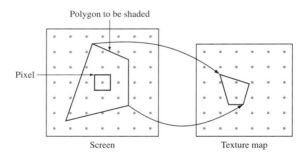

Figure 13.29　Texture mapping.

3D texture mapping is easier to use with 3D objects than 2D texture mapping. It is a function that maps the object's spatial coordinates into 3D texture space and uses 3D textures. Thus, no matter what the object's shape is, the texture on its surface is consistent. This is useful for modeling materials such as wood and marble. Due to the space required to store a 3D array of pixels, procedural textures can be used. However, they are difficult to antialias. Texture mapping can contain other surface properties besides color to increase the illusion of complexity. For example, surface normal perturbations could be stored in the texture map (bump mapping) to enable the simulation of wrinkled surfaces.

EXAMPLE 13.1　　**Use a shading model.**

Apply the shading model given by Eq. (13.25) to the scene of the two boxes shown in Figure 13.12*a*. Assume that the visible surfaces have been identified as shown in Figure 13.12*c*. Use $n = 100$ for the specular reflection component.

SOLUTION   Let us assume that the point light source is placed at infinity, that is, coincident with the viewing eye. Table 13.2 can be derived to enable us to use Eq. (13.25). The table shows the components of the vectors of each face in the VCS shown in Figure 13.12a.

Table 13.2   Shading vectors.

Face	$\hat{n}$	$\hat{l}$	$\hat{r}$	$\hat{v}$
F_1	(0, 0, 1)	(0, 0, −1)	(0, 0, 1)	(0, 0, 1)
F_2	(1, 0, 0)	(0, 0, −1)	(0, 0, 1)	(0, 0, 1)
F_3	(0, 1, 0)	(0, 0, −1)	(0, 0, 1)	(0, 0, 1)
F_4	(0, 0, 1)	(0, 0, −1)	(0, 0, 1)	(0, 0, 1)
F_5	(1, 0, 0)	(0, 0, −1)	(0, 0, 1)	(0, 0, 1)
F_6	(0, 1, 0)	(0, 0, −1)	(0, 0, 1)	(0, 0, 1)

Substituting the values from Table 13.2 into Eq. (13.25) gives

Face F_1: $\quad I_P = I_a K_a + I_L[-K_d + (K_s)^{100}]$

Face F_2: $\quad I_P = I_a K_a + I_L(K_s)^{100}$

Face F_3: $\quad I_P = I_a K_a + I_L(K_s)^{100}$

Face F_4: $\quad I_P = I_a K_a + I_L[-K_d + (K_s)^{100}]$

Face F_5: $\quad I_P = I_a K_a + I_L(K_s)^{100}$

Face F_6: $\quad I_P = I_a K_a + I_L(K_s)^{100}$

The intensities I_a and I_L are chosen based on the maximum intensity a pixel may have. The coefficients K_a, K_d, and K_s are chosen based on experimental measurements. The preceding equations assume constant shading. Notice that the shading of F_1 and F_4, and F_2, F_3, F_5, and F_6 are equal. This will make F_2 and F_3 and F_5 and F_6 indistinguishable in the shaded image. A solution to this problem would be to use Gourand or Phong shading. The reader is encouraged to calculate both of them as an exercise and compare intensities (see the end-of-chapter problems).

13.6.4 Shading Solids

Shading algorithms of solids can be developed based on exact solid representation schemes (B-rep and CSG) or on some approximations of these schemes (faceted B-rep). A wide variety of shading algorithms of solids exist and are directly related to the representation scheme utilized. Shading algorithms are implemented into special-purpose tiling engines (hardware) to speed their execution.

13.6.4.1   Ray-Tracing Shading Algorithm for CSG

The ray-tracing algorithm described in Section 13.5.1 is considered a natural tool to create shaded images. To create a shaded image, cast one ray per pixel into the scene, and identify the visible surface S_1 (Figure 13.18) in this pixel. Compute the surface normal at the visible point t_1. Use Eq. (13.25) to calculate the pixel intensity value. Processing all pixels this way produces a raster-type picture of the scene.

Special effects and additional realism are possible by adding transparency and shadowing. Transparent surfaces may be modeled with or without refraction. Nonrefractive transparency is easy because the ray remains one straight line (no refraction by going through solid). With transparency, more than one surface may be visible at pixel. If surface S_1 (Figure 13.18) transmits any light, then the surface normal at point t_2 on surface S_2 must be computed to calculate the intensity at the point. The net intensity at the given pixel can be calculated using Eq. (13.30). If S_2 transmits light, then S_3 must be processed in a similar way, and so forth.

To model shadows, the following procedure is executed for each point light source:

1. Cast a second ray connecting the visible point t_1 with the point light source as shown in Figure 13.30.

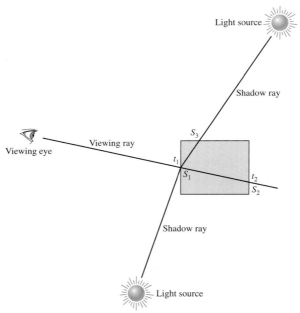

Figure 13.30   Modeling shadows.

2. If the ray intersects any surface (S_3) between the visible (relative to the viewing eye) point and the light source, then the point is in shadow.
3. If the surface that shadows (S_3) is transparent, then attenuate the intensity of the light that

passes through and add it to the intensity of the pixel at point t_1.

13.6.4.2   z-Buffer Algorithm for B-rep and CSG

z-buffer (also known as depth-buffer) algorithms that operate on B-reps, especially polygonal approximations, are simple and well known. The basic algorithm defines two arrays, the depth array $z(x, y)$ and the intensity array $I(x, y)$, to store the depth and intensity information of the screen pixels located by (x, y) values.

Considering Figure 13.31, a z-buffer algorithm for B-rep can be described as follows. Initialize the $z(x, y)$ and $I(x, y)$ arrays with a large number and background intensity respectively. For each face of the solid, check the distance d between each point on the face (P_1 and P_2) and the viewing eye. If d is smaller than the z value in the proper element of the $z(x, y)$ array, write d onto $z(x, y)$ and compute the intensity using Eq. (13.25). Update the intensity buffer by writing the intensity value onto $I(x, y)$. At the end of the scan, the intensity buffer contains the correct image values. Typically, the frame buffer of the graphics display is used to store the intensity array. Thus, the display can be updated incrementally, whenever a new intensity for a pixel is computed, by writing the new value onto the frame buffer.

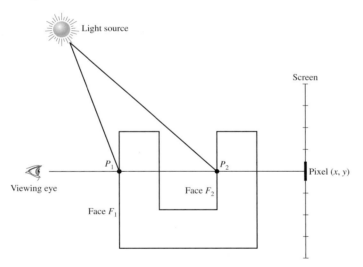

Figure 13.31   z-buffer shading algorithm for B-rep.

To apply the previously described z-buffer algorithm for B-rep to CSG models, it must be modified to reflect the fact that the actual solid faces are not available. Instead, the primitives' faces must be scanned to determine the elements of the $z(x, y)$ and $I(x, y)$ arrays. Scanning the primitives' faces yields a superset of the points needed by the z-buffer algorithm. Exploiting the generate-and-classify paradigm based on the point-membership classification, we can discard the points of the superset that are not on the actual solid faces.

The z-buffer algorithm for CSG models is described as follows. Initialize the $z(x, y)$ and $I(x, y)$ arrays as in the B-rep procedure. For each face of each primitive of the solid, check the distance d between each point on the face and the viewing eye. If d is smaller than the z value in the proper element of $z(x, y)$, then classify the point with respect to the solid (by traversing the CSG tree). If the point is classified as on the solid, then write d onto $z(x, y)$, and compute the intensity and write it onto $I(x, y)$. Otherwise, if the point is classified as in or out of the solid, discard it and move to the next point.

13.7 Colors

The use of colors in CAD/CAM has two main objectives. Colors can be used in geometric construction. In this case, various entities can be assigned different colors to distinguish them. Color is one of the two main ingredients (the second being texture) of shaded images. In some engineering applications such as finite element analysis, colors can be used effectively to display contour images such as stress or heat-flux contours.

Black-and-white raster displays provide achromatic colors, while color displays provide chromatic color. Achromatic colors are described as black, various levels of gray (dark or light gray), and white. The only attribute of achromatic light is its intensity, or amount. A scalar value between 0 (as black) and 1 (as white) is usually associated with the intensity. Thus, a medium gray is assigned a value of 0.5. For multiple-plane displays, different levels (scale) of gray can be produced. For example, 256 (2^8) different levels of gray (intensities) per pixel can be produced for an eight–plane display. The pixel value V_i is related to the intensity level I_i by the following equation:

$$V_i = \left(\frac{I_i}{C}\right)^{1/\gamma}$$

(13.31)

The values C and γ depend on the display in use. If the raster display has no lookup table, V_i (for example, 00010111 in an eight–plane display) is placed directly in the proper pixel. If there is a table, i is placed in the pixel and V_i is placed in an entry i of the table. Use of the lookup table in this manner is called gamma-correction, after the exponent in Eq. (13.31).

Chromatic colors produce more pleasing effects on the human vision system than achromatic colors. However, they are more complex to study and generate. Color is created by taking advantage of the fundamental trichromacy of the human eye. Three different-colored images are combined additively at photoreceptors in the eye to form a single perceived image whose color is a combination of the three prime colors. Red, green, and blue colors are typically used to produce the wide range of desired colors.

Color descriptions and specifications generally include three properties: hue, saturation, and brightness. Hue associates a color with some position in the color spectrum. Red, green, and yellow are hue names. Saturation describes the vividness or purity of a color, or it describes how diluted the color is by white light. Pure spectral colors are fully saturated colors, and grays are desaturated colors. Brightness is related to the intensity, value, or lightness of the color.

There exists a wealth of studies and methods of how for specifying and measuring colors. Some methods are subjective, such as the Munsell and pigment-mixing methods. The Munsell method is widely used and is based on visually comparing unknown colors against a set of standard colors. The pigment-mixing method is used by artists. Other methods used in physics are objective and treat visible light with a given wavelength as electromagnetic energy with a spectral energy distribution.

Our primary interest in this section is not to review these studies and methods, but to describe some existing color models so application programs can choose the desired colors properly. We will also show how some of these models can be converted to red, green, and blue since most of the commonly used raster displays demand three digital values, specifying an intensity for each of the colors.

13.7.1 Color Models

A color model or space is a 3D color coordinate system that specifies of colors within some color range. Each displayable color is represented by a point in a color model. There are quite a number of color models available. Some popular models are discussed here. These models are based on the red, green, and blue (RGB) primaries. For any one of these models, coordinates are translated into three voltage values in order to control the display. This process is shown in Figure 13.32, which summarizes the sequence of transformation for some models. The gamma correction is performed to obtain a linear relationship between digital RGB values and the intensity of light emitted by the display.

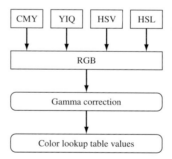

Figure 13.32   Transformation of a color model to RGB.

13.7.1.1   RGB Model

The RGB color space uses a cartesian coordinate system as shown in Figure 13.33a. Any point (color) in the space is obtained from the three RGB primaries; that is, color is additive. The main diagonal of the cube is the locus of equal amounts of each primary and therefore represents the gray scale or levels. In the RGB model, black is at the origin and is represented by (0, 0, 0), and white is represented by (1, 1, 1). In the RGB model, the lowest intensity (0 for each color) produces the black color and the maximum intensity (1 for each color) produces the white color.

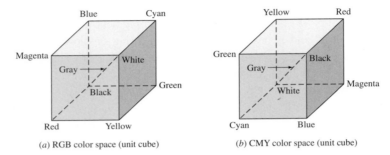

(a) RGB color space (unit cube)   (b) CMY color space (unit cube)

Figure 13.33   Two color models.

13.7.1.2   CMY Model

CMY (cyan, magenta, yellow) model shown in Figure 10.33b is the complement of the RGB model. The cyan, magenta, and yellow colors are the complements of the red, green, and blue, respectively. The white is at the origin (0, 0, 0) of the model and the black is at point (1, 1, 1), which is opposite to the RGB model. The CMY model is considered a subtractive model because the primary colors subtract some color from white light. For example, a red color is obtained by subtracting a cyan color from the white light (instead of adding magenta and yellow).

The conversion from CMY to RGB is achieved by the following equation:

$$V_i = \left(\frac{I_i}{C}\right)^{1/\gamma} \tag{13.32}$$

The unit column vector represents white in the RGB model or black in the CMY model.

PROBLEMS

Part I: Theory

13.1 Sketch the minimax boxes for the following tangent polygon. What are your conclusions?

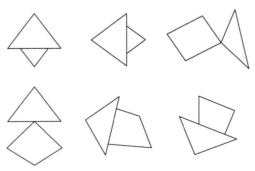

13.2 Find the sum, $\sum \Delta \theta$, of the angles subtended by the edges of the following polygons with respect to point P. What conclusions can you make?

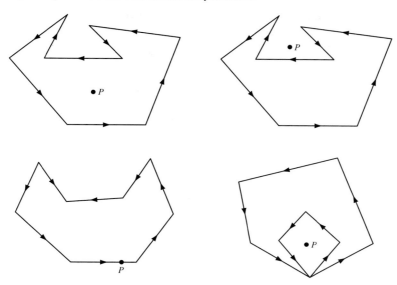

13.3 Find the face and priority lists of the following scenes. Assume that back faces have been removed to simplify the problem.

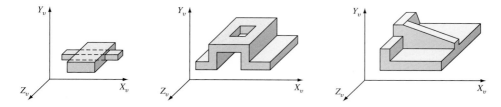

13.4 Apply the area-oriented algorithm to the scenes of Problem 13.3.

13.5 Find the ray/sphere, ray/cone, and ray/torus intersections and the corresponding bounds tests. Use Eqs. (9.70) to (9.72). For the ray/torus intersection, a fourth-order polynomial in t results which has a closed-form solution.

13.6 Solve Example 13.1 for both Gourand and Phong shading. How do you ensure that edges are clearly visible in the resulting shading images?

13.7 Apply solid shading techniques to the following models.

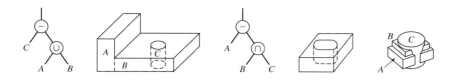

Part II: Laboratory

Use your in-house CAD/CAM system to answer the questions in this part.

13.8 Investigate the shading, lighting, shadows, texture, and transparencies offered by your CAD system. Apply the concepts covered in this chapter to them.

Part III: Programming

Write the code so that the user can input values to control the output. The program should display the results. Use OpenGL with C or C++, or use Java 3D.

13.9 Write a program that sorts a list of vertices of a polygon and produces its bounding box.

13.10 Write a program to classify a ray for a ray-tracing algorithm against a primitive and against a solid.

13.11 Write a program that would compute box enclosures around primitives in the SCS. Extend the program to combine the enclosures at tree nodes for the three operators $\cup*$, $\cap*$, and $-*$.

13.12 Write a program to implement Eq. (13.25) for the case of parallel (orthographic) view with a point light source coincident with the viewing eye.

13.13 Write a program to implement ray tracing to generate shaded images.

13.14 Write a program to implement a z–buffer to generate shaded images.

Computer Animation

GOAL

Understand and master the concepts of motion and animation and their importance to geometric modeling, and understand the related algorithms.

OBJECTIVES

After reading this chapter, you should understand the following concepts:

- Need for motion and animation
- Computer animation
- Frame-buffer animation
- Real-time playback
- Keyframe technique
- Simulation approach
- Animation-related problems
- Animation of articulated bodies

CHAPTER HEADLINES

14.1 Introduction

Animation is the process in which the illusion of movement is achieved by creating and displaying a sequence of images with elements that appear to have motion. The illusion of movement can be achieved in various ways. The most obvious way is to change the locations of various elements of various images in the sequence. Other ways include transforming an object to another (metamorphosis), changing the color of an object, or changing light intensities.

The world of entertainment has used both conventional (manual) and computer animation in producing animated cartoons, movies, logos, and ads. The emphasis of conventional animation is usually on the artistic aspects and looks of the images in the animation sequence. The animator draws a sequence of pictures which, if animated, produces the illusion of the proper motion. The use of computer animation in entertainment has introduced more realistic images and more complicated motions than what manual animation can offer. Also, it has enabled animators to incorporate physical laws into animation.

Animation is viewed as a valuable extension of modeling and simulation in the world of science and engineering. It is a very useful visualization aid for many modeling and simulation applications, especially those for which large amounts of scientific data and results are generated. Visualizing this data is important to its correct interpretation and understanding. In engineering problems that involve simulation, animation may be considered a discretization of the time domain, generating a sequence of the appropriate images at discrete time values, and displaying the sequence to restore time continuity so that the development of events over time can be better understood.

14.2 Conventional Animation

The goal of conventional animation is to produce cartoon-animated films. It is used primarily to animate 2D scenes. Its extension to 3D animation is usually difficult and time-consuming. However, an understanding of conventional animation is important to computer animation. Most of the terminology and concepts utilized by animation software originates with conventional animation.

An animated film tells a story. At its conception, the story is described by a synopsis or summary. The scenario of the story is developed next. The scenario is the detailed text of the story without any cinematographic references. A storyboard is then developed based on the scenario. The storyboard is a film in an outline form. It is a set of drawings resembling a comic strip which indicates the key sequences of the film scenes. These key sequences (also called keyframes) form the basis for the animated film. The following steps are utilized in conventional animation:

1. **Create keyframes.** Animators draw keyframes which correspond to the movement of the film characters and the timing required. An animator is usually assigned to one specific

character. Animators are skilled individuals who understand human and animal motions thoroughly and who have good imagination.

2. **Create inbetweens.** Interpolation between any two keyframes must be made to produce smooth animation. This interpolation process, which is done manually, and the resulting frames are known as inbetweening and inbetweens, respectively. Assistant animators draw the main artistry of inbetweens, and inbetweeners draw the remaining figures. For the smoothest animation, 24 frames must be drawn for every second of animation. If the movement in a certain scene of the animation is to take N seconds, the inbetweener must draw $24N$ frames of the same scene to complete this movement.

3. **Perform line testing,** the drawings (line drawings thus far) of the keyframes and inbetweens are photocopied onto transparent acetate (cel) and are filmed under a rostrum camera to test the quality of the movements produced.

4. **Paint frames.** After any modifications arising from the line tests, cells are painted to introduce color, which results in a color film. Painting also gives the animated characters a sense of solidity. Static backgrounds have to be painted also.

5. **Produce the film.** The final photography under the rostrum camera is carried out on color films or videotapes. Soundtracks of voice, music, and effects (such as thunderstorming) are added to the film. Figure 14.1 shows these steps with an illustration of a runner.

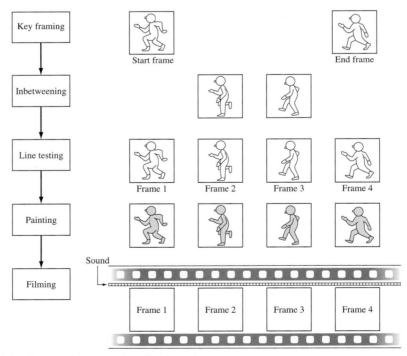

Figure 14.1   Steps of conventional animation.

14.3 Computer Animation

Conventional animation, even for small cartoon films, is a very labor-intensive process involving hundreds, perhaps thousands, of hours. The majority of the time is spent in making all the drawings of the characters and the backgrounds for both the keyframes and the inbetweens. Computer animation is a viable solution to alleviate the shortcomings of conventional animation. The science of computer animation has advanced enough to meet the demands of modeling and simulation in engineering applications.

Two classes of computer animation can be distinguished: entertainment and engineering. The role of computers in the first class can be identified from Figure 14.1. Computer graphics techniques can be used to generate the drawings of keyframes and inbetweens. The drawings of keyframes can be created with an interactive graphics editor, can be digitized, or can be produced by programming. The inbetweens can be completely calculated by the computer by interpolation along complex paths of motion. The use of layers in generating these drawings is very helpful. Drawings that are shared by more than one frame are stored on separate layers and shared by all the frames by overlaying the proper layers.

Shading techniques can be used to paint the drawings of the various frames. Not only can these techniques add visual realism to the animation, but they can also provide a tenfold reduction in preparation time.

Filming an animated film can be assisted by computer. If filming is done with a camera, its movements can be controlled by the computer, or virtual cameras can be completely programmed. If a film is recorded with a video recorder, the computer can control the recorder.

Modeled animation (sometimes called 3D animation), on the other hand, is an entirely different medium. It opens up the possibility of utilizing the available computer graphics and CAD techniques to create scenes, movements, and images which are difficult to achieve by conventional means. In particular, accurate representations of objects, as well as smooth and complex 3D movements, are facilitated. After the storyboard is prepared for a film, modeled animation is achieved as follows:

1. **Create geometric description.** In order to allow the complete and general 3D animation of drawn objects, they should be described as geometric models. A high degree of realism can be added to these models after their images are generated with shading attributes such as color, texture, reflectance, translucency, and so forth.

2. **Generate frames.** With animation software, we can use geometric transformations to create frames, both keyframes and inbetweens.

3. **Perform line testing.** After all frames are generated, their corresponding images are generated by shading them. These images can be animated and displayed on the graphics display to test the movements in real time.

4. **Record the animation sequence.** When all the frames are satisfactory, they are recorded frame by frame onto video recorders or films.

Figure 14.2 shows the steps of computer animation. The example of a slider-crank mechanism is used to illustrate these steps.

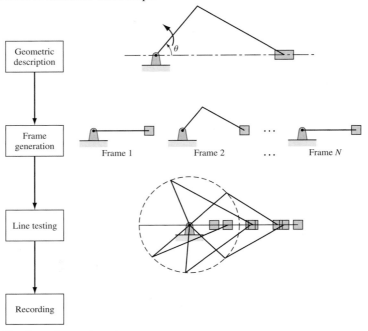

Figure 14.2   Engineering animation.

14.4 Engineering Animation

Animation has been utilized in CAD/CAM applications primarily for visualization purposes. NC tool paths can be verified by displaying the tool as it moves on the surfaces to be machined. Design engineers can use animation to detect interferences during assembly processes. Similarly, mechanisms engineers can display motions to verify kinematic constraints.

Animation as described here is mainly an extension of geometric modeling. A geometric model is usually created first. Some of the geometric parameters of the model are changed according to geometrical and/or analytical procedures. The values of these parameters are used to generate the various frame for animation. The slider-crank mechanism shown in Figure 14.2 is an example. The angle θ of the crank is incrementally changed within its range (0 to 360°) to produce the various frames for animation.

Engineering animation can be thought of in another way, that is, as an extension of analytical modeling and simulation. In this context, animation becomes an effective way of analyzing the large amount of numerical data that results from simulation. By visualizing the data in a continuous animated fashion (with freeze capabilities), engineers and scientists can track precisely the development of various phenomena.

An engineering animation system needs to meet the following requirements:

1. **Exact representation and display of data.** By data, we mean objects or numerical results of simulation. Data must be displayable in an image (shaded) form for better visualization and understanding. Raster displays are usually used to display animation (sometimes called raster animation).

2. **High speed and automatic production of animation.** In order to use animation as an aid to communication between engineers and designers, animated frames must be produced quickly. A real-time animation system is ideal. In addition, engineering animation must be performed automatically because engineers and designers who produce the animation frames are not professional animators.

14.5 Animation Types

In terms of the time it needed to generate animation, there are three types of computer animation. They are frame-buffer animation, real-time playback, and real-time animation.

14.5.1 Frame-Buffer Animation

Frame buffers offer a variety of techniques for limited animation. These techniques, if used properly, can provide the illusion of animation for a large class of applications. They may not be suitable for engineering animation, though, because they are all based on static images, in the sense that the contents of the frame buffer (pixel memory) do not change during animation. Dynamics are added by modifying the pattern in which pixels are read from memory and the way in which their contents (bits) are interpreted to provide color information using a color lookup table. There are three types of frame-buffer animation: color lookup table, zoom-pan-scroll, and crossbar animation.

Color lookup table animation is achieved by color cycling, alternate color, or bit-plane extraction. In color cycling, the entries in the lookup table are rotated. This gives the illusion of motion as colors appear to flow across the screen. Alternate color animation paints more than one image into the frame buffer memory using different pixel values for each image. Setting the lookup table entries to background for all but one of the images allows a single image to be viewed. Changing the subset of visible colors alternately displays one image and then another. Bit-plane extraction partitions the frame buffer into bit planes and assigns each image to a separate set of these planes. Loading the lookup table to ignore all but the bit planes corresponding to a particular image allows images to be displayed separately. Using this technique, it is possible to display more than one image simultaneously to mix them or overlay one with another using a priority scheme.

Figure 14.3 shows an example of displaying a pendulum's motion using cycling animation. The figure shows how the frame buffer could be loaded; the numbers indicate the pixel values placed in each region of the buffer. Table 14.1 shows how the lookup table is loaded at each step of animation to display the pendulum motion. The idea is to display all but one of the pendulums at the background color 1. The motion effects result by cycling the contents of the lookup table.

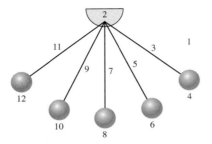

Figure 14.3 Pendulum animation.

Table 14.1 Lookup table for pendulum motion (blank entries mean white).

Entry number	Animation steps with corresponding colors loaded in lookup table								
	1	2	3	4	5	6	7	8	9
1									
2	B	B	B	B	B	B	B	B	B
3	B								B
4	R								R
5		B						B	
6		R						R	
7			B				B		
8			R				R		
9				B		B			
10				R		R			
11					B				
12					R				

In zoom-pan-scroll animation, the frame buffer is logically divided into different regions, each containing a separate, low-resolution image. Figure 14.4 illustrates the case in which four 256×256 images and sixteen 128×128 images are stored within a 512×512 frame buffer. By setting horizontal and vertical zoom factors of 2 with a 512×512 viewport, the display can cycle through the four images (Figure 14.4a) by successively setting the window location to the upper left-hand pixel of each image. One advantage of this approach over lookup table animation is that each pixel maintains its full color range instead of sacrificing palette size for enhanced dynamics. However, it sacrifices the spatial resolution of the image. There is clearly a trade-off between spatial resolution and intensity resolution.

Crossbar animation is achieved by partitioning bits within pixels. In other words, crossbar animation consists of routing any of the bits from pixel memory to any of the input lines in the lookup table. With a 32-bit frame buffer, we can store four 8-bit images, eight 4-bit images, sixteen 2-bit images, or even thirty-two 1-bit images. Figure 14.5 shows the partition of the buffer to obtain four images. Crossbar animation coupled with color table animation and zoom-pan-scroll is a particularly effective combination.

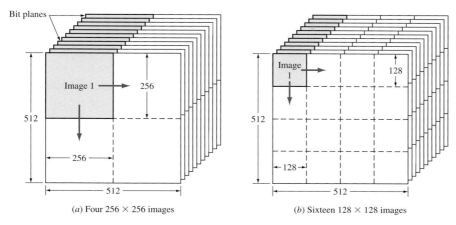

(a) Four 256 × 256 images   (b) Sixteen 128 × 128 images

Figure 14.4   Zoom-pan-scroll animation.

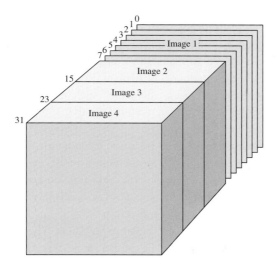

Figure 14.5   Crossbar animation.

14.5.2 Real-Time Playback

This is the most popular form of animation. Frames are generated frame by frame in advance at non–real-time rates, and then saved in a file. A real-time program can display the frames later to create the animation. In this case, animation is real-time in the sense of playing back the sequence of frames. This is the type of animation that CAD systems offer. A designer generates the geometry of each frame, one at a time, and saves (records) them in a file. Later, the designer uses a CAD function to display the file contents sequentially.

One example is animating the assembly steps of an assembly model. In this example, each frame captures the assembly parts (components) in different locations and orientations relative to each other. Other useful animation applications are found in nuclear weapons simulations, car-crash simulations, fluid dynamics (flame propagation), meteorology, chemical and molecular modeling, CT scans of parts of the human body (head, neck, spine, chest, liver, kidney, and stomach), ultrasound scans, trajectory planning and obstacle avoidance in robotics, and verification of NC tool paths. The animation of tool paths shows a cutting tool moving along its cutting paths in contact with the workpiece to determine any gouging problems.

Real-time playback is used where it is generally impossible to generate frames at the rates required for real-time presentation. This occurs in applications which require intensive calculations or in images that must display a certain degree of realism. Frames must be displayed at the rate of 24 frames per second for smooth animation, and at the rate of 30 frames per second for flicker-free display. Therefore, if the time it takes to calculate any frame in the animation sequence is greater than $1/24$ s, the frames must be recorded and then played back at the rate of 24 frames/second.

14.5.3 Real-Time Animation

Real-time computer animation is limited by the capabilities of the computer and data rates. Several minutes may be required to calculate a single frame in number-crunching–oriented applications, or in images with a great deal of realism. Real-time animation is possible with the development and advancement of special hardware such as array and graphics processors. Real-time animation is not possible for calculation-intensive applications running on multiuser computers with high-resolution raster displays.

Many engineering animations require intensive calculations. To simulate the evolution of a thunderstorm, for example, software for a simulation model based on flow equations that describe thunderstorm dynamics is used to generate data at given time values. The amount of the data generated is too large to study in its numerical form. Its display in an animated form enables a thorough study of how gusts and tornadoes develop.

Another example is the animation of facial expressions. These animations are based on nonlinear models of muscles which can be pulled or squeezed to create the deformable topology of faces. These animations could facilitate teaching lipreading to the hearing-impaired and help determine preoperatively the mobility that would remain after facial surgery.

14.6 Animation Techniques

The essence of the computer animation problem is determining the proper sequence of frames based on a given animation model. The animation model could differ depending on the application at hand. In entertainment, the model is simply a given set of keyframes and a set of motion (spatial and temporal) constraints. The solution to the animation problem in this case is the automatic inbetweening (generation of the inbetweens). In engineering, the model is an equilibrium equation with the proper boundary and/or initial conditions. The solutions to this equation at certain time values yield the desired sequence of frames.

The solution to the animation problem is directly related to the animation model. If the model is based on shape information only (see Section 14.7), the frames are generally fast to generate, but the quality of animation (how smooth and realistic the resulting motions are) usually suffers. If the model is based on physical laws (see Section 14.8) in addition to shape information, more computer time is needed to generate the frames, but the quality of the animation is better. Existing animation techniques offer various degrees of compromise between the exactness of the animation model and the time required to generate the frames.

14.7 Keyframe Technique

This technique is an extension of keyframing used in conventional animation. Given a set of keyframes, the technique automatically determines the correspondences between them, and then generates all the frames (inbetweens) between them by some form of interpolation. The correspondence between the geometry of the keyframes is a simple process if they have the same number of points and curves. All that is needed to establish correspondence in this case is to ask the animator to input the geometry of the keyframes in the same order.

The other case, where the keyframes do not have the same number of points and curves, requires preprocessing to make the number equal for all the keyframes. Different preprocessing methods are used depending on whether only the number of points, only the number of curves, or both are different. Preprocessing methods are not covered in this book. We will always assume that all frames have the same number of points and curves.

There are two fundamental approaches to keyframe animation. The first is called image-based keyframe animation (also called shape interpolation). In this technique, the inbetweens are obtained by interpolating the keyframe images themselves. Linear and/or nonlinear interpolation algorithms exist. Linear algorithms, in general, produce undesirable effects such as lack of smooth motion, discontinuities in motion speed, distortions in rotations, and contortions in the generated frames.

The second approach is called parametric keyframe animation (or key-transformation animation). Here, better images can be produced by interpolating the parameters that describe the keyframe images, that is, the parameters of the geometric model itself. In a parameter model, the animator creates keyframes by specifying the parameter values. These values are interpolated and the inbetweens are individually constructed from the interpolated values. The remainder of this section describes some algorithms used for image-based keyframe animation.

14.7.1 Skeleton Algorithm

The idea behind the skeleton algorithm stems from the manual inbetweening. Instead of using the images themselves as the basis for inbetweening, skeletons of the figures can be used. A skeleton, or stick figure, is a simple image of the original one composed of only the key points and curves that describe the form of movement required. This allows the animator to create many keyframes consisting of skeletons only. These keyframe skeletons are then interpolated by the computer to create the inbetween skeletons.

Details can be added to both keyframe and inbetween skeletons according to a single model. The inbetweens created this way are much better because the keyframes are similar. Skeletons can be defined by curves or four-sided polygons. Figure 14.6 shows an example of skeleton animation.

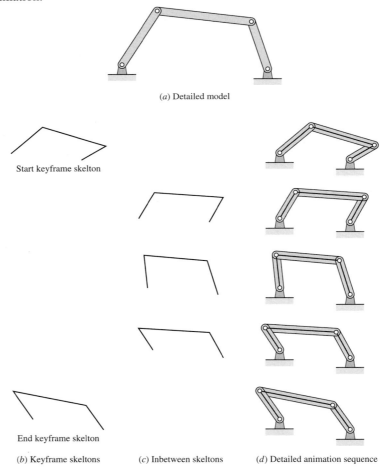

(a) Detailed model

Start keyframe skelton

End keyframe skelton

(b) Keyframe skeltons   (c) Inbetween skeltons   (d) Detailed animation sequence

Figure 14.6   Skeleton animation.

14.7.2 The Path of Motion and P-Curves

The keyframe technique is based solely on linear interpolation of the shape of the object to be animated. It does not consider the dynamics or movements of the object while the inbetweens are generated. The time is considered only when selecting the keyframes. If the motion is uniform, the positions of the keyframes in the time space become less critical for generating smooth animation. If abrupt motion or dynamics exist, unsmooth animation can result. The ideal solution is, of course, to develop and solve the dynamic equilibrium equations that describe the motion.

The use of path description and P-curves provides the animator with the information about the motion and its dynamics and dynamics that is needed to define the keyframes. A **P-curve** defines both the trajectory of a point in space and its location in time. Thus, the curve provides both spatial and temporal information about the motion. Figure 14.7 shows the P-curve for a person who walks along two adjacent walls to go from point A to point C. Both motion trajectory $[y = f(x)]$ and motion curves $[x = x(t), y = y(t)]$ are necessary to describe the motion. The person accelerates in the X direction until time t_1, continues with almost uniform motion (constant speed) until time t_2, and then accelerates in the Y direction to reach point C. The P-curve combines both Figures 14.7a and b. It has the shape of the trajectory, but a trail of symbols is used to indicate the path. These symbols are equally spaced in time.

The dynamics of a motion are represented on its P-curve by the local density of the symbols as shown in Figure 14.7c. The animator can use the density as a guide to choose the locations of the keyframes and to decide on the necessary number of inbetweens so that the resulting animation sequence is smooth and looks natural.

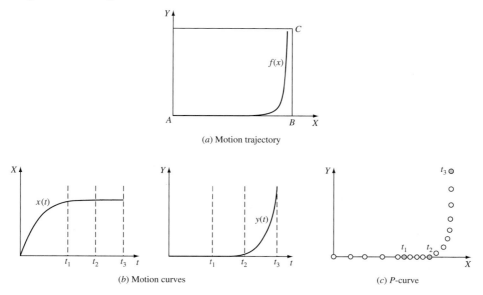

(a) Motion trajectory

(b) Motion curves

(c) P-curve

Figure 14.7   Animation via P-curve.

14.7.3 Inbetweening Utilizing Moving Point Constraints

This technique allows the animator more control over the inbetweening process than the previous techniques. The animator can specify, in addition to a set of keyframes, a set of constraints called moving points. **Moving points** are curves varying in space and time which constrain both the trajectory and the dynamics (path and speed) of certain points on the keyframes similar to P-curves. The set of keyframes and moving points form a constraint or patch specification of the desired dynamics.

Figure 14.8 shows an example of a patch network. The network is formed from the animator's input. It consists of an ordered set of keyframes $\{K_1, K_2,\dots K_{n-1}, K_n\}$ which define the shape of the object to be animated at the animator-specified times $\{t_1, t_2,\dots, t_{n-1}, t_n\}$, and a set of moving points $\{M1, M2,\dots, M_{q-1}, M_q\}$. Each keyframe can be considered a static shape positioned at a fixed point in time. Thus, each keyframe acts as a constraint in the motion sequence.

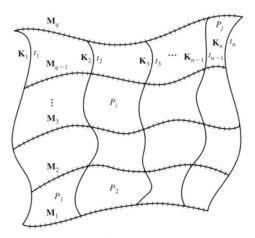

Figure 14.8   Animation via patch network.

The patch network is subdivided into patches $P_1, P_2, \dots, P_j$. Any patch P_i is defined by four boundary curves; two are static boundaries derived from the two bounding keyframes, and two are dynamic boundaries derived from the two bounding moving points. The inbetweening of the patch network can thus be reduced to the sum of inbetweening the individual patches. Figure 14.9 shows a single patch geometry. The patch is described by the two parameters u and t; u is the parametric space variable and t is the parametric time variable. The parameters are normalized to the interval $[0, 1]$. The two static boundaries are $\mathbf{P}(u, 0)$ and $\mathbf{P}(u, 1)$ which are parametric curves describing the geometry of keyframes. The two dynamic boundaries are $\mathbf{P}(0, t)$ and $\mathbf{P}(1, t)$. The corners of the patch are the endpoints of the keyframe curves. The subscript associated with each corner denotes its parametric values. For example, P_{00} is the point $P(0, 0)$.

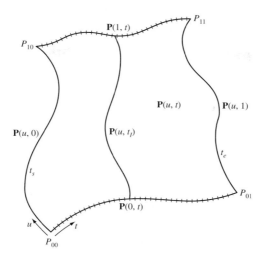

Figure 14.9   Patch specification.

A betweening algorithm based on this technique is to find a parametric function $\mathbf{P}(u, t)$ for each patch in the patch network. A patch time interval $[t_s, t_e]$ should be normalized to the interval $[0, 1]$ to facilitate calculations. (t_s and t_e are the start and end times of the patch, respectively.) If an inbetween frame $\mathbf{P}(u, t_I)$ (refer to Figure 14.9) is to be generated at time t_I, the corresponding normalized time is given by:

$$t_n = \frac{t_I - t_s}{t_e - t_s} \tag{14.1}$$

If the functions $\mathbf{P}(u, t)$ are available for all the patches, the inbetween frame $\mathbf{P}(u, t_I)$ is obtained by evaluating all the patches overlapping it by holding the time variable constant at t_I.

Three inbetweening algorithms are available. They are linear space inbetweening, cubic metric space inbetweening, and Coons patch inbetweening. The linear space inbetweening (Miura) algorithm first associates a linear time function called a basic curve with each keyframe $\mathbf{K}_0$ and $\mathbf{K}_1$ of a given patch, as shown in Figure 14.10. Each time function passes through the two end points of its corresponding keyframe. To generate an inbetween frame $\mathbf{P}(u, t)$ at time t, the basis curve $\mathbf{L}_t$ (line in this case) is obtained by connecting the two points $\mathbf{M}_0(t)$ and $\mathbf{M}_1(t)$. The inbetween frame is defined as

$$\mathbf{P}(u, t) = (1 - t)\mathbf{A}_0 + t\mathbf{A}_1 \tag{14.2}$$

that is, the inbetween frame (or the static curve) interpolated for time t is the time-weighted average of the term $\mathbf{A}_0$ and $\mathbf{A}_1$. $\mathbf{A}_0$ is derived from the initial static shape $\mathbf{K}_0$ at $t = 0$, and $\mathbf{A}_1$ is derived from the final static curve $\mathbf{K}_1$ at $t = 1$. The term $\mathbf{A}_0$ is the transformation of $\mathbf{K}_0$ by a transformation matrix $[T_0]$, and $\mathbf{A}_1$ is the transformation of $\mathbf{K}_1$ by $[T_1]$. Thus,

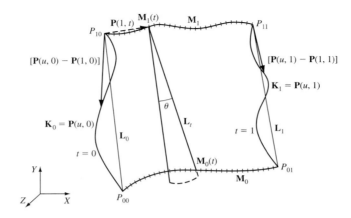

Figure 14.10   Linear space inbetweening algorithm.

$$\mathbf{A}_0 = [T_0][\mathbf{P}(u, 0) - \mathbf{P}(1, 0)] \tag{14.3}$$

and

$$\mathbf{A}_1 = [T_1][\mathbf{P}(u, 1) - \mathbf{P}(1, 1)] \tag{14.4}$$

The vector $\mathbf{P}(1, 0)$ is subtracted from the keyframe $\mathbf{P}(u, 0)$ in Eq. (14.3) to transform the origin of the XYZ coordinate system to point P_{10} to facilitate calculating $[T_0]$. Similarly, P_{11} is used as the point of rotation of the keyframe $\mathbf{P}(u, 1)$.

The matrix $[T_0]$ is found based on the argument that the matrix that transforms the basis curve $\mathbf{L}_0$ at $t = 0$ to the basis curve $\mathbf{L}_t$ at time t should be the same matrix that transforms $\mathbf{K}_0$ to $\mathbf{P}(u, t)$. As seen in Figure 14.10, $[T_0]$ involves rotation, translation, and scaling. Therefore, $[T_0]$ rotates, translates, and scales $\mathbf{K}_0$ to a new orientation, position, and size, but with the same shape. Similarly $[T_0]$ maps $\mathbf{K}_1$ to $\mathbf{A}_1$.

To find $[T_0]$, the basis curve $\mathbf{L}_0$ is rotated the angle θ shown in Figure 14.10 about the point P_{10}, translated along the vector $\mathbf{P}(1, t)$, and its length scaled from $\mathbf{L}_0$ to $\mathbf{L}_t$. Assuming a 2D case for simplicity, we can write:

$$[T_0] = s \begin{bmatrix} \cos\theta & -\sin\theta & l \\ \sin\theta & \cos\theta & m \\ 0 & 0 & 1 \end{bmatrix} \tag{14.5}$$

where l, m, and s are the translation in the X direction, the Y direction, and scaling factor, respectively. These variables are given by:

$$\cos\theta = \frac{\mathbf{L}_0 \cdot \mathbf{L}_t}{|\mathbf{L}_0|\ |\mathbf{L}_t|} \tag{14.6}$$

$$\sin\theta = \frac{\left| \mathbf{L}_0 \times \mathbf{L}_t \right|}{\left\| \mathbf{L}_0 \right| \left| \mathbf{L}_t \right\|} \tag{14.7}$$

$$l = x(1, t) \tag{14.8}$$
$$m = y(1, t)$$

$$s = \frac{\left| \mathbf{L}_t \right|}{\left| \mathbf{L}_0 \right|} \tag{14.9}$$

Substituting the end points of $\mathbf{L}_0$ and $\mathbf{L}_1$ into Eqs. (14.6) to (14.9) and substituting the results into Eq. (14.5), we obtain:

$$[T_0] = \frac{1}{\left| \mathbf{L}_0 \right|^2} \begin{bmatrix} r_1 & -r_2 & l \\ r_2 & r_1 & m \\ 0 & 0 & 1 \end{bmatrix} \tag{14.10}$$

where

$$r_1 = (x_{00} - x_{10})(x_{0t} - x_{1t}) + (y_{00} - y_{10})(y_{0t} - y_{1t}) \tag{14.11}$$

$$r_2 = (y_{00} - y_{10})(x_{0t} - x_{1t}) - (x_{00} - x_{10})(y_{0t} - y_{1t}) \tag{14.12}$$

$$\left| \mathbf{L}_0 \right|^2 = (x_{00} - x_{10})^2 + (y_{00} - y_{10})^2 \tag{14.13}$$

Similarly, the matrix $[T_1]$ can be evaluated as

$$[T_1] = \frac{1}{\left| \mathbf{L}_1 \right|^2} \begin{bmatrix} b_1 & -b_2 & n \\ b_2 & b_1 & q \\ 0 & 0 & 1 \end{bmatrix} \tag{14.14}$$

where

$$b_1 = (x_{01} - x_{11})(x_{0t} - x_{1t}) + (y_{01} - y_{11})(y_{0t} - y_{1t}) \tag{14.15}$$

$$b_2 = (y_{01} - y_{11})(x_{0t} - x_{1t}) - (x_{01} - x_{11})(y_{0t} - y_{1t}) \tag{14.16}$$

$$\left| \mathbf{L}_1 \right|^2 = (x_{01} - x_{11})^2 + (y_{01} - y_{11})^2 \tag{14.17}$$

The interpolation algorithm is applied to each patch in the network. The linear space inbetweening algorithm suffers from slope discontinuity, that is, speed discontinuity along the boundaries of the patches.

The cubic metric space inbetweening algorithm is very similar to the Miura algorithm, but it is designed to rectify its cross-boundary derivative discontinuities. The cubic metric space algorithm uses Hermite cubic splines as basis curves, as shown in Figure 14.11. The cubic splines $\mathbf{C}_0$ and $\mathbf{C}_1$ are the basis vectors of keyframes $\mathbf{K}_0$ and $\mathbf{K}_1$, respectively. Each spline is defined by two end points and two end slopes. At time t, which corresponds to an inbetween frame $\mathbf{P}(u, t)$, a similar spline curve $\mathbf{C}_t$ is defined. Its end points are the positions of the moving points at time t, and its end slopes are the time-weighted average of the slopes at the corresponding end points of $\mathbf{K}_0$ and $\mathbf{K}_1$.

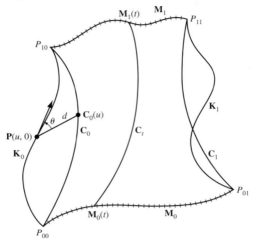

Figure 14.11   Linear space inbetweening algorithm.

In a cubic metric space, each point on a keyframe has a corresponding point on its basis curve, that is, both have the same value of the u parameters as shown in Figure 14.11. The distance d between $\mathbf{P}(u, 0)$ and $\mathbf{C}_0(u)$ and the angle θ between the tangent to $\mathbf{K}_0$ and the line joining the two points determine the transformation matrix $[T_0]$. Similarly, the matrix $[T_1]$ can be determined using $\mathbf{K}_1$ and $\mathbf{C}_1$. $[T_0]$ and $[T_1]$ transform $\mathbf{K}_0$ and $\mathbf{K}_1$ to $\mathbf{P}(u, t)$ such that $\mathbf{P}(u, t)$ matches the positions and slopes of $\mathbf{C}_t$ at its end points. The details of developing $[T_0]$ and $[T_1]$ are not covered here.

The third algorithm is the Coons patch inbetweening algorithm. It is based on the Coons patch representation that is shown in Figure 14.9, and it has the advantage of controlling the normal derivatives across patch boundaries. This ensures slope continuities along boundaries between adjacent patches. A linear Coons patch is usually used for inbetweening. To find the inbetween frame $\mathbf{P}(u, t)$ at time t, this time value is substituted into the patch equation (see Chapter 7).

The three inbetweening algorithms just described are general enough to handle keyframes made from any curves. However, the Coons patch inbetweening algorithm is the best of the three. It does not suffer from derivative discontinuities as does the Miura algorithm, nor does it require the animator to input slope values at the end points of keyframes as does the cubic metric space algorithm.

In addition, the average execution time required by the Coons algorithm for inbetweening is a little higher than that required by the Miura algorithm and much less than that required by the cubic space algorithm.

14.8 Simulation Technique

This approach (also called behavioral or algorithmic animation) is based on the physical laws that control the motion or the dynamic behavior of the object to be animated. The physical laws that describe the motion are developed and solved, thus allowing the motion to be described algorithmically. A frame is obtained by substituting a given time value into the solution, and the corresponding position and configuration of the object are calculated. A sequence of frames can be calculated and animated for a given time interval.

The simulation technique of animation attempts to combine characteristics of objects and environments traditionally modeled in graphics (such as shape shading and illumination) with physical laws that require the physical properties of objects (such as mass and inertia) and environment (such as friction and wind pressure).

The simulation approach to animation is versatile, produces realistic animation quality, and eliminates the unnatural and jerky motion that may result from keyframe animation. However, finding the physical laws that control the motion may be difficult, and solving them (in most cases numerical solutions are utilized) for a large sequence may be too expensive.

14.9 Animation Problems

The successful generation and display of an animation sequence depend on various factors such as the nature of objects being animated (whether they are fuzzy or not), the dynamic model used, the time interval between frames in the animation sequence (i.e., time sampling rate), interaction of the animated object with its environment (e.g., collision detection), the enhancement calculations used to generate the frames (such as shading, transparency, shadowing, and so forth), and the hardware characteristics of graphics displays (resolution, refresh rate, and so forth). This section briefly describes some of the common problems encountered in animation.

Two of the common problems in animation which control the frame rate calculations (number of frames computed and displayed per second) are frame-to-frame flicker and frame-to-frame discontinuity. Flicker is the blinking effect of a graphics display caused by the blank period between erasing and generating the contents of its pixels. It takes time to erase and redraw a screen image. To avoid frame flicker, a frame rate of 30 Hz must be used. This is the same rate used to scan graphics displays for flicker-free images. With such high a rate, only playback animation is possible if the animation sequence is to be flicker free.

Frame-to-frame discontinuity results from motion sampling in the time domain. If the time sampling rate is not adequate, discontinuous and jerky motion results. Changes in position of image elements from one frame to the next should be gradual. Frame-to-frame discontinuities are unnoticeable and blend into smooth, realistic motion at a time-sampling rate of 24 frames/second.

Two other problems that affect the quality of animation are spatial aliasing and temporal aliasing. Spatial aliasing is related to the resolution of the graphics display and results in the staircase effect (jagged edges). Many spatial antialiasing techniques are available to solve this problem. Temporal aliasing is related to the discretization of the time domain into discrete time values at which frames are calculated. Temporal antialiasing is not as common as spatial antialiasing. However , it is desired when motion blur is to be modeled into computer animation.

Algorithms for spatial antialiasing can be extended to temporal antialiasing. For example, supersampling algorithms with filtering can be used. In this case, each pixel intensity might be determined in four or more different locations, rather than just one. A filter may then be applied to derive the actual intensity of the output pixel. Supersampling is applied to each moving image (frame), and then filtering is applied to each resulting intensity function to "multiply expose" each output picture.

Different filter types can be used to achieve different effects of motion blur in the image. For example, the standard box filter tends to create an image in which objects are fainter at their extremes in the direction of motion, where they cover pixels for a shorter duration. Gaussian and triangular filters exaggerate this effect.

When several objects are animated at once in a scene, the problem of detecting and controlling object interactions is encountered. Some animation systems require the animator to visually inspect the scene for object interaction and respond in the proper fashion. This is a time-consuming and difficult process even for keyframe systems, where the user defines the motion explicitly. It is even more difficult for algorithmic animation where motion is obtained by solving the physical laws of motion.

Collision detection has been studied extensively in the fields of CAD/CAM and robotics. Some algorithms are more general than what computer animation may require. An existing algorithm is designed to test the interpenetration of surfaces that model flexible objects. If surfaces are modeled as triangular patches, collision between two surfaces is detected by testing for penetration of each vertex point through the planes of any triangle not including that vertex.

Surfaces are assumed to be initially separate. For each time step of animation, the positions of the points at the beginning and the end of the time step are compared to see if any point went through a triangle during that time step. If so, collision has occurred. The mathematical details of the algorithm are not covered here. Another algorithm to test for the collision between convex polyhedral solids is available. It is based on the Cyrus-Beck clipping algorithm.

Once a collision is detected, a response to it is necessary. Keyframe animation systems can follow a predetermined set of rules about the motion of objects immediately following the collision. Animation systems using dynamic simulation inherently must respond to collisions automatically and realistically. The response is usually based on the conservation of linear and angular momentum. Surface friction and elasticity of the colliding objects can be considered in the momentum equations. Details of the collision response are not covered here.

PROBLEMS

Part I: Theory

14.1 Each frame in an animation sequence takes five minutes of computer time to generate. Assuming 24 frames per second of animation, how long does it take to produce one minute of animation?

14.2 A four-bar linkage is shown on the right. Find the two extreme positions of the linkage. If these two positions are used as the start and end keyframes in keyframe animation

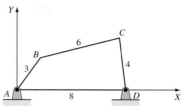

a. Find the motion trajectory, motion curves, and P-curves of points B and C if the P-curves animation method is used. Assume constant angular acceleration for links AB and CD of 4 rad/s^2 and 6 rad/s^2, respectively. Also, find an appropriate number of inbetweens based on the P-curves. Generate these inbetweens.

b. Using the same number and positions of the inbetweens as in (a), generate them using the linear space inbetweening algorithm.

c. Repeat (b) but for the linear Coons patch inbetweening algorithm.

d. Compare the inbetweens generated in (a), (b), and (c).

Part II: Laboratory

Use your in-house CAD/CAM system to create and animate the following objects.

14.3 A human heart.

14.4 A four–cylinder car engine.

14.5 Choose your favorite object(s), system(s), or application(s) and animate them.

14.6 Animate the following systems:

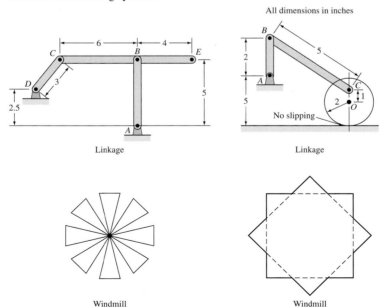

Linkage   Linkage

Windmill   Windmill

Part III: Programming

Write the code so that the user can input values to control the output. The program should display the results. Use OpenGL with C or C++, or use Java 3D.

14.7 Harmonic motion.

14.8 A bouncing ball.

14.9 A pendulum motion.

14.10 A yoyo motion.

14.11 Inserting a disk into a shaft.

14.12 Your favorite application.

Product Design
and Development

This part covers some concepts that are utilized in product development and design. These concepts are supported by most CAD/CAM systems. This part should help CAD designers to realize the ROI (return on investment) after spending time creating intricate geometric models of complex parts and assemblies. To achieve this goal, we accomplish the following objectives:

1. Perform and analyze volume calculations **(Chapter 15)**
2. Create and analyze asssembly models **(Chapter 16)**
3. Perform and evaluate FEM/FEA **(Chapter 17)**
4. Move CAD data from one CAD system to another **(Chapter 18)**
5. Work as teams in a collaborative design environment **(Chapter 19)**

575

Mass Properties

GOAL

Understand and master geometric and mass properties, how to calculate them, the concept of numerical integration, and how to use CAD systems to calculate mass properties.

OBJECTIVES

After reading this chapter, you should understand the following concepts:

- Importance of mass properties in CAD design

- Calculate curve and contour lengths

- Calculate areas

- Calculate volumes

- Calculate centroids

- Calculate inertia properties

- Use Gauss quadrature

- Use mass properties in part design

CHAPTER HEADLINES

15.1   Introduction

Mass property calculation was one of the first engineering applications to be implemented in CAD/CAM systems. This is perhaps due to the strong dependence of these calculations on the geometry and topology of objects. These calculations typically involve masses, centroids (centers of gravity), and inertial properties (moments of inertia). They form the basis for the study and analysis of both rigid and deformable body mechanics (statics and dynamics). For various objects, one can create their geometric models first, and then use them to calculate their mass properties, which can later be used for analysis.

Mass property calculations usually involve evaluating various integrals. Exact evaluation of these integrals is only possible for simple shapes. For complex shapes, approximate methods are usually used to evaluate these integrals. These methods have the important property that they monotonically converge to the exact solution which is, of course, not known. Mass property algorithms that utilize these methods are fully automatic and require no additional input except mass attributes, such as the density of the model.

15.2   Geometric Properties

In this section, we develop the equations needed to calculate geometric properties, specifically length, area, surface area, and volume. These properties form the basis for mass property calculations. As seen in this section, length, area, and volume are formulated as single, double, and triple integrals, respectively.

15.2.1   Curve Length

Calculating the length of a given curve between two end points is useful in many applications. For example, in mechanism analysis, we might be interested in calculating the length of a given locus in space.

To calculate the length of a spatial curve between two points P_1 and P_2, consider the curve shown in Figure 15.1. Given an incremental length ΔL of the curve, the curve's total length between P_1 and P_2 can be given by the following integral:

$$L = \int_{P_1}^{P_2} dL \tag{15.1}$$

Curves are usually represented in a parametric form as discussed in Chapter 6 and given by Eq. (6.3). If the length element ΔL is bounded by the points P_i and P_{i+1} as shown in Figure 15.1, ΔL can be approximated by the length of the vector connecting the two points, that is,

$$\Delta L = \left| \mathbf{P}_{i+1} - \mathbf{P}_i \right| \tag{15.2}$$

or

$$\Delta L = \sqrt{(x_{i+1} - x_i)^2 + (y_{i+1} - y_i)^2 + (z_{i+1} - z_i)^2} \tag{15.3}$$

or

$$\Delta L = \sqrt{(\Delta x)^2 + (\Delta y)^2 + (\Delta z)^2} \qquad (15.4)$$

Dividing both sides by Δu and taking the limit when u approaches zero, we get

$$\lim_{u \to 0} \frac{\Delta L}{\Delta u} = \lim_{u \to 0} \sqrt{\left(\frac{\Delta x}{\Delta u}\right)^2 + \left(\frac{\Delta y}{\Delta u}\right)^2 + \left(\frac{\Delta z}{\Delta u}\right)^2} \qquad (15.5)$$

or

$$\frac{dL}{du} = \sqrt{\left(\frac{dx}{du}\right)^2 + \left(\frac{dy}{du}\right)^2 + \left(\frac{dz}{du}\right)^2} \qquad (15.6)$$

or

$$dL = \sqrt{x'^2 + y'^2 + z'^2}\, du \qquad (15.7)$$

Substituting Eq. (15.7) into Eq. (15.1) gives

$$L = \int_{u_1}^{u_2} \sqrt{x'^2 + y'^2 + z'^2}\, du \qquad (15.8)$$

or

$$L = \int_{u_1}^{u_2} \sqrt{\mathbf{P}' \cdot \mathbf{P}'}\, du \qquad (15.9)$$

Equation (15.9) gives the exact length of a curve segment bounded by the parametric values u_1 and u_2 as the integral, with respect to u, of the square root of the dot product of the tangent vector of the curve. Equation (15.9) requires that the curve is C^1 continuous. It applies to both open and closed curves.

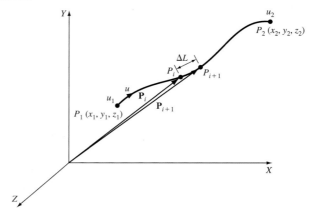

Figure 15.1   Curve length.

15.2.2 Cross-Sectional Area

A cross-sectional area is a planar region bounded by a closed boundary. The boundary consists of a set of C^1 continuous curves connected together. Thus, the boundary is piecewise continuous. There are three properties associated with a planar region: the length of its contour (boundary), its area, and its centroid.

Figure 15.2a shows a region R that is oriented generally in space. The region's plane coincides with the $X_L Y_L$ plane of the $X_L Y_L Z_L$ local coordinate system. The mapping between this local system and the XYZ coordinate system (MCS) shown in the figure can be achieved by using the mapping matrix given by Eq. (12.81).

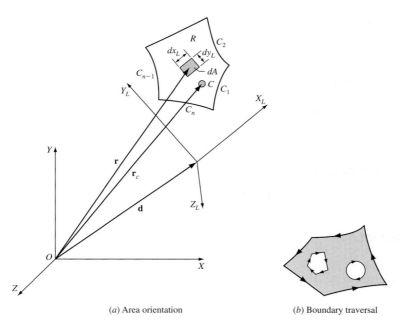

(a) Area orientation   (b) Boundary traversal

Figure 15.2   Cross-sectional area.

The region is bounded by the curves C_1, C_2, ..., C_{n-1}, and C_n. The length of the contour is given by the sum of the lengths of C_1, C_2, ..., and C_n, that is,

$$L = \sum_{i=1}^{n} L_i \tag{15.10}$$

where L_i is the length of curve C_i, which can be calculated using Eq. (15.9).

To calculate the area A of the region R, consider an area element dA of sides dx_L and dy_L as shown in Figure 15.2a. Integrating over the region gives:

$$A = \iint_R dA = \iint_R dx_L\, dy_L \tag{15.11}$$

The evaluation of the integral in Eq. (15.11) requires mapping this equation from the local coordinate system of the region R to the global system (MCS) because the parametric equations describing the boundary curves are stored in the region's geometric database with respect to the latter. Utilizing the matrix $[T]$ given by Eq. (12.81), we can write:

$$\begin{bmatrix} x_L \\ y_L \\ z_L \\ 1 \end{bmatrix} = [T]\begin{bmatrix} x \\ y \\ z \\ 1 \end{bmatrix} \tag{15.12}$$

The matrix $[T]$ can be evaluated using three noncollinear points as discussed in Chapter 2. End points of curves C_i can be used for this purpose. Expanding Eq. (15.12) using the notation of Eq. (12.81), we obtain:

$$\begin{bmatrix} x_L \\ y_L \\ z_L \\ 1 \end{bmatrix} = \begin{bmatrix} r_{11}x + r_{12}y + r_{13}z + x_d \\ r_{21}x + r_{22}y + r_{23}z + y_d \\ 0 \\ 1 \end{bmatrix} \tag{15.13}$$

The z_L coordinate is shown as zero in Eq. (15.13) because the region R lies in the X_LY_L plane. The differential elements dx_L and dy_L can be written using Eq. (15.13) as:

$$dx_L = r_{11}dx + r_{12}dy + r_{13}dz \tag{15.14}$$

$$dy_L = r_{21}dx + r_{22}dy + r_{23}dz \tag{15.15}$$

Substituting Eqs. (15.14) and (15.15) into Eq. (15.11) and reducing, the area A becomes:

$$A = \alpha_1\iint_R dx\, dx + \alpha_2\iint_R dy\, dy + \alpha_3\iint_R dz\, dz + \alpha_4\iint_R dx\, dy + \alpha_5\iint_R dx\, dz + \alpha_6\iint_R dy\, dz \tag{15.16}$$

where

$$\alpha_1 = r_{11}r_{21} \tag{15.17}$$

$$\alpha_2 = r_{12}r_{22} \tag{15.18}$$

$$\alpha_3 = r_{13}r_{23} \tag{15.19}$$

$$\alpha_4 = r_{11}r_{22} + r_{12}r_{21} \tag{15.20}$$

$$\alpha_5 = r_{11}r_{23} + r_{13}r_{21} \tag{15.21}$$

$$\alpha_6 = r_{12}r_{23} + r_{13}r_{22} \tag{15.22}$$

It is beneficial to change the double integrals shown in Eq. (15.16) into line integrals that can be evaluated over the boundary of the region R. This is achieved by using Green's theorem, which can be stated as follows: Given two functions $f(x, y)$ and $g(x, y)$ over a closed region R, we can write:

$$\iint_R \left(\frac{\partial f}{\partial x} + \frac{\partial g}{\partial y}\right) dx\, dy = \oint_B (f\, dy - g\, dx) \tag{15.23}$$

The f and g in Eq. (15.23) are continuous and single-valued functions of x and y. The line integral is taken in the positive direction around the boundary B. The functions f and g are usually not unique. There are three choices for both f and g for each integral of Eq. (15.16). For example, Green's theorem can be used for the $\iint_R dx\, dy$ with $f = x$ and $g = 0$, $f = 0$ and $g = y$, or $f = x/2$ and $g = y/2$. Applying Green's theorem to Eq. (15.16) using $f = x$ and $g = 0$, $f = y$ and $g = 0$, $f = z$ and $g = 0$, $f = x$ and $g = 0$, $f = 0$ and $g = z$, and $f = y$ and $g = 0$ for the six integrals respectively, we obtain:

$$A = \oint_B (\alpha_1 x - \alpha_5 z)\, dx + \oint_B (\alpha_2 y + \alpha_4 x)\, dy + \oint_B (\alpha_3 z + \alpha_6 y)\, dz \tag{15.24}$$

Considering the curves C_i of the boundary B, Eq (15.24) becomes:

$$A = \sum_{i=1}^{n} \left[\int_{C_i} (\alpha_1 x - \alpha_5 z)\, dx + \int_{C_i} (\alpha_2 y + \alpha_4 x)\, dy + \int_{C_i} (\alpha_3 z + \alpha_6 y)\, dz \right] \tag{15.25}$$

The boundary B must be traversed such that the interior lies to the left, that is, traversed in a counterclockwise direction for an exterior boundary and clockwise for hole boundaries (Figure 15.2b). The traversal direction affects the limits of the integrals (see Example 15.2). Assuming that each curve C_i is represented parametrically, then $x = x(u)$, $y = y(u)$, and $z = z(u)$. This gives

$$dx = x'\, du$$
$$dy = y'\, du \tag{15.26}$$
$$dz = z'\, du$$

Substituting Eq. (15.26) into Eq. (15.25) gives

$$A = \sum_{i=1}^{n} \left[\int_{u_{i1}}^{u_{i2}} (\alpha_1 x - \alpha_5 z) x'\, du + \int_{u_{i1}}^{u_{i2}} (\alpha_2 y + \alpha_4 x) y'\, du + \int_{u_{i1}}^{u_{i2}} (\alpha_3 z + \alpha_6 y) z'\, du \right] \tag{15.27}$$

where u_{i1} and u_{i2} are the limits of the parameter u for the ith curve.

The terms in Eq. (15.27) are reduced significantly depending on the relative orientation of the $X_L Y_L Z_L$ coordinate system with respect to the MCS. For example, if both systems are identical, then $[T]$ becomes an identity matrix. As a result, $\alpha_1 = \alpha_2 = \alpha_3 = \alpha_5 = \alpha_6 = 0$, $\alpha_4 = 1$, and Eq. (15.27) reduces to:

$$A = \sum_{i=1}^{n} \int_{u_{i1}}^{u_{i2}} xy' \, du \tag{15.28}$$

The centroid of the region R is obtained by equating the moment of the entire area A lumped at its centroid C (see Figure 15.2a) to the sum of the moments of the element areas of the region. The moments are taken with respect to the origin O of the MCS. Figure 15.2a shows an element area dA located by the position vector $\mathbf{r}_c$ from the origin O. The centroid of the region is located by the vector $\mathbf{r}_c$. Equating the two moments, we get:

$$\mathbf{r}_c = \frac{\iint_R \mathbf{r} \, dA}{\iint_R dA} = \frac{\iint_R \mathbf{r} \, dx_L \, dy_L}{A} \tag{15.29}$$

which, in scalar form, gives:

$$x_c = \frac{\iint_R x \, dx_L \, dy_L}{A} \qquad y_c = \frac{\iint_R y \, dx_L \, dy_L}{A} \qquad z_c = \frac{\iint_R z \, dx_L \, dy_L}{A} \tag{15.30}$$

where x_c, y_c, and z_c are the MCS coordinates of the centroid. The integrals $\iint_R x \, dx_L \, dy_L$, $\iint_R y \, dx_L \, dy_L$, and $\iint_R z \, dx_L \, dy_L$ are sometimes called the first moments of the area with respect to the YZ plane, the XZ plane, and the XY plane, respectively.

Substituting Eq. (15.14) and (15.15) into Eq. (15.30), using Green's theorem to change the resulting double integrals into line integrals, and reducing the results in a similar way to the area calculations, Eq. (15.30) becomes:

$$x_c = \frac{1}{A} \sum_{i=1}^{n} \left[\int_{u_{i1}}^{u_{i2}} \left(\frac{\alpha_1 x^2}{2} - \alpha_5 xz \right) x' \, du + \int_{u_{i1}}^{u_{i2}} \left(\alpha_2 xy + \frac{\alpha_4 x^2}{2} \right) y' \, du + \int_{u_{i1}}^{u_{i2}} (\alpha_3 xz + \alpha_6 xy) z' \, du \right]$$

$$y_c = \frac{1}{A} \sum_{i=1}^{n} \left[\int_{u_{i1}}^{u_{i2}} (\alpha_1 xy - \alpha_5 yz) x' \, du + \int_{u_{i1}}^{u_{i2}} \left(\frac{\alpha_2 y^2}{2} + \alpha_4 xy \right) y' \, du + \int_{u_{i1}}^{u_{i2}} \left(\alpha_3 yz + \frac{\alpha_6 y^2}{2} \right) z' \, du \right] \tag{15.31}$$

$$z_c = \frac{1}{A} \sum_{i=1}^{n} \left[\int_{u_{i1}}^{u_{i2}} \left(\alpha_1 xz - \frac{\alpha_5 z^2}{2} \right) x' \, du + \int_{u_{i1}}^{u_{i2}} (\alpha_2 yz + \alpha_4 xz) y' \, du + \int_{u_{i1}}^{u_{i2}} \left(\frac{\alpha_3 z^2}{2} + \alpha_6 yz \right) z' \, du \right]$$

It is worth noting that x, y, and z variables are used in Eq. (15.30) and not x_L, y_L, and 0 to facilitate the formulation of the centroid calculations. If the coordinates of the centroid C are needed relative to the $X_L Y_L Z_L$ coordinate system, they can be mapped from the MCS to the local coordinate system as explained in Section 15.3.5.

The preceding formulations apply to a singly connected region, that is, a region with only one outside closed boundary. They can be extended to multiply connected regions. A multiply connected region is a region with holes inside it, that is, it has one outside boundary with more than one inside boundary. In this case, the net cross-sectional area of a multiply connected region is given by:

$$A_m = A - \sum_{j=1}^{m} A_{hj} \tag{15.32}$$

where A_m is the net area, A is the area of the singly connected (that is, excluding the holes) region bounded by the outside boundary, and A_{hj} is the area of the jth hole. Each of the areas A and any A_{hj} can be calculated using Eq. (15.27). The method by which to obtain the centroid of a multiply connected region is covered in Section 15.6.

15.2.3 Surface Area

The surface area A_s of a bounded surface, shown in Figure 15.3, can be formulated in a similar way to the cross-sectional area. The major difference is that A_s is not planar in general as in the case of a cylindrical, spherical, B-spline, or a Bezier surface.

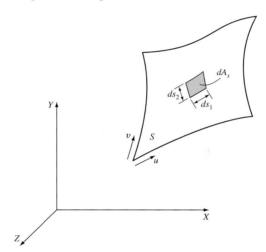

Figure 15.3   Surface area.

Figure 15.3 shows a surface S and a surface area element dA_s. The surface area A_s of the surface is given by:

$$A_s = \int_S dA_s \tag{15.33}$$

To evaluate the integral in Eq. (15.33), dA_s, which is a rectangle, must be written as the product of the lengths of two perpendicular sides. These sides are taken along the u and v direction as ds_1 and ds_2 (Figure 15.3). This is a convenient choice because surfaces are represented in parametric form in CAD/CAM systems. Thus, Eq. (15.33) can be written as:

$$A_s = \iint_S ds_1 \, ds_2 \tag{15.34}$$

Using the first fundamental quadric form of a surface given by Eq. (7.13), and observing that ds_1 is taken along the u direction, that is, $dv = 0$, and ds_2 is taken along the v direction (that is, $du = 0$), we can write:

$$ds_1 = \sqrt{\mathbf{P}_u \cdot \mathbf{P}_u} \, du \tag{15.35}$$

and

$$ds_2 = \sqrt{\mathbf{P}_v \cdot \mathbf{P}_v} \, dv \tag{15.36}$$

Substituting Eqs. (15.35) and (15.36) into Eq. (15.34), we get:

$$A_s = \int_{v_1}^{v_2} \int_{u_1}^{u_2} \sqrt{(\mathbf{P}_u \cdot \mathbf{P}_u)(\mathbf{P}_v \cdot \mathbf{P}_v)} \, du \, dv \tag{15.37}$$

Equation (15.37) can also be written in terms of the Jacobian $[J] = \|\mathbf{P}_u \times \mathbf{P}_v\|$ (see Problem 15.5 at the end of the chapter.)

For an object consisting of multiple surfaces, its total surface area is equal to the sum of its individual surface areas, that is,

$$A_s = \sum_{i=1}^{n} A_{si} \tag{15.38}$$

Notice that the contribution of holes in an object to its surface area is positive as shown by Eq. (15.38). (Their contribution to the object's volume is negative).

The centroid of a surface can be found using Eq. (15.29). Substituting $dA_s = ds_1 ds_2$ for dA in Eq. (15.29), Eq. (15.30) becomes:

$$x_c = \frac{\int_{v_1}^{v_2} \int_{u_1}^{u_2} xK \, du \, dv}{A_s} \qquad y_c = \frac{\int_{v_1}^{v_2} \int_{u_1}^{u_2} yK \, du \, dv}{A_s} \qquad z_c = \frac{\int_{v_1}^{v_2} \int_{u_1}^{u_2} zK \, du \, dv}{A_s} \tag{15.39}$$

where

$$K = \sqrt{(\mathbf{P}_u \cdot \mathbf{P}_u)(\mathbf{P}_v \cdot \mathbf{P}_v)} \tag{15.40}$$

15.2.4 Volume

Figure 15.4 shows an object whose volume is V. The volume V can be expressed as a triple integral by integrating the volume element dV, that is,

$$V = \iiint_V dV = \iiint_V dx \, dy \, dz \tag{15.41}$$

The volume integral of Eq. (15.41) can be changed into a surface integral using the Gauss divergence theorem, which can be expressed as follows:

$$\iiint_V \nabla \cdot \mathbf{F} \, dV = \iint_S \mathbf{F} \cdot \hat{\mathbf{n}} \, dS \tag{15.42}$$

where

$$\nabla = \frac{\partial}{\partial x}\hat{\mathbf{i}} + \frac{\partial}{\partial y}\hat{\mathbf{j}} + \frac{\partial}{\partial z}\hat{\mathbf{k}} \tag{15.43}$$

and $\hat{\mathbf{n}}$ is the unit normal vector of the surface of the body.

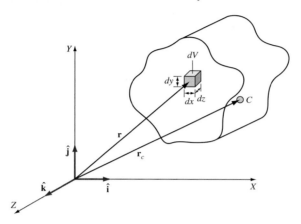

Figure 15.4   Volume.

The vector function $\mathbf{F}$ used in Eq. (15.42) must be C^1 continuous over the closed regular surface S and its interior V. To convert Eq. (15.29) to a surface integral, we have to find a vector function $\mathbf{F}$ which satisfies the condition $\nabla \cdot \mathbf{F} = 1$. As in Green's theorem, the vector function $\mathbf{F}$ is not unique. Some candidate functions are $x\hat{\mathbf{i}}$, $y\hat{\mathbf{j}}$, $z\hat{\mathbf{k}}$, or $(x/3)\hat{\mathbf{i}} + (y/3)\hat{\mathbf{j}} + (z/3)\hat{\mathbf{k}}$. Choosing one of the first three candidates is better than the last one because it can eliminate unnecessary surface integrals. For example, choosing $\mathbf{F} = z\hat{\mathbf{k}}$ eliminates integrals over surfaces with normals perpendicular to the Z axis. Using this choice, the volume can be written as:

$$V = \iiint_V dV = \iint_S z(\hat{\mathbf{k}} \cdot \hat{\mathbf{n}}) \, dA_s \tag{15.44}$$

Substituting Eqs. (15.35) and (15.36) for dA_s into Eq. (15.44), we obtain:

$$V = \iiint_V dV = \int_{v_1}^{v_2} \int_{u_1}^{u_2} z(\hat{\mathbf{k}} \cdot \hat{\mathbf{n}}) \sqrt{(\mathbf{P}_u \cdot \mathbf{P}_u)(\mathbf{P}_v \cdot \mathbf{P}_v)} \, du \, dv \tag{15.45}$$

If the object is closed by multiple surfaces, its volume becomes:

$$V = \sum_{i=1}^{n} \int_{v_{i1}}^{v_{i2}} \int_{u_{i1}}^{u_{i2}} z(\hat{\mathbf{k}} \cdot \hat{\mathbf{n}}_i) \sqrt{(\mathbf{P}_{ui} \cdot \mathbf{P}_{ui})(\mathbf{P}_{vi} \cdot \mathbf{P}_{vi})} \, du \, dv \tag{15.46}$$

Equation (15.46) applies to a singly connected object (no holes in its interior). The volume V_m of a multiply connected object (with holes) is given by:

$$V_m = V - \sum_{j=1}^{m} V_{hj} \tag{15.47}$$

where V_{hj} is the volume of hole j.

The centroid of the object is located by the vector $\mathbf{r}_c$ shown in Figure 15.4. This vector is given by [similar to Eq. (15.29)]:

$$\mathbf{r}_c = \frac{\iiint_V \mathbf{r}\, dV}{\iiint_V dV} = \frac{\iiint_V \mathbf{r}\, dV}{V} \tag{15.48}$$

or

$$x_c = \frac{1}{V}\iiint_V x\, dV \qquad y_c = \frac{1}{V}\iiint_V y\, dV \qquad z_c = \frac{1}{V}\iiint_V z\, dV \tag{15.49}$$

Using the Gauss divergence theorem with $\mathbf{F} = xz\hat{\mathbf{k}}$, $\mathbf{F} = yz\hat{\mathbf{k}}$, and $\mathbf{F} = (z^2/2)\hat{\mathbf{k}}$ for x_c, y_c, and z_c, respectively, Eq. (15.49) becomes

$$x_c = \frac{1}{V}\iint_S xz(\hat{\mathbf{k}}\cdot\hat{\mathbf{n}})\, dA_s$$

$$y_c = \frac{1}{V}\iint_S yz(\hat{\mathbf{k}}\cdot\hat{\mathbf{n}})\, dA_s \tag{15.50}$$

$$z_c = \frac{1}{V}\iint_S \frac{z^2}{2}(\hat{\mathbf{k}}\cdot\hat{\mathbf{n}})\, dA_s$$

Using $dA_s = ds_1 ds_2$ together with Eqs. (15.35), (15.36), and (15.40), Eq. (15.50) becomes:

$$x_c = \frac{1}{V}\int_{v_1}^{v_2}\int_{u_1}^{u_2} xzK(\hat{\mathbf{k}}\cdot\hat{\mathbf{n}})\, du\, dv$$

$$y_c = \frac{1}{V}\int_{v_1}^{v_2}\int_{u_1}^{u_2} yzK(\hat{\mathbf{k}}\cdot\hat{\mathbf{n}})\, du\, dv \tag{15.51}$$

$$z_c = \frac{1}{V}\int_{v_1}^{v_2}\int_{u_1}^{u_2} \frac{z^2}{2}K(\hat{\mathbf{k}}\cdot\hat{\mathbf{n}})\, du\, dv$$

If the object is closed by multiple surfaces, the integrals in Eq. (15.51) are summed over these surfaces to give:

$$x_c = \frac{1}{V}\sum_{i=1}^{n}\int_{v_{i1}}^{v_{i2}}\int_{u_{i1}}^{u_{i2}} xzK_i(\hat{\mathbf{k}}\cdot\hat{\mathbf{n}}_i)\, du\, dv$$

$$y_c = \frac{1}{V}\sum_{i=1}^{n}\int_{v_{i1}}^{v_{i2}}\int_{u_{i1}}^{u_{i2}} yzK_i(\hat{\mathbf{k}}\cdot\hat{\mathbf{n}}_i)\, du\, dv \tag{15.52}$$

$$z_c = \frac{1}{V} \sum_{i=1}^{n} \int_{v_{i1}}^{v_{i2}} \int_{u_{i1}}^{u_{i2}} \frac{z^2}{2} K_i(\hat{\mathbf{k}} \cdot \hat{\mathbf{n}}_i) \, du \, dv$$

15.3 Mass Properties

The mass properties of an object are a set of useful properties used in various engineering applications. These properties include mass, centroid, first moments, and second moments of inertia. The main difference between mass and geometric properties is the inclusion of the density of the object material in the former. Formally, an object can have a centroid (of its volume), a center of mass (of its mass), and a center of gravity (of its weight) that may differ from each other if the acceleration of gravity g and/or the density ρ of the object material is not constant.

In this chapter, we assume that g and ρ are constants. Therefore the three centers (of volume, of mass, and of weight) coincide and equal the centroid (of the volume) of the object. This assumption implies that objects of interest are homogeneous and are always close to the surface of the earth.

15.3.1 Mass

The mass of an object can be formulated in a way similar to formulating its volume. If we replace the volume element dV shown in Figure 15.4 by a mass element, we can write:

$$dm = \rho \, dV \tag{15.53}$$

Integrating Eq. (15.53) over the distributed mass of the object gives:

$$m = \iiint_m \rho \, dV \tag{15.54}$$

Assuming the density ρ to be uniform (constant), Eq. (15.54) becomes:

$$m = \rho \iiint_V dV = \rho V \tag{15.55}$$

As Eq. (15.55) shows, once the volume of an object is calculated, it is multiplied by its density to obtain its mass.

15.3.2 Centroid

Equation (15.48) can be used to find the center of mass of an object by replacing the volume V by the mass m, that is,

$$\mathbf{r}_c = \frac{\iiint_m \mathbf{r} \, dm}{m} \tag{15.56}$$

If we substitute ρV for m in Eq. (15.56), Eq. (15.48) results, and the center of mass is coincident with the center of volume. Therefore, Eqs. (15.49) to (15.52) apply to the center of mass (centroid).

15.3.3 First Moments of Inertia

The first moment of an area, volume, or mass is a mathematical property that appears in various calculations. It is defined as the moment of an object property (area, volume, or mass) with respect to a given plane. For a lumped mass, the first moment of the mass about a given plane is equal to the product of the mass and its perpendicular distance from the plane. Using this definition, the first moments of a distributed mass of an object with respect to the XY, XZ, and YZ planes are given by, respectively:

$$M_{xy} = \iiint_m z\,dm = \rho \iiint_V z\,dV \qquad (15.57)$$

$$M_{xz} = \iiint_m y\,dm = \rho \iiint_V y\,dV \qquad (15.58)$$

$$M_{yz} = \iiint_m x\,dm = \rho \iiint_V x\,dV \qquad (15.59)$$

We can easily recognize that these expressions for first moments appear in the centroid Eq. (15.49). Substituting Eq. (15.37) into Eqs. (15.57) to (15.59) gives:

$$M_{xy} = \rho V z_c = m z_c \qquad (15.60)$$

$$M_{xz} = \rho V y_c = m y_c \qquad (15.61)$$

$$M_{yz} = \rho V x_c = m x_c \qquad (15.62)$$

As Eqs. (15.60) to (15.62) show, the first moments are by-products of the volume and centroid calculations.

15.3.4 Second Moments and Products of Inertia

The second moment of inertia of a lumped mass about a given axis is the product of the mass and the square of the perpendicular distance between the mass and the axis. The second moments of inertia of a distributed mass about the X, Y, and Z axes can be written as:

$$I_{xx} = \iiint_m (y^2 + z^2)\,dm = \rho \iiint_V (y^2 + z^2)\,dV \qquad (15.63)$$

$$I_{yy} = \iiint_m (x^2 + z^2)\,dm = \rho \iiint_V (x^2 + z^2)\,dV \qquad (15.64)$$

$$I_{zz} = \int\int\int_m (x^2 + y^2)\, dm = \rho \int\int\int_V (x^2 + y^2)\, dV \tag{15.65}$$

The physical interpretation of a second moment of inertia of an object about an axis is that it represents the resistance of the object to any rotation about the axis (examples: the torque equation $T = I\alpha$, and the stress equation $\sigma = Mc/I$).

The inertia integrals can be changed to surface integrals via the Gauss divergence theorem. Using $\mathbf{F} = (y^2 z + z^3/3)\hat{\mathbf{k}}$, $\mathbf{F} = (x^2 z + z^3/3)\hat{\mathbf{k}}$, and $\mathbf{F} = (x^2 z + y^2 z)\hat{\mathbf{k}}$ for I_{xx}, I_{yy}, and I_{zz}, respectively, Eqs. (15.63) to (15.65) become:

$$I_{xx} = \rho \int\int_S z\left(y^2 + \frac{z^2}{3}\right)(\hat{\mathbf{k}} \cdot \hat{\mathbf{n}})\, dA_s \tag{15.66}$$

$$I_{yy} = \rho \int\int_S z\left(x^2 + \frac{z^2}{3}\right)(\hat{\mathbf{k}} \cdot \hat{\mathbf{n}})\, dA_s \tag{15.67}$$

$$I_{zz} = \rho \int\int_S z(x^2 + y^2)(\hat{\mathbf{k}} \cdot \hat{\mathbf{n}})\, dA_s \tag{15.68}$$

Using $dA_s = ds_1 ds_2$ together with Eqs. (15.35), (15.36), and (15.40), Eqs. (15.66) to (15.68) become:

$$I_{xx} = \rho \int_{v_1}^{v_2} \int_{u_1}^{u_2} zK\left(y^2 + \frac{z^2}{3}\right)(\hat{\mathbf{k}} \cdot \hat{\mathbf{n}})\, du\, dv \tag{15.69}$$

$$I_{yy} = \rho \int_{v_1}^{v_2} \int_{u_1}^{u_2} zK\left(x^2 + \frac{z^2}{3}\right)(\hat{\mathbf{k}} \cdot \hat{\mathbf{n}})\, du\, dv \tag{15.70}$$

$$I_{zz} = \rho \int_{v_1}^{v_2} \int_{u_1}^{u_2} zK(x^2 + y^2)(\hat{\mathbf{k}} \cdot \hat{\mathbf{n}})\, du\, dv \tag{15.71}$$

If the object is closed by multiple surfaces, the integrals in Eqs. (15.69) to (15.71) are summed over these surfaces to give:

$$I_{xx} = \rho \sum_{i=1}^n \int_{v_{i1}}^{v_{i2}} \int_{u_{i1}}^{u_{i2}} zK_i\left(y^2 + \frac{z^2}{3}\right)(\hat{\mathbf{k}} \cdot \hat{\mathbf{n}}_i)\, du\, dv \tag{15.72}$$

$$I_{yy} = \rho \sum_{i=1}^n \int_{v_{i1}}^{v_{i2}} \int_{u_{i1}}^{u_{i2}} zK_i\left(x^2 + \frac{z^2}{3}\right)(\hat{\mathbf{k}} \cdot \hat{\mathbf{n}})\, du\, dv \tag{15.73}$$

$$I_{zz} = \rho \sum_{i=1}^n \int_{v_{i1}}^{v_{i2}} \int_{u_{i1}}^{u_{i2}} zK_i(x^2 + y^2)(\hat{\mathbf{k}} \cdot \hat{\mathbf{n}})\, du\, dv \tag{15.74}$$

Like second moments of inertia, products of inertia are useful and are defined by the following equations:

$$I_{xy} = \iiint_m xy \, dm = \rho \iiint_V xy \, dV \tag{15.75}$$

$$I_{xz} = \iiint_m xz \, dm = \rho \iiint_V xz \, dV \tag{15.76}$$

$$I_{yz} = \iiint_m yz \, dm = \rho \iiint_V yz \, dV \tag{15.77}$$

Following a similar approach to that used in the second moments of inertia, Eqs. (15.75) to (15.77) can be rewritten as:

$$I_{xy} = \rho \int_{v_1}^{v_2} \int_{u_1}^{u_2} xyzK(\hat{\mathbf{k}} \cdot \hat{\mathbf{n}}) \, du \, dv \tag{15.78}$$

$$I_{xz} = \rho \int_{v_1}^{v_2} \int_{u_1}^{u_2} \frac{xz^2}{2} K(\hat{\mathbf{k}} \cdot \hat{\mathbf{n}}) \, du \, dv \tag{15.79}$$

$$I_{yz} = \rho \int_{v_1}^{v_2} \int_{u_1}^{u_2} \frac{yz^2}{2} K(\hat{\mathbf{k}} \cdot \hat{\mathbf{n}}) \, du \, dv \tag{15.80}$$

and for multiple surface objects;

$$I_{xy} = \rho \sum_{i=1}^{n} \int_{v_{i1}}^{v_{i2}} \int_{u_{i1}}^{u_{i2}} xyzK_i(\hat{\mathbf{k}} \cdot \hat{\mathbf{n}}_i) \, du \, dv \tag{15.81}$$

$$I_{xz} = \rho \sum_{i=1}^{n} \int_{v_{i1}}^{v_{i2}} \int_{u_{i1}}^{u_{i2}} \frac{xz^2}{2} K_i(\hat{\mathbf{k}} \cdot \hat{\mathbf{n}}_i) \, du \, dv \tag{15.82}$$

$$I_{yz} = \rho \sum_{i=1}^{n} \int_{v_{i1}}^{v_{i2}} \int_{u_{i1}}^{u_{i2}} \frac{yz^2}{2} K_i(\hat{\mathbf{k}} \cdot \hat{\mathbf{n}}_i) \, du \, dv \tag{15.83}$$

15.3.5 Property Mapping

The volume and mass properties already presented are formulated with respect to a given *XYZ* coordinate system. For a geometric model, this system is the MCS of the model database. The MCS is a convenient system because all the curve, surface, or solid equations and other geometric information are stored with respect to this system. If mass properties are to be calculated with respect to other systems, these properties must be mapped from the MCS to the other systems. In the following discussions, let us assume that mass properties are already available in the MCS (*XYZ*) and we need to map them to a given WCS ($X_W Y_W Z_W$).

The coordinates of the centroid (x_c, y_c, z_c) can be mapped from the MCS to the WCS using the following equation:

$$\mathbf{r}_{cW} = [T]\mathbf{r}_c \tag{15.84}$$

or

$$\begin{bmatrix} x_{cW} \\ y_{cW} \\ z_{cW} \\ 1 \end{bmatrix} = [T] \begin{bmatrix} x_c \\ y_c \\ z_c \\ 1 \end{bmatrix} \tag{15.85}$$

where $[T]$ is the general mapping matrix given by Eq. (12.81).

Once the centroid is mapped, the first moments M_{xy}, M_{xz}, and M_{yz} are automatically mapped as seen from Eqs. (15.60) to (15.62). These equations become:

$$\begin{bmatrix} M_{xyW} \\ M_{xzW} \\ M_{yzW} \\ 1 \end{bmatrix} = m \begin{bmatrix} x_{cW} \\ y_{cW} \\ z_{cW} \\ 1 \end{bmatrix} = m[T] \begin{bmatrix} x_c \\ y_c \\ z_c \\ 1 \end{bmatrix} \tag{15.86}$$

The second moments and products of inertia of an object with respect to its MCS can be mapped to compute the moments of inertia of the object about any arbitrary axis. The axis may or may not pass through the origin of the MCS. Figure 15.5 shows some examples. The axis AA passes through the origin O, and its direction in space is defined by the unit vector $\hat{\mathbf{n}}$. By definition, $I_{aa} = \iiint_m b^2 dm$, where b is the perpendicular distance from dm to AA. If the position of dm is located using $\mathbf{r}$, then $b = r\sin\theta = |\hat{\mathbf{n}} \times \mathbf{r}|$. Hence, I_{aa} can be expressed as:

$$I_{aa} = \iiint |\hat{\mathbf{n}} \times \mathbf{r}|^2 dm = \iiint (\hat{\mathbf{n}} \times \mathbf{r}) \cdot (\hat{\mathbf{n}} \times \mathbf{r}) \, dm \tag{15.87}$$

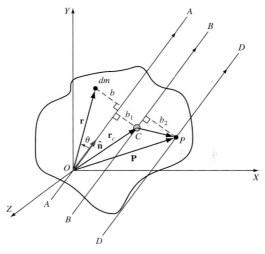

Figure 15.5   Moments of inertia about an arbitrary axis.

Using $\hat{\mathbf{n}} = n_x\hat{\mathbf{i}} + n_y\hat{\mathbf{j}} + n_z\hat{\mathbf{k}}$, $\mathbf{r} = x\hat{\mathbf{i}} + y\hat{\mathbf{j}} + z\hat{\mathbf{k}}$, and Eqs. (15.63) to (15.65) and (15.75) to (15.77), Eq. (15.87) becomes:

$$I_{aa} = I_{xx}n_x^2 + I_{yy}n_y^2 + I_{zz}n_z^2 - 2I_{xy}n_xn_y - 2I_{xz}n_xn_z - 2I_{yz}n_yn_z \tag{15.88}$$

If the moment of inertia about a centroidal axis BB (Figure 15.5) that is parallel to axis AA is to be computed, the parallel axis theorem can be used to give:

$$I_{bb} = I_{aa} - mb_1^2 \tag{15.89}$$

Substituting $\left|\hat{\mathbf{n}} \times \mathbf{r}_c\right|$ for b_1, Eq. (15.89) becomes:

$$I_{bb} = I_{aa} - m(\hat{\mathbf{n}} \times \mathbf{r}_c) \cdot (\hat{\mathbf{n}} \times \mathbf{r}_c) \tag{15.90}$$

which can be reduced to:

$$I_{bb} = I_{aa} - m[(n_yz_c - n_zy_c)^2 + (n_zx_c - n_xz_c)^2 + (n_xy_c - n_yx_c)^2] \tag{15.91}$$

The moment of inertia about any axis DD (Figure 15.5) that is parallel to AA but passes through a general point P can be computed using the parallel axis theorem again, that is,

$$I_{dd} = I_{bb} + mb_2^2 \tag{15.92}$$

The distance b_2 is equal to $\left|\hat{\mathbf{n}} \times \overline{\mathbf{CP}}\right| = \left|\hat{\mathbf{n}} \times (\mathbf{P} - \mathbf{r}_c)\right|$. Hence Eq. (15.91) becomes:

$$I_{dd} = I_{bb} + m\{[n_y(z - z_c) - n_z(y - y_c)]^2$$
$$+ [n_z(x - x_c) - n_x(z - z_c)]^2 + [n_x(y - y_c) - n_y(x - x_c)]^2\} \tag{15.93}$$

The moment of inertia of an object about axes parallel to the MCS but passing through a given point in space can be determined using the parallel axis theorem. Figure 15.6 shows three parallel coordinate systems: the MCS, a centroidal system $(X_cY_cZ_c)$, and a WCS that has an origin at point P. Let us assume that the moments and products of inertia have been calculated with respect to the MCS. Applying the parallel axis theorem twice, we can write:

$$\begin{aligned}
(I_{xx})_C &= I_{xx} - m(y_c^2 + z_c^2) \\
(I_{yy})_C &= I_{yy} - m(x_c^2 + z_c^2) \\
(I_{zz})_C &= I_{zz} - (x_c^2 + y_c^2) \\
(I_{xy})_C &= I_{xy} - mx_cy_c \\
(I_{xz})_C &= I_{xz} - mx_cz_c \\
(I_{yz})_C &= I_{yz} - my_cz_c
\end{aligned} \tag{15.94}$$

and

$$(I_{xx})_W = (I_{xx})_C + m[(y-y_c)^2 + (z-z_c)^2]$$

$$(I_{yy})_W = (I_{yy})_C + m[(x-x_c)^2 + (z-z_c)^2]$$

$$(I_{zz})_W = (I_{zz})_C + m[(x-x_c)^2 + (y-y_c)^2] \qquad (15.95)$$

$$(I_{xy})_W = (I_{xy})_C + m(x-x_c)(y-y_c)$$

$$(I_{xz})_W = (I_{xz})_C + m(x-x_c)(z-z_c)$$

$$(I_{yz})_W = (I_{yz})_C + m(y-y_c)(z-z_c)$$

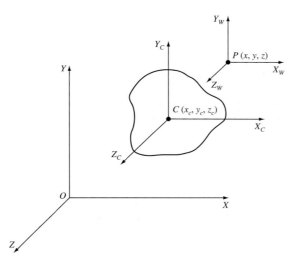

Figure 15.6 Moments of inertia about axes parallel to the MCS.

Principal moments of inertia are useful properties to determine for an object. Their corresponding axes are the principal axes of inertia. These axes are characterized such that products of inertia with respect to them are zeros. Thus, the principal inertia tensor is a diagonal matrix whose diagonal elements are the principal moments of inertia I_x, I_y, and I_z. These moments are the roots of the following cubic equation:

$$I^3 - (I_{xx} + I_{yy} + I_{zz})I^2 + (I_{xx}I_{yy} + I_{yy}I_{zz} + I_{zz}I_{xx} - I_{xy}^2 - I_{yz}^2 - I_{xz}^2)I$$

$$- (I_{xx}I_{yy}I_{zz} - 2I_{xy}I_{yz}I_{xz} - I_{xx}I_{yz}^2 - I_{yy}I_{xz}^2 - I_{zz}I_{xy}^2) = 0 \qquad (15.96)$$

The determination of the directions of the principal axes of inertia are not discussed here. It involves angular momentum equations. Readers interested on how to derive Eq. (15.96) or how to find these directions of the principal axes should refer to books on statics and dynamics subjects.

The radius of gyration R_g of an object with respect to an axis is one of the inertial properties of the object. It is defined by the following equation:

$$R_g = \sqrt{\frac{I}{m}} \qquad\qquad (15.97)$$

where I is the moment of inertia with respect to the same axis of R_g.

15.4 Properties Evaluation

Geometric and mass properties of an object have been formulated in the previous two sections. These properties and the equations that describe them are listed in Table 15.1. All these equations are integral equations in the parametric space. Some are line integrals to be evaluated over curves or edges of the object. In this case, the integrand is a function of the parameter u. Others are surface integrals to be evaluated over surfaces or faces of the object, in which case the integrand is a function of the parameters u and v. These surface integrals can be further reduced to line integrals for analytic surfaces.

Table 15.1   Summary of geometric and mass properties.

Property type	Property	Equation number
Geometric	Curve length	(15.9)
	Cross-sectional area:	
	Area	(15.27)
	Centroid	(15.31)
	Surface area:	
	Area	(15.37)
	Centroid	(15.39)
	Volume:	
	Volume	(15.46)
	Centroid	(15.52)
Mass	Mass:	
	Mass	(15.55)
	Center of mass	(15.52)
	First moments of inertia	(15.60) to (15.62)
	Second moments of inertia	(15.72) to (15.74)
	Products of inertia	(15.81) to (15.83)

The equations listed in Table 15.1 are applicable to all curves, surfaces, and solids covered in Chapters 6,7, and 9, respectively. For many curves, such as lines, circles, and conics, and for many surfaces such as spherical, conic, and bicubic surfaces, the integrals shown in the equations can be integrated exactly to generate the final expressions of geometric and mass properties in closed form.

Over some surfaces like B-spline surfaces or curved surfaces with irregular boundaries, the surface integrals for geometric and mass properties cannot be evaluated exactly for various reasons. First, the parametric equation of the surface may not allow the conversion of the double integration over a parametric domain into a set of line integrals directly. In this case, numerical integration such as Gauss quadrature can be used to approximately evaluate the integral.

Second, the analytic parametric equation of the boundary over which the line integral is to be evaluated may not be available. In this case, an interpolation function for the boundary of the parametric domain must be chosen even when simple surfaces such as arbitrarily oriented cylinders intersect each other. Third, the final line integral may only be possible to evaluate numerically.

Numerical integration (sometimes called numerical quadrature) is a major subject in itself. Interested readers should refer to books and other literature on numerical analysis. In this section, we review two methods: Newton-Cotes quadrature and Gauss quadrature. These methods are numerically efficient. Mass property equations whose numbers are listed in Table 15.1 involve either 1D (in u) or 2D (in u and v) integrals. These two types of integrals take the following general forms:

$$I = \int_{u_1}^{u_2} f(u)\, du \tag{15.98}$$

$$I = \int_{v_1}^{v_2} \int_{u_1}^{u_2} f(u, v)\, du\, dv \tag{15.99}$$

Let us apply the two methods to Eq. (15.98). To evaluate the integral in Eq. (15.98) numerically, the integral is approximated by a polynomial and a remainder, that is,

$$I = I_a + R \tag{15.100}$$

where I_a is the approximate value of I, and R is the remainder. I_a usually takes the form of a polynomial, and R is the source of error or approximation in the numerical evaluation.

In Newton-Cotes integration, it is assumed that the sampling points of $f(u)$ are spaced at equal distances in the interval $[u_1, u_2]$. If we use $(n + 1)$ sampling points, we can define:

$$u_0 = u_1 \qquad u_n = u_2 \qquad h = \frac{u_2 - u_1}{n} \tag{15.101}$$

and the integration formula for the Newton-Cotes method can be written as:

$$I = \int_{u_1}^{u_2} f(u)\, du = (u_2 - u_1) \sum_{i=0}^{n} C_i^n f_i + R \tag{15.102}$$

where C_i^n are the Newton-Cotes constants and $f_i = f(u_i) = f(u_0 + ih)$. The cases $n = 1$ and $n = 2$ are the well-known trapezoidal and Simpson rules shown in Figure 15.7. The constants for these two rules are given by:

$n = 1$;

$$C_0^1 = C_1^1 = \tfrac{1}{2}$$ (15.103)

$n = 2$;

$$C_0^2 = \tfrac{1}{6} \qquad C_1^2 = \tfrac{4}{6} \qquad C_2^2 = \tfrac{1}{6}$$ (15.104)

Therefore, Eq. (15.101) becomes for the trapezoidal rule (Figure 15.7a):

$$I = \int_{u_1}^{u_2} f(u)\, du \approx \frac{u_2 - u_1}{2}(f_0 + f_1) = \frac{h}{2}(f_0 + f_1)$$ (15.105)

and for the Simpson rule (Figure 15.7b):

$$I = \int_{u_1}^{u_2} f(u)\, du \approx \frac{u_2 - u_1}{6}(f_0 + 4f_1 + f_2)$$

$$= \frac{h}{3}(f_0 + 4f_1 + f_2)$$ (15.106)

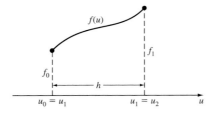

(a) Trapezoidal rule {two points at $u = u_1$ and u_2}

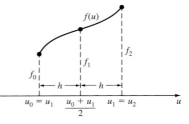

(b) Simpson rule {three points at $u = u_1$, $(u_1 + u_2)/2$, and u_2}

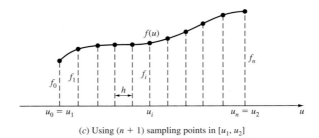

(c) Using $(n + 1)$ sampling points in $[u_1, u_2]$

Figure 15.7   Sampling points for trapezoidal and Simpson rules.

The minimum number of sampling points in the interval $[u_1, u_2]$ are two and three for the trapezoidal and Simpson rules, respectively. The two methods can still be applied if $(n + 1)$ sampling points are used in the interval $[u_1, u_2]$ as shown in Figure 15.7c. In this case, apply each method repetitively using two sampling points for the former and three for the latter, and then add to obtain:

   trapezoidal rule:

$$I = \int_{u_1}^{u_2} f(u)\, du \approx h(\tfrac{1}{2}f_0 + f_1 + f_2 + \cdots + f_{n-1} + \tfrac{1}{2}f_n) \tag{15.107}$$

   Simpson rule:

$$I = \int_{u_1}^{u_2} f(u)\, du \approx \frac{h}{3}(f_0 + 4f_1 + 2f_2 + 4f_3 + \cdots + 2f_{n-2} + 4f_{n-1} + f_n) \tag{15.108}$$

for the Simpson rule, n must be an even number.

The error E introduced by Newton-Cotes integration is equal to the remainder R [Eq. (15.100)]. An upper bound on the error E is given by:

   trapezoidal rule:

$$E < \frac{(u_2 - u_1)^3}{10}\left(\frac{d^2 f}{du^2}\right)_{u_2} \tag{15.109}$$

   Simpson rule:

$$E < \frac{(u_2 - u_1)^5}{1000}\left(\frac{d^4 f}{du^4}\right)_{u_2} \tag{15.110}$$

The actual errors can be significantly less than these upper bounds. They mainly depend on the function $f(u)$ that is integrated. It should be emphasized here that actual errors include, in addition to the remainder, the truncation and round-off errors that result during computations. The Simpson rule is usually more accurate than trapezoidal rule. If $f(u)$ is linear, both produce the exact integration. If $f(u)$ is parabolic, Simpson rule is still exact.

The extension of the trapezoidal and Simpson rules to two dimensions is straightforward. In this case the sampling points are obtained by dividing the intervals $[u_1, u_2]$ and $[v_1, v_2]$. The details are not covered here.

Both the trapezoidal and Simpson rules use equally spaced sampling points. These methods are effective when measurements of an unknown function to be integrated have been taken at equal intervals. However, in the integration of geometric-related equations such as mass property calculations, a function is called to evaluate the function $f(u)$ at given points. These points may be chosen anywhere in the interval $[u_1, u_2]$. Therefore, it seems natural to optimize the positions of the sampling points to improve the accuracy of numerical integration. **Gauss quadrature** is a numerical integration method in which both the positions of the sampling points and the associated weights have been optimized.

The basic formula in Gauss numerical integration is:

$$I = \int_{u_1}^{u_2} f(u)\, du = \sum_{i=1}^{n} V_i f_i + R \tag{15.111}$$

where V_i and u_i are the weight factors and the sampling points, respectively. V_i and u_i are derived from a Legendre polynomial of order $n + 1$, and are given by:

$$V_i = \frac{u_2 - u_1}{2} W_i \tag{15.112}$$

$$u_i = \frac{u_2 - u_1}{2} C_i + \frac{u_1 + u_2}{2} \tag{15.113}$$

C_i and W_i are the locations (coordinates in the sampling space) of sampling points and the weights associated with them, respectively. They have been published for values of $n = 1$ to $n = 16$. Table 15.2 shows values for $n = 1$ to 6 for C_i and W_i (to 15 decimal places).

Substituting Eqs. (15.112) and (15.113) into Eq. (15.111) and dropping R, we get:

$$I = \int_{u_1}^{u_2} f(u)\, du \approx \frac{u_2 - u_1}{2} \sum_{i=1}^{n} W_i f\left(\frac{u_2 - u_1}{2} C_i + \frac{u_1 + u_2}{2} \right) \tag{15.114}$$

Table 15.2   Gauss quadrature data.

Number of sampling points, n	Location C_i	Weight W_i
1	0.000000000000000	2.000000000000000
2	± 0.577350269189626	1.000000000000000
3	± 0.774596669241483	0.555555555555556
	0.000000000000000	0.888888888888889
4	± 0.861136311594053	0.347854845137454
	± 0.339981043584856	0.652145154862546
5	± 0.906179845938664	0.236926885056189
	± 0.538469310105683	0.478628670499366
	0.000000000000000	0.568888888888889
6	± 0.932469514203152	0.171324492379170
	± 0.661209386466265	0.360761573048139
	± 0.238619186083197	0.467913934572691

Gauss quadrature produces exact results for integrating polynomials according to the following rule: n sampling points yield exact results for polynomials of degree $\leq 2n - 1$. Thus, three-point ($n = 3$) Gauss quadrature should be adequate for mass property calculations. Figure 15.8 shows this case of three sampling points.

If the interval $[u_1, u_2]$ over which the integral is to be evaluated is large and the function $f(u)$ changes significantly over the interval, the interval can be divided into subintervals as shown in Figure 15.8b. Each subinterval could be chosen to have a width Δu of 1 or 2. Gauss quadrature is applied to each subinterval and the results are added. This is similar to the case of Simpson rule shown in Figure 15.7c.

In summary, in Newton-Cotes formulas, we use $(n + 1)$ equally spaced sampling points (as shown in Figure 15.7a and 15.7b), and polynomials of order at most n are integrated exactly. In Gauss quadrature n unequally spaced sampling points (Figure 15.8) are required to integrate a polynomial of order at most $(2n - 1)$. Polynomials of orders less than n and $(2n - 1)$ for the two methods, respectively, would also be integrated exactly. For both methods, if a nonpolynomial function is integrated, an error will result, the magnitude of which depends on how well the polynomial matches the function. In general, Gauss quadrature is usually more accurate than Newton-Cotes methods.

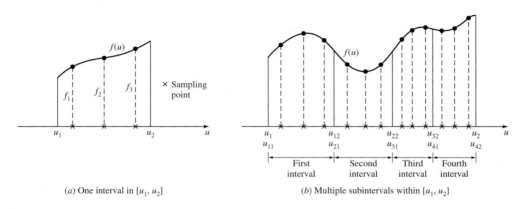

(a) One interval in $[u_1, u_2]$ (b) Multiple subintervals within $[u_1, u_2]$

Figure 15.8   Three-point Gauss quadrature.

The extension of Gauss quadrature to evaluate double integrals can be accomplished without much difficulty. Analogous to Eq. (15.111), the integral in Eq. (15.99) can be written as:

$$I = \int_{v_1}^{v_2} \int_{u_1}^{u_2} f(u, v)\, du\, dv = \sum_{i=1}^{n} \sum_{j=1}^{n} V_i V_j f_{ij} + R \tag{15.115}$$

where:

$$V_j = \frac{v_2 - v_1}{2} W_j \tag{15.116}$$

$$v_j = \frac{v_2 - v_1}{2} C_j + \frac{v_1 + v_2}{2} \tag{15.117}$$

and V_i and u_i are given by Eqs. (15.112) and (15.113), respectively. Thus, Eq. (15.115) can be written as :

$$I = \int_{v_1}^{v_2} \int_{u_1}^{u_2} f(u,\, v)\, du\, dv \approx \frac{(u_2 - u_1)(v_2 - v_1)}{4}$$

$$\times \sum_{i=1}^{n} \sum_{j=1}^{n} W_i W_j f\left(\frac{u_2 - u_1}{2}C_i + \frac{u_1 + u_2}{2},\, \frac{v_2 - v_1}{2}C_j + \frac{v_1 + v_2}{2}\right)$$

(15.118)

The sampling points for 2D Gauss quadrature is shown in Figure 15.9.

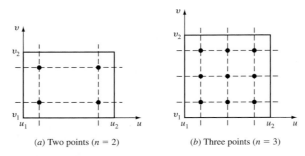

(a) Two points ($n = 2$)   (b) Three points ($n = 3$)

Figure 15.9   Sampling points for 2D Gauss quadrature.

Other integration schemes are available. One particular scheme is the Monte Carlo method discussed in Chapter 20. What kind of integration scheme should be used? It seems that Gauss quadrature with three sampling points is attractive for mass property calculations based on the formulation presented in this chapter.

EXAMPLE 15.1   **Calculate the length of a B-spline curve.**
Find the length of the B-spline curve developed in Example 6.19 in Chapter 6. The curve and its control points are shown in Figure 15.10.

SOLUTION   The B-spline curve equation is given by:

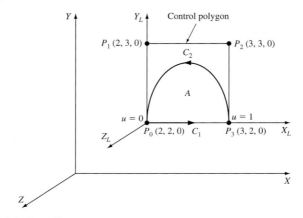

Figure 15.10   A cubic B-spline curve.

$$\mathbf{P}(u) = \mathbf{P}_0(1-u)^3 + 3\mathbf{P}_1 u(1-u)^2 + 3\mathbf{P}_2 u^2(1-u) + \mathbf{P}_3 u^3, \qquad 0 \le u \le 1 \qquad (15.119)$$

where $\mathbf{P}_0$, $\mathbf{P}_1$, $\mathbf{P}_2$, and $\mathbf{P}_3$ are shown in Figure 15.10. Using Eq. (15.119), the tangent vector is

$$\mathbf{P}' = (3\mathbf{P}_3 - 9\mathbf{P}_2 + 9\mathbf{P}_1 - 3\mathbf{P}_0)u^2 + (6\mathbf{P}_2 - 12\mathbf{P}_1 + 6\mathbf{P}_0)u + 3(\mathbf{P}_1 - \mathbf{P}_0) \qquad (15.120)$$

$$\begin{bmatrix} x' \\ y' \\ z' \end{bmatrix} = \begin{bmatrix} 6u - 6u^2 \\ 3 - 6u \\ 0 \end{bmatrix} \qquad (15.121)$$

Substituting Eq. (15.121) into Eq. (15.8) and reducing, the length of the B-spline is

$$L = 3\int_0^1 \sqrt{4u^4 - 8u^3 + 8u^2 - 4u + 1}\, du \qquad (15.122)$$

The integral in Eq. (15.122) can be evaluated numerically using Gauss quadrature with $f(u)$ being the square root function. Using two sampling points, we can write

$$L \approx \frac{3}{2}\sum_{i=1}^{2} W_i f_i = \frac{3}{2}(W_1 f_1 + W_2 f_2) \qquad (15.123)$$

Using Eq. (15.113) and Table 15.2, we calculate u_1 and u_2 as 0.211324865 and 0.788675134. Calculating f_1 and f_2 and knowing that $W_1 = W_2 = 1$, Eq. (15.123) gives

$$L \approx \frac{3}{2}(0.666666667 + 0.666666665) = 1.999999998$$

Using three sampling points, the length L is given by

$$L \approx \frac{3}{2}\sum_{i=1}^{3} W_i f_i = \frac{3}{2}(W_1 f_1 + W_2 f_2 + W_3 f_3) \qquad (15.124)$$

Note that u_1, u_2, and u_3 are calculated to be 0.112701665, 0.5, and 0.887298334 respectively. Thus,

$$L \approx \frac{3}{2}(0.444444445 + 0.444444444 + 0.444444444) = 2.0$$

Notice that because the B-spline is symmetric about an axis passing through the point $u = 0.5$ and the weights are symmetric, the resulting f_i are also symmetric. The reader is encouraged to apply the trapezoidal and Simpson rules and compare their accuracies with that of Gauss quadrature.

EXAMPLE 15.2   **Calculate the area and centroid of a bounded region.**

Calculate the area (and its centroid) bounded by the B-spline curve of Example 15.1 and the X_L axis, as shown in Figure 15.10.

SOLUTION   Let us assume the $X_L Y_L Z_L$ coordinate system has the same orientation as the MCS as shown in Figure 15.10. Thus, Eq. (15.28) applies and gives:

$$A = \sum_{i=1}^{2} \int_{u_{i1}}^{u_{i2}} xy' du \qquad (15.125)$$

The boundary consists of two curves: C_1 which is a straight line and C_2 which is the B-spline. Using Eq. (6.11) and (6.13), we can write for C_1:

$$\mathbf{P}(u) = \mathbf{P}_0 + u(\mathbf{P}_3 - \mathbf{P}_0), \qquad 0 \le u \le 1 \qquad (15.126)$$

$$\mathbf{P}' = \mathbf{P}_3 - \mathbf{P}_0 \qquad (15.127)$$

Substituting Eqs. (15.119), (15.120), (15.126), and (15.127) into Eq. (15.125) and reducing, we get:

$$A = \int_0^1 [x_0 + u(x_3 - x_0)](y_3 - y_0)\, du$$
$$+ \int_1^0 x_0(1-u)^3 + 3x_1 u(1-u)^2 + 3x_2 u^2(1-u) + x_3 u^2](3-6u)\, du \qquad (15.128)$$

The first term on the right-hand side of Eq. (15.128) is zero because the coordinates y_3 and y_0 are equal. The limits of the second integral reflect the fact that the curves C_i of the boundary must be traversed in a counterclockwise order so that the area usually lies to the left of any boundary curve. Substituting the values for the x coordinates and reducing, Eq. (15.128) becomes:

$$A = 3 \int_1^0 (4u^4 - 8u^3 + 3u^2 - 4u + 2)\, du \qquad (15.129)$$

Exact integration of Eq. (15.129) gives $A = 0.6$. Let us compare this result with two- and three-sampling-points Gauss quadrature. For two points, we get

$$A \approx -\tfrac{1}{2}(W_1 f_1 + W_2 f_2) \qquad (15.130)$$

Note that u_1 and u_2 are given by 0.211324865 and 0.788675134, respectively. Calculating f_1 and f_2 and substituting into Eq. (15.130), we get:

$$A \approx -\frac{3}{2}\,(1.221153452\text{-}1.665597891) = 0.666666659$$

For three-point integration, the area is

$$A \approx -\tfrac{3}{2}(W_1 f_1 + W_2 f_2 + W_3 f_3) \qquad (15.131)$$

Note that u_1, u_2, and u_3 are 0.112701665, 0.5, and 0.887298334, respectively. Equation (15.131) gives

$$A \approx -\frac{3}{2}\,(0.875828708 + 0 - 1.275828689) = 0.599999971$$

This area should be exact, that is, 0.6 because, as mentioned earlier, using $n = 3$ should yield exact results for polynomials of degree ≤ 5. The polynomial in this example [Eq. (15.129)] is of degree 4. The difference is due to truncation and round-off errors. On the other hand, the accuracy of the two-point integration is not as good as expected.

The centroid of the area is given by Eq. (15.31), which reduces to:

$$x_c = \frac{1}{A} \sum_{i=1}^{n} \int_{u_{i1}}^{u_{i2}} \frac{x^2}{2} y' \, du$$

$$y_c = \frac{1}{A} \sum_{i=1}^{n} \int_{u_{i1}}^{u_{i2}} xyy' \, du \qquad (15.132)$$

$$z_c = \frac{1}{A} \sum_{i=1}^{n} \int_{u_{i1}}^{u_{i2}} xzy' \, du$$

$z_c = 0$ because the area lies in the XY plane. Similar to reducing Eq. (15.128), Eq. (15.132) reduces to:

$$x_c = \frac{3}{2A} \int_1^0 (-8u^7 + 28u^6 - 30u^5 + 25u^4 - 32u^3 + 12u^2 - 8u + 4) \, du \qquad (15.133)$$

$$y_c = \frac{3}{A} \int_1^0 (-12u^6 + 36u^5 - 25u^4 + 5u^3 - 12u^2 - 2u + 4) \, du \qquad (15.134)$$

Exact integration of Eqs. (15.133) and (15.134) gives $x_c = 2.5$ and $y_c = 2.321428571$. Notice that due to symmetry of the area with respect to the axis given by the equation $x_L = 0.5$ (or $x = 2.5$), the centroid lies on this axis. Thus $x_c = 2.5$ is correct as expected.

Let us compare these exact results with three-sampling-points Gauss quadrature. Applying the same Gauss data we needed for calculating A to Eqs. (15.133) and (15.134), we obtain:

$$x_c = -\frac{3}{4A}(1.782523371 + 0 - 3.782523365)$$

$$= -\frac{3 \times -1.999999994}{4 \times 0.599999971} = 2.500000113$$

$$y_c = -\frac{3}{2A}(2.014406029 + 0 - 2.934406026)$$

$$= -\frac{3 \times -0.919999997}{2 \times 0.599999971} = 2.300000104$$

In summary, the exact area A is 0.6 and the exact centroid is (2.5, 2.321428571, 0) with respect to the MCS (XYZ system). And three-sampling-points Gauss quadrature gives excellent results.

EXAMPLE 15.3   **Calculate the surface area of a bounded region.**

A plane passes through the three points $P_0(1, 2, 3)$, $P_1(2, 4, 5)$, and $P_2(4, 2, 3)$. Find the surface area that is bounded by the parametric domain $u = [0, 1]$ and $v = [0, 1]$.

SOLUTION   The plane equation is given by Eq. (7.25), that is,

$$\mathbf{P}(u, v) = \mathbf{P}_0 + u(\mathbf{P}_1 - \mathbf{P}_0) + v(\mathbf{P}_2 - \mathbf{P}_0), \qquad \begin{cases} 0 \le u \le 1 \\ 0 \le v \le 1 \end{cases} \qquad (15.135)$$

Equation (15.135) gives

$$\mathbf{P}_u = \mathbf{P}_1 - \mathbf{P}_0 \qquad \mathbf{P}_v = \mathbf{P}_2 - \mathbf{P}_0 \tag{15.136}$$

Substituting into Eq. (15.37), we get

$$A_s = C \int_0^1 \int_0^1 du\, dv = C \tag{15.137}$$

where $C = \sqrt{[(x_1 - x_0)^2 + (y_1 - y_0)^2 + (z_1 - z_0)^2][(x_2 - x_0)^2 + (y_2 - y_0)^2 + (z_2 - z_0)^2]} = 9$

Thus, $A_s = 9$.

Let us use Gauss quadrature with two sampling points. Equation (15.117) gives

$$A_s = \frac{C}{4} \sum_{i=1}^{2} \sum_{j=1}^{2} W_i W_j f_{ij} = \frac{C}{4}(W_1 W_1 f_{11} + W_1 W_2 f_{12} + W_2 W_1 f_{21} + W_2 W_2 f_{22}) \tag{15.138}$$

From Eq. (15.137), $f(u, v) = 1$. Thus, $f_{11} = f_{12} = f_{21} = f_{22} = 1$. Therefore, Eq. (15.138) gives:

$$A_s = \frac{C}{4}\ (1.000000000 + 1.000000000 + 1.000000000 + 1.000000000) = C = 9$$

which is the exact answer as expected.

15.5 Properties of Composite Objects

Frequently an object can be divided into two or more simple subobjects whose integral properties can be readily or easily obtained. For example, a 3D object may be decomposed into few 2½D subobjects. The original object is referred to as a composite object.

When an object can be divided into a number of simple objects, the volume of the object is the sum of the volumes ot its subobjects (if subobjects such as holes are removed, their corresponding volumes are subtracted), that is,

$$V = \sum_{i=1}^{n} V_i \tag{15.139}$$

where V, V_i, and n are the volume of the object, the volume of ith subobject, and the number of subobjects, respectively. The centroid of the object is given by:

$$\mathbf{r}_c = \frac{\displaystyle\sum_{i=1}^{n} \mathbf{r}_{ci} V_i}{\displaystyle\sum_{i=1}^{n} V_i} = \frac{\displaystyle\sum_{i=1}^{n} \mathbf{r}_{ci} V_i}{V} \tag{15.140}$$

or

$$x_c = \frac{\displaystyle\sum_{i=1}^{n} x_{ci}V_i}{V} \qquad y_c = \frac{\displaystyle\sum_{i=1}^{n} y_{ci}V_i}{V} \qquad z_c = \frac{\displaystyle\sum_{i=1}^{n} z_{ci}V_i}{V} \qquad (15.141)$$

where r_{ci} is the position vector of the centroid of the ith subobject with respect to the MCS. The first moments, second moments, and products of inertia are given by:

$$M_{xy} = \sum_{i=1}^{n} (M_{xy})_i \qquad M_{xz} = \sum_{i=1}^{n} (M_{xz})_i \qquad M_{yz} = \sum_{i=1}^{n} (M_{yz})_i \qquad (15.142)$$

$$I_{xx} = \sum_{i=1}^{n} (I_{xx})_i \qquad I_{yy} = \sum_{i=1}^{n} (I_{yy})_i \qquad I_{zz} = \sum_{i=1}^{n} (I_{zz})_i \qquad (15.143)$$

$$I_{xy} = \sum_{i=1}^{n} (I_{xy})_i \qquad I_{xz} = \sum_{i=1}^{n} (I_{xz})_i \qquad I_{yz} = \sum_{i=1}^{n} (I_{yz})_i \qquad (15.144)$$

In Eqs. (15.142) and (15.144), the sums are algebraic sums because first moments and products of inertia may be positive or negative.

15.6 Mass Properties on CAD/CAM Systems

CAD/CAM systems typically calculate the mass properties formulated in Sections 15.2 and 15.3. Several mass properties commands and/or modifiers exist on CAD/CAM systems to support calculations for 2D planar areas, 2½D objects, and 3D objects.

In using mass property commands, it is the user's responsibility to ensure the correctness of units of the density so the proper units for the mass and inertial properties are produced. It is also the user's responsibility to be aware of the errors that may occur in the calculations as discussed in this chapter and to try to reduce these errors as much as the error analysis permits.

CAD systems provide users with options to control the calculations of the mass properties of CAD models. These options include the choice of measurement units for length (mm or inches) and angles (radians or degrees), the choice of decimal places (two, three, four, and so forth), and the choice of density units (lb/in^3 or g/mm^3). The screenshot on the right shows an example of these options.

EXAMPLE 15.4   **Calculate the mass properties of a uniform-thickness object.**

Calculate the mass properties of the solid model shown in Figure 15.11. Use a density of $\rho = 1.0$ lb/in^3.

SOLUTION   The model shown is a 2½D model with uniform thickness of 2 inches. Figure 15.11 shows the model geometry and its mass properties as calculated by SolidWorks.

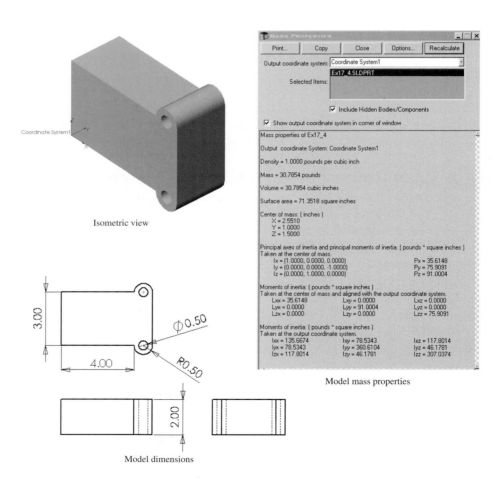

Isometric view

Model mass properties

Model dimensions

Figure 15.11   Mass properties of a uniform-thickness model.

EXAMPLE 15.5   **Calculate the mass properties of an axisymmetric object.**

Figure 15.12 shows a cross section (lies in the *YZ* plane) that is rotated about an axis parallel to the *X* axis at a distance $y = -2$. The cross section is rotated an angle of 120°. The beginning and ending angles of rotation are 30° and 150°, respectively. Calculate the mass properties of the resulting model. Use a density of $\rho = 1.0 \text{ lb/in}^3$.

SOLUTION   The model shown is an axisymmetric 2½D model. Figure 15.12 shows the model geometry and its mass properties as calculated by SolidWorks.

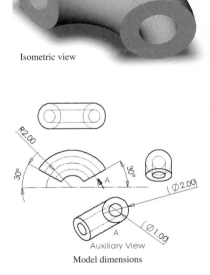

Isometric view

Model dimensions

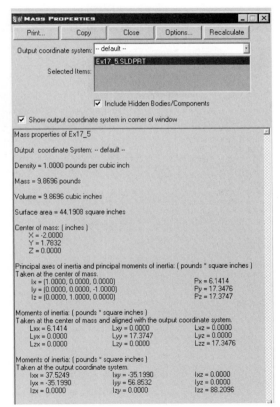

Model mass properties

Figure 15.12   Mass properties of an axisymmetric model.

| EXAMPLE 15.6 | **Calculate the mass properties of a 3D object.** |

Figure 15.13 shows a 3D model of density 1.0 lb/in³. Calculate the mass properties of the model.

| SOLUTION | The model shown is a general 3D model. Figure 15.13 shows the model geometry and its mass properties as calculated by SolidWorks.

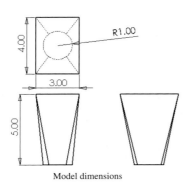

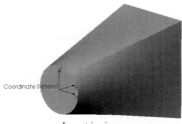

Isometric view

Model dimensions

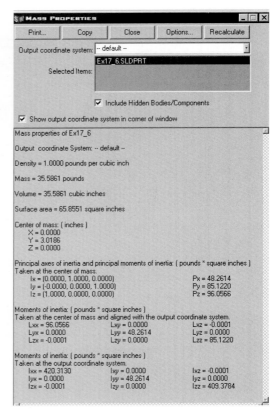

Model mass properties

Figure 15.13 Mass properties of a 3D model.

EXAMPLE 15.7	**Calculate the mass properties of a composite object.**

Figure 15.14 shows a composite 2½D model. Calculate the mass properties of the object assuming a density of 1 lb/in^3.

SOLUTION	The model shown is a composite 2½D model. Figure 15.14 shows the

model geometry and its mass properties as calculated by SolidWorks.

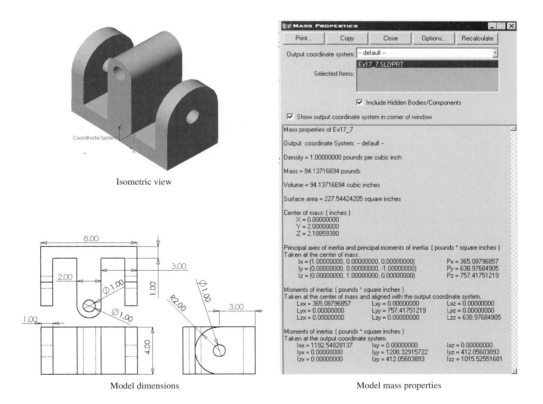

Figure 15.14   Mass properties of a composite 2½D model.

Problems

Part I: Theory

15.1 Derive Eq. (15.31) in detail.

15.2 Derive Eq. (15.88) in detail.

15.3 How would Eqs. (15.88), (15.91), and (15.92) change if the arbitrary axis is defined by two points instead of a point and a unit vector **n** ?

15.4 How does Eq. (15.31) reduce for the following cases:

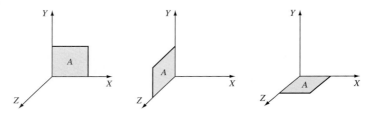

15.5 Prove that Eq. (15.37) can be rewritten as:

$$A_s = \int_v^{v_2} \int_{u_1}^{u_2} |J| \, du \, dv$$

where $|J|$ is the determinant of the jacobian $[J]$ which is given by the equation

$$[J] = \|P_u \times P_v\| = \begin{vmatrix} \hat{i} & \hat{j} & \hat{k} \\ \dfrac{\partial x}{\partial u} & \dfrac{\partial y}{\partial u} & \dfrac{\partial z}{\partial u} \\ \dfrac{\partial x}{\partial v} & \dfrac{\partial y}{\partial v} & \dfrac{\partial z}{\partial v} \end{vmatrix}$$

Hint: Expand $(P_u \cdot P_u)(P_v \cdot P_v)$ in Eq. (15.37) and expand $|J|$. Then use the fact that $P_u \cdot P_v = 0$ to prove the equality.

15.6 Derive the principal moments of inertia of an object given its moments about a coordinate system.

15.7 Use Eq. (15.9) to find the curve length for all analytic and synthetic curves covered in Chapter 6. Which are exact and which are approximate?

15.8 Use equations listed in Table 15.1 to find properties of objects made of surfaces covered in Chapter 7. Which are exact and which are approximate?

15.9 Evaluate the following integrals using the trapezoidal rule, Simpson rule, and Gauss quadrature (with one, two, and three sampling points). Solve the integrals exactly and compare.

a. $\displaystyle\int_0^1 (4 + 5u^2)\, du$

b. $\displaystyle\int_0^{\pi/2} \cos u\, du$

c. $\displaystyle\int_{-1}^1 (8u^3 + 5u^2)\, du$

d. $\displaystyle\int_0^2 u e^{2u}\, du$

e. $\displaystyle\int_0^6 \left(\frac{u^2}{4} - 4u + 16\right) du$

f. $\displaystyle\int_0^1 u^5\, du$

g. $\displaystyle\int_{-1}^2 (u + 2 - u^2)\, du$

15.10 Evaluate the following integrals using Gauss quadrature with two and three sampling points. Solve the integrals exactly and compare the results.

a. $\displaystyle\int_0^1 \int_0^1 (3u^3 v + 4u^2 v^3)\, du\, dv$

b. $\displaystyle\int_0^{\pi} \int_0^2 r^2 \sin\theta\, dr\, d\theta$

(This is the first moment of half a circle of radius 2 about the X axis.)

15.11 A cubic spline curve in the XY plane of the MCS passes through the origin and is described by the following equation:

$$\mathbf{P}(u) = \begin{bmatrix} 0.03u^3 + 0.3u^2 + 1.5u \\ -0.05u^3 + 1.5u \\ 0 \end{bmatrix}, \quad 0 \le u \le 1$$

Calculate the curve length, the area under the spline, and its centroid. Compare the exact results with the three-sampling-points Gauss quadrature.

15.12 A sphere with a radius R and a center at (x_0, y_0, z_0) is described by the following equation:

$$\mathbf{P}(u, v) = \begin{bmatrix} x_0 + R\cos u \cos v \\ y_0 + R\cos u \sin v \\ z_0 + R\sin u \end{bmatrix}, \quad \begin{cases} -\pi/2 \le u \le \pi/2 \\ 0 \le v \le 2\pi \end{cases}$$

For $R = 1$ and center at $(1, 1, 1)$, calculate the surface area and the centroid exactly. Use the three-sampling-points Gauss quadrature. Compare the results.

15.13 Given a volume element $dV = dx\, dy\, dz$, prove that $dV = dx\, dy\, dz = |J|\, du\, dv\, dw$ where $|J|$ is given by the following equation:

$$[J] = \begin{bmatrix} \dfrac{\partial x}{\partial u} & \dfrac{\partial y}{\partial u} & \dfrac{\partial z}{\partial u} \\[2ex] \dfrac{\partial x}{\partial v} & \dfrac{\partial y}{\partial v} & \dfrac{\partial z}{\partial v} \\[2ex] \dfrac{\partial x}{\partial w} & \dfrac{\partial y}{\partial w} & \dfrac{\partial z}{\partial w} \end{bmatrix}$$

Part II: Laboratory

Use your in-house CAD/CAM system to answer the questions in this part.

15.14 Calculate the mass properties of models from Chapters 6, 7, and 9.

Part III: Programming

Write the code so that the user can input values to control the output. The program should display the results. Use OpenGL with C or C++, or use Java 3D.

15.15 Write various computer programs that implement the equations listed in Table 15.1. Use exact solutions if possible, or use Gauss quadrature with three sampling points.

Assembly Modeling

**OAL**
Understand and master assemblies, the different approaches to creating them, assembly analysis, and how to use CAD systems to create assembly models.

◉BJECTIVES

After reading this chapter, you should understand the following concepts:

- Differences between part and assembly modeling
- Mating conditions
- Bottom-up assembly modeling approach
- Top-down assembly modeling approach
- WCS and mate methods to assemble parts
- Managing assemblies
- Working with subassemblies
- Assembly analysis

CHAPTER HEADLINES

16.1   Introduction

In most engineering designs, the product of interest is a composition of parts, formed into an assembly. Modeling and representing assemblies as well as analyzing assemblies are all relevant issues to geometric modeling and the CAD/CAM technology. They form the focus of this chapter. Parts and/or subassemblies of a given product can be modeled separately, most often by different members of the design team, on a CAD/CAM system. Instances (copies) of these parts can then be merged into a base part or a host to generate the assembly model.

Assembly modeling is considered an extension of part modeling. CAD/CAM systems act as assembly modelers, in addition to being geometric modelers. These systems provide users with an assembly mode to create assemblies. (Other modes are part and drawing modes.) After a designer creates all the individual parts of an assembly, the designer uses them to create the assembly, as we discuss in this chapter.

Assembly modeling raises two modeling issues that do not exist at the part modeling level: hierarchy and mating. These two issues distinguish assembly modeling from part modeling. Individual parts and subassemblies must be assembled in the right hierarchy (sequence), which is captured (stored) in an assembly tree for each assembly or product. The assembly tree may not be unique, as there may be more than one sequence to create the same assembly.

Mating conditions are used to determine the mating (spatial relationships and orientations) between the assembly parts. For example, the axes of a shaft and a hole may have to be lined up, in which case a concentric mating condition is required. Two faces may have to be planar, in which case a planar (coplanar) condition is used.

16.2   Assembly Modeling

An **assembly** is a collection of independent parts. It is important to understand the nature and the structure of dependencies between parts in an assembly to be able to model the assembly properly. In order to determine, for example, whether a part can be moved and which other parts will move with it, the assembly model must include the spatial positions and hierarchical relationships among the parts, and the assembly (attachment) relationships (mating conditions) between parts.

Figure 16.1 shows how an assembly model can be created using a CAD system. Designers first create the individual parts. They can also analyze the parts separately. Once the parts design is complete, designers can proceed to create the assembly and analyze it. Creating the assembly from its parts requires specifying the spatial and mating relationships between the parts. Assembly analysis may include interference checking, mass properties, kinematic and dynamic analysis, and finite element analysis. CAD systems establish a link between an assembly and its individual parts such that designers need only change individual parts for design modification, and the system updates the assembly model automatically.

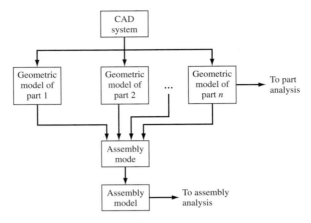

Figure 16.1   Creating an assembly model.

16.3 Assembly Tree

The most natural way to represent the hierarchical relationships between the various parts of an assembly is an assembly tree, as shown in Figure 16.2. An assembly is divided into several subassemblies at different levels (shown in Figure 16.2, as the tree depths). Each subassembly at depth $(n - 1)$ is composed of various parts. The leaves of the tree represent individual parts or subassemblies. The nodes of the tree represent parts and/or subassemblies, and its root represents the assembly itself. The assembly is located at the top of the tree at depth 0 or at the highest hierarchy n of the assembly sequence.

Figure 16.3 shows an electric clutch assembly. The clutch consists of three main elements: the field, the rotor, and the armature. The field and the coil are held stationary, and the rotor is driven by the electric motor through the gear, pinion, and rotor shaft. The armature is attached to the load shaft through the hub. The load shaft carries the load (not shown) to be overcome by the clutch. The assembly tree for this clutch is shown in Figure 16.4. The tree represents an assembly sequence by which the clutch assembly can be produced. The assembly tree is not unique; other valid assembly sequences can be generated.

16.4 Assembly Planning

Assembly planning is a key to creating successful assemblies, especially the large ones that are typically encountered in practice. The important issue is not only creating the assembly, but also updating it in the future when design changes are made to the individual parts. These updates should be done automatically and correctly. As with the models of individual parts, an assembly model should be fully parametric and flexible. This means that the relations between the assembly parts should be easy to change and update. When a designer changes some of the assembly parameters, the others should update accordingly.

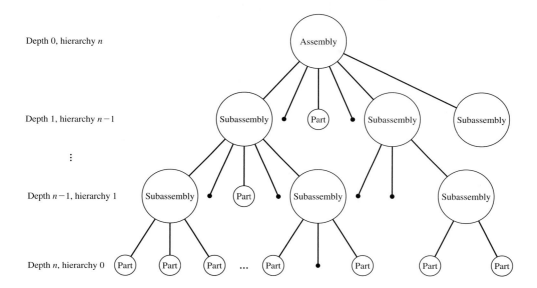

Figure 16.2   Assembly tree.

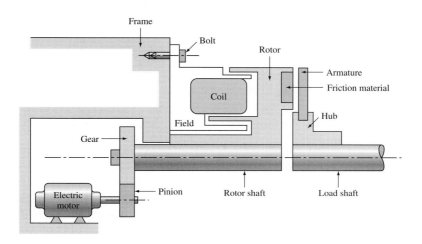

Figure 16.3   Electric clutch assembly.

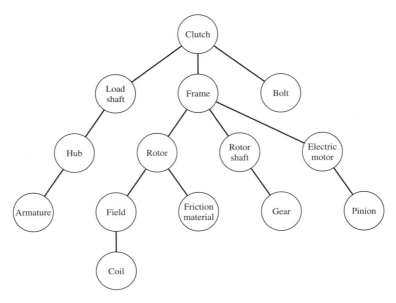

Figure 16.4   Electric clutch assembly.

Before we begin to create an assembly, we should consider the following issues:

1. **Identify the dependencies between the components of an assembly.** Assembly components are individual parts or entire subassemblies that, in turn, consist of other subassemblies and/or parts, as shown in Figure 16.2. Identifying these dependencies helps us to determine the best approach to create the assembly: bottom-up, top-down, or both. We discuss these two approaches in detail in the coming sections. Consider, for example, the assembly of two blocks and a bolt that holds them together. The bolt size depends on the hole size in each block. Thus, the best approach to create this assembly is to create the two blocks as separate parts, and merge them into a blank assembly model; this is the bottom-up approach. We then create the bolt in the assembly model to relate its size to that of the holes; this is the top-down approach.

2. **Identify the dependencies between the features of each part.** These dependencies include symmetry and geometric arrays (patterns). Understanding these dependencies helps us to optimize the creation of assemblies. Consider the creation of a part that has a rim with four holes. We can create the four holes as separate features in the rim, or we create one hole and use it as a base feature in a circular array (pattern) command. The latter approach is better because it creates a dependency between the base hole and the other three. When we use the rim part in an assembly, we can make one change, the size (diameter) of the base hole, and the rest of the rim part is updated automatically.

3. **Analyze the order of assembling the parts.** This order determines the ease of the assembly process on the shop floor. It also determines the cost of creating the assembly. The order also affects the manufacturing processes of the parts. For example, if an assembly requires a part with a flat, blind hole, the hole must be milled with a flat-end mill-cutting tool and cannot be drilled.

How much planning ahead should we consider before beginning creating an assembly? The more the better. Thorough planning avoids many impossible impasses later on. While planning may not seem that important for simple assemblies that consist of a small number of parts, this is not the case in large assemblies that contain thousands or tens of thousands of parts, such as airplanes and automobiles. Think of assembly planning as an important activity of the conceptual design phase of a product, and not as an afterthought process of parts modeling.

16.5 Mating Conditions

Individual parts of an assembly are usually created separately using a CAD/CAM system and then merged (assembled) together, using a `merge` or `insert` command, to form the assembly. Each part has its own database with its own MCS. Typically, the user selects one of the parts as a base part (host) and merges the other parts into it. Alternatively, the user can begin with a blank part as the host. The MCS of the host becomes the global coordinate system, that is, the MCS of the assembly. A part MCS becomes a local coordinate system for this part.

The final correct position of each part in the assembly is obtained by locating and orienting its MCS properly with respect to the global coordinate system of the assembly. Figure 16.5 shows an example. The XYZ is the global coordinate system of the database of the assembly model. Its origin O is the (0, 0, 0) point. The $X_1Y_1Z_1$, $X_2Y_2Z_2$, $X_3Y_3Z_3$ and $X_4Y_4Z_4$ are local coordinate systems of four parts that make the assembly. Their origins O_1, O_2, O_3, and O_4 are located properly relative to the assembly origin O, and their orientations relative to the XYZ coordinate system reflect the proper orientations of the parts in their assembly.

Locating and orienting parts in their assembly is achieved by specifying mating conditions among them. These conditions specify the spatial relationships among the parts. Mating conditions can be provided interactively with ease because they use simple geometric entities such as faces and centerlines. For example, a mating condition can consist of planar faces butting up against one another, or requiring centerlines of individual parts to be collinear.

The most common mating conditions are `coincident`, `concentric`, `tangent`, `coplanar`, `parallel faces`, and `perpendicular faces`. Some CAD systems may use these names, different names, and/or additional mating conditions. CAD users should consult with their respective systems.

When using mating conditions to assemble two parts, keep in mind that there are six degrees of freedom in E^3: three translations along the axes of the assembly MCS and three rotations about the same axes. A part should be fully constrained to allow the creation of the correct assembly. Mating conditions between two parts may require, for example, a coincident condition of two faces and two points on the faces. If we mate the two faces using the coincident condition only, they can still rotate relative to each other.

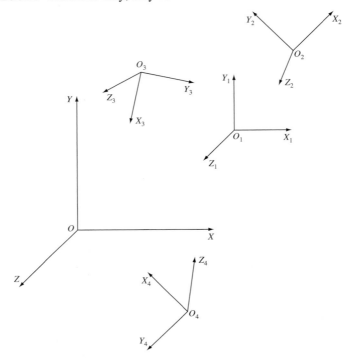

Figure 16.5   Assembling parts.

The coincident mating condition is applied between two planar faces, or between a planar face and a cylindrical face (shaft). This condition is illustrated in Figure 16.6. Part 1 and Part 2 have the MCSs $X_1 Y_1 Z_1$ and $X_2 Y_2 Z_2$, respectively. The hatched faces are the faces to be mated. Each face is specified by its unit normal vector and any one point on the face with respect to the part MCS. The planar face of Part 1 is specified by the unit normal $\hat{\mathbf{n}}_1$ and by the point P_1 with respect to the $X_1 Y_1 Z_1$ coordinate system. Similarly, the planar face of Part 2 is specified by $\hat{\mathbf{n}}_2$ and P_2 with respect to the $X_2 Y_2 Z_2$ coordinate system. The coincident condition is satisfied by forcing $\hat{\mathbf{n}}_1$ and $\hat{\mathbf{n}}_2$ to be opposite each other, and the two faces touch each other such that P_1 and P_2 are coincident.

The concentric mating condition holds between two cylindrical faces: a shaft cylindrical face and a hole cylindrical face, as shown in Figure 16.7. The concentric mating condition is achieved by forcing the shaft and hole axes to be collinear. Each axis is specified by two points. The hole axis is specified by the two points P_1 and P_2 defined with respect to the $X_1 Y_1 Z_1$ MCS. Similarly, the shaft axis is specified by the two points P_3 and P_4 with respect to the $X_2 Y_2 Z_2$ MCS.

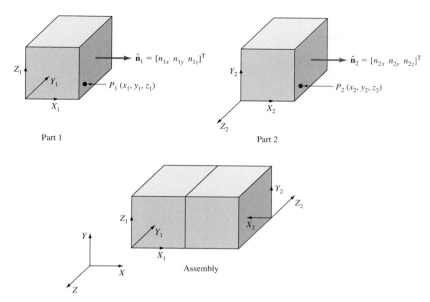

Figure 16.6 Coincident mating condition.

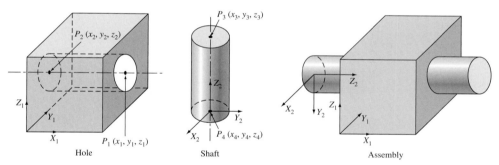

Figure 16.7 Concentric mating condition.

The tangent mating condition is applicable between two planar/cylindrical or cylindrical/cylindrical faces, as shown in Figure 16.8. The tangent mating condition is achieved by forcing a cylindrical face to be tangent to a planar (flat) face. The difference between the tangent and coincident mating conditions is that the former uses at least one cylindrical face, while the latter uses two planar (flat) faces. Figure 16.8 shows the tangent mating between cylindrical and planar faces.

The coplanar mating condition holds between two planar faces when they lie in the same plane. This condition is illustrated in Figure 16.9. It is similar to the coincident condition except that the points P_1 and P_2 are chosen to lie on the two edges to mate. The coplanar condition is the complement (opposite) of the coincident condition and is satisfied by forcing the two normals $\hat{\mathbf{n}}_1$ and $\hat{\mathbf{n}}_2$ to be in the same direction.

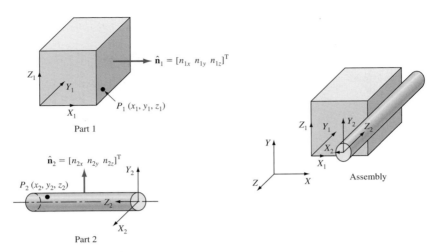

Figure 16.8 Tangent **mating condition.**

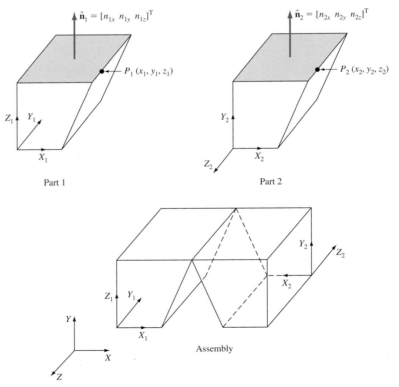

Figure 16.9 Coplanar **mating condition.**

The coincident, concentric, tangent, and coplanar mating conditions as described allow both rotational and translational freedom of movement between the mating parts. In the coincident mating condition shown in Figure 16.6, Part 2 can slide on Part 1 or rotate relative to it after the two faces designated by the two normals $\hat{\mathbf{n}}_1$ and $\hat{\mathbf{n}}_2$ are mated. Similarly, in the concentric condition shown in Figure 16.7, the shaft can slide and/or rotate inside the hole. These additional movements can be eliminated by adding more constraints to each of these mating conditions.

The parallel faces mating condition is similar to the coincident mating condition except that the two mating faces are not in contact with one another. This condition is achieved by forcing the normals of the two faces to be parallel and opposite. If we separate the two mating faces shown in Figure 16.6, we have a parallel faces mating condition. The perpendicular faces mating condition is specified by requiring two faces to be perpendicular to each other. The two shaded faces of Part 1 and Part 2 shown in Figure 16.8 are made perpendicular to each other by forcing their normals $\hat{\mathbf{n}}_1$ and $\hat{\mathbf{n}}_2$ to be perpendicular.

16.6 Bottom-Up Assembly Approach

Three assembly approaches exist: bottom-up, top-down, or a combination of both. Each approach has its advantages and disadvantages. This section covers the first approach. Section 16.7 covers the second approach.

The bottom-up approach is most common as it is the traditional and most logical approach. In this approach, we create the individual parts independently, insert them into an assembly, and use the mating conditions to locate and orient them in the assembly as required by the assembly design.

The assembly modeling process itself begins with creating a blank assembly model (file) using the assembly mode of a CAD system. We import (insert) the assembly parts into this model, one at a time. The first part we insert is known as the base part or the host, on top of which other parts are assembled. We use the proper mating conditions to place and orient each inserted part correctly in the assembly model.

When we insert parts into the assembly, we insert copies of the parts. These copies are known as instances. We can use multiple instances of any part if the assembly requires it. The CAD software maintains a link (known as a pointer in database and programming jargon) between each instance and its original part. If we change the original part, we can change all its instances in an assembly by simply updating the assembly. These links (call them assembly links) between an assembly and its individual parts is the most important fundamental concept behind assembly modeling. As a matter of fact, an assembly link is bidirectional; we change the part and update the assembly, or we change the instance in the assembly and update the part.

How do we specify assembly constraints among different parts in the bottom-up approach? We specify them among the instances of the parts in the assembly model itself. We cannot specify them otherwise because each part has its own file. These constraints are required to maintain the correct proportions among the assembly parts (see Example 16.3).

The bottom-up approach has some advantages. It is the preferred technique if the parts have already been constructed, as in the case of off-the-shelf parts. It also allows designers to focus on the individual parts. It also makes it easier and simpler to maintain the relationships and regeneration behavior of parts than in the top-down approach.

EXAMPLE 16.1   **Use the bottom-up assembly approach.**
         Use the bottom-up approach to create the block-plate-pin assembly model shown below.

SOLUTION   The block-plate-pin assembly shown below has three parts: block, plate, and pin. The pin holds the other two parts together.

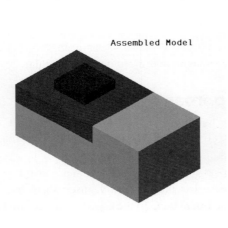

Assembled Model

Exploded View

We create the three individual parts of the assembly first and save each one in its own file. Parts dimensions are shown at the end. Here is how to create the assembly model.

Modeling strategy:
1. Log in and start the CAD/CAM system.
2. Select the assembly mode.
3. Open a new assembly file.
4. Use an `Insert => Component => From File` command (or its equivalent) to insert the block.
5. Repeat Step 4 for the plate. The assembly state, after Steps 1 to 5, is shown on the right.
6. Mate and align the plate with the block. Use a `coincident` mating condition between `Face 1` and `Face 2`. Also position the two instances close together using a `Closest` alignment command.
7. Repeat Step 6 for `Face 3` and `Face 4`.

Face 4   Face 2

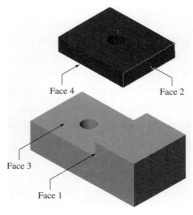

Face 3

Face 1

8. Mate the holes in the block and the plate. Use a `concentric` mating condition. The assembly state, after Steps 6 to 8, is shown below.

9. Insert the pin into the assembly using an `Insert` command or its equivalent.

10. Mate the pin with its hole. Use a `concentric` mating condition between the two. Also, use a `coincident` mating condition between `Face 5` and `Face 6` as shown below.

11. Save all the files and exit the CAD system.

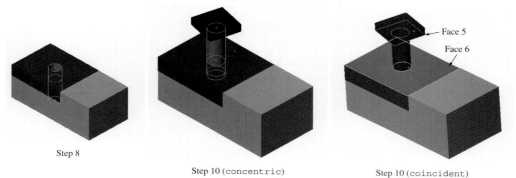

Face 5

Face 6

Step 8   Step 10 (concentric)   Step 10 (coincident)

Example 16.1 discussion:

Instead of using an `Insert` command as shown in Step 4, some CAD systems, such as SolidWorks, allow you to drag components from their files and drop them onto the assembly file. Both the assembly and components files must be open.

The assembly of each component requires enough mating conditions to eliminate the undesired translations (movements) and rotations. For example, Steps 6 and 7 use two `coincident` mating conditions and four faces to fully constrain the relative movements between the block and the plate. Similarly, Step 10 uses `concentric` and `coincident` mating conditions to mate the pin in its hole.

The assembly tree shows the state of each component whether it is fully constrained or not in the assembly model. Some CAD systems, such as SolidWorks, use the prefix *f* or "–" in a component name in the assembly tree to indicate its state. *f* means the component is fixed, that is, it does not move. There is always one fixed component in each assembly. It is the base component or the host. It is the first component that you insert. "–" means that the component is floating, that is, it needs more mating conditions to fully constrain it. A fully constrained component does not have a prefix in its name. The screenshot on the right shows the assembly tree for this example.

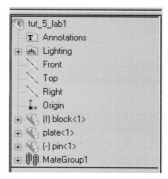

CAD systems allow you to move and rotate components to check their available degrees of freedom after applying mating conditions. For example, the pin can only rotate about its axis.

The parts dimensions are shown in the forthcoming screenshot. Use them for the hands-on exercise of this example.

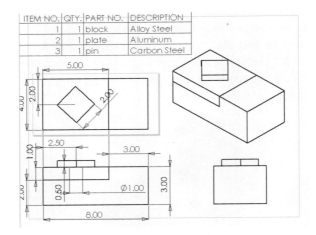

ITEM NO.	QTY.	PART NO.	DESCRIPTION
1	1	block	Alloy Steel
2	1	plate	Aluminum
3	1	pin	Carbon Steel

Example 16.1 hands-on exercise:

1. Apply `move` and `rotate` commands to the three parts to check their assembly states.
2. Can you find other mating conditions that you can use to create the assembly?
3. Use subassemblies to create the assembly. Create a subassembly of the block and the plate. Save the subassembly in a file called *sub1*. Open a new assembly file. Insert *sub1* and the pin into it.
4. Create a slot feature in the assembly without editing the parts or subassemblies files. Do it in the assembly itself. Use a command such as `Insert => Assembly feature => Cut` or its equivalent. The screenshot on the right shows the cut. The slot is one inch wide.

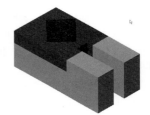

EXAMPLE 16.2   **Use multiple instances of a part in an assembly.**

       Use the bottom-up approach to create the assembly model of the universal joint shown in the forthcoming screenshot.

SOLUTION   The universal-joint assembly shown on the following page has seven parts: two yokes, a center block, a main pin, a pin, and two bushings to hold the pin in place. The pin goes through the main pin in its center hole. This locks the main pin in place. We create only five individual parts of the assembly first and save each one in its own file. We use two instances of each of the yoke and bushing parts during the assembly process. Here is how to create the assembly model.

Modeling strategy:

1. Log in and start the CAD/CAM system.

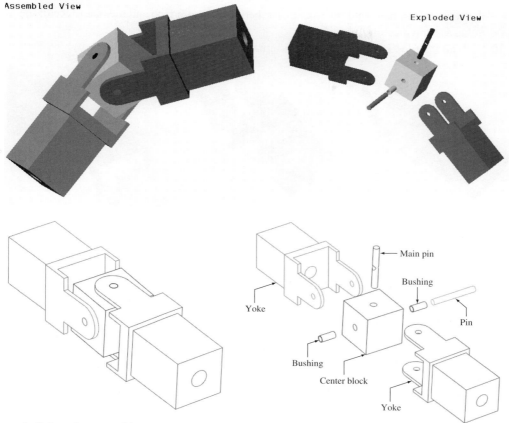

Assembled View

Exploded View

Main pin

Bushing

Pin

Yoke

Bushing

Center block

Yoke

2. Select the assembly mode.

3. Open a new assembly file.

4. Use an `Insert => Component => From File` command (its equivalent, or drag and drop) to insert one yoke and the center block. Apply a `concentric` mate between the holes in the yoke forks and the center block. The assembly state is shown on the following page.

5. Insert the main pin into the assembly. Apply two `concentric` mates: one between the main pin and a hole of the center block, and one between the main pin again and the other hole of the center block. The screenshot that follows shows the result of this step.

6. Insert the second yoke and use a `concentric` mate between the yoke hole and the hole of the center block, as shown.

7. Insert one bushing and apply two mates: `concentric` between the bushing and the pin hole in the center block, and `coincident` between the bushing and the outer face of the center block.

8. Repeat Step 7 for the other bushing.

9. Insert the pin into the assembly and apply two mates: `concentric` between the pin and the center block hole, and `coincident` between the pin face and the outer face of the yoke, as shown below.

10. Save all the files and exit the CAD system.

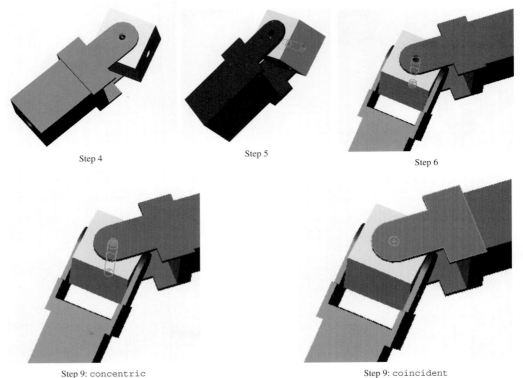

Step 4   Step 5

Step 6

Step 9: `concentric`   Step 9: `coincident`

Example 16.2 discussion:

We only create five parts and save them in their own files. We use two instances of the yoke and two instances of the bushing to create the assembly. The difference between the instances of each part comes in their orientation and location in the assembly model.

Example 16.2 hands-on exercise:

1. Find other mating conditions to create the assembly. Compare with the conditions of this example.

2. It has been identified that the main pin is a critical component of the universal joint. We need to increase the diameter of the main pin and the corresponding hole in the center block through which it passes by approximately 10%, that is, from 0.45 to 0.5 in for better reliability of the universal joint. Apply the following constrain relationship between these two parameters to update the features simultaneously.

Main_Pin Dia = CenterBlock_Hole_Dia through which the main pin passes.

EXAMPLE 16.3	**Create and apply assembly constraints.**

Show an example that uses assembly constraints.

SOLUTION	When you merge parts into an assembly model, you merge an instance

(copy) of the part into the assembly. If you change the parent (original) part, the assembly gets updated automatically by replacing the old instance with the new one. When you change one part, you run the risk of the assembly getting out of proportion. To avoid this problem, you can add constraints (equations) to the assembly to relate individual parts together.

Create the following bolt (*part1.prt*) and nut (*part2.prt*) parts.

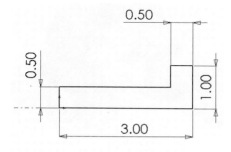

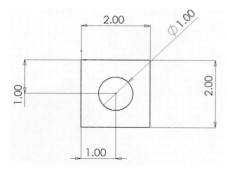
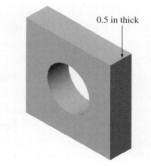

0.5 in thick

Assemble the bolt and the nut to create the assembly shown on the right. To add a relationship, click Tools (from the main toolbar) => Equations (or its equivalent). An Equations dialog box appears. Click Add to create a new relationship. A New Equation dialog box appears. Now select (double-click) the nut sketch from the assembly navigation tree, which has the dimension you need to constrain. Now the sketch and the associated dimensions appear on the graphical window. Select the Diameter dimension from the sketch, that is, ∅1.00. Now this dimension will be represented in the Equations

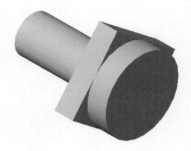

dialog box as `D3@Skecth1@hole` (this parameter name will differ depending on the construction steps). The following screenshots show the equation.

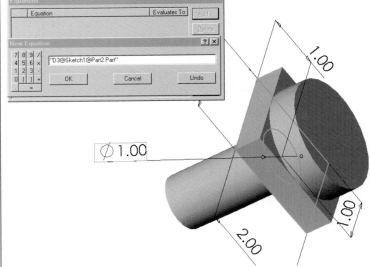

Now click on the Equals button on the `New Equation` dialog box shown above. Accept the selection of the Dimension dialog box in the feature navigation graphical area on the left. Double-click on the *part1.prt* sketch to activate the sketch in the graphical window. Before selecting the radius of the shaft of the bolt, update the equation as `D3@Skecth1@part2.prt=2*`. Now click on the half diameter dimension of the bolt shaft. The equation will now appear as follows.

`D3@Skecth1@part2.prt = 2 * D2@Sketch1@part1.prt`

The results are shown in the screenshot on the right.

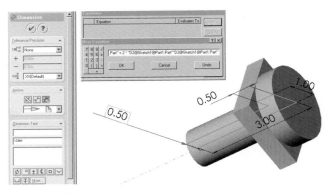

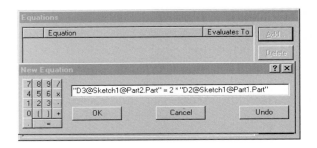

To finish creating the constraint equation, click OK in the `New Equation` dialog box. The new equation/relationship appears as shown on the right. You may edit it by selecting the equation and clicking `Edit All`.

Now whenever you edit the dimension of the bolt, within the assembly (by double-clicking on the bolt shaft sketch in the navigation tree) or outside (by opening and editing *part1.prt*), the nut hole diameter also gets updated to fit the bolt shaft. Try changing the bolt radius to 0.25 in and see the effect on the nut hole diameter.

Example 16.3 discussion:

The addition of assembly constraints (equations) between the parts of an assembly makes the assembly "intelligent" and adds important design intent to it. The assembly becomes intelligent as its updates check for nonsense results such as a smaller hole diameter than the diameter of the shaft that mates with it. The design intent captured and expressed by the constraints guarantees that the assembly performs as designed.

Example 16.3 hands-on exercise:

Add two new constraints to the assembly. The first constraint requires that the diameter of the bolt head is equal to the length of the nut side. The second constraint specifies that the nut thickness is equal to 20% of the bolt length (excluding the thickness of its head) — it is 0.5/2.5 in this example.

16.7 Top-Down Assembly Approach

The bottom-up approach covered in Section 16.6 appeals to small assemblies consisting of, for example, a hundred or maybe a thousand components. The top-down approach, while good for any size assembly, is ideal for large assemblies consisting of tens of thousands of components. It provides an effective tool and a well-organized approach to managing the design of large assemblies. It allows a project leader to break up product specifications, assign work teams, and enforce downstream design changes at a high level.

The top-down assembly approach fosters a systems engineering approach to product design, in which the assembly layout communicates design criteria to subsystem developers, including suppliers. This tight control allows distributed design teams to work concurrently within a common product framework. It also allows detailed design to begin while the assembly layout is being finalized.

The top-down approach lends itself well to the conceptual design phase. It captures the design intent of a product in the early design stages at a high level of abstraction. After all, assembly design does not always require detailed design of constituent parts and subassemblies. This allows designers to validate different design concepts before implementing them. The top-down approach also allows designers to practice the what-if design scenarios with ease.

The top-down assembly approach begins with an assembly layout sketch (also known as assembly sketch or skeleton model). The layout serves as the behind-the-scenes backbone of the assembly. The layout defines components in the context of an assembly. These components are "empty" as they do not have any external references to actual parts and subassemblies files yet. The assembly layout sketch defines skeletal, space claim, and other physical properties that may be used to define the geometry of and the relationships between components (parts or subassemblies).

The space claim is the most important property of an assembly layout because the layout shows where each assembly component belongs. When a designer lays down the skeletons of all the assembly components in the layout, the designer can clearly see any interferences, clearances, or overlapping between them. The designer can change the locations of the components, relative to each other, in the layout to better meet product design requirements.

The process of creating an assembly using the top-down approach is as follows:

1. Log in and start the CAD/CAM system.

2. Select the assembly mode.

3. Open a new assembly file.

4. Create a sketch in which various entities represent components (parts or subassemblies) in the assembly. Indicate a tentative location for each component, capturing the overall design intent of the assembly.

5. Create more sketches if needed. A complex assembly may require multiple sketches, similar to creating complex features of individual parts. For example, we may use the Front, Top, and Right sketch planes to create the assembly layout.

6. Use the sketch(es) to define the component size, skeletal shape, and location within the assembly.

7. Edit the assembly sketch(es) as needed until the assembly layout is finalized. For example, change the locations or the sizes of some components.

8. Create the individual parts and save them in their respective files. Make sure that each part references the assembly sketch(es).

9. Evaluate the assembly after the parts are fully constructed. If needed, modify the assembly sketch, and update the assembly and parts. Such an update is an automatic process.

10. Save the assembly file and exit the CAD system.

The top-down approach has many advantages. The major advantage is that if we change the layout sketch, the assembly and its parts are automatically updated upon exiting the sketch. We make all the changes quickly in one place, the assembly layout sketch.

The assembly layout sketch does not have to be the master plan for the design. For example, if we had a model of an engine with some fixed pulley locations, we could make the circles that represent the pulleys in the layout sketch coincident with the known locations of the pulleys in the model. If we change the locations of the pulleys in the engine model, the assembly model will update automatically.

EXAMPLE 16.4	**Use one assembly layout sketch.**

Use the top-down approach to create the pulley-belt assembly model shown below.

SOLUTION   We create the pulley-belt assembly shown on the right. The layout sketch shown below is the top-level assembly plan. The assembly will be driven by this sketch. We will be able to modify the parameters in the layout sketch to change the assembly, instead of modifying the part and rebuilding the assembly, as we do in the bottom-up assembly approach. Here are the detailed steps.

Modeling strategy:

1. Log in and start the CAD/CAM system.
2. Select the assembly mode.
3. Open a new assembly file.
4. Create the assembly layout sketch shown below. Define the dimensions as shown in the layout. The lines connecting the two pulleys are tangent to the pulleys. (Apply appropriate constraints to make the lines tangent.) Save the assembly file as *topDown.asm*.
5. Open a new part file, construct the large pulley part, and save it as *pulley1.prt* and exit. Use 3.0 inches as the pulley diameter and 0.5 inch as its thickness, as shown below.
6. Open a new part file, construct the small pulley part, and save it as *pulley2.prt* and exit. Use 1.0 inch as the pulley diameter and 0.5 inch as its thickness, as shown below.
7. Open a new part file, construct the belt part, and save it as *belt.prt* and exit. Use 5.0×0.5 in rectangular section and 0.1 in as its thickness, as shown below.

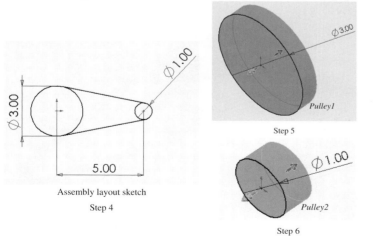

Assembly layout sketch

Step 4

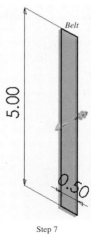

Step 5

Step 6

Step 7

8. Open the assembly file *topDown.asm* and insert the following components: *pulley1.prt*, *pulley2.prt*, and two instances of *belt.prt*. Now apply various mates that are needed to assemble the components as shown below. (Fully constrain the system with no relative movements.)

9. Relate the reference dimensions of the individual components to the corresponding dimensions in the assembly layout sketch. We make the radii of the small and large pulleys, and the belt length equal to the corresponding variables in the sketch, as shown below. Thus,

pulley1 diameter = Layout Sketch Big Circle Diameter

pulley2 diameter = Layout Sketch Small Circle Diameter

length of the belt = Layout Sketch Distance between the Circles

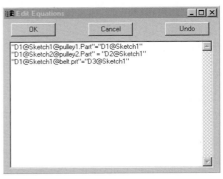

Step 8   Step 9

10. Now you may change the parameters in the layout sketch to see the effect in the individual components and in the assembly. Change the big pulley diameter to 5.0 inches and the distance between pulley centers to 10.0 inches. Regenerate (update) the assembly to obtain the results shown below.

11. Save all the files and exit the CAD system.

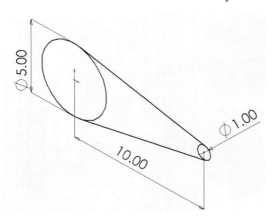

Example 16.4 discussion:

The assembly uses four instances, one of each of the small and the large pulleys and two of the belt. Step 9 relates the key dimensions of the instances to those of the assembly layout sketch. When we change the dimensions of the layout sketch and update (regenerate) the assembly as we do in Step 10, the dimensions of the instances, and their corresponding parts, change automatically.

Example 16.4 hands-on exercise:

Add a constraint to the assembly that requires the diameter of the large pulley to be twice that of the small pulley. Change the dimensions of the diameters and update the assembly. Do the assembly updates satisfy all the assembly constraints (four of them)?

EXAMPLE 16.5   **Use multiple assembly layout sketches.**
Use the top-down approach to create the block-plate-pin assembly model of Example 16.1.

SOLUTION   We have used the bottom-up approach to create the assembly model shown on the right. We redo the example here using the top-down approach. Unlike the assembly of Example 16.4, this assembly requires three sketches to fully define and constrain the assembly layout sketch. Here are the detailed steps.

Modeling strategy:

1. Log in and start the CAD/CAM system.
2. Select the assembly mode.
3. Open a new assembly file.
4. Select the Front sketch plane and create the following sketch that shows the front view of the assembly. We call this sketch *sketch1*. We label the dimensions as shown below. We use these labels as parameters in the assembly constraint equations in later steps.
5. Select the Top sketch plane and create the following sketch that shows the top view of the assembly. We call this sketch *sketch2*. We label the dimensions as shown below. We use these labels as parameters in the assembly constraint equations in later steps.

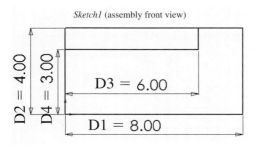

Step 4

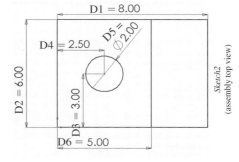

Step 5

6. Select the `Right` sketch plane and create the following sketch that shows the right view of the assembly. We call this sketch *sketch3*. We label the dimensions as shown below. We use these labels as parameters in the assembly constraint equations in later steps. The isometric view of the three assembly sketches is shown below.

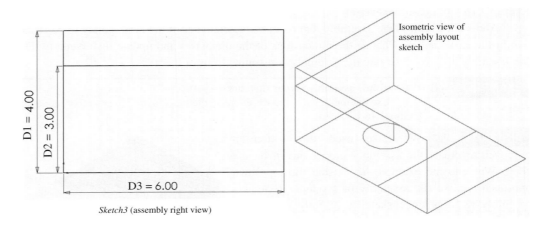

Sketch3 (assembly right view)

7. Associate the assembly layout sketch with the components' parameters. The following 18 equations relate the components' parameters to those of the layout sketch. Note that, you may not define a LHS (left-hand side) parameter using more than one layout sketch parameter. For example, if a component sketch parameter *D1@Sketch1@block.prt* is controlled by a layout sketch parameter *D1@Sketch1*, and in another view the same component parameter is controlled by another sketch parameter, say *D2@Sketch2*, then we may not be able to consider both cases of controlling the LHS parameter. We have to decide on selecting only one option. In other words we cannot define the LHS parameter more than once. The components' parameters shown in the equations below are shown in the following additional screenshots.

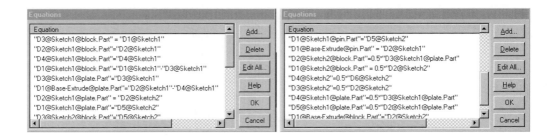

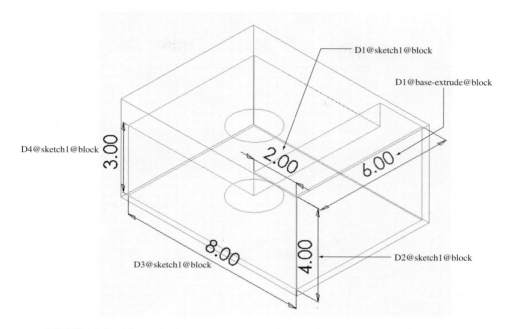

D1@sketch1@block
D1@base-extrude@block
D4@sketch1@block
3.00
2.00
6.00
8.00
4.00
D3@sketch1@block
D2@sketch1@block

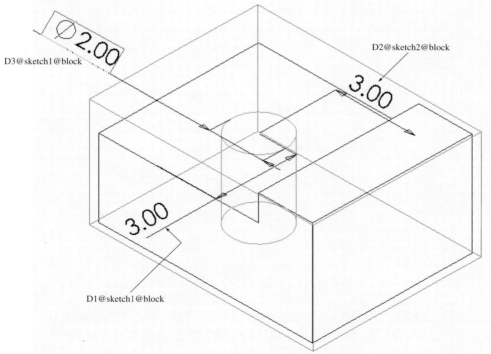

⌀2.00
D3@sketch1@block
D2@sketch2@block
3.00
3.00
D1@sketch1@block

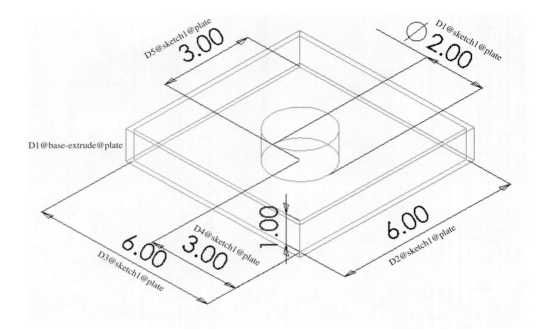

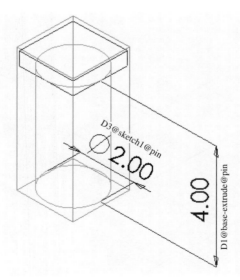

8. In the current assembly we can modify most of the dimensions of the assembly using the
   Front and the Top layout sketches. The Right layout sketch is not used, since that will
   result in redefining the LHS, which is not allowed.

9. Modify the front view layout sketch as shown below and regenerate the assembly.

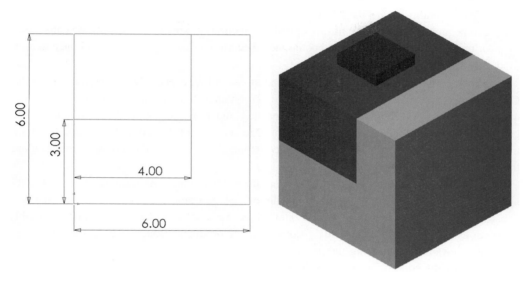

10. Modify the top view layout sketch as shown below and regenerate the assembly.

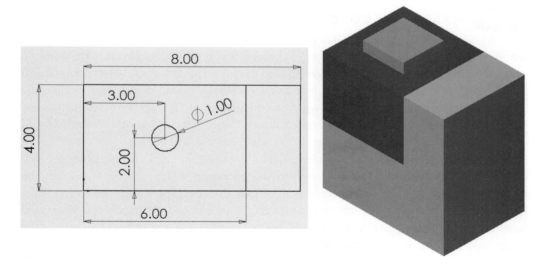

Example 16.5 discussion:

The top-down approach seems more complex for this assembly than the bottom-up approach. We have to create three layout sketches and define 18 constraint equations.

Example 16.5 hands-on exercise:

Find a different set of constraint equations to define the assembly.

16.8 Testing Mating Conditions

Mating conditions are the key to the successful creation of assemblies. These conditions can become complex at times as illustrated in Sections 16.6 and 16.7. Components could over- or under-mate. An under-mated component becomes a floating component in the assembly modeling space.

How can we ensure that we have created an assembly, especially a large one, correctly? CAD systems provide a `Move Component` command(s) or its equivalent to answer this question. With this command, a designer can drag and move any assembly component. The component can only move along its current free degrees of freedom. For example, if the translation or rotation about the X axis is not constrained, the component moves or rotates along this axis.

Why is it important to test the mating conditions? For simulation reasons. We may use the assembly model to create an animation file of the assembly sequence for the assembly department. We also create exploded views, as we explain later, to produce assembly instructions that customers can use in do-it-yourself products. Both the animation files and exploded views require moving the parts in space along their free degrees of freedom. Finally, testing the mating conditions provides the opportunity to ensure that an assembly functions as it is supposed to according to its design.

16.9 Assembly Load Options

When we insert a component into an assembly, a reference (pointer) between the assembly and the component is created, that is, the component information is loaded into the assembly model. CAD systems provide two options to load the information: *fully* or *partially loaded*. In the *fully loaded* mode, all the component information is fully loaded in memory. The *partially loaded* (sometimes known as *lightweight*) mode loads only a subset of the component information in memory. The remaining information is loaded if the component is selected or if it is changed by the current editing session.

The loading option affects the performance of large assemblies during the active assembly session. The *partially loaded* option improves the performance significantly. It performs the assembly activities and commands a lot more quickly than the *fully loaded* option. For small assemblies, there is no significant benefits to using *partially loaded* components. We recommend using the *partially loaded* option for large assemblies as a start. As you begin selecting components to work with, they get resolved one by one, thus resolving components on demand.

16.10 Managing Assemblies

After we insert components into an assembly, we can perform many operations to manage them. We can hide, freeze (suppress), copy, or delete them. When hiding a component, we make it invisible from the display screen; however, it still gets updated when we update the assembly model. Freezing a component does the opposite to hiding it; when we freeze it, it is still visible on the screen, but it is not updated. Copying components is useful for building symmetric assemblies.

CAD systems provide mate property managers to manage mating conditions. A manager allows a designer to assign mating conditions, group them, copy them, or delete them. Some systems provide smart mates that predict mate intent. For example, if a designer drops a shaft on a hole, the CAD system can display a `concentric` mate symbol to guide the designer to use the correct mate.

16.11 Working with Subassemblies

When an assembly is a component of another assembly, it is known as a subassembly. We can nest subassemblies in multiple levels in the assembly tree to reflect the product hierarchy, as shown in Figure 16.2.

There are three alternatives for creating subassemblies using CAD systems:

• Create an assembly document (file) as a separate operation. Then insert it as a component into a higher-level assembly model. This makes it a subassembly of the higher-level assembly.

• Insert a new, empty subassembly at any hierarchical level of an open assembly model (file), then add components to it.

• Form a subassembly by selecting a group of existing components from an assembly.

As we develop assemblies, we may have a need to edit their subassemblies. We can work with subassemblies in a variety of ways:

• Dissolve a subassembly into individual components.

• Edit the subassembly structure by moving its components up or down the assembly tree or to a different tree branch. Simply drag the components from the assembly tree manager and drop where appropriate in other tree locations (levels).

• Move components in and out of subassemblies to the parent assembly.

16.12 Inference of Position and Orientation

The inference of the position (location) and orientation of a part in an assembly from mating conditions requires computing its 4×4 homogeneous transformation matrix from these conditions. This matrix relates the part's local coordinate system (part MCS) to the assembly's global coordinate system (assembly MCS). In reference to Figure 16.5, the location of `Part 1` is represented by the vector $\mathbf{OO}_1$ connecting the original O of the assembly MCS to the origin O_1 of the $X_1Y_1Z_1$ MCS of the part. The orientation is represented by the rotation matrix between the two systems. The transformation matrix can be written as [see Eq. (12.81)]:

$$[T] = \begin{bmatrix} m_x & q_y & r_z & \vdots & x \\ m_y & q_y & r_y & \vdots & y \\ m_z & q_z & r_z & \vdots & z \\ \hdashline 0 & 0 & 0 & \vdots & 1 \end{bmatrix} \qquad (16.1)$$

This matrix has 12 variables (nine rotational and three translation elements) that must be determined from the mating conditions. For an assembly of N parts, and choosing one of them as a host, $N - 1$ transformation matrices have to be computed. Therefore, the variables to solve for simultaneously are the $12(N-1)$ elements of these matrices. Before we present the general solution, we cover an easier version first; we call it the WCS method.

16.12.1 WCS Method

The simplest method for specifying the location and orientation of each part in an assembly is to provide the 4×4 homogeneous transformation matrix $[T]$ directly, instead of inferring it from mating conditions. This method provides us with a good understanding of the basics that we need in Section 16.12.2. The matrix transforms the coordinates of the geometric entities of the part from its MCS to the assembly MCS.

One way for the user to provide the transformation matrix interactively is by defining a WCS in the assembly model such that its origin and orientation match the final location and orientation of the MCS of the part that we need to insert into the assembly. We then write the $[T]$ matrix, given by Eq. (16.1), that converts the coordinates from the part MCS (now the WCS) to the assembly MCS. When we apply $[T]$ to all the part geometry, we effectively insert (merge) the part into the assembly.

As a WCS is completely defined by specifying its X and Y axes or its XY plane, the proper WCS used to merge a part into its assembly can be defined such that its XY plane coincides with the XY plane of the part MCS. Figure 16.10 shows an example. The assembly consists of two parts, A and B. Three instances of Part B are used in the assembly. The user first creates the two parts with the MCS of each part as shown in Figure 16.10a. To create the assembly, we insert an instance of Part A into a blank assembly file. We use the instance as the assembly base. It is usually beneficial to assign a separate layer for each instance for ease of managing the assembly.

To merge the instance of B on top of A, the $X_1Y_1Z_1$ WCS is defined by the user as shown in Figure 16.10b, and then the instance is merged. Similarly, the $X_2Y_2Z_2$ and $X_3Y_3Z_3$ WCSs are defined and shown in Figure 16.10b. The transformation matrices to merge the three instances of B into A are given by, respectively:

$$[T_1] = \begin{bmatrix} 1 & 0 & 0 & \vdots & 1.5 \\ 0 & 1 & 0 & \vdots & 1.5 \\ 0 & 0 & 1 & \vdots & 3 \\ \hdashline 0 & 0 & 0 & \vdots & 1 \end{bmatrix} \qquad (16.2)$$

$$[T_2] = \begin{bmatrix} 0 & 1 & 0 & \vdots & 2 \\ -1 & 0 & 0 & \vdots & 0 \\ 0 & 0 & 1 & \vdots & 1 \\ \cdots & \cdots & \cdots & & \cdots \\ 0 & 0 & 0 & \vdots & 1 \end{bmatrix} \tag{16.3}$$

$$[T_3] = \begin{bmatrix} 1 & 0 & 0 & \vdots & 6 \\ 0 & 1 & 0 & \vdots & 1.5 \\ 0 & 0 & 1 & \vdots & 1 \\ \cdots & \cdots & \cdots & & \cdots \\ 0 & 0 & 0 & \vdots & 1 \end{bmatrix} \tag{16.4}$$

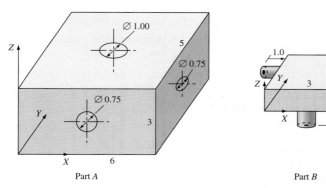

(a) Individual parts of the assembly

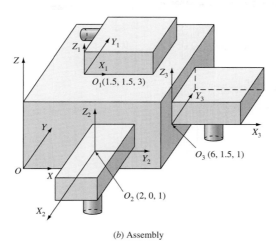

(b) Assembly

Figure 16.10   Assembly creation via the WCS method.

16.12.2 Mate Method

In a typical assembly, the mating conditions between two components are not enough by themselves to completely constrain the two components. An intertwinement of mating conditions usually exist among all of the parts. In general, a group of parts must be solved simultaneously to find the transformation matrix $[T]$. The mating conditions along with the properties of $[T]$ provide the necessary equations to solve for the $12(N-1)$ variables. The number of equations is always equal to or greater than the number of variables. Therefore, the method of solution must account for the number of redundant equations, and it must eliminate these equations from the system of equations to be solved.

Before we discuss the details of possible methods of solution, we present the development of constraint equations from mating conditions. We discuss the three basic mating conditions: coincident, concentric, and coplanar. For the coincident condition shown in Figure 16.6, each face where the two parts mate (butt up against one another) is specified by a unit normal and a point described in the MCS of its corresponding part. Let $[T_1]$ and $[T_2]$ be the transformation matrices from the $X_1Y_1Z_1$ and $X_2Y_2Z_2$ coordinate systems, respectively, to the MCS of the assembly. The unit normals and the two points specifying the mating conditions can be expressed in terms of the MCS (XYZ) system as follows:

$$
\begin{bmatrix} n_{1x}^a \\ n_{1y}^a \\ n_{1z}^a \\ 0 \end{bmatrix} = [T_1] \begin{bmatrix} n_{1x} \\ n_{1y} \\ n_{1z} \\ 0 \end{bmatrix}
\tag{16.5}
$$

$$
\begin{bmatrix} x_1^a \\ y_1^a \\ z_1^a \\ 1 \end{bmatrix} = [T_1] \begin{bmatrix} x_1 \\ y_1 \\ z_1 \\ 1 \end{bmatrix}
\tag{16.6}
$$

$$
\begin{bmatrix} n_{2x}^a \\ n_{2y}^a \\ n_{2z}^a \\ 0 \end{bmatrix} = [T_2] \begin{bmatrix} n_{2x} \\ n_{2y} \\ n_{2z} \\ 0 \end{bmatrix}
\tag{16.7}
$$

and

$$\begin{bmatrix} x_2^a \\ y_2^a \\ z_2^a \\ 1 \end{bmatrix} = [T_2] \begin{bmatrix} x_2 \\ y_2 \\ z_2 \\ 1 \end{bmatrix} \quad (16.8)$$

In the preceding equations, the superscript a indicates assembly. The `coincident` condition requires that the directions of the two unit normals must be equal and opposite, and the two points must lie in the same plane at which the two faces mate. The `coincident` condition requires four equations that can be expressed as follows:

$$n_{1x}^a = -n_{2x}^a \quad (16.9)$$

$$n_{1y}^a = -n_{2y}^a \quad (16.10)$$

$$n_{1z}^a = -n_{2z}^a \quad (16.11)$$

$$[n_{1x}^a \quad n_{1y}^a \quad n_{1z}^a \quad 0] \left\{ \begin{bmatrix} x_1^a \\ y_1^a \\ z_1^a \\ 1 \end{bmatrix} - \begin{bmatrix} x_2^a \\ y_2^a \\ z_2^a \\ 1 \end{bmatrix} \right\} = 0 \quad (16.12)$$

The `concentric` condition requires the centerlines of the shaft and the hole to be collinear as shown in Figure 16.7. The equation of the centerline of, say, the hole can be written as:

$$\frac{x - x_1^a}{x_2^a - x_1^a} = \frac{y - y_1^a}{y_2^a - y_1^a} = \frac{z - z_1^a}{z_2^a - z_1^a} \quad (16.13)$$

If the shaft axis is collinear with the hole centerline, points P_3 and P_4 defining the axis should satisfy Eq. (16.13). These points must first be transformed using $[T_2]$ to the MCS coordinates. The constraint equations required for each `concentric` condition can be written as:

$$\frac{x_3^a - x_1^a}{x_2^a - x_1^a} = \frac{y_3^a - y_1^a}{y_2^a - y_1^a} = \frac{z_3^a - z_1^a}{z_2^a - z_1^a} \quad (16.14)$$

and

$$\frac{x_4^a - x_1^a}{x_2^a - x_1^a} = \frac{y_4^a - y_1^a}{y_2^a - y_1^a} = \frac{z_4^a - z_1^a}{z_2^a - z_1^a} \quad (16.15)$$

Each of Eqs. (16.14) and (16.15) yield three combinations of equations resulting in six equations for each `concentric` condition. In general, two of these equations are redundant because Eqs. (16.14) and (16.15) each yield only two independent equations instead of three. However, it is necessary to carry all three to cover the case where the centerline passing through points P_1 and P_2 is parallel to any of the assembly MCS axes. For example, if the centerline is parallel to the X axis as shown in Figure 16.7, Eq. (16.14) becomes:

$$\frac{x_3^a - x_1^a}{x_2^a - x_1^a} = \frac{y_3^a - y_1^a}{0} = \frac{z_3^a - z_1^a}{0} \tag{16.16}$$

Equation (16.16) gives the following two equations only:

$$(y_3^a - y_1^a)(x_2^a - x_1^a) = 0 \tag{16.17}$$

and

$$(z_3^a - z_1^a)(x_2^a - x_1^a) = 0 \tag{16.18}$$

Hence, it can be seen that all three equations must be carried so that at least two independent equations can be written for all cases, although this introduces redundancy in the system of equations.

The constraint equations for the coplanar condition are the same as for the coincident condition except that the two unit normals are in the same direction as shown in Figure 16.9. Thus Eqs. (16.9) to (16.12) can be used after replacing the minus sign in Eqs. (16.9) to (16.11) with a plus sign.

One last constraint equation can be written, and it applies to all free rotating parts in the assembly such as bolts, pins, and shafts. A free rotating part is defined here as a part which rotates freely about a centerline axis. The rotation of these parts usually does not alter the appearance of the assembly, thus the name free rotation. There are infinite possible orientations of a free rotating part.

Figure 16.11 shows a bolt with its $X_1 Y_1 Z_1$ MCS oriented in the assembly XYZ MCS. The bolt can rotate freely about its X_1 axis any angle ϕ. As Figure 16.11 shows, the orientation of the Y_1 and the Z_1 axes is insignificant to the assembly. If this free orientation is not constrained, the calculations of the transformation matrix from the mating conditions will diverge. As long as the angle ϕ is arbitrary, it can be set to zero, that is,

$$\phi = 0 \tag{16.19}$$

Equation (16.19) represents the constraint equation associated with free rotating parts. In reference to Figure 16.11, Eq. (16.19) is equivalent to the two equations

$$a = b = 0 \tag{16.20}$$

where a and b are the components of the unit vectors (along the Y_1 and Z_1 axes) in the Z_1 and Y_1 directions respectively. If the X_1 axis is coincident with the X axis of the assembly MCS, Eq. (16.20) can be written in terms of the elements of the matrix $[T]$ given by Eq. (16.1) as:

$$q_z = r_y = 0 \tag{16.21}$$

For a part rotating freely about its Y_1 or Z_1 axis, we can write, respectively:

$$m_z = r_x = 0 \tag{16.22}$$

$$m_y = q_x = 0 \tag{16.23}$$

With all the constraint equations derived for the various mating conditions, we now calculate the total number of equations and unknowns that can be used to infer the position and orientation of a part from mating conditions. For each `coincident` condition, 16 equations can be written: 12 are provided by Eqs. (16.5) to (16.8), and the other four are Eqs. (16.9) to (16.12).

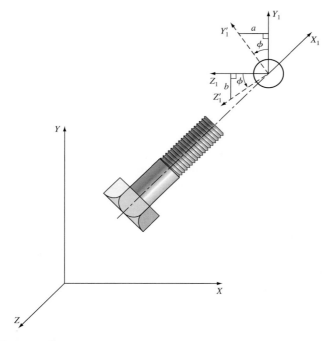

Figure 16.11   Free rotating part.

For each `concentric` conditions, 18 equations can be written: 12 are provided by Eqs. (16.5) to (16.8), and the other six are Eqs. (16.14) and (16.15). The `coplanar` condition provides the same number of equations as the `coincident` condition.

For each free rotating part, two equations [Eq. (16.21) and (16.22) or (16.23)] are available. In addition, the properties of the transformation matrix [T] [Eq. (16.1)] provide six equations: three from the unit vector length property, and three from the orthogonality property. These can be written as:

$$m_x^2 + m_y^2 + m_z^2 = 1 \tag{16.24}$$

$$q_x^2 + q_y^2 + q_z^2 = 1 \tag{16.25}$$

$$m_x q_x + m_y q_y + m_z q_z = 0 \tag{16.26}$$

$$r_x = m_y q_z - m_z q_y \tag{16.27}$$

$$r_y = m_z q_x - m_x q_z \tag{16.28}$$

$$r_z = m_x q_y - m_y q_x \tag{16.29}$$

The unit length requirement of (r_x, r_y, r_z) and its orthogonality with the other two unit vectors is satisfied automatically by Eqs. (16.27) to (16.29).

For an assembly of N parts, the total number of equations that can be written is given by:

$$M = 6(N-1) + 16\text{NA} + 16\text{NC} + 18\text{NF} + 2\text{NR} \tag{16.30}$$

where *NA, NC, NF,* and *NR* are, respectively, the number of `coincident`, `coplanar`, `concentric`, and free-rotation conditions. The number of variables is:

$$V = 12(N-1) + 12(\text{NA} + \text{NC} + \text{NF}) \tag{16.31}$$

The first term of Eq. (16.31) represents the elements of the $(N-1)$ transformation matrices, and the second term is the number of variables introduced by the `coincident`, `coplanar`, and `concentric` conditions. Each condition introduces 12 variables: six components of the unit normal vectors $\hat{\mathbf{n}}_1$ and $\hat{\mathbf{n}}_2$, and six coordinates of the two points P_1 and P_2 (or P_3 and P_4 for the `coplanar` condition). Equations (16.30) and (16.31) show that the number of equations and the number of variables are not equal. In general, the former is always equal to or greater than the latter.

The representation of the transformation matrix by Eq. (16.1) increases both the number of variables and the number of equations, which in turn makes finding a solution an expensive proposition. In an effort to reduce the number of variables, Eq. (16.1) may be rewritten using the following approach. The position of the part is still given by the coordinates x, y, and z. The orientation can be described by a sequence of rotation about the X, Y, or Z axes instead of using the components of the unit vectors as given by the nine elements of the matrix.

Let us assume that the MCS of a part can be oriented properly in its assembly by three rotations about the axes of the assembly MCS in the following order: α about the Z axis, β about the Y axis, and γ about the X axis. Thus, we can rewrite Eq. (16.1) as:

$$[T] = \begin{bmatrix} 1 & 0 & 0 & x \\ 0 & 1 & 0 & y \\ 0 & 0 & 1 & z \\ 0 & 0 & 0 & 1 \end{bmatrix} \begin{bmatrix} & & & 0 \\ & [R] & & 0 \\ & & & 0 \\ 0 & 0 & 0 & 1 \end{bmatrix} \tag{16.32}$$

where $[R]$ is given by Eq. (12.68).

Using Eq. (16.32) instead of Eq. (16.1) reduces the number of variables per matrix by half; from 12 to 6 (α, β, γ, x, y, z). Consequently, six equations [(16.24) to (16.29)] are eliminated. In addition, only one constraint equation [(16.19)] is used for free rotation.

To reduce the number of equations and variables even further, we can eliminate the twelve variables (the global description of unit normals and points) given by Eqs. (16.5) to (16.8) by considering them known in terms of the elements of the transformation matrix. Thus, Eqs. (16.30) and (16.31) become, respectively:

$$M = 4NA + 4NC + 6NF + NR \tag{16.33}$$

and

$$V = 6(N-1) \tag{16.34}$$

To simplify the implementation of the free rotation condition, we can define the Z axis as the axis of free rotation. Therefore the angle ϕ in Eq. (16.19) is about this axis. If the free rotating axis is the Y axis of the part MCS, multiply the local coordinates of all the part points and unit vectors by $[R_x(90)]$. This rotation matrix rotates the Y axis 90° about the X axis (see Figure 16.11). The new axis of rotation becomes the Z axis. If the part rotates freely about its X axis, then multiply all points and vectors by $[R_Y(-90)]$ to change the axis of rotation to the Z axis. In an interactive system, before assigning mating conditions the user is prompted to select the free rotating parts displayed on the screen.

16.12.3 Solving the Mate Equations

We now discuss the solution of the system of equations that results from applying the mating conditions. This system is nonlinear due to the trigonometric functions that appear in the transformation matrix $[T]$. Since the number of equations is equal to or exceeds the number of variables, a method is needed to remove the redundant equations. The method discussed here utilizes the least-squares technique to eliminate redundancy first, followed by using the Newton-Raphson iteration method to solve the resulting set of independent equations. The Newton-Raphson method for n nonlinear equations in n variables can be written as:

$$\mathbf{X}_{k+1} = \mathbf{X}_k + [J(\mathbf{X}_k)]^{-1}\mathbf{R}_k \tag{16.35}$$

where $\mathbf{X}_k$ is the solution vector at the kth iteration. $[J(\mathbf{X}_k)]$ is the Jacobian matrix, and $\mathbf{R}_k$ is the residual vector, both of which are evaluated at the current solution vector $\mathbf{X}_k$.

When redundancy exists, the inverse of the Jacobian may not exist because the Jacobian itself may not be square and/or it may be singular. The following procedure can be used to solve for n variables ($\mathbf{X} = [x_1 \ x_2 \ ... \ x_n]^T$) using the following m equations:

$$f_1(x_1, x_2, ..., x_n) = 0$$
$$f_2(x_1, x_2, ..., x_n) = 0$$
$$\vdots \tag{16.36}$$
$$f_m(x_1, x_2, ..., x_n) = 0$$

In a vector form, Eq. (16.36) becomes

$$\mathbf{F(X)} = \mathbf{0} \tag{16.37}$$

To write Eq. (16.36) in Newton-Raphson iterative form, let us assume that a solution $\mathbf{X}_i$ exits at step i and the solution at step $i + 1$ is $\mathbf{X}_{i+1}$ such that

$$\mathbf{X}_{i+1} = \mathbf{X}_i + \Delta\mathbf{X}_i \tag{16.38}$$

Linearizing Eq. (16.36) about $\mathbf{X}_i$ gives:

$$\mathbf{F}_{i+1} = \mathbf{F}_i + \frac{\partial \mathbf{F}(\mathbf{X}_i)}{\partial \mathbf{X}} \Delta \mathbf{X}_i \tag{16.39}$$

If $\mathbf{X}_{i+1}$ is the solution, then Eq. (16.37) holds, that is, $\mathbf{F}_{i+1} = \mathbf{0}$. Thus, Eq. (16.39) becomes:

$$\frac{\partial \mathbf{F}(\mathbf{X}_i)}{\partial \mathbf{X}} \Delta \mathbf{X}_i = -\mathbf{F}_i \tag{16.40}$$

Expanding Eq. (16.40) gives:

$$
\begin{bmatrix}
\dfrac{\partial f_1}{\partial x_1} & \dfrac{\partial f_1}{\partial x_2} & \cdots & \dfrac{\partial f_1}{\partial x_n} \\[2mm]
\dfrac{\partial f_2}{\partial x_1} & \dfrac{\partial f_2}{\partial x_2} & \cdots & \dfrac{\partial f_2}{\partial x_n} \\[2mm]
& \vdots & & \\[2mm]
\dfrac{\partial f_m}{\partial x_1} & \dfrac{\partial f_m}{\partial x_2} & \cdots & \dfrac{\partial f_m}{\partial x_n}
\end{bmatrix}_i
\begin{bmatrix}
\Delta x_1 \\[2mm] \Delta x_2 \\[2mm] \vdots \\[2mm] \Delta x_n
\end{bmatrix}_i
= -
\begin{bmatrix}
f_1(x_1, x_2, \ldots, x_n) \\[2mm]
f_2(x_1, x_2, \ldots, x_n) \\[2mm]
\vdots \\[2mm]
f_m(x_1, x_2, \ldots, x_n)
\end{bmatrix}_i
\tag{16.41}
$$

or

$$[J]_i \Delta \mathbf{X}_i = \mathbf{R}_i \tag{16.42}$$

where $[J]_i = [J(\mathbf{X}_i)]$, $\Delta \mathbf{X}_i$, and $\mathbf{R}_i$ are the Jacobian matrix, the incremental solution, and the residual vector at iteration i respectively. The Jacobian $[J(\mathbf{X}_i)]$ is non-square of size $m \times n$.

The least-squares method may be used to solve Eq. (16.42) for $\Delta \mathbf{X}_i$. The method is based on multiplying both sides of Eq. (16.42) by $[J]_i^T$, that is,

$$[J]_i^T [J]_i \Delta \mathbf{X}_i = [J]_i^T \mathbf{R}_i \tag{16.43}$$

Solving this equation for $\Delta \mathbf{X}_i$ gives:

$$\Delta \mathbf{X}_i = [J^T J]_i^{-1} [J]_i^T \mathbf{R}_i \tag{16.44}$$

The algorithm to solve for $\Delta \mathbf{X}_i$ can be described as follows: An initial guess $\mathbf{X}_0$ is made. The Jacobian $[J]_0$ and the residual vector $\mathbf{R}_0$ are computed. Next, Eq. (16.44) is used to calculate $\Delta \mathbf{X}_0$. Lastly, Eq. (16.38) is used to compute $\mathbf{X}_1$. These steps are repeated to obtain $\Delta \mathbf{X}_1$ and $\mathbf{X}_2$, $\Delta \mathbf{X}_2$ and $\mathbf{X}_3$, ..., and $\Delta \mathbf{X}_{n-1}$ and $\mathbf{X}_n$. Convergence is achieved when the elements of the residual vector $\mathbf{R}$ or the incremental solution $\Delta \mathbf{X}$ approaches zero.

EXAMPLE 16.6   **Use the mate method to infer position and location.**

Figure 16.12 shows a pin and a block with their MCSs. The pin is to be assembled into the hole in the block. Use the mate method to find the location and orientation of the pin in its assembly. Use the block MCS as the assembly MCS.

SOLUTION   The assembly consists of two parts: the pin and the block. The location and orientation of the $X_1Y_1Z_1$ MCS of the pin must be found relative to the assembly XYZ system. By inspection, it is obvious that the origin P_3 of the $X_1Y_1Z_1$ system must coincide with point P_1 in the assembled position, and its orientation is the same as that of the XYZ system. We now see how to reach the same conclusion using the mate method.

As seen from Figure 16.12, there is one `coincident` condition, one `concentric` condition, and one `free-rotation` condition. In addition, the number of parts N is equal to 2. Substituting this information into Eqs. (16.33) and (16.34) gives $M = 10$ and $V = 6$. The six variables are α, β, γ, x, y, and z where x, y, and z are the coordinates of P_3 measured in the XYZ system. Using Figure 16.12, we can write:

$$\hat{\mathbf{n}}_1 = [0 \quad 0 \quad -1 \quad 0]^T \qquad \hat{\mathbf{n}}_2 = [0 \quad 0 \quad 1 \quad 0]^T$$

$$\mathbf{P}_1 = [3 \quad 2 \quad 2 \quad 1]^T \qquad \mathbf{P}_2 = [3 \quad 2 \quad 0 \quad 1]^T$$

$$\mathbf{P}_3 = [0 \quad 0 \quad 0 \quad 1]^T \qquad \mathbf{P}_4 = [0 \quad 0 \quad -3 \quad 1]^T$$

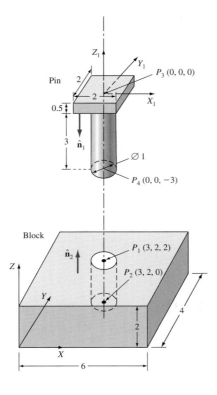

Figure 16.12   Block-pin assembly.

The normal $\hat{\mathbf{n}}_2$ and the points P_1 and P_2 are already expressed in terms of the assembly MCS. To transform $\mathbf{n}_1$, P_3, and P_4, use Eqs. (16.5) and (16.6) with Eq. (16.32) to obtain:

$$\hat{\mathbf{n}}_1^a = [-s\beta \quad c\beta s\gamma \quad -c\beta c\gamma \quad 0]^T$$
$$\mathbf{P}_3^a = [x \quad y \quad z \quad 1]^T$$
$$\mathbf{P}_4^a = [x - 3s\beta \quad y + 3c\beta s\gamma \quad z - 3c\beta c\gamma \quad 1]^T$$

The following constraint equations can be written:
The `coincident` condition [use Eqs. (16.9) to (16.12)]:

$$s\beta = 0$$
$$c\beta s\gamma = 0$$
$$c\beta c\gamma - 1 = 0$$
$$c\beta s\gamma(y - 2) - s\beta(x - 3) - c\beta c\gamma(z - 2) = 0$$

The `concentric` condition [use Eqs. (16.14) and (16.15)]:

$$x - 3 = 0$$
$$y - 2 = 0$$
$$x - 3s\beta - 3 = 0$$
$$y + 3c\beta s\gamma - 2 = 0$$

The free rotation of the pin about the Z axis of the assembly MCS gives the condition:

$$\alpha = 0$$

Note that the `concentric` condition gives four constraint equations instead of six because the X_1 and the Y_1 axes are parallel to the X and Y axes of the assembly MCS, respectively. Comparing the preceding nine constraint equations with Eq. (16.36), we find

$$\mathbf{X} = [\alpha \; \beta \; \gamma \; x \; y \; z]^T$$

$$f_1 = s\beta$$
$$f_2 = c\beta s\gamma$$
$$f_3 = c\beta c\gamma - 1$$
$$f_4 = c\beta s\gamma(y - 2) - s\beta(x - 3) - c\beta c\gamma(z - 2)$$
$$f_5 = x - 3$$
$$f_6 = y - 2$$
$$f_7 = x - 3s\beta - 3$$
$$f_8 = y + 3c\beta s\gamma - 2$$
$$f_9 = \alpha$$

The Jacobian matrix becomes:

$$[J] = \begin{bmatrix} 0 & c\beta & 0 & 0 & 0 & 0 \\ 0 & -s\beta s\gamma & c\beta c\gamma & 0 & 0 & 0 \\ 0 & -s\beta c\gamma & -c\beta s\gamma & 0 & 0 & 0 \\ 0 & \begin{matrix}-s\beta s\gamma(y-2)\\ -c\beta(x-3)\\ +s\beta c\gamma(z-2)\end{matrix} & \begin{matrix}c\beta c\gamma(y-2)\\ +c\beta s\gamma(z-2)\end{matrix} & -s\beta & c\beta s\gamma & -c\beta c\gamma \\ 0 & 0 & 0 & 1 & 0 & 0 \\ 0 & 0 & 0 & 0 & 1 & 0 \\ 0 & -3c\beta & 0 & 1 & 0 & 0 \\ 0 & -3s\beta s\gamma & 3c\beta c\gamma & 0 & 1 & 0 \\ 1 & 0 & 0 & 0 & 0 & 0 \end{bmatrix}$$

To find the solution, let us assume the initial guess

$$\mathbf{X}_0 = [0\,0\,0\,0\,0\,0\,0\,0\,0]^T$$

Substituting this initial guess into the above matrix gives $[J]_0$ as

$$[J]_0 = \begin{bmatrix} 0 & 1 & 0 & 0 & 0 & 0 \\ 0 & 0 & 1 & 0 & 0 & 0 \\ 0 & 0 & 0 & 0 & 0 & 0 \\ 0 & 3 & -2 & 0 & 0 & -1 \\ 0 & 0 & 0 & 1 & 0 & 0 \\ 0 & 0 & 0 & 0 & 1 & 0 \\ 0 & -3 & 0 & 1 & 0 & 0 \\ 0 & 0 & 3 & 0 & 1 & 0 \\ 1 & 0 & 0 & 0 & 0 & 0 \end{bmatrix}$$

then:

$$[J]_0^T = \begin{bmatrix} 0 & 0 & 0 & 0 & 0 & 0 & 0 & 0 & 1 \\ 1 & 0 & 0 & 3 & 0 & 0 & -3 & 0 & 0 \\ 0 & 1 & 0 & -2 & 0 & 0 & 0 & 3 & 0 \\ 0 & 0 & 0 & 0 & 1 & 0 & 1 & 0 & 0 \\ 0 & 0 & 0 & 0 & 0 & 1 & 0 & 1 & 0 \\ 0 & 0 & 0 & -1 & 0 & 0 & 0 & 0 & 0 \end{bmatrix}$$

$$[J^T J]_0 = \begin{bmatrix} 1 & 0 & 0 & 0 & 0 & 0 \\ 0 & 19 & -6 & -3 & 0 & -3 \\ 0 & -6 & 14 & 0 & 3 & 2 \\ 0 & -3 & 0 & 2 & 0 & 0 \\ 0 & 0 & 3 & 0 & 2 & 0 \\ 0 & -3 & 2 & 0 & 0 & 1 \end{bmatrix}$$

Notice that $[J^T J]$ is always symmetric. Also,

$$[J^T J]_0^{-1} = \begin{bmatrix} 1 & 0 & 0 & 0 & 0 & 0 \\ 0 & 0.1818 & 0 & 0.2727 & 0 & 0.5455 \\ 0 & 0 & 0.1818 & 0 & -0.2727 & -0.3636 \\ 0 & 0.2727 & 0 & 0.9091 & 0 & 0.8182 \\ 0 & 0 & -0.2727 & 0 & 0.9091 & 0.5455 \\ 0 & 0.5455 & -0.3636 & 0.8182 & 0.5455 & 3.3636 \end{bmatrix}$$

$$[J^T J]_0^{-1} [J]_0^T = \begin{bmatrix} 0 & 0 & 0 & 0 & 0 & 0 & 0 & 0 & 1 \\ 0.1818 & 0 & 0 & 0 & 0.2727 & 0 & -0.2727 & 0 & 0 \\ 0 & 0.1818 & 0 & 0 & 0 & -0.2727 & 0 & 0.2727 & 0 \\ 0.2727 & 0 & 0 & 0 & 0.9091 & 0 & 0.0909 & 0 & 0 \\ 0 & -0.2727 & 0 & 0 & 0 & 0.9091 & 0 & 0.0909 & 0 \\ 0.5455 & -0.3636 & 0 & -1 & 0.8182 & 0.5455 & -0.8182 & -0.5455 & 0 \end{bmatrix}$$

To calculate the residual vector $\mathbf{R}_0$, substitute the initial guess $\mathbf{X}_0$ into the nine f functions to obtain:

$$\mathbf{R}_0 = \begin{bmatrix} 0 & 0 & 0 & -2 & 3 & 2 & 3 & 2 & 0 \end{bmatrix}^T$$

Using Eq. (16.44), we obtain:

$$\Delta\mathbf{X}_0 = \begin{bmatrix} 0 & 0 & 0 & 3 & 2 & 2 \end{bmatrix}^T$$

and Eq. (16.38) gives:

$$\mathbf{X}_1 = \begin{bmatrix} 0 & 0 & 0 & 3 & 2 & 2 \end{bmatrix}^T$$

We now use $\mathbf{X}_1$ as an initial guess for the second iteration. Before proceeding, if we calculate $\mathbf{R}_1$, we find it to be $\mathbf{0}$. Therefore $\Delta\mathbf{X}_1 = \mathbf{0}$, and $\mathbf{X}_2 = \mathbf{X}_1$. Thus, $\mathbf{X}_1$ is the solution. Actually, it is the exact solution.

Substituting the solution $\mathbf{X}_1$ into Eq. (16.32) gives the transformation matrix as:

$$[T] = \left[\begin{array}{ccc|c} 1 & 0 & 0 & 3 \\ 0 & 1 & 0 & 2 \\ 0 & 0 & 1 & 2 \\ \hline 0 & 0 & 0 & 1 \end{array}\right]$$

Multiplying the local coordinates (pin MCS coordinates) of each point of the pin by the preceding $[T]$ matrix transforms them to the assembly MCS. Notice that this is the same transformation matrix that would be used if the WCS method were to be used to create the assembly.

EXAMPLE 16.7   **Use the mate method to infer position and location.**
Calculate the locations and orientations of the instances of the assembly shown in Figure 16.10.

SOLUTION   This assembly consists of four parts, that is, $N = 4$. Each instance of part B requires one `coincident` condition, one `concentric` condition, and one free-rotation condition to be assembled. Therefore, the number of equations is $M = 4 \times 3 + 6 \times 3 + 1 \times 3 = 33$, and $V = 6(4 - 1) = 18$. The 18 variables are the three rotations (α, β, γ) and the three coordinates (x, y, z) of the origin of each of the $X_1Y_1Z_1$, $X_2Y_2Z_2$, and $X_3Y_3Z_3$ coordinate systems.

The constraint equations for the `coincident`, `concentric`, and free-rotation conditions can be obtained in a similar fashion to Example 16.6. However, the free-rotation condition differs for the three instances. It is $\alpha = 0$, $\beta = 0$, and $\gamma = 0$ for the instance with origins O_1, O_2, and O_3, respectively. For the `concentric` condition, four equations instead of six can be written for each instance. This brings M down to 27 from 33.

The vector of variables to solve for are

$$\mathbf{X} = [\alpha_1 \quad \beta_1 \quad \gamma_1 \quad x_{01} \quad y_{01} \quad z_{01} \quad \alpha_2 \quad \beta_2 \quad \gamma_2 \quad x_{02} \quad y_{02} \quad z_{02} \quad \alpha_3 \quad \beta_3 \quad \gamma_3 \quad x_{03} \quad y_{03} \quad z_{03}]^T$$

Because the three instances are not interrelated, Eq. (16.42) can be written as:

$$\begin{bmatrix} [J_{01}] & 0 & 0 \\ 0 & [J_{02}] & 0 \\ 0 & 0 & [J_{03}] \end{bmatrix} \begin{bmatrix} \Delta\mathbf{X}_{01} \\ \Delta\mathbf{X}_{02} \\ \Delta\mathbf{X}_{03} \end{bmatrix} = \begin{bmatrix} \mathbf{R}_{01} \\ \mathbf{R}_{02} \\ \mathbf{R}_{03} \end{bmatrix}$$

This is a system of 27 equations in 18 variables ($[J_0]$ has 27 rows and 18 columns). However, the system is decoupled as shown, and we can solve for the location and orientation of each instance independently from the other two instances. Thus, we can write:

$$[J_{01}] \Delta\mathbf{X}_{01} = \mathbf{R}_{01} \qquad \text{for instance with origin } O_1$$
$$[J_{02}] \Delta\mathbf{X}_{02} = \mathbf{R}_{02} \qquad \text{for instance with origin } O_2$$
$$[J_{03}] \Delta\mathbf{X}_{03} = \mathbf{R}_{03} \qquad \text{for instance with origin } O_3$$

From an algorithmic and software development point of view, solving the 27×18 system is preferred because the software may not have the intelligence to judge if the system of equations for an assembly is decoupled or not. The software may also be developed in such a way as to allow the user to assemble two parts at a time instead of assembling all the parts at once by specifying all the mating conditions simultaneously. The solution to this example is given by Eqs. (16.2) to (16.4). The reader is encouraged to work out the details to obtain the solution.

16.13 Assembly Analysis

CAD systems provide various tools to analyze assemblies once they are created. Here is a list of assembly analysis activities:

1. **Generate assembly drawings.** An assembly drawing is no different from a part drawing. This drawing may include the standard four views or just an isometric view. The procedure for creating an assembly drawing is the same as a part drawing; here we obviously use the assembly model. The screenshot on the right shows an example of an assembly drawing.

ITEM NO.	QTY.	PART NO.	DESCRIPTION
1	1	block	Alloy Steel
2	1	plate	Aluminum
3	1	pin	Carbon Steel

2. **Generate a parts list.** The parts list is also known as bill of materials (BOM). This list is a table that shows the part names and how many instances of each part are used in the assembly. The above screenshot shows an example. The BOM is usually inserted in the assembly drawing via a command sequence such as `Insert => Bill Of Materials =>` select a template `=> Open =>` select the default template `=>` input information. SolidWorks uses an `Excel` sheet as the BOM file.

3. **Generate an exploded view.** This is an isometric (ISO) view that shows the parts of an assembly displayed apart from each other. The normal ISO view is known as a collapsed (assembly) view. The screenshot on the right shows an example. We may create an exploded view by specifying the direction and distance to which each part will move while exploding. Parts can be exploded in the directions of their permissible degrees of freedom after applying the mating conditions. An exploded view may consist of one or more `explode` steps. Each step shows the assembly component(s) displaced into a new location. All the steps of the exploded view are stored in the view itself. Upon displaying the view, the steps are displayed sequentially to give the illusion of animation, or assembly sequence. The `explode` steps may be edited after they are created, or new steps may be added. Finally, an exploded view may be collapsed to return it to its assembled state.

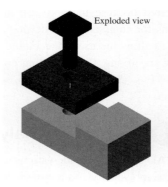

Exploded view

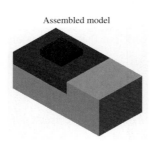

Assembled model

4. **Generate sectional views.** This is similar to generating sectional or other types of views of individual parts, as discussed in Chapter 4.

5. **Perform interference checking.** Allows you to check if any parts of an assembly pierce each other or not. If an interference is detected between two parts, the CAD system displays the interference volume to allow users to examine and rectify/eliminate it. The user may also request the CAD system to report overlapping or tangent edges/faces as interference if needed. Here is an example of interference detection using SolidWorks. Click this sequence: `Tools => Interference Detection =>` select the components to be checked for interference `=> Check`. The system will indicate the interference volume, if any, between the components. Click the `Treat coincidence as interference` checkbox if you want coincident entities (faces, edges, or vertices that touch or overlap) to be reported as interferences. Otherwise, touching or overlapping entities are ignored.

6. **Perform collision detection.** We detect the collision of a component with other components of an assembly, by moving or rotating it along its degrees of freedom. Collision implies that the component cannot be assembled correctly into the assembly, or it cannot move or rotate freely in its final position in the assembly.

7. **Perform mass property calculations.** We can calculate the mass properties of an entire assembly as we do for a single part (see Chapter 15). The centroid or the inertia properties of an entire assembly or subassembly may be important in some applications.

EXAMPLE 16.8 | **Interference checking.**
Perform an interference detection on the universal joint of Example 16.2.

SOLUTION Open the assembly of the universal joint file, and test for interference between parts. Click this SolidWorks sequence (or its equivalent on your CAD system): `Tools => Interference Detection =>` select the entire assembly to be checked (select the top-level assembly component from the feature manager tree) for interference `=> Check`. The system will indicate the interference volume, if any, between the components. Click the `Treat coincidence as interference` checkbox if you want coincident entities (faces, edges, or vertices that touch or overlap) to be reported as interferences. Otherwise, touching or overlapping entities are ignored. The screenshots show the results.

Example 16.8 discussion:

The preceding screenshots show that there is an interference between the center block and one yoke.

Example 16.8 hands-on exercise:

1. Modify the dimensions of the parts files of the universal joint to clear the interference volumes.
2. Get the assembly statistics of the universal joint. These statistics provide the part count, unique parts, subassemblies, unique subassemblies, and so forth.

16.14 Tutorials

This section shows how to create assemblies from subassemblies. In this tutorial, we assume that the parts have been already created. We also assume that you have opened an assembly file as well as the individual parts files.

16.14.1 Gillette MACH3 Razor Assembly

The MACH3 razor consists of three parts: the razor, the cartridge, and the holder. Here are the assembly steps.

1. **Insert the parts.** Drag all the parts and insert them into the assembly file as shown below.

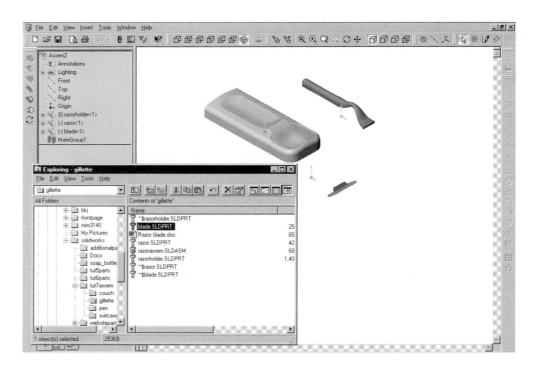

2. **Create the razor-cartridge subassembly.** Apply the `coincident` mate between the four edges shown below.

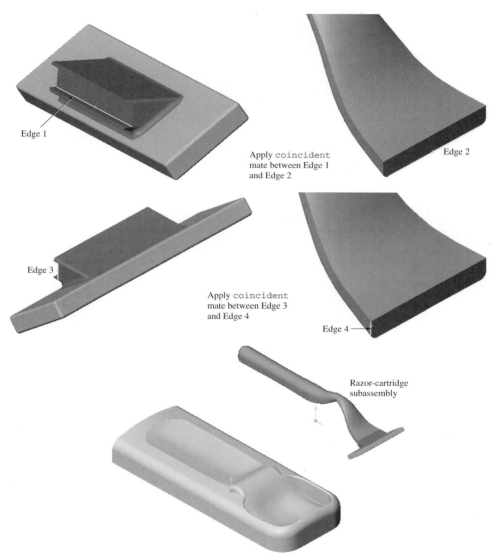

Edge 1

Apply `coincident` mate between Edge 1 and Edge 2

Edge 2

Edge 3

Apply `coincident` mate between Edge 3 and Edge 4

Edge 4

Razor-cartridge subassembly

3. **Move the razor-cartridge subassembly.** We now assemble the razor-cartridge subassembly and the razor holder. Move the razor-cartridge subassembly a distance of 2 mm along the *Y* axis (vertical axis shown on the right) of the assembly to make the subassembly touch the razor holder, as shown on the right.

4. **Create the final assembly.** Use the `distance` constraint and the `parallel` mate to finish the assembly as shown below.

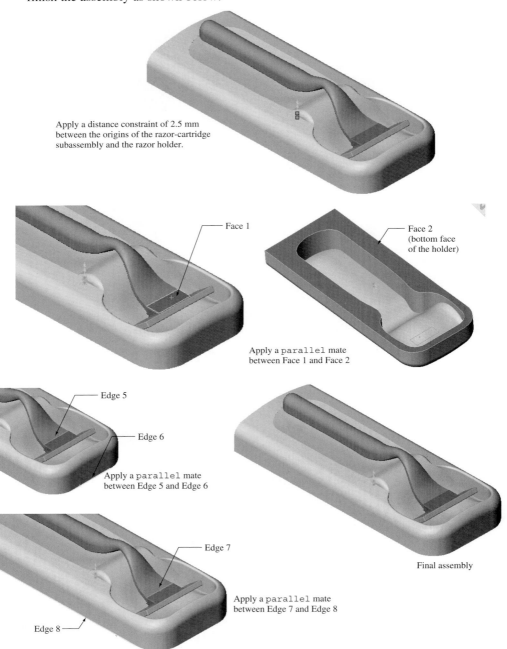

Apply a distance constraint of 2.5 mm between the origins of the razor-cartridge subassembly and the razor holder.

Face 1

Face 2 (bottom face of the holder)

Apply a `parallel` mate between Face 1 and Face 2

Edge 5

Edge 6

Apply a `parallel` mate between Edge 5 and Edge 6

Edge 7

Final assembly

Apply a `parallel` mate between Edge 7 and Edge 8

Edge 8

PROBLEMS

Part I: Theory

16.1 Generate the assembly tree for the shaft system shown below.

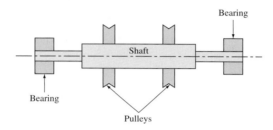

16.2 Generate the assembly tree for the driving unit of a small centrifugal fan shown below.

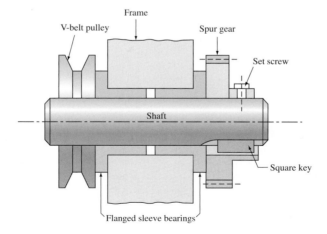

16.3 Generate the assembly tree for the screw jack shown below.

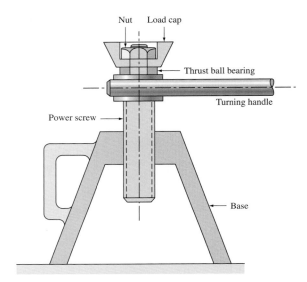

16.4 Generate the assembly tree for the block-plate-pin assembly shown below (see also Examples 16.1 and 16.5).

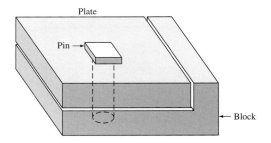

16.5 Generate the assembly tree for the pressure sensor shown below.

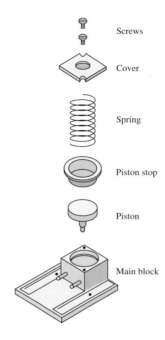

Screws

Cover

Spring

Piston stop

Piston

Main block

16.6 Generate the assembly tree for the household fan shown below.

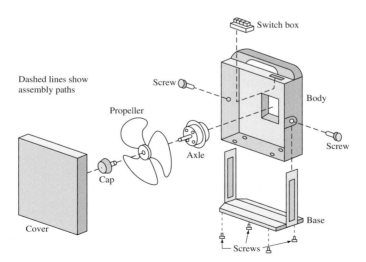

Switch box

Dashed lines show
assembly paths

Screw

Body

Propeller

Screw

Axle

Cap

Base

Cover

Screws

16.7   Generate the assembly tree for the axisymmetric mechanical assembly shown below.

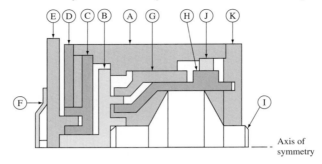

16.8   Generate the assembly tree for the bell shown below.

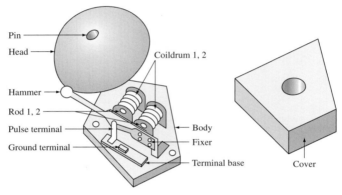

16.9   Generate the assembly tree for the three-pin power plug shown below.

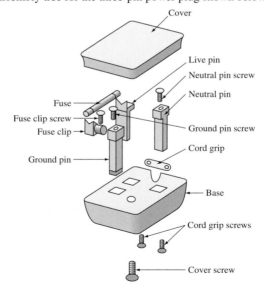

16.10 Find the mating conditions in Problems 16.1 to 16.9.

16.11 Use the WCS and mate methods to assemble the following parts:

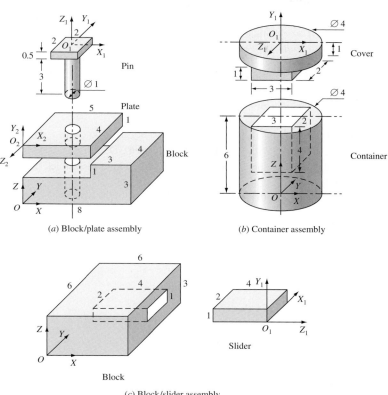

(a) Block/plate assembly   (b) Container assembly

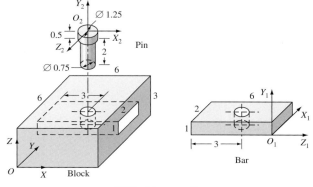

(c) Block/slider assembly

(d) Block/bar assembly

Part II: Laboratory

Use your in-house CAD/CAM system to answer the questions in this part.

16.12 Use the bottom-up approach to create the assemblies shown in Problems 16.1 to 16.9 and 16.11. Obtain drawings of individual parts as well as the assembly drawing. Obtain exploded views as well. Also obtain isometric views with hidden-line removal.

16.13 Repeat Problem 16.12, but using the top-down approach.

16.14 Obtain shaded images of the assemblies you create in Problem 16.12. For each assembly obtain two images, one of the collapsed view and the other of the exploded view.

Part III: Programming

Write the code so that the user can input values to control the output. The program should display the results. Use OpenGL with C or C++, or use Java 3D.

16.15 Implement the WCS method. Test the program for simple assemblies (assemblies of two to four parts).

16.16 Same as Problem 16.15, but using the mate method with the least-squares solution.

Finite Element Method

OAL

Understand finite element modeling and analysis, how to calculate stresses and strains, and how to use CAD systems to help perform FEM/FEA.

BJECTIVES

After reading this chapter, you should understand the following concepts:

- Why the finite element method
- Procedure of the finite element method
- FEA
- FEM
- Preprocessors: mesh generation
- Postprocessors: results display
- Understanding the results
- How CAD systems facilitate both FEM and FEA

CHAPTER HEADLINES

17.1 Introduction

Finite element modeling (FEM) and analysis (FEA) is one of the most popular mechanical engineering applications offered by existing CAD/CAM systems. This is attributed to the fact that the finite element method is perhaps the most popular numerical technique for solving engineering problems. The method is general enough to handle any complex shape or geometry (problem domain), any material properties, any boundary conditions, and any loading conditions.

The generality of the finite element method fits the analysis requirements of today's complex engineering systems and designs, where closed-form solutions of governing equilibrium equations are usually not available. In addition, it is an efficient design tool by which designers can perform parametric design studies by considering various design cases (different shapes, materials, loads, boundary conditions, and so forth), analyzing them, and choosing the optimum design.

The **finite element method** (also meaning FEA in this book) is a numerical analysis technique for obtaining approximate solutions to a wide variety of engineering problems. The method is based on dividing a complex shape into small elements, solving the equilibrium equations at hand for each element, and then assembling the elements' results to obtain the solution to the original problem. The shape division, and the choice of the element and the analysis types, are among the important decisions for the success of the method.

The interpretation of the results of FEA requires a good understanding of the principles of engineering such as linear/nonlinear mechanics, heat transfer, fluid mechanics, dynamics, and so forth. Using an FEM/FEA package that is offered by a CAD system is not a substitute to such understanding. An industrial team typically consists of a senior analyst who leads young engineers through the FEM/FEA process and who guides the what-if parametric studies.

17.2 Finite Element Procedure

The solution of a continuum problem by the finite element method usually follows an orderly, step-by-step process. The following steps show in general how the finite element method works. These steps will become more understandable when the FEA is covered in more detail later in this chapter.

1. **Create the finite elements.** The essence of the finite element method is to divide a continuum (the problem domain) into quasidisjoint nonoverlapping elements. This is achieved by replacing the continuum by a set of key points, called nodes, which when connected properly produce the elements. The collection of nodes and elements forms the finite element mesh.

   A variety of element shapes and types are available. The analyst or designer can mix element types to solve one problem. The number of nodes and elements that can be used in a problem is a matter of engineering judgment. As a general rule, the larger the number of nodes and elements, the more accurate the finite element solution, but also the more expensive the solution is; more memory space is needed to store the finite element model, and more computer time is needed to obtain the solution.

Figure 17.1 shows an example of creating a finite element mesh for a cantilever beam. The beam is made out of steel and supports a concentrated load at its free end. Figure 17.1c shows two types (four-node and six-node) of a quadrilateral (element shape) element.

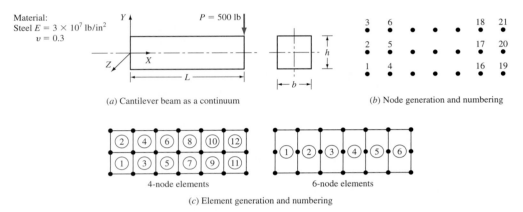

(a) Cantilever beam as a continuum

(b) Node generation and numbering

4-node elements

6-node elements

(c) Element generation and numbering

Figure 17.1   Finite element mesh of a beam.

2. **Approximate the solution within an element.** The variation of the unknown (called field variable) in the problem is approximated within each element by a polynomial. The field variable may be scalar (e.g., temperature) or a vector (e.g., horizontal and vertical displacements). Polynomials are usually used to approximate the solution over an element domain because they are easy to integrate and differentiate. The degree of the polynomial depends on the number of nodes per element, the number of unknowns (components of field variable) at each node, and certain continuity requirements along element boundaries.

3. **Develop element matrices and equations.** The finite element formulation presented in the next section involves transforming the governing equilibrium equations from the continuum domain to the element domain. Once the nodes and material properties of a given element are defined, its corresponding matrices (stiffness matrix, mass matrix, and so forth) and equations can be derived. Four methods are available to derive element matrices and equations: the direct method, the variational method, the weighted residual method, and the energy method. In this chapter, we cover the second (suitable for solid mechanics problems) and the third (suitable for thermalfluids problems) methods.

4. **Generate the global system matrix equation.** The individual element matrices are added together by summing the equilibrium equations of the elements to obtain the global matrices and the system of algebraic equations. Before solving this system, it must be

modified by applying the boundary conditions. If boundary conditions are not applied, wrong results are obtained, or a singular system of equations may result.

5. **Solve for the unknowns at the nodes.** The global system of algebraic equations is solved via Gauss elimination methods to provide the values of the field variables at the nodes of the finite element mesh. Values of field variables and their derivatives at the nodes form the complete finite element solution of the original continuum problem. Values at other points inside the continuum other than the nodes can be obtained although it is not customarily done.

6. **Interpret the results.** The final step is to analyze the solution and results obtained from the previous step to make design decisions. The correct interpretation of these results requires a sound background in both engineering and FEA. This is why treating FEM/FEA packages as a black box is usually not recommended and is, in fact, considered dangerous.

In the context of the preceding steps, it is clear that there are various critical decisions that practitioners of the finite element method have to make, for example, the type of analysis, the number of nodes, the degrees of freedom (components of the field variable) at each node, the element shape and type, the material type, the external loads, the boundary conditions, and finally the interpretation of the results. Making these decisions becomes more obvious after we discuss the preceding steps in more detail while covering FEA.

17.3 Finite Element Analysis

Finite element analysis is based on the following premise: Instead of solving the governing differential (equilibrium) equations directly, the finite element method solves an integral form of these equations. The solution of such an integral form is approximate. In obtaining such a solution, the finite element method leaves the differential operator intact and approximates the solutions space (by using interpolation functions). In contrast, the finite difference method amounts to a finite-difference approximation of the differential operator while keeping the solution space intact.

FEA begins with the differential equations of a system and ends with solving them approximately. It goes through a number of steps in between. It converts the differential equations into integral equations by using the variational approach or the weighted residual method. Next, it divides the problem domain into elements and develops the element equations. It assembles the element equations to obtain the global system matrix equation. The boundary conditions and the external loads are applied to this system before solving. The results of the solution are available at the nodes of the elements. Finite element analysts can display them in a graphical form to analyze them, to make design decisions and recommendations. The remaining sections of the chapter cover the mathematical details and the formulation of these steps.

CAD systems have FEM/FEA packages built into them. The packages are integrated with CAD databases to automate both FEM and FEA. These systems also provide designers with easy-to-use interfaces to display results in the form of curves, contours, or animation.

17.4 Development of Integral Equations

We cover two popular approaches to deriving integral equations which form the basis of the finite element solution. These approaches are the variational principle and the method of weighted residuals. In both approaches, we begin with the governing differential equations.

17.4.1 Variational Approach

In this approach, a variational statement (called a functional Π) is derived from the differential equation of the problem. Let us assume that the continuum problem at hand can be described in general by the following equation:

$$L(\phi) - f = 0 \tag{17.1}$$

over the continuum domain D and subject to the boundary conditions (b.c.)

$$\phi = \Phi \qquad \text{on } B \tag{17.2}$$

where ϕ is the dependent variable to solve for, f is a known function of the independent variable, L is a linear or nonlinear differential (ordinary or partial) operator, and B is the boundary of the domain D.

To derive the functional Π that corresponds to Eqs. (17.1) and (17.2), we seek an integral $I(\phi)$ whose first variation with respect to ϕ vanishes. The rationale behind this step is well explained and documented in books on variational calculus (calculus of variations) and on the finite element method. Readers should refer to these books to convince themselves. Multiplying Eq. (17.1) by the first variation of ϕ, $\delta\phi$, and integrating over the domain D, we get

$$\int_D \delta\phi[L(\phi) - f] \, dD = 0 \tag{17.3}$$

The objective now is to manipulate Eq. (17.3) in a way that allows the variational operator δ to be moved outside the integral sign, that is, Eq. (17.3) becomes

$$\delta \int_D [L^*(\phi) - f^*] \, dD = 0 \tag{17.4}$$

where $L^*(\phi)$ is the new operator resulting from the original operator $L(\phi)$. When Eq. (17.4) is reached, the integral [call it $I(\phi)$] in the equation is the desired functional Π because its first variation [as given by Eq. (17.4)] is zero, that is, Eq. (17.4) can be written as:

$$\delta[I(\phi)] = \delta\Pi = 0 \tag{17.5}$$

where

$$\Pi = I(\phi) = \int_D [L^*(\phi) - f^*] \, dD \tag{17.6}$$

There is no general consistent set of rules that can be followed while manipulating Eq. (17.1) to obtain Eq. (17.6). However, it is always the case that the boundary conditions given by Eq. (17.2) must be invoked during the manipulation process. Integration by parts is usually useful for one- and 2D problems as seen from the examples that follow. In some cases, the Gauss divergence theorem [Eq. (15.42)] may be useful.

While Eq. (17.1) is a good start from which to derive Eq. (17.6) for field problems (heat transfer and so forth), it is efficient for solid mechanics problems to write the functional Π directly using the variational principle (equivalent to minimum potential energy) or Hamilton principle.

EXAMPLE 17.1	**Derive the integral equation of a bar.**

Derive the functional Π for the bar shown in the figure on the right. The bar has a length L and a variable cross-sectional area A. It is loaded with a concentrated axial load P at its free end. The axial displacement at any point along the bar is governed by the following differential equation:

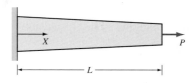

$$AEu'' = 0 \qquad (17.7)$$

where E is the modulus of elasticity of the bar material, and $u'' = d^2u/dx^2$.

SOLUTION	This example illustrates how a functional Π is derived from a differential

equation by using integration by parts. Care must be given to the boundary conditions during the derivation as shown below. In terms of Eq. (17.1), Eq. (17.7) shows that $\phi = u$, $L = AE\, d^2/dx^2$, and $f = 0$. The domain D of this example is 1D. It is the length L, that is, $D = \{x: x \in [0, L]\}$. Utilizing Eq. (17.3), following is how Eqs. (17.4) and (17.6) are derived for the bar.

Multiplying Eq. (17.7) by δu and integrating over the bar length, we can write

$$\int_0^L AEu'' \, \delta u \, dx = 0 \qquad (17.8)$$

Using the equality $u'' \, dx = du'$ and integrating by parts, we get

$$AEu' \, \delta u \Big|_0^L - \int_0^L AEu' \, d(\delta u) = 0 \qquad (17.9)$$

where we used the integration by parts rule:

$$\int_0^L U \, dV = UV \Big|_0^L - \int_0^L V \, dU \qquad (17.10)$$

where $U = EA \, \delta u$ and $V = u'$.

The first term in Eq. (17.9) can be reduced using the boundary conditions. At $x = 0$, $u = 0$; thus $\delta u = 0$. At $x = L$, $AEu' = P$. This condition is derived from axial loading where the stress $\sigma = P/A = E\varepsilon = Eu'$. The second term in Eq. (17.9) can be reduced if we interchange δ and d/dx in order to write:

$$d(\delta u) = \frac{d(\delta u)}{dx} dx = \delta\left(\frac{du}{dx}\right) dx = \delta(u') \, dx \qquad (17.11)$$

Substituting the boundary conditions and Eq. (17.11) into Eq. (17.9), we get

$$\int_0^L AEu' \, \delta u' \, dx - P \, \delta u_L = 0 \tag{17.12}$$

where u_L is the axial displacement at $x = L$. Rewriting $u' \, \delta u'$ as $\frac{1}{2}\delta u'^2$ (similar to $du^2 = 2u \, du$),

$$\int_0^L \delta\left(\frac{AE}{2}u'^2 dx\right) - \delta(Pu_L) = 0 \tag{17.13}$$

Interchanging δ and $\int$, we get:

$$\delta\left(\int_0^L \frac{AE}{2}u'^2 dx - Pu_L\right) = 0 \tag{17.14}$$

Comparing Eq. (17.4) and (17.14), the functional Π that corresponds to the differential Eq. (17.7) is

$$\Pi = \frac{1}{2}\int_0^L AEu'^2 dx - Pu_L \tag{17.15}$$

EXAMPLE 17.2   **Derive the integral equation of a cantilever beam.**

Derive the functional Π for the cantilever shown in the figure on the right. The beam has a length L and a cross-sectional moment of inertia I. It is loaded with a transverse load P at its free end. The beam's lateral deflection is given by the following equation:

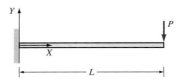

$$EIy'''' = 0 \tag{17.16}$$

where E is the modulus of elasticity of the bar material, and $y'''' = d^4y/dx^4$.

SOLUTION   In terms of Eq. (17.1), Eq. (17.16) shows that $\phi = y$, $L = EI \, d^4/dx^4$, and $f = 0$. The domain D of this example is 1D. It is the length L, that is, $D = \{x: x \in [0, L]\}$. Utilizing Eq. (17.3), following is how Eqs. (17.4) and (17.6) are derived for the cantilever.

Multiplying Eq. (17.16) by δy and integrating over the beam length, we get

$$\int_0^L EIy'''' \, \delta y \, dx = 0 \tag{17.17}$$

Integrating by parts ($U = EI \, \delta y$ and $V = y'''$), Eq. (17.17) becomes

$$EIy''' \, \delta y \Big|_0^L - \int_0^L EIy''' \, d(\delta y) = 0 \tag{17.18}$$

Let us use the boundary conditions of the beam to reduce the first term in Eq. (17.18). At $x = 0$, $y = 0$; thus $\delta y = 0$. At $x = L$, $EIy''' = V = -P$. This condition states that the shear force V at the free end (given by EIy''' from the beam theory) is equal to minus the applied load P. Similar to Eq. (17.11), $d(\delta y) = \delta(y') \, dx$. Thus, Eq. (17.18) becomes:

$$\int_0^L EIy''' \, \delta y' \, dx + P \, \delta y_L = 0 \tag{17.19}$$

where y_L is the beam deflection at the free end.

Using $dy'' = y''' \, dx$ and integrating by parts ($U = EI \, \delta y'$ and $V = y''$), the first term of Eq. (17.19) reduces to:

$$\int_0^L EIy''' \, \delta y' \, dx = EIy'' \, \delta y' \Big|_0^L - \int_0^L EIy'' \, d(\delta y') \tag{17.20}$$

At $x = 0$, $y' = 0$; thus $\delta y' = 0$ (slope at fixed end = 0). At $x = L$, $EIy'' = 0$. This condition states that the bending moment (given by EIy'' from the beam theory) at the free end is zero. Using these boundary conditions, and rewriting $y'' \, d(\delta y')$ as $\frac{1}{2} \delta y''^2 \, dx$, Eq. (17.20) becomes:

$$\int_0^L EIy''' \, \delta y' \, dx = -\int_0^L \frac{EI}{2} \, \delta y''^2 \, dx \tag{17.21}$$

Substituting Eq. (17.21) into Eq. (17.19), we get:

$$\delta\left(\int_0^L \frac{EI}{2} y''^2 \, dx - Py_L\right) = 0 \tag{17.22}$$

which gives

$$\Pi = \frac{1}{2}\int_0^L EIy''^2 \, dx - Py_L \tag{17.23}$$

EXAMPLE 17.3   **Derive the integral equation for an infinite slab.**

Derive the functional Π for the infinite slab shown in the figure on the right. The slab has a width L, and is subjected to a a constant uniform heat flux input q, as shown below. The temperature T inside the slab is governed by the following heat conduction equation:

$$KT'' = 0 \tag{17.24}$$

where K is the thermal conductivity and $T'' = d^2T/dx^2$.

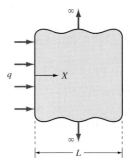

SOLUTION   In terms of Eq. (17.1), Eq. (17.24) shows that $\phi = T$, $L = K \, d^2/dx^2$, and $f = 0$. The domain D of this example is 1D. It is the length L, that is, $D = \{x : x \in [0, L]\}$. Utilizing Eq. (17.3), following is how Eqs. (17.4) and (17.6) are derived for the slab.

Multiplying Eq. (17.24) by T and integrating over the slab width gives

$$\int_0^L KT'' \, \delta T \, dx = 0 \tag{17.25}$$

Equation (17.25) is identical in form to Eq. (17.8). The same mathematical manipulation used to obtain Eq. (17.15) can be followed here. Thus, Eq. (17.25) gives

$$KT' \, \delta T \big|_0^L - \int_0^L KT' \, d(\delta T) = 0 \tag{17.26}$$

At $x = 0$, $KT' = -q$ (which is Fourier's Law). At $x = L$, $T = T_L$ (assuming that the slab is insulated at $x = L$ to keep T_L constant); thus $\delta T = 0$. After manipulating Eq. (17.26), the functional Π is

$$\Pi = \frac{1}{2} \int_0^L KT'^2 \, dx - qT_0 \tag{17.27}$$

where T_0 is the temperature at $x = 0$.

| **EXAMPLE 17.4** | **Derive the integral equation for a permeable media.** |

Derive the functional Π for the permeable medium shown in the figure on the right. The media has a cross-sectional area A. It also has a constant fluid flow V across the end at $x = 0$, as shown. The pressure P inside the media is given by the following equation:

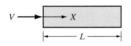

$$AgP'' = 0 \tag{17.28}$$

where g is the coefficient of permeambility and $P'' = d^2P/dx^2$.

| **SOLUTION** | In terms of Eq. (17.1), Eq. (17.28) shows that $\phi = P$, $L = Ag \, d^2/dx^2$, and |

$f = 0$. The domain D of this example is 1D. It is the length L, that is, $D = \{x: x \in [0, L]\}$. Utilizing Eq. (17.3), following is how Eqs. (17.4) and (17.6) are derived for the media.

Multiplying Eq. (17.28) δP and integrating over the media length, we get

$$\int_0^L AgP'' \, \delta P \, dx = 0 \tag{17.29}$$

Again, this equation is identical in form to Eq. (17.8). Integrating it gives

$$AgP' \, \delta P \big|_0^L - \int_0^L AgP' \, d(\delta P) = 0 \tag{17.30}$$

At $x = 0$, $AgP' = -V$ (fluid flow is proportional to pressure gradient P'). At $x = L$, $P = P_L$ (assuming that pressure at $x = L$ is kept constant). Thus, Eq. (17.30) gives:

$$\Pi = \frac{1}{2} \int_0^L AgP'^2 \, dx - VP_0 \tag{17.31}$$

Example 17.4 discussion:

The development of the functional Π in this example, as in Examples 17.1 to 17.3, shows that applying the boundary conditions correctly is crucial to their correct development of Π. Boundary conditions that are functions of derivatives of the dependent variable (at $x = 0$, AgP' in this example) are so-called natural boundary conditions. Conditions that are specified directly on the dependent variable (at $x = 0$, $P = P_0$, and at $x = L$, $P = P_L$ in this example) are referred to as geometric or rigid boundary conditions. It is only the geometric boundary conditions that are applied to the global system of equations mentioned in Step 4 in Section 17.2.

EXAMPLE 17.5   **Derive the integral equation for a plane stress problem.**

Derive the functional Π for a plane stress problem over the rectangular domain shown in the figure on the right. The concentrated loads P_1 and P_2 are applied at the center distances in the X and Y directions, respectively.

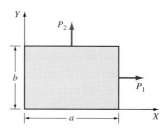

SOLUTION   Examples 17.1 to 17.4 are 1D; the functional Π for each problem is a 1D integral over the length of the problem domain. This example and Example 17.6 use 2D domains that result in 2D integrals for the functional Π.

For plane stress continuum problems, the following five equations can be written:

1. Displacement vector: $\qquad d = [u \quad v]^T$ (17.32)

where u and v are the displacement components in the X and Y directions, respectively.

2. Strain vector: $\varepsilon = [\varepsilon_x \quad \varepsilon_y \quad \gamma_{xy}]^T = \left[\dfrac{\partial u}{\partial x} \quad \dfrac{\partial v}{\partial y} \quad \dfrac{\partial u}{\partial y} + \dfrac{\partial v}{\partial x}\right]^T$ (17.33)

where ε_x, ε_y, and γ_{xy} are the normal strain in the X direction, the normal strain in the Y direction, and the shear strain, respectively.

In addition, assuming an isotropic material, the normal strain ε_z in the Z direction is given by

$$\varepsilon_z = \frac{v}{1-v}(\varepsilon_x + \varepsilon_y)$$ (17.34)

where v is the Poisson's ratio.

3. Stress vector: $\qquad \sigma = [\sigma_x \quad \sigma_y \quad \tau_{xy}]^T$ (17.35)

where σ_x, σ_y, τ_{xy} are the normal stress in the X direction, the normal stress in the Y direction, and the shear stress, respectively. For plane stress, $\sigma_z = \tau_{xz} = \tau_{yz} = 0$.

4. Hook's Law: $\qquad \sigma = [D]\varepsilon$ (17.36)

where $[D]$ is the material stiffness matrix and is given by

$$[D] = \frac{E}{1-v^2}\begin{bmatrix} 1 & v & 0 \\ v & 1 & 0 \\ 0 & 0 & \dfrac{1-v}{2} \end{bmatrix}$$ (17.37)

where E is the modulus of elasticity.

5. Static equilibrium:

$$\frac{\partial \sigma_x}{\partial x} + \frac{\partial \tau_{xy}}{\partial y} = 0$$

$$\frac{\partial \tau_{xy}}{\partial x} + \frac{\partial \sigma_y}{\partial y} = 0$$

(17.38)

The virtual work principle is utilized to derive the functional Π for this problem instead of using the differential equation [Eq. (17.38)]. This principle states that if a body in equilibrium is subjected to a small virtual displacement, it remains in equilibrium and the virtual work done by the internal forces is equal to the virtual work done by the externally applied forces, that is,

$$W_{int} = W_{ext}$$

(17.39)

For an elastic body, W_{int} results from the internal stresses and any other internal and body forces. For our problem, Eq. (17.39) becomes

$$\int_V \delta \varepsilon^T \sigma \, dV = P_1 \, \delta u_1 + P_2 \, \delta v_2$$

(17.40)

where u_1 and v_2 are the horizontal and vertical displacement components under P_1 and P_2, respectively.

The virtual work expression given by the left-hand side of Eq. (17.40) is general and applies to any solid mechanics problem. Equation (17.40) can be recast as:

$$\int_V \delta \varepsilon^T \sigma \, dV - P_1 \, \delta u_1 - P_2 \, \delta v_2 = 0$$

(17.41)

Our goal now is to find the functional Π from Eq. (17.41) by reducing it to the form given by Eq. (17.5). Substituting Eq. (17.36) into the first term in Eq. (17.41) gives

$$\int_V \delta \varepsilon^T \sigma \, dV = \int_V \delta \varepsilon^T [D] \varepsilon \, dV$$

(17.42)

For linear elastic material, the matrix $[D]$ is symmetric, and $\delta \varepsilon^T [D] \varepsilon$ can be reduced to $\frac{1}{2} \delta(\varepsilon^T [D] \varepsilon)$. To verify this result for our problem, we expand $\delta \varepsilon^T [D] \varepsilon$ to get:

$$\delta \varepsilon^T [D] \varepsilon = [\delta \varepsilon_x \quad \delta \varepsilon_y \quad \delta \gamma_{xy}][D] \begin{bmatrix} \varepsilon_x \\ \varepsilon_y \\ \gamma_{xy} \end{bmatrix}$$

(17.43)

Expanding Eq. (17.43) and reducing, we get:

$$\delta \varepsilon^T [D] \varepsilon = \frac{E}{1 - v^2} \left[\varepsilon_x \delta \varepsilon_x + v(\varepsilon_x \delta \varepsilon_y + \varepsilon_y \delta \varepsilon_x) + \varepsilon_y \delta \varepsilon_y + \frac{1 - v}{2} \gamma_{xy} \delta \gamma_{xy} \right]$$

$$= \frac{E}{2(1 - v^2)} \left[\delta \varepsilon_x^2 + 2v \, \delta(\varepsilon_x \varepsilon_y) + \delta \varepsilon_y^2 + \frac{1 - v}{2} \delta \gamma_{xy}^2 \right]$$

$$= \frac{E}{2(1 - v^2)} \delta \left(\varepsilon_x^2 + 2v \varepsilon_x \varepsilon_y + \varepsilon_y^2 + \frac{1 - v}{2} \gamma_{xy}^2 \right)$$

$$= \frac{1}{2}\delta(\varepsilon^T[D]\varepsilon) \tag{17.44}$$

Substituting Eq. (17.44) into Eq. (17.41) gives:

$$\frac{1}{2}\delta\int_V \varepsilon^T[D]\varepsilon\,dV - \delta(P_1 u_1) - \delta(P_2 v_2) = 0 \tag{17.45}$$

where $P_1\,\delta u_1$ and $P_2\,\delta v_2$ in Eq. (17.41) are written as shown in Eq. (17.45) because P_1 and P_2 are constants (or independent of displacements). Factoring the operator δ, Eq. (17.45) becomes:

$$\delta\left(\frac{1}{2}\int_V \varepsilon^T[D]\varepsilon\,dV - P_1 u_1 - P_2 v_2\right) = 0 \tag{17.46}$$

Therefore:

$$\Pi = \frac{1}{2}\int_V \varepsilon^T[D]\varepsilon\,dV - P_1 u_1 - P_2 v_2 \tag{17.47}$$

or

$$\Pi = \frac{1}{2}\int_V \varepsilon^T \sigma\,dV - P_1 u_1 - P_2 v_2 \tag{17.48}$$

Equation (17.47) or (17.48) is also recognized as the potential (strain) energy equation of the continuum solid under study. The integral term is the strain energy due to deformation, and the other two terms are the work done by the external loads P_1 and P_2. For plane problems, $dV = t\,dA$ where t is the thickness of the solid. Thus Eq. (17.48) becomes

$$\Pi = \frac{1}{2}t\int_A \varepsilon^T \sigma\,dA - P_1 u_1 - P_2 v_2 \tag{17.49}$$

Notice that the preceding approach to deriving Π is general and applicable to a wide variety of 2D and 3D solid mechanics problems.

EXAMPLE 17.6   **Derive the integral equation for a heat conduction problem.**

Derive the functional Π for a steady-state heat conduction problem over the rectangular domain shown in the figure to the right. The domain is subject to heat fluxes q_1 and q_2 in the X and Y directions, respectively. The material is isotropic and homogeneous with a thermal conductivity of K. The differential equation is given by:

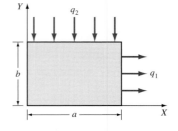

$$K\frac{\partial^2 T}{\partial x^2} + K\frac{\partial^2 T}{\partial y^2} = 0 \tag{17.50}$$

SOLUTION   Equation (17.50) is the familiar Laplace equation. Many physical problems are governed by the Laplace equation. In this example, we follow a general procedure to derive Π for systems governed by the Laplace equation. This equation can be written as

$$\nabla^2 \phi = 0 \tag{17.51}$$

where the 2D Laplace operator is

$$\nabla^2 = \frac{\partial^2}{\partial x^2} + \frac{\partial^2}{\partial y^2} \tag{17.52}$$

For Eq. (17.50), ϕ is obviously the temperature T. Substituting Eq. (17.51) into Eq. (17.3), we obtain

$$\int_A \nabla^2 \phi \, \delta\phi \, dA = 0 \tag{17.53}$$

Recognizing that $\nabla^2 \phi = \nabla \cdot \nabla \phi$, where $\nabla = \frac{\partial}{\partial x}\hat{\mathbf{i}} + \frac{\partial}{\partial y}\hat{\mathbf{j}}$, we can write

$$\nabla \cdot (\delta\phi \nabla \phi) = \nabla(\delta\phi) \cdot \nabla\phi + \delta\phi \nabla \cdot \nabla\phi$$

$$= \nabla(\delta\phi) \cdot \nabla\phi + \delta\phi \nabla^2\phi \tag{17.54}$$

or

$$\delta\phi \nabla^2\phi = \nabla \cdot (\delta\phi \nabla\phi) - \nabla(\delta\phi) \cdot \nabla\phi \tag{17.55}$$

Substituting Eq. (17.55) into the integral equation (17.53) gives

$$\int_A \nabla \cdot (\delta\phi \nabla\phi) \, dA - \int_A \nabla(\delta\phi) \cdot \nabla\phi \, dA = 0 \tag{17.56}$$

Applying the Gauss divergence theorem to the first term of Eq. (17.56) to invoke the boundary conditions, we get

$$\oint_B \hat{\mathbf{n}} \cdot \delta\phi \nabla\phi \, dS - \int_A \nabla(\delta\phi) \cdot \nabla\phi \, dA = 0 \tag{17.57}$$

The first term of Eq. (17.57) can be reduced further if we recognize that $\delta\phi$ is scalar and $\hat{\mathbf{n}} \cdot \nabla\phi = \frac{\partial\phi}{\partial n}$ where $\frac{\partial\phi}{\partial n}$ is the derivative normal to the boundary. For the second term, we realize that $\nabla(\delta\phi) = \delta\nabla\phi$ and $\nabla(\delta\phi).\nabla\phi = \delta\nabla\phi.\nabla\phi = \frac{1}{2}\delta(\nabla\phi.\nabla\phi)$. Thus, Eq. (17.57) becomes

$$\frac{1}{2}\delta \int_A \nabla\phi \cdot \nabla\phi \, dA - \oint_B \delta\phi \frac{\partial\phi}{\partial n} \, dS = 0 \tag{17.58}$$

The second term in Eq. (17.58) is the natural boundary conditions. Once this term is known for a given problem, the functional Π is easily obtained from Eq. (17.58). For example, if all natural boundary conditions are zeros, Eq. (17.58) gives

$$\Pi = \int_A \nabla\phi \cdot \nabla\phi \, dA = \int_A \left[\left(\frac{\partial\phi}{\partial x} \right)^2 + \left(\frac{\partial\phi}{\partial y} \right)^2 \right] dA \tag{17.59}$$

Let us apply Eq. (17.58) to the heat conduction problem described by Eq. (17.50). Assuming that the sides $x = 0$ and $y = 0$ are insulated, then the temperatures at these two sides are constants or $\delta T = 0$. Thus, we can write

$$\oint_B \delta T \frac{\partial T}{\partial n} \, dS = \int_0^b \delta T \frac{\partial T}{\partial x} \, dy + \int_0^a \delta T \frac{\partial T}{\partial y} \, dx \tag{17.60}$$

Using Fourier's Law, $K\left(\frac{\partial T}{\partial n} \right) = -q_n$, Eq. (17.60) reduces to

$$\oint_B \delta T \frac{\partial T}{\partial n} dS = \frac{1}{K} \int_0^b \delta T(-q_1) \, dy + \frac{1}{K} \int_0^a \delta T(q_2) \, dx$$

$$= -\frac{1}{K} \int_0^b \delta(q_1 T) \, dy + \frac{1}{K} \int_0^a \delta(q_2 T) \, dx \qquad (17.61)$$

Substituting Eq. (17.61) into Eq. (17.58) gives

$$\delta \left(\frac{1}{2} \int_A K \nabla \phi \cdot \nabla \phi \, dA + \int_0^b q_1 T \, dy - \int_0^a q_2 T \, dx \right) = 0 \qquad (17.62)$$

which gives

$$\Pi = \frac{1}{2} \int_0^b \int_0^a K \left[\left(\frac{\partial T}{\partial x} \right)^2 + \left(\frac{\partial T}{\partial y} \right)^2 \right] dx \, dy + \int_0^b q_1 T \, dy - \int_0^a q_2 T \, dx \qquad (17.63)$$

Notice that the preceding development can easily be extended to 3D domains. In this case, the 3D Laplace operator is

$$\nabla = \frac{\partial^2}{\partial x^2} + \frac{\partial^2}{\partial y^2} + \frac{\partial^2}{\partial z^2}$$

17.4.2   Method of Weighted Residuals

In some applications such as transient (time-dependent) heat conduction and some types of flow in fluid mechanics, variational principles may not exist or are unknown. However, the governing differential equations and the boundary conditions (natural and geometric) are known. In these applications, the method of weighted residuals is used instead of the variational principle to derive the integral equation which is equivalent to the equation $\delta \Pi = 0$ when the variational principle is used.

The method of weighted residuals is a numerical technique for obtaining approximate solutions to differential equations. Its application involves two steps. First, an approximate solution that satisfies the differential equation and its geometric boundary conditions is chosen. This approximate solution may be, for example, a polynomial with unknown coefficients.

When this solution is substituted into the differential equation and boundary conditions, an error or a residual results. This residual is chosen to vanish in some average sense over the entire solution domain. This leads to the integral equation which corresponds to the original differential equation.

In the second step of the method of weighted residuals, the integral equation is solved for the unknown coefficients, thus producing the solution.

To solve Eq. (17.1) subject to Eq. (17.2) using the method of weighted residuals, we begin by assuming an approximate solution ϕ_a such that

$$\phi \approx \phi_a = \sum_{i=1}^n c_i g_i \qquad (17.64)$$

where c_i are unknown coefficients and g_i are known assumed functions of the independent variable(s).

Substituting Eq. (17.64) into Eqs. (17.1) and (17.2) gives

$$L(\phi_a) - f \neq 0 \tag{17.65}$$

or

$$R = L(\phi_a) - f \tag{17.66}$$

where R is the residual or error. To minimize the residual R over the entire domain, we form a weighted average which should vanish over the domain; then

$$\int_D R W_i \, dD = \int_D [L(\phi_a) - f] W_i \, dD = 0, \qquad i = 1, 2, \ldots, n \tag{17.67}$$

There is a weighting factor W_i associated with each term of Eq. (17.64). These factors may be chosen according to different criteria. Galerkin's method, for example, uses the known functions g_i in Eq. (17.64) as W_i. Thus, Eq. (17.67) for Galerkin's method becomes:

$$\int_D R g_i \, dD = \int_D [L(\phi_a) - f] g_i \, dD = 0, \qquad i = 1, 2, \ldots, n \tag{17.68}$$

Equation (17.69) can be solved for the coefficients c_i.

Equation (17.68) is the integral equation for Galerkin's method, as Eq. (17.5) is the integral equation for the variational principle. Either equation can form the core of the finite element formulation.

EXAMPLE 17.7   **Derive the integral equation using Galerkin's method.**

Solve the following differential equation using Galerkin's method:

$$y'' + y = -x, \qquad 0 \le x \le 1$$

using these boundary conditions: $y(0) = y(1) = 0$

SOLUTION   The above boundary conditions are only geometric. Let us assume the following approximate solution (which satisfies the above geometric boundary conditions):

$$y(x) \approx y_a(x) = c_1 x(1 - x^2) + c_2 x^2(1 - x)$$

$$= c_1 g_1(x) + c_2 g_2(x)$$

$$= \sum_{i=1}^{2} c_i g_i(x)$$

Notice that $y_a(x)$ satisfies the b.c. Using the differential equation, we can write

$$R = y_a'' + y_a + x$$

$$= (-5x - x^3) c_1 + (2 - 6x + x^2 - x^3) c_2 + x$$

Using Eq. (17.68), we can write

$$\int_0^1 R g_1 \, dx = 0 \qquad \int_0^1 R g_2 \, dx = 0$$

or

and
$$\int_0^1 x(1-x^2)R\,dx = 0$$

$$\int_0^1 x^2(1-x)R\,dx = 0$$

Substituting R into these two equations, performing the integrations, and reducing give:

$$0.7238c_1 + 0.2738c_2 = 0.1333$$

$$0.2738c_1 + 0.1238c_2 = 0.05$$

Solving these two equations together gives $c_1 = 0.19211$ and $c_2 = -0.0210$. Thus

$$y(x) \approx y_a(x) = 0.19211x(1-x^2) - 0.0210x^2(1-x), \qquad 0 \le x \le 1$$

Note: Compare this approximate solution with the exact solution:

$$y(x) = \frac{\sin x}{\sin 1} - x$$

EXAMPLE 17.8   **Compare the variational method and Galerkin's method.**
Use Galerkin's method to derive the integral equation of the cantilever beam described by Eq. (17.16) in Example 17.2.

SOLUTION   Using Eq. (17.64), let us assume an approximate beam deflection in the following form

$$y_a(x) = \sum_{i=1}^{n} c_i g_i(x)$$

Using the above equation together with Eqs. (17.16) and (17.67), the beam integral equation is

$$\int_0^L EIy_a''''g_i(x)\,dx = 0, \qquad i = 1, 2, \ldots, n$$

Integrating by parts, this equation gives:

$$EIy_a''' g_i(x)\Big|_0^L - \int_0^L EIy_a''' g_i'(x)\,dx = 0, \qquad i = 1, 2, \ldots, n$$

At $x = L$, $EIy_a''' = V = -P$. At $x = 0$, $y_a(0) = 0$; thus $g_i(0) = 0$ becomes $c_i \ne 0$. Substituting these b.c. and integrating by parts again give:

$$-Pg_i(L) - EIy_a'' g_i'(x)\Big|_0^L + \int_0^L EIy_a'' g_i''(x)\,dx = 0, \qquad i = 1, 2, \ldots, n$$

The second term in the above equation vanishes because the slope at the fixed end is equal to 0; that is $y_a'(0) = 0$ which gives $g_i'(0) = 0$, and because $EIy_a''(L) = 0$. Thus, the above integral equation becomes:

$$\int_0^L EIy_a'' g_i''(x)\,dx - Pg_i(L) = 0, \qquad i = 1, 2, \ldots, n$$

This equation and the functional Π given by Eq. (17.23) give identical finite element equations as presented later in this chapter.

17.5 Finite Elements

The FEA begins by approximating the continuum under study by an assemblage of discrete finite elements, as presented in Section 17.2. The elements are interconnected at the nodal points (nodes) on the element boundaries. Therefore, the integral equation, Eq. (17.5) or (17.68), which is written for the continuum domain, is changed into a sum, over the number of elements, of integral equations each of which is written for the element domain. Equation (17.5) or (17.68) takes the following general form:

$$\int_D H(\phi)\, dD = 0 \tag{17.69}$$

The function $H(\phi)$ is a general function of the unknown variable ϕ. If the continuum domain D is divided into element domains D^e, then

$$D \approx \sum_{e=1}^{m} D^e \tag{17.70}$$

$$\phi \approx \sum_{e=1}^{m} \phi^e \tag{17.71}$$

where m is the total number of elements that form the continuum domain, and ϕ^e is the unknown variable within the element domain. Equation (17.69) can be rewritten at the element level as

$$\int_D H(\phi)\, dD = \sum_{e=1}^{m} \int_{D^e} H^e(\phi^e)\, dD^e = 0 \tag{17.72}$$

Equation (17.72) forms the basis from which to derive element equations and matrices.

Finite elements are characterized by several features. An element is completely described by its shape, the number and type (interior or exterior) of nodes, nodal variables, and the type of shape function. These element features control the behavior of the element in solving problems; that is, they control the accuracy and convergence of the finite element solution to the exact solution (which in many problems does not exist).

17.5.1 Element Shape

Most finite elements are geometrically simple to meet the fundamental premise of the finite element method that a continuum of an arbitrary shape can be accurately modeled by an assemblage of elements. This premise also implies that element dimensionality is the same as the continuum dimensionality. For 1D elements, there is one independent variable, and elements are line segments as shown in Figure 17.2a. The number of nodes per element depends on the nodal variables (degrees of freedom) and the continuity requirements between the elements.

At first glance, 1D elements may not seem necessary because 1D problems are usually governed by differential equations whose solutions can be obtained by other analytic or numeric techniques. However, these elements are useful in modeling 2D and 3D problems where part of the problem is 1D. For example, a cantilever beam with a spring attached to its free end. While the beam can be modeled by 2D elements, the spring is modeled as a 1D element.

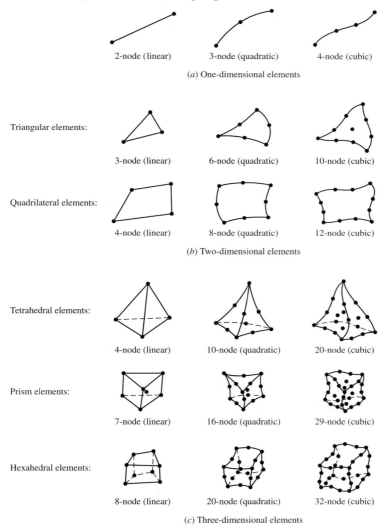

Figure 17.2   **Element types and dimensionality.**

Two popular 1D elements in solid mechanics are the truss and beam elements. The truss element has two to four nodes with one variable (axial displacement) per node. This element can only support tension or compression (no bending). The beam element also has two, three, or four nodes with two variables (beam deflection and slope) per node.

Figure 17.2*b* shows common 2D elements. Historically, triangular elements were developed first because they were easy to develop and formulate by hand. The 3-node linear triangular element is the simplest 2D element. The 10-node triangular element has nine nodes on its boundary (called exterior nodes) and one node inside the boundary (called interior node).

A quadrilateral element has a minimum of four nodes and as many as twelve nodes. In addition to modeling plane stress and strain problems, 2D elements can be used to model axisymmetric problems. In this case, an element represents the cross section of an axisymmetric shape whose thickness is given by the length of its arc segment. In general, 2D elements can model 2D and 2½D objects (continuums).

3D elements, shown in Figure 17.2*c*, are usually the 3D counterparts of 2D elements. These elements can be used to discretize 3D objects (continuums). Creating (and visualizing) 3D finite element meshes is usually a labor-intensive and error-prone process. Thus, using pre-processors and automatic mesh generation algorithms are beneficial in discretizing 3D objects.

The exterior nodes of any element shown in Figure 17.2 can be divided into two types: corner and midside nodes. Corner nodes are the minimum required nodes to define the element shape and, as the name implies, are located at the corners of the element. Midside nodes are added to improve the element accuracy or to meet continuity requirements between elements; they are located along the sides of the element.

The various elements supported by a particular FEA package are known as the element library. The higher the number of elements in the library, the more versatile the package is; that is, it can handle a large number of problems. In addition, many of these packages provide their users with the ability to add their own customized elements to the library. This is useful in a research and development environment. The element library should form an important criterion in evaluating existing FEA packages.

17.5.2 Element Nodal Degrees of Freedom

Continuum discretization into elements and element definition by nodes is effectively a replacement of the continuum by a set of nodes. The continuous unknown variable(s) in the governing integral equation(s) [Eq. (17.5) or (17.68)] becomes discrete unknown values at the nodes. These values are referred to as the nodal variables or nodal degrees of freedom. These degrees of freedom depend on the variable in the integral equation [Eq. (17.5) or (17.68)] and on the continuity requirements between elements.

The specific meaning of nodal degrees of freedom stems from the problem at hand. A specific element can be used in a variety of problems such as solid mechanics, heat transfer, or fluid mechanics as long as the associated integral equation is the same and has the same characteristics. For example, one degree of freedom at a node can represent an axial displacement for uniaxial loading problem (Example 17.1), a temperature for heat conduction problem (Example 17.3), or a pressure for fluid flow problem (Example 17.4).

17.5.3 Element Design

One of the most important decisions in the FEA is the selection of the element type and the nodal degrees of freedom. For practitioners and users of the finite element method, this means choosing the proper element with the proper number of nodes from the available element library (see Figure 17.2). The correlation between the integral equation governing the problem and the number of element nodes enables one to make the proper decision.

The ability to formulate the individual element equations from the integral equation [Eq. (17.5) or (17.68)] and the privilege to assemble these equations to obtain the global system of equations rely on the assumption that the interpolation polynomial within each element (ϕ^e) must satisfy certain requirements. These requirements stem from the need to ensure that Eq. (17.72) holds and that the approximate solution converges to the correct solution when we use an increasing number of small elements, that is, when we refine the element mesh. This type of convergence (the finer the mesh, the better the solution) is known as monotonic convergence.

The assurance of monotonic convergence as the element size decreases is controlled primarily by the highest derivative that appears in the element integral equation which is identical in form to the continuum integral equation [see Eq. (17.72)]. Let us assume that the element integral equation contains derivatives up to the rth order. For monotonic convergence, the assumed element interpolation polynomial must meet the following two requirements:

1. **Compatibility requirement.** C^{r-1} continuity must hold at element boundaries; that is, the field variable and any of its derivatives up to one order less than the highest-order derivative appearing in the integral equation [Eq. (17.5) or (17.68)] must be continuous at the interelement boundary. Elements that satisfy this requirement are desirable and are known as compatible or conforming elements. If elements are nonconforming, a patch test can be used to test their validity. Nonconforming elements are not discussed here. Physically, compatibility assures that no gaps occur between elements when the assemblage is loaded.

2. **Completeness requirement.** C^r continuity must hold within an element; that is, all uniform states of ϕ and its derivatives up to the highest order appearing in the integral equation should be represented in ϕ^e when, in the limit, the element size shrinks to zero. Physically, when the finite element mesh has a very large number of elements, the element domain D^e becomes so small that ϕ^e and its related derivatives are approximately constant in the domain. This condition is achieved by having a constant in ϕ^e and its derivatives. For example if $\phi^e = \alpha_1 + \alpha_2 x + \alpha_3 x^2$, then α_1 and α_2 are the uniform state of ϕ^e and its first derivative, respectively. As the element size goes to zero, that is, as x goes to zero, ϕ^e and $d\phi^e/dx$ go to α_1 and α_2, respectively, that is, constants (uniform states). Thus, the completeness requirement is achieved by having all the terms in the polynomial ϕ^e up to the order that provides a constant (at least) value of the rth derivative. Therefore, the polynomial should be of at least order r. In one dimension, any polynomial is complete, and a polynomial of order $r = n$ may be written as:

$$\phi^e(x) = \sum_{i=1}^{L} \alpha_i x^{i-1}$$
(17.73)

where $L = n + 1$.

In two dimensions, a complete n^{th} order polynomial is given by

$$\phi^e(x, y) = \sum_{k=1}^{L} \alpha_k x^i y^j, \qquad i + j \le n$$
(17.74)

where $L = (n + 1)(n + 2)/2$. The well-known Pascal triangle (Figure 17.3a) helps identify the term of a complete polynomial.

In three dimensions, a complete n^{th} order polynomial is given by:

$$\phi^e(x, y, z) = \sum_{l=1}^{L} \alpha_l x^i y^j z^k, \qquad i + j + k \le n$$
(17.75)

where $L = (n + 1)(n + 2)(n + 3)/6$. The coefficients α_i in Eqs. (17.73), (17.74), or (17.75) are unknown constants called the generalized coordinates of the element.

If an interpolation polynomial is incomplete (has fewer terms than L shown above), or if it has additional terms (more terms than L), then it is desirable but not necessary that the polynomial contain the appropriate terms to preserve symmetry as shown in Figure 17.3. When symmetry is preserved, the polynomial is said to have geometric invariance, that is, it is independent of the origin and orientation of the coordinate system. In addition to selecting the symmetric terms, geometric invariance requires that an element shape be unbiased, that is, does not have a preferred direction. For example, if we use triangular elements other than isosceles, rectangular elements of aspect ratio (ratio between the two dimensions of the rectangle) not equal to 1, or an element with a different number of nodes for each of its sides, a preferred direction already exists. This is why geometric invariance is fortunately a desirable but not necessary requirement.

If the interpolation polynomial of a finite element satisfies both the compatibility and the completeness requirements, how can we investigate the nonmonotonic convergence of the FEA to the exact solution, which may not be known? This question can also be rephrased as follows: How can we refine a finite element mesh to study the monotonic convergence of FEA? There are three conditions by which a mesh can be refined in a regular fashion: (1) elements can be made as small as possible without any gaps or other discontinuities between them; (2) all coarser (previous) meshes must be contained in the refined ones; and (3) the types of elements (the number of nodes per element) must remain the same for all meshes. This means that the interpolation polynomial remains unchanged during the process of mesh refinement.

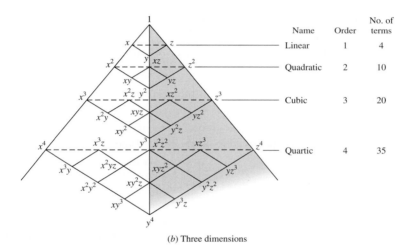

Name	Order	No. of terms
Linear	1	3
Quadratic	2	6
Cubic	3	10
Quartic	4	15
Quintic	5	21
Hexadic	6	28
Septic	7	36

(a) Two dimensions

Name	Order	No. of terms
Linear	1	4
Quadratic	2	10
Cubic	3	20
Quartic	4	35

(b) Three dimensions

Figure 17.3   Array of terms in complete polynomials.

Having decided on the order and the number of terms of the interpolation polynomial of a finite element, how can we use this information to design (or choose an element from a library) an element, that is, determine its number of nodes and the degrees of freedom per node? Applying the completeness requirement, we can determine the order of the interpolation polynomial and its number of terms, that is, the number of the generalized coordinates n_G. Applying the compatibility requirement, we can determine the degrees of freedom per node (n_F) which equal the field variable ϕ^e and any of its derivatives up to the rth derivative. Finally, the number of nodes per element must equal n_G/n_F, that is, the total degrees of freedom at the nodes of an element must be equal to n_G so that we can have enough equations to determine the generalized coordinates α_i as seen in the next section.

The preceding procedure to design or choose a finite element is iterative in nature. For example, n_G/n_F may have to be rounded to the nearest integer or the geometry of the element may impose a minimum number of nodes. A linear triangular element requires three corner nodes to define its geometry, while a linear quadrilateral element requires four corner nodes. Once the number of nodes n_G/n_F is changed, then the interpolation polynomial (that is, the number of α_i) must be adjusted accordingly. The excess of the number of nodes from the corner nodes is distributed between the element sides in a way that satisfies compatibility at the interelement boundary. At the first iteration of the procedure, emphasis is usually put on satisfying the compatibility requirement at the element corner nodes.

The design of finite elements becomes more complex as the compatibility requirements increase. C^0 continuity elements, that is, elements that require continuity of the field variable only, are much easier to design than C^1 continuity (continuity of ϕ and its first derivative) elements. This is why nonconforming elements are sometimes used. While these nonconforming elements preserve continuity of ϕ, they may violate its slope (or higher derivatives) continuity between elements, though naturally not at the nodes where such continuity is imposed.

EXAMPLE 17.9   **Design 1D finite elements.**

Design finite elements for the systems whose integral equations are derived in Examples 17.1 to 17.4.

SOLUTION   1D elements are used for all the domains described in Examples 17.1 to 17.4 because these domains are 1D. The functionals given by Eqs. (17.15), (17.27), and (17.31) are identical in form. Therefore, the same finite element can be used to discretize their corresponding domains. The highest derivative in any of the three equations is a first derivative. Thus, a linear interpolation polynomial is sufficient to satisfy the completeness requirement, and can be written, using Eq. (17.73), as:

$$\phi^e = \alpha_1 + \alpha_2 x$$

The compatibility requirement is satisfied by requiring continuity of ϕ^e at the interelement boundary. Because 1D elements are connected at nodes only (no side-by-side boundaries), then specified nodal values of ϕ^e cause the elements to conform. This also implies that one degree of freedom per node ($n_F = 1$) is sufficient. With $n_G = 2$ (unknown α_1 and α_2) and $n_F = 1$, a two-node element is sufficient to provide monotonic convergence of the FEA of the three systems under discussion.

Figure 17.4a shows the element and its degrees of freedom. Notice that $\phi = u$, T, or P. In solid mechanics applications, this element is known as a truss or bar element. The two-node element is known as a linear element because ϕ^e is linear as written in the preceding equation. It is also known as a constant-strain element because the strain within the element domain is $\varepsilon_x = d\phi/dx = \alpha_2$.

In using this element to obtain a problem solution, the element length L_e must be sufficiently small to yield good results. For faster-converging elements (which means using larger-size elements), we can use quadratic or cubic elements. The quadratic element requires three nodes (two corner nodes and one midside node) because it uses the second-order polynomial $\phi^e = \alpha_1 + \alpha_2 x + \alpha_3 x^2$. This polynomial provides a linear function ($\alpha_2 + 2\alpha_3 x$) of the slope $d\phi^e/dx$.

Similarly, the cubic element requires four nodes (two corner and two midside), uses the third-order polynomial $\phi^e = \alpha_1 + \alpha_2 x + \alpha_3 x^2 + \alpha_4 x^3$, and provides a quadratic function $(\alpha_2 + 2\alpha_3 x + 3\alpha_4 x^2)$ of the slope $d\phi^e/dx$.

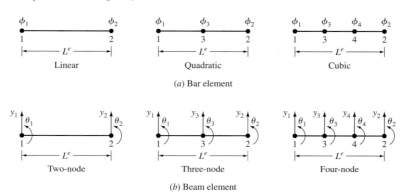

(a) Bar element

(b) Beam element

Figure 17.4   1D elements.

The functional Π given by Eq. (17.23) has a second derivative as its highest derivative. Thus, interelement compatibility is achieved by using a deflection y and slope $y' = \theta$ as degrees of freedom at each node. Completeness requires that the interpolation polynomial must be at least quadratic. Realizing that a minimum of two nodes are required to define an element geometrically, and with two degrees of freedom per node, a cubic polynomial $(y^e = \alpha_1 + \alpha_2 x + \alpha_3 x^2 + \alpha_4 x^3)$ of the beam deflection meets the compatibility, completeness, and geometric requirements of the element. Figure 17.4b shows the two-node beam element. Within the element, the polynomial provides a quadratic function of the slope y' and a linear function of the curvature y''. Three-node and four-node beam elements which use quintic and septic polynomials, respectively, are also shown in Figure 17.4b.

The numbering system of element nodes shown in Figure 17.4 is worth noting. The corner nodes are numbered first, followed by the midside nodes. This system is especially useful, particular when we cover the isoparametric evaluation in Section 17.12.

EXAMPLE 17.10   **Design 2D finite elements.**
Design finite elements for the systems whose integral equations are derived in Examples 17.5 and 17.6.

SOLUTION   The two functionals given by Eqs. (17.47) or (17.63) have the same form. Therefore, the element requirements for both the plane stress and heat conduction problems are the same. The finite elements for these problems are 2D, and so are their interpolation polynomials. Completeness requires at least a linear function which results in a constant-strain element for the plane stress problem and a constant-heat-flux element for the steady state heat conduction problem. Compatibility requires the continuity of ϕ at the nodes and the element sides. Thus, one degree of freedom per node is sufficient.

Let us now consider a triangular and a quadrilateral element. A triangular element requires a minimum of three nodes to define its geometry. Using the Pascal triangle shown in Figure 17.3a, the interpolation polynomial $\phi^e = \alpha_1 + \alpha_2 x + \alpha_3 y$ is the required one. For plane stress, the displacement has two components u and v which must be applied to each node. Thus, two interpolation polynomials, $u^e = \alpha_1 + \alpha_2 x + \alpha_3 y$ and $v^e = \beta_1 + \beta_2 x + \beta_3 y$, are required for each component.

For a heat conduction problem, the polynomial is $T^e = \alpha_1 + \alpha_2 x + \alpha_3 y$. Figure 17.5a shows the three-node linear triangle element. Higher-order elements such as quadratic and cubi elements (Figure 17.5a) can be used to improve the convergence of the FEA.

The complete quadratic polynomial is given by $\phi^e = \alpha_1 + \alpha_2 x + \alpha_3 y + \alpha_4 x^2 + \alpha_5 xy + \alpha_6 y^2$. With six unknown α values, a six-node element is used. Similarly, a ten-node element is required to support the complete cubic polynomial given by

$$\phi^e = \alpha_1 + \alpha_2 x + \alpha_3 y + \alpha_4 x^2 + \alpha_5 xy + \alpha_6 y^2 + \alpha_7 x^3 + \alpha_8 x^2 y + \alpha_9 xy^2 + \alpha_{10} y^3$$

A quadrilateral element requires a minimum of four nodes to define its geometry, as shown in Figure 17.5b. With three generalized coordinates (α_1, α_2, and α_3) in the linear polynomial, a minimum of four nodes, and one degree of freedom per node, we have three unknowns (α_1, α_2, and α_3) and four equations (one for each node). To resolve this discrepancy, we must add an additional term to the linear interpolation polynomial. This term must maintain the linearity of the polynomial in both x and y, as well as its geometric invariance (symmetry). The xy term in the quadratic row in the Pascal triangle is the one that meets these two criteria. Therefore, the interpolation polynomial for the linear quadrilateral element is given by $\phi^e = \alpha_1 + \alpha_2 x + \alpha_3 y + \alpha_4 xy$. This argument can be extended to quadratic and cubic elements. The polynomial for an eight-node quadrilateral element is given by

$$\phi^e = \alpha_1 + \alpha_2 x + \alpha_3 y + \alpha_4 x^2 + \alpha_5 xy + \alpha_6 y^2 + \alpha_7 x^2 y + \alpha_8 xy^2,$$

and for a twelve-node cubic element by

$$\phi^e = \alpha_1 + \alpha_2 x + \phi^e \alpha_3 y + \alpha_4 x^2 + \alpha_5 xy + \alpha_6 y^2 + \alpha_7 x^3 + \alpha_8 x^2 y + \alpha_9 xy^2 + \alpha_{10} y^3 + \alpha_{11} x^3 y + \alpha_{12} xy^3$$

The cubic terms $\alpha_{11} x^3 y$ and $\alpha_{12} xy^3$ are chosen to maintain the symmetry of ϕ^e (see Figure 17.3a). If the term $x^2 y^2$ were to be chosen from the Pascal triangle and symmetry maintained, the element would have to have eleven nodes only.

17.5.4 Element Shape Function

Thus far we have seen how a finite element can be designed to meet the monotonic convergence criteria (compatibility and completeness) and how a field variable can be represented within an element as a polynomial whose coefficients are the generalized coordinates of the element. In this section, we present how these coefficients are expressed in terms of the nodal values of the field variable as well as the coordinates of the nodes. This enables us later to express the element and global equations in terms of the nodal values which we need to solve for. It also enables us to express the element matrices in terms of the element geometry.

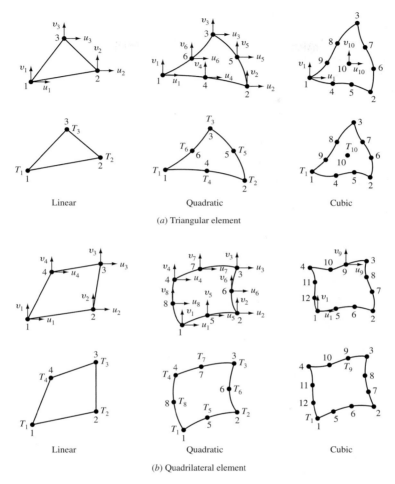

(a) Triangular element

(b) Quadrilateral element

Figure 17.5 2D elements.

To illustrate the procedure, let us consider linear triangular and linear quadrilateral elements with one degree of freedom per node. The field variable within a triangular element is given by:

$$\phi^e(x, y) = \alpha_1 + \alpha_2 x + \alpha_3 y \tag{17.76}$$

Let us assume that the field variable ϕ assumes the values ϕ_1, ϕ_2, and ϕ_3 at the element nodes 1, 2, and 3, respectively, as shown in Figure 17.6a. These nodal values must satisfy Eq. (17.76). Substituting the values with the nodal coordinates into Eq. (17.76) gives:

$$\phi_1 = \alpha_1 + \alpha_2 x_1 + \alpha_3 y_1$$

$$\phi_2 = \alpha_1 + \alpha_2 x_2 + \alpha_3 y_2 \qquad (17.77)$$

$$\phi_3 = \alpha_1 + \alpha_2 x_3 + \alpha_3 y_3$$

or, in matrix notation,

$$\phi = [G]\alpha \qquad (17.78)$$

where

$$\phi = [\phi_1 \quad \phi_2 \quad \phi_3]^T \qquad (17.79)$$

$$[G] = \begin{bmatrix} 1 & x_1 & y_1 \\ 1 & x_2 & y_2 \\ 1 & x_3 & y_3 \end{bmatrix} \qquad (17.80)$$

$$\alpha = [\alpha_1 \quad \alpha_2 \quad \alpha_3]^T \qquad (17.81)$$

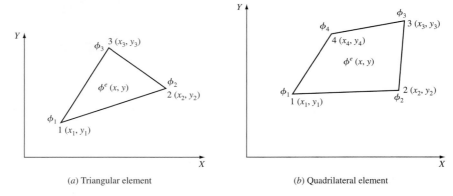

(a) Triangular element   (b) Quadrilateral element

Figure 17.6   Linear 2D elements.

Solving Eq. (17.78) for the generalized coordinates gives

$$\alpha = [G]^{-1}\phi \qquad (17.82)$$

Equation (17.76) can be written in a vector form as:

$$\phi^e = [P]\alpha \qquad (17.83)$$

where $[P] = [1 \ x \ y]$. Substituting Eq. (17.82) into (17.83), we get:

$$\phi^e = [P][G]^{-1}\phi = [N]\phi \qquad (17.84)$$

where

$$[N] = [P][G]^{-1} = [N_1 \quad N_2 \quad N_3] \qquad (17.85)$$

Thus, Eq. (17.84) can be expanded to give

$$\phi^e = N_1\phi_1 + N_2\phi_2 + N_3\phi_3 = \sum_{i=1}^{3} N_i\phi_i \tag{17.86}$$

where $N_i = N_i(x, y)$.

The matrix $[N]$ is known as the element shape function, and N_i is the nodal shape function associated with node i. The procedure followed to obtain Eq. (17.86) from Eq. (17.76) is general and applicable to other elements. The original interpolation polynomial as given by Eq. (17.76) should not be confused with the nodal shape functions N_i. The former applies to the whole element domain, whereas the latter refer to individual nodes, and collectively they represent the field variable behavior. Based on Eq. (17.86), one can think of the nodal shape functions N_i as interpolation functions that spread the values of the field variables at the element nodes (ϕ_1, ϕ_2, ϕ_3) back to the interior of the element.

An important property of N_i is that it must have a value of unity at node i and zero at any other node. We can prove this property as follows. At point (x_1, y_1) we have

$$\phi^e(x_1, y_1) = \phi_1 \tag{17.87}$$

Comparing Eq. (17.87) and Eq. (17.86) shows that $N_1(x_1, y_1) = 1$ and $N_2(x_1, y_1) = N_3(x_1, y_1) = 0$. Another property is that N_i has an identical form to ϕ with the difference in the coefficients (see Problem 17.9 at the end of the chapter). These two properties are useful in deriving shape functions by inspection as seen later in the isoparametric evaluation.

Let us apply the procedure just described to the linear quadrilateral element shown in Figure 17.6b. The interpolation polynomial is given by:

$$\phi(x, y) = \alpha_1 + \alpha_2 x + \alpha_3 y + \alpha_4 xy \tag{17.88}$$

Substituting the element nodal values into Eq. (17.88), similar equations to (17.77) to (17.86) can be developed. In this case:

$$\phi = [\phi_1 \quad \phi_2 \quad \phi_3 \quad \phi_4]^T \tag{17.89}$$

$$[G] = \begin{bmatrix} 1 & x_1 & y_1 & x_1y_1 \\ 1 & x_2 & y_2 & x_2y_2 \\ 1 & x_3 & y_3 & x_3y_3 \\ 1 & x_4 & y_4 & x_4y_4 \end{bmatrix} \tag{17.90}$$

$$\alpha = [\alpha_1 \quad \alpha_2 \quad \alpha_3 \quad \alpha_4]^T \tag{17.91}$$

$$[P] = [1 \quad x \quad y \quad xy] \tag{17.92}$$

$$[N] = [N_1 \quad N_2 \quad N_3 \quad N_4] \tag{17.93}$$

$$\phi^e = N_1\phi_1 + N_2\phi_2 + N_3\phi_3 + N_4\phi_4 = \sum_{i=1}^{4} N_i\phi_i \qquad (17.94)$$

The preceding equations can easily be generalized for n-node elements. For example the shape function and the interpolation polynomial can be written as:

$$[N] = [N_1 \quad N_2 \quad \cdots \quad N_n] \qquad (17.95)$$

$$\phi^e = N_1\phi_1 + N_2\phi_2 + \cdots + N_n\phi_n = \sum_{i=1}^{n} N_i\phi_i \qquad (17.96)$$

The procedure just described for deriving element shape functions is straightforward. However, its main disadvantages are

1. the inverse of the matrix $[G]$ may not exist for all element orientations in the global XYZ coordinate system, and

2. even if $[G]^{-1}$ exists, the computational effort to obtain it is not trivial, especially for higher-order elements such as cubic ones. This leads to finding N_i by inspection, which relies on the use of element local coordinate systems called natural coordinates. In this book, we refer to them as the parametric coordinates, as an extension of the parametric spaces used in geometric modeling.

17.5.5 Element Equations Using the Variational Approach

The variational finite element formulation is equivalent to the Ritz method used to solve differential equations with some differences not discussed here. The finite element solution to a problem involves choosing the unknown nodal values (ϕ_i) of the field variable ϕ to make the functional Π stationary or minimum. This is equivalent to minimizing the potential energy of a system in solid mechanics problems. To make Π, given by Eq. (17.6), stationary with respect to the nodal values of ϕ, we require (from variational calculus) that:

$$\delta\Pi = \delta I(\phi) = 0 \qquad (17.97)$$

Applying the discretization Eq. (17.72), Eq. (17.97) becomes

$$\delta\Pi = \sum_{i=1}^{m} \delta\Pi^e = \sum_{i=1}^{m} \delta I^e(\phi^e) = 0 \qquad (17.98)$$

where the variation of I^e is taken only with respect to the nodal values associated with the element e. Equation (17.98) implies

$$\begin{bmatrix} \dfrac{\partial I^e}{\partial \phi_1} \\[2mm] \dfrac{\partial I^e}{\partial \phi_2} \\[2mm] \vdots \\[2mm] \dfrac{\partial I^e}{\partial \phi_n} \end{bmatrix} = \begin{bmatrix} 0 \\ 0 \\ \vdots \\ 0 \end{bmatrix} \tag{17.99}$$

where n is the number of nodes of the element. In vector form, Eq. (17.99) is

$$\left\{ \frac{\partial I^e}{\partial \phi_i} \right\} = \{0\}, \qquad i = 1, 2, \dots, n \tag{17.100}$$

Equation (17.99) or (17.100) provides the necessary element equations that characterize the behavior of element e. The development of the element matrices involves substituting Eq. (17.96) into the element functional Π^e or I^e, and then differentiating the result to obtain Eq. (17.99). This procedure is illustrated in the following examples.

EXAMPLE 17.11   **Derive 1D element equations using the variational approach.**

Use the variational approach to derive the element equations for the systems whose integral equations are derived in Examples 17.1 to 17.4.

SOLUTION   We have designed two-node, three-node, and four-node elements for these systems. We use the two-node element in this example to illustrate how to derive the element equations and matrices. Using Eq. (17.96) with $n = 2$, we get

$$\phi^e = N_1 \phi_1 + N_2 \phi_2 \tag{17.101}$$

Applying Eq. (17.72) to the functional Π given by Eq. (17.15), we write

$$\Pi^e = I^e = \frac{1}{2} \int_0^{L^e} A E u^{e'^2} \, dx - P u_L^e \tag{17.102}$$

where L^e is the element length. Substituting Eq. (17.101) into Eq. (17.102), and realizing that $\phi^e = u^e$, we obtain

$$I^e = \frac{1}{2} \int_0^{L^e} A E (N_1' u_1 + N_2' u_2)^2 \, dx - P(N_1 u_1 + N_2 u_2)_L \tag{17.103}$$

Applying the stationarity condition given by Eq. (17.99), Eq. (17.103) gives

$$\int_0^{L^e} A E (N_1' u_1 + N_2' u_2) N_1' \, dx - P N_{1L} = 0 \tag{17.104}$$

and

$$\int_0^{L^e} A E (N_1' u_1 + N_2' u_2) N_2' \, dx - P N_{2L} = 0 \tag{17.105}$$

Because $N_i = 1$ at node i and zero elsewhere, the second term in Eqs. (17.104) and (17.105) is equal to zero except for the boundary element adjacent to the bar end at $x = L$. Thus, for any other element, Eqs. (17.104) and (17.105) can be written in a matrix form as:

$$\begin{bmatrix} k_{11} & k_{12} \\ k_{21} & k_{22} \end{bmatrix}^e \begin{bmatrix} u_1 \\ u_2 \end{bmatrix}^e = \begin{bmatrix} 0 \\ 0 \end{bmatrix} \qquad \text{or} \qquad [K]^e \mathbf{U}^e = \mathbf{0} \tag{17.106}$$

where any element k_{ij} is given by

$$k_{ij} = \int_0^{L^e} AEN_i' \, N_j' \, dx \tag{17.107}$$

Equation (17.106) is the element equation. $[K]^e$ is known as the element stiffness matrix, and $\mathbf{U}^e$ is the element displacement vector. Notice that $[K]^e$ is symmetric, as can easily be seen from Eq. (17.107). In general, many engineering applications result in a symmetric positive definite stiffness matrix. If we know the form of N_i and perform the integral of Eq. (17.107), $[K]^e$ would be completely evaluated. Isoparametric evaluation of $[K]^e$ is discussed in Section 17.12.

For the boundary element adjacent to the end $x = L$, $N_{1L} = 0$ and $N_{2L} = 1$. Thus, this element equation is

$$\begin{bmatrix} k_{11} & k_{12} \\ k_{21} & k_{22} \end{bmatrix}^e \begin{bmatrix} u_1 \\ u_2 \end{bmatrix}^e = \begin{bmatrix} 0 \\ P \end{bmatrix} \tag{17.108}$$

The element equations and matrices can be obtained for the heat conduction and fluid flow systems following the same approach. Using $\phi^e = T^e$ in Eq. (17.101) together with Eq. (17.27), the element equation is identical to Eq. (17.106), but with replacing u by T. In this case, the element matrix $[K]^e$ is known as the element conductivity matrix, and any element k_{ij} is given by

$$k_{ij} = \int_0^{L^e} KN_i' \, N_j' \, dx \tag{17.109}$$

The equation of the boundary element at $x = 0$ is

$$\begin{bmatrix} k_{11} & k_{12} \\ k_{21} & k_{22} \end{bmatrix}^e \begin{bmatrix} T_1 \\ T_2 \end{bmatrix}^e = \begin{bmatrix} q \\ 0 \end{bmatrix} \tag{17.110}$$

Similarly for the fluid flow system, $\phi^e = P^e$, and the element matrix is the flow matrix whose elements are given by

$$k_{ij} = \int_0^{L^e} AgN_i' \, N_j' \, dx \tag{17.111}$$

The equation of the boundary element at $x = 0$ is

$$\begin{bmatrix} k_{11} & k_{12} \\ k_{21} & k_{22} \end{bmatrix}^e \begin{bmatrix} P_1 \\ P_2 \end{bmatrix}^e = \begin{bmatrix} V \\ 0 \end{bmatrix} \tag{17.112}$$

We now develop the element equations and matrices and matrices for beam bending. Here, we have two degrees of freedom (y and θ) per node. From Example 17.9, the beam deflection for a two-node element is given by

$$y^e = \alpha_1 + \alpha_2 x + \alpha_3 x^2 + \alpha_4 x^3 \tag{17.113}$$

Differentiating Eq. (17.113), the beam slope is

$$\theta^e = \frac{dy^e}{dx} = \alpha_2 + 2\alpha_3 x + 3\alpha_4 x^2 \tag{17.114}$$

Substituting the nodal values (y_1, θ_1, y_2, θ_2) into Eqs. (17.113) and (17.114) and reducing, Eq. (17.113) can be written as

$$y^e = N_1 y_1 + N_2 \theta_1 + N_3 y_2 + N_4 \theta_2 \tag{17.115}$$

Although we can continue with Eq. (17.115) to develop the element equations and matrices, we stop the development here and leave it to interested readers. The beam element that utilizes the polynomial given by Eq. (17.113) is C^1 continuous and is sometimes known as the standard-type beam element. Its nodal shape functions cannot easily be obtained by inspection for isoparametric evaluation.

We present here another approach that treats both y and θ as independent variables. The merit of this approach is that it leads to beam elements that are C^0 continuous and whose shape functions are easier to find by inspection.

Using the beam slope $y' = \theta$, Eq. (17.23) can be written as:

$$\Pi = \frac{1}{2} \int_0^L EI\theta'^2 \, dx - Py_L \tag{17.116}$$

Equation (17.116) is identical in form to the equations of the other systems we discussed. We can write an interpolation polynomial (linear for two-node elements) for both y and θ as follows:

$$y^e = N_1 y_1 + N_2 y_2 \tag{17.117}$$

$$\theta^e = N_1 \theta_1 + N_2 \theta_2 \tag{17.118}$$

Notice that N_1 and N_2 in Eqs. (17.117) and (17.118) are the same because they are functions of the element geometry only.

Applying Eq. (17.72) to Eq. (17.116) and using Eqs. (17.117) and (17.118) in the result, we obtain

$$\Pi^e = I^e = \frac{1}{2} \int_0^{L^e} EI(N_1' \theta_1 + N_2' \theta_2)^2 \, dx - P(N_1 y_1 + N_2 y_2)_L \tag{17.119}$$

Following the same procedure we used to obtain Eqs. (17.106) to (17.108), Eq. (17.119) gives

$$\begin{bmatrix} k_{11} & k_{12} \\ k_{21} & k_{22} \end{bmatrix}^e \begin{bmatrix} \theta_1 \\ \theta_2 \end{bmatrix}^e = \begin{bmatrix} 0 \\ 0 \end{bmatrix} \tag{17.120}$$

where

$$k_{ij} = \int_0^{L^e} EIN_i' N_j' \, dx \tag{17.121}$$

Equation (17.108) applies to the boundary beam element at $x = L$ after replacing u_1 and u_2 by θ_1 and θ_2, respectively.

Equation (17.120) does not include the displacements y_1 and y_2 at the element nodes. This is because Eq. (17.116) does not include them. Terms including y_1 and y_2 appear in Eq. (17.116) if the strain energy due to the beam shear deformation is included, and/or if external distributed loads and moments are added (see Problem 17.11 at the end of the chapter). In this case, the matrix equation [Eq. (17.120)] changes from 2×2 to 4×4, and the element displacement vector $\mathbf{U}^e$ becomes $[y_1 \; \theta_1 \; y_2 \; \theta_2]^{eT}$. Using Eq. (17.120), the finite element solution gives the slopes θ at the nodes. To obtain the nodal displacements, we use $\theta = dy/dx$, or $\Delta y = \theta \, \Delta x$, or $y_{i+1} = y_i + \theta_i L^e$.

Readers can easily extend the procedure presented in this example to three-node and four-node elements. Refer to Problem 17.10 at the end of the chapter.

EXAMPLE 17.12   **Derive 2D element equations using the variational approach.**
Use the variational approach to derive the element equations for the systems whose integral equations are derived in Examples 17.5 and 17.6.

SOLUTION   We have designed triangular and quadrilateral elements of various numbers of nodes for these systems. Let us derive the element equations and matrices for a 3-node triangular element. The two independent displacement components u and v for the plane stress problem can be written using Eq. (17.86) as follows:

$$u^e = N_1 u_1 + N_2 u_2 + N_3 u_3 \tag{17.122}$$

$$v^e = N_1 v_1 + N_2 v_2 + N_3 v_3 \tag{17.123}$$

which can be written in vector form as:

$$\begin{bmatrix} u \\ v \end{bmatrix}^e = \begin{bmatrix} N_1 & 0 & N_2 & 0 & N_3 & 0 \\ 0 & N_1 & 0 & N_2 & 0 & N_3 \end{bmatrix} \begin{bmatrix} u_1 \\ v_1 \\ u_2 \\ v_2 \\ u_3 \\ v_3 \end{bmatrix}^e = [N]\mathbf{U}^e \tag{17.124}$$

Substituting Eqs. (17.122) and (17.123) into Eq. (17.33) and rearranging, we obtain

$$\varepsilon = [B]\mathbf{U}^e \tag{17.125}$$

where

$$[B] = \begin{bmatrix} N_{1,x} & 0 & N_{2,x} & 0 & N_{3,x} & 0 \\ 0 & N_{1,y} & 0 & N_{2,y} & 0 & N_{3,y} \\ N_{1,y} & N_{1,x} & N_{2,y} & N_{2,x} & N_{3,y} & N_{3,x} \end{bmatrix} \tag{17.126}$$

and $\mathbf{U}^e$ is the element displacement vector shown in Eq. (17.124). $N_{i,x}$ and $N_{i,y}$ means the partial derivative of N_i of node i with respect to x and y, respectively.

Applying Eq. (17.72) to Eq. (17.47) and using Eq. (E17.125), we obtain

$$\Pi^e = I^e = \frac{1}{2}\int_{V^e} \mathbf{U}^{eT}[B]^T[D][B]\mathbf{U}^e \, dV$$

$$-P_1[N_1 \;\; 0 \;\; N_2 \;\; 0 \;\; N_3]_1 \mathbf{U}^e$$

$$-P_2[0 \;\; N_1 \;\; 0 \;\; N_2 \;\; 0 \;\; N_3]_2 \mathbf{U}^e \tag{17.127}$$

where $[N_1 \; 0 \; ...]_1$ and $[0 \; N_1 \; 0 \; ...]_2$ are evaluated at the points where the load P_1 and P_2 are applied, respectively. These terms contribute as loads in the appropriate element equation as shown in Example 17.11.

Applying the stationarity condition given by Eq. (17.99), six independent element equations can be written in a matrix form as

$$[K]^e\mathbf{U}^e = \mathbf{0} \tag{17.128}$$

where the element stiffness matrix $[K]^e$ is given by

$$[K]^e = \int_{V^e}[B]^T[D][B] \, dV$$

$$= t\int_{A^e}[B]^T[D][B] \, dA \tag{17.129}$$

The $[K]^e$ matrix of Eq. (17.129) is symmetric and positive definite. Readers are encouraged to expand Eq. (17.129) to obtain the individual elements of $[K]^e$.

Elements whose nodes carry the applied loads P_1 and P_2 have a nonzero right-hand side in Eq. (17.128).

For 2D problems, more than one element may share, for example, the node that carries P_1. In such a case, P_1 can be divided equally between the elements. There is a better approach for handling applied loads than this that we will discuss when we cover the assembly process and the application of external loads.

For the heat conduction problem, we assume the temperature polynomial to be

$$T^e = N_1T_1 + N_2T_2 + N_3T_3 = [N]\mathbf{T}^e \tag{17.130}$$

where $[N] = [N_1 \; N_2 \; N_3]$ and $\mathbf{T}^e = [T_1 \; T_2 \; T_3]^T$. To simplify manipulations, Eq. (17.63) can be rewritten as

$$\Pi = \frac{1}{2}\int_0^b\int_0^a \mathbf{T'}^T[K_T]\mathbf{T'} \, dx \, dy + \int_0^b q_1T \, dy - \int_0^a q_2T \, dx \tag{17.131}$$

where

$$\mathbf{T}' = \begin{bmatrix} \dfrac{\partial T}{\partial x} & \dfrac{\partial T}{\partial y} \end{bmatrix}^T$$

and $[K_T]$ given by:

$$[K_T] = \begin{bmatrix} K & 0 \\ 0 & K \end{bmatrix} \tag{17.132}$$

The first term in Eq. (17.131) is identical in form to that of Eq. (17.47). Using Eqs. (17.72) and (17.130), Eq. (17.131) gives:

$$\Pi^e = I^e = \frac{1}{2} \int_0^{b^e} \int_0^{a^e} \mathbf{T}^{eT} [B]^T [K_T][B] \mathbf{T}^e \, dx \, dy + \int_0^{b^e} q_1 [N] \mathbf{T}^e \, dy - \int_0^{a^e} q_2 [N] \mathbf{T}^e \, dx \tag{17.133}$$

where

$$[B] = \begin{bmatrix} N_{1,x} & N_{2,x} & N_{3,x} \\ N_{1,y} & N_{2,y} & N_{3,y} \end{bmatrix} \tag{17.134}$$

Similar to the plane stress problem, applying the stationarity condition given by Eq. (17.99) to Eq. (17.133), we obtain the element equations as

$$[K]^e \mathbf{T}^e - \mathbf{Q}_1 + \mathbf{Q}_2 = \mathbf{0} \tag{17.135}$$

where $[K]^e$ is the element conductivity matrix,

$$[K]^e = \int_0^{b^e} \int_0^{a^e} [B]^T [K][B] \, dx \, dy \tag{17.136}$$

and $\mathbf{Q}_1$ and $\mathbf{Q}_2$ are heat flow vectors,

$$\mathbf{Q}_1 = \int_0^{b^e} q_1 [N]^T \, dy \tag{17.137}$$

$$\mathbf{Q}_2 = \int_0^{a^e} q_2 [N]^T \, dx \tag{17.138}$$

17.5.6 Element Equations Using Weighted Residuals

Galerkin's method described by Eqs. (17.64) and (17.68) forms the basis of Galerkin's finite element method. Using the interpolation polynomial given by Eq. (17.96) in place of Eq. (17.64) and applying the discretization equation (17.72) to Eq. (17.68), the element equations based on Galerkin's finite element method are given by

$$\int_{D^e} [L(\phi^e) - f^e] N_i \, dD = 0, \qquad i = 1, 2, \dots, n \tag{17.139}$$

where n is the number of element nodes.

Before deciding on the explicit form of the shape function N_i to meet the compatibility and completeness requirements, Eq. (17.139) is always integrated by parts (or by using the Gauss divergence theorem), as we did in deriving a functional Π to lower the highest-order derivative appearing in this equation. This, in turn, reduces the order of the element continuity (C^0, C^1, etc.) and, therefore, the order of its interpolation polynomial. There are close similarities between manipulating Eq. (17.139) to develop the element equations and the variational approach as illustrated in Example 17.13.

EXAMPLE 17.13	**Derive 1D element equations using Galerkins's method.**

Use Galerkin's method to derive the element equations for the cantilever beam of Examples 17.2.

SOLUTION   We have derived the integral equation of the beam using Galerkin's method in Example 17.8. That equation can be utilized to develop the standard-type beam element. In this example, as we did in Example 17.11, we use the beam slope θ. From Example 17.8, the approximate solution and the beam integral equation are given in terms of the beam deflection y as follows:

$$y_a = \sum_{i=1}^{n} c_i g_i(x) \tag{17.140}$$

and

$$\int_0^L EIy_a'' g_i''(x)\, dx - Pg_i(L) = 0, \qquad i = 1, 2, \ldots, n \tag{17.141}$$

Differentiating Eq. (17.141), the beam approximate slope is

$$y_a' = \theta_a = \sum_{i=1}^{n} c_i g_i'(x) = \sum_{i=1}^{n} c_i g_{i\theta}(x) \tag{17.142}$$

Using Eq. (17.142) in Eq. (17.141) by substituting $y_a'' = \theta_a'$ and $g_i''(x) = g_{i\theta}'(x)$, we obtain

$$\int_0^L EI\theta_a' g_{i\theta}'(x)\, dx - Pg_i(L) = 0, \qquad i = 1, 2, \ldots, n \tag{17.143}$$

If we replace Eqs. (17.140) and (17.142) by Eqs. (17.117) and (17.118), and we write Eq. (17.143) for an element domain, we obtain

$$\int_0^{L^e} EI(N_1' \theta_1 + N_2' \theta_2)N_i' - PN_{iL} = 0, \qquad i = 1, 2, \ldots, n \tag{17.144}$$

The element equations given by Eq. (17.144) are identical to those obtained by applying the stationarity condition to Eq. (17.119). Therefore, the element matrices are identical to those obtained via the variational approach.

17.6 Assembly of Element Equations

The discretization process given by Eq. (17.72) results in a matrix equation for each element in the finite element mesh. To obtain the finite element solution, we must combine these equations into one matrix equation. This process of combining element equations is known as the assembly process, and the matrix equation is known as the global or total matrix equation. This global matrix equation is a system of algebraic equations.

The assembly process and its procedure for constructing the system (continuum) equations from the element equations are the same regardless of the type of problem being analyzed, the complexity of the system of elements, or the mixture of element types that make the mesh.

An element matrix equation cannot be solved by itself for the unknown nodal values at the element nodes because it does not represent element equilibrium. No internal reactions are included in the element equation. And because these reactions always appear in pairs of equal and opposite magnitudes, assembly of the elements eliminates these reactions in principle (they are not included in element equations) and makes the global system of equations a valid system to solve for the nodal values.

In addition, the assembly process is valid because of the compatibility requirement at the nodes which is reinforced during the element design and the choice of the interpolation polynomial.

So that we can discuss the general assembly procedure and the algorithm to carry out the assembly process, the matrix equation of any element can be written in the following form:

$$[K]^e \phi^e = \mathbf{R}^e \tag{17.145}$$

where $[K]^e$ is the element matrix, ϕ^e is the vector of the element nodal values, and $\mathbf{R}^e$ is the element load vector.

From the previous examples, $[K]^e$ may represent stiffness, conductivity, or flow matrix; ϕ^e may represent displacements, temperatures, or pressures; $\mathbf{R}^e$ includes the effect of external and internal (body forces, heat sources, etc.) loads. Realizing that $[K]^e \phi^e - \mathbf{R}^e$ is equivalent to the element integral in Eq. (17.72), Eq. (17.145) can be rewritten as

$$\int_D H(\phi)\, dD = \sum_{e=1}^{m} ([K]^e \phi^e - \mathbf{R}^e) = 0 \tag{17.146}$$

or

$$\sum_{e=1}^{m} [K]^e \phi^e = \sum_{e=1}^{m} \mathbf{R}^e \tag{17.147}$$

or

$$[K]\phi = \mathbf{R} \tag{17.148}$$

where

$$[K] = \sum_{e=1}^{m} [K]^e \tag{17.149}$$

$$\phi = [\phi_1 \quad \phi_2 \quad \cdots \quad \phi_n]^T \tag{17.150}$$

$$\mathbf{R} = \sum_{e=1}^{m} \mathbf{R}^e \tag{17.151}$$

$[K]$, ϕ, and $\mathbf{R}$ are the global (assembled) system matrix, global nodal values, and global load vector, respectively. The size of $[K]$, ϕ, and $\mathbf{R}$ is determined from the number of nodes N multiplied by the number of degrees of freedom per node n_F. The matrix $[K]$ is $n \times n$ where $n = N \times n_F$, and both ϕ and $\mathbf{R}$ have n elements.

The assembly procedure and its related algorithm is based on accumulating (adding) the contributions of elements attached to a given node to the equations, in the global system, that correspond to the degrees of freedom of this node.

To illustrate this procedure, let us consider the three-node, two-element mesh shown in Figure 17.7. There are two numbering schemes that are required to assemble element matrices: global and local, as shown in Figure 17.7a. The global numbering scheme is established to identify the mesh nodes and elements. This scheme can be created manually or algorithmically. The local numbering scheme is used to generate the element equations before assembly.

The relationship between the two schemes is established at the user input level of mesh data. For example, when a user inputs nodes 2 and 3 as the nodes of element ② in Figure 17.7a, node 2 becomes its local node 1 and node 3 becomes its local node 2.

Utilizing the global numbering scheme, the element equations of the two elements shown in Figure 17.7a can be written as follows:

$$\begin{bmatrix} k_{11}^{①} & k_{12}^{①} \\ k_{21}^{①} & k_{22}^{①} \end{bmatrix} \begin{bmatrix} \phi_1 \\ \phi_2 \end{bmatrix} = \begin{bmatrix} R_1^{①} \\ R_2^{①} \end{bmatrix} \tag{17.152}$$

$$\begin{bmatrix} k_{11}^{②} & k_{12}^{②} \\ k_{21}^{②} & k_{22}^{②} \end{bmatrix} \begin{bmatrix} \phi_2 \\ \phi_3 \end{bmatrix} = \begin{bmatrix} R_1^{②} \\ R_2^{②} \end{bmatrix} \tag{17.153}$$

Each row in an element equation represents an equation in a given direction, that is, an equation for a given degree of freedom. For example, the equation $k_{11}^{①}\phi_1 + k_{12}^{①}\phi_2 = R_1^{①}$ from the first element matrix Eq. (17.152) is associated with the degree of freedom of global node 1, that is, ϕ_1.

Global node 2 has two contributions, an equation from each element ① and ②. When these two equations are added algebraically, the total node equation results, which can be placed in the global system of equations as shown in Figure 17.7b.

Each element in the element matrix is identified by a row and a column. Each corresponds to a nodal degree of freedom. For example, the element $k_{12}^{②}$ is identified by the row of ϕ_2 and the column of ϕ_3. The row and column must be preserved while $k_{12}^{②}$ is placed in the global matrix.

Therefore, the assembly procedure is equivalent to stretching and placing the element equations in the global system of equations and adding the overlapping elements in $[K]$ and $\mathbf{R}$, as illustrated in Figure 17.7b.

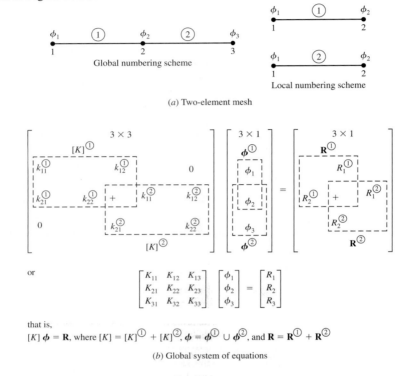

(a) Two-element mesh

that is,

$[K]\,\boldsymbol{\phi} = \mathbf{R}$, where $[K] = [K]^{①} + [K]^{②}$, $\boldsymbol{\phi} = \boldsymbol{\phi}^{①} \cup \boldsymbol{\phi}^{②}$, and $\mathbf{R} = \mathbf{R}^{①} + \mathbf{R}^{②}$

(b) Global system of equations

Figure 17.7   Assembly of element equations.

Due to the stretching effect during the assembly process, the global numbering scheme controls directly the locations of nonzero elements of the global matrix (the matrix bandwidth). A **bandwidth** is the width of nonzero elements of a matrix measured from the diagonal. Nodal numbering must be done to reduce the bandwidth.

There are various algorithms for optimal nodal numbering for bandwidth reduction. The general rule is that the smaller the difference between the maximum and minimum global node numbers of any element, the smaller the bandwidth of the resulting global matrix, and consequently the closer to optimum the global numbering scheme.

On this basis, the numbering scheme shown in Figure 17.1b, and again in Figure 17.8a, is optimal, while the numbering scheme of the same mesh shown in Figure 17.8b is not. In Figure 17.8a, the maximum difference between the numbers of any two nodes of any element is 4 for a four-node element and 5 for a six-node element. In Figure 17.8b, the difference jumps to 8 and 15 respectively.

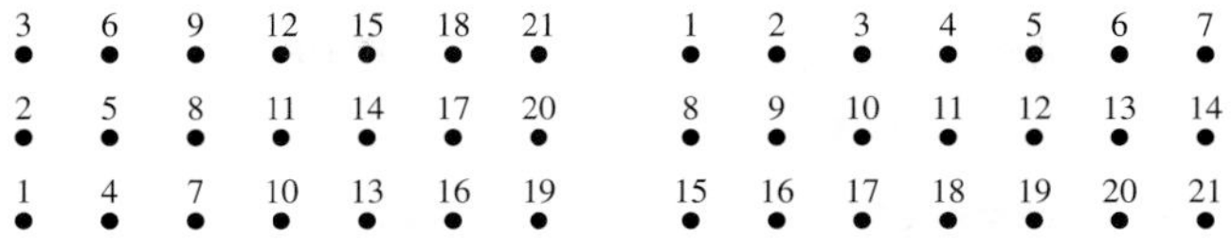

3	6	9	12	15	18	21		1	2	3	4	5	6	7
•	•	•	•	•	•	•		•	•	•	•	•	•	•
2	5	8	11	14	17	20		8	9	10	11	12	13	14
•	•	•	•	•	•	•		•	•	•	•	•	•	•
1	4	7	10	13	16	19		15	16	17	18	19	20	21
•	•	•	•	•	•	•		•	•	•	•	•	•	•

(a) Optimal scheme   (b) Nonoptimal scheme

Figure 17.8   Numbering schemes of mesh nodes.

17.7 Applying Boundary Conditions

The global system of equations resulting from the assembly process and given by Eq. (17.148) cannot be solved for the nodal values ϕ of the field variable until they have been modified to account for the geometric boundary conditions of the problem at hand. The natural boundary conditions are already included in the element equations. Element matrices $[K]^e$ and the global (assembled) matrix $[K]$ are always singular before applying the boundary conditions, that is, their inverse cannot be found because their determinants are zero.

Before the geometric boundary conditions are applied, the system of equations is not completely defined, similar to the inability to find constants in solutions of differential equations if enough boundary conditions are not available. In structural applications, having fixed nodes prevents the structure from moving in space as a rigid body when external loads are applied.

Applying zero boundary conditions amounts to eliminating the row and the column from the global system of equations that correspond to the zero degree of freedom. For example, if

$$
\begin{bmatrix}
k_{11} & k_{12} & \cdots & & k_{1n} \\
& k_{22} & \cdots & & k_{2n} \\
& & & & \vdots \\
\text{Symmetric} & k_{mm} & \cdots & k_{mn} \\
& & & & \vdots \\
& & & & k_{nn}
\end{bmatrix}
\begin{bmatrix}
\phi_1 \\ \phi_2 \\ \vdots \\ \phi_m \\ \vdots \\ \phi_n
\end{bmatrix}
=
\begin{bmatrix}
R_1 \\ R_2 \\ \vdots \\ R_m \\ \vdots \\ R_n
\end{bmatrix}
\tag{17.154}
$$

and $\phi_1 = \phi_m = 0$ as boundary conditions, applying these conditions to Eq. (17.154) eliminates the first row and column (for $\phi_1 = 0$) and the mth row and column (for $\phi_m = 0$). Thus, the number of simultaneous equations reduces to $(n - 2)$ from n, and Eq. (17.154) becomes

$$
\begin{bmatrix}
k_{22} & k_{23} & \cdots & & k_{2n} \\
& k_{33} & \cdots & & k_{3n} \\
& & & & \vdots \\
\text{Symmetric} & k_{m-1} & \cdots & k_{m-1,n} \\
& & k_{m+1} & \cdots & k_{m+1,n} \\
& & & & \vdots \\
& & & & k_{n-1}
\end{bmatrix}
\begin{bmatrix}
\phi_2 \\ \phi_3 \\ \vdots \\ \phi_{m-1} \\ \phi_{m+1} \\ \vdots \\ \phi_n
\end{bmatrix}
=
\begin{bmatrix}
R_2 \\ R_3 \\ \vdots \\ R_{m-1} \\ R_{m+1} \\ \vdots \\ R_n
\end{bmatrix}
\tag{17.155}
$$

Applying nonzero geometric boundary conditions is possible and usually requires rearranging Eq. (17.154); it is not discussed here.

17.8 Nodal External Applied Loads

Equation (17.155) represents the original continuum problem after applying the finite element method to it. The continuum is replaced by a set of nodes. Therefore, any continuum-related properties (such as mass, damping, internal heat source, etc.) or external applied loads must be lumped at the nodes.

Two types of external loads exist: concentrated and distributed. Examples of the former include concentrated forces, moments, and so forth. Examples of the latter include distributed forces, moments, heat fluxes, fluid fluxes, and so forth.

To apply concentrated loads to a finite element model, all that is needed is to create a node at the point of application of each load. In the cantilever beam example shown in Figure 17.1, node number 21 carries the load P of 500 lb, or P is applied to the model through node 21. In some cases, the load might be split between more than one node. In other cases, distributed loads may be lumped at various nodes. This is a matter of judgement left to the FEA analyst.

Distributed loads are automatically applied through the proper equations of boundary elements. Equations (17.137) and (17.138) show an example. In this case, the user would have to input the intensity of the distributed load and the nodes they are to be applied to or lumped at.

During developing element equations and assembling them, external applied (distributed or concentrated) loads may be neglected until the development of the global load vector $\mathbf{R}$ is performed. This makes all element equations similar and eliminates having to identify boundary and interior elements. The development of the global load vector can be based on the nodes and the boundary segments they are applied to.

17.9 Finite Element Solution

Thus far we have discussed in detail the derivation of the finite element equations given by the global system equation [Eq. (17.155)]. For practical design and FEA problems, this system may have hundreds or thousands of degrees of freedom. The effective solution of such a large number of simultaneous equations not only determines the cost of the analysis, but it also controls the accuracy of the solution. In nonlinear analysis, the convergence to the correct solution is also controlled by the method of solution.

For linear static or steady-state analysis, Gauss elimination is the common numerical method of solution. After applying the boundary conditions and the external applied loads to Eq. (17.148), the nodal values can be obtained as:

$$\phi = [K]^{-1}\mathbf{R} \qquad\qquad (17.156)$$

where $[K]^{-1}$ is the inverse of the global matrix $[K]$. Many implementation factors are usually considered to speed up the inversion process. For example, a skyline is sometimes used to avoid wasting time dealing with zero elements in [K]. A **skyline** is a piecewise line that separates the zero elements from the nonzero elements of [K].

For linear dynamic or transient analysis, numerical time integration is used. There are various integration schemes that keep the cost of the solution down by speeding its convergence. Experienced FEA analysts realize that the time increment (time step) is the most sensitive factor that controls the convergence of dynamic analysis.

In nonlinear static analysis, iterative solution methods are utilized to solve for the system response. The Newton-Raphson method is the method commonly used in the solution. The solution is incremental in nature. The external load is applied to the continuum in increments (load increments or steps), and the continuum response is calculated incrementally.

The load step is crucial for the solution to converge and requires experience to choose. Two types of nonlinearities may exist in a problem: geometric and material. A geometrically nonlinear problem is one in which large deformation occurs, and a materially nonlinear problem is one in which material properties are nonlinear.

Nonlinear dynamic analysis is the most complex and sensitive analysis an FEA analyst may face. Here the interaction between the load step and the time step depends to a large extent on the nonlinearity at hand, and it changes from one problem to another. The solution is usually obtained by trial and error. Readers who are interested in knowing more about the solution methods should consult finite element books.

17.10 Dynamic and Nonlinear Analyses

The material presented in this chapter thus far has been related to linear static FEA because this is the simplest and most widely used analysis. However, other analyses can be viewed as (nontrivial) extensions of all the concepts presented here. The steps of the finite element method presented here are essentially the same. The major difference comes in finding the integral equation and the numerical methods used to solve the global systems of equations.

Equation (17.148) represents linear static problems. Linear dynamic problems can be represented by:

$$[M]\ddot{\phi} + [C]\dot{\phi} + [K]\phi = \mathbf{R} \qquad (17.157)$$

where [M], [C], [K], $\ddot{\phi}$, $\dot{\phi}$ and ϕ are, respectively, the system mass matrix, damping matrix, stiffness matrix, acceleration vector, velocity vector, and displacement vector.

In solid mechanics applications, the first term in Eq. (17.157) is the system inertia forces, the second term is the dissipation or frictional forces, and the third term is the spring forces. [M] and [C] are, for many applications, symmetric and positive definite. They can be formulated in a similar way to the stiffness matrix [K].

When numerical integration schemes are applied to solve Eq. (17.157), the equation takes the following form:

$$[\bar{K}]\phi_{n+1} = \mathbf{R}_{n+1} \qquad (17.158)$$

where $[\bar{K}]$ is an effective matrix given by:

$$[\bar{K}] = \alpha[M] + \beta[C] + \gamma[K] \tag{17.159}$$

where α, β, and γ are constants determined by the time numerical integration scheme. R_{n+1} is the load vector at step $(n + 1)$. It is also a function of the original load vector **R** and other terms which result from the integration scheme.

For nonlinear static analysis, Eq. (17.148) becomes:

$$[K(\phi)]\phi = \mathbf{R} \tag{17.160}$$

and for nonlinear dynamic analysis, the equation is

$$[M]\ddot{\phi} + [C(\dot{\phi})]\dot{\phi} + [K(\phi)]\phi = \mathbf{R} \tag{17.161}$$

Utilizing the Newton-Raphson method to solve Eqs. (17.160) and (17.161), an initial guess of ϕ (usually **0**) for Eq. (17.160), and of $\dot{\phi}$ (also **0**) and $\ddot{\phi}$ (also **0**) additionally for Eq. (17.161), is required.

17.11 Accuracy of Finite Element Solutions

The finite element method is an approximate method and thus introduces errors to problem solutions. There are two sources of errors of approximations: the discretization process and the element shape functions. The type and number of elements employed in a finite element mesh as well as the order of the element shape function control the quality of finite element solutions.

To reduce the errors of approximation and therefore improve the solution quality, we can either refine the mesh or increase the order of the interpolation polynomial. The convergence criteria discussed in Section 17.5.3 are used, and the approximation error is controlled by refining the mesh: as the continuum is divided into smaller elements for each analysis iteration, the error decreases.

This refining method of controlling the quality of the approximation is called the *h*-version (because the size of the elements is generally denoted by *h*) of the finite element method. In the *h*-version, solution improvement is achieved by an orderly sequence of uniform mesh refinements, and the shape functions remain constant and constructed from polynomials of low order (linear, quadratic, or cubic).

The other method of improving the quality of a finite element solution is called the *p*-version (because the polynomial degree of shape functions is generally denoted by *p*). Instead of changing the mesh, the degree of the polynomial describing the elements is changed. In the *p*-version, the finite element mesh is fixed, and the solution improvement is achieved by increasing the polynomial order of the shape function. Polynomial order can change between one and eight.

Although it has been established theoretically and by examples that for a given problem, *h*- or *p*-version alone converges to the same solution, it has also been shown that the *p*-version is more efficient. With the proper mesh design, *p*-version can achieve the near-optimum rates of convergence. Meshes for *p*-version are usually different than those for *h*-version.

Meshes for *p*-version are often simpler than those for *h*-version, partly because *p*-version meshes can be graded so that the element size decreases in geometric progression toward points of interest such as stress concentration in elasticity. In some cases, the meshes in *p*-version are so simple that complex mesh generators are not needed. Because *p*-version elements are generally large, mapping techniques based on the blending function method are used so that curves such as circles, ellipses, hyperbolas, and parabolas are represented exactly.

17.12 Isoparametric Evaluation of Element Matrices

Element matrices have been expressed as integrals, over the element domain, of the shape functions and other element properties. The evaluation of these matrices is a crucial step in FEA because it directly affects the cost and accuracy of the FEA. While shape functions and element integrals can be evaluated in the global coordinate system, such an approach is not very efficient. It is not always possible to find $[G]^{-1}$ as discussed in Section 17.5.4, nor it is efficient to compute $[G]^{-1}$ if it exists.

Isoparametric evaluation of element matrices offers a very attractive alternative. In addition to avoiding the inversion problem, it enables the creation of curve-sided elements, which means that a less number of elements could be used to represent curved boundaries. Consequently, the size and cost of the finite element model is reduced.

The idea behind the isoparametric evaluation is similar to that of parametric representation in geometric modeling. Element geometry is mapped from the cartesian space to the parametric space, where element geometric shape is simple. In the parametric space, 1D elements are always line segments of length 2; 2D elements are squares of 2×2 sides; and 3D elements are cubes of $2 \times 2 \times 2$ sides.

Figure 17.9 shows some isoparametric elements. Due to the simple shapes of elements in the parametric spaces, element shape functions can be easily constructed in this space by inspection, and numerical integration to evaluate element matrices can be easily performed.

17.12.1 Element Mapping

The isoparametric evaluation of a finite element requires mapping both its geometric shape and its interpolation polynomial from the cartesian space to the parametric space. Instead of interpreting Eq. (17.96) to be over the element cartesian domain, we use it over the element parametric domain. Thus, N_i changes from functions of the cartesian space to functions of the element parametric space.

For one, two, and three dimensions, N_i becomes $N_i(u)$, $N_i(u, v)$, and $N_i(u, v, w)$, respectively. To map the element geometric shape, we can think of the coordinates (x, y, z) of any point within the element in a similar way to field variable ϕ. Thus, Eq. (17.96) can be rewritten to interpolate the coordinates as follows:

$$x = \sum_{i=1}^{m} F_i x_i \qquad y = \sum_{i=1}^{m} F_i y_i \qquad z = \sum_{i=1}^{m} F_i z_i \qquad (17.162)$$

where F_i is the shape function that interpolates the element geometry, (x_i, y_i, z_i) are the cartesian coordinates of node i, and m is the number of nodes that interpolates the geometry.

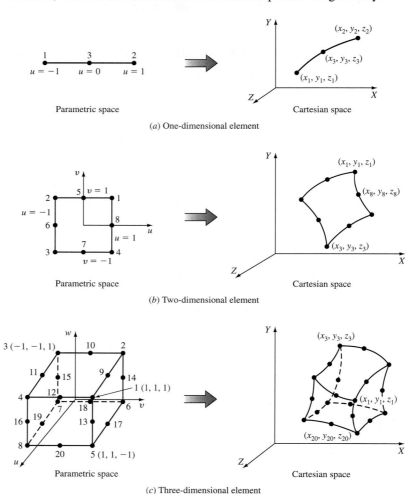

Figure 17.9   Sample isoparametric elements.

In relation to the number of nodes n that interpolates the field variable, superparametric elements occur when $m > n$, subparametric elements occur when $m < n$, and isoparametric elements occur when $m = n$ (consequently $F_i = N_i$). In this section, we consider isoparametric elements only, for which Eq. (17.162) changes to:

$$x = \sum_{i=1}^{n} N_i x_i \qquad y = \sum_{i=1}^{n} N_i y_i \qquad z = \sum_{i=1}^{n} N_i z_i \qquad (17.163)$$

Equation (17.163) enables having elements with curved sides because the element sides are fitted between the nodes. In standard-type element formulation, this is not possible because the element sides are always straight regardless of any midside nodes.

When writing Eq. (17.163), we assume that the mapping between the parametric and cartesian spaces is unique (one-to-one mapping), that is, each point in one space has one and only one corresponding point in the other space. If the mapping is not unique, we may expect violent and undesirable distortions in the global space that may fold the curved element back upon itself. A method for checking for nonuniqueness and violent distortion is by evaluating the Jacobian $[J]$ of the element and checking the sign of its determinant $|J|$. If the sign of $|J|$ does not change for all the elements in the solution domain (mesh), acceptable mapping is guaranteed.

An important consideration in isoparametric evaluation is to ensure that element shape functions established normally in the parametric space preserve the convergence conditions (compatibility and completeness) in the global space. Without proofs, it turns out that if N_i meets these conditions in the parametric space, they are automatically met in the global space. Therefore, the element design discussed in Section 17.5.3 for the global space applies for the parametric one.

17.12.2 Shape Functions by Inspection

Element shape functions for isoparametric elements can be established in the element parametric space by inspection based on the insight gained about them in Section 17.5.4. The fundamental property of a shape function N_i is that its value is unity at node i and zero at all other nodes. Using this property, N_i corresponding to a specific nodal layout could be obtained in a systematic manner by inspection as follows. First, we construct the shape functions N_i corresponding to the corner (basic) nodes of the element. The addition of another node (interior or midside) results in an additional shape function and a correction to be applied to the already existing shape functions.

Let us illustrate this concept for 1D elements (see Figure 17.9a). The origin of the parametric coordinate system is taken as the midpoint of the element as shown. If the element has only the two corner nodes 1 and 2, their shape functions are linear and can be written as:

$$N_1 = \frac{1}{2}(1-u) \qquad N_2 = \frac{1}{2}(1+u) \qquad (17.164)$$

Notice that N_1 and N_2 satisfy the fundamental property of a shape function.

If we add node 3 at the center ($u = 0$), its shape function becomes (by inspection)

$$N_3 = 1 - u^2 \qquad (17.165)$$

With the addition of node 3, does Eq. (17.164) still define valid element shape functions? At node 3, $u = 0$, which gives, by substituting in Eq. (17.164), $N_1 = N_2 = \frac{1}{2}$. This violates the fundamental property. To make N_1 or N_2 equal zero at node 3, we simply modify Eq. (17.164) to give $N_1 = \frac{1}{2}(1-u) - \frac{1}{2}$ and $N_2 = \frac{1}{2}(1+u) - \frac{1}{2}$. This new N_1 and N_2 still violate the fundamental property at (now) nodes 1 and 2. The proper solution is:

$$N_1 = \frac{1}{2}(1-u) - \frac{1}{2}N_3 \qquad N_2 = \frac{1}{2}(1+u) - \frac{1}{2}N_3 \qquad (17.166)$$

For a four-node element, nodes 3 and 4 are located at $u = -\frac{1}{3}$ and $u = \frac{1}{3}$, respectively. Following the same concept, we can write

$$N_3 = \frac{9}{16}(1-u^2)(1-3u) \qquad (17.167)$$

$$N_4 = \frac{9}{16}(1-u^2)(1+3u) \qquad (17.168)$$

and the correct N_1 and N_2 can be written as

$$N_1 = \frac{1}{2}(1-u) - \frac{2}{3}N_3 - \frac{1}{3}N_4 \qquad (17.169)$$

$$N_2 = \frac{1}{2}(1+u) - \frac{1}{3}N_3 - \frac{2}{3}N_4 \qquad (17.170)$$

This systematic generation of shape functions for 1D elements is shown in Figure 17.10.

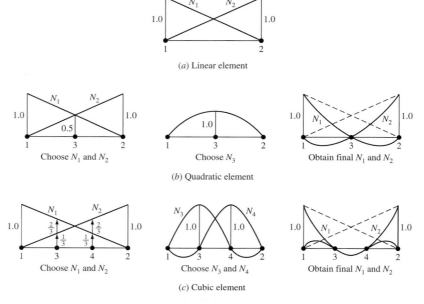

Figure 17.10 Isoparametric shape functions for 1D elements.

The preceding concept can be extended to 2D elements. Considering the quadrilateral element shown in Figure 17.9b, we can write for the corner nodes:

$$N_1 = \frac{1}{4}(1 + u)(1 + v) \tag{17.171}$$

$$N_2 = \frac{1}{4}(1 - u)(1 + v) \tag{17.172}$$

$$N_3 = \frac{1}{4}(1 - u)(1 - v) \tag{17.173}$$

$$N_4 = \frac{1}{4}(1 + u)(1 - v) \tag{17.174}$$

and for the midside nodes, we have

$$N_5 = \frac{1}{4}(1 - u^2)(1 + v) \tag{17.175}$$

$$N_6 = \frac{1}{4}(1 - v^2)(1 - u) \tag{17.176}$$

$$N_7 = \frac{1}{4}(1 - u^2)(1 - v) \tag{17.177}$$

$$N_8 = \frac{1}{4}(1 - v^2)(1 + u) \tag{17.178}$$

For each present midside node, the shape function of each surrounding corner node must by corrected by subtracting half of the shape function of this midside node. For example, if node 5 is present, then $\frac{1}{2}N_5$ is subtracted from N_1 and N_2 given in the preceding equations, that is, node 1 has $N_1 - \frac{1}{2}N_5$ and node 2 has $N_2 - \frac{1}{2}N_5$. If, in addition, node 8 is present, then node 1 has $N_1 - \frac{1}{2}N_5 - \frac{1}{2}N_8$ and node 4 has $N_4 - \frac{1}{2}N_8$.

Similarly, we can write the following equations for the 3D element shown in Figure 17.9c:

$$N_1 = \frac{1}{8}(1 + u)(1 + v)(1 + w) \tag{17.179}$$

$$N_9 = \frac{1}{4}(1 - u^2)(1 + v)(1 + w) \tag{17.180}$$

Equations for N_2 to N_8 are similar to Eq. (17.179) with the proper sign change, and equations for N_{10} to N_{20} are similar to Eq. (17.180).

17.12.3 Evaluation of Element Matrices

In order to perform the isoparametric evaluation of element matrices, we should be able to evaluate the integrals that appear in the element equations. In general, for 2D problems these integrals take the form

$$\int_{A^e} f\left(\phi^e, \frac{d\phi^e}{dx}, \frac{d\phi^e}{dy}, \ldots\right) dx \, dy \tag{17.181}$$

Thus, based on Eq. (17.181), we should express the derivatives of ϕ^e in the parametric space and transform the integral domain to the parametric space. To map the derivatives, Eq. (17.96) gives:

$$\frac{\partial \phi^e}{\partial x} = \sum_{i=1}^{n} \frac{\partial N_i}{\partial x} \phi_i \qquad \frac{\partial \phi^e}{\partial y} = \sum_{i=1}^{n} \frac{\partial N_i}{\partial y} \phi_i \tag{17.182}$$

To express $\partial N_i/\partial x$ and $\partial N_i/\partial y$ in terms of u and v, we can write [notice that $N_i = N_i(u, v)$]

$$\frac{\partial N_i}{\partial u} = \frac{\partial N_i}{\partial x}\frac{\partial x}{\partial u} + \frac{\partial N_i}{\partial y}\frac{\partial y}{\partial u} \tag{17.183}$$

$$\frac{\partial N_i}{\partial v} = \frac{\partial N_i}{\partial x}\frac{\partial x}{\partial v} + \frac{\partial N_i}{\partial y}\frac{\partial y}{\partial v} \tag{17.184}$$

or

$$\begin{bmatrix} \dfrac{\partial N_i}{\partial u} \\[2ex] \dfrac{\partial N_i}{\partial v} \end{bmatrix} = \begin{bmatrix} \dfrac{\partial x}{\partial u} & \dfrac{\partial y}{\partial u} \\[2ex] \dfrac{\partial x}{\partial v} & \dfrac{\partial y}{\partial v} \end{bmatrix} \begin{bmatrix} \dfrac{\partial N_i}{\partial x} \\[2ex] \dfrac{\partial N_i}{\partial y} \end{bmatrix} \tag{17.185}$$

Using Eq. (17.163), Eq. (17.185) becomes

$$\begin{bmatrix} \dfrac{\partial N_i}{\partial u} \\[2ex] \dfrac{\partial N_i}{\partial v} \end{bmatrix} = \begin{bmatrix} \displaystyle\sum_{i=1}^{n} \dfrac{\partial N_i}{\partial u} x_i & \displaystyle\sum_{i=1}^{n} \dfrac{\partial N_i}{\partial u} y_i \\ \displaystyle\sum_{i=1}^{n} \dfrac{\partial N_i}{\partial v} x_i & \displaystyle\sum_{i=1}^{n} \dfrac{\partial N_i}{\partial v} y_i \end{bmatrix} \begin{bmatrix} \dfrac{\partial N_i}{\partial x} \\[2ex] \dfrac{\partial N_i}{\partial y} \end{bmatrix} = [J] \begin{bmatrix} \dfrac{\partial N_i}{\partial x} \\[2ex] \dfrac{\partial N_i}{\partial y} \end{bmatrix} \tag{17.186}$$

where $[J]$ is known as the Jacobian, that is,

$$[J] = \begin{vmatrix} \displaystyle\sum_{i=1}^{n} \dfrac{\partial N_i}{\partial u} x_i & \displaystyle\sum_{i=1}^{n} \dfrac{\partial N_i}{\partial u} y_i \\ \displaystyle\sum_{i=1}^{n} \dfrac{\partial N_i}{\partial v} x_i & \displaystyle\sum_{i=1}^{n} \dfrac{\partial N_i}{\partial v} y_i \end{vmatrix} = \begin{bmatrix} \dfrac{\partial N_1}{\partial u} & \dfrac{\partial N_2}{\partial u} & \cdots & \dfrac{\partial N_n}{\partial u} \\[2ex] \dfrac{\partial N_1}{\partial v} & \dfrac{\partial N_2}{\partial v} & \cdots & \dfrac{\partial N_n}{\partial v} \end{bmatrix} \begin{bmatrix} x_1 & y_1 \\ x_2 & y_2 \\ \vdots & \vdots \\ x_n & y_n \end{bmatrix} \tag{17.187}$$

Inverting Eq. (17.186), the global derivatives are given by:

$$\begin{bmatrix} \dfrac{\partial N_i}{\partial x} \\ \dfrac{\partial N_i}{\partial y} \end{bmatrix} = [J]^{-1} \begin{bmatrix} \dfrac{\partial N_i}{\partial u} \\ \dfrac{\partial N_i}{\partial v} \end{bmatrix} \tag{17.188}$$

With Eq. (17.188), $\partial\phi^e/\partial x$ and $\partial\phi^e/\partial y$ can be expressed in the parametric spaces as follows. Equation (17.182) can be rewritten as:

$$\begin{bmatrix} \dfrac{\partial\phi^e}{\partial x} \\ \dfrac{\partial\phi^e}{\partial y} \end{bmatrix} = \begin{bmatrix} \dfrac{\partial N_1}{\partial x} & \dfrac{\partial N_2}{\partial x} & \cdots & \dfrac{\partial N_n}{\partial x} \\ \dfrac{\partial N_1}{\partial y} & \dfrac{\partial N_2}{\partial y} & \cdots & \dfrac{\partial N_n}{\partial y} \end{bmatrix} \begin{bmatrix} \phi_1 \\ \phi_2 \\ \vdots \\ \phi_n \end{bmatrix} \tag{17.189}$$

Using Eq. (17.188), Eq. (17.189) becomes:

$$\begin{bmatrix} \dfrac{\partial\phi^e}{\partial x} \\ \dfrac{\partial\phi^e}{\partial y} \end{bmatrix} = [J]^{-1} \begin{bmatrix} \dfrac{\partial N_1}{\partial u} & \dfrac{\partial N_2}{\partial u} & \cdots & \dfrac{\partial N_n}{\partial u} \\ \dfrac{\partial N_1}{\partial v} & \dfrac{\partial N_2}{\partial v} & \cdots & \dfrac{\partial N_n}{\partial v} \end{bmatrix} \begin{bmatrix} \phi_1 \\ \phi_2 \\ \vdots \\ \phi_n \end{bmatrix} \tag{17.190}$$

To complete the evaluation of the integral, we need to express the area element $dx\,dy$ in terms of $du\,dv$. As shown in Chapter 15, we write

$$dx\,dy = |J|\,du\,dv \tag{17.191}$$

For 1D elements,

$$dx = \frac{dx}{du}\,du = \sum_{i=1}^{n} \frac{dN_i}{du}\,du = |J|\,du$$

For 3D elements readers can follow a similar development ([J] becomes 3×3).

When we express $dx\,dy$ in terms of $du\,dv$, the integral in Eq. (17.181) reduces to

$$\int_{-1}^{1} \int_{-1}^{1} g(u,\,v)\,du\,dv \tag{17.192}$$

The integral given by Eq. (17.192) can be evaluated numerically using Gauss quadrature as explained in Chapter 15. Because the integration limits always go from -1 to 1 in the element parametric space, Eqs. (15.112) and (15.113) becomes:

$$V_i = W_i \tag{17.193}$$

$$u_i = C_i \tag{17.194}$$

Equations (15.114) and (15.118) can be adjusted accordingly.

As discussed in Chapter 15, sampling points are required to use Gauss quadrature to evaluate integrals numerically. The number of sampling points used to evaluate element integrals is known as the order of quadrature in the finite element literature. The practical rule is that it is desirable to keep the order of quadrature as low as possible to reduce the cost of the FEA and minimize computational errors.

There is a lower limit on the number of sampling points because as the element size decreases, the integrand becomes constant and the integral becomes the area or the volume of the element. Thus, all the discussions presented in Chapter 15 regarding the choice of the number of sampling points apply here. Table 17.1 shows the reliable integration order and the reduced order used in practice.

17.13 Finite Element Modeling

Thus far, we have presented the finite element theory in order to provide readers with a clear understanding of the requirements of FEA, the information typically needed to perform one, and how to interpret the results. While much of the theory presented has been codified in commercial finite element codes available for engineers and designers to use, most of the burden of finite element modeling and of analyzing the results lies on the engineers and designers themselves. A typical finite element model is comprised of nodes, degrees of freedom, boundary conditions, elements, material properties, externally applied loads, and analysis type. Engineers must choose all these modeling attributes carefully to obtain the right FEA results.

In practice, practitioners must first decide on the mesh layout, that is, the number of nodes and elements. Zones of expected abrupt changes in the field variable (such as stress concentration around holes) require more dense numbers of nodes and elements than zones where gradual changes occur. This is known as mesh gradation.

After the mesh layout is chosen, practitioners must choose the type of analysis (static/dynamic, linear/nonlinear, beam theory, plane stress, 3D, etc.), the type (deflection, rotation, temperature, flux, etc.) and number of degrees of freedom at each node, the boundary conditions, the element information (type, number of nodes per element, and Gauss quadrature order), material properties, and lumping external loads at the nodes.

Once the finite element model is defined by choosing the parameters of the corresponding mesh, it must be input to the code that performs the FEA. A review of most existing FEA commercial codes reveals that they require their users to provide the data of the finite element model in a data file with a specific format. A data file consists of records (rows of data in the file); each record consists of 80 spaces (to hold 80 characters of data). Each record is divided into various fields (a field consists of a certain number of characters), and each field holds the value of a specific variable.

Apart from the specific detailed format of a certain data file, such a file usually has five major sections: control, nodal, element, material, and loading sections. While the latter four sections are self explanatory, the control section includes information such as the problem description (heading), the total number of nodes, the type of analysis, and so forth.

Table 17.1   Gauss quadrature order

Dimension	Number of nodes	Element shape	Reliable Gauss quadrature order	Reduced order used in practice
One	2		1	Same
	3		2	Same
	4		3	Same
Two	4		2×2	Same
	8		3×3	2×2
	12		4×4	3×3
Three	8		$2 \times 2 \times 2$	Same
	20		$3 \times 3 \times 3$	$2 \times 2 \times 2$
	32		$4 \times 4 \times 4$	$3 \times 3 \times 3$

The output from FEA codes is primarily in a numerical form. It usually consists of the nodal values of the field variable and its derivatives. For example, in solid mechanics problems the output is nodal displacements and element stresses. In heat transfer problems, the output is nodal temperatures and element heat fluxes. Graphical outputs are usually more informative in providing trends of continuum behavior. Thus, curves and contours of the field variable can be plotted and displayed. Also, deformed shapes can be displayed superposed on undeformed shapes. Vibration modes can also be displayed.

As FEM is a very labor-intensive task, it has been the target of automation. Powerful pre- and postprocessors exist that can perform most of the FEM activities and tasks. The classification *pre* and *post* is relative to the FEA phase, as shown in Figure 17.11. Preprocessors, with the help of the user, can generate data files automatically. Users no longer need to know the exact format of the data file required by an FEA code.

After the user generates a finite element model using a certain preprocessor, a single command can produce the data file with the required format. Most preprocessors support most of the existing commercial FEA codes. If not, a preprocessor can be interfaced with a given FEA code. This task usually requires good knowledge of the database structure of the preprocessor and the format of the data file of the FEA code.

Postprocessors are usually automatic and do not require user assistance. They usually process the numerical results and display them in the desired form requested by the user. Postprocessor commands are usually simple to use.

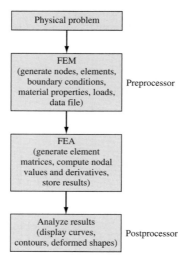

Figure 17.11   Finite element pre- and postprocessors.

17.14 Mesh Generation

Mesh generation forms the backbone of FEM. Mesh generation refers to the generation of nodal coordinates and elements. It also includes the automatic numbering of nodes and elements based on a minimal amount of user-supplied data. Automatic mesh generation reduces errors and saves a great deal of user time, therefore reducing the FEA cost.

Before the existence of preprocessors, finite element meshes were generated manually. In manual mesh generation, the analyst discretizes the simplified geometry of the object to be studied, that is, the geometric model of the object, into nodes and elements. Nodes are defined by specifying their coordinates, while element connectivity (connecting nodes) defines the elements. Manual meshing is inefficient and error prone, and meshing data can grow rapidly and become confusing for complex objects—especially 3D ones.

Preprocessors provide a wide variety of algorithms, schemes, and methods for mesh generation. They have various levels of automation and different user-input requirements. We classify mesh generation into semiautomatic and fullyautomatic. The latter is taken to mean a method in which only the shape (both geometry and topology) of the object to be meshed and the mesh attributes (mesh density, element type, boundary conditions, loads, etc.) are required as input. Any other method that may require additional input, such as subdividing the object into subdomains or regions, is a semiautomatic one.

The most important criterion in mesh generation is to ensure the validity and correctness of the resulting mesh. It is important to list the requirements that make a mesh valid, that is, produces the correct FEA results. Some of the requirements are listed here (some are necessary while others are desirable):

1. **Nodal locations.** Nodes must lie inside or on the boundaries of the geometric model to be meshed. Nodes that are very close to the boundaries must be pulled to lie on them to accurately mesh the model. Some generation methods offset (shrink) the model boundary by a small amount ε, generate the nodes based on the offset boundary, and then pull the boundary nodes to the original boundary of the model.

2. **Element type and shape.** It is desirable if various elements (large element library) are available to provide users with the required flexibility to meet the compatibility and completeness requirements.

3. **Mesh gradation.** This usually refers to mesh smoothing and density control. Most often, objects on which FEA is performed may have holes or sharp corners. It is usually required that mesh density (number of nodes and elements) is increased around these regions to capture the rapid change (e.g., stress variation around holes and sharp corners) of the field variable. These regions are known as transition regions.

4. **Mesh conversion.** It may be desirable to convert a mesh of a given type of element to another mesh of a different element type. In 2D meshes, for example, it is always possible to convert a triangular element into three quadrilateral elements (a tetrahedron can be subdivided into four hexahedra) or to combine two triangular elements to produce a

quadrilateral element. A quadrilateral-element mesh may be converted into a triangular-element mesh by splitting each quadrilateral into two triangles. Mesh conversion must be done with care as poorly formed elements (especially in 3D) may result.

5. **Element aspect ratio.** For geometric invariance as discussed earlier, it is important to keep the aspect ratio of any element close to 1, that is, all sides of an element are equal in length.

6. **Mesh geometry and topology.** As the object to be meshed has geometry and topology, so should its mesh. Mesh geometry refers to the coordinates of nodal points and connectivity information of elements. Mesh topology refers to mesh orientation relative to object topology. Object topology always determines the mesh topology.

7. **Compatibility with representation schemes.** A mesh generation method is inherently related to the geometric model to be meshed. Solid models support fully automatic mesh generation.

8. **Cost effectiveness.** The time it takes to generate a mesh and the time it takes to perform the FEA are crucial. To reduce both, it is important that the mesh generation method optimizes the mesh and minimizes the number of nodes and elements that comprise the mesh and yet meets the conversion requirements.

17.15 Tutorials

17.15.1 Perform FEM/FEA on a CAD System

This tutorial shows how to use a CAD system to perform a static linear analysis of the beam shown on the right. The beam material is steel (isotropic elastic material). We load the beam with a concentrated load. We use two different load cases that use different boundary conditions. In the first case we have a cantilever beam loaded at its free end. In the second case, we clamp the beam down on one end, place a simple support on the other end, and load the beam at its center. All dimensions are in inches.

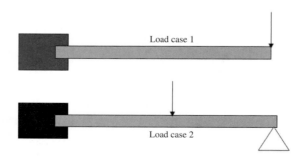

FEM/FEA steps:

1. **Create the beam model.** The beam model is an extrusion with a thickness of 4 in. Sketch the beam cross section in the Top sketch plane as shown below.

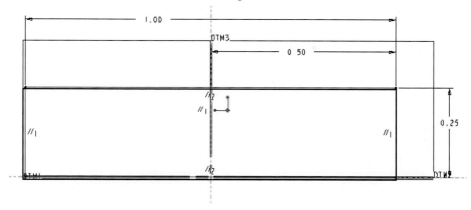

2. **Start the FEM/FEA application.** CAD systems provide the FEM/FEA application as a built-in module. Use your CAD system applications menu to invoke the module.

3. **Enter the material properties.** Select steel from the materials library of the FEM/FEA application. Enter the steel properties—use $E = 3.0 \times 10^7$ for Young's Modulus [units are psi (pounds per square inch)], and $v = 0.33$ for Poisson's ratio which is dimensionless.

4. **Apply the boundary conditions.** We must constrain the fixed (clamped) end of the beam. Select one of the beam end faces and fix it. We fix the left end in this tutorial. Find the proper constraint menu on your FEM/FEA application and fix this end.

5. **Apply the external load.** Select the right edge of the beam and apply a concentrated force of 10 lb in the $-Y$ direction as shown the upcoming screenshot. Use the name LoadCase1 for this load.

6. **Autogenerate the mesh.** Use an element offered by your FEM/FEA application. The application generates the nodes and elements mesh automatically as shown below.

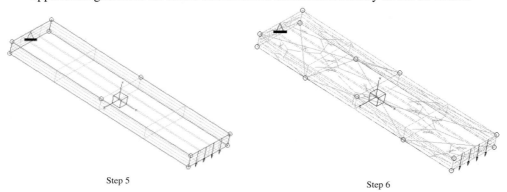

Step 5   Step 6

7. **Run the analysis.** Invoke the FEA in your application to run the analysis. Watch for errors while running the analysis. Your CAD system should provide a log file that you can examine to trace the analysis steps and read errors if any occur.

8. **Create results file.** This step allows you to save the FEA results set in a file, so you can display them in the next step.

9. **Display the results.** You can display the nodal displacements and element stresses in two modes: contours or graphs, as shown in the following screenshots.

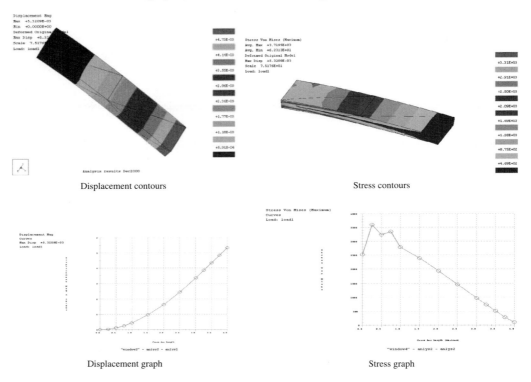

Displacement contours   Stress contours

Displacement graph   Stress graph

10. **Animate beam displacement.** Create an animation of the beam deflection. This is an easy task to achieve. The CAD system displays the beam deflecting incrementally from the original zero-deflection state to the maximum-deflection state under the applied load.

11. **Delete the load case 1.** Now delete the load case 1 to prepare for using load case 2.

12. **Change the boundary conditions at the free end.** In load case 2, the free end becomes simply supported. This allows only rotation about the Z axis of the beam MCS. Change the boundary conditions at the free end to reflect this constraint.

13. **Apply the external load.** Create and select the middle line on the top face of the beam. Apply a concentrated force of 10 lb in the $-Y$ direction as shown below. Use the name LoadCase2 for this load.

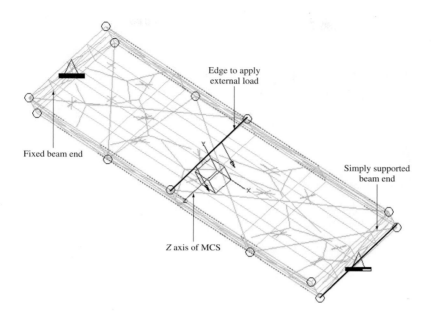

14. **Repeat steps 7 to 10.** The following screenshots show the FEA results for load case 2.

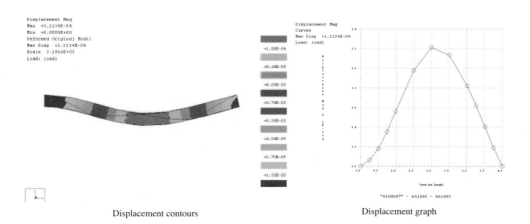

Displacement contours   Displacement graph

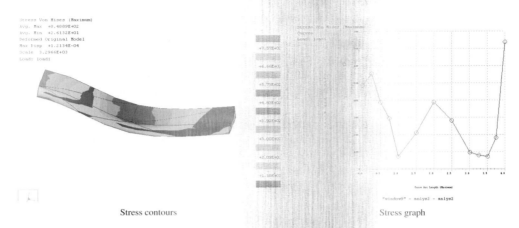

Stress contours   Stress graph

Tutorial 17.15.1 discussion:

This tutorial illustrates the typical FEM/FEA activities that CAD systems support. As is evident here, the designer must understand all the theoretical concepts covered in this chapter to be able to peform FEM/FEA effectively on CAD systems.

The two load cases use the same mesh because the beam shape did not change from one case to the other. In both cases, we only change the external loads values, location (free end or middle edge), and type (e.g., concentrated or distributed). In this example, we have to change the boundary conditions also. Typically, this is not done for different load cases.

The mesh has linear tetrahedral elements. Each element has four nodes. Tetrahedral elements are the easiest to generate. They are easier to generate than, for example, hexahedral elements. This is why CAD systems have them as the default element type.

Once a FEM model is developed, it is easy to run parametric studies. We only change the load to generate a new load case, run the analysis, display the results, analyze them, and make design decisions. We repeat this cycle for as many load cases as we want to study. This is a great tool to support parametric design studies to investigate the what-if questions.

Tutorial 17.15.1 hands-on exercise:

1. Change the beam into a simply-supported beam on both ends. Run the FEA using the same data given in the tutorial.
2. Now change the beam to have fixed (clamped) ends on both sides. Run the FEA analysis again using the same data given in the tutorial.
3. Compare the four sets of results for the four sets of end supports: fixed-free, fixed-simply, simply-simply, and fixed-fixed.
4. Which beam is the strongest, that is, which beam has the least maximum deflection and maximum stress? Does the beam that has the least minimum deflection have the least minimum stress also? Explain your results.
5. For which of these four types of beams can you find the exact solution from the beam theory? Compare the exact results with the FEA results.

PROBLEMS

Part I: Theory

17.1 Figure 17.12 shows various 1D systems with their boundary conditions. Derive the functional Π for each system.

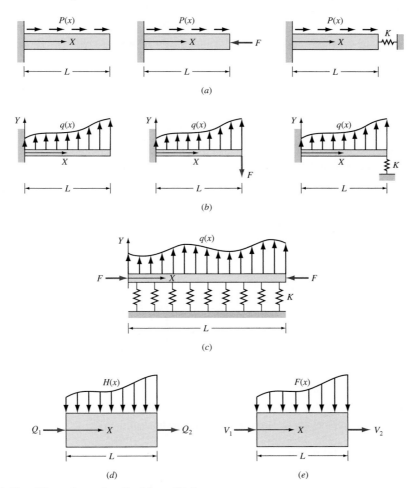

Figure 17.12   1D systems for Problem 17.1.

Following are the equilibrium equations of the systems shown in Figure 17.12:

a. A bar under a distributed axial load $P = P(x)$ (see Figure 17.12a):

$$AEu'' + P(x) = 0$$

b. A cantilever beam under a distributed load of density $q(x)$ per unit length (see Figure 17.12b):

$$EI y'''' - q(x) = 0$$

c. A beam on an elastic foundation of stiffness density $K(x)$ per unit length and subjected to a distributed load of density $q(x)$ per unit length (see Figure 17.12c):

$$EI y'''' + K(x)y - q(x) = 0$$

d. A region of a cross-sectional area A with steady-state uniaxial heat flow Q and a source heat flux $H(x)$ (see Figure 17.12d):

$$AKT'' + H(x) = 0$$

e. A steady-state flow through a region with a cross-sectional area A, a fluid flow V across the ends of the region, and a fluid flux $F(x)$ per unit length distributed along the length of the region (see Figure 17.12e):

$$AgP'' + F(x) = 0$$

17.2 The accompanying figure shows two 2D systems with their boundary conditions. Derive the functional Π for each system. Following are the equilibrium equations of these systems.

a. Plane stress problem over a rectangular domain as shown in Figure 17.13a, where $q_1(y)$ and $q_2(x)$ are the distributed load densities.

b. A steady-state heat conduction over a rectangular domain (see Figure 17.13b), where Q_1 and Q_2 are heat flows, $H(x)$ is a heat flux, and Q is the internal heat generation (heat source):

$$K\frac{\partial^2 T}{\partial x^2} + K\frac{\partial^2 T}{\partial y^2} + Q = 0$$

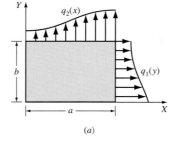

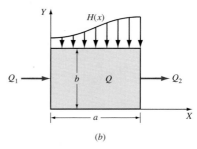

(a)   (b)

17.3 Starting with Eq. (17.50) in Example 17.6, rederive Eq. (17.63) using integration by parts.

17.4 Use Galerkin's method to derive the integral equations of the systems of Examples 17.1, 17.3, and 17.6.

17.5 Equation (17.48) in Example 17.5 applies to 3D solid mechanics problems. Knowing that the displacement components are u, v, and w in the X, Y, and Z directions, respectively, design 3D linear, quadratic, and cubic elements. Write the interpolation polynomial for each element.

17.6 The 3D counterpart of Eq. (17.50) in Example 17.6 is given as

$$K\frac{\partial^2 T}{\partial x^2} + K\frac{\partial^2 T}{\partial y^2} + K\frac{\partial^2 T}{\partial z^2} = 0$$

Derive the corresponding functional Π [*Hint:* modify Eq. (17.50).] Design finite elements for the problem and write the interpolation polynomial for each element.

17.7 The differential equilibrium equation for bending of an isotropic thin plate of constant thickness t and subject to a distributed load of intensity q per unit area is given by:

$$\frac{\partial^4 w}{\partial x^4} + 2\frac{\partial^4 w}{\partial x^2 \partial y^2} + \frac{\partial^4 w}{\partial y^4} + \frac{12(1-v^2)}{Et^3}q = 0$$

where w is the plate lateral deflection in the Z direction and the plate lies in the XY plane as shown below. Derive the following corresponding functional Π:

$$\Pi = \frac{1}{2}\iint_A \frac{Et^3}{12(1-v^2)}\left\{\left(\frac{\partial^2 w}{\partial x^2} + \frac{\partial^2 w}{\partial y^2}\right)^2 - 2(1-v)\left[\frac{\partial^2 w}{\partial x^2}\frac{\partial^2 w}{\partial y^2} - \left(\frac{\partial^2 w}{\partial x \partial y}\right)^2\right] + qw\right\}dxdy$$

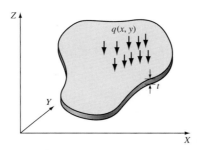

17.8 The 8-node rectangular element shown on the right has been designed for a given problem. Write the corresponding interpolation polynomial for the field variable ϕ assuming the element is a C^0 element [i.e., one degree of freedom (ϕ) per node]. Investigating the order and the terms of the polynomial, how does ϕ change in the problem domain? Does the polynomial have geometric invariance? Explain your answer.

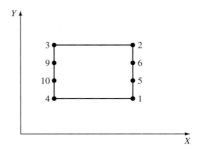

    Note: The node numbering follows the isoparametric element numbering scheme.

17.9 Derive the explicit form of N_i (in terms of nodal coordinates) used in Eq. (17.86) for the linear triangular element, and Eq. (17.94) for the linear quadrilateral element.

17.10 Following Example 17.11, develop the element equations and matrices for three-node and four-node elements for the systems of Examples 17.1 to 17.4.

17.11 The functional Π for a beam loaded with a distributed load of intensity $p(x)$ and a distributed moment of intensity $m(x)$, and including the shear deformation is given by

$$\Pi = \frac{1}{2}\int_0^L EI\theta'^2 dx + \frac{1}{2}\int_0^L GAS(y' - \theta)^2 dx - \int_0^L Py\, dx - \int_0^L m\theta\, dx$$

Derive the element equations and matrices for two-node, three-node, and four-node elements. G, A, and S are the shear modulus, the cross-sectional area, and the shear correction factor, respectively.

17.12 Write the polynomials and all relevant equations for quadratic and cubic triangular elements. Follow Example 17.12. Repeat for linear, quadratic, and cubic quadrilateral elements.

17.13 Derive the element equations and matrices for differential equations of Problems 17.5, 17.6, and 17.7 using the variational approach. Use 3D linear, quadratic, and cubic tetrahedral and hexahedral elements.

17.14 Utilizing the integral equations developed in problem 17.4, derive the corresponding element equations and matrices. Compare the results with those of Example 17.11.

17.15 Assuming each continuum in Examples 17.1 to 17.4 is modeled by three two-node elements, write the element equations for each one and assemble them into a global system of equations.

17.16 Assuming each continuum in Examples 17.5 and 17.6 is modeled by two four-node rectangular elements, write the element equations for each one and assemble them into a global system of equations.

17.17 Apply the geometric boundary conditions to each of the global systems of equations that results in Problems 17.15 and 17.16. Write the final systems of equations.

17.18 Apply the external loads to the global systems of equations that resulted in Problem 17.17.

17.19 Develop the Jacobian [J] for 3D elements for isoparametric evaluation.

Part II: Laboratory

Use your in-house CAD/CAM system to answer the questions in this part.

17.20 Utilizing the FEM package provided by your CAD system, generate the meshes and the FEA data files for the geometric models shown in the problems at the ends of Chapters 7, 8, and 9.

17.21 Perform finite element modeling and analysis on the following models.

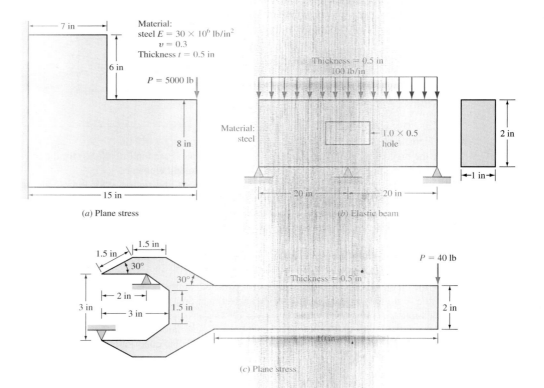

(a) Plane stress

(b) Elastic beam

(c) Plane stress

Part III: Programming

Write the code so that the user can input values to control the output. The program should display the results. Use OpenGL with C or C++, or use Java 3D.

17.22 Write a program that generates a finite element mesh for a four-sided, 2D region using the transfinite mapping method.

17.23 Same as Problem 17.22 but for a 3D region with six surfaces.

Product Data Exchange

GOAL

Understand data exchange standards including IGES and STEP, how translators work, why we need preprocessors and postprocessors, and how to test translators to ensure the correct data conversion.

OBJECTIVES

After reading this chapter, you should understand the following concepts:

- Necessary evil of CAD/CAM data translation

- Data exchange format

- IGES standard

- STEP standard

- ACIS format

- DXF *de facto* standard

- File structure and format

- Testing and verification

CHAPTER HEADLINES

18.1 Introduction

Fundamental incompatibilities among entity representations greatly complicate exchanging modeling data among CAD/CAM systems. Even simple geometric entities such as circular arcs are represented by incompatible forms in many systems. Some systems use NURBS to represent them, as we cover in Chapter 8, while others use the usual parametric representation covered in Chapter 6.

The transfer of data between dissimilar CAD/CAM systems must embrace the complete description of a product stored in its database. Four types of modeling data make up this description. These are shape, nonshape, design, and manufacturing data. Shape data consists of both geometrical and topological information. Nonshape data includes shaded images and measuring units of the database. Design data includes FEM/FEA. Manufacturing data includes tolerancing and bill of materials.

Where similar CAD/CAM systems are operated by both parties, no difficulty of exchange exists because the files that store modeling data are compatible. However, many dissimilar CAD/CAM systems are in existence, and here data communication problems arise.

Realizing the importance of product data exchange among different CAD/CAM systems, all CAD/CAM vendors and many organizations have been collaborating to set exchange standards to make it manageable for users of CAD/CAM systems to communicate their product information effectively.

The evolution of the exchange standards mimics the evolution of the CAD/CAM technology itself. Older standards such as IGES focus on geometric data, while newer standards such as STEP embrace the four types of product data just described. Other de facto standards such as ACIS and AutoCAD DXF have become popular due to their pervasive use in industry. We cover all these standards in this chapter.

18.2 Types of Translators

Having acknowledged the need to exchange modeling data, how can we solve the problem of data exchange? Two solutions exist: direct and indirect. The direct solution produces direct translators. It entails translating the modeling data directly from one CAD/CAM system format to another, usually in one step. This solution converts the data (database) format from one native format to another. It requires a knowledge of both native formats.

On the other hand, the indirect solution produces indirect translators. It is more general and adopts the philosophy of creating a neutral database structure (also called a neutral file) which is independent of any existing or future CAD/CAM system. This structure acts as an intermediary and a focal point of communication among the dissimilar database structures of CAD/CAM systems. This solution converts native formats to a neutral format that all CAD/CAM systems can interpret and understand.

Figure 18.1 shows how both solutions work. Direct translators convert data directly in one step. They are typically written by computer service companies that specialize in CAD/CAM database conversion. Direct translators are considered dedicated translation programs, two of which link a system pair as indicated by the dual direction arrows shown in Figure 18.1. For example, two translators are needed to transfer data between `System 1` and `System 2`; one from `System 1` to `System 2` and the other from `System 2` to `System 1`.

Indirect translators utilize some neutral file format, which reflects the neutral database structure. Each translation system has its own pair of translators to translate data to and from the neutral format. The translator that converts data from the native format of a given CAD/CAM system to the neutral format is called a preprocessor, while the translator that does the opposite translation is known as a postprocessor.

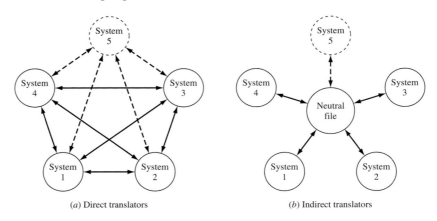

(a) Direct translators   (b) Indirect translators

Figure 18.1   Types of CAD/CAM data translators.

Each type of translator has its advantages and disadvantages. Direct translators provide a satisfactory solution when only a small number of systems are involved, but as this number increases the number of translator programs that need to be written becomes prohibitive. In general, if modeling data is to be transferred between all possible pairs of n CAD/CAM systems, then the total number of translators, N, that must be written is given by:

$$N = 2\binom{n}{2} = n(n-1) \tag{18.1}$$

where

$$\binom{n}{2} = \frac{n!}{2(n-2)!}$$

The symbol "!" in Eq. (18.1) indicates factorial. The coefficient 2 of the left-hand side of the same equation reflects the fact that two translators must be written for each pair of systems. Adding one system to the existing n systems would require writing $2n$ additional translators. In Figure 18.1a, eight additional translators (shown dashed) are needed to accommodate the addition of `System 5`.

On the plus side, direct translators run more quickly than the indirect ones, and the data files they produce are smaller than the neutral files created by indirect translators.

As for indirect translators, they do not suffer from the increasing number of programs that have to be written as in the case of direct ones. For the case of n systems, the total number of indirect translators that must be written is given by:

$$N = 2n \qquad\qquad (18.2)$$

Equation (18.2) applies for $n > 3$. For $n = 3$, Eqs. (18.1) and (18.2) give the same number of translators. For $n > 3$, the required number of translators based on the neutral format philosophy is less than those based on direct conversion.

Adding one system (System 5 shown in Figure 18.1b) to n existing systems would only require writing two additional translators (shown dashed in the figure) regardless of the value of n. Moreover, indirect translator philosophy provides stable communication between CAD/CAM systems, protects against system obsolescence, and eliminates dependence on a single system supplier.

A side benefit of neutral files is that they can potentially be archived. Some companies, in the aerospace industry for example, need to keep CAD/CAM databases for 20 to 50 years. Indirect translators based on a standard neutral file format are now the common practice, while direct translators are seldom used. The remainder of this chapter presents material related to indirect translators only.

18.3 IGES

IGES (Initial Graphics Exchange Specification) is the first standard exchange format developed to address the concept of communicating product data among dissimilar CAD/CAM systems. IGES is the ANSI Standard Y14.26M. IGES has gone through various revisions since its inception. Currently, it supports solid modeling, including both B-rep and CSG schemes.

IGES defines a neutral database, in the form of a neutral file format, which describes an "IGES model" of modeling data of a given product. The IGES model can be read and interpreted by dissimilar CAD/CAM systems. Therefore, the corresponding product data can be exchanged among these systems. IGES describes the possible entities that can be used to build an IGES model, the necessary parameters (data) for the definition of model entities, and the possible relationships and associativities between model entities.

Like most CAD/CAM systems, an IGES model is based on the concept of entities. The fundamental unit of information in the model, and consequently in the IGES file, is the entity; all product data is expressed as a list of predefined entities. Each entity defined by IGES is assigned a specific entity type number to refer to it in the IGES file. Entity numbers 1 through 599 and 700 through 5000 are allocated for specific assignments. Entity type numbers 600 through 699 and 10000 through 99999 are for implementor-defined entities (via macro definitions). Entity type numbers 5001 through 9999 are reserved for macro entities.

IGES has three data types: geometric, annotation, and structure. The latter two are nongeometric data types. Geometric entities define the product shape and include curves, surfaces, and solids. Nongeometric entities provide views and drawings of the model to enrich its representation and include annotation and structure entities. Annotation entities include various types of dimensions (linear, angular, ordinate), centerlines, notes, general labels, symbols, and cross-hatching. Structure entities include views, drawings, attributes (such as line and text fonts, colors, and layers), properties (e.g., mass properties), subfigures and external cross reference entities (for surfaces and assemblies), symbols (e.g., mechanical and electrical symbols), and macros (to define parametric parts).

With the information structure (geometry, annotation, and structure) of geometric models captured, the remainder of data representation consists primarily of specifying the data and parameters of typical geometric entities such as curves, surfaces, and solids, and nongeometric entities such as annotation. Due to the similarities between IGES representations of various entities, only selected entity types are covered in some detail in this chapter.

18.3.1 Geometric Entities

IGES uses two distinct but related cartesian coordinate systems to represent geometric entity types. These are the MCS and WCS introduced in Chapter 2. IGES refers to the WCS as the definition space. The WCS plays a simplifying role in representing planar entities. In such a case, the *XY* plane of the WCS is taken as the entity plane, and therefore only *x* and *y* coordinates relative to the WCS are needed to represent the entity. To complete the representation, a transformation matrix is assigned (via a pointer) to the entity as one of its parameters to map its description from WCS to MCS. This matrix itself is defined in IGES as entity type 124. Each geometric entity type in IGES has such a matrix.

If an entity is directly described relative to the MCS, then no transformation is required. This is achieved in IGES by setting the value of the matrix pointer to zero to prevent unnecessary processing. As a general rule, all geometric entity types in IGES are defined in terms of a WCS and a transformation matrix. The case when MCS and WCS are identical is triggered by a zero value of the matrix pointer.

Directionality is important to exchanging curves, especially parametric curves. Within IGES, all curves are directed. Each curve has a starting point, an ending point, and a parameter *u*. The information may not be enough to uniquely define the curve, as in the case of a circular or conic arc. Thus, some entity types refer to a "counterclockwise direction" with respect to a WCS. In IGES, the definition of this direction is based on an observer positioned along the positive *Z* axis of the WCS and looking down upon the *XY* plane of the WCS.

IGES reserves entity numbers 100 to 199 inclusive for its geometric entities. Sample entity type numbers used by IGES are shown in Table 18.1.

Table 18.1   IGES geometric entities

Entity number	Entity description	Entity number	Entity description
100	Circular arc	132	Connect point
102	Composite curve	136	Finite element
104	Conic arc	138	Nodal display and rotation
106	Copious data	140	Offset surface
108	Plane	142	Curve on a parametric surface
110	Line	144	Trimmed parametric surface
112	Parametric spline curve	146	Nodal results
114	Parametric spline surface	148	Element results
116	Point	150	Block
118	Ruled surface	152	Right angular wedge
120	Surface of revolution	154	Right circular cylinder
122	Tabulated cylinder	156	Right circular cone
124	Transformation matrix	158	Sphere
125	Flash	160	Torus
126	Rational B-spline curve	162	Solid of revolution
128	Rational B-spline surface	164	Solid of linear extrusion
130	Offset curve	186	Ellipsoid

Specifications and descriptions of entities, including geometric entities, in IGES follow one pattern. Each entity has two main types of data: directory data and parameter data. The former is the entity type number, and the latter are the parameters required to uniquely and completely define the entity. In addition, IGES specifies other parameters related to entity attributes and to IGES file structure.

18.3.2 Annotation Entities

Drafting data are represented in IGES via its annotation entities. Many IGES annotation entities are constructed by using other basic entities that IGES defines, such as copious data (centerline, section, and witness line), leader (arrow), and a general note.

An annotation entity may be defined in the modeling space (WCS) or in the drawing space (a given drawing). If a dimension is inserted by the user in model mode, then it requires a transformation matrix pointer when it is translated into IGES.

Table 18.2 shows some IGES annotation entities.

Table 18.2   IGES annotation entities

Entity number	Entity description	Entity number	Entity description
202	Angular dimension	216	Linear dimension
206	Diameter dimension	218	Ordinate dimension
208	Flag note	220	Point dimension
210	General label	222	Radius dimension
212	General note	228	General symbol
214	Leader (arrow)	230	Sectional area

18.3.3 Structure Entities

The previous two sections show how geometric and drafting data can be represented in IGES. Product definition includes much more information. IGES permits a valuable set of product data to be represented via its structure entities. These entities include associativity, drawing, view, external reference, property, subfigure, macro, and attribute entities. Attributes include line fonts, text fonts, and color definition. Table 18.3 shows IGES structure entities.

Table 18.3   IGES structure entities

Entity number	Entity description	Entity number	Entity description
302	Associativity definition	406	Property
304	Line font definition	408	Singular subfigure instance
306	Macro definition	410	View
308	Subfigure definition	412	Rect. array subfig. instance
310	Text font definition	414	Circular array subfig. instance
312	Text display template	416	External reference
314	Color definition	418	Nodal load/constraint
320	Network subfigure definition	420	Network subfigure instance
402	Associativity instance	600 – 699	Macro instance (user defined)
404	Drawing	10000 – 99999	Macro instance (user defined)

The associativity definition entity (type number 302) allows IGES to define a special relationship (called associativity schema) between various entities of a given model. The collection of entities that are related to each other via the associativity schema is called a class. Two kinds of associativities are permitted within IGES. Predefined associativities have (form) number 1 to 5000, and the second kind is implementor-defined and has numbers 5001 to 9999. Each time an associativity relation is needed in the IGES file, an associativity instance entity (type number 402) is used.

The external reference entity (type number 416) enables IGES files to relate to each other. This entity provides a link between an entity in one file and the definition or a logically related entity in another file. Three forms of external reference entity are defined. Form 0 is used when a single definition from the referenced file, which may contain a collection of definitions, is desired. Form 1 is used when the entire file is to be instanced, which is the case where the referenced file contains a complete subassembly. Form 3 is used when an entity in one file refers to another entity in a separate file. This is the case when each sheet of a drawing is a separate file and, for example, a flange on one sheet mates with a flange on another sheet.

The property entity (type number 406) in IGES contains numerical and textural data. Due to the wide range of properties, each one is assigned a form number and each form number may contain different property types (ptypes). For example, form number 11 contains tabular data that is organized under n ptypes. ptypes 1, 2, 3, 4, as an example, refer to Young's modulus, Poisson's ratio, shear modulus, and material matrix, respectively. There are 17 form numbers that can be specified with the property entity.

The macro capability in IGES (type number 306) allows the family of parts and/or entities grouped by the user for special purposes to be exchanged. Macros can only define a "new" entity in terms of the entities supported by IGES. This capability allows the extension of IGES beyond its common entity subset by utilizing a formal mechanism which is a part of IGES itself. A "new" entity can only be defined once in an IGES file, but it can be referenced as many times in the file as needed by using the macro instance entity (type number 600 – 699 or 10000 – 99999). This number is referred to in the macro definition entity as the entity type ID.

18.3.4 File Structure and Format

A typical CAD/CAM system which supports IGES usually provides its users with two IGES commands. One command enables the user to create an IGES file of a given model residing in the system, while the other allows the user to read an existing IGES file of a model into the system.

An IGES file consists of a sequence of records. Depending on the chosen file format, the record length can be fixed or variable. There are two different formats to represent IGES data in a file: ASCII and binary. The ASCII form has two format types: fixed 80-character record (line) length format, and compressed format. The binary form consists of bytes representing the data. Both the compressed ASCII and binary formats are aimed at reducing the IGES file size. We only cover the fixed 80-character length format here.

The file is divided into sections. Within each section, the records are labeled and numbered. IGES data is written in columns 1 through 72 inclusive of each record. Column 73 stores the section identification character. Columns 74 through 80 are specified for the section sequence number of each record.

An IGES file has a specific structure. It consists of six sections which must appear in the following order: Flag section (optional), Start section (S), Global section (G), Directory Entry (D) section, Parameter Data (P) section, and Terminal section (T), as shown in Figure 18.2.

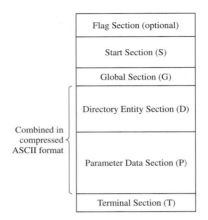

Figure 18.2   IGES file structure.

Figure 18.2 shows the section code, also called the identification character (Column 73 of each record) for the IGES file sections. These codes are S, G, D, P, and T. The Flag section does not have a code. The **Flag** section is used only with compressed ASCII and binary format. It is a single record (line) that precedes the Start section in the IGES file with the character "C" in Column 73 to identify the file as compressed ASCII. The compressed ASCII form is intended to be simply converted to and from the regular ASCII form. In the Binary file format, the Flag section is called the Binary information section, and the first byte (eight bits) of this section has the ASCII letter "B" as the file identifier.

The **Start** section is a human-readable introduction to the file. It is commonly described as a "prologue" to the IGES file. This section includes user-relevant information, such as the name of the CAD/CAM system generating the IGES file, and a brief description of the product being converted. IGES does not specify how this section could be used.

The **Global** section includes information describing the preprocessor and information needed by the postprocessor to interpret the file. Some of the parameters that are specified in this section are the characters used as delimiters between individual entries and between records (usually commas and semicolons, respectively), the name of the IGES file, the vendor and software version of the sending CAD/CAM system, the number of significant digits in the representation of integers and single- and the double-precision floating-point numbers on the sending systems, the date and time of file generation, model space scale, model units, the minimum resolution and maximum coordinate values, and the name and organization of the author of the IGES file.

The **Directory Entry** section is a list of all the entities defined in the IGES file together with certain attributes associated with them. The entry for each entity occupies two 80-character records, which are divided into a total of 20 8-character fields. The first and the eleventh (beginning of the second record of any given entity) fields contain the entity type number (Tables 18.1 to 18.3). The second field contains a pointer to the parameter data entry for the entity in the Parameter Data section. The pointer of an entity is simply its sequence number in the Directory Entry section. Some of the entity attributes specified in this section are line font, layer number, transformation matrix, line weight, and color.

The **Parameter Data** section contains the actual data defining each entity listed in the Directory Entry section. For example, a straight line entity is defined by the six coordinates of its two end points. While each entity always has two records in the Directory Entry section, the number of records needed for each entity in the Parameter Data section varies from one entity to another (minimum is one record) and depends on the amount of data. Parameter data are placed in free format in columns 1 through 64. The parameter delimiter (usually a comma) is used to separate parameters, and the record delimiter (usually a semicolon) is used to terminate the list of parameters. Both delimiters are specified in the Global section of the IGES file. Column 65 is left blank. Columns 66 to 72 on all Parameter Data records contain the entity pointer specified in the first record of the entity in the Data Entry section.

The **Terminate** section contains a single record which specifies the number of records in each of the four preceding sections for checking purposes.

18.4 STEP

STEP (STandard for Exchange of Product data) is an exchange for product data in support of industrial automation. "Product data" is interpreted to be more general than the "product definition data" which forms the core philosophy of IGES. "Product data" encompasses data relevant to the entire life cycle of a product such as design, manufacturing, quality assurance, testing, and support.

STEP is an ISO standard. It has absorbed PDES (Product Data Exchange Standard) which was an ANSI standard. Before merging the two, PDES was a U.S. project. In June 1985, the IGES Steering Committee voted that PDES should represent U.S. interest in the STEP effort.

In order to support industrial automation, STEP files are fully interpretable by computer. For example, tolerance information would be carried in a form directly interpretable by a computer rather than a computerized text form which requires human intervention to interpret. In addition, this information would be associated with those entities in the model affected by the tolerance. Thus, the general emphasis of STEP is to eliminate the human presence from the "product data" exchange; that is, to obviate the use of engineering drawings and other paper documents as necessary means of passing information between different product phases that may be performed on similar or dissimilar CAD/CAM systems.

There is a fundamental difference in philosophy between exchanging data in IGES and in STEP. The central unit of data exchange in the IGES model is the entity, while the central unit of data exchange in the STEP model is the application, which contains various types of entities. Therefore, when data is exchanged between systems, it is done in terms of "application" units. This approach maintains all the meaningful associativities and relationships between the application entities which make industrial automation possible.

The goal of STEP is to represent all product information, in a common data format, throughout a product's entire life cycle. The data being transferred is geometry (e.g., curves, surfaces, solids), but also analysis, manufacturing, implementation, and testing procedures. Essentially, STEP is a common structure, operating as a template, for sharing data among multiple users, across all functional areas.

18.4.1 Architecture

To achieve the STEP philosophy, product data is exchanged by STEP according to "discipline models" or "schemas." Both the sender who originates the discipline model and the receiver of the model must be aware of the meaning of the discipline model being exchanged in order to recover the correct meaning of data in the exchange. Discipline models are standardized and defined by STEP to be interpreted and used by another computer. Examples are the Mechanical Products discipline model, Electrical Products discipline model, and AEC discipline model. Thus, the concept of discipline models makes STEP flexible to accommodate any future models and application areas when they become available.

STEP uses a three-layer architecture, as shown in Figure 18.3 that forms the core of STEP structure. The three layers are the application layer, the logical layer, and the physical layer. The application layer is the interface between the user and STEP. It contains all the descriptions and information of various application areas. The purpose of the logical layer is to provide a consistent, computer-independent description of the data constructs that contain the information to be exchanged. The physical layer deals with the data structures and data format for the exchange file itself. The main goal here is to establish and maintain efficiency in the file size and processing time.

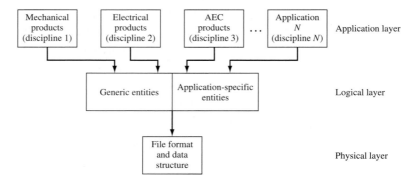

Figure 18.3   STEP three-layer architecture.

18.4.2 Implementation

STEP is built on a data exchange language, called EXPRESS, to formally describe a model and the file format that stores it. EXPRESS stores both the model data and semantics.

The basic unit in EXPRESS is the entity. An **entity** is a collection of data, constraints, and operations. The operations work on data. A set of entities make up a model. The realtionships (semantics) between model entities are carried over, and maintained by STEP, from the native CAD database of the model.

EXPRESS follows the object oriented (OO) paradigm covered in Chapter 5. It first defines an object schema, and then defines instances of such object. Consider the example of a cricle. EXPRESS defines the circle (the object) schema as its center and radius. the center is a point, and the radius is a real number. EXPRESS defines a point schema as three numbers for the x, y, and z coordinates. The circle center is an instance of the point object.

There are several sections within STEP, called Application Protocols (APs), which are built into a common data model. These APs include definitions not only of typical geometry and drafting elements, but also of data types and processes for specific industries such as automotive, aerospace, shipbuilding, electronics, plant construction and maintenance. Sample APs are AP203 (STEP format to save solid models) and AP210 (STEP format to save electronics). AP203 is further divided into classes for defining wireframe geometry, surfaces and solid modeling.

One of the latest significant developments in STEP is the recent agreement to provide mapping to XML (eXtensible markup language). This new technology is rapidly becoming the preferred method for complex data access on the Web. The flexibility and growing availability of commercial Web/XML tools with STEP greatly increases the sharing of information across disciplines, with universal access.

XML lends itself well to STEP. XML is a standard format for data representation. It complements XHTML (eXtensible hypertext markup langauge) which is a standard for data presentation. XML defines data schema in a DTD (document type definition) file.

18.5 ACIS and DXF

ACIS is used as a kernel in a number of commercial CAD/CAM systems. Spatial, the maker of ACIS, provides a translator for these systems use. Spatial's translator allows the exchange of solid, surface, and wireframe data via a variety of neutral and native formats, including IGES, STEP, Pro/E, SolidWorks, CATIA, Parasolid (PS), Unigraphics (UG), and Inventor. These major systems, therefore, offer these formats in their translation menus.

DXF (Data eXchange Format) is a *de facto* standard due to its popularity. DXF is an AutoCAD format. AutoDesk Inc., the maker of AutoCAD, publishes, supports, and maintains it. DXF 3D is a format that translates CAD models (part files), while DXF/DWG is a format that translates drawing files. DXF/DWG does not and cannot translate part files. DXF files come in two formats: ASCII and binary. The ASCII version is the most widely used in industry.

A DXF file consists of four sections: Header, Tables, Blocks, and Entities. The Header section includes the AutoCAD system settings such as dimension style and layers. The Tables section includes line styles and user-defined coordinate systems. The Blocks section includes drawing blocks (instances). The Entities section includes entity definition and data.

18.6 Processors

A standard, such as IGES or STEP, in itself is just a document describing what should go into a data file. Interested developers (CAD/CAM vendors or companies specialized in database transfer) must interpret, understand, and implement the standard into programs, often called processors or translators.

The processors translate from their systems to the standard format and vice versa. The software that translates from the native file format of a given CAD/CAM system to a standard format is called a preprocessor. The software that translates in the opposite way (from a standard to a CAD/CAM system) is called a postprocessor. The user interface to access these processors usually takes the form of simple commands, accompanied by proper dialogues.

Figure 18.4 shows file exchange using a translator. The source system is the originating or sending CAD/CAM system, and the target system is the receiving one. The archival database is a side benefit of using standards. Such archived databases could be kept for as long as needed. If system B in Figure 18.4 becomes the source and system A becomes the target, the processors reverse positions.

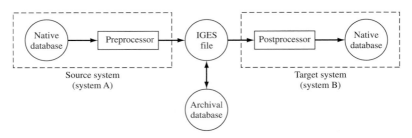

Figure 18.4   Pre- and postprocessors of a translator.

18.6.1   Design and Implementation

Designing and writing processors is a significant challenge. A typical database might contain many instances of many entity types. Many of these entity types involve complex mathematics and complex data structures. Problems in writing a processor relate to the definition and format of the standard itself. Some of these problems are:

1. **Entity set.** Any standard (IGES or STEP) does not and cannot contain a real superset of entities which are found in all of today's CAD/CAM systems. The standard may contain an entity which has no equivalence on a specific CAD/CAM system. Or, the system may contain an entity for which no standard entity exists. A processor could either ignore translating the entity, or translate it into a similar one, destroying its original meaning.

2. **Format.** While a standard allows exchanging complex structures and relationships, its format must be processible by a wide range of different computer systems and therefore

can only use simple data formats and management methods known to these systems and, in the meantime, independent of any system specifics.

3. **Limitations of individual CAD/CAM systems.** These limitations are based on specific systems and are related to things such as model size, model space, and data precision.

The designing of processors, with all the preceding problems in mind, divides into the following steps:

1. **Analyze and tabulate entity characteristics.** This step involves the study of the entity mathematical representations utilized by both the standard and the CAD/CAM system. In many cases, an entity can be represented by a number of nearly, but not completely, equivalent methods.

2. **Define conversion algorithms.** Step 1 clearly provides the information required to design the proper conversion algorithms to convert an entity to and from the standard.

3. **Develop a complete specification of the processors.** Steps 1 and 2 form the core of the design process of processors. Once completed, other specifications of the processors must be developed. These include the standard revision that the processor ought to support, the subset of the standard entities it can support, and the user interface of the processor.

4. **Design verification procedures.** Careful verification of processors is very important because processors operate at the interface between different organizations and vendors. Processors must be verified by constructing test data, running it through the processors, and comparing the actual results with those expected. Ideally, two sets of test data would be required: a set for implementations to use during processor development, and a more comprehensive set for final processor verification. In addition, more customized tests for specific user requirements can also be developed in a collaboration between users and implementors of the standard.

18.6.2 Testing and Verification

A newly developed processor must be carefully tested before it is used in a production environment. For example, there is an IGES Test, Evaluate, and Support Committee whose function is to provide test data. There is an IGES test library prepared by the committee which allows testing of the basic implementation of an entity. However, the library does not allow the checking of the variations that occur in production data due to numerical and computational errors. These variations must be tested by implementors and users themselves.

Verification of the results of a processor is a time-intensive task. In most cases, it is not sufficient to check converted models visually; more comprehensive tests are needed. The common methods of testing are shown in Figure 18.5. They are:

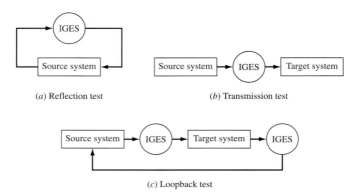

(a) Reflection test   (b) Transmission test

(c) Loopback test

Figure 18.5   Methods of verification of processors.

1. **Reflection test**. In this test, a neutral file created by a translator preprocessor is read by its own postprocessor to create a native file of the translated model. This test is used to establish that a translator's processors could read and write common entities, making them symmetric.

2. **Transmission test.** Here a neutral file of a model created by the preprocessor of a source system is transferred to a target system whose postprocessor is used to re-create the model on the target system. This test essentially determines the capabilities of the preprocessor and the postprocessor of the source and the target systems, respectively.

3. **Loopback test.** In this test, a neutral file created by the source system is read by the target system which, in turn, creates another neutral file and then transfers this file back to the source system to read it. This test checks the pre- and postprocessors of both the source and the target systems.

18.6.3 Error Handling

Error handling and reporting when processing a neutral file is important. There are two major error sources when processing IGES files: programming errors in the processor, and misinterpretation of the standard itself. These sources apply to both pre- and postprocessors. The way a processor reports these errors, and the information given with these reports, determine whether the correction of an error becomes a laborious task or not. The preprocessor should report the entity type, number of unprocessed entries, reasons for unprocessing, and other relevant database information of these unprocessed entities.

On the other hand, the postprocessor should report the number of unprocessed entities, their types, their forms, their record numbers in the Directory Entry and Parameter Data sections, and the reasons for unprocessing. It should also report any invalid or missing data encountered in reading neutral files, especially those that were edited.

 Tutorials

18.7.1 Create IGES, STEP, ACIS, and DXF Files

In this tutorial we use four CAD data exchange formats: IGES, STEP, ACIS, and DXF. We export the native CAD file of the following cube part to neutral files using these formats so that we can import them to other CAD systems. We keep the part and the drawing as simple as possible to limit the size of the neutral files because they tend to be large, even for these simple files.

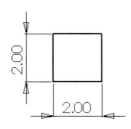

The neutral files are ASCII files. We open them in a text editor, such as the Notepad, and annotate them, using the C language comment-statement style, for presentation purposes in this tutorial. We generally recommend against manual editing of these files in practice.

Follow these steps to create the model and its neutral files:

1. **Create a simple cubic block.** Create the block as an extrusion, or as a primitive, depending on your CAD system.

2. **Create a drawing of the block.** Select the drawing mode of your CAD system, define a drawing with one view, and dimension it as shown above.

3. **Create the IGES file.** Simply save the part file as an IGES file. Click this sequence, or its equivalent, on your CAD system: `File => Save As => Type a file name =>` Select the file type as `IGES (*.igs) => OK`. We open the file in a text editor and remove some repetitive records from it to reduce its length. Here is the resulting IGES file:

```
📄 DataExCubeModied.igs - Notepad                                                    _ □ ✕
File  Edit  Format  View  Help
Solidworks IGES FILE using analytic representation for surfaces      S     1  //Start section (S)|

1H,,1H;,17HDataExCube.SLDPRT,52HC:\AcademicStuff\CSEG240-F03\MassProp\DaG         1  //Global section (G)
taExCube.IGS,41HSolidworks 2003 by Solidworks Corporation,11HVersion 5.3G         2
,32,308,15,308,15,17HDataExCube.SLDPRT,1.,1,2HIN,50,0.125,13H031103.1702G         3
36,1E-008,19684.6456692913,11HNishit Shah,,11,0,;                        G         4
                                                                            //blank lines are not part of file

        314        1        0        0        0         00000200D     1  //Directory Entity section (D)
        314        0        8        1        0               0D     2  //only partial results are shown
        128        2        0        0        0         01010000D     3  //25 entites are shown
        128        0        0        2        0               0D     4  //Each entity has two records (lines)
        126        4        0        0        0         01010500D     5
        126        0        0        2        0               0D     6
        110        6        0        0        0         01010000D     7
        110        0        0        1        0               0D     8
        126        7        0        0        0         01010500D     9
        126        0        0        2        0               0D    10
        110        9        0        0        0         01010000D    11
        110        0        0        1        0               0D    12
        126       10        0        0        0         01010500D    13
        126        0        0        2        0               0D    14
        110       12        0        0        0         01010000D    15
        110        0        0        1        0               0D    16
        126       13        0        0        0         01010500D    17
        126        0        0        2        0               0D    18
        110       15        0        0        0         01010000D    19
        110        0        0        1        0               0D    20
        102       16        0        0        0         01010500D    21
        102        0        0        1        0               0D    22
        102       17        0        0        0         01010000D    23
        102        0        0        1        0               0D    24
        142       18        0        0        0         00010500D    25
        142        0        0        1        0               0D    26
        144       19        0        0        0         00000000D    27
        144        0       -1        1        0               0D    28
        128       20        0        0        0         01010000D    29
        128        0        0        2        0               0D    30
        126       22        0        0        0         01010500D    31
        126        0        0        2        0               0D    32
        110       24        0        0        0         01010000D    33
        110        0        0        1        0               0D    34
        126       25        0        0        0         01010500D    35
        126        0        0        2        0               0D    36
        110       27        0        0        0         01010000D    37
        110        0        0        1        0               0D    38
        126       28        0        0        0         01010500D    39
        126        0        0        2        0               0D    40
        110       30        0        0        0         01010000D    41
        110        0        0        1        0               0D    42
        126       31        0        0        0         01010500D    43
        126        0        0        2        0               0D    44
        110       33        0        0        0         01010000D    45
        110        0        0        1        0               0D    46
        102       34        0        0        0         01010500D    47
        102        0        0        1        0               0D    48
        102       35        0        0        0         01010000D    49
        102        0        0        1        0               0D    50

314,75.2941176470588,75.2941176470588,75.2941176470588,;              1P     1  //Parameter Data section (P)
128,1,1,1,0,0,1,0,0,0.,0.,1.,1.,0.,0.,1.,1.,1.,1.,1.,0.,              3P     2  //only partial results are shown
0.,-2.,0.,0.,0.,0.,2.,-2.,0.,2.,0.,0.,1.,0.,1.;                       3P     3  //25 entites are shown
126,1,1,0,1,0,0,0.,1.,1.,1.,1.,1.,0.,0.,1.,1.,0.,0.,1.,0.,            5P     4  //They match entities in the D section
0.,1.;                                                                5P     5  //Each entity has one record (line)
110,0.,0.,0.,0.,2.,0.;                                                7P     6
126,1,1,1,0,1,0,0.,1.,1.,1.,1.,1.,1.,0.,0.,1.,0.,0.,1.,0.,            9P     7
0.,1.;                                                                9P     8
110,0.,2.,0.,0.,2.,-2.;                                              11P     9
126,1,1,1,0,1,0,0.,1.,1.,1.,1.,0.,1.,0.,0.,0.,0.,0.,1.,0.,           13P    10
0.,1.;                                                               13P    11
110,0.,2.,-2.,0.,0.,-2.;                                            15P    12
126,1,1,0,1,0,0.,1.,1.,1.,1.,0.,0.,0.,1.,0.,0.,0.,1.,0.,            17P    13
0.,1.;                                                               17P    14
110,0.,0.,-2.,0.,0.,0.;                                             19P    15
102,4,5,9,13,17;                                                    21P    16
102,4,7,11,15,19;                                                   23P    17
142,1,3,21,23,1;                                                    25P    18
144,3,1,0,25;                                                       27P    19
128,1,1,1,0,0,1,0,0,0.,1.,1.,0.,0.,1.,1.,1.,1.,1.,0.,              29P    20
2.,-2.,0.,2.,0.,2.,2.,-2.,2.,2.,0.,0.,1.,0.,1.;                     29P    21
126,1,1,0,1,0,0.,1.,1.,1.,1.,1.,0.,0.,1.,1.,0.,0.,1.,0.,            31P    22
0.,1.;                                                               31P    23
110,0.,2.,0.,2.,2.,0.;                                              33P    24
126,1,1,0,1,0,0.,1.,1.,1.,1.,1.,1.,0.,0.,1.,0.,0.,1.,0.,            35P    25
0.,1.;                                                               35P    26
110,2.,2.,0.,2.,2.,-2.;                                             37P    27
126,1,1,1,0,1,0,0.,1.,1.,1.,1.,0.,1.,0.,0.,0.,0.,0.,1.,0.,           39P    28
0.,1.;                                                               39P    29
110,2.,2.,-2.,0.,2.,-2.;                                            41P    30
126,1,1,0,1,0,0.,1.,1.,1.,0.,0.,0.,1.,0.,0.,0.,0.,1.,0.,            43P    31
0.,1.;                                                               43P    32
110,0.,2.,-2.,0.,2.,0.;                                             45P    33
102,4,31,35,39,43;                                                  47P    34
102,4,33,37,41,45;                                                  49P    35

S      1G     4D   158P    109                                        T     1  //Terminate section (T)
```

4. **Create the STEP file.** Simply save the part file as a STEP file. Click this sequence, or its equivalent, on your CAD system: File => Save As => Type a file name => Select the file type as STEP (*.step) => OK. We edit the STEP file to remove some repetitive records from it to reduce its length. Here is the resulting STEP file.

```
DataExCubeModified.STEP - Notepad
File  Edit  Format  View  Help
ISO-10303-21;
HEADER;
FILE_DESCRIPTION (( 'STEP AP214' ),
    '1' );
FILE_NAME ('DataExCube.STEP',
    '2003-11-03T22:09:50',
    ( 'Nishit' ),
    ( 'Self' ),
    'SwSTEP 2.0',
    'Solidworks 2002296',
    '' );
FILE_SCHEMA (( 'AUTOMOTIVE_DESIGN' ));
ENDSEC;

DATA;
#1 = VERTEX_POINT ( 'NONE', #152 ) ;
#2 = CARTESIAN_POINT ( 'NONE', ( 2.000000000000000000, 0.000000000000000000, 0.000000000000000000 ) ) ;
#3 = COLOUR_RGB ( '',0.7529411764705882200, 0.7529411764705882200, 0.7529411764705882200 ) ;
#4 = FILL_AREA_STYLE_COLOUR ( '', #3 ) ;
#5 = FILL_AREA_STYLE ('',( #4 ) ) ;
#6 = SURFACE_STYLE_FILL_AREA ( #5 ) ;
#7 = SURFACE_SIDE_STYLE ('',( #6 ) ) ;
#8 = SURFACE_STYLE_USAGE ( .BOTH., #7 ) ;
#9 = PRESENTATION_STYLE_ASSIGNMENT (( #8 ) ) ;
#10 = STYLED_ITEM ( 'NONE', ( #9 ), #118 ) ;
#11 = PRESENTATION_LAYER_ASSIGNMENT ( '', '', ( #10 ) ) ;
#12 = MECHANICAL_DESIGN_GEOMETRIC_PRESENTATION_REPRESENTATION ( '', ( #10 ), #20 ) ;
#13 = DIMENSIONAL_EXPONENTS ( 1.000000000000000000, 0.000000000000000000, 0.000000000000000000, 0.000000000000000000, 
0.000000000000000000, 0.000000000000000000, 0.000000000000000000 ) ;
#14 = LENGTH_MEASURE_WITH_UNIT ( LENGTH_MEASURE( 0.02539999999999999900 ), #16 );
#15 =( CONVERSION_BASED_UNIT ( 'INCH', #14 ) LENGTH_UNIT ( ) NAMED_UNIT ( #13 ) );
#16 =( LENGTH_UNIT ( ) NAMED_UNIT ( * ) SI_UNIT ( $, .METRE. ) );
#17 =( NAMED_UNIT ( * ) PLANE_ANGLE_UNIT ( ) SI_UNIT ( $, .RADIAN. ) );
#18 =( NAMED_UNIT ( * ) SI_UNIT ( $, .STERADIAN. ) SOLID_ANGLE_UNIT ( ) );
#19 = UNCERTAINTY_MEASURE_WITH_UNIT (LENGTH_MEASURE( 1.000000000000000100E-005 ), #15, 'distance_accuracy_value', 
'NONE');
#20 =( GEOMETRIC_REPRESENTATION_CONTEXT ( 3 ) GLOBAL_UNCERTAINTY_ASSIGNED_CONTEXT ( ( #19 ) )
GLOBAL_UNIT_ASSIGNED_CONTEXT ( ( #15, #17, 18 ) ) REPRESENTATION_CONTEXT ( 'NONE', 'WORKASPACE' ) );
#21 = CARTESIAN_POINT ( 'NONE', ( 2.000000000000000000, 2.000000000000000000, -2.000000000000000000 ) ) ;
#22 = CARTESIAN_POINT ( 'NONE', ( 0.000000000000000000, 0.000000000000000000, -2.000000000000000000 ) ) ;
#23 = LINE ( 'NONE', #24, #25 ) ;
#24 = CARTESIAN_POINT ( 'NONE', ( 0.000000000000000000, 0.000000000000000000, 0.000000000000000000 ) ) ;
#25 = VECTOR ( 'NONE', #26, 39.37007874015748100 ) ;
#26 = DIRECTION ( 'NONE', ( 1.000000000000000000, 0.000000000000000000, 0.000000000000000000 ) ) ;
#27 = LINE ( 'NONE', #28, #29 ) ;
#28 = CARTESIAN_POINT ( 'NONE', ( 0.000000000000000000, 2.000000000000000000, -2.000000000000000000 ) ) ;
#29 = VECTOR ( 'NONE', #30, 39.37007874015748100 ) ;
#30 = DIRECTION ( 'NONE', ( 0.000000000000000000, 0.000000000000000000, 1.000000000000000000 ) ) ;
#31 = LINE ( 'NONE', #32, #33 ) ;
#32 = CARTESIAN_POINT ( 'NONE', ( 0.000000000000000000, 0.000000000000000000, -2.000000000000000000 ) ) ;
#33 = VECTOR ( 'NONE', #36, 39.37007874015748100 ) ;
#34 = ORIENTED_EDGE ( 'NONE', *, *, #191, .F. ) ;
#35 = ADVANCED_FACE ( 'NONE', ( #157 ), #158, .T. ) ;
#36 = DIRECTION ( 'NONE', ( 0.000000000000000000, 0.000000000000000000, 1.000000000000000000 ) ) ;
#37 = FACE_OUTER_BOUND ( 'NONE', #126, .T. ) ;
#38 = PLANE ( 'NONE', #39 ) ;
#39 = AXIS2_PLACEMENT_3D ( 'NONE', #40, #41, #42 ) ;
#40 = CARTESIAN_POINT ( 'NONE', ( 2.000000000000000000, 0.000000000000000000, -2.000000000000000000 ) ) ;
#41 = DIRECTION ( 'NONE', ( 1.000000000000000000, 0.000000000000000000, 0.000000000000000000 ) ) ;
#42 = DIRECTION ( 'NONE', ( 0.000000000000000000, 0.000000000000000000, -1.000000000000000000 ) ) ;
#43 = LINE ( 'NONE', #44, #45 ) ;
#44 = CARTESIAN_POINT ( 'NONE', ( 2.000000000000000000, 0.000000000000000000, -2.000000000000000000 ) ) ;
#45 = VECTOR ( 'NONE', #46, 39.37007874015748100 ) ;
#46 = DIRECTION ( 'NONE', ( 0.000000000000000000, 1.000000000000000000, 0.000000000000000000 ) ) ;
#47 = CARTESIAN_POINT ( 'NONE', ( 2.000000000000000000, 2.000000000000000000, 0.000000000000000000 ) ) ;
#48 = FACE_OUTER_BOUND ( 'NONE', #125, .T. ) ;
#49 = PLANE ( 'NONE', #50 ) ;
#50 = AXIS2_PLACEMENT_3D ( 'NONE', #51, #52, #53 ) ;
#51 = CARTESIAN_POINT ( 'NONE', ( 0.000000000000000000, 2.000000000000000000, -2.000000000000000000 ) ) ;
#52 = DIRECTION ( 'NONE', ( 0.000000000000000000, 1.000000000000000000, 0.000000000000000000 ) ) ;
#53 = DIRECTION ( 'NONE', ( 0.000000000000000000, 0.000000000000000000, 1.000000000000000000 ) ) ;
#54 = LINE ( 'NONE', #55, #56 ) ;
#55 = CARTESIAN_POINT ( 'NONE', ( 2.000000000000000000, 0.000000000000000000, 0.000000000000000000 ) ) ;
#56 = VECTOR ( 'NONE', #57, 39.37007874015748100 ) ;
#57 = DIRECTION ( 'NONE', ( 0.000000000000000000, 1.000000000000000000, 0.000000000000000000 ) ) ;
#58 = LINE ( 'NONE', #59, #60 ) ;
#59 = CARTESIAN_POINT ( 'NONE', ( 0.000000000000000000, 0.000000000000000000, -2.000000000000000000 ) ) ;
#60 = VECTOR ( 'NONE', #61, 39.37007874015748100 ) ;
#61 = DIRECTION ( 'NONE', ( 0.000000000000000000, -1.000000000000000000, 0.000000000000000000 ) ) ;
#62 = FACE_OUTER_BOUND ( 'NONE', #136, .T. ) ;
#63 = PLANE ( 'NONE', #64 ) ;
#64 = AXIS2_PLACEMENT_3D ( 'NONE', #65, #66, #67 ) ;
#65 = CARTESIAN_POINT ( 'NONE', ( 0.000000000000000000, 2.000000000000000000, 0.000000000000000000 ) ) ;
#66 = DIRECTION ( 'NONE', ( 0.000000000000000000, 0.000000000000000000, -1.000000000000000000 ) ) ;
#67 = DIRECTION ( 'NONE', ( -1.000000000000000000, 0.000000000000000000, 0.000000000000000000 ) ) ;
#68 = LINE ( 'NONE', #69, #70 ) ;
#69 = CARTESIAN_POINT ( 'NONE', ( 2.000000000000000000, 2.000000000000000000, -2.000000000000000000 ) ) ;
//removed records #70 to #193
ENDSEC;
END-ISO-10303-21;
```

5. **Create the ACIS file.** Simply save the part file as an ACIS file. Click this sequence, or its equivalent, on your CAD system: `File => Save As =>` Type a file name `=>` Select the file type as `ACIS (*.sat) => OK`. We did not have to edit the ACIS file as it is not excessively long. Here is the originally generated ACIS file.

```
DataExCubeModified.SAT - Notepad
File Edit Format View Help
700 0 1 0
37 SolidWorks(2002296)-Sat-Convertor-2.0 13 ACIS 7.0.3 NT 24 Mon Nov 03 17:13:45 2003
1 9.9999999999999995e-007 1e-010
-0 body $1 -1 $-1 $2 $-1 $-1 #
-1 name_attrib-gen-attrib $-1 -1 $-1 $-1 $0 keep keep_kept ignore copy @10 DataExCube #
-2 lump $3 -1 $-1 $-1 $4 $0 #
-3 rgb_color-st-attrib $-1 -1 $-1 $-1 $2 0.75294117647058822 0.75294117647058822 0.75294117647058822 #
-4 shell $-1 -1 $-1 $-1 $5 $-1 $2 #
-5 face $-1 -1 $-1 $6 $7 $4 $-1 $8 forward single #
-6 face $-1 -1 $-1 $9 $10 $4 $-1 $11 forward single #
-7 loop $-1 -1 $-1 $-1 $12 $5 #
-8 plane-surface $-1 -1 $-1 0 0 -50.799999999999997 0 -1 0 0 0 -1000 forward_v I I I I #
-9 face $-1 -1 $-1 $13 $14 $4 $-1 $15 forward single #
-10 loop $-1 -1 $-1 $-1 $16 $6 #
-11 plane-surface $-1 -1 $-1 0 50.799999999999997 -50.799999999999997 0 0 -1 -1000 0 0 forward_v I I I I #
-12 coedge $-1 -1 $-1 $17 $18 $19 $20 reversed $7 $-1 #
-13 face $-1 -1 $-1 $21 $22 $4 $-1 $23 forward single #
-14 loop $-1 -1 $-1 $-1 $24 $9 #
-15 plane-surface $-1 -1 $-1 50.799999999999997 0 -50.799999999999997 1 0 0 0 0 -1000 forward_v I I I I #
-16 coedge $-1 -1 $-1 $25 $26 $27 $28 reversed $10 $-1 #
-17 coedge $-1 -1 $-1 $29 $12 $30 $31 reversed $7 $-1 #
-18 coedge $-1 -1 $-1 $12 $29 $32 $33 forward $7 $-1 #
-19 coedge $-1 -1 $-1 $34 $35 $12 $20 forward $36 $-1 #
-20 edge $-1 -1 $-1 $37 0 $38 0.050799999999999998 $19 $39 forward @7 unknown #
-21 face $-1 -1 $-1 $40 $36 $4 $-1 $41 reversed single #
-22 loop $-1 -1 $-1 $-1 $42 $13 #
-23 plane-surface $-1 -1 $-1 0 0 -50.799999999999997 -1 0 0 0 0 1000 forward_v I I I I #
-24 coedge $-1 -1 $-1 $32 $43 $34 $44 reversed $14 $-1 #
-25 coedge $-1 -1 $-1 $45 $16 $46 $47 reversed $10 $-1 #
-26 coedge $-1 -1 $-1 $16 $45 $29 $48 forward $10 $-1 #
-27 coedge $-1 -1 $-1 $30 $49 $16 $28 forward $22 $-1 #
-28 edge $-1 -1 $-1 $50 -0.050799999999999998 $51 0 $27 $52 forward @7 unknown #
-29 coedge $-1 -1 $-1 $18 $17 $26 $48 forward $7 $-1 #
-30 coedge $-1 -1 $-1 $42 $27 $17 $31 forward $22 $-1 #
-31 edge $-1 -1 $-1 $51 0 $37 0.050799999999999998 $30 $53 forward @7 unknown #
-32 coedge $-1 -1 $-1 $54 $24 $18 $33 reversed $14 $-1 #
-33 coedge $-1 -1 $-1 $55 0 $38 0.050799999999999998 $18 $56 forward @7 unknown #
-34 coedge $-1 -1 $-1 $57 $19 $24 $44 forward $36 $-1 #
-35 coedge $-1 -1 $-1 $19 $57 $42 $58 forward $36 $-1 #
-36 loop $-1 -1 $-1 $-1 $35 $21 #
-37 vertex $-1 -1 $-1 $20 $59 #
-38 vertex $-1 -1 $-1 $33 $60 #
-39 straight-curve $-1 -1 $-1 0 0 0 1000 0 0 I I #
-40 face $-1 -1 $-1 $-1 $61 $4 $-1 $62 forward single #
-41 plane-surface $-1 -1 $-1 0 50.799999999999997 0 0 0 -1 -1000 0 0 forward_v I I I I #
-42 coedge $-1 -1 $-1 $49 $30 $35 $58 reversed $22 $-1 #
-43 coedge $-1 -1 $-1 $24 $54 $63 $64 forward $14 $-1 #
-44 edge $-1 -1 $-1 $38 0 $65 0.050799999999999998 $34 $66 forward @7 unknown #
-45 coedge $-1 -1 $-1 $26 $25 $54 $67 reversed $10 $-1 #
-46 coedge $-1 -1 $-1 $68 $63 $25 $47 forward $61 $-1 #
-47 edge $-1 -1 $-1 $69 -0.050799999999999998 $50 0 $46 $70 forward @7 unknown #
-48 edge $-1 -1 $-1 $51 0 $55 0.050799999999999998 $29 $71 forward @7 unknown #
-49 coedge $-1 -1 $-1 $27 $42 $68 $72 reversed $22 $-1 #
-50 vertex $-1 -1 $-1 $72 $73 #
-51 vertex $-1 -1 $-1 $31 $74 #
-52 straight-curve $-1 -1 $-1 0 0 -50.799999999999997 0 -1000 0 I I #
-53 straight-curve $-1 -1 $-1 0 0 -50.799999999999997 0 0 1000 I I #
-54 coedge $-1 -1 $-1 $43 $32 $45 $67 forward $14 $-1 #
-55 vertex $-1 -1 $-1 $33 $75 #
-56 straight-curve $-1 -1 $-1 50.799999999999997 0 -50.799999999999997 0 0 1000 I I #
-57 coedge $-1 -1 $-1 $35 $34 $76 $77 forward $36 $-1 #
-58 edge $-1 -1 $-1 $78 -0.050799999999999998 $37 0 $35 $79 forward @7 unknown #
-59 point $-1 -1 $-1 0 0 0 #
-60 point $-1 -1 $-1 50.799999999999997 0 0 #
-61 loop $-1 -1 $-1 $-1 $76 $40 #
-62 plane-surface $-1 -1 $-1 0 50.799999999999997 -50.799999999999997 0 1 0 0 0 1000 forward_v I I I I #
-63 coedge $-1 -1 $-1 $46 $76 $43 $64 reversed $61 $-1 #
-64 edge $-1 -1 $-1 $69 0 $65 0.050799999999999998 $43 $80 forward @7 unknown #
-65 vertex $-1 -1 $-1 $64 $81 #
-66 straight-curve $-1 -1 $-1 50.799999999999997 0 0 0 1000 0 I I #
-67 edge $-1 -1 $-1 $55 0 $69 0.050799999999999998 $54 $82 forward @7 unknown #
-68 coedge $-1 -1 $-1 $76 $46 $49 $72 forward $61 $-1 #
-69 vertex $-1 -1 $-1 $47 $83 #
-70 straight-curve $-1 -1 $-1 0 50.799999999999997 -50.799999999999997 -1000 0 0 I I #
-71 straight-curve $-1 -1 $-1 0 50.799999999999997 0 0 1000 0 0 I I #
-72 edge $-1 -1 $-1 $50 0 $78 0.050799999999999998 $68 $84 forward @7 unknown #
-73 point $-1 -1 $-1 0 50.799999999999997 -50.799999999999997 #
-74 point $-1 -1 $-1 0 0 -50.799999999999997 #
-75 point $-1 -1 $-1 50.799999999999997 0 -50.799999999999997 #
-76 coedge $-1 -1 $-1 $63 $68 $57 $77 reversed $61 $-1 #
-77 edge $-1 -1 $-1 $65 -0.050799999999999998 $78 0 $57 $85 forward @7 unknown #
-78 vertex $-1 -1 $-1 $72 $86 #
-79 straight-curve $-1 -1 $-1 0 0 0 0 -1000 0 I I #
-80 straight-curve $-1 -1 $-1 50.799999999999997 50.799999999999997 -50.799999999999997 0 0 1000 I I #
-81 point $-1 -1 $-1 50.799999999999997 50.799999999999997 0 #
-82 straight-curve $-1 -1 $-1 50.799999999999997 0 -50.799999999999997 0 1000 0 I I #
-83 point $-1 -1 $-1 50.799999999999997 50.799999999999997 -50.799999999999997 #
-84 straight-curve $-1 -1 $-1 0 50.799999999999997 -50.799999999999997 0 0 1000 I I #
-85 straight-curve $-1 -1 $-1 0 50.799999999999997 0 -1000 0 0 I I #
-86 point $-1 -1 $-1 0 50.799999999999997 0 #
End-of-ACIS-data
```

6. **Create the DXF file.** Simply save the drawing file as a DXF file. Click this sequence, or its equivalent, on your CAD system: `File => Save As =>` Type a file name `=>` Select the file type as `DXF (*.dxf) => OK`. DXF format is very verbose. The DXF file for the simple drawing of this tutorial is 85 pages long when printed. We only show its first page in the screenshot below as an illustration of the format.

```
  0
SECTION
  2
HEADER
  9
$ACADVER
  1
AC1012
  9
$DWGCODEPAGE
  3
ansi_1252
  9
$INSBASE
 10
0.0
 20
0.0
 30
0.0
  9
$EXTMIN
 10
0.0
 20
0.0
 30
0.0
  9
$EXTMAX
 10
11.0
 20
8.5
 30
0.0
  9
$LIMMIN
 10
0.0
 20
0.0
  9
$LIMMAX
 10
0.0
 20
0.0
  9
$ORTHOMODE
 70
        0
  9
$REGENMODE
 70
        1
  9
$FILLMODE
 70
        1
  9
$QTEXTMODE
 70
        0
  9
$MIRRTEXT
 70
        1
  9
$DRAGMODE
 70
        2
  9
$LTSCALE
 40
1.0
  9
$OSMODE
 70
        0
  9
$ATTMODE
 70
        1
  9
```

Tutorial 18.7.1 discussion:

We create neutral files of the part using IGES, STEP, and ACIS. We create a neutral file of the part drawing using DXF.

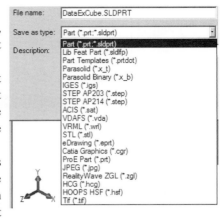

We edit the neutral files to show all the different sections of the file of each exchange standard, and yet keep the file length to a minimum. Relate the file information and sections to the material covered in the chapter for each format.

Check the Save As popup menu, or its equivalent, of your CAD system to find out the possible existing exchange format. The screenshot on the right shows an example. It shows some direct translations to Parasolid, Pro/E, and HOOPS. It also shows indirect translations to IGES, STEP, and ACIS.

Tutorial 18.7.1 hands-on exercise:

1. Read one of the neutral files back into your CAD system. Use the STEP file. Edit the part to add the fillet shown in the screenshot on the right. Use a fillet radius of one inch.
2. Perform the three processor tests covered in Section 18.6 and shown in Figure 18.5.

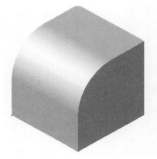

P ROBLEMS

Part I: Theory

18.1 Discuss the requirements of product data exchange between dissimilar CAD/CAM systems.
18.2 Describe the IGES methodology.
18.3 Compare the various testing methods for processors. Which test is the most comprehensive? Why?
18.4 Describe the STEP methodology.
18.5 Compare IGES and STEP.

Part II: Laboratory

Use your in-house CAD/CAM system to answer the questions in this part.

18.6 Generate IGES and STEP files for the geometric models of Chapters 6 and 7. Prepare files to demonstrate the exchange of the geometric, annotation, and structure entities.
18.7 For the IGES and STEP files you generate in Problem 18.6, perform the reflection test by reading the files back into your system. Did you get back the exact original models? What are your comments?

18.8 If you can use more than one CAD/CAM system, perform the other types of processor tests using the IGES and STEP files you generate in Problem 18.6. What are your observations?

18.9 Create a parametric model of a block with a hole. Export the model using IGES. Imoprt the IGES file back into your CAD/CAM system. Try to change the dimensions of the block and the hole in the sketch mode. What happens? Hint: IGES does not export parametrics and constraints.

18.10 Repeat Problem 18.9, but for STEP.

18.11 Carry out Tutorial 18.7.1 on your CAD system.Compare the sizes of the neutral files of the different exchange formats. Which format produces the smallest file size?

Collaborative Design

**OAL**
Understand the difference between traditional and collaborative design, become familiar with collaborative design concepts, understand its requirements, and know what tools are needed to perform collaborative design.

OBJECTIVES

After reading this chapter, you should understand the following concepts:

- Synchronous and asynchronous communication
- Distributed computing model
- Instant messaging
- Virtual reality
- Collaborative design principles
- Collaboration approaches
- Collaboration tools
- Collaborative design software

CHAPTER HEADLINES

19.1   Introduction

Collaborative design is an interactive process of real-time communication between members of a design team who are physically in different locations. During this process, several engineers or a team of designers are simultaneously involved to agree or disagree on design issues and address them during the early phases of product design. Concurrent engineering is considered the grander philosophy that motivates collaborative design as a key ingredient.

Collaborative design fosters the ability to conceptually design a product in a distributed manner by involving supply-chain and other members who have the expertise to accomplish the design task at hand. It has the advantages over traditional design, of increasing productivity, shortening product development lifecycle, and improving competitiveness.

The key advantage of collaborative over traditional design is real-time communication among team members regardless of where they are. Collaborative design is the closest thing to face-to-face meetings between team members in the same room. Collaborative design is more than just exchanging documents between designers on time. When there is real collaboration in design, there is a common design goal to achieve in real time. All design partners make effective use of each other's insight and talent. Teams must be stimulated in creativity, and collaboration must be organized both synchronously and asynchronously.

Collaborative design systems and software tools utilize various concepts of distributed computing such as the Internet, intranets, extranets, the client/server model of communication, instant messaging, and virtual reality. We cover these concepts here briefly.

19.2   Distributed Computing

Distributed computing is the pervasive computing model today. It is based on the client/server model that is central to the Internet's distributed network hardware/communication and services. The client/server model uses two computer programs installed on two separate, but networked, computers at different locations. The **client** program (or client, for short), installed on one computer (known as the client computer) communicates with the **server** program (or server, for short) installed on the other computer, as shown in Figure 19.1. The client requests services from the server. While the server is running, it acts as a daemon. A **daemon** is a process that awaits (listens to) client requests. When it receives a request from a client, it grants the request and sends a response back to the client. A common example of client/server communication is a Web browser, acting as a client, requesting a Web page from a Web server.

Multiple clients can access a server concurrently as shown on the right. This increases the demand (known as the server load, traffic, or hits) on the server, and may slow its response significantly. In such a case, the server computer (also known as the host computer or host machine, or host for short) must be upgraded or

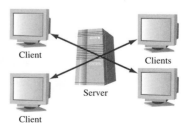

replaced with a faster computer. Balancing a server load to improve its response time is a ery interesting and practical problem, but it is beyond the scope of this book.

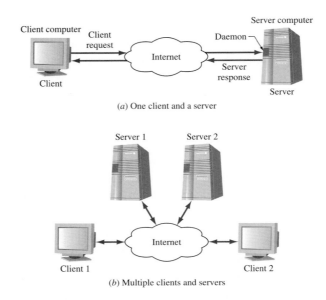

(a) One client and a server

(b) Multiple clients and servers

Figure 19.1   Client/server model of computing.

19.3   Intranets and Extranets

The underlying concept for the Internet, intranets, and extranets is the same. They are all networks using the same hardware and software. The differences occur in their geographical domains and in the level of security they use. Figure 19.2 shows the three types of networks. An **intranet** is a network that is contained within an organization; that is, it is an internal or private network. Organizations use intranets to share information and computing resources among employees in different departments and widespread offices.

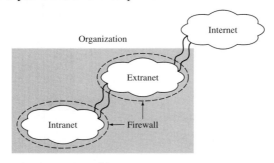

Figure 19.2   The Internet, extranet and intranet.

As with the Internet, intranets connect heterogeneous computer hardware, including server computers, workstations, PCs, and so forth. Because intranets provide company-wide (intracompany) access to the internal private information of the entire company, they must be protected from outside intruders. Thus, it is typical for companies to install firewalls around their intranets to prevent unauthorized access. A **firewall** is a collection of related computer programs installed on a server computer that acts as a gateway to the intranet. The firewall intercepts each incoming message or request to determine whether to forward it to its destination.

An **extranet** is an organization network that allows access to outside networks such as the Internet. It allows users outside a company to communicate with its employees, and vice versa. They can access the company's internal information and resources as well. An extranet facilitates intercompany relationships. Extranets typically link companies and businesses with their customers, suppliers, vendors, partners, and other businesses over the Internet. Extranets provide secure and private links via advances in network security and the use of firewalls. An extranet may be viewed as an intermediate network between the Internet and the intranet.

19.4 Instant Messaging

Collaborative design communication is based on Web communication concepts. The Web is all about communication in an easy and convenient way, anytime, anywhere. Communication via the Web has two modes: synchronous or asynchronous. **Synchronous communication** is an activity in which the two communicating persons send messages to (or receive them from) each other *simultaneously, in real time*. Examples include instant messaging (IM) and chatting. **Asynchronous communication** is an activity in which the two communicating persons send messages to (or receive them from) each other *independently, at different times*. Examples include e-mail and newsgroups.

Chatting on the Web is fun and easy to do. It enables us to keep and stay in touch. In addition to chatting for fun, the use of Web chatting is unlimited. For example, business sales people who are constantly on the road can use it to talk to their customers and clients. Company executives can hold meetings from remote locations. Professors and instructors can hold office hours with their students online, by setting a specific time where students can post questions and ask for help. Members of design teams can use chatting as a mode of communication during activities of collaborative design.

Three types of content exist for Web chatting. In text-only chatting, short text messages are exchanged. These messages usually use abbreviations, or acronyms. In instant messaging, graphics and text content is used, where text messages are augmented with colors and graphics in the form of what is known as emoticons. The third type of chatting includes both audio and video content. This type is known as Webcasting. Existing collaborative design tools make use of instant messaging, audio, and video communications.

Web chatting is a distributed application that is based on the client/server model of communication. Web chatting software can be client- or Web-based. In the client-based type, the user needs to download and install a chatting client. Examples include America Online Instant

Messenger (AIM), Yahoo Messenger (YM), and Microsoft Messenger (MSN Messenger). In the Web-based type, users chat through Web pages. Existing collaborative design tools make use of client-based Web chatting.

19.5 Virtual Reality Modeling Language

Virtual reality modeling language (VRML) lends itself well to collaborative design and CAD/CAM as a useful communication tool for 3D models. Section 2.16 describes how to use VRML for viewing purposes. **VRML** is a graphics systems that intermediates communication between output devices (graphics displays) and the application model (CAD part file).

Collaborative design tools convert a CAD file into a VRML file that can be rendered in a Web browser to serve as the center of discussion between team members during a real-time design session. The resulting VRML file does not require a CAD system. It can be displayed in a Web browser that has a VRML plug-in installed. Team members can rotate the VRML model and mark it up in real time.

VRML uses most of the computer graphics concepts covered in Part III of this book. It uses a 3D space to describe and render the modeling scene or the CAD model. It uses orientations, viewpoints, and transformations to move objects in space. It uses polygons and surfaces to describe the objects. It uses different types of light sources (such as ambient light, directional lights, spotlights, and point lights) to render objects. It also uses shading techniques (flat, Gourand, and Phong) and ray-tracing algorithms to add realism to models. It also uses texture mapping.

VRML files use the *.wrl* (for world) extension as the file type. A VRML file stores a scene graph that describes a 3D scene. VRML resembles to a great extent Java 3D and uses the object concepts we cover in Chapter 5. The building block of a VRML file is a node. Nodes make up the scene. Nodes have the parent/child relationship. VRML uses the instance concept when a scene uses the same basic shape (primitive) multiple times.

Different types of VRML nodes exist. Appearance nodes describe the colors, shading and lighting in the scene. Geometry nodes describe the shapes used in the scene. The shapes that VRML uses are the solids primitives we cover in Chapter 9. More specifically, VRML uses a box, cone, cylinder, sphere, and extrusion. Grouping nodes describe geometric transformations.

All the VRML concepts are made transparent to designers during collaborative design sessions because the collaboration software takes care of it. This allows designers to focus only on the design task at hand.

19.6 Traditional Design

Traditional design is a serialized process as shown in Figure 1.1 in Chapter 1. It begins with the design need and ends with the documentation of the design. In between come activities such as developing design requirements and specifications, generating model geometry, performing modeling and simulation, carrying detailed analysis and design optimization, and evaluation.

While concepts such as DFA (design for assembly), DFM (design for manufacturing), DFX (design for anything), and concurrent engineering have helped to improve and shorten the design process, it still largely remains a serialized activity.

After design is complete, the manufacturing phase begins with the process planning as shown in Figure 1.1. The point to make here is that design dictates manufacturing. Bad design decisions affect manufacturing cost and product quality directly. If a wrong decision is made at any stage it will affect the decisions in the following stages. It is well known that the process of revision of the design is slow and expensive.

Moreover, design is performed at many sites by many subcontractors. Each of them uses different tools and different programs. It is difficult to maintain multiple tools and to train users on multiple systems. Thus, consistent and cross-discipline representations are needed to approach a design from a systems perspective.

Also, due to design complexity and lack of collaboration tools, the teams only interact at preset milestones instead of continuously throughout the process. At these milestones, specifications of design are exchanged over the wall. Downstream functions are largely ignored due to the complexity and lack of integrated tool sets.

The main problem with the traditional design approach is the lack of collaborative and continuous communication between design teams and members on a daily, or even hourly, basis leading up to crucial deadlines. It is one thing to send an e-mail message with an attachment trying to convey a design idea, and it is another thing to talk about it with team members. We all agree that the latter approach is much more effective in getting the point across.

19.7 Collaborative Design

A collaborative design environment provides a decision-support and resource-management system that interconnects computational resources (CAD tools, design tools, resource planning, project management tools, etc.) and users. It also provides a communication infrastructure to enable distinct resources in different geographical locations and users to seamlessly collaborate and solve problems. By linking these tools within such an environment, they become available to users, regardless of where the tools and/or users are located.

The main difference between collaborative and traditional designs is that the former make the communication feasible and seamless anytime. We expect that it will take time for the philosophy of collaborative design to catch on and become a mainstream daily design tool and activity.

19.8 Collaboration Principles

The central issue in collaborative design is centralization and damage control. Should there be a centralized depository of design documents, and who should own the documents? If a miscommunication occurs during collaboration, how can we contain it? Due to this issue and others, there exist collaboration principles that help guide the development of collaborative design systems. They all relate to design knowledge and information, and how to deal with it. Here is a list of some principles:

1. **Ownership control**. Each group that generates design information should own and control it. This implies that the group is responsible for the updates and maintenance of the information. After a collaborative design session is over, the group should organize and store the information and stand by ready for any other changes that are deemed necessary by future collaboration.

2. **Distribution control.** Only the required information should be provided by group owners, and only when needed. This help prevents "choking" other designers by overwhelming them with too much design knowledge when they do not need it.

3. **Damage control.** This principle relates to information "defects." What happens if wrong information (known as information defects) is generated inadvertently during a session of collaboration? These defects should not be propagated through future sessions. These defects should be brought to the attention of its owners to correct as soon as possible to prevent extensive future damage.

4. **Access control.** Information should be provided only to those who need it. Information that does not concern a design group should not be given to it to avoid confusing it. This is different from distribution control. Distribution and access control complement each other. After we identify a design group that should get design information (access control here), we apply the principle of distribution control to this design information.

5. **Guidelines control.** Designers should not feel constrained during collaborative design sessions. It is imperative to the success of collaboration that designers feel free, within reasonable limits, to share and exchange design information with other designers, especially those from other organizations. In this regard, "soft" design guidelines should benefit most cases and should avoid difficult exception cases. On the other hand, organizations should avoid imposing "hard" design constraints such as "it must be done this way, and only this way." The use of published design standards as a guideline is a good example of soft control. An example of hard control is when a company requires a designer to seek approval or permission before releasing design information during a collaboration session.

19.9 Collaboration Approaches

Engineering design is a large, diverse discipline that can be viewed differently by different groups and teams. This depends largely on the point of view of design that is of interest to a group or team. For example, design may be viewed at a macro level, where a systems approach is used to guide design development and evaluate different design alternatives. On the other hand, design may be viewed at a micro level, where detailed analysis and computations are used to evaluate alternatives and make design decisions.

Engineering design approaches directly affect collaborative design tools and software systems. Different aspects of the engineering design process are dealt with differently in collaborative design. Collaborative design can be classified into three approaches. Each approach focuses on different aspects of engineering design and provides collaboration support for understanding and implementing these aspects. The three approaches are:

1. **Collaboration based on design methodologies.** This approach focuses on generating formalized design methodologies by investigating how the technical design decisions are made. The design process models are often implied in these design methodologies and theories. These models include the systematic design model, axiomatic design model, Quality function deployment (QFD), general design theory, and so forth. Basically, these design methodologies provide the guidelines for making technical decisions more consciously and systematically.

Collaborative design tools that belong to this approach would implement some of these design models in tools and software systems. Designers can focus on this aspect of design during a collaboration session by using these tools and systems.

2. **Collaboration based on design workflow.** This approach views the design process as a workflow with task dependencies and information exchange. The methodologies and techniques of this approach come mainly from research in business operation and project management. From this aspect, engineering design is viewed as an information-driven process among design activities.

The design organization is viewed as a stochastic processing network in which engineering resources are workstations that perform different tasks. The design tasks themselves are jobs that flow among the workstations. With workstations and tasks in mind, a set of techniques to manipulate the design activities are developed to aid during collaborative design sessions. Such techniques include signal flow diagrams, design structure matrices, and design process networks.

3. **Collaboration based on CAD/CAM/CAE.** Several methodologies from CAD, CAM, and CAE (computer-aided engineering) areas view collaborative design as individuals and groups accessing data and sharing design information. The design process is specified accordingly as the management of the product data at different abstraction levels. During this management, the technological, scientific, and interdisciplinary dependencies of the design information could be established, handled, and maintained by a product model— the CAD model. Information systems are built around these models to support the storage and processing of various types of interest to designers. These systems use IT (information technologies) and CAD/CAM/CAE techniques.

PLM (product lifecycle management) systems are built based on this IT approach. These systems provide designers with powerful collaborative design tools and systems. PLM is covered Chapter 23.

Which of the three collaboration approaches is best to use? This question is meaningless because they all serve different needs of the design process. We can safely say, though, that the tools and systems that are built based on the third approach are far more advanced and more powerful than those tools and systems that are built based on the other two approaches. This is because PLM systems have been in development by CAD/CAM vendors for quite some time.

19.10 Collaboration Tools

Collaborative design is a relatively new field. While there is a wealth of theories and concepts, technological advancement has not caught up yet. Until recently, only a small set of tools that somehow support collaborative design activities has been developed. For example, some tools for collaborative model annotation and visualization via the Internet have been available. They utilize techniques such as shared cameras and telepointers that allow designers, at different geographical locations, to annotate and visualize models on computer screens during collaboration sessions.

However, such tools are primarily visual. They focus on inspection and/or design-change management. They do not support real modeling activities. In other words, they are valuable assistants for teamwork, but not for real CAD modeling on CAD systems. Some more recent research is focusing on the possibility of enhancing existing CAD systems with collaborative facilities, with the goal of providing real modeling facilities to participants in a collaborative session.

The two main obstacles to developing modeling-based collaborative design tools and systems (to go beyond just visualization tools and systems) are concurrency and synchronization. The real challenge is developing a collaborative modeling system that offers (or hooks up to) all facilities of CAD/CAM/CAE modeling systems to its users, while at the same time providing them with the necessary coordination mechanisms that guarantee an effective collaboration.

Solutions must be provided to the critical problems of concurrency and synchronization that characterize collaborative design environments. Concurrency involves the management of different processes while trying to access and manipulate the same data simultaneously. Synchronization involves propagating evolving data among users of a distributed application, in order to keep their data consistent.

The Internet and its related technologies have been a major driving force in the recent development and advancement of solving the concurrency and synchronization problems. A new breed of collaborative tools and systems now enable interactive mechanical design sessions over the Internet. These are primarily PLM and related systems.

Computing technologies, networks, and information management products are key ingredients to the implementation of collaborative design tools and systems. Collaborative design software is a relatively new meld of networking and CAD. With this software, you can leverage the Internet, intranets, and/or extranets to extend access and interaction privileges to engineering data that exists in different locations. For example, some collaborating engineers may require access to assemblies, parts, and drawings (but not a full-blown design suite) of other engineers.

One of the popular schemes for building and developing suites of collaborative design tools and systems is to add them to existing CAD/CAM/CAE systems as add-ons. These add-on systems are usually third-party modules integrated into an application suite or a system. For example, I-DEAS Web Access is an add-on for EDS I-DEAS Master Series CAD/CAM system that allows data interaction. With I-DEAS Web Access, users can measure and mark up data and get at selected property information within a secure and managed data-sharing environment that ensures data concurrency, synchronization, and accuracy while limiting data proliferation.

19.11 Collaborative Design Systems

Collaborative modeling systems are distributed multiuser systems that allow both concurrent and synchronized correspondence, aimed at supporting engineering teams and coordinating their modeling activities. These systems turn design into an interactive process, instead of an iterative asynchronous process where product data are sent back and forth among several team members in an offline fashion.

Some commercial collaborative design systems are available and can be used as add-ons to CAD/CAM/CAE systems. Companies that offer collaborative design software include Adaptive Media, Centric Software, CoCreate, RealityWave, and EDS. Readers are encouraged to visit the Websites of these companies to learn more about these systems.

These collaborative design systems and others attempt to address the same issues: how to support collaborative design sessions and activities, and how to solve the two problems of concurrency and synchronization. Therefore, it is not surprising that these systems share similar characteristics. Table 19.1 summarizes them. Readers can use this table for evaluation purposes of collaborative design systems.

Table 19.1   Evaluation criteria for collaborative design systems.

Criterion	Criterion description
Operating systems (OSs) and platforms	All collaborative design systems support PC platforms. In addition, some support Unix platforms. Supported OSs include Windows, Linux, and Unix.
Supported file formats	File formats has to do with the CAD/CAM/CAE system that a collaborative design system is added to. The collaborative system must support the file formats of the CAD system to access part files and make them available during collaboration sessions. The popular supported file formats include IGES, STEP, VRML, STL, and DWG.

Table 19.1   Evaluation criteria for collaborative design systems. (Continued)

Criterion	Criterion description
Available collaboration tools	There is a common superset of collaboration tools that are found among collaborative design systems. Not all existing systems support all the tools listed here. They all support a subset of them. The superset of collaboration tools includes bookmark views, redline marking, markup, last view, live chat, URL links, video/audio conferencing, instant messaging, annotation, behavioral modeling, NetMeeting, view part, interferences, and simultaneous collaboration.
Supported CAD/CAM/CAE system	Different collaborative design systems support different CAD/CAM/CAE systems where they are used as add-ons to these systems. Among the CAD/CAM/CAE systems that are supported are Pro/E, SolidWorks, AutoCAD, CADKEY, I-DEAS, Unigraphics, Solid Edge, CATIA, MicroStation, and Solid Designer. In addition to supporting CAD/CAM/CAE systems, some collaborative design systems support office applications (e.g., word processors).
Data transfer methods	CAD/CAM/CAE and other modeling data must be transferred over the networks from one location to another to support collaboration during a design session. The two methods that exist are download or streaming. Download is an offline method where designers must wait for the entire file to download from the server to their computer (client) after requesting the file. This method is slow. It is similar to downloading music from the Internet over a slow modem. The streaming method is a fast, real-time method by which file content is streamed continuously from the server to the clients, similar to streaming audio/video over the Internet.
Pricing	The pricing structure of collaborative design systems is based on the number of permanent seats (users) and additional occasional viewers. All systems charge a price per seat. If it happens that an additional seat is required for a viewer, a separate charge is incurred. This is, effectively, an on-demand pricing. The viewer only views the session, but has limited, if any, participation during the collaborative design session.

Problems

Part I: Theory

19.1 What is collaborative design? What is its key benefit? What is its key advantage over traditional design?

19.2 Describe the client/server model of distributed computing.

19.3 What is the difference between the Internet, intranets, and extranets?

19.4 In what way does VRML benefit collaborative design?

19.5 What are the two main drawbacks of traditional design?

19.6 What is the difference between collaborative design and traditional design?

19.7 List and explain the collaboration principles.

19.8 List and explain the collaboration approaches.

19.9 What are the two main obstacles that hinder the development of collaborative design systems? Explain each one.

19.10 List and explain the criteria that are used to evaluate collaborative design systems.

Part II: Laboratory

Use your in-house CAD/CAM system to answer the questions in this part.

19.11 Find out if your CAD/CAM system has modules that support collaborative design. If you find some, apply the criteria of Table 19.1 to them. Also use them in an actual collaboration session with some other designers. Write and submit a full report.

19.12 Perform a Web search about existing collaborative design systems. Summarize each system by applying Table 19.1 to it. Submit a written report that describes each system briefly.

19.13 You are a team member of your capstone design project, or another project. Each member on your team produces a design document (file), be it a CAD part, a report, or an image. The team members can communicate only via computers. The modes of communication are using hardware media (e.g., CDs, zip disks, etc.), upload/download files to a common design working area (directory) via FTP, e-mail, instant messaging, voice over IP (Internet Protocol), and a PDM (Product Data Management) system. (See Chapter 23 for PDM.) The main goal of communication is that members of the design team share, mark up, and discuss the design documents among themselves. After communicating and sharing the files using each communication mode, answer the following questions:

1. Which is the best (fastest, easiest, most convenient, most accessible) mode of communication? Why?

2. How easy is it to communicate?

3. Apply the collaboration principles of Section 19.8 of this chapter to your collaboration experience. Be specific.

4. Which of the collaboration approaches of Section 19.9 of this chapter do you think you have used? Why?

5. Write a follow-up evaluation of your collaboration process. Suggest improvements. Be specific.

Product Manufacturing and Management

This part covers some concepts that are utilized in product management and manufacturing. These concepts are supported by most CAD/CAM systems. This part should help CAD designers and CAM engineers to realize the ROI (return on investment) after spending time creating intricate geometric models of complex parts and assemblies. To achieve this goal, we accomplish the following objectives:

1. Apply tolerances to parts **(Chapter 20)**
2. Plan part production **(Chapter 21)**
3. Perform and evaluate NC programming **(Chapter 22)**
4. Manage part and product data through all phases **(Chapter 23)**

Engineering Tolerances

GOAL

Understand the importance and role of tolerances in manufacturing, the different types of tolerances, part inspection and its relation to tolerances, and how CAD systems implement and use tolerances.

OBJECTIVES

After reading this chapter, you should understand the following concepts:

- Tolerance concepts: limits and fits, accumulation, cost, and quality

- Conventional tolerances

- Geometric tolerances

- Datums

- ANSI GD&T Y14.5M standards

- Inspection gages

- Tolerance stackup analysis

- Tolerance synthesis

CHAPTER HEADLINES

20.1 Introduction

Manufacturing parts to exact dimensions, such as exactly 12 in long, is impossible from practical experience. A variety of physical limitations on manufacturing processes (such as cutting conditions, hardware accuracy, software accuracy, and the skills of machine operators) as well as material properties contribute to limiting the precision with which we can manufacture parts.

To account for this variability of dimensions (due to manufacturing) at the design phase, we assign a tolerance to each suitable dimension of the part (not every dimension requires a tolerance). The tolerance produces a range of acceptable values for the dimension. If a part size and shape are not within the maximum and minimum limits defined by the part tolerances, the part is not acceptable; it is rejected during inspection and becomes scrap.

20.2 Need for Tolerances

The assignment of actual values to the tolerance limits has a major influence on the overall cost and quality of an assembly or a product. If the tolerances are too small (tight), the individual parts will cost more to make. If the tolerances are too large (loose), an unacceptable percentage of assemblies may be scrapped (rejected) or require rework.

In addition to manufacturing cost considerations, tolerances are usually specified to meet the functional requirements of assemblies. In order for mating features (faces) of mating parts to fit together and operate properly, each part must be manufactured within these tolerance limits. For example, sliding parts such as journals and pistons must be made so that they are capable of moving relative to other parts but without so much freedom that they will not function properly. On the other hand, keys, gears on shafts, and other similar members mounted by press or shrink fit are toleranced so that the desired interference is maintained without being so large as to make the assembly impossible or the resulting stresses too high.

Tolerancing is an important and essential element of mass production and interchangeable manufacturing, by which parts can be made in widely separated locations and then brought together for assembly. Tolerancing also makes it possible for spare parts to replace broken or worn ones in existing products successfully. Without interchangeable manufacturing, modern industry could not exist, and without effective size control by the engineer, interchangeable manufacturing could not be achieved.

Tolerancing information is essential for part design and manufacturing. Design engineers need tolerance analysis to distribute allowances among related design dimensions, to check design results, or to design assemblies. Production engineers need tolerances for process planning, assembly operations, part inspection, and other production activities.

In spite of the importance of tolerances in manufacturing, some design engineers do not give serious consideration to them. Moreover, often there is a natural reluctance to change a proven design to reduce manufacturing cost. A thorough understanding of the tolerancing theory should rectify these problems. In addition, the availability of good tolerance software should enable engineers to apply the tolerancing theory in a convenient and easy manner.

20.3 Conventional Tolerances

Conventional and geometric tolerances make up the tolerancing theory. Geometric tolerances are covered later in this chapter. This section introduces the basic concepts, standards, and common practices regarding the conventional tolerances of mechanical parts. We begin by introducing the ANSI (American National Standards Institute) definitions of tolerance terms.

Nominal size is the designation used for the purpose of general identification. It is the dimension that results from design calculations such as stress or heat transfer analysis. It is usually expressed in common fractions or decimals. For example, a pipe may be designated as a $3\frac{1}{2}$- or 3.5-inch-diameter pipe.

Basic size is the theoretical size from which limits of size are derived by the application of allowances and tolerances. It is the decimal equivalent of the nominal size. The number of decimal places determines the precision or accuracy required. For example, if the nominal size of a pipe is $3\frac{1}{2}$ in and we require accuracy to three decimal places, its basic size becomes 3.500 in. We should point out that the sizes of 3.5, 3.50, and 3.500 are three different sizes when we consider manufacturing and producing these dimensions in the machine shop.

Actual size is the measured size of the finished part. It is a combination of the basic size and its assigned tolerance. It is the size that results from producing the part, including all the manufacturing imperfections. The actual size must lie within the maximum and minimum allowable values for the size (known as limits); otherwise the part is rejected during inspection, therefore becoming scrap.

Tolerance is the total amount by which a dimension may vary. It is used to determine the permissible limits (maximum and minimum) of the dimensions. Tolerance can be expressed in either of two ways. A **bilateral tolerance** is specified as plus or minus deviation from the basic size, for example, 1.750 ± 0.002 in. A **unilateral tolerance** is a tolerance in which variation is permitted only on one side of the basic size, for example, $1.750^{+0.004}_{-0.000}$ or $1.750^{+0.000}_{-0.003}$.

Allowance is the difference between the maximum material limits of mating parts. It is the minimum clearance (positive allowance) or maximum interference (negative allowance) between mating parts. It is also known as the tightest fit between mating parts—the smallest clearance or the largest interference. Allowance is an algebraic value; it could be a positive or negative value. Allowance is used to determine the type of fit (clearance, transition, or interference), as we discuss later.

20.3.1 Clearance Fit Tolerance Calculations

Conventional tolerances are almost always applied to shaft-hole pairs of dimensions. Here is how we apply the preceding definitions to a particular dimension. Let us consider a shaft transmitting a load through a gear mounted on it. After the designer has performed the stress calculations, a minimal size of the shaft diameter is obtained—2 in. This is the nominal size of the shaft. It is also the nominal size of the hole in the gear, as shown in Figure 20.1a.

Assuming that allowance and tolerances are specified to four decimal places, the precision required is then the same, and the basic size is expressed as 2.0000 in as shown in Figure 20.1b.

From the functional and assembly requirements between the gear and the shaft (power is transmitted between the two via a key), a clearance fit between the two parts is adequate. Let us assume a tolerance h for the hole in the gear of value 0.0030 in, a shaft tolerance s of 0.0030 in, and an allowance a of 0.0030 in. The variables h, s, and a each define a tolerance zone.

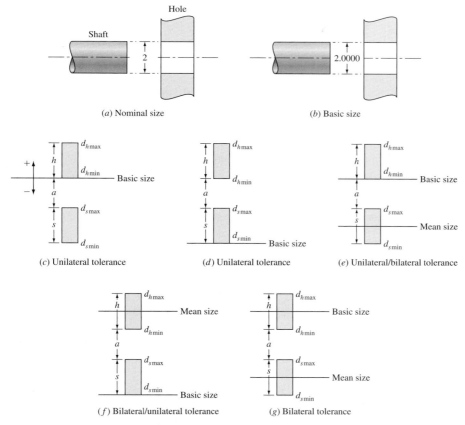

Figure 20.1　Clearance fit calculations.

In order to determine the diameters of the gear hole and the shaft given the values for h, s, and a, let us consider all the possible variations in the values of these dimensions relative to the basic size. Ten possibilities exist. Five of them are shown as bar diagrams in Figure 20.1c to 20.1g. The other five possibilities are complements (reverse the location of the bar adjacent to the basic-size datum) of those shown in the figure. The shaded bars are the tolerance zones.

The relative locations of these zones with respect to the basic size and the mean size determine the type of tolerance, and consequently the hole and shaft diameters. To determine the maximum and minimum diameters for the hole (d_{hmax} and d_{hmin}) and the shaft (d_{smax} and d_{smin}), a tolerance zone (h or s) is allocated relative to the basic-size datum, followed by the allowance a and the other tolerance zone. From the definition of allowance (minimum clearance), h, s, and *a* do not overlap in the case of clearance fit. In Figure 20.1*e*, the hole has unilateral tolerance while the shaft has a bilateral tolerance. Figure 20.1*f* shows the opposite case.

The calculations of the toleranced dimension from the bar diagrams are simple. For the five possibilities shown in Figure 20.1*c* to *g*, the hole dimensions are, respectively, $2.0000^{+0.0030}_{-0.0000}$, $2.0060^{+0.0030}_{-0.0000}$, $2.0000^{+0.0030}_{-0.0000}$, $2.0060^{+0.0015}_{-0.0015}$, and $2.0000^{+0.0015}_{-0.0015}$. The shaft dimensions are calculated in a similar fashion — simply add *h*, *s*, and *a* algebraically to the basic size. Observe that values above the basic-size line are positive and those below the basic-size line are negative. Thus the five toleranced values for Figure 20.1*c* to *g* are, respectively, $1.9970^{+0.0000}_{-0.0030}$, $2.0000^{+0.0030}_{-0.0000}$, $1.9955^{+0.0015}_{-0.0015}$, $2.0000^{+0.0030}_{-0.0000}$, and $1.9955^{+0.0015}_{-0.0015}$.

Using these dimensions for both the hole and the shaft, the reader can easily verify that the minimum clearance is equal to the allowance *a*. The other five cases that are not shown in Figure 20.1 are obtained from the cases shown in the figure by flipping the tolerance zone adjacent to the basic-size datum to the other side of the datum; except for the case shown in Figure 20.1*g*, the basic-size and mean-size datums are interchanged. The toleranced dimensions of the hole for the complements of possibilities of Figure 20.1*c* to *g* are, respectively, $2.0000^{+0.0000}_{-0.0030}$, $2.0030^{+0.0030}_{-0.0000}$, $2.0000^{+0.0000}_{-0.0030}$, $2.0045^{+0.0015}_{-0.0015}$, $2.0045^{+0.0030}_{-0.0000}$. The toleranced dimensions of the shafts for the complements of possibilities of Figure 20.1*c* to *g* are, respectively, $1.9955^{+0.0000}_{-0.0030}$, $2.0000^{+0.0000}_{-0.0030}$, $1.9925^{+0.0015}_{-0.0015}$, $2.0000^{+0.0000}_{-0.0030}$, and $2.0000^{+0.0015}_{-0.0015}$.

Out of the ten possibilities for the toleranced dimensions of both the shaft and the hole, what is the best possibility? Practice shows that drawings made with unilateral tolerances are usually easier to check than those made with bilateral tolerances. Thus, the six possibilities that result from Figures 20.1*e* to *g* and their complements are eliminated. In addition, it is usually easier to machine shafts than holes to any desired size. This eliminates the possibility shown in Figure 20.1*d* and its complement. The complement possibility of Figure 20.1*c* is also eliminated as it is practically easier for the machinist to aim at a minimum hole diameter equal to the basic size than at a diameter equal to the basic size minus *h*. The clearance fit for possibility of Figure 20.1*c* is shown in Figure 20.2.

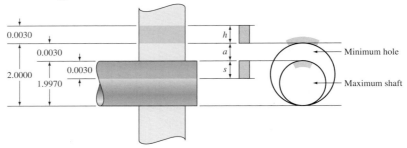

Figure 20.2   Clearance fit of Figure 20.1*c* (figure not drawn to scale).

In some situations, the bilateral method of tolerancing is very appropriate. Examples include the location of features when the variation from the basic size is equally critical in both directions, welded assemblies, and large tolerances. For large tolerances, it is sometimes more convenient to give the mean dimension and the variation each way.

20.3.2 Interference Fit Tolerance Calculations

Figures 20.3 and 20.4 show how to apply the clearance fit calculations shown in Figures 20.1 and 20.2 to an interference fit. The key difference is that h, s, and a overlap based on the definition of allowance a (maximum interference) in the case of interference fit. Here, we use 0.0008 in, 0.0008 in, and 0.002 in for h, s, and a, respectively. For the five possibilities shown in Figure 20.3c to g, the hole dimensions are, respectively, $2.0000^{+0.0008}_{-0.0000}$, $1.9996^{+0.0000}_{-0.0008}$, $2.0000^{+0.0008}_{-0.0000}$, $1.9992^{+0.0004}_{-0.0004}$, and $2.0000^{+0.0004}_{-0.0004}$. The shaft dimensions are calculated in a similar fashion—simply add h, s, and a algebraically to the basic size. Observe that values above the basic-size line are positive and those below the basic size line are negative. The five toleranced values for Figure 20.1c to g are, respectively, $2.0012^{+0.0008}_{-0.0000}$, $2.0000^{+0.0008}_{-0.0000}$, $2.0016^{+0.0004}_{-0.0004}$, $2.0000^{+0.0008}_{-0.0000}$, and $2.0012^{+0.0004}_{-0.0004}$.

Using these dimensions for both the shaft and the hole, the reader can easily verify that the maximum interference is equal to the allowance a. The preferred interference fit of Figure 20.3c is shown in Figure 20.4.

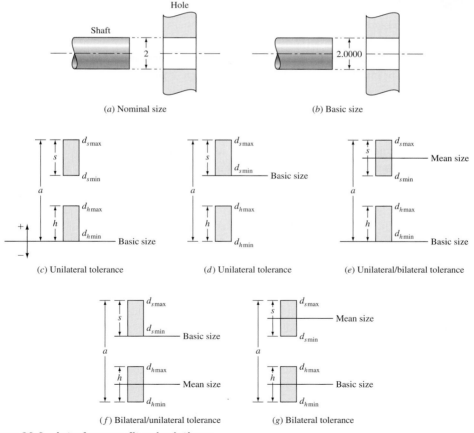

(a) Nominal size   (b) Basic size

(c) Unilateral tolerance   (d) Unilateral tolerance   (e) Unilateral/bilateral tolerance

(f) Bilateral/unilateral tolerance   (g) Bilateral tolerance

Figure 20.3   Interference fit calculations.

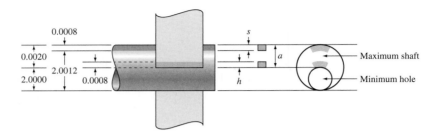

Figure 20.4   Interference fit of Figure 20.3*c* (figure not drawn to scale).

20.4 Fits and Limits

Fits between mating parts specify the range of tightness or looseness that may result from the application of specific tolerances to mating parts. In the previous sections, the limits (the maximum and minimum diameters) of the shaft and the hole are determined on the basis of the fit type (clearance or interference) given a hole tolerance h, a shaft tolerance s, and an allowance a. As seen, the type of fit is crucial in interpreting the allowance a (minimum clearance or maximum interference), which in turn determines the limits. The differences between the hole and the shaft limits determine the range of tightness or looseness.

Fits between mating parts can be identified as cylindrical or location fits. Cylindrical fits apply when an internal member fits in an external member as a shaft in a hole. Location fits are intended to determine only the location of the mating parts. Both cylindrical and location fits can be divided into three types: clearance fits, interference fits, and transition fits.

In clearance fit, one part is always loose relative to the other; that is, a shaft is loose in a hole or two stationary parts can be freely assembled or disassembled. In interference fit, one part is forced tight into the other during assembly and an internal pressure between the two results. Interference fits which can transmit torques or forces between mating parts (e.g., a shaft and a pulley) are usually referred to as force fits. Location interference fits are used when accuracy of location is important and for parts requiring alignment. A transition fit is a fit which may result in either clearance or interference condition.

When utilizing a given fit to calculate the limits and toleranced dimensions of mating parts, the designer must assign the basic size found from design calculations to either the shaft or the hole as seen in the previous section. The choice determines the basis or the system for calculating the limits and toleranced dimensions. There exist two systems: the basic hole system and the basic shaft system. Both systems assume unilateral tolerances h and s for the hole and the shaft, respectively.

In the basic hole system, the minimum hole is taken as the basic size, and the allowance and tolerances are applied accordingly. Figure 20.1c shows an example of using the basic hole system to apply tolerances and calculate limits.

In the basic shaft system, the maximum shaft is taken as the basic size, and the allowance and tolerances are applied accordingly. The complement of Figure 20.1d would show an example of using the basic shaft system to apply tolerances and calculate limits.

The basic hole system is the most widely used and recommended system in practice. This is due to manufacturing considerations. It is usually easier to machine shafts to any desired size than holes. Holes are often produced by using standard reamers, broaches, and other standard tools; and standard plug gages are used to check the actual sizes. The basic shaft system should not be used unless there is a reason for it. For example, it is advantageous when several parts having different fits, but one basic size, are mounted on a single shaft. Typically, the textile industry uses the basic shaft system.

20.4.1 MMC and LMC

Assuming we can calculate the limits of hole and shaft sizes for a given fit, how can we show them on an engineering drawing? There are two approaches of dimensioning a drawing; one is based on the maximum material condition (MMC), and the other is based on the least material condition (LMC). The MMC is defined as the condition where the maximum amount of material is preserved in the shaft-hole system. The allowances for clearance and interference fits have been defined to maintain the MMC. The LMC is the opposite of the MMC. It is the condition where the least amount of material is preserved in the shaft-hole system. Figure 20.5 shows how the MMC and LMC are used to show limits or dimensions for the clearance fit shown in Figure 20.2 and the interference fit shown in Figure 20.4.

Figure 20.5 shows that the dimensions of the maximum shaft and minimum hole represent the MMC and consequently the tightest fit, while the dimensions of the minimum shaft and maximum hole represent the LMC and, therefore, the loosest fit. Thus, the MMC produces the least possibility of assembly because it produces the most dangerous condition (tightest fit). However, once assembled, the resulting assembly best meets its functional requirement. The LMC, on the other hand, produces the best possibility of assembly because it produces the least dangerous condition (loosest fit). However, once assembled, the resulting assembly least meets its functional requirement. The MMC is usually preferred over the LMC.

Dimensioning based on MMC has another advantage. During manufacturing, the machinist aims at the principal dimension which is the one shown above the dimensioning line, that is, the maximum shaft or minimum hole in the MMC. Should the machinist, through error, produce an oversized hole or an undersized shaft, the parts might still be acceptable, providing the dimensions are within the limits specified by the drawing. In effect, the MMC reduces the amount of scrap, which is a valuable economic gain.

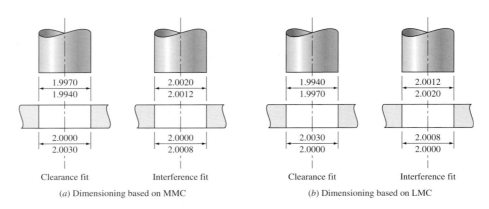

Clearance fit — 1.9970 / 1.9940 (top), 2.0000 / 2.0030 (bottom)
Interference fit — 2.0020 / 2.0012 (top), 2.0000 / 2.0008 (bottom)
Clearance fit — 1.9940 / 1.9970 (top), 2.0030 / 2.0000 (bottom)
Interference fit — 2.0012 / 2.0020 (top), 2.0008 / 2.0000 (bottom)

(*a*) Dimensioning based on MMC   (*b*) Dimensioning based on LMC

Figure 20.5   Dimensioning according to MMC and LMC.

20.4.2   ANSI Calculations

In order to achieve interchangeable manufacturing, the calculations of fits, tolerances, and limits have to be standardized. Starting with the functional requirements of the mating parts, the designer can choose a suitable fit. Tolerance zones, allowances, and limits can then be evaluated, and the dimensioning limits are calculated. We present the calculations of tolerances and limits in the two standards systems: ANSI and ISO (International Standards Organization).

ANSI has established eight classes (types) of cylindrical fit that specify the amount of allowance a, the hole tolerance h, and the shaft tolerance s as functions of the basic size (diameter) d as shown in Table 20.1. With a basic size d and a desired fit, the designer can use Table 20.1 to calculate a, h, and s, which in turn can be used to calculate the toleranced dimensions and limits as discussed in the previous sections and shown in Figures 20.1 to 20.5.

Table 20.1   ANSI classification of cylindrical fits.

Class of fit	Type	Description	Allowance (clearance) a	Allowance (interference) a	Hole tolerance h	Shaft tolerance s
1	Clearance fit	Loose fit	$0.0025\sqrt[3]{d^2}$		$0.0025\sqrt[3]{d}$ [a]	$0.0025\sqrt[3]{d}$
2		Free fit	$0.0014\sqrt[3]{d^2}$		$0.0013\sqrt[3]{d}$	$0.0013\sqrt[3]{d}$
3		Medium fit	$0.0009\sqrt[3]{d^2}$		$0.0008\sqrt[3]{d}$	$0.0008\sqrt[3]{d}$
4	Transition fit	Snug fit	0		$0.0006\sqrt[3]{d}$	$0.0004\sqrt[3]{d}$
5		Wringing fit		0	$0.0006\sqrt[3]{d}$	$0.0004\sqrt[3]{d}$

Table 20.1   ANSI classification of cylindrical fits. (Continued)

Class of fit	Type	Description	Allowance (clearance) a	Allowance (interference) a	Hole tolerance h	Shaft tolerance s
6	Interference fit	Tight fit		$0.00025d$	$0.0006\,\sqrt[3]{d}$	$0.0006\,\sqrt[3]{d}$
7		Medium force fit		$0.0005d$	$0.0006\,\sqrt[3]{d}$	$0.0006\,\sqrt[3]{d}$
8		Heavy force (shrink) fit		$0.001d$	$0.0006\,\sqrt[3]{d}$	$0.0006\,\sqrt[3]{d}$

a. d = basic size (diameter)

Given h, s, and a, and assuming unilateral tolerances and using the basic hole system, the limits of a cylindrical fit can be written as (see Figures 20.1 to 20.5);

$$d_{h\,\text{min}} = d \tag{20.1}$$

$$d_{h\,\text{max}} = d + h \tag{20.2}$$

$$d_{s\,\text{max}} = d - a \tag{20.3}$$

$$d_{s\,\text{min}} = d - a - s \tag{20.4}$$

Equations (20.1) to (20.4) apply for clearance, transition, or interference fit. In Eqs. (20.3) and (20.4), the allowance a is an algebraic value; it is positive for clearance fits and negative for interference fits. The tolerances h and s are always positive values. Subtracting Eq. (20.3) from (20.1) gives

$$a = d_{h\,\text{min}} - d_{s\,\text{max}} \tag{20.5}$$

Equation (20.5) agrees with the definition of the allowance a. It also shows that a is positive for clearance fits and negative for interference fits.

While Table 20.1 and Eqs. (20.1) to (20.4) can be used to calculate hole and shaft sizes, it is useful in practice to transform this information into tables that designers can look up to determine limits and toleranced dimensions. ANSI and ISO systems exist and are equivalent.

ANSI rearranges Table 20.1 to incorporate the ISO system, and extends it to include location fits. As a result, there exist five standard classes of fits (with their ISO equivalents) with several grades under each class. A grade of a fit is determined by the permissible variation in its relative looseness or tightness; it depends on quality of machining or workmanship. The five classes are running or sliding fits RC, location clearance fits LC, location transition fits LT, location interference fits LN, and force or shrink fits FN. The classes, grades, and their equivalent ISO symbols are shown in Table 20.2.

Table 20.2   ANSI fits and their ISO equivalents.

Class	ANSI symbol Grade	ISO symbol Hole	ISO symbol Shaft	Class	ANSI symbol Grade	ISO symbol Hole	ISO symbol Shaft
Running or sliding fits RC	RC1	H5	g4	Location transition fits LT	LT1	H7	js6
	RC2	H6	g5		LT2	H8	js7
	RC3	H7	f6		LT3	H7	k6
	RC4	H8	f7		LT4	H8	k7
	RC5	H8	e7		LT5	H7	n6
	RC6	H9	e8		LT6	H7	n7
	RC7	H9	d8	Location interference fits LN	LN1	H6	n5
	RC8	H10	c9		LN2	H7	p6
	RC9	H11	c11		LN3	H7	r6
Location clearance fits LC	LC1	H6	h5	Force or shrink fits FN	FN1	H6	n6
	LC2	H7	h6		FN2	H7	s6
	LC3	H8	h7		FN3	H7	t6
	LC4	H10	h9		FN4	H7	u6
	LC5	H7	g6		FN5	H8	x7
	LC6	H9	f8				
	LC7	H10	e9				
	LC8	H10	d9				
	LC9	H11	c10				
	LC10	H12	c12				
	LC11	H13	c13				

ANSI applies the equations listed in Table 20.1 to all the fits listed in Table 20.2 for a range of basic sizes. The result is a set of tables that designers can use to find a, h, and s for a given fit and a given size, thus simplifying tolerance calculations greatly for designers and practicing engineers. These tables are listed in Appendix C.

The goal of using the ANSI tolerance tables listed in Appendix C is to find a, h, and s. We substitute them into Eqs. (20.1) to (20.4) to find the hole and shaft limits. The tables have three columns for each fit listed in Table 20.2: Limit, Hole, and Shaft. The Limit column has a slightly different heading, depending on the type of fit—Limit clearance or Limit interference.

Here is how we can use these tables. For clearance fit, the allowance a is the smallest number in the Limit column of the tables, and for interference fit it is the largest number in the Limit column. This selection is consistent with the requirement that a maintains the MMC. The nonzero tolerance limit (the other is always zero due to use of the basic hole system) in the

Hole column of the tables is h. The shaft tolerance s can be obtained from the Shaft column of the tables using the following equation:

$$s = |L_{min} + a| \tag{20.6}$$

where L_{min} is the algebraic minimum limit in the Shaft column, and a is the algebraic allowance. In Eq. (20.6), use a as a positive number for clearance and as a negative number for interference. $|\cdot|$ denotes the absolute value.

In summary, the procedure to find hole and shaft limits based on ANSI standards is

1. Decide on a fit type (clearance, transition, or interference) based on the design's functional requirements.
2. Use Table 20.1 to find the ANSI symbol for the selected fit.
3. Use ANSI tolerance tables (Appendix C) to calculate a, h, and s.
4. Substitute a, h, and s into Eqs. (20.1) to (20.4) to calculate the hole and shaft limits.

EXAMPLE 20.1	**Use ANSI tolerance calculations.**

Calculate the tolerance zones h and s, the allowance a, and the limits for two fits of classes RC2 and FN4 if the basic size is 3.0000 in. Use ANSI calculations.

SOLUTION   Table 20.2 shows that RC2 and FN4 are clearance and interference fits, respectively. Using the ANSI tolerance tables in Appendix C, we obtain, for RC2, $a = 0.0004$ in, $h = 0.0007$ in, and $s = |-0.0009 + 0.0004| = 0.0005$ in. Using Eqs. (20.1) to (20.4) gives these limits: $d_{hmax} = 3.0007$, $d_{hmin} = 3.0000$, $d_{smax} = 2.9996$, and $d_{smin} = 2.9991$ in.

For the interference fit FN4, similar calculations can be performed. Using the proper ANSI tables, we obtain $a = -0.0047$ in, h = 0.0012 in, and $s = |0.0040 - 0.0047| = 0.0007$ in. Using Eqs. (20.1) to (20.4) gives these limits: $d_{hmax} = 3.0012$, $d_{hmin} = 3.0000$, $d_{smax} = 3,0047$, and $d_{smin} = 3,0040$ in.

20.4.3 ISO Calculations

ISO symbols can also be utilized to determine tolerances and limits as follows. The capital letter H is used to describe the classes of holes, and small letters are used to describe the classes of shafts. The grade within a class is described by a number called the tolerance grade number or IT number.

For example, H7/f6 (see Table 20.2) represents a clearance fit where the hole has class H and IT number 7, and the shaft has class f and IT number 6. For each shaft class, there is an amount of tolerance called fundamental deviation δF. For each IT number, there is another amount of tolerance; call it Δd. ISO tolerance tables (see Appendix C) lists δ_F and Δd for various classes and IT numbers, for a practical range of basic sizes. The allowance a, the hole tolerance zone h, and the shaft tolerance zone s can be related to δ_F and Δd as follows:

$$h = \Delta d_h \tag{20.7}$$

$$s = \Delta d_s \tag{20.8}$$

$$a = \begin{cases} \delta_F & \text{clearance} \\ \delta_F + \Delta d_s & \text{interference} \end{cases} \tag{20.9}$$

where subscripts h and s denote hole and shaft, respectively.

Once we have h, s, and a, we use Eqs. (20.1) to (20.4) to calculate the limits. In effect, the difference between ANSI and ISO calculations boils down to the method and tables that we must use to calculate h, s, and a.

In summary, the procedure to find hole and shaft limits based on ISO standards is

1. Decide on a fit type (clearance, transition, or interference) based on the design functional requirements.
2. Use Table 20.1 to find the ISO symbol for the selected fit.
3. Use the hole IT number and the ISO tolerance tables (Appendix C) to calculate Δd_h.
4. Use the shaft IT number and the ISO tolerance tables (Appendix C) to calculate Δd_s.
5. Use the shaft class and the ISO tolerance tables (Appendix C) to calculate δ_F.
6. Use Eqs. (20.7) to (20.9) to calculate h, s, and a.
7. Substitute a, h, and s into Eqs. (20.1) to (20.4) to calculate the hole and shaft limits.

EXAMPLE 20.2 | **Use ISO tolerance calculations. Redo Example 20.1.**
Calculate the tolerance zones h and s, the allowance a, and the limits for two fits of classes RC2 and FN4 if the basic size is 3.0000 in. Use ISO calculations.

SOLUTION   Using the proper ISO tables gives, for RC2, $\Delta d_h = 0.0007$ in for IT number 6, $\Delta d_s = 0.0005$ in for IT number 5, and $\delta_F = -0.0004$ in for class g. Using Eqs. (20.7) to (20.9) gives $h = 0.0007$, $s = 0.0005$, and $a = 0.0004$ in, which agree with ANSI calculations.

Using the proper ISO tables again gives, for FN4, $\Delta d_h = 0.0012$ in for IT number 7, $\Delta d_s = 0.0007$ in for IT number 6, and $\delta_F = 0.0040$ in for class u. Using Eqs. (20.7) to (20.9) gives $h = 0.0012$, $s = 0.0007$, and $a = 0.0047$ in, which agree with ANSI calculations.

Example 20.2 discussion:

We have used ISO calculations to redo Example 20.1, to illustrate the similarities between the two methods. The use of one method over the other is a matter of personal preference although the ANSI method seems to be more straightforward—no need to calculate $\Delta d_h, \Delta d_s$, or δ_F.

20.5 Tolerance Accumulation

Placing toleranced dimensions on an engineering drawing is not a random process; it requires careful consideration of the effect of one tolerance on another. Whenever more than one tolerance in a given direction (for example, horizontal or vertical) affect the location of a given surface of a part, tolerances are cumulative. Noncumulative tolerances and dimensioning are always preferred because cumulative ones may result in a higher percentage of rejection (scrap) during inspection. Accumulation (stackup) of tolerances usually occurs statistically.

Figure 20.6 shows three possibilities for dimensioning the same drawing. Figure 20.6a shows each dimension in the horizontal direction with its associated tolerance. This practice of dimensioning should be discouraged. It is confusing to the machinist who will produce the part. The machinist cannot produce the four toleranced dimensions; only any three can be controlled, and the fourth is an outcome. Therefore, the machinist must make a decision on which three dimensions to aim for. If the left-out dimension is crucial to the function of the part, the rejection rate of the part during inspection will most likely be high.

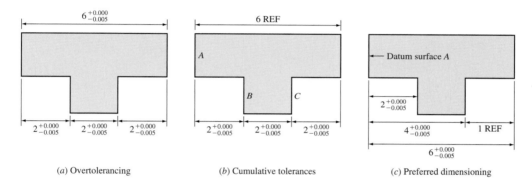

(a) Overtolerancing   (b) Cumulative tolerances   (c) Preferred dimensioning

Figure 20.6   Tolerance accumulation.

Figure 20.6b shows a possible solution to the problem of overtolerancing (or superfluous dimensioning). Here, only the three necessary dimensions are given with their associated tolerances. The overall length is just a reference and, therefore, is shown without tolerance as well as being marked REF. With this solution the designer assumes that the overall length is not important to the functional requirements (assembly) of the part.

While Figure 20.6b may seem an acceptable solution, it has two disadvantages. First, tolerances are cumulative. For example, the middle dimension is controlled by two tolerances, and any of the side dimensions is controlled by three tolerances. Second, three surfaces (A, B, and C) are reference surfaces (datums) from which dimensions are measured. Thus, the machinist must carefully machine these surfaces even if the machining quality of these surfaces is immaterial to the part's function. This may increase the machining cost of the part.

Figure 20.6c shows the preferred solution for dimensioning the part. Here only one reference surface (A) is used as a datum and only three toleranced dimensions are shown. Therefore, as a general rule, it is best to dimension surfaces so that each surface is affected by only one dimension. This can be achieved by measuring all the dimensions from a single datum surface, as shown in Figure 20.6c.

20.6 Tolerance-Cost Relationship

Tolerance is a key factor in determining the cost of a part. The relationship between tolerances and manufacturing cost is shown in Figure 20.7. The manufacturing (total) cost is divided into machining and scrap cost. The machining cost is the cost of first producing the part. This cost consists of labor, overhead, gages, tools, jigs and fixtures, inspection, and so forth. The scrap cost is the cost encountered due to rejecting some parts that fall outside the specified tolerance range, and/or due to repairing some of these parts. The cost of parts which are produced by multiple manufacturing processes (Figure 20.7b) is the sum of costs encountered in each process. An example of multiple processes is a cylinder that may need rough turning, finish turning, and grinding.

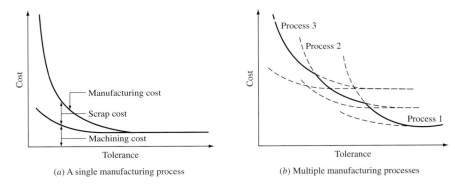

(a) A single manufacturing process   (b) Multiple manufacturing processes

Figure 20.7   Effect of tolerances on manufacturing cost.

As expected, and shown in Figure 20.7, the tighter the tolerance, the more expensive it is to manufacture a part. This trend provides the fundamental rule in selecting tolerances by designers at the design phase; tolerances should be as large as possible as long as they meet the functional and assembly requirements of the part. It may be worthwhile to change designs to relax tolerance requirements for cost purposes. Larger tolerances result in using less skilled machinists (less expensive labor), lower inspection cost, and reduced scrapping of material.

It is important for designers to be aware of manufacturing accuracies attainable by various manufacturing processes. To assist designers in relating tolerances to machining processes, Table 20.3, developed by ANSI, relates tolerance grades (IT number) to machining processes. The range of grades for each process accounts for the conditions (old, new, well maintained, etc.) of the machine and the level of skills the machine operator has. Having selected a tolerance based on the functional requirements of a part, the designer can determine the tolerance grade for a given size using tolerance grade tables (Table 20.2). With the tolerance grade known, the designer can determine the proper machining process by using Table 20.3.

Table 20.3   Relationship between tolerances and machining.

Machining process	Tolerance grade (IT number)									
	4	5	6	7	8	9	10	11	12	13
Lapping and honing	X	X								
Cylindrical grinding		X	X	X						
Surface grinding		X	X	X	X					
Diamond turning		X	X	X						
Diamond boring		X	X	X						
Broaching		X	X	X	X					
Powder metal—sizes			X	X	X					
Reaming			X	X	X	X	X			
Turning				X	X	X	X	X	X	X
Powder metal—sintered			X	X	X	X				
Boring					X	X	X	X	X	X
Milling							X	X	X	X
Planning and shaping							X	X	X	X
Drilling							X	X	X	X
Punching							X	X	X	X
Die casting								X	X	X

20.7 Surface Quality

Surface quality is an important factor that affects the performance of mating parts relative to each other as well as the choice of the manufacturing processes. Tolerances and surface quality are interrelated in the sense that they are both a direct outcome of manufacturing processes. A manufacturing process, such as lapping and honing, that produces small tolerances (small tolerance grades) also produces smooth surfaces. Therefore, in specifying tolerances, a designer should consider the requirements of surface finish in addition to functional and assembly requirements.

For example, an interference fit made on rough surfaces may have a reduced contact area which results in a subsequent reduction of the interference force between the mating parts. Higher surface quality results in higher production costs. Thus, designers normally leave a surface as rough as is feasible.

Surface finish can be evaluated quantitatively by using various measures. The most popular measures are surface roughness and waviness. Figure 20.8a shows that a close look at a surface texture shows that it is not absolutely smooth. Instead, it consists of fine irregularities superposed on larger or wavelike variations as shown in Figure 20.8b. The measure of the irregularities over a sampling length A (Figure 20.8a) is defined as the surface roughness, whereas the measure of the large variations over a wavelength B defines the waviness of the surface.

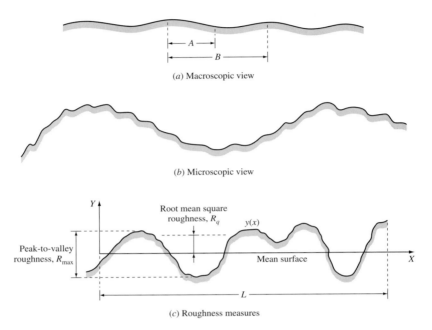

(a) Macroscopic view

(b) Microscopic view

(c) Roughness measures

Figure 20.8   Surface quality.

There are three methods for calculating the surface roughness R of a surface. Let us define an imaginary mean surface (Figure 20.8c) such that the total variations (measured by the sum of the areas between the mean surface and the profile of the actual surface) above the mean surface are equal to the total variations beneath it. The roughness average R_a measures the average of the absolute displacement (variation) relative to the mean surface:

$$R_a = \frac{1}{L} \int_0^L |y|\, dx \tag{20.10}$$

where $|y|$ is the absolute value of the roughness function $y(x)$. The roughness average R_a is also known as the arithmetic average (AA). It is usually measured using a planimeter to calculate the area above and below the mean surface.

R_a values are usually expressed as micrometers (μm) or microinches (μ in). The value of R_a can vary quite considerably without affecting the surface function. Table 20.4 gives the ANSI ranges of R_a produced by common manufacturing processes. The ISO roughness code is also shown in the table.

Table 20.4   ANSI surface roughness.

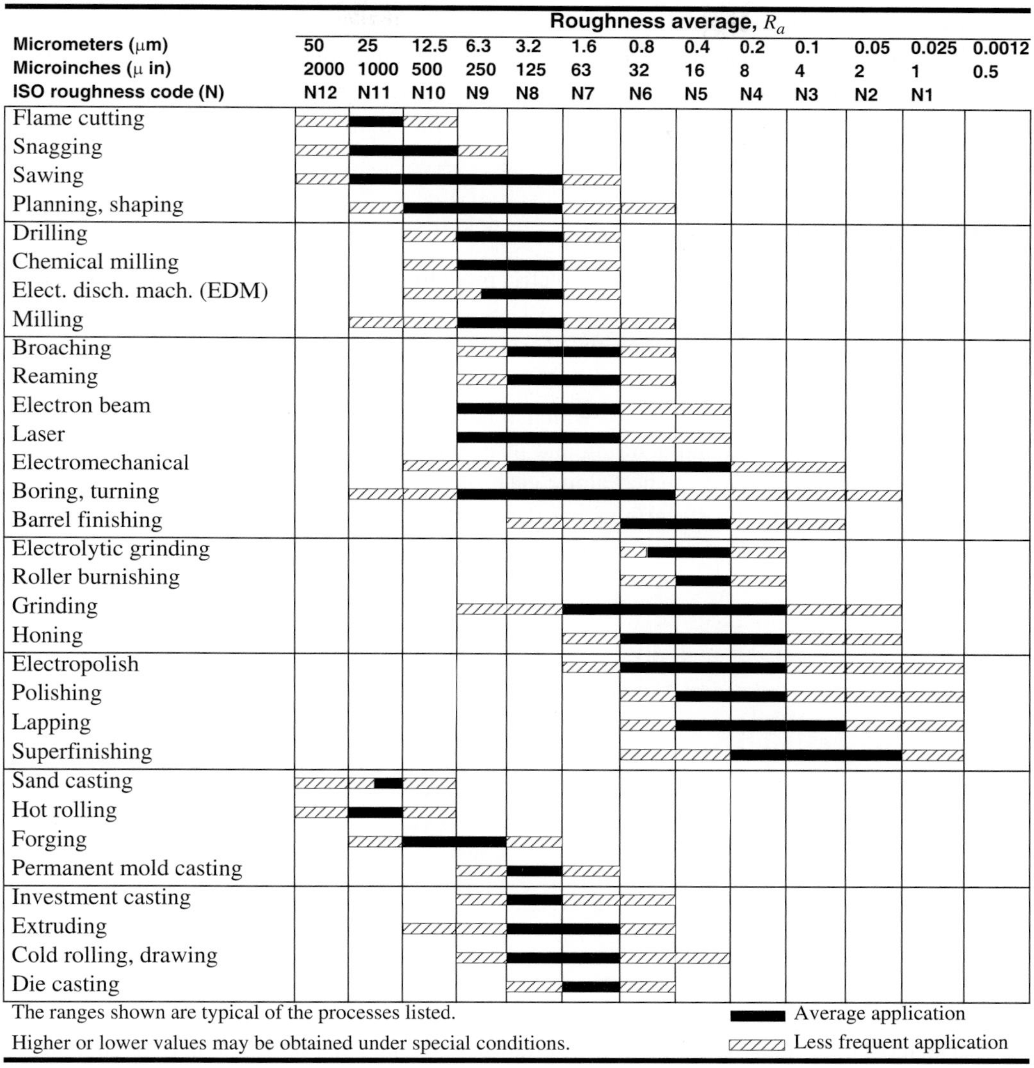

	Roughness average, R_a												
Micrometers (μm)	50	25	12.5	6.3	3.2	1.6	0.8	0.4	0.2	0.1	0.05	0.025	0.0012
Microinches (μ in)	2000	1000	500	250	125	63	32	16	8	4	2	1	0.5
ISO roughness code (N)	N12	N11	N10	N9	N8	N7	N6	N5	N4	N3	N2	N1	

Processes listed (rows):

Flame cutting
Snagging
Sawing
Planning, shaping

Drilling
Chemical milling
Elect. disch. mach. (EDM)
Milling

Broaching
Reaming
Electron beam
Laser
Electromechanical
Boring, turning
Barrel finishing

Electrolytic grinding
Roller burnishing
Grinding
Honing

Electropolish
Polishing
Lapping
Superfinishing

Sand casting
Hot rolling
Forging
Permanent mold casting

Investment casting
Extruding
Cold rolling, drawing
Die casting

The ranges shown are typical of the processes listed.
Higher or lower values may be obtained under special conditions.

■ Average application
▨ Less frequent application

Another measure of surface roughness is given by the rms (root mean square) value R_q (Figure 20.8c) given by:

$$R_q^2 = \frac{1}{L} \int_0^L y^2 \, dx \tag{20.11}$$

The rms method is still an averaging method.

The third measure of roughness is given by the maximum peak-to-valley height R_{max} (Figure 20.8c). Sometimes, R_{max} is evaluated at various locations over the length of the surface, and an average value is calculated.

Various methods exist to determine the wavelength and consequently the waviness of a surface. These methods concentrate on counting the frequency rather than the amplitude of the surface profile. Some of these methods use concepts from Fourier analysis to determine average wavelengths. These methods are not covered here.

20.8 Datums

The concept of datums is used in geometric modeling as we cover in Section 2.8. It is also an important concept in geometric tolerances. The use of datums in geometric tolerancing helps in manufacturing because they define the workpiece surfaces that the machinist must use to measure part dimensions from. The machinist uses these surfaces as references.

A **datum** is a "theoretical plane" which acts as a master reference for locating features (surfaces) of a part during manufacturing and inspection. A datum may be a point, a line, or a plane. A datum plane may be created along each of the three $(X, Y, \text{and } Z)$ axes as shown in Figure 20.9. The three mutually orthogonal datum planes form what is sometimes called the datum reference system (also known as the master datum system).

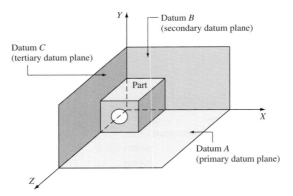

Figure 20.9   Datum reference system.

The most important plane to part measurements is termed datum A (the horizontal plane in Figure 20.9) and is known as the primary datum. The second most important is datum B (known as the secondary datum), and the least important is datum C (tertiary datum). The designer usually decides the priority of importance of the three datum planes based on the functional and assembly requirements of the part.

A distinction must be made between the datum plane and the actual surface of the part that is in contact with the plane. Such a surface is known as the datum feature. For example, the bottom surface of the part shown in Figure 20.9 which is in contact with the primary datum plane is called the primary datum feature. Similarly, the rear and left surfaces of the part are called the secondary datum feature and the tertiary datum feature, respectively.

While the datum features are actual physical surfaces of a part, the datum planes represent surfaces of machine tables, fixtures, inspection devices, and gages. From this point of view, we can explain the physical interpretation and implication of datum priority as follows: In reference to Figure 20.9, the part is pushed onto Datum A until contact is established. Then, the part is pushed onto Datum B while in contact with Datum A, that is, slide it on A until contact with B is achieved. Finally, contact with Datum C is established in a similar way while maintaining the contact with A and B simultaneously. This order of establishing contact with the datums defines datum priority during inspection.

Figure 20.9 shows datum planes and datum features as perfect planes. In practice, these planes are not perfect and have surface irregularities because they are manufactured. If we assume that datum planes are manufactured with much higher precision than datum features, one can assume a perfect datum plane and an irregular datum feature as shown in Figure 20.10.

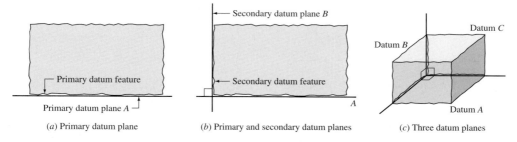

(a) Primary datum plane   (b) Primary and secondary datum planes   (c) Three datum planes

Figure 20.10   **Contact between datum surfaces and datum planes.**

The primary datum plane A, shown in Figure 20.10, contacts the actual primary datum feature of the part at three or more noncollinear points. The secondary datum plane B contacts the secondary datum feature of the part at two or more distinct points not on the same normal to A, and the tertiary datum plane contacts the tertiary datum feature of the part at one or more points. It is clear now that the order or precedence of the datums is important, since the actual datum feature may sit differently against the datum plane depending on the number of points which are required to make contact.

In defining the datum reference system shown in Figure 20.10, the contact points between the datum features and the datum planes are not precisely located. It is assumed that the surface finish of the parts is so good that the precise contact points are unimportant. For less regular surfaces such as those of castings and forgings, spot and line datums are required for precise datum definition.

Functionally, a primary datum feature might be replaced by three-point contact, or by one point and one line contact as shown in Figure 20.11. These replacements are known as datum targets. Figure 20.11b shows how the primary datum A is specified using datum targets. The letter *A* in the circular symbols refers to the datum name. The integer on the right (2 in the figure) specifies the number of targets used to define the datum while the integer on the left (1 or 2 in the figure) reflects the sequence of the corresponding target.

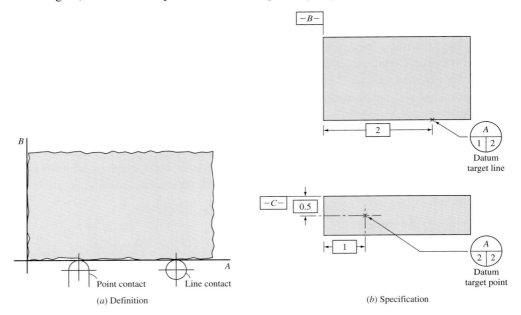

Figure 20.11   Datum targets.

Figure 20.12 shows the interpretation of datum targets in terms of gages and placement procedure for inspection. Datum A is the primary datum and is defined by two target points (two points are sufficient for 2D while three points are needed for 3D). Datum *B* is the secondary datum, thus requiring one target point. Figures 20.12*b* and *c* show how the part is placed and inspected. The part is pushed onto the gage in such a way that surface F_1 is in contact first with the target points of *A* and then surface F_2 is brought into contact with the target point of *B*. If F_1 and F_2 touch the gage in the three target points simultaneously, it passes inspection; otherwise it is rejected.

Other datum reference systems may be established using cylindrical datums. Datum reference systems may also be formed by combinations of cylindrical and planar datums. Precedence is again important. Figure 20.13 shows examples.

The specification of datum features includes definition of the criteria for applying material conditions (MMC, LMC, or RFS—regardless of feature size) to the datum. These conditions specify how datums are established while inspecting the part. Material conditions are used with surfaces that have sizes and may be used as datums such, as holes. If the surface of a hole is specified as a datum at the MMC, then the surface of the minimum hole should be used during inspection. If, on the other hand, the RFS is used, then the effect of the size of the hole is eliminated while establishing the corresponding datum during inspection.

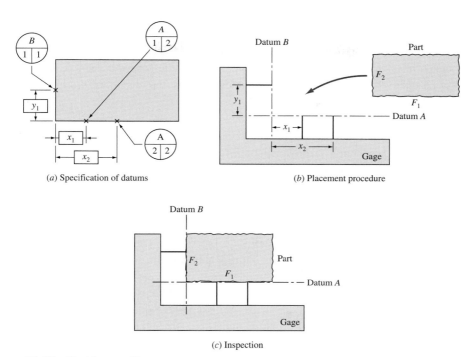

(a) Specification of datums

(b) Placement procedure

(c) Inspection

Figure 20.12   Part inspection.

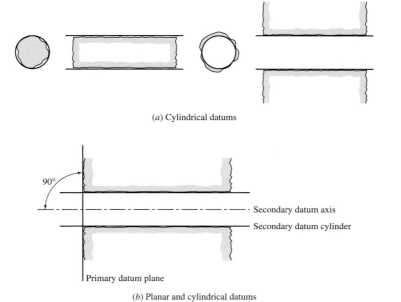

(a) Cylindrical datums

90°

Secondary datum axis

Secondary datum cylinder

Primary datum plane

(b) Planar and cylindrical datums

Figure 20.13   Cylindrical datums.

Datum features should be actual physical surfaces of parts from which measurements can be made. Abstract concepts such as centerlines, center planes, or axes of symmetry should not be used as datum features. While these concepts are useful for line drawings, they do not have any associated physical resemblance such as actual surfaces, edges, or corners. Datum features should not be taken through a row of holes, either, because the centers of the holes can shift anywhere within the tolerance zones.

Designers should develop the habit of establishing datums. The proper choice of datums can make a design easier to read and interpret and easier to manufacture and inspect. Proper datums can reduce production errors, which in turn can reduce the number of scrap parts. Therefore, proper datums can be very cost-effective. The number of datums needed to control positions on a part depends on the function of the part and how the designer wishes it to be made. A single datum is usually needed. Sometimes, two or three datums may be required.

20.9 Geometric Tolerances

Conventional tolerances are known as traditional or coordinate tolerancing methods. They typically control the variability of linear dimensions that describe location, size, and angle as shown in Figure 20.14. Conventional tolerancing is also known as tolerancing of perfect form because it does not address variability of shape. For example, the tolerances ΔC and ΔE shown in Figure 20.14 do not convey any information about the perpendicularity of the horizontal and vertical planes shown in the figure.

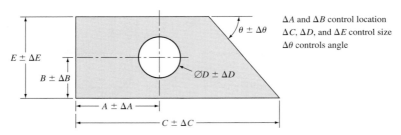

Figure 20.14   Control of linear dimensions.

Conventional tolerancing methods have three main shortcomings. First, they are incapable of controlling all aspects of the shape of a part. In addition to control of location, size, and angle, control is also needed for the form (shape) of part features, such as straightness, flatness, parallelism, or angularity of specific portions of the part.

Second, conventional tolerancing does not use the datum concept, which is important for manufacturing and inspecting the part. It does not explicitly specify datums or their precedence. Datums are usually implied from the way the part drawings are dimensioned. For example, Figure 20.14 implies that the bottom horizontal and the left vertical planes are used as datums. However, which plane is more important than the other (precedence or datums) cannot be determined. In some cases, implied datums are too ambiguous to identify easily.

Third, extending the conventional tolerancing methods to control locations (i.e., 2D or 3D control) described in rectangular coordinates or angular dimensions introduces awkward situations. Consider, for example, the control of the location of the center of the hole shown in Figure 20.15. Specifying a tolerance of ±0.005 in on the coordinates of the center (2.000, 2.000) results in the 0.010×0.010 in square tolerance zone shown in Figure 20.15b. While the designer might think that he/she is controlling the location of the center of the hole within a 0.010 in boundary, the center could actually vary across the diagonal of the square tolerance zone, yielding a maximum tolerance of 0.014 in instead of 0.010 in.

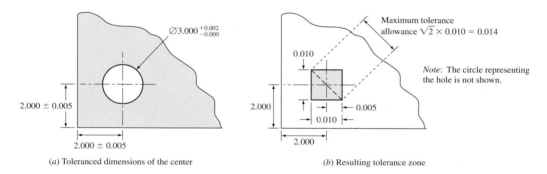

(a) Toleranced dimensions of the center   (b) Resulting tolerance zone

Figure 20.15   Two-dimensional tolerance zones.

Figure 20.16 shows similar results for angular tolerance zones. Thus, conventional tolerancing gives more freedom in the diagonal directions and may unnecessarily constrain errors in the horizontal (or radial) and vertical (angular) directions.

Determining tolerance zones of locations, for example, of holes, which are interrelated is more complex due to the combined effects of individual tolerance zones. This discrepancy indicates that a better tolerancing system to control locations is needed. In the new system, it seems natural to force the tolerance zone of the hole center to become circular instead of square, rectangular, or polar. This new tolerancing system is known as geometric tolerancing. It provides a more functional and better-controlled location, size, angle, and form of parts.

20.9.1   Types of Geometric Tolerances

Many books and references have been written on the subject of geometric dimensioning and tolerancing (GD&T). Samples include *Dimensioning and Tolerancing Handbook* by Bruce A. Wilson, Genium Publishing Corporation (ISBN 0-931690-80-3); Chapter 13 in *Principles of Technical Drawings* by F.E. Giesecke et al., MacMillian, 1992; and the famous *ANSI Y14.5M-1994* standards on GD&T.

Four types of geometric tolerances exist: size, location, form, and orientation. Size tolerance (as in conventional tolerancing) controls the size (lengths and/or diameters) of part features as shown in Figure 20.14. Location tolerance (as in conventional tolerancing) controls position and concentricity of various features as shown in Figure 20.14. Form tolerance controls the shape of individual features such as the flatness or straightness of a single surface. Orientation tolerance controls the shape of related features such as the parallelism or the perpendicularity of two surfaces. It consists of angulairty, perpendicularity, and parallelism tolerances.

Geometric tolerancing permits explicit definition of datums, with clear specification of the datum precedence in relation to each tolerance specification.

The two distinct concepts that geometric tolerancing introduces are the use of datums and form and orientation tolerances; conventional tolerancing lacks both. Both concepts provide good control of the the shape of the various features of a part. Such control credits geometric tolerancing for reducing the scrap percentage during part inspection.

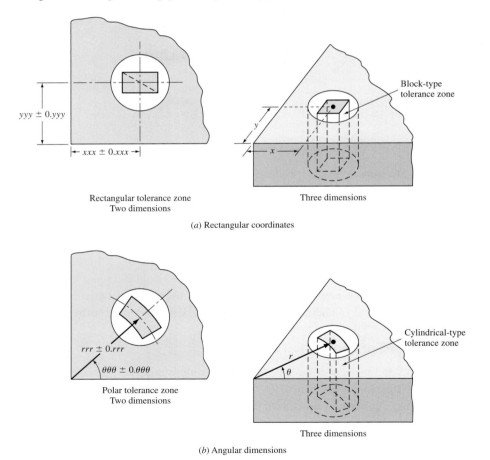

Figure 20.16   Three-dimensional tolerance zones.

20.9.2 ANSI and ISO Symbols

In order to communicate geometric tolerances with their related information effectively and concisely on engineering drawings, ANSI and ISO have developed a set of standards and symbology for GD&T. The ANSI Y14.5M-1982 standards incorporate most of the symbology and features of the international ISO 1101 standards that are significant to GD&T.

The 1982 version of ANSI Y14.5M includes the ISO symbology related to datum targets as well as methods for specifying limits and fits. ISO in turn has accepted the projected tolerance zone, the three-plane datum concept, the multiple datum principle, and the method of designating datum targets from the ANSI Y14.5M. In spite of this mutual exchange, differences (especially in symbology) between the two standards still exist.

Geometric tolerances are usually expressed on engineering drawings by displaying symbols called feature control symbols. These symbols may contain basic sizes, tolerances, datums, ANSI symbols (see Table 20.5), and ANSI modifying symbols (Table 20.6). Table 20.5 shows the ANSI symbols that we should use to specify a geometric tolerance on a part feature. These symbols apply only to form tolerances. The other two types of geometric tolerances (size and location) use the standards from conventional tolerancing. Table 20.6 shows the ANSI modifying symbols that, together with symbols from Table 20.5, provide complete specification of any type of geometric tolerance: size, location, form, or orientation.

Table 20.5   ANSI symbols for geometric tolerancing.

Feature	Tolerance type	Description	ANSI symbol
Individual features	Form	Straightness	—
		Flatness	▱
		Circularity (roundness)	○
		Cylindricity	⌭
Individual or related feature	Profile	Profile of a line	⌒
		Profile of a surface	⌓
Related features	Orientation	Angularity	∠
		Perpendicularity	⊥
		Parallelism	//
	Location	Position	⊕
		Concentricity	◎
	Runout	Circular runout	↗
		Total runout	↗↗

The symbols shown in Tables 20.5 and 20.6 serve as building blocks to create geometric tolerances and add them to engineering drawings. We assemble elements from both tables to create what we call "feature control symbols." A feature control symbol completely specifies a geometric tolerance on a part feature. Samples are shown in Figure 20.17.

Table 20.6   ANSI modifying symbols.

Modifying term	ANSI symbol	Modifying term	ANSI symbol
At maximum material condition	Ⓜ	Spherical diameter	SØ
Regardless of feature size	Ⓢ	Radius	R
At least material condition	Ⓛ	Spherical radius	SR
Projected tolerance zone	Ⓟ	Reference	( )
Diameter	Ø	Arc length	⌒

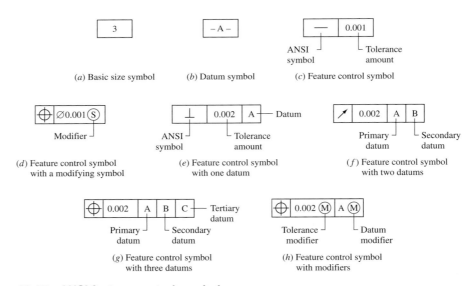

Figure 20.17   ANSI feature control symbols.

20.9.3 Size Tolerances

A size tolerance (known as feature size tolerance) is the same as a conventional tolerance. A feature of size, such as a hole or a slot, may be toleranced similar to conventional tolerancing by direct attachment of the size tolerance (T_s) to the basic size, or by specifying the upper and lower limits of size as shown in Figure 20.18.

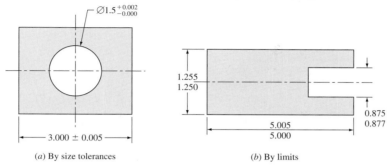

(a) By size tolerances   (b) By limits

Figure 20.18   SIze tolerances.

20.9.4 Location Tolerances

This type of geometric tolerancing is designed to eliminate the undesired tolerance zones (see Figures 20.15 and 20.16) that result from using conventional tolerancing methods to control the locations of features. There are two types of location tolerances (see Table 20.5): position and concentricity. Positional tolerancing is the most widely used and is known as the true position tolerancing.

In positional tolerancing, tolerances are specified on the actual position and not on its coordinates; thus the name true position tolerances. Figure 20.19 converts Figure 20.15 from conventional tolerances to true position tolerances. The resulting circular tolerance zone shows that the tolerance on the center of the hole is indeed 0.010 in, and the axis of the hole must lie within the cylindrical tolerance zone, which has a circle of diameter equal to the tolerance value (0.010 in) and a height equal to the hole depth.

In true position tolerancing, the true position of a feature is the basic location of a point, line, or plane of a feature with respect to a datum or other feature. It is specified using basic dimensions and enclosed in boxes as shown in Figure 20.19.

For cylindrical features, the true position tolerance is the diameter of the cylindrical tolerance zone within which the feature axis must lie as shown in Figure 20.19. Figure 20.19 also illustrates how symbols are used in geometric tolerancing. The boxed 2.000 indicates the basic size, and the feature control symbol expresses a positional tolerance of 0.010 inch on the hole center under MMC by using the modifier Ⓜ in the symbol.

The Ⓜ modifier means that the 0.010 in tolerance circle applies only when the hole is at MMC, that is, when it has its minimum diameter. This means that the center of the MMC (minimum) hole of diameter 3.000 inches is allowed to depart from the true position shown in Figure 20.19a by half of the amount 0.010 in in any angular direction about the true position, with respect to datums A and B.

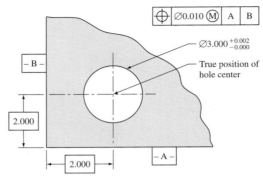

(a) Size and position tolerances

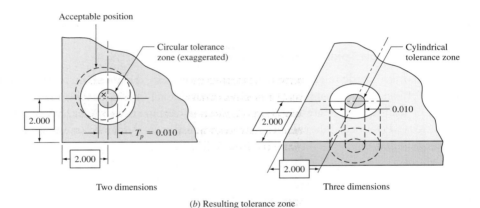

(b) Resulting tolerance zone

Figure 20.19   Location Tolerances.

During part inspection, the (M) modifier means that the part must accept a gage pin (i.e., a shaft) of diameter d_s and make contact with both datums A and B as shown in Figure 20.20. From Figure 20.20, the part must satisfy the following equation in order to pass inspection:

$$d_s = d_{h\min} - T_p \tag{20.12}$$

where T_p is the true position tolerance (0.010 in Figure 20.19b).

Figure 20.19a shows that two values of tolerances are needed to fully describe the hole. A size tolerance on the hole diameter is given as 0.002 in, and a true position tolerance of 0.010 in on the position of its center is specified. The true position and feature size tolerances (T_p and T_s) may be interdependent or independent, depending on the material condition attached to the former. The MMC or LMC makes T_p and T_s interdependent, while the RFS condition makes them independent of each other.

Consider the example shown in Figure 20.19a, which uses the $\textcircled{M}$ modifier in the feature control symbol that specifies T_p. If the hole is at MMC (smallest size), T_p is not affected, but if the hole is larger, the available T_p is larger.

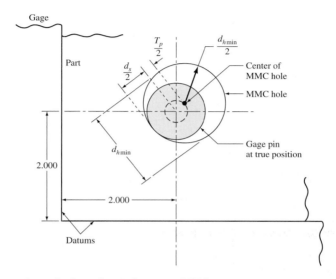

Figure 20.20   Inspection of a location tolerance: MMC case.

As shown in Figure 20.21a, if the hole is exactly 3.000 in in diameter (MMC, or smallest size), its center location may vary from 1.995 in to 2.005 in; that is, within the T_p zone of 0.010. The MMC hole can receive a gage pin of 2.990 in diameter [Eq. (20.12)] located at the true position (2.000, 2.000). At the extreme position of the hole to the right, the outer side of the hole contacts the outer side of the pin. At the extreme position of the hole to the left, the inner side of the hole contacts the inner side of the pin.

If the hole is 3.002 inch diameter, that is, maximum size, it will be acceptable by the gage pin in both extreme positions (right or left) if the proper sides (inner or outer) of both the hole and the pin are in contact as shown by the solid circles in Figure 20.21b. As seen from Figure 20.21b, the hole center location may vary from 1.994 in to 2.006 in, which are outside the specified positional tolerance permitted by T_p alone.

In effect, when the hole is not at MMC, the positional tolerance is greater than T_p and reaches $T_p + T_s$ at the maximum hole diameter. Thus, when the hole is not at MMC, a greater positional tolerance becomes available. Therefore, it has become common practice for both manufacturing and inspection to assume that positional tolerance applies to MMC and that greater positional tolerance becomes permissible when the part is not at MMC.

If, for any reason, the hole center must be arbitrarily held within the tolerance zone specified by T_p, the modifying symbol $\textcircled{S}$ (see Table 20.6) should replace the symbol $\textcircled{M}$ in the feature control symbol shown in Figure 20.19a. For this RFS case, the hole size and its center location are shown by the dashed circle in Figure 20.21b. The oversize tolerance zone of $T_p + T_s$ diameter is no longer permissible since the modifier (S) restricts its diameter to T_p. In this case, the ordinary fixed gage (shown in Figure 20.20) can no longer be used to inspect the part for the location of the hole. Other methods (perhaps more expensive) must be used to inspect the hole if its center is to be restricted to the tolerance zone of the diameter T_p.

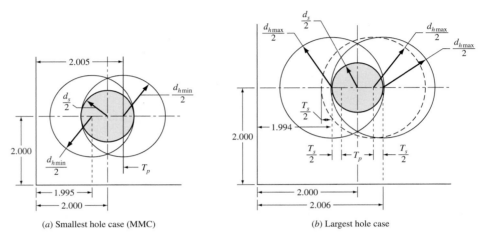

(a) Smallest hole case (MMC)   (b) Largest hole case

Figure 20.21   Inspection of a location tolerance: RFS case.

In addition to solving the problem of the square/rectangular tolerance zones that result from using conventional tolerancing methods, positional tolerancing also eliminates the accumulation of tolerances even in a chain of dimensions as shown in Figure 20.22. This is because positional tolerances are usually measured relative to the true positions, which are specified by basic sizes only. For example, the true position of the middle hole shown in Figure 20.22 is measured from datums A and B by 2.000 and 1.000 in, respectively. This true position is independent of the tolerances of the left hole.

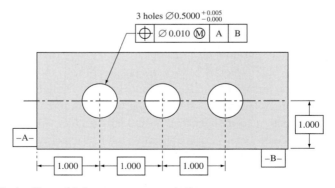

Figure 20.22   Elimination of tolerance accumulation.

After positional tolerance, the second type of location tolerance is concentricity tolerance. It is defined as a relationship between the axis of the toleranced feature(s) and the axis of a datum feature(s). It specifies the collinearity of axes of adjacent features as shown in Figure 20.23. Concentricity tolerance is used less often than positional tolerance because it is difficult to locate actual centers. It is usually preferred to replace this type of tolerance with the runout tolerance (Table 20.5). The axis of the concentricity cylindrical tolerance zone is taken to be the axis of the datum feature(s) as shown in Figure 20.23.

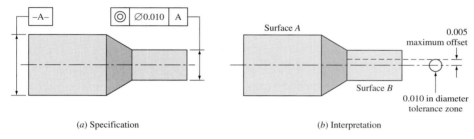

(a) Specification   (b) Interpretation

Figure 20.23   Concentricity tolerance.

EXAMPLE 20.3 | **Apply positional tolerances to a part.**

A designer has designed the plate shown in the following figure with its nominal dimensions. A tolerance of 0.005 in is acceptable for drilling the holes. For successful assembling of the plate, a tolerance of 0.002 in is allowed on the position of the center of any hole, and another tolerance of 0.001 in is allowed on the axis of any hole. Re-sketch the figure shown below to show these tolerances. Sketch the resulting tolerance zones.

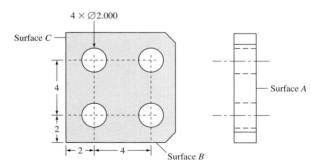

SOLUTION | The tolerance 0.005 in resulting from drilling is a size tolerance on any hole size; that is, $T_s = 0.005$ in. There is a positional tolerance $T_p = 0.001$ in on the location of any hole center. This tolerance requires datums. From the nominal dimensions shown in the figure, surfaces A (bottom of the plate), B, and C are proper datums. These surfaces would have to be machined and prepared properly before the holes are drilled.

The third tolerance on the axis of any hole can be thought of as a form tolerance; that is, $T_F = 0.001$ in. If we measure this tolerance with respect to surface A of the plate, T_F would become a perpendicularity tolerance between the axis of any hole and the bottom surface of the plate. Figure 20.24a shows the specifications of the three types of tolerances (T_s, T_p, and T_F), assuming MMC and unilateral tolerance for T_s. Figure 20.24b shows the two tolerance zones in 2D and 3D that result from T_p and T_F. Figure 20.24b shows one of the many possible tolerance zone patterns.

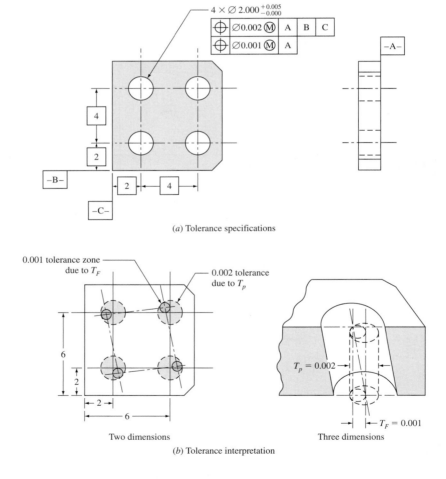

(a) Tolerance specifications

(b) Tolerance interpretation

Figure 20.24   Positional tolerancing of a plate.

EXAMPLE 20.4 | **Assembly of toleranced parts.**

Two parts such as the one shown below are to be assembled by a bolt whose diameter ranges from 0.250 to 0.260 in. The designer, using positional tolerancing, specifies tolerances on the hole where the bolt is to be inserted, as shown in the figure. Assuming that the parts should be in contact with the datums during assembly, is it possible to assemble the two parts with the bolt? Why? How can you solve the problem?

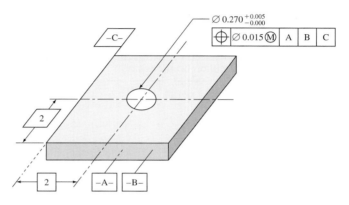

SOLUTION | Using Eq. (20.12) with MMC, we write

$$d_{smax} = d_{hmin} - T_p$$

substituting $d_{smax} = 0.260$ in and $d_{hmin} = 0.270$ in into this equation gives $T_p = 0.010$ in. The specified tolerance T_p is 0.015 in, which means that it is not possible to assemble the two parts with the bolt with the given tolerances. For successful assembly, Eq. (20.12) implies that $T_p < d_{hmin} - d_{smax}$. In this example, T_p should be less than 0.010 in.

20.9.5 Form Tolerances

Form tolerances (T_F), together with location tolerances, provide the designer with complete tolerancing theory capable of controlling most aspects of part geometry and, therefore, its manufacturing. Form tolerances control the variabilities in the shapes of part features. Table 20.5 shows the various form tolerances. Some of them apply to individual features. They are straightness, flatness, roundness, cylindricity, and profile.

Other form tolerances apply to related features. They control the variability of one feature relative to another. They are angularity, perpendicularity, parallelism, runout, and profile. The specifications on a drawing and the meaning of each form tolerance are summarized in Table 20.7.

20.9.6 Orientation Tolerances

According to ANSI definition shown in Table 20.5, an orientation tolerance specifies the orientation of a feature relative to another feature in the same part. An orientation tolerance is made up of three tolerances: angularity ($\angle$), perpendicularity ($\perp$), and parallelism ($//$). Each of the three orientation tolerances controls a shape aspect of a feature, and specifies its own tolerance zone.

Table 20.7 shows how to specify and interpret orientation tolerances. An angularity tolerance specifies the amount of deviation from a nominal angle. Perpendicularity and parallelism tolerances specify the amount of deviation from 90° and 0°, respectively.

Table 20.7   ANSI form tolerances.

Form tolerance	Specification on a drawing	Interpretation
Straightness		
Flatness		
Roundness (circularity)		
Cylindricity		

Table 20.7   ANSI form tolerances. (Continued)

Form tolerance	Specification on a drawing	Interpretation
Profile		Control profile of a line
Angularity		
Perpendicularity		Control perpendicularity of two planes
		Control perpendicularity of an axis and a plane

Table 20.7 ANSI form tolerances. (Continued)

Form tolerance	Specification on a drawing	Interpretation
Parallelism		Control parallelism of two planes
		Control parallelism of two axes
		Control parallelism of an axis and a plane
Runout		Control circular runout
		Control total runout

The table is self-explanatory for all forms. The runout tolerance deserves some comments. It is a measure of deviation from perfect form. It is defined by a tolerance zone contained between two surfaces of revolution with respect to a specified datum axis, or between two planes normal to the axis. It is determined by rotating the form 360° about its axis. It can be found by reading the net change on a dial as shown in the table.

For circular runout, the dial is fixed in a certain location; that is, it measures a circular element only. For total runout, the dial moves along the entire surface while the form is rotating. The runout is measured in FIM (Full Indicator Movement) units of the dial. Runout tolerance is a composite tolerance which incorporates variations in straightness, roundness, and parallelism.

The tolerance zone of the runout tolerance is used to express the zone within which the feature surface must remain when the part is rotated about the given axis. The axis is established by one long cylinder, two widely separated coaxial cylindrical surfaces, or one cylindrical surface and a plane face at a right angle to it as shown in Figure 20.25.

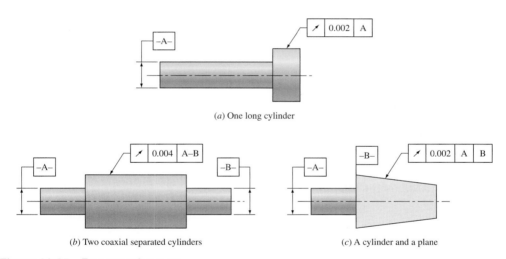

(a) One long cylinder

(b) Two coaxial separated cylinders

(c) A cylinder and a plane

Figure 20.25 Runout tolerance.

20.9.7 ANSI Metric Tolerance Example

Figure 20.26 shows an example to illustrate the use of geometric tolerancing. The interpretation of the various tolerances shown is left as an exercise to the reader (refer to the problems at end of the chapter).

20.9.8 Use of Form Tolerances

Several form tolerances may be applied to a feature. The resulting tolerance zones may be interrelated, and careful interpretation of the interrelationship is required. For example, one tolerance form may imply the other. In addition, the location of form tolerance zones is usually not specified in the form tolerance itself, but in the related positional or size tolerance. In general, all form tolerance zones must lie within the size or positional tolerance zones which determine their location. In Figure 20.27, for example, the parallelism tolerance zone lies somewhere within the size tolerance zone. Similarly, the flatness tolerance zone must lie within the size tolerance zone, but neither its location nor orientation is precisely fixed.

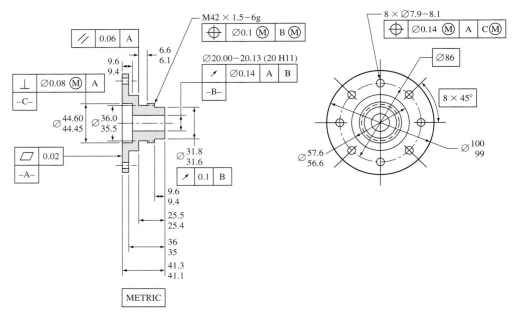

Figure 20.26   ANSI metric tolerance example.

20.10 Tolerance Practices in Drafting

Both conventional and geometric tolerancing methods have been covered thus far. While conventional tolerances are easy to understand and use, they are ambiguous in most cases and do not allow control of form (shape). On the other hand, geometric tolerances are more specific and control both location and form, but assigning and interpreting them usually require more care. One would expect the wide use of geometric tolerances in industry and no or little use of conventional tolerances. Practices in GD&T show a reasonable mixture of conventional and geometric tolerances. In spite of the distinct advantages of geometric tolerances, conventional tolerancing has been retained in many companies where tolerance requirements are less stringent.

For many parts in which assembly conditions are not critical, or where known process capability ensures satisfactory results, it is an accepted practice to indicate linear and angular dimensions from feature to feature without preferring either as a datum from which the other is measured. This results in greater flexibility for production and inspection (quality) departments, which may reduce costs. In practice, geometric tolerancing is applied only when real advantages result, or when specific needs and functional requirements are demanded.

It should be emphasized here that geometric tolerancing is quite crucial to achieve the automation of design and manufacturing demanded from the CAD/CAM technology. Unlike conventional tolerances, geometric tolerances are unambiguous and can be modeled geometrically and used for part manufacturing and inspection. Thus, geometric tolerancing provides full automation of GD&T, while conventional tolerancing provides computer-assisted GD&T because human interpretation is always required.

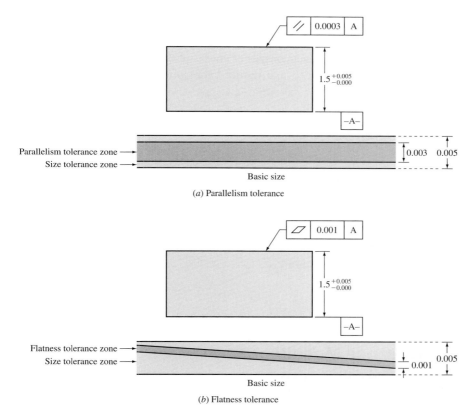

Figure 20.27   Use of several form tolerances.

20.11 Tolerance Practices in Manufacturing

When the assembly drawing and the detailed drawings are complete, the dimensional information is used by the checker (a person who checks the validity and compatibility of dimensions) to ensure correct assembly of the parts and compliance with functional requirements. Such checking is initially applied to the ideal (nominal) part. Then, the checker may perform tolerance analysis, in which the statistical properties of the manufacturing processes are taken into account.

In addition to their use in tolerance analysis, drawings are widely used to plan manufacturing operations (process planning), order and process material, design toolings and jigs and fixtures (clamps), machine parts, inspect product quality, assemble and test finished products, and guide post-manufacturing tasks such as shipping, installation, operation, and maintenance. We cover process planning in more detail in Chapter 21.

Product inspection consists of in-process inspection and final inspection. The former is usually carried out using simple gaging techniques involving go/no-go, gages as shown in Figure 20.28. The use of these gages is usually more effective if the MMC is utilized in GD&T. The final inspection of components usually requires checking every dimension and note on the drawing. Final inspection of a product consists of checks of overall dimensioning, inspection of functional dimensions and tolerances, and testing for required functions. In addition to other methods of inspection, a coordinate measuring machine (CMM) may be used. A control file is produced from a part program similar to that used in NC, and the results of the measurements may be used in analysis programs.

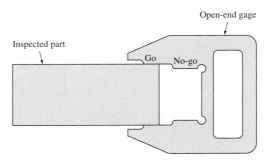

Figure 20.28   Go/no-go inspection gage.

20.12   Tolerance Modeling

Tolerances define a class of parts that are topologically identical to their corresponding nominal part and that are interchangeable in assembly operations and functionally equivalent to each other. Such classes are referred to as variational classes. Mathematically, a variational class is modeled by a set of solid models which contain the solid model of the nominal part. Thus, variational classes may be modeled as a collection of r-sets (see Chapter 9). The precise nature of such a collection is largely unknown.

Geometrically, tolerance specifications are representations of variational classes. They can be thought of as entities of computational nature; that is, they yield valid solid models if computed. Consider the part shown in Figure 20.29. Only conventional tolerances are shown for simplicity. Three entities are needed to store the tolerances $\pm \Delta A$, $\pm \Delta B$, and $\pm \Delta D$.

When these entities are evaluated relative to the nominal part, two extreme or limiting models can be defined. One extreme model corresponds to the MMC and is called the MMC model, while the other is the LMC model and corresponds to the LMC. These are shown by the solid and dashed boundaries, respectively, in Figure 20.29b. In this example, the variational class is all valid solid models (i.e., r-sets) contained between, and including, the MMC and LMC parts.

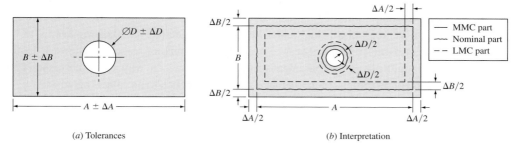

(a) Tolerances   (b) Interpretation

Figure 20.29   Variational classes of a part.

Readers should be cautious when extending the interpretation shown in Figure 20.29 to more complex parts or to include geometric tolerances. The difficulty usually arises when tolerance zones overlap or when datums are implicitly assumed as in conventional tolerances. In Figure 20.29, we divide the tolerances ΔA and ΔB equally around the nominal part. If, however, we assume that the bottom and left faces of the part are implicit datums, all ΔA and ΔB would be applied to the right and the top faces respectively. This would result in a different variational class.

Readers are encouraged to change the part shown in Figure 20.29 into an L bracket shape, assign dimensions and tolerances to it (without overdimensioning), and attempt to interpret the tolerance geometrically.

There are two approaches to generating variational classes from tolerance specifications: parametrization and nonparametrization. Parametrization is similar to instancing or generating a family of parts. As shown in Figure 20.29a, one would parametrize the nominal part into parameters A, B, and D. These parameters represent a family of parts. The limiting parts correspond to the tolerance limits as shown in Figure 20.29b. Assigning specific values to these parameters results in a specific part of the family. In B-rep, parametrization is achieved by assigning parameters to the vertices of the part. In CSG, it is achieved by assigning variables to the parameters of the primitives that make the part.

Variational geometry is a parametric approach to generating variational classes. In conventional solid modelers, dimensions are subordinate to geometry. The designer creates the exact geometry of a nominal part, and dimensions are derived from the geometry. Variational geometry enables the reversal of this relationship. The designer creates the part topology and a set of dimensions (parameters) from which exact geometry is derived.

Therefore, dimensions can be changed directly, and the new model is created automatically. Thus conventional tolerances can be represented as allowable ranges (limits) of the explicitly represented dimensions. A disadvantage of the variational geometry approach is that there is no obvious way to deal with geometric tolerances. The details of variational geometry are not covered here. Interested readers should refer to the literature.

Offsetting is a nonparametric approach to generate variational classes. In this approach, the boundary of the nominal part is offset by the amount of the specified tolerances to generate the limiting parts. One can think of Figure 20.29*b* as offsetting the nominal part once outward and once inward to generate the MMC and LMC parts, respectively. Offsetting seems to be more appealing than parametrization because it catches the spirit of tolerance zones. It is not always easy or straightforward to parametrize the part in such a way as to correspond to specified tolerances. In offsetting, each tolerance zone corresponds to an offsetting zone.

The offsetting theory in geometric modeling is not covered here. But it should be mentioned that offsetting supports bilateral size tolerances of equal value (e.g., ±0.005). Both of these values must not be null. Unilateral tolerance specification or unequal bilateral limits must be redefined to have equal values. For example, a size tolerance of $'^{+0.0015}_{-0.0005}$ must be redefined to become ±0.0010 for the offsetting theory to apply. However, this may not be acceptable in practice.

How can a variational class be displayed on a graphics display? This question may seem trivial at first. To appreciate the question, let us assume that displaying the three parts as shown in Figure 20.29*b* is an acceptable display. It provides users with the nominal part display together with the limiting acceptable parts. Even with this simplification, the dimensions of the two limiting parts relative to the nominal part form an ill-conditioned display problem. Tolerances are usually in the order of hundredths or thousandths of the nominal dimensions. Therefore, the display of the three parts appears coincident on the screen. Magnifying local zones of the display provides a partial solution.

20.13 Tolerance Representation

Tolerancing by constraining configuration and position parameters of primitive half-spaces is ideal for conventional tolerancing. Conventional GD&T can be interpreted as a variancing scheme based on the specification of limit (±) constraints on independent configuration and position parameters of primitive half-spaces. While configuration parameters are easy to deal with (just assign them), position parameters must be carefully assigned because such assignment determines the dimensioning scheme of the part.

For example, consider the part shown in Figure 20.30*a*. The part is a 2D solid (rectangle with a hole) consisting of a block (rectangle) minus a cylinder (circle). The block is defined by the intersection of four primitive planar half-spaces P_1 to P_4, and the cylinder is defined by a cylindrical half-space C_1. Planar half-spaces P_1 to P_4 do not have configuration parameters, while half-space C_1 has the diameter D as its configuration parameter. If the dimension scheme is to be as shown, then half-spaces P_1 and P_3 have zero positional (translation is zero) parameters, P_2 has X translation of A, P_4 has Y translation of B, and C_1 has X translation and Y translation of C and E, respectively. Thus, this dimensioning scheme requires the positional parameters A, B, C, and E.

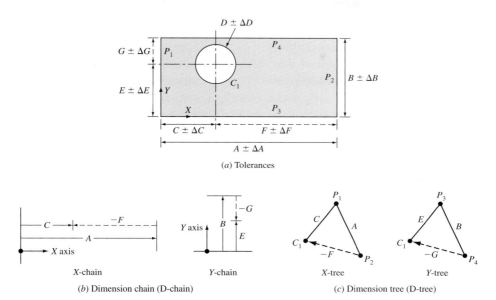

(a) Tolerances

X-chain   Y-chain   X-tree   Y-tree

(b) Dimension chain (D-chain)   (c) Dimension tree (D-tree)

Figure 20.30   Tolerance representation.

Two other dimensioning schemes are possible for this part. In one of them, we can locate the hole by the distances F and G (shown dashed) instead of C and E, respectively. This scheme would then require the positional parameters A, B, F, and G. In the other scheme, the hole is located by C and E while P_2 and P_4 are located by F and G, respectively, thus requiring the positional parameters C, E, F, and G. Notice that A, C, and F cannot coexist, otherwise overdimensioning would result; only any two parameters are allowed. Similarly, any two parameters of B, E, and G are permitted to avoid overdimensioning in the Y direction.

The positional parameters which determine the locations of primitive half-spaces (part features) can be defined systematically and algorithmically by introducing the concept of dimension chains. A dimension chain (D-chain) defines a location relative to another location. Thus, a D-chain is a directed chain with positive and negative dimensions. A positive dimension is one that is measured in the positive direction of the corresponding axis, A negative dimension is measured in the negative direction of an axis.

There is one D-chain per axis of measurement. Figure 20.30b shows the two D-chains for the X and Y axes. The X-chains for three dimensioning schemes discussed above are A and C, A and $-F$, and C and F. The Y-chains are B and E, B and $-G$, and E and G.

D-chains could be used to further develop the concept of dimension trees (D-trees). D-trees are more useful algorithmically than D-chains. Their traversal can be used to detect existing overdimensioning problems of a given dimensioning scheme, as well as to perform tolerance stack analysis as we show later in this chapter. A **D-tree** is a tree whose nodes (interior or exterior) are half-spaces and whose branches are directed (positive or negative) dimensions.

Like D-chains, there is one D-tree per axis of dimensioning. Figure 20.30c shows the X-tree and the Y-tree for the dimensioning scheme shown in Figure 20.30a. Two branches per tree exist. The trees of the other two dimensioning schemes are also shown. The trees for both schemes still have the nodes P_1, P_2, and C_1 for the X-tree, and P_3, P_4, and C_1 for the Y-tree. However, the branches correspond to the D-chains mentioned in the previous paragraph.

Two important rules for manipulating D-chains and D-trees exist. The first rule is related to null dimensions if they occur. Null dimensions have no effect on a D-chain or a D-tree. Thus, the D-chain A + null − F − null is equivalent to the chain $A − F$.

The second rule is related to condensing a D-chain or a D-tree. A component dimension and its immediate successor cancel one another if and only if they are identical and opposite in direction. The D-chain $A − A$ is null, while the D-chain $A + B − B$ is equivalent to A. However, the D-chain $A + B − A$ is not equivalent to B.

How can D-trees be used to detect overdimensioning? If the dimensioning scheme of a part results in a dimensioning graph and not a tree, then the scheme is inappropriate. Thus, for properly dimensioned parts there is a unique path between any two nodes of such graphs. For example, if we specify the dimensions A, C, and F simultaneously in the X direction, and B, E, and G in the Y direction, the X-tree and Y-tree shown in Figure 20.30c become closed (have cycles or loops), thus changing from trees to graphs. Thus, overdimensioning occurs if any of the part dimensioning results in graphs and not trees.

While the preceding discussion is limited to translations of half-spaces, it can be extended in a similar fashion to rotations of the half-spaces as well, resulting in angular D-chains and D-trees. May these chains and trees be decomposed into several independent chains and trees (one per direction), or are they able to detect overdimensioning? The answers to these questions are left as an exercise to the reader.

In the context of D-chains or D-trees, conventional tolerances are applied to the branches (dimensions) of these chains or trees. The nominal value of a D-chain can be found by applying "+" or "−" as arithmetic operators to the nominal values of the components of dimensions. The tolerance values, however, would require tolerance stack analysis to evaluate.

EXAMPLE 20.5   **Sketch D-chains and D-trees.**
                    Sketch the corresponding D-chains and D-tree of the ANSI example part shown in Figure 20.26.

SOLUTION   The part shown in Figure 20.26 is axisymmetric. Two dimensioning axes can be identified: the horizontal and the radial. Dimensions in the radial axis are the diameters of the various cylindrical faces shown in the Figure. A D-chain or D-tree for this axis is useless (no chain effect exists) because the diameters themselves are not interrelated, although they collectively affect the various radial thicknesses of the part.

Figure 20.31a shows the half-spaces dimensioned in the horizontal direction with P_1, P_2, and so forth. Assuming bilateral tolerances, the nominal dimensions of the limits shown in Figure 20.26 can be easily calculated. Figure 20.31b shows the D-chain, and Figure 20.31c shows the corresponding D-tree assuming that the horizontal axis originates from the location of half-space P_1 and is positive to the right. It is obvious that no overdimensioning problems exist because the resulting D-tree is a tree and not a graph.

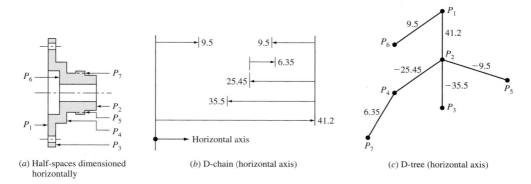

(a) Half-spaces dimensioned horizontally

(b) D-chain (horizontal axis)

(c) D-tree (horizontal axis)

Figure 20.31   Tolerance representation.

20.14 Tolerance Analysis

With all tolerances assigned to the various components (parts) of an assembly, the designer must make sure that the combined effect or accumulation of all these tolerances (tolerance stackup) does not cause an inoperable or malfunctioning assembly. Analysis of tolerances and their stackup is important because tolerance assignments are usually done on a part-by-part basis. **Tolerance analysis** is defined as the process of checking the tolerances to verify that all the design constraints are met. Tolerance analysis is sometimes known as design assurance.

The objective of tolerance analysis is to determine the variability of any quantity that is a function of product dimensions, and are called design functions. Most often, these qualities are themselves dimensions. Product dimensions and variables that control the behavior of a design function are called design function variables.

The variability of design functions is used to assess the suitability of a particular tolerance specification. Figure 20.32 shows an example of a design function for the case of two blocks assembled into a slot. The design function F is the clearance between the two blocks, and is a function of the dimensions of the slot and the two blocks. A tolerance specification for these dimensions is satisfactory if it prevents F from being less than zero.

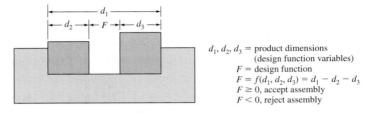

d_1, d_2, d_3 = product dimensions
                (design function variables)
F = design function
$F = f(d_1, d_2, d_3) = d_1 - d_2 - d_3$
$F \geq 0$, accept assembly
$F < 0$, reject assembly

Figure 20.32   Tolerance representation.

The formulation of tolerance analysis can be stated as follows. Given a set of tolerances $\{T\} = \{T_1, T_2, ..., T_n\}$ on a set of dimensions $\{d\} = \{d_1, d_2, ..., d_n\}$, and a set of design constraints $\{C\} = \{C_1, C_2, ..., C_m\}$, is $\{T\}$ satisfactory? Constraints could be functional requirements of an assembly, manufacturing costs, and so forth. The dimensions in the set $\{d\}$ include both nominal dimensions $\{d_N]$ and their tolerances $\{T\}$, that is,

$$\{d\} = \{d_N\} + \{T\} \tag{20.13}$$

To assess tolerance suitability, we formulate a design function in terms of $\{d\}$, as follows:

$$F = f(\{d\}) = f(d_1, d_2, ..., d_n) \tag{20.14}$$

The variability of F due to variability in $\{d\}$ is determined (using methods described in the discussion that follows). If F satisfies $\{C\}$ all the time, $\{T\}$ is satisfactory and the assembly is accepted. If not, $\{T\}$ is unsatisfactory and the assembly is rejected. Design functions are often complex, and their formulation forms the hardest part of tolerance analysis and can be time consuming.

Tolerance analysis methods can be divided into two types. In the simpler type, dimensions have conventional tolerances, and the result of tolerance analysis is the nominal value of the design function (F_N) and its upper (F_{max}) and lower (F_{min}) limits. This type of analysis is sometimes called worst-case analysis. This means that all possible combinations of in-tolerance parts must result in an assembly that satisfies the design constraints.

The upper and lower limits of the design function represent the worst possible combination of the tolerances of the design function variables. However, the likelihood of worst-case combination of these tolerances in any particular product is very low. Therefore, worst-case tolerance analysis is very conservative.

The other type of tolerance analysis is performed on a statistical basis. Tolerance analysis methods of this type allow statistical tolerances and output a statistical distribution for the design function. This allows for more realistic analysis. Manufacturing costs are reduced by loosening up the tolerances and accepting a calculated risk that the design constraints $\{C\}$ may not be satisfied 100% of the time.

By assuming a probability distribution for each toleranced dimension, it is possible to determine the likelihood that the specified design limits will be exceeded. Effectively, a reject rate is determined for the assembly. A nonzero reject rate may be preferable to an increase in individual part manufacturing costs due to tight tolerances. Both the worst-case and statistical approaches are used in practice.

20.14.1 Worst-Case Arithmetic Method

The arithmetic tolerance method is the worst-case analysis method. It uses the limits of dimensions to carry out the tolerance calculations. The actual or expected distribution of dimensions is not taken into account. All manufactured parts are interchangeable since the maximum values are used. Arithmetic tolerances require greater manufacturing accuracy. It is used in job shop production (very few parts are produced) and in cases where totally or 100% interchangeable assembly is required.

Let us assume a closed-loop (meaning the resulting dimension is obtained by adding and/or subtracting the given dimensions) dimension set $\{d\}$ of n elements such that the design function (resultant dimension) F is obtained by adding the first m elements (called increasing dimensions) and subtracting the last $(n - m)$ elements (called decreasing dimensions). Using this method, all tolerance information about F is obtained by adding and/or subtracting the corresponding information about the individual dimensions. Thus, we can write:

Nominal dimension:

$$F_N = \sum_{i=1}^{m} d_{iN} - \sum_{i=m+1}^{n} d_{iN} \tag{20.15}$$

Maximum dimension:

$$F_{max} = \sum_{i=1}^{m} d_{i\,max} - \sum_{i=m+1}^{n} d_{i\,min} \tag{20.16}$$

Minimum dimension:

$$F_{min} = \sum_{i=1}^{m} d_{i\,min} - \sum_{i=m+1}^{n} d_{i\,max} \tag{20.17}$$

Tolerance on F:

$$T_F = F_{max} - F_{min} = \sum_{i=1}^{m} T_i + \sum_{i=m+1}^{n} T_i = \sum_{i=1}^{n} T_i \tag{20.18}$$

Upper tolerance on F:

$$T_{uF} = F_{max} - F_N = \sum_{i=1}^{m} (d_{i\,max} - d_{iN}) - \sum_{i=m+1}^{n} (d_{i\,min} - d_{iN})$$

$$= \sum_{i=1}^{m} T_{ui} - \sum_{i=m+1}^{n} T_{Li} \qquad (20.19)$$

Lower tolerance on F:

$$T_{LF} = F_{min} - F_N = \sum_{i=1}^{m} (d_{imin} - d_{iN}) - \sum_{i=m+1}^{n} (d_{imax} - d_{iN})$$

$$= \sum_{i=1}^{m} T_{Li} - \sum_{i=m+1}^{n} T_{ui} \qquad (20.20)$$

where T_{ui} and T_{Li} are the upper and lower tolerances on dimension d_i, respectively. For unilateral tolerances, one of these variables is zero.

EXAMPLE 20.6 **Arithmetic tolerance analysis.**

Figure 20.33a shows a part with assigned dimensions and tolerances. Use the arithmetic method to calculate the tolerance information for the axial dimension design function F of the outside surface shown.

SOLUTION Figures 20.33b and c shows the D-chain and D-tree of the dimensions of the part design. It is obvious that the design function F is affected by the dimensions in its chain. The dimension d_6 is independent of the chain and, therefore, is not expected to affect F in the tolerance analysis. There are five dimensions in the chain (d_1 to d_5) excluding F; two of them are increasing (positive) dimensions (d_1 and d_3) and three are decreasing (negative) dimensions (d_2, d_4, and d_5).

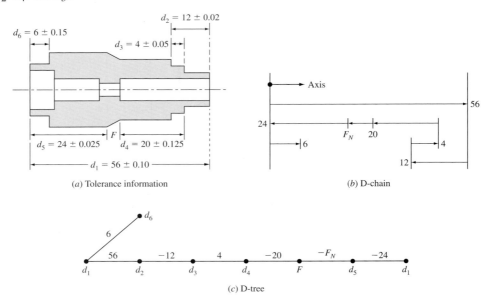

(a) Tolerance information

(b) D-chain

(c) D-tree

Figure 20.33   Tolerance analysis of a part.

Substituting the tolerance information shown in Figure 20.32a into Eqs. (20.15) to (20.20), we obtain:

$$F_N = (56 + 4) - (12 + 20 + 24) = 4 \text{ in}$$

$$F_{max} = (56.1 + 4.05) - (11.98 + 19.875 + 23.975) = 4.32 \text{ in}$$

$$F_{min} = (55.9 + 3.95) - (12.02 + 20.125 + 24.025) = 3.68 \text{ in}$$

$$T_F = 4.32 - 3.68 = 0.64 \text{ in}$$

$$T_{uF} = (0.10 + 0.05) - (-0.02 - 0.125 - 0.025) = 0.32 \text{ in}$$

$$T_{LF} = (-0.10 - 0.05) - (0.02 + 0.125 + 0.025) = -0.32 \text{ in}$$

20.14.2 Worst-Case Statistical Method

This method, like the arithmetic method, uses the limits of dimensions to perform tolerance analysis. However, unlike the arithmetic method, it takes into consideration the fact that dimensions of parts of an assembly follow a probabilistic distribution curve. Consequently, the frequency distribution curve of the dimensions of the final assembly follow a probabilistic distribution curve. Typically, the probabilistic distribution curve is assumed to be a normal distribution curve. This method is used in both batch and mass production. It allows for variabilities in manufacturing conditions such as tool wear, machine conditions, and random errors. It increases the manufacturing efficiency by increasing tolerance limits and, therefore, reducing the required accuracy of manufacturing.

This method is applied to a closed-loop dimension set $\{d\}$ with each element d_i of the set having a probability distribution curve. The design function F is obtained in the same way as in the arithmetic method. The tolerance information about F [similar to Eqs. (20.15) to (20.20)] can be obtained statistically as follows.

A normal distribution is considered the basis of the analysis. Parameters relating other distributions to the normal distribution are shown in Table 20.8. Figure 20.34 shows the parameters of a distribution curve for one of the elements of the dimension set.

Table 20.8   Probabilistic distribution curves.

Distribution	Normal	Uniform	Quasi-uniform	Triangle	Left skew	Right skew
Shape						
α	0	0	0	0	−0.26	0.26
K	1.0	1.73	1 − 1.5	1.22	1.17	1.17

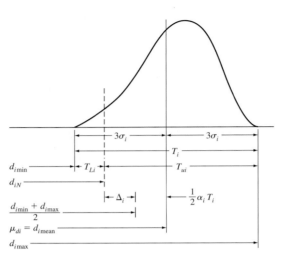

Figure 20.34   Probabilistic distribution curve of a dimension.

When the elements in the dimension set become large enough, the distribution of the design function F (the resulting dimension) will be asymptotically normal and independent of the distributions of the individual dimension. Thus we can write:

$$F_N = \sum_{i=1}^{m} d_{iN} - \sum_{i=m+1}^{n} d_{iN} \tag{20.21}$$

$$T_F = \sqrt{\sum_{i=1}^{n} K^2 T_i^2} \tag{20.22}$$

$$\sigma_F = \sqrt{\sum_{i=1}^{n} \sigma_i^2} = \frac{T_F}{6} \tag{20.23}$$

$$T_{uF} = \Delta + \frac{1}{2}T_F \tag{20.24}$$

$$T_{LF} = \Delta - \frac{1}{2}T_F \tag{20.25}$$

$$\Delta = \sum_{i=1}^{m} \left(\Delta_i + \frac{1}{2}\alpha_i T_i\right) - \sum_{i=m+1}^{n} \left(\Delta_i + \frac{1}{2}\alpha_i T_i\right) \tag{20.26}$$

The $\sigma = T_F/6$ used in Eq. (20.23) is based on assuming a range of 6σ for the distribution curve (3σ on each side of the mean tolerance) as shown in Figure 20.34. Equations (20.24) to

(20.26) can be viewed as dividing T_F into the upper and lower limits T_{uF} and T_{LF}, respectively. If bilateral equal tolerance limits are assumed, then $T_{uF} = \frac{1}{2}T_F$ and $T_{LF} = -\frac{1}{2}T_F$.

EXAMPLE 20.7   **Statistical tolerance analysis.**

Use the worst-case statistical method to calculate the tolerance information of F for Example 20.6. Assume a normal distribution curve.

SOLUTION   The use of Eqs. (20.21) to (20.23) is straightforward and gives $F_N = 4$ in, $T_F = 0.341$ in, and $\sigma_F = 0.0568$ in. To calculate T_{uF} and T_{LF}, notice that $\alpha_i = 0$ for the normal distribution curve. For worst-case analysis, d_{iN} shown in Figure 20.34 is either d_{imin} or d_{imax}. In either case, it is shown from the figure that $\Delta_i = (d_{imax} - d_{imin})/2 = T_i/2$. Thus Eq. (20.26) becomes:

$$\Delta = \sum_{i=1}^{m} \frac{T_i}{2} - \sum_{i=m+1}^{n} \frac{T_i}{2} = \frac{1}{2}(0.2 + 0.1) - \frac{1}{2}(0.04 + 0.25 + 0.05) = -0.02 \text{ in}$$

Substituting into Eqs. (20.24) and (20.25) gives T_{uF} and T_{LF} as 0.151 in and −0.190 in respectively.

EXAMPLE 20.8   **Statistical tolerance analysis.**

The assembly shown in the figure below consists of three parts: A, B, and C. A and B must "just" fit into A when put end-to-end. The three parts have a 0.1 percent tolerance of their basic dimensions. The designer responsible for this design must answer the following questions:

1. Is interference likely to happen in the assembly?
2. If it does, find the percentage of assemblies that must be rejected due to interference.

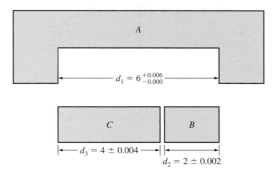

SOLUTION   The preceding figure shows the three dimensions of d_1, d_2, and d_3 that control the assembly behavior. Let us consider d_1 as the design function F and d_2 and d_3 as the design variables. To check for interference, consider the given tolerance on d_1, $^{+0.006}_{-0.000}$, as a design constraint. Thus, the tolerance analysis problem is formulated as follows: Given $\{T\} = \{\pm 0.002, \pm 0.004\}$ on $\{d_N\} = \{2, 4\}$, and given $\{C\} = \{+0.006, -0.000\}$, and assuming $F = d_1 = d_2 + d_3$, will interference occur?

The answer to the preceding question entails finding the tolerance T_F on F and comparing it with $\{C\}$. Using Eq. (20.22) with $K_i = 1$ gives:

$$T_F = \sqrt{(0.004)^2 + (0.008)^2} = 0.0089 \text{ in}$$

This means that $\{T_F\} > \{C\}$, which means that interference is bound to happen.

To find the rejection percentage due to interference, let us find the mean μ_F and the standard deviation σ_F of the sample F. The means μ_B and μ_C of the samples B and C are 2 and 4, respectively. The mean μ_F is calculated as $\mu_B + \mu_C = 6$. Using Eq. (20.23) gives $\sigma_F = 0.0015$. Thus the probability density function $f(z)$ of the design function F, assuming normal distribution, is given as:

$$f(z) = \frac{1}{\sqrt{2\pi}} e^{-z^2/2}$$

This is a standard normal distribution with z as the normalized variable, that is, $z = (x - \mu_F)/\sigma_F$ where x is any dimension. To find the probability of rejection, we substitute the mean value of d_1, which is $(d_{1max} + d_{1min})/2 = 6.003$ for x to obtain $z = 2$. Using probability tables for a standard normal distribution gives a probability of 0.0228 for $z = 2$. Thus the rejection percentage of the assembly at hand due to interference is 2.28%.

Note: Readers should be familiar with population combinations in statistics. For a population $R = X_1 \pm X_2 \pm \dots \pm X_n$, the mean and standard deviation are

$$\mu_R = \mu_1 \pm \mu_2 \pm \dots \pm \mu_n$$

$$\sigma_R = \sqrt{\sigma_1^2 + \sigma_2^2 + \dots + \sigma_n^2}$$

20.14.3 Monte Carlo Simulation Method

The previous two methods are only applicable to conventional tolerances—mainly for closed-loop dimensional sets with linear design functions. When these functions become more complex or nonlinear, applying these methods becomes less obvious if not impossible. Consider the simple example of a box that has sides of lengths a, b, and c with 1% tolerances on each dimension. To calculate the resulting tolerances on a diagonal of the box, the design function F is the length of the diagonal and given by the relation:

$$F = \sqrt{a^2 + b^2 + c^2} \tag{20.27}$$

To calculate the tolerances on the diagonal using the worst-case arithmetic method, we reduce (or increase) each dimension by 1%, which gives a tolerance on the diagonal of $\pm 0.01 \sqrt{a^2 + b^2 + c^2}$. To use the worst-case statistical method is less obvious and may require linearizing Eq. (20.27).

One may conclude that it is not always possible to find a design function to perform tolerance analysis. Or even if one is found, it may be difficult to use. In addition, geometric tolerances must be considered in the tolerance analysis. To bring these tolerances into the analysis, 2D (such as area) or 3D (such as volume) design functions may have to be formulated instead of the 1D (dimension) functions used in the previous methods.

These 2D and 3D functions can be written first in terms of nominal dimensions and then perturbed using the geometric tolerances. While this approach enables including geometric tolerances in the tolerance analysis, it still requires a designed function which it may not be possible to find.

One of the methods that seems useful in performing tolerance analysis using geometric tolerances without a need for an explicit design function is the Monte Carlo simulation method. The idea of the Monte Carlo method stems from the manufacturing practice wherein a prototype of a given assembly is built and assembled to test its tolerances and functionality. The Monte Carlo method achieves the same result (within the reliability limits of the method) without the time and cost of part manufacturing.

The Monte Carlo method is applicable whether the design variables (these are the design function variables used in the previous two methods) are linear or not. Either a worst-case statistical analysis (using tolerance limits which give d_{imax} and d_{imin} as shown in Figure 20.34) or just a statistical analysis (using tolerance values between tolerance limits which give d_{iN} as shown in Figure 20.34) can be performed.

The method operates by generating (with computer random generation) a large sample of assembly instances of the assembly model to be analyzed. Each assembly instance consists of parts' instances, each of which corresponds to a set of dimensions that are generated randomly using the statistical distributions assumed for nominal dimensions (Figure 20.34). Each generated assembly instance is checked to determine whether it meets the specified design constraints or not. Thus a statistical distribution can be generated for each design constraint. Probabilities of accepting or rejecting assemblies can, therefore, be estimated.

An algorithm based on the Monte Carlo method and implemented into a solid modeler can be described as follows:

1. Generate a candidate instance of an as-manufactured part using a normal (or other) distribution random number generator to perturb the vertices of the part within the specified size tolerance zone.
2. Check if the part instance meets the specified form tolerances. This is needed because form tolerances may be tighter than size tolerances, and because normal distribution may, in rare cases, generate perturbations with standard deviations beyond the size tolerance zone.
3. If one or more of the vertices of the part instance are found outside the zone of form tolerance, the assembly instance is rejected. If all vertices are inside the zone, the assembly instance may be accepted.
4. Repeat steps 1 to 3 for all other parts in the assembly.
5. Use the solid modeler to create the assembly instance using all the instances created in step 4. These instances are positioned relative to datums established by part features.

6. Check if the assembly instance from step 5 satisfies the design constraints. If it does, the assembly is accepted, otherwise it is rejected.

7. Repeat steps 1 to 6 as many times as is needed by the desired sample size (number of assembly instances) for calculating the statistics. The larger the sample size the better, and the more confidence we have in the results.

In theory, the potential accuracy of the Monte Carlo method is unlimited. In practice, the sample size required to obtain reasonable estimates can be prohibitive. For a large, complex assembly, the dimensionality of the sample space is so great that a Monte Carlo analysis, even one based on a large number of assembly instances, does not carry a high degree of reliability. Nevertheless, a Monte Carlo analysis may act to focus attention on potential problem areas.

Generating an instance of a part (step 1) depends primarily on whether the solid modeler is a B-rep or CSG. For B-rep models, where faces, edges, and vertices are stored, size and form tolerances are applied as variations to the surface equations of the part faces. Position tolerances are applied to features of position. New edges and vertices are computed at the intersections of the varianced faces.

For CSG models, since part geometry is stored in unevaluated form as an ordered sequence of features (primitive or compound half-spaces), each feature is perturbed first using proper tolerances, then combined to generate a unique instance of the part. This maintains features as the main building blocks of the assembly and reduces the instancing task to that of simply perturbing features individually.

The random generation of vertices is achieved by random perturbation of the coordinates of the vertices as follows. Given a nominal location d_N of a vertex with a size tolerance $T_s(d_N \pm T_s/2)$, the vertex is moved an amount x according to the size tolerance. Figure 20.34 can be used to interpret this movement. Assuming a normal distribution for x, and using $\sigma = T_s/6$, the density distribution function of the variational component x is given by:

$$f(x) = \frac{1}{\sigma\sqrt{2\pi}} e^{-x^2/2\sigma^2}, \qquad \sigma > 0 \qquad (20.28)$$

If vertices are perturbed as mentioned in step 1 of the preceding algorithm, the solid model must be triangulated before the algorithm begins. This may be unacceptable. In this case, tolerances may be applied directly to equations of surfaces in B-rep or of half-spaces in CSG.

20.14.4 Other Methods

Other methods for tolerance analysis exist. Optimization methods do treat the design function and the design constraints as an optimization problem, while the design variables are viewed as the decision variables for optimization. The design function may be nonlinear. Linear programming methods linearize the design function and design constraint equations, and they solve tolerance analysis using linear programming. After the linear programming problem is solved, a sensitivity analysis can be performed to determine the relative contribution of each of the tolerances.

The Taylor series method and the quadrature method are other statistical methods, and they approximate the probability density function of the design function without generating the large number of samples required by the Monte Carlo method.

Tolerance charts are a semigraphical, spreadsheet-like method which can be used to formulate design functions and perform tolerance analysis on the resulting functions. This method deals only with one dimensional problems.

Bjorke's method formulates design functions and performs their statistical analysis. The method uses the concept of tolerance chain and can deal with general 3D problems. However, it cannot accommodate design functions that are not dimensions. Bjorke assumes that the design function and the dimensions have a beta (β) distribution.

20.15 Tolerance Synthesis

In tolerance analysis, tolerances on dimensions are known and the tolerance on a resultant dimension is to be determined. In terms of a design function, the variability of the function is to be obtained. **Tolerance synthesis** (also called tolerance distribution) is the inverse problem; the allowable tolerance, or allowable variations in the design function, on a dimension are known and the tolerances on other dimensions are to be determined. In terms of Figure 20.35, given the tolerance on A, $\pm\Delta A$, find the tolerances on B, C, and D such that their tolerance analysis produces back the tolerance $\pm\Delta A$ on A.

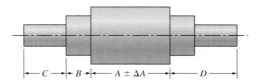

Figure 20.35   Tolerance synthesis.

There are two criteria to distribute a tolerance: the equal tolerance criterion and the equal precision criterion. The former simply distributes the tolerance equally among dimensions that affect the given tolerance. While it is easy to compute manually, it is not accurate based on design and manufacturing experience. The latter is more suitable to real designs, but it involves more complicated calculations.

When the nominal dimensions are less than or equal to 500 mm, there is a relation between the standard tolerance and precision in ISO:

$$T_i = aI_i = a(C\sqrt[3]{d_{iav}} + 0.001 d_{iav}) \tag{20.29}$$

where C is a constant, a is the precision coefficient for different IT grades, d_{iav} is the sectioned standard length value, and I_i is the fundamental tolerance unit.

The first term in Eq. (20.29) accounts for manufacturing errors, and is a statistical result of machine shop data. The second term includes temperature change error and measurement error caused by dimension change. It is very small and is neglected. The length d_{iav} is used instead of the actual dimension d_i to reduce the number of constants in the equation as shown. There are tables that relate d_{iav} to d_i. Equation (20.29) forms the basis of two methods for tolerance synthesis as discussed Sections 20.15.1 and 20.15.2.

20.15.1 Arithmetic Method

Neglecting the second term in Eq. (20.29) gives:

$$T_i = aI_i = aC\sqrt[3]{d_{iav}} \tag{20.30}$$

Substituting Eq. (20.30) into Eq. (20.18) gives:

$$a = \frac{T_F}{C \displaystyle\sum_{i=1}^{n} \sqrt[3]{d_{iav}}} \tag{20.31}$$

substituting a given by Eq. (20.31) back into (20.30) gives the tolerance T_i on each dimension in the dimension set.

The tolerance T_i can be divided into upper and lower limits in various ways. For a shaft-based dimension, $T_{ui} = 0$, $T_{Li} = -T_i$. For a hole-based dimension, $T_{ui} = T_i$, $T_{Li} = 0$. For any other less critical dimension, we have

For an increasing dimension:

$$T_{u\mathrm{adj}} = T_{uF} + \sum_{m+1}^{n} T_{Li} - \sum_{i=1,\,i\neq\mathrm{adj}}^{m} T_{ui} \tag{20.32}$$

$$T_{L\mathrm{adj}} = T_{LF} + \sum_{m+1}^{n} T_{ui} - \sum_{i=1,\,i\neq\mathrm{adj}}^{m} T_{Li} \tag{20.33}$$

For a decreasing dimension:

$$T_{u\mathrm{adj}} = -T_{uF} + \sum_{i=1}^{m} T_{Li} - \sum_{i=m+1,\,i\neq\mathrm{adj}}^{n} T_{ui} \tag{20.34}$$

$$T_{L\mathrm{adj}} = -T_{LF} + \sum_{i=1}^{m} T_{ui} - \sum_{i=m+1,\,i\neq\mathrm{adj}}^{n} T_{Li} \tag{20.35}$$

20.15.2 Statistical Method

Substituting Eq. (20.30) into Eq. (20.22) gives

$$a = \frac{T_F}{\sqrt{\sum_{i=1}^{n} K_i^2 \left(C \sqrt[3]{d_{iav}}\right)^2}} \tag{20.36}$$

Substituting Eq. (20.36) into Eq. (20.30) produces the tolerance T_i on each dimension in the dimension set.

The division of T_i into upper and lower limits for a shaft- or hole-based dimension is the same as in the arithmetic method. Modifications of the tolerance of other dimensions are given by (assuming $\alpha_F = 0$):

$$T_{adj} = \frac{1}{K_{adj}} \sqrt{T_F^2 - \sum_{i=1}^{n} K_i^2 T_i^2} \tag{20.37}$$

For an increasing dimension:

$$\Delta_{adj} = \Delta_F + \sum_{m+1}^{n} \left(\Delta_i + \frac{1}{2}\alpha_i T_i\right) - \sum_{i=1, i \neq adj}^{m} \left(\Delta_i + \frac{1}{2}\alpha_i T_i\right) - \frac{1}{2}\alpha_{adj} T_{adj} \tag{20.38}$$

For a decreasing dimension:

$$\Delta_{adj} = \Delta_F + \sum_{i=1}^{m} \left(\Delta_i + \frac{1}{2}\alpha_i T_i\right) - \sum_{i=m+1, i \neq adj}^{n} \left(\Delta_i + \frac{1}{2}\alpha_i T_i\right) - \frac{1}{2}\alpha_{adj} T_{adj} \tag{20.39}$$

where

$$\Delta_i = \frac{1}{2}(\Delta_{ui} + \Delta_{Li}) \tag{20.40}$$

Then, the adjusted dimension is

$$d_{adj} + \Delta_{adj} \pm \frac{T_{adj}}{2} \tag{20.41}$$

where d_{adj} is d_i to be adjusted.

20.15.3 Taguchi Method

As in the case of tolerance analysis, the preceding two methods for tolerance synthesis are limited in their application. Some other methods are based on mathematical programming. Another method is the Taguchi method, which is based on the principles of experimental design.

The Taguchi method not only determines tolerances but also determines the ideal nominal values for the dimensions. This is referred to as dimension centering. The method finds the nominal dimensions that allow the largest, lowest-cost tolerances to be assigned.

It selects dimensions and tolerances with regard to their effect on a single design function. The method uses fractional factorial experiments to find the nominal dimensions and tolerances which maximize the so-called signal-to-noise (S/N) ratio. The signal is a measure of how close the design function is to its desired nominal value. The noise is a measure of the variability of the design function caused by tolerances.

The main disadvantage of the Taguchi method is its inability to handle more than one design function. Finding one design function for a product may not be practical at all.

20.16 Tutorials

20.16.1 Create Conventional Tolerances

In this tutorial we create the following drawing with different types of conventional tolerances. All dimensions are in inches.

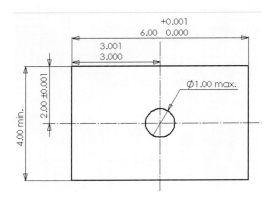

Modeling strategy:

1. **Create the part.** We create a block with the dimensions shown in the screenshot below. The block is an extrusion with a thickness of 1 in. Select the `Front` sketch plane. Create the sketch shown below. Then extrude it a distance of 1 in. Save the part in a file.

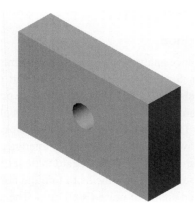

2. **Create a drawing.** While the part is active on the screen, open a new drawing file. Add a front view to the drawing as shown below. Add centerlines to the hole. Also add dimensions as shown below.

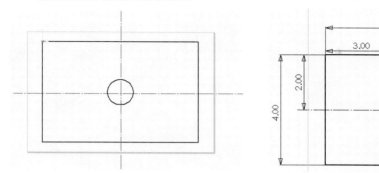

3. **Add conventional tolerances.** Activate the dimensioning dialog box or menu on your CAD system. Locate the conventional tolerances menu. The screenshot below shows an example of a tolerance menu. Use it to add the tolerances shown below.

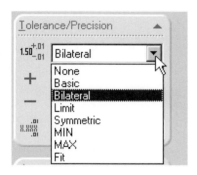
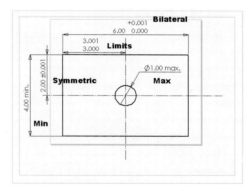

Tutorial 20.6.1 discussion:

The tutorial shows how to create different styles of conventional tolerance specifications, as shown in the screenshots of step 3. However, it is not recommended in practice to mix various styles of tolerances in one drawing. We should use only one style for consistency, and to avoid potential confusion in the machine shop during production.

The screenshot shown on the right in step 3 shows the different styles that are shown in the screenshot shown on the left in step 3.

Tutorial 20.6.1 hands-on exercise:

1. **Use a consistent tolerance style per drawing.** Modify the tolerances of the tutorial to use only one tolerance style—bilateral style as shown on the left of the forthcoming screenshot.

2. **Understand tolerance accumulation.** The forthcoming screenshot shown on the right represents a drawing of a chamfered block. Is this the best way of dimensioning? Do you

have any suggestions for improvement? Make the necessary changes to incorporate your suggestions. Hint: Use an ordinate dimensioning scheme to eliminate tolerance accumulation.

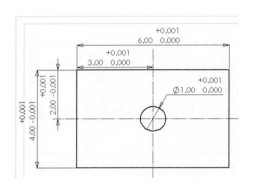

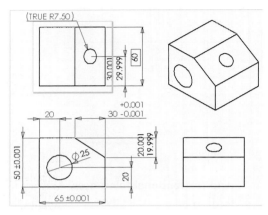

20.16.2 Create Geometric Tolerances

In this tutorial we create the drawing shown on the right with different geometric tolerances. This tutorial uses the same part as Tutorial 20.16.1. All dimensions are in inches.

Modeling strategy:

1. **Create the part.** Follow step 1 of Tutorial 20.16.1.
2. **Create a drawing.** Follow step 2 of Tutorial 20.16.1.

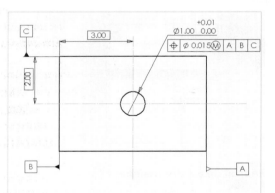

3. **Create datums.** Activate the dimensioning dialog box or menu on your CAD system. Locate the geometric tolerances menu. The forthcoming screenshots in Step 4 show an example of a tolerance menu. Use it to create the datums shown on the right.

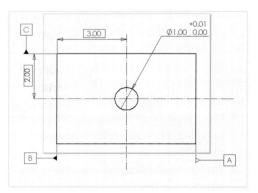

4. **Add geometric tolerances.** Activate the dimensioning dialog box or menu on your CAD system. Locate the geometric tolerances menu. The following screenshots show example menus. Use them to add the tolerances shown below.

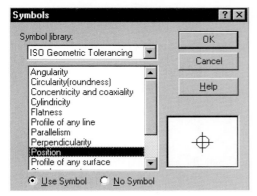

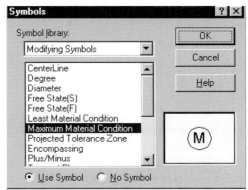

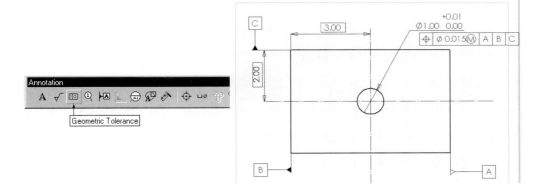

Tutorial 20.6.2 discussion:

The tutorial shows how to create geometric tolerances. The screenshot shown on the left in step 4 shows form tolerances. The screenshot shown on the right in step 4 shows ANSI modifying symbols.

Tutorial 20.6.2 hands-on exercise:

1. **Add a perpendicular geometric tolerance.** Modify Tutorial 20.6.2 to include the perpendicular tolerance shown in the screenshot on the right.

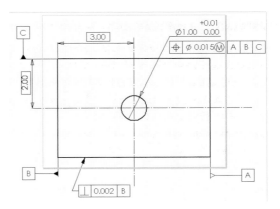

2. **Add more geometric tolerances.** The
   screenshot on the right shows more
   geometric tolerances. Create a new
   drawing and add them.

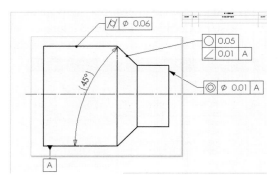

3. **Add tolerances to a hole and a slot.**
   The following screenshot shows a part with a hole and a slot . The required datums and
   tolerances to manufacture the part are shown. Create the part and a drawing with the two
   views as shown.

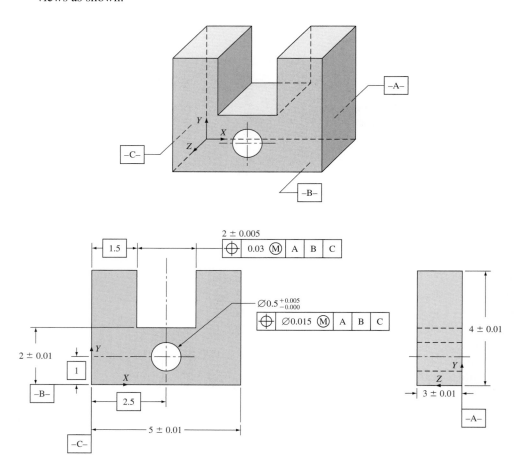

Problems

Part I: Theory

20.1 Describe and sketch the various available dimensioning schemes.

20.2 Find the basic rules that ANSI requires in dimensioning any drawing.

20.3 Cylindrical fits are fits between shafts and holes. For a nominal diameter $d = 1.7500$ in, use the basic hole system to calculate h, s, a, and the hole and shaft limits for class fit 2, 4, 6, and 7. Use Table 20.1 in your calculations. Follow Figure 20.1 and sketch the bar diagrams for each fit.

20.4 Calculate the tolerance zones h and s, the allowance a, and the limits for the following fits:

RC8; basic size = 2.5000 in

LC3; basic size = 7.0000 in

LT4; basic size =10.0000 in

LN2; basic size =12.0000 in

FN2; basic size = 3.0000 in

20.5 Interpret the following positional tolerances. Sketch the tolerance zone(s) for each case:

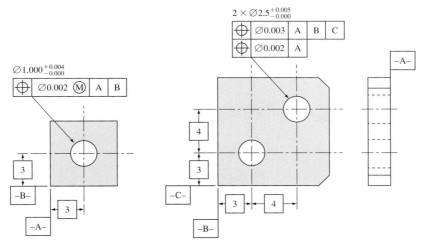

20.6 Two parts of the one shown on the right are to be assembled with bolts. Derive the positional tolerance T_p on the hole center to satisfy the following fits: clearance fit H8/f7, transition fit H7/k6, and interference fit H7/s6. Sketch the tolerance zone for each fit.

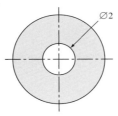

20.7   Interpret the various geometric tolerances used in the following parts.

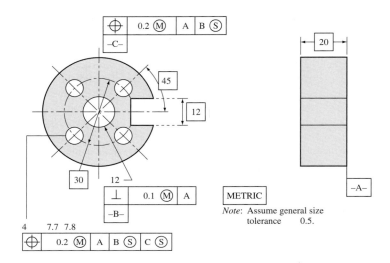

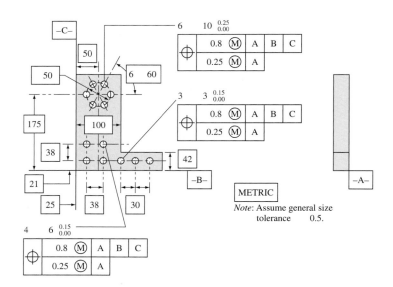

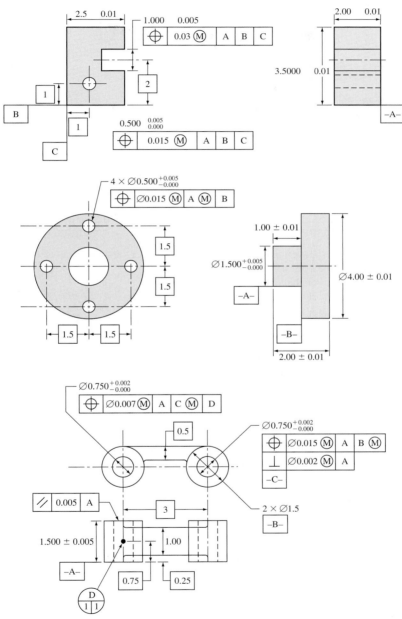

20.8 Sketch the D-chains and D-trees for the models shown in Problems 20.5 and 20.7. Are there any overdimensioning cases? Why?

20.9 Identify the design function(s) F in the following part designs. Then use the worst-case arithmetic method of tolerance analysis to calculate the tolerance information for F.

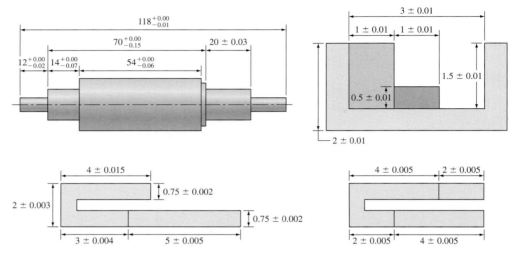

20.10 Use the worst-case statistical method of tolerance analysis to calculate the tolerance information for F for parts shown in Problem 20.9.

20.11 A shaft and a hole have nominal diameters of 1 in as shown on the right.

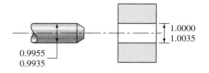

   a. Will the assembly have a clearance less than 0.006 in?

   b. What is the percentage of assembly rejection due to a clearance less than 0.006 in?

Part II: Laboratory

Use your in-house CAD/CAM system to answer the questions in this part.

20.12 Find the dimension commands and their respective modifiers which are supported by your CAD/CAM systems. What types of tolerances (conventional or geometric) do they support?

20.13 Create the toleranced views of the parts shown in Problems 20.5, 20.7, 20.9, and 20.11. Create the parts first, then drawings, and finally views in the drawings. Assume any missing dimensions.

20.14 Does your system have tolerance analysis or synthesis commands? If it does, what type (arithmetic, statistical, Monte Carlo, etc.)?

Part III: Programming

Use OpenGL with C or C++, or use Java 3D.

20.15 Write a program that can read a text string of a dimension and its tolerance, and then separate the basic dimension and the tolerance into two numerical values.

20.16 Write a program that implements the worst-case arithmetic method for tolerance analysis.

20.17 Repeat Problem 20.16 but for worst-case statistical method.

20.18 Write a program that can randomly generate coordinates of a point (vertex). Assume a size tolerance zone T_s on the point and study the statistical behavior of the point.

Process Planning

OAL

Understand the interface between design and manufacturing and between CAD and CAM, the importance of process planning, the basics of CAPP, and the benefits of CAPP.

BJECTIVES

After reading this chapter, you should understand the following concepts:

- Manual process planning
- Essentials of CAPP
- Input and output of CAPP systems
- CAPP models
- Architecture of CAPP systems
- CAPP approaches: variant, generative, and hybrid
- CAPP benefits
- CAPP software

CHAPTER HEADLINES

21.1 Introduction

Process planning is a common task in discrete part manufacturing. It is the activity responsible for the conversion of design data to manufacturing or work instructions. More specifically, **process planning** is defined as the activity that translates part design specifications from an engineering drawing into the manufacturing instructions required to convert a part from a rough to a finished state.

Process planners must first evaluate design data and specifications such as geometric features, dimensional sizes, tolerances, surface finish, and materials in order to select an appropriate sequence of processing operations and specific machines. Operation details such as setup and cut planning, stock preparation, jigs and fixtures planning, speeds, feeds, tooling, and assembly steps are then determined, and standard times and costs are calculated.

The outcome of the design evaluation and the preparation of operation details is a process plan to produce the part. This plan is then documented as a job routing (or operation) sheet, a cost estimate, and coded instructions for NC machine tools.

Process planning, as viewed by some, represents the link between engineering design and shop floor manufacturing. It is a major determinant of manufacturing cost and profitability. The gap between CAD and CAM can be shortened considerably by developing better systems for process planning.

The difficulty in developing a reliable process planning system is due to the qualitative nature of many of the decisions involved in addition to the many variables that the process planner must consider. In this respect, the process planning phase in manufacturing resembles the conceptual design phase in engineering design.

21.2 Activities

Before we discuss the various approaches to process planning, it is useful to examine the activities included in process planning. These are material selection and order, process selection, process sequencing, machine selection, design toolings and jigs and fixtures (clamps), intermediate surface determination, fixture selection, machining parameters selection, cost and time estimation, process plan preparation, NC tool path generation, inspecting product quality, and assembly and testing of finished products.

Post-manufacturing tasks such as shipping, installation, operation, and maintenance may also be considered as part of process planning. The activities of process planning are interrelated. This makes a single, universally applicable process plan for all part manufacturing unlikely to be attained, or even attempted.

The process planner has the task of breaking down the manufacturing of components into individual process steps. The planner specifies the machines on which the components are to be made, and the functional requirements of tools, jigs and fixtures, and inspection gages. For each process step, the planner identifies the process datum(s) which determines the way in which the part is to be set up.

Process dimensions and tolerances are also defined. In the early stages of machining of castings, forgings, or sheet metal parts, spot or line datums are often used since the surfaces of these parts are usually irregular. In general, faces, edges, and vertices of a part may be used as datums.

The process plan also includes a description of in-process inspection requirements for each process step. These requirements are deduced from the part dimensioning and tolerancing. From known information about process capabilities, the process planner decides on which dimensions are to be checked, and the frequency and nature of the checks.

The clamping method, besides the machining processes, must be taken into account in assessing the process capability to achieve a given tolerance. In the design of clamps, the resulting distortion and its effect on critical dimensions must be considered. Where distortion is unavoidable, general tolerances are specified, or the maximum permissible loads which may be applied to produce the correct shape are given. These loads should be acceptable to the designer as safe loads to be applied on the assembly.

If the process planner decides that the design requirements are too expensive or impossible to meet with existing equipment, the planner may notify the responsible designer with suggested changes. This usually involves a need to relax close tolerances, or a recommendation to use standard sizes. In many companies, process planning is brought into the design cycle earlier to reduce the need for this iterative process. Therefore, tentative process planning would have already taken place before the formal planning process begins.

Process plans must also include instructions for assembly. In planning a subassembly, dimensional allowances may be required to provide for final machining of the subassembly to true size to remove assembly-induced distortion. Both the function of the assembly and the resulting distortion have to be considered in determining assembly forces for interference fits, torques for bolted assemblies, and so forth.

A process plan is documented in multiple documents including routing, process, cost, time, setup, and inspection sheets. Each sheet provides the details of the corresponding activity. For example, a setup sheet provides the steps required to set up a machine tool. Multiple setup sheets may be needed, one per operation.

The detailed process plan and its sheets are crucial to the successful production of the part. When these sheets are distributed to the various production workstations in different work centers and cells, they lose their coherence. Machinists would not know anything about the part; only what they see on the sheets. Thus these sheets must be detailed enough to stand on their own.

21.3 Process Plan Development

Process planning begins with an examination of an engineering part drawing and ends with manufacturing process plans and instructions that are based on a knowledge of manufacturing processes and machine capabilities, tooling, materials, related costs, and shop practices.

Process planning depends heavily on the personal experience and background of the process planner for plans which are feasible, low cost, and consistent with other plans for similar parts.

If the part to be produced belongs to an existing family of parts, the process planning usually involves recalling existing process plans for similar parts and modifying them to create a plan for the new part. Process plans are usually classified and stored; thus they can be retrieved.

If the part to be produced is new, the planner may have to generate a new plan. The development of a new plan is highly subjective, labor intensive, time-consuming, tedious, and often boring. Furthermore, it requires personnel who are well trained and experienced in manufacturing shop floor practices.

Two levels of process planning can be identified: high-level planning and low-level planning. During the high-level planning, the planner identifies the machinable features (surfaces) of the part, groups them into setups, and orders these setups. Each setup is listed in the order in which it is to be done, the features to be cut in each of the setups, and the tools for cutting each feature.

The low-level planning includes specifying the details of performing each step that results from the first level such as choosing machines, cutting conditions (speed and feed), types of fixtures, cost and time estimates, and so forth.

Most often, process planners follow more or less a consistent set of steps to develop process plans for new products. These steps involve primarily stock preparation, plan generation, and specification of manufacturing parameters. The detailed steps can be listed as follows:

1. **Get oriented.** Investigate the engineering drawing to identify the basic structure and the potential difficulties of the part to be produced. The planner reads the design data and specifications on the drawing and checks for any major problems: Can the part be clamped or does it fit between the jaws of a vice? Is the part too long and thin in certain directions so that it will bend when clamped? and so on.

2. **Recognize the outer envelope of the finished part.** With the help of the engineering drawing, recognize the outer, or bounding, envelope of the part. The recognition includes both the geometric shape and the surface finish of the envelope. The recognition of the envelope helps to determine the optimal stock shape from which to produce the finished part; "optimal" means the shape that results in the least amount of material waste and/or removal to convert the stock into the finished part.

3. **Choose the optimal stock.** Based on the above Step 2, choose the stock shape, dimensions, surface finish, and material. Standard stock shapes exist and are shown in Figure 21.1. The dimensions of a stock are typically about 1/4 in larger than the finished part dimensions. The finish of the sides of the stock could be rough (saw-cut) or smooth (rolled or machined).

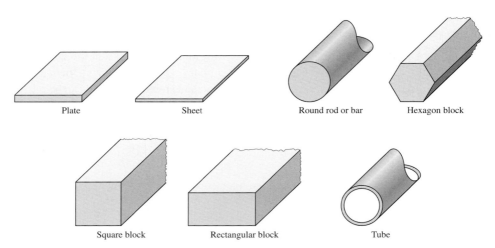

Plate   Sheet   Round rod or bar   Hexagon block

Square block   Rectangular block   Tube

Figure 21.1   Standard stock shapes.

4. **Recognize part features.** Identify and list the features and subfeatures that are subtracted from the part's outer envelope. Sample examples of typical features and subfeatures encountered in manufacturing are shown in Figures 21.2 and 21.3, respectively. Features are the individual geometric shapes that are cut into the stock to form the part. Although the shapes of features are simple by themselves, they can be combined in many ways to form more complex shapes. Shapes that make useful features reflect the shape of the tools, the movements of the machine, and the paths they can cut through the workpiece (stock).

5. **Choose stock preparation plan.** Outline all methods that can convert the stock chosen in step 3 into the part shape with the minimum waste of material. Machine the part features into the stock, one by one, until the final shape is obtained. The plan begins by squaring off the sides of the stock in a particular order, to provide reference surfaces for the machining operations that follow. This order is stored in what is sometimes called a squaring graph. Some of the squaring rules utilized to develop squaring graphs are: machine a largest side first; machine a medium side second; machine the remaining sides in any order. Figure 21.4 shows the squaring graph of a block. Observe that when branching happens at the same level as for setups C and D, the setups can be done in any order.

6. **Consider alternative methods for producing each feature.** There may be many ways of making each part feature recognized in Step 4. For example, to make an angle, a sine table, sine bar, angle plate, or angle cutter might be used. Some ways of producing a feature are better than others, so the alternatives generated must be ordered according to preference. The preference is usually determined based partly on judgment and partly on strategy. Judgment is related to the particular characteristics of the part. The planner may ask questions such as: If I make an angle using the sine bar, can the part be clamped firmly enough? Strategy, on the other hand, determines how trade-offs are to be made during

manufacturing. For example, the planner may ask questions such as: Which is more important for the part, cost or accuracy? Is it worth the risk of breaking a tool to discover a faster way to make the part? Is the part a prototype or is it used in mass production?

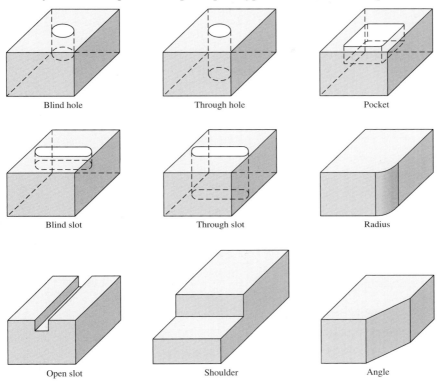

Figure 21.2   Sample manufacturing features.

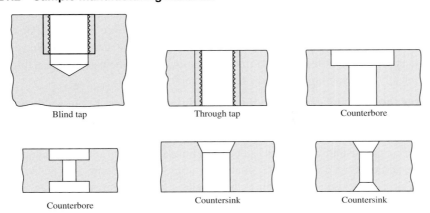

Figure 21.3   Sample manufacturing subfeatures.

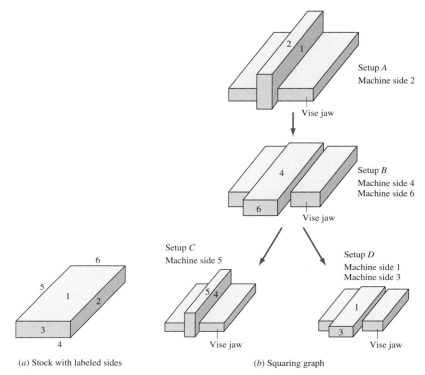

(a) Stock with labeled sides   (b) Squaring graph

Figure 21.4   Stock squaring graph.

7. **Generate a plan by exploring feature interactions.** Utilizing the outcomes of step 6, generate a plan to produce the part. The plan should consider feature interactions that happen when machining operations of one collection of features affect or destroy the way in which other features can be machined. Feature interactions have several different causes. Most commonly they result from clamping or reference problems. For example, producing one feature may destroy the clamping surfaces needed to grip the workpiece while cutting another feature. The final generation plan of the part may be expressed graphically in what may be called a feature interaction graph.

8. **Merge the squaring graph with the feature interaction graph.** Merge the two graphs by combining their overlapping steps so that we can get a compact sequence. The more overlap, the better and more concise the final plan. For example, we can use a setup step to produce a part feature in the same step, thus eliminating a step from the interaction graph. The resulting merged graph is the final plan outline and represents the final ordering for the desired process plan. We refer to this resulting graph as the process plan graph.

9. **Make a final check on the plan.** Verify the process plan by checking that the setups are actually feasible, that the clamps are not in the way of the tools, and so forth.

10. **Elaborate the process plan.** The preceding nine steps represent the high-level planning of the process plan. This step represents the low-level planning. Now, the planner generates more details for producing each individual feature, choosing feeds and speeds, estimating costs and standard times, and so forth. The planner verifies all these details and releases the process plan to the various departments for execution.

EXAMPLE 21.1   **Develop a process plan.**

Figure 21.5a shows a typical part. The part has a slot and two tapped holes with countersinks (on side 4). Develop the plan graph and the high-level process plan that can successfully produce the part. Show the squaring graph and the interaction graph as well.

SOLUTION   We follow the preceding ten steps here to develop the plan graph and the process plan. Figure 21.5a shows the part with five features: a slot in the center, two holes, and two angles α and β. The part also has four subfeatures: two taps and two countersinks. The appropriate stock from which to produce this part is a block of dimensions $1.75 \times 0.50 \times 3.25$ in (following the rule of 1/4 in larger than the finished part dimensions). Assuming the stock block to be saw-cut, Figure 21.4 shows the correct squaring graph.

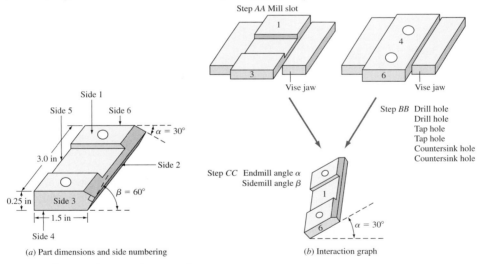

(a) Part dimensions and side numbering   (b) Interaction graph

Figure 21.5   Feature interaction graph.

Studying the interactions of the five part features reveals some interesting observations. The two holes interact with the angle α, and the slot interacts with the angle β. In both cases, making the angle first will cause difficulty in gripping the part to make the other feature. The restriction that this interaction puts on the plan is that the two holes must be made before angle α, and the slot before angle β. Thus, one would cut the slot and drill the two holes before cutting the two angles.

Because there are no tolerance specifications on the part dimensions, the process planner must assume them to be tight and recommend the machining methods accordingly. In situations like this, the planner usually requests a single roughing and finishing pass on the machine. If tolerances were specified, the planner could take advantage of this information to make the plan more efficient by reducing the number of machining steps (by requesting, say, a single roughing pass only on the machine and eliminating the finishing pass).

Figure 21.5*b* shows the resulting interaction graph for this part. The graph shows the machining sequence in which features are cut (subtracted) from the stock to produce the finished part. Each step in the sequence shows the machining activities that must be achieved. Observe that the order of carrying out steps *AA* and *BB* is unimportant.

We now merge the squaring graph of Figure 21.4*b* with the interaction graph of Figure 21.5*b* to obtain the final plan graph. Investigating both graphs reveals that setup *D* of the squaring graph overlaps with the setup requirement of step *AA* in the interaction graph. Therefore, these two setups can be combined. This implies that while setting up the stock to machine its sides 1 and 3, we mill the slot afterwards. This results in saving one unnecessary setup.

Figure 21.6 shows the plan graph to produce the part from the rectangular saw-cut stock. Figure 21.6 can be translated into the following high-level process plan:

```
 1  SET UP SIDE 2            10  SIDEMILL SIDE 3
 2  ENDMILL SIDE 2           11  ENDMILL SLOT
 3  SET UP SIDES 4 AND 6     12  SET UP SIDE 4
 4  ENDMILL SIDE 4           13  DRILL HOLES
 5  SIDEMILL SIDE 6          14  TAP HOLES
 6  SET UP SIDE 5            15  COUNTERSINK HOLES
 7  ENDMILL SIDE 5           16  SET UP ANGLES α AND β
 8  SET UP SIDED 1 AND 3     17  ENDMILL ANGLE α
 9  ENDMILL SIDE 1           18  SIDEMILL ANGLE β
```

The last step to fully develop the process plan is to perform the low-level planning by adding the required details for each operation and rewriting the plan utilizing customary process planning codes known to process planners. To be able to specify the proper cutting conditions (feeds and speeds), the planner must know the material of the part. Aluminum or steel can be used. As part of the low-level planning, the planner can add the cost and time estimates as well as the final inspection of the finished part.

21.4 CAPP

Process planning can be done manually or with the aid of a computer. Manual process planning has some disadvantages. It is closely tied to the personal experience and knowledge of the planner of production facilities, equipment, their capabilities, processes, and tooling; this results in inconsistent plans. Manual process planning is time-consuming, and slow. Moreover, it is slow in responding to changes in product design and production.

Addressing these issues leads to the need for computerized systems that will allow process planning to be performed either entirely or partially by a computer. CAPP (computer-aided process planning) helps to simplify and improve the activity of process planning, provide the user with optimum process plans in a quick consistent fashion, and achieve more effective use of manufacturing resources.

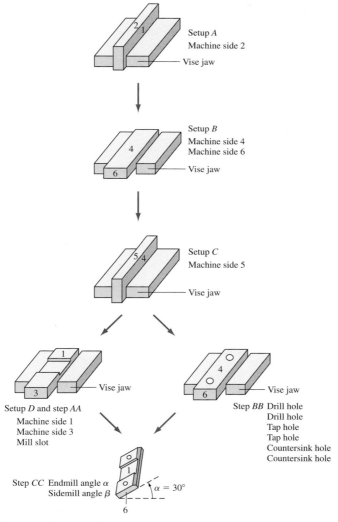

Figure 21.6 Process plan graph.

The input to a CAPP system includes a part description and a production size. The output is a description of the process plan which includes four sets of information. General information includes part name, number, class, and drawing. This set associates the plan to the part. This is useful in reusing the process plan in the future for similar parts.

Typical documents that are produced from a CAPP system include method sheets, routing sheets, and tool kit sheets. Method sheets are typical manufacturing instructions at an operation level (low level). They give details of the exact method an operation must follow, including things such as speeds, feeds, and tooling. Routing sheets give details of the work centers through which the part passes and may include details such as time value, and tooling (high-level information). Toolkit sheets act as a listing from which tools can be ordered and/or designed prior to the beginning of production.

Process structure information includes setups of different manufacturing processes, machine and tool positions, and required cuts to be performed on the stock (workpiece).

Operation information includes operation name and number, name of production department (work center), workstation name and type, part drawing before and after the operation, specifications of both the part and tool fixtures, cut sequence, standard operation time, and control program. The operation information belongs to the low-level plan we covered in Section 21.3.

Cut information includes cut description in words, cut number, cutting tool type and code, machining parameters such as speed and feed, and the time it takes to finish the cut. This information is repeated for each cut. Cut information belongs to the low-level plan.

21.5 CAPP Benefits

The benefits of implementing and using a CAPP system are

1. Reduction in process planning effort
2. Saving in direct labor
3. Saving in material
4. Saving in scrap
5. Saving in tooling
6. Reduction in work-in process
7. Reduced process planning lead time
8. Greater process plan consistency
9. Improved cost estimating procedures and fewer calculation errors
10. More complete and detailed process plans
11. Improved production scheduling and capacity utilization
12. Improved ability to introduce new manufacturing technology and rapid updates
13. Process plans to utilize the improved technology

21.6 CAPP Model

CAPP works at the interface between CAD and CAM. It takes CAD data, converts it to production data, and feeds the latter to a production system. Figure 21.7 shows a CAPP model based on this interface concept. The CAPP model utilizes the flow shown in the figure to convert CAD data into production data.

After the CAD model is created, it is prepared for transfer into the CAPP model. This preparation step is performed by a preprocessor, and it could involve producing an IGES or STEP file that the CAPP model can read. This step is necessary because both models are independent of each other. CAD data also needs to be prepared to obtain the proper product definition as required by the CAPP model.

The CAPP model applies its knowledge and rules to the prepared CAD data to produce its output, the process plan.

The CAPP model performs the necessary postprocessing operations on its output to produce output that production and scheduling systems can read and utilize in their own activities.

Figure 21.7 shows that the components of the CAPP model are independent of both the CAD and production systems. Thus the model requires two conversion steps: one to convert CAD data, and the other to convert the CAPP output itself.

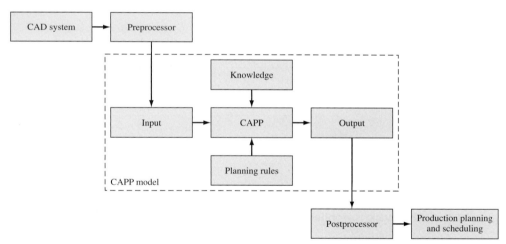

Figure 21.7   CAPP model.

21.7 Architecture of a CAPP System

The CAPP model shown in Figure 21.7 forms the basis for building CAPP systems and software. Figure 21.8 shows the architecture of a CAPP system. It consists of four modules. The preprocessor prepares the CAD data for the machining module.

The machining module receives the processed CAD data and uses its machining knowledge base to select the required machining operations and processes. It also generates all the low-level information for each process such as feeds, cutting speeds, cutting conditions, and so forth.

The constraint creation module selects the appropriate processing constraints for the data. It uses the concepts discussed in Section 21.3. For example, it decides on the part envelope, stock shape, features, and subfeatures. The constraint module uses a manufacturing knowledge base that has data and rules similar to those of human process planners.

The constraint application module takes the stock shape and features as input and checks them for feasibility and potential conflicts. It uses the concepts of squaring (Figure 21.4) and interaction (Figure 21.5) graphs. It searches all possible solutions to find an optimal plan. It may request additional information from the manufacturing knowledge base. The outcome from this module is the process plan graph (Figure 21.6).

The process plan generation module generates both the high-level and low-level plans discussed in Section 21.3. It generates all the plan sheets that are sent to various production entities. Production managers and supervisors receive the sheets and delegate them to the appropriate production personnel. For example, tooling sheets are sent to the tooling department. Each production stop or work center receives a routing sheet that shows the next stop where the work-in-progress part should be sent.

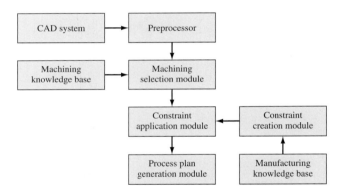

Figure 21.8 CAPP system architecture.

21.8 CAPP Approaches

Two CAPP approaches, and therefore two types of CAPP systems, exist: variant and generative. A third approach combines these two approaches; it is known as the hybrid or semi-generative. Each approach has advantages and disadvantages, as we discuss here.

21.8.1 Variant CAPP

This approach is also known as the retrieval approach. Variant CAPP has evolved out of the traditional manual process planning method. A process plan for a new part is created by identifying and retrieving an existing plan for a similar part, followed by making the necessary modifications to adapt it to the new part.

Variant CAPP is based on the group technology (GT) classification and coding approach used to identify a larger number of part attributes or parameters. These attributes allow the CAPP system to select a baseline process plan for the part family and accomplish about 90% of the planning work. The planner adds the remaining 10% of the planning by modifying the baseline plan. Baseline process plans, stored in CAPP databases, are manually entered after standard plans are developed based on the accumulated experience and knowledge of multiple planners and manufacturing engineers.

The system logic involved in establishing a variant CAPP is relatively straightforward. It involves matching a code with a pre-established process plan maintained in the system. The initial challenge is in developing the GT classification and coding structure for the part families and in manually developing a standard baseline process plan for each part family.

The use of a variant CAPP system has two modes: new and old. If the part under consideration does not belong to any existing part family in the CAPP system, a new process plan must be developed manually and then entered into the system database to create a new baseline process plan for future use.

If the part matches an existing family, the process plan engineer retrieves the baseline (master) process plan and modifies it.

One of the disadvantages of variant CAPP is that the quality of the process plan still depends on the knowledge and background of a process planner. However, this method has some advantages. The investment, in hardware and software, required to acquire a variant CAPP system is not high. Once acquired, the system offers a shorter development time and lower manpower consumption to develop process plans. Moreover, the system is very reliable and reasonable in real production environments for small- and medium-sized companies.

21.8.2 Generative CAPP

The generative process planning approach is viewed as the automated approach to process planning. Unlike the variant approach, the generative approach does not require assistance from the user to generate a process plan. It usually accepts geometric and manufacturing data from the user, and utilizes computerized searches and decision logics to develop the part process plan automatically. Generative process planning systems do not require or store predefined master process plans. Instead, the system automatically generates a unique plan for a part every time the part is ordered and released for manufacturing.

In generative CAPP, process plans are generated by means of decision logic, formulas, technology algorithms, and geometry-based data that are built or fed as input to the system. Generally, the format of input to CAPP systems can be divided into two categories. In text input format, the user answers questions in an interactive mode (interactive input). In graphic input format, data is obtained from CAD models (interface input).

The first key to implementing a generative system is the development of decision rules appropriate for the parts to be processed. These decision rules are specified using decision trees, computer languages involving logic-type statements such as if-then, if-then-else, or artificial intelligence approaches with object-oriented programming. The nature of the parts affects the complexity of the decision rules for generative planning and ultimately the degree of success in implementing the generative CAPP system.

The second key to implementing generative CAPP is the available data related to the part to drive the planning. Simple forms of generative CAPP systems may be driven by GT codes.

A pure generative system can produce a complete process plan from part classification and other design data which does not require any further modification or manual interaction.

Some approaches that are used in generative process planning are decision trees, decision tables, axiomatic approach, artificial intelligence–based approach, and expert systems. All these approaches generate a plan. In generating such a plan, an initial state of the part (stock) must be defined in order to reach the final state (goal or the finished part). The path taken between the two states defines the process plan or the sequence of processes.

Two types of planning are available: forward and backward planning. Using forward planning, we begin with the stock as the initial state and remove part features until the finished part, the final state, is obtained. Backward planning uses the reverse procedure. We begin with the finished part as the initial state, and our goal is to fill it with part features until the stock is obtained. In backward planning, a drilling process fills a hole instead of its conventional meaning in forward planning.

Forward and backward planning are not as similar as they may seem, and they affect the programming of a corresponding system significantly. The requirements (preconditions) and the results (postconditions) of a setup in forward planning are the results and requirements, respectively, of the setup in backward planning.

While forward planning seems natural, why does one need backward planning? Forward planning suffers from conditioning problems; the result of a setup affects the next setup. In backward planning, conditioning problems are eliminated because setups are selected to satisfy the initial requirements only.

The generative approach has all the advantages that the variant approach has. However, it has the additional advantages that it is fully automatic and that an up-to-date process plan is generated each time a part is ordered. However, a generative system would require major revisions in decision logic if new equipment or processing capabilities became available. In addition, the development of the system in the first place is a formidable task.

21.8.3 Hybrid CAPP

This approach is an interim approach used when problems in building purely generative process planning systems occur. This approach can be characterized as an advanced application of variant technology employing generative-type features. Systems using this method should co-operate with process planners who possess technological knowledge. The planner's responsibility is the interpretation of decision data and/or a working drawing. The approach can be implemented in three ways:

1. The variant approach can be used to develop the general process plan, then the generative method can be used to modify it.
2. The generative approach can be used to create as much of the process plan as possible, then the variant approach can be used to fill in the details.
3. Process planners can select either generative mode for complicated part features or variant mode for fast process plan generation.

21.9 CAPP Systems

Many CAPP systems exist. They use the three approaches covered in Section 21.7. They are written in various programming languages such as C++, LISP, Turbo Pascal, Fortran, Prolog, and dBASE. Sample variant systems are CAMSS, GLM, and PerMIA. Sample generative systems are AFR, ALPS, AMOPPS, CROPS, ESTPAR, FBD, FCAPP/SM, GEOPDE, GFAS, GLM, IKOOP, KAPLAN, K-base, MCOES, PART, ROBEX and RATE, and TVCAPP. Sample hybrid systems are CADEXCAP, COMPLAN, RDCAPP, SMT, and TAMCAM.

PROBLEMS

Part I: Theory

21.1 What is process planning? Why do we need it?

21.2 Describe the activities that make up process planning.

21.3 Summarize the ten steps of process plan development.

21.4 What is the input and output to CAPP?

21.5 List and describe the documents that a CAPP system generates.

21.6 What are the elements of a CAPP model?

21.7 Describe the modules and knowledge bases of a CAPP system.

21.8 List the CAPP approaches with their advantages and disadvantages.

21.9 How does the variant CAPP approach work?

21.10 How does the generative CAPP approach work?

21.11 What are the differences between forward and backward process planning? Why do we use both?

21.12 What are the different ways we can use a hybrid CAPP system?

21.13 What are the benefits of CAPP?

21.14 Develop the squaring graph, the interaction graph, and the plan graph of the parts whose geometry are shown in problems in Chapters 6, 7, 9, and 10. Write the sequence of steps (using high-level planning format) of the resulting process plan for each part.

21.15 This problem illustrates how to handle process planning with incomplete information due to lack of information itself or due to a lack of experience. Consider yourself a process planner with very little hands-on experience with titanium. How can you develop a process plan for the following titanium part.

Chapter 21

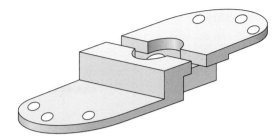

21.16 Search the various resources and collect the existing commercial CAPP systems. Classify them into variant, generative, and hybrid systems. What are the documents that each one produces (method sheets, routing sheets, tool kit sheets, etc.). For the variant systems, what is the GT classifying and code system used? For the generative system, what methods and decision logic are used?

Part II: Laboratory

Use your in-house CAD/CAM system to answer the questions in this part.

21.17 Using a process planning package, generate process plans for the parts shown in the problems in Chapters 6, 7, 9, and 10.

Part Programming

OAL

Understand the basics of machine tools and their programming, part programming and its basics, the fundamentals of coding languages and structure, and the use of CAD/CAM systems for generating and verifying toolpaths.

OBJECTIVES

After reading this chapter, you should understand the following concepts:

- Importance of part programming
- Working at the interface between CAD and CAM
- CAM requirements of CAD data
- Fundamentals of machine tools: axes, table motion, and programming
- NC programming: toolpaths, zero radius programming, and tool offsets
- APT programming: geometric and motion statements
- Toolpath generation
- Toolpath verification

CHAPTER HEADLINES

22.1 Introduction

One of the outcomes of process planning is the method sheet that lists the manufacturing instructions for various machine tools to produce the finished part. The information in these sheets can be utilized to program machine tools to produce the part. We cover the basics of part programming. Readers should realize that the topic of part programming deserves a book by itself, and they should refer to existing books for further information.

The basic premise of part programming is that the machining steps of a part are captured in a program (called a **part program**) that the control system (controller) of the machine tool reads and executes to produce the finished part. This enables the part to be machined automatically without a need for human operators. Part programming utilizes the NC (numerical control) technology. Part programming is considered the turning point of the metal-cutting industry to meet the demands of intricate designs, with tight tolerances, which could never have been machined using the manual approach.

The NC technology is based on controlling the motion of the drives of the machine tool as well as the motion of the cutting tool via an NC part program. An **NC program** is a set of statements (instructions) which can be interpreted and executed by the machine controller and converted into signals that move the machine spindles and drives.

NC software is available in two modes: integrated and standalone. In the integrated mode, NC software is included as an add-on package or module of a CAD/CAM system. Standalone NC software may be used independent of a CAD/CAM system, or may be interfaced with it via IGES or STEP. Examples of standalone NC software include MASTERCAM and SURFCAM.

22.2 Integrating CAD, NC, and CAM

The need to increase productivity and survivability in an increasingly competitive market requires the integration between CAD and CAM, and prompts the notion of CIM (computer-integrated manufacturing). The original tenet of CIM thinking is to establish a common and central database to hold the great bulk of data necessary to run a company.

The technologies that would take part in the two-way communication with the central database include CAD, CAM, FMS (flexible manufacturing systems), production management, and automated assembly. The achievement of a true CIM system is a challenging task because the CAD and CAM data are heterogeneous and incompatible. For example, the geometry data stored in the CAD database would be of no use to production management, and vice versa; data related to production management is of no use to the design department.

To meet the challenge to have a centrally integrated CIM database, it is more attractive and practical to link multiple standalone CAD and CAM databases to create a "linked" CIM system. The linked CIM system provides quick and cost-effective means of achieving integration between CAD and CAM. The linked CIM system would rely heavily on communication software to link the various involved hardware (mainly computers).

There are many problems to face in terms of the compatibility of different computer systems and hardware, the compatibility of computer languages, and the compatibility of software packages. The Internet, intrants, and extranets have solved many of the networking problems. Communication protocols at the software level such as IGES, STEP, and MAP (manufacturing automation protocol) have achieved similar success.

A crucial step in NC programming is the effective use of CAD information by the NC programming software. This is sometimes referred to as linking CAD to NC. To assure the successful linking between CAD and NC, the geometric entities of the part must be ordered properly and NC information be put on separate layers. Imposing these requirements on the generation of engineering drawings enables the automatic generation of NC programs. These requirements form a CAM constraint that must be considered at the CAD phase to avoid having to scrap engineering drawings entirely and create new ones to develop NC part programs.

22.3 Preparing CAD Data for NC

Sometimes, an engineering drawing that looks fine from a drafting or CAD point of view is a garbled mess when transferred to an NC software package. Although some NC packages include translation routines that automatically format drawings, some do not, and others require the NC programmer to prepare the drawing for programming. For a successful electronic linking of CAD to NC, designers should keep the following set of guidelines in mind when entering CAD data that will eventually be used for NC programming:

1. **Part data must be accurate.** The coordinates must be accurate before the drawing is sent to the NC package. The accuracy of the data is usually within the round-off and truncation errors of the computer running the CAD software. For the most part, the default accuracy is enough. This accuracy is particularly important when compared to part tolerances. Consider, for example, storing the dimension 10 ± 0.0005 in a database. If the nominal size is stored as 10.0003 instead of 10.0000, then the tolerance becomes meaningless. Many CAD/CAM systems let their users specify the accuracy of the database (single or double precision) and the number of decimal places. It is also possible to write programs to check the accuracy of numbers stored in databases.

2. **Part data should be toleranced properly.** If parts are drawn using nominal dimensions, they must be edited to actual sizes before data is sent for NC programming.

3. **Drawing tolerances should match manufacturing practice.** Certain dimensions and tolerances are easier to program than others because of the accuracy limits of NC controllers and machine tools. Bilateral tolerances are usually easier to program. Preferred dimensioning tolerance notation usually specifies the nominal size plus or minus a tolerance since compensation can be made for cutter wear, material deflection, and tooling setup inaccuracies.

4. **Separate drawing annotation from drawing data.** Place annotation (notes and dimensions) and data (geometrical entities) on different layers. If annotation information is on the same layer with part geometry, the NC package might reject the part data file.

5. **Establish standards for layer assignments.** The layers available on all CAD/CAM systems can be employed to separate information, such as dimensions, notes, and material lists, within a drawing. By adopting a layer structure which controls the placement of data, subsequent isolation of geometry for NC part programming purposes can be effected in an easy and rapid manner.

6. **Profile entities must connect**. A typical NC machine tool will cut continuously only within 0.0001 in. It is, therefore, imperative that CAD databases meet or exceed machine tool accuracy.

7. **Avoid overlapping drawing entities.** If an arc or line is copied over an existing one, the resulting toolpath will be unusable.

8. **Dimensionality requirement must be specified.** If a design calls for milling operations which require Z-axis motion, the CAD package used to create the part database must have 3D capabilities. However, 2D-only packages would be sufficient for some applications such as routing.

9. **Input part geometry in manufacturing order.** Geometric entities should be input in the order they are connected to each other so that they form a continuous unidirectional (clockwise or counterclockwise) path that the cutting tool might follow. However, requiring designers to enter geometry in manufacturing order could severely curtail creativity and design productivity. Therefore, most existing NC packages reorder geometry automatically or semiautomatically. Part profile (geometric entities) is usually reordered according to coincident end points.

22.4 Machine Tools

An NC part program consists of a combination of machine tool code and machine-specific instructions. It contains geometric information about part geometry and motion statements to move the cutting tool. Cutting speed, feedrate, and auxiliary functions (coolant on and off, spindle direction, etc.) are programmed according to surface finish and tolerance requirements.

Part programmers who prepare part programs for NC machine tools must understand the coordinate systems of these machine tools and how they work. This section serves this purpose.

22.4.1 Description of Machine Tools

Figure 22.1 shows a schematic of an NC machine tool which typically consists of the MCU (machine control unit) and the machine tool itself. The MCU (also known as the controller unit, or controller for short) is considered the brain of the machine. It reads the part program and controls the machine tool operations. After reading the part program, the MCU decodes it to provide commands and instructions to the various control loops of the machine axes of motion.

The MCU performs two functions: reads the part program and controls the machine tool. It consists of two units, one for each function. The DPU (data processing unit) reads and decodes the part program statements, processes the decoded information, and provides data to the CLU (control loop unit).

The CLU receives the data from the DPU and converts it to control signals. The data usually provides the control information such as the new required position of each axis, its direction of motion and velocity, and auxiliary control signals to relays. The CLU also instructs the DPU to read new instructions from the part program when needed, controls the drives attached to the machine leadscrews, and receives feedback signals on the actual position and velocity of the axes of the machine. Each axis of motion of the machine tool has its own leadscrew, control signals, and feedback signals.

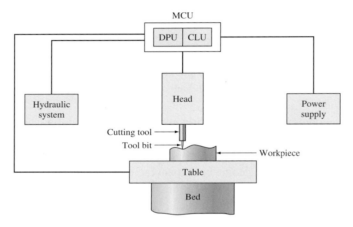

Figure 22.1   Schematic of an NC machine tool.

22.4.2   Motions and Axes of Machine Tools

A workpiece is machined to the finished shape by allowing a relative motion between the workpiece and the cutting tool. Such relative motion can be provided by holding the workpiece stationary and moving the cutting tool, as in drilling, or by moving the cutting tool and the workpiece simultaneously, as in milling and turning.

Regardless of whether the tool or the workpiece moves, each motion adds to the versatility of the machine tool and requires its own axis of motion. A machine tool may also have more than one head to provide additional axes of motion, one axis per head motion.

In NC machine tools, each axis of motion is equipped with a driving device to replace the handwheel of the conventional machine. The type of a driving device is selected mainly according to the power requirements of the machine. A driving device may be a dc motor, a hydraulic actuator, or a stepper motor.

An **axis of motion** is defined as an axis where relative motion between the cutting tool and workpiece occurs. This movement is achieved by the motion (slides) of the machine tool table. The primary three axes of motion are referred to as the *X*, *Y*, and *Z* axes and form the machine tool *XYZ* coordinate system. Figure 22.2 shows the coordinate system and the axes of motion of a typical machine tool.

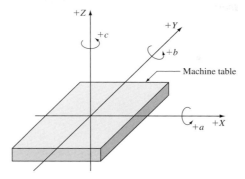

Figure 22.2   Axes of motion of a machine tool.

The *XYZ* system is a right-hand system, and the location of its origin may be fixed or adjustable. The positive directions of the axes are defined by the manufacturer of the machine tool. However, it is a convention that the positive direction of the *Z* axis moves the cutting tool away from the workpiece and the machine table. This orients the *XY* plane of the *XYZ* system along the table plane of the machine table as shown in Figure 22.2.

In addition to the primary slide motions in the *X*, *Y*, and *Z* directions, secondary slide motions may exist and may be labeled *U*, *V*, and *W*. Rotary motions around axes parallel to *X*, *Y*, and *Z* may also exist and are designated *a*, *b*, and *c*, respectively, as shown in Figure 22.2. These notations are EIA (Electronics Industry Association) standards.

It is conventional that machine tools are designated by the number of axes of motion they can provide to control the tool position and orientation. Most often, we hear terminology such as $2\frac{1}{2}$-axis, 3-axis, or 5-axis milling machines. These axes refer to the possible axes of motion that the machine tool can control simultaneously. We should not correlate them to spatial degrees of freedom to avoid confusion and misinterpretation.

If the machine tool can simultaneously control the tool along only two axes, it is classified as a 2-axis machine. In this machine, the tool is parallel and independently controlled along the third axis. Geometrically, this means that the machine tool controller can guide the cutting tool along a 2D contour with only independent movement specified along the third axis. Figure 22.3a shows an example of a 2-axis machine. Here the *Z*-axis control plane is parallel to the *XY* plane.

If the tool can be controlled to follow an inclined *Z*-axis control plane, we have a $2\frac{1}{2}$-axis machine as shown in Figure 22.3b. In a 3-axis machine, the tool is controlled along the three axes (*X*, *Y*, and *Z*) simultaneously, but the tool orientation does not change with the tool motion as shown in Figure 22.3c.

If the tool axis orientation varies with the tool motion in 3D, we have a multi-axis orientation machine (4-, 5-, or 6-axis). A 6-axis machine, for example, is capable of moving the tool simultaneously along each primary axis and, in the meantime, simultaneously rotating it about each primary axis. Figure 22.3d shows a multi-axis machine tool. Up to 9-axis machine tools are commercially available.

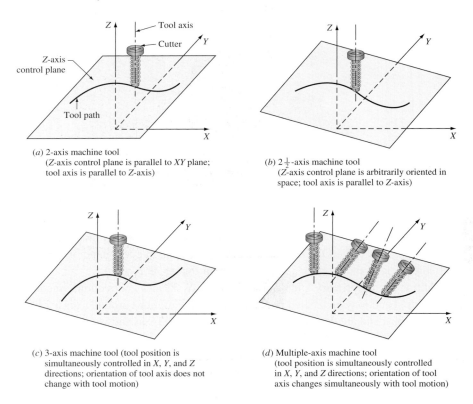

(a) 2-axis machine tool
(Z-axis control plane is parallel to XY plane; tool axis is parallel to Z-axis)

(b) $2\frac{1}{2}$-axis machine tool
(Z-axis control plane is arbitrarily oriented in space; tool axis is parallel to Z-axis)

(c) 3-axis machine tool (tool position is simultaneously controlled in X, Y, and Z directions; orientation of tool axis does not change with tool motion)

(d) Multiple-axis machine tool (tool position is simultaneously controlled in X, Y, and Z directions; orientation of tool axis changes simultaneously with tool motion)

Figure 22.3　Types of machine tools based on motion axes.

22.4.3　Point-to-Point and Continuous-Path Machining

Machine tools can perform two types of machining: point-to-point (PTP) and continuous path. The main features of each type are discussed in this section.

PTP machining is considered the simplest type of machining. An obvious example is drilling a hole. In PTP machining, the cutting tool performs operations on the workpiece at specific points. The tool is not always in contact with the workpiece throughout its motion or its path. The exact path the tool takes in moving from point to point is immaterial assuming, of course, that the time required is reasonable, and the tool does not collide with either the workpiece or the holding fixture. Straight-line toolpaths between points are common.

Drilling holes shown in Figure 22.4a provides an example of PTP machining. During drilling, the drill moves to a position directly over the hole to be drilled. Once in position, the drill is lowered at predetermined speed and feed. After drilling a hole, the drill moves out of the hole at a rapid retract rate. The drill moves to another point and the cycle is repeated.

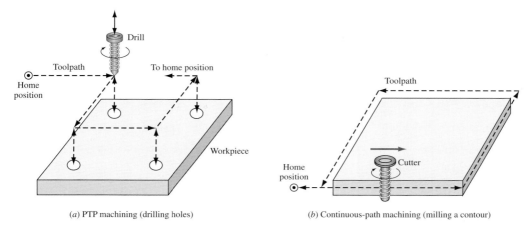

(a) PTP machining (drilling holes)   (b) Continuous-path machining (milling a contour)

Figure 22.4   PTP and continuous-path machining.

Unlike PTP machining, the cutting tool is always in contact with the workpiece in continuous-path machining. Therefore, the workpiece is being affected throughout the entire movement. The entire travel of the cutting tool must therefore be controlled to a high degree of accuracy both for position and velocity. Milling operations provide an example of continuous-path machining. Figure 22.4b shows contour milling. NC milling machines are very popular. This is attributed to their versatility. NC milling machines can also be used for PTP machining. NC lathes provide another example of continuous-path machining for NC turning.

22.4.4 NC, CNC, and DNC

Considering the structure of the controller unit, machine tools are categorized as NC, CNC (computer numerical control), and DNC (distributed numerical control) machines. In an NC machine tool, the DPU is a tape reader as shown in Figure 22.5a. Each time the NC program needs to be executed to machine a new part, the punched tape must be read by the tape reader. In a CNC machine tool, the DPU is an onboard computer (a PC). The computer provides ROM and display screen for the machine tool as shown in Figure 22.5b. The computer reads and stores the NC program like any other file. The NC programmer can have the program stored on a floppy disk or a CD.

The CNC concept has a few drawbacks. If the same NC program is used on various machine tools, it has to be loaded (read) separately into each machine, a nonproductive and repetitive task. More importantly, CNC systems are limited in the area of feedback. It is necessary for some sort of feedback of information to be generated by the CNC system to track such variables as machine downtimes, work-in-process, production rates, scrap rates, and other production data.

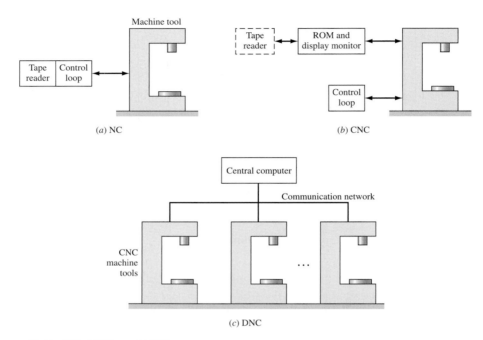

Figure 22.5   NC, CNC, and DNC.

The DNC system, shown in Figure 22.5c, has been created for this purpose. A DNC system consists of a central computer to which a group of CNC machine tools are connected via a communication network. The communication in a DNC system is usually achieved using a standard protocol such as TCP/IP or MAP.

The central computer has various functions. It stores the NC programs. It also downloads these programs to any number of CNC machine tools in the network, thus avoiding reloading of the same program. It also performs the feedback task described previously (data collection, processing, and reporting) as well as providing communications between the various components of the DNC system.

In some DNC systems, various hierarchical levels of computers and networks may exist between the CNC machine tools and the central computer. The intermediate computers (sometimes called satellite computers) provide various levels of local control for different CNC machines that belong to the DNC system.

The major advantage of a DNC system is that the system can be centrally monitored. This is important when dealing with different operators, in different shifts, working on different machines.

22.5 NC Programming

NC programming begins by getting an NC programmer to program the part. Using their machining and NC knowledge, part programmers plan the required toolpath(s) with the proper cutting parameters. A **toolpath** is the path that the cutting tool must follow from its home (park) position to machine the part and come back to home position. The path is usually repeated more than once (each time is called a pass) for a given number of passes to finish the part. A **pass** is one cycle in which the cutting tool traverses the toolpath once. Each pass, the tool is fed further into the stock to remove more material to eventually change the stock into the finished part.

Once the path is planned, NC programmers utilize their NC knowledge to generate the details of the path. This includes calculating the x, y, and z coordinates of the necessary points of the path, writing the NC program using the syntax of one of the programming languages, verifying the generated toolpath, postprocessing the program, and finally using the program to produce the actual part. The calculation of the point coordinates of the path involves the geometry of the part profile (boundaries) and requires trigonometric and geometric skills of the programmer if the toolpath is generated manually.

Programming languages such as APT are considered high-level languages. They provide the programmer with geometric statements that utilize the calculated coordinates and machining statements that control the motion of the cutting tool as well as the cutting conditions. NC controllers are usually designed to accept low-level languages utilizing G-code and M-code. Therefore, APT programs are usually postprocessed to generate the low-level programs.

Before using an NC program in production, it is usually necessary to verify it for possible programming mistakes. Therefore, the toolpath is either plotted or displayed on a monitor to be checked visually. Once verified, the NC program is ready for use in production to cut the part.

If NC programming is used in an integrated CAD/CAM environment, the toolpath generation, verification, and postprocessing are performed by the NC package which is part of the overall software. Utilizing the CAD database with all its geometrical information, the coordinates required to define the path are automatically evaluated.

The NC package also provides programmers with tool libraries that they can use to define various tools with the proper tool geometric parameters. For example, the programmer can define a drill or a cutting mill. Once a tool is defined, a toolpath can be generated using the tool and the part geometry.

When a toolpath is generated, it can be verified on the graphics display. This usually involves the display of an animation sequence that shows the tool moving along its generated path, which is superimposed on the part geometry. Colors and shaded images are usually used to enhance the visualization process. If mistakes are spotted, the toolpath can be modified accordingly and reverified.

Once accepted, time and cost estimates can be calculated. In addition, the NC package generates the APT program automatically from the toolpath information stored in the part database. Finally, the NC package postprocesses the APT program to translate it to the proper format (G- and M-code) that a particular NC controller requires.

22.6 Basics of NC Programming

NC programming deals with the control of machine tools with all their related components such as the machine table and coolant system, cutting tools, jigs and fixtures, workpieces, and so forth. NC programming languages have semantics and syntax to follow. Before we discuss the details and syntax of some of these languages, we review some of the concepts and basics typically employed in writing NC programs.

22.6.1 Machine Tool Coordinate System

Most existing machine tools use cartesian coordinate systems. The origin and orientation of these systems is usually provided by the manufacturers with the machine's documentation. Typically, the Z axis is the tool axis. It is important for the NC programmer to ensure that the orientation of the coordinate system where the toolpath is described geometrically is identical to that of the machine tool that will read and execute the corresponding NC program. If the toolpath is generated from a CAD database, the MCS, or a chosen WCS, must have the same orientation as the machine tool.

22.6.2 Mathematics of ToolPaths

In order for the cutting tool to produce the required finished part, its toolpath must guide it properly. This is especially important when different machinable surfaces meet or intersect. Transitional points must be calculated properly. Figure 22.6 shows some examples. If programming is done manually, the programmer must calculate the location $(\Delta x, \Delta y)$ of the center of the cutter at the intersecting edge of the two surfaces A and B.

The calculation of Δx and Δy for various cases requires various mathematical backgrounds, including trigonometry, angular relationships, analytic geometry, and cutter geometry. Actually, many common cases for Δx, Δy have been calculated and tabulated for use by NC programmers.

If the toolpath is generated from a CAD database, all of these calculations have been incorporated into the NC software and are performed without user intervention.

22.6.3 Machining Forces

It is the responsibility of the part programmers to choose the proper feeds and speeds for the various machining processes. As the dimensional accuracy of the workpiece relates to the programming accuracy, the efficiency of the machining process relates to the feeds and speeds. The feeds and speeds control the surface finish of the resulting part. They also control the machining forces that affect the cutting tool during machining. Higher feeds and speeds increase the surface roughness of the part. Excessive machining forces due to higher feeds and speeds may result in damage to the tool bit or broken cutting tools.

There are two common types of materials for cutting tools: high speed steel (HSS) and carbide. Carbide tool bits usually stand higher machining forces than HSS tool bits; thus they can stand higher feeds and speeds. However, they are more expensive. There are formulas and tables that help the programmer to calculate and/or choose the cutting speed of the cutting tool, the rate of metal removal, and the horsepower for each machining process (drilling, milling, turning, etc.)

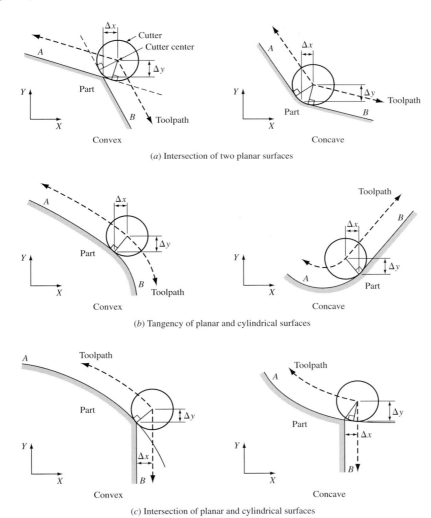

(a) Intersection of two planar surfaces

(b) Tangency of planar and cylindrical surfaces

(c) Intersection of planar and cylindrical surfaces

Figure 22.6   Toolpath calculations.

22.6.4   Cutter Programming

There are two methods for generating and programming a toolpath. The first method (less popular) is called the cutter center programming or "programming the tool." It considers the actual diameter of the cutter during the toolpath generation. In this method, the program guides the cutter around the part contour. The cutter will have to follow the path at a set distance away from the part, at each point on the path, corresponding to the cutter radius. This method is error prone, and the resulting program is valid for only one cutter diameter.

The second method (more popular) is used with NC machine tools that have cutter compensation features. This method is called zero-radius programming or "programming the part." In this method, the program still guides the zero-radius cutter around the part contour. By specifying a tool offset equal to the cutter diameter, the machine tool compensates for the diameter while executing the program by using the proper tool offset (cutter radius in this case). This method of programming is easier than cutter center programming. It is also more flexible; it allows the use of cutters of different diameters.

22.6.5 Tool Offsets

Tool offsets are useful features provided by NC machines to make part programs more flexible and easier to create. In Section 22.6.4, we discussed the cutter diameter compensation. Tool length compensation is also possible, and it allows a program to be written to perform milling, drilling, tapping, or boring without presetting the tool to a specific length.

Other useful tool offsets are those used for the flexible positioning of holding fixtures or parts, and multiple part machining. Generally, in addition to the machine and program origins we discussed previously, we have the fixture origin and the part origin. If a fixture holding the part is located in a certain position on the machine table, the part must be located properly relative to the fixture origin.

The part origin (identical to the program origin) is therefore offset from the machine origin. Tool offset can be used to compensate for this offset during program execution. Similarly, if more than one part (similar or different) is set up on a machine table, they can be machined one after the other without stopping or interrupting the machine tool by using tool offsets to compensate for the various origins of the various parts.

Figure 22.7 shows how four holes can be drilled in four workpieces in one machine setup. After drilling the hole in workpiece WP_1, we offset the tool a distance of T_y in the Y direction to line up the drill bit with the center of the hole of WP_2 to drill the hole. Similarly, we use two more tool offsets T_x and $-T_y$ to drill the holes in WP_3 and WP_4, respectively. The three tool offsets are programmed into the NC program that drills the four holes. After the machinist sets up the four workpieces (WP_1 to WP_4) on the machine table at once, the NC program runs once and drills the four holes.

22.6.6 Programming Steps

A part program uses some or all of the following principles:

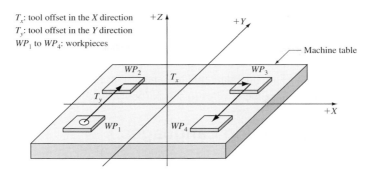

T_x: tool offset in the X direction
T_y: tool offset in the Y direction
WP_1 to WP_4: workpieces

Figure 22.7   Tool offsets.

1. **Absolute programming.** In this mode of programming, an origin for each axis of motion of the machine tool has to be selected before starting the program. The intersection of the X, Y, and Z axes of motion defines the machine origin, which must be coincident with the part program origin. Once the origin is selected, all locations are defined with respect to this origin, and all motions have to be started with respect to this origin.

2. **Incremental programming.** Unlike absolute programming, it is not necessary to select an origin in incremental programming. All motions are specified from the immediate last position of the tool. Thus, motions are described in increments from a given position. The immediate advantage of incremental programming is that the programmed motion directly matches the actual motion of the tool because both occur from the last position of the cutter.

3. **Rapid positioning.** In PTP machining, rapid positioning of the programmed machine slides to a required location is desired to increase the machine's productivity. Such motion is useful in drilling and boring when the tool is approaching the part or moving between holes. The tool or the machine table always travels at the highest machine speed (e.g., 10 in/second). It should be mentioned here that regardless of the actual element of the machine that provides the motion, it is always the tool that moves relative to the part in any NC program.

4. **Linear interpolation.** This feature is programmed using a certain programming code. It produces straight-line motions by allowing the corresponding machine slides to move at different speeds independent of each other.

5. **Circular interpolation.** To achieve this motion, the relative positions and velocities of two machine slides are initiated and maintained on a constantly changing basis, but starting and stopping at the same time. A special code is used in the NC program to indicate circular interpolation. Both linear and circular interpolations are used to guide the tool along linear and circular paths, respectively.

6. **Canned cycles.** These can be defined as standard subroutines or procedures stored in the "library" of the machine tool. Any NC program can call and use any of these subroutines.

Examples include subroutines for drilling, tapping, boring, turning, and threading. The word "canned" indicates that the subroutine is prewritten and stored, while the word "cycle" indicates that the subroutine is used over and over in a relative (cyclic) fashion. There are two types of canned cycles; fixed and variable canned cycles. The former implies that subroutines are not adaptable to a specific user's needs, while the latter allow the programmer to write canned cycles to suit particular applications.

22.7 NC Programming Languages

A part programming language allows a part programmer to take the geometry of a part from an engineering drawing (in manual programming) or from a CAD database (when using a NC package) and describe the path to be taken by a cutter to produce the part. Most of these languages have English-like syntax and statements.

The English-like statements are composed of a combination of vocabulary words, symbolic names, and data (normally obtained from a drawing or a CAD database), these being separated by punctuation. These statements are designed to require the minimal amount of data input from the programmer, leaving the computer to generate the vast amount of tool offsets and other required information.

There is a handful of part programming languages. Most of them handle PTP and continuous-path machining, particularly milling. In many languages, turning is treated as a sub-set of milling with few deletions or additions. Some languages are developed specifically for turning operations, such as CINTURN II.

Some of the popular and powerful part programming languages are APT (Automatically Programmed Tools) and APT-like (ADAPT, EXAPT, UNIAPT, MINIAPT. etc.), COMPACT II, SPLIT, PROMPT, and CINTURN II. Most of these languages were developed first for mainframe computers. They were later modified and adapted to run on minicomputers and microcomputers.

The statements of a part programming language can be divided into eight groups:

1. **Language features.** These features are similar to those found in programming languages such as C. Part programmers can define variables and arrays and geometric entities in a part program.
2. **Geometric statements.** These statements follow the same description of curves and surfaces covered in this book. In a CAD environment, geometric information required by these statements is generated interactively when the user selects the surface to be machined when using NC software. Geometric statements must precede the motion statements in a part program.
3. **Tool statements.** Tool shape (geometry), tool-axis orientation, and tool-to-part tolerance can be defined via these statements. Tool shapes are usually stored in a tool library maintained by NC software. A user can choose a specific tool and enter specific values for the tool variables, or the user can define a new tool and add it to the library.

4. **Motion statements.** These statements guide the tool in its motion. They provide information for type of machining (PTP or continuous path), direction of cutting, speed and feedrate, and so forth.
5. **Arithmetic statements.** As in C, addition, subtraction, and so forth, as well as functions (square root, sine, etc.) are available.
6. **Repetitive programming.** Statements that provide looping, branching, coordinate copying, and coordinate transformations are provided. Moreover, macro facilities enable programmers to deal with repetitive programming more effectively.
7. **Output facilities.** A part program forms what is called the CL-data, CL for cutter location. The CL-data is stored in a file called the CL-file. A computer printout of this file is known as the CL-printout. The CL-file is usually stored in ASCII format. In an attempt to reduce the file size and speed its postprocessing, the part program can be stored in a binary format. In this case, the file is known as BCL-file, with the "B" for binary. Other output facilities include listing the part program with syntax error diagnostics.
8. **Postprocessor statements.** The CL-file (or BCL-file) is written in the programming language syntax. This syntax is a high-level form which the machine tool controller is incapable of reading and interpreting. Therefore, the CL-file must be postprocessed by a postprocessor. Postprocessors are hardware dependent. A postprocessor is written for a specific controller and machine combination. Many controllers accept G-codes and M-codes, which describe types of movements of the machine tool, as well as F-codes, which describe feedrate.

22.8 G-Code and M-Code

Part programming can be done manually or by using computers to automate the part programming process. Following are the common methods that are used to create an NC part program:

1. **Manual part programming.** The NC programmer writes the code manually and stores it in a file, similar to C, C++, and Java programming.
2. **Computer-assisted part programming.** The NC programmer uses a standalone NC software such as MasterCAM to generate NC programs.
3. **Part programming using CAD/CAM systems.** The NC programmer uses the NC package of a CAD/CAM system to generate NC programs. This approach is similar to approach 2. However, the benefit here is that the programmer uses the CAD database of the part directly; no need to translate it and import it into a CAM program through IGES or STEP.
4. **Manual data input.** The NC programmer uses the controller of a machine tool to input the NC data directly. The programmer can use the controller to verify the toolpath after data entry to ensure its correctness before using it in actual machining. The controller also allows the programmer to save the data and its program.

Irrespective of the method that is used to create the NC part program, the instruction statements in the program must be in a basic code, that is, a low-level machine language understood by the machine controller (MCU). A conversion process (postprocessing) may be necessary when high-level languages are used to write the NC part program.

The instructional statements of an NC program written in basic code consists of code words followed by data. The general syntax of a statement takes this form:

```
codeWord data
```

Existing code words are:

- **N-word (sequence number).** Specifies line numbers in the NC program.
- **G-word (preparatory word).** Defines what to do on the geometry that follows.
- **x, y, z (coordinates).** Define points for the related G-words.
- **F-word (feedrate).** Specifies feedrate for cutting tool.
- **S-word (spindle speed).** Specifies the speed of the spindle of the machine tool.
- **T-word (tool selection).** Specifies a tool from the tool library of the machine tool.
- **M-word (miscellaneous command).** Specifies instructions related to the machine tool.
- **EOB symbol (end of block).** Specifies the end of an NC program.

G-words and M-words are important in any NC program. G-words prepare the MCU for the instructions and data contained in the statement. More than one G-word may be needed to prepare the MCU for the movement. The most commonly used G-words are listed in Table 22.1.

M-words are used to specify miscellaneous or auxiliary functions that are available on the machine tool. Examples of such functions are spindle rotation, stopping the spindle for a tool change, or turning the cutting fluid on or off. The most commonly used M-words are listed Table 22.2.

Table 22.1   G-code

G-word	Description	G-word	Description
G00	Rapid point-to-point movement	G20	Input data in inches
G01	Linear motion between points	G21	Input data in millimeters
G02	Clockwise circular motion	G28	Go to reference point
G03	Counterclockwise circular motion	G90	Absolute coordinates
G04	Tool dwell	G91	Incremental coordinates
G10	Tool offset	G94	Feed/minute in milling and drilling
G17	Selection of XY plane	G95	Feed/revolution in milling and drilling
G18	Selection of XZ plane	G98	Feed/minute in turning
G19	Selection of YZ plane	G99	Feed/revolution in turning

Table 22.2   M-code

M-word	Description	M-word	Description
M02	End of program and machine stop	M09	Turn off cutting fluid
M03	Clockwise spindle start (CSS)	M10	Automatic clamping of fixtures
M04	Counterclockwise spindle start (CCSS)	M11	Automatic unclamping of fixtures
M05	Spindle stop	M13	CSS and turn on cutting fluid
M06	Tool change	M14	CCSS and turn on cutting fluid
M07	Turn cutting fluid on; flood mode	M17	Turn off spindle and cutting fluid
M08	Turn cutting fluid on; mist mode	M19	Turn off spindle at oriented position

EXAMPLE 22.1   **Programming in G-code and M-code.**

Figure 22.8 shows the NU logo and the toolpath required to engrave it in a rectangular plate. We use a milling operation to cut the plate. The plate thickness is 0.500 in. The letter U is lowered relative to the letter N for an artistic look.

Write an NC program to mill the logo. Use a feed rate of 10 in/min.

SOLUTION   We write the NC program in G-code and M-code. Figure 22.8 shows the two letters. The toolpath to mill each letter is shown by the white centerline; the arrows indicate the motion direction of the cutting tool (mill).

Figure 22.8 shows the geometry required to cut each letter. The coordinates of the top left corner of each letter are shown, along with the width and height of each letter.

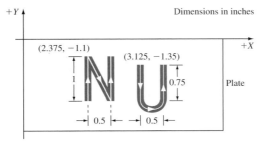

Figure 22.8   Geometry and toolpath of the NU logo.

Based on the geometry and the toolpath shown in Figure 22.8, the NC program is listed below, annotated in C comment style.

```
//cut the letter N
N01 G00 X2.3750 Y-2.1000 Z0.2000 M13//go to bottom left corner of N
N02 G01 Z-0.0625 FI0.0000         //cut the letter depth
N02 G01 Y-1.1000                  //cut the left leg of N
N03 G01 X2.8750 Y-2. 1000         //cut the diagonal side of N
```

```
N04 G01 Y -1.1000                    //cut the right leg of N
N05 G00 Z0.2000                      //clear the workpiece
//cut the letter U
N06 G00 X3.1250 Y-1.3500 Z0.2000       //go to top left corner of U
N07 G01 Z-0.0625 F10.0000              //cut the letter depth
N08 G01 Y-2.1000                       //cut the left leg of U
N09 G03 X3.6250 Y-2.1000 I3.3750 J-2.0500 //cut the bottom arc of U
N10 G01 Y-1.3500                       //cut the right leg of U
N11 G00 Z0.2000                        //clear the workpiece
N12 M17                          //turn off spindle and cutting fluid
```

Example 22.1 discussion:

The statements of the NC program are numbered sequentially using the N-word for ease of reference. The program has two main sections, one to cut each letter. The geometric data used in the G-code can easily be traced back to Figure 22.8. Observe the positive and negative directions of the X, Y, and Z axes.

Statement N01 instructs the cutting tool to go to the bottom left corner of the letter N, but at $z = 0.2000$ in, that is, above the top plane of the workpiece. We must do this to avoid damaging the workpiece because G00 is a rapid movement of the tool from its previous location. Statement N01 also turns on the spindle rotation and the cutting fluid in preparation for for drilling.

Statement N02 cuts a depth of 0.0625 in from the workpiece at a feed rate of 10.0 in/min. The tool does not move in the horizontal plane during this cut; it simply spins in location after the cut.

Statements N03 and N04 cut the letter N. During any cut, the previous x, y, or z coordinates are used unless they are specified explicitly in the G-code of the new statement.

Statement N05 moves the cutting tool up to the location $z = 0.2000$ to clear the workpiece because the tool must move rapidly, according to statement N06, to get ready to cut the letter U.

Statements N07 to N11 are similar to those we that we just explained for the letter N. They engrave the letter U in the workpiece.

Statement N12 stops the machine spindle and shuts off the cutting fluid as the drilling operation is completed.

Statements N01 and N12 show that we can write M-code next to G-code on the same line or on separate lines.

Example 22.1 hands-on exercise:

1. Reverse the directions of the toolpaths for N and U that are shown in Figure 22.8. Write the NC program, using the G-code and M-code, based on the new toolpath.
2. Repeat the example for your favorite logo.

22.9 APT Programming

In this section, we review the APT syntax and show examples of how to develop a part program to illustrate the many concepts previously discussed. In this review, we present the minimum set of syntax. Not all the modifiers that accompany many of the statements are shown.

22.9.1 Geometric Statements

A geometric statement in APT takes the following form:

```
variable = geometric entity/geometric data to define the entity
```

`variable` is a user-chosen variable. `geometric entity` is an APT reserved word.

Sample geometric statements are:

```
P1 = POINT/3.0,  2.0,  -1.0
P2 = POINT/6.0,  5.0,  3.0
L1 = LINE/P1,  P2
C1 = CIRCLE/CENTER,  P1,  RADIUS,  1.5
```

The first two statements define two points `P1` and `P2` with coordinates (3, 2, −1) and (6, 5, 3), respectively. The third statement creates a line `L1` between `P1` and `P2`, and the last statement creates a circle with `P1` as a center and a radius of 1.5. Other modifiers exist to define points (e.g., intersection modifier) and lines (e.g., parallel, perpendicular, tangent, etc.).

22.9.2 Motion Statements

A motion statement in APT takes the form:

```
motion statement/motion data
```

Three of the useful motion statements are

```
FROM/P1
GO TO/3.0,  4.0,  -2.0
GODLTA/2.0,  -1.0,  0.0
```

The first statement instructs the tool to move from its current position to point `P1`. The second statement is an absolute programming statement where the tool must move to the point defined by $x = 3$, $y = 4.0$, and $z = −2.0$ relative to the machine tool origin. The third statement represents incremental programming, where the tool is to move from its current position to a new position defined by $\Delta x = 2.0$, $\Delta y = −1.0$, and $\Delta z = 0$.

These three statements are useful in PTP machining in particular. A tool needs more guidance, and therefore instructions, for continuous-path machining. A motion statement for continuous-path machining requires three surfaces to guide the tool motion: part, drive, and check surfaces as shown in Figure 22.9.

The part surface is defined as the surface that controls the tool motion along the tool axis. It controls the depth of the machining operation.

The drive surface controls the tool motion perpendicular to the toolpath, that is, the direction in which the tool is moving. Alternatively, the drive surface is always tangent to the toolpath or the direction of the tool motion.

The check surface terminates the tool motion. The part surface shown in Figure 22.9 usually intersects both the drive and check surfaces. However, the drive and check surfaces may or may not intersect.

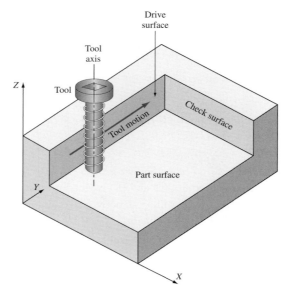

Figure 22.9 `Tool guiding surfaces for APT continuous-path programming.`

APT provides a GO statement using these three surfaces with the following syntax.

      `GO/TO, drive surface, TO, part surface, TO, check surface`

The `TO` in the preceding statement is a modifier. Other modifiers are `ON` and `PAST`. These modifiers control the relative position of the cutter with respect to the surface following the modifier. Figure 22.10 shows these meanings graphically.

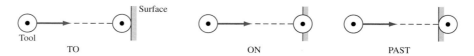

Figure 22.10 `TO, ON,` and `PAST` **APT modifiers.**

A tool statement describes the tool to be used in machining. It takes the form:

      `CUTTER/d`

where `d` is the cutter diameter. Tool description is required so that the APT program can use the tool offset (`d`/2) to compute the coordinates of the toolpath.

22.9.3 Auxiliary Statements

APT has auxiliary statements to control the operations of machine tools during machining. The general syntax of an auxiliary statement is:

auxiliary statement/description data

Sample statements are MCHIN, PARTNO, COOLNT, RAPID, RETRACT, STOP, and FINI. The use of some of these statements is shown in the following examples.

EXAMPLE 22.2   **Write a drilling program in APT.**

Figure 22.11a shows a part with thickness of 1.000 in. Two holes in the positions shown are to be drilled in the part. Write an APT program to drill the two holes. Use the tool home position shown, a drilling speed of 500 rpm and a feedrate of 3.55 in/min. The machine that performs the drilling is number 5 and its controller is coded as *DRILL*.

SOLUTION   This example illustrates the use APT in PTP machining. Utilizing the home position shown, the toolpath is shown in Figure 22.11b.

In this example, the machine coordinate system (X_m, Y_m, Z_m) is coincident with the part coordinate system (X_p, Y_p, Z_p). In addition, the machine origin, the part origin, and the program origin are identical and located at the left bottom corner of the part.

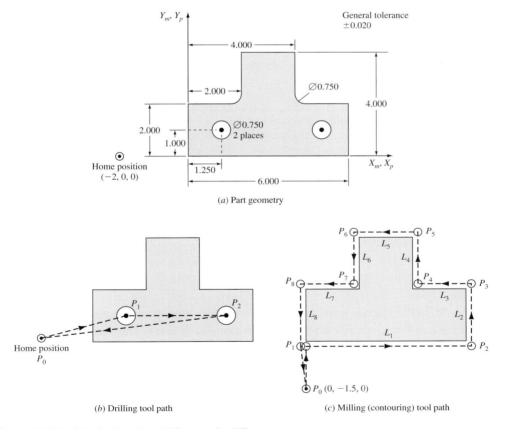

(a) Part geometry

(b) Drilling tool path

(c) Milling (contouring) tool path

Figure 22.11   Toolpaths for drilling and milling a part.

Using this information, the APT program may look as follows:

```
PARTNO  PTP-EXAMPLE                    FROM/P0
MACHIN/DRILL,5                         GO TO/P1
CLPRNT                                 GODLTA/0,0, -1.5
CUTTER/0.750                           GODLTA/0,0,  1.5
P0 = POINT/-2.0,  0.0,  0.0            GO TO/P2
P1 = POINT/1.25,  1.0,  0.0            GODLTA/0,0, -1.5
P2 = POINT/4.75,  1.0,  0.0            GODLTA/0,0,  1.5
SPINDL/500                             GO TO/P0
FEDRAT/3.55                            COOLNT/OFF
COOLNT/ON                              FINI
```

Example 22.2 discussion:

The NC APT program starts the cutting tool (drill) from the home position, drills the hole whose center is at P_1 first, then drills the second hole before going back home. This program is the CL-file to drill the four holes. It must be postprocessed to generate the G-code and M-code before downloading into the controller of the machine tool.

Example 22.2 hands-on exercise:

Reverse the direction of the toolpath that is shown in Figure 22.11b. Write the APT program based on the new toolpath.

EXAMPLE 22.3   **Write a milling program in APT.**

Write an APT program to mill the contour of the part shown in Figure 22.11a. Use an end-mill cutter with diameter of 0.5 in. Assuming the part is made of steel, use a cutting speed of 580 rpm and feedrate of 2.30 in/min. The machine that performs the milling is coded as *MILL5*.

SOLUTION   This is an example of continuous-path machining. Figure 22.11c shows the toolpath, assuming a home position at $P_0(0, -1, 0)$.

Example 22.3 discussion:

To properly guide the tool on its path in continuous-path machining, one could imagine guiding the tool with a remote control around the part profile, thus making the proper left and right turns.

The NC APT program starts the cutting tool (mill) from the home position and mills the part sides in a counterclockwise direction before going back home. This program is the CL-file to mill the part sides. It must be postprocessed to generate the G-code and M-code before downloading into the controller of the machine tool.

Example 22.3 hands-on exercise:

Reverse the directions of the toolpath that is shown in Figure 22.11b. Write the APT program based on the new toolpath.

The APT program is as follows:

```
PARTNO  MILLING-EXAMPLE
MACHINE/MILL5
CLPRNT
INTOL/0.002
OUTOL/0.002
CUTTER/0.750
P0 = POINT/0, -1.5, 0
P1 = POINT/0, 0, 0
P2 = POINT/6.0, 0, 0
P3 = POINT/6.0,2.0,0
P4 = POINT/4.0,2.0,0
P5 = POINT/4.0,4.0,0
P6 = POINT/2.0,4.0,0
P7 = POINT/2.0, 2.0,0
P8 = POINT/0, 2.0,0
L1 = LINE/P1,P2
L2 = LINE/P2,P3
L3 = LINE/P3,P4
L4 = LINE/P4,P5
L5 = LINE/P5,P6
L6 = LINE/P6,P7

L7 = LINE/P7,P8
L8 = LINE/P8,P1
PL1 = PLANE/0.0, -1.5,
6.0, 0,-1.5, 0, 2.0, -1.5
SPINDL/580
FEDRAT/2.30
COOLNT/ON
FROM/P0
GO/TO, L1, TO, PL1, TO, L8
GORGT/L1, PAST, L2
GOLFT/L2, PAST, L3
GOLFT/L3, TO, L4
GORGT/L4, PAST, L5
GOLFT/L5, PAST, L6
GOLFT/L6, TO, L7
GOLRGT/L7, PAST, L8
GOLFT/L8, PAST, L1
RAPID
GO TO/P0
COOLNT/OFF
FINI
```

22.10 Toolpath Generation

Four steps can be identified for successful NC machining of parts utilizing CAD/CAM databases. These are recognition of machined surfaces (feature recognition), toolpath generation, toolpath verification, and collision detection. Automatic feature recognition for manufacturing is a crucial step in integrating CAD and CAM. We have touched on it in process planning.

Collision detection is concerned with finding out if the cutting tool and its assembly (tool, shank, etc.) may collide with other components of the manufacturing environment such as the workpiece itself, jigs, fixtures, spindles, and so forth. CAD/CAM systems can perform collision checking and detection as covered in Chapter 16. If the intersection between components is not null, collision occurs and the components have to be rearranged.

This section is devoted primarily to generating toolpaths. Readers interested in more details on feature recognition and collision detection and avoidance should consult the literature.

In cutting complex surfaces, one is faced with the problems of cutter offset, accuracy, and cutter interference with the workpiece (tool gouging). Customarily, ball-end (or ball-nosed) cutters are used in these circumstances, and the calculation of cutter offsets is achieved by finding the directions of normals on the surface. Having determined the directions of normals, the tool is then offset by the radius of the cutter along the normal vectors over the surface.

Before calculating cutter offsets, we must generate the toolpath. To mill a surface parametrically, cutter location points are generated using the surface definition and machining parameters. These points are stored in the CL-file. The ball-end cutter tip is directed to move in straight-line segments to each point. The fewer the points, the more rough the resulting machined surface. However, large point files should be avoided to minimize storage, computation, and milling time.

For example, if we choose a step size of 0.1 in both the u and v directions of a parametric surface, and assuming both u and v have a range from 0 to 1, 11 points are created along each of the u and v directions. Thus, 121 points are needed to mill the surface. The problem with this step definition is that it does not take into account areas of local flattening, where fewer points are required. In addition, it does not take tolerance specification on the surface into account.

Out of the many existing algorithms to generate toolpaths for surfaces, we present the following algorithm. The algorithm applies to 3-axis machining. The algorithm subdivides the surface into parametric space curves (in the u or v direction). Each curve is further approximated by a sequence of line segments. The toolpaths for these segments are calculated for ball-end cutters.

The algorithm attempts to minimize the number of points evaluated on the surface to generate the toolpath and yet stay within the specified tolerance. The key issues in the algorithm are how to approximate the parametric space curves and how to determine the toolpaths.

Figure 22.12a shows a parametric surface. We assume the machining direction of the surface to be the u direction. Thus, the tool moves along curves of $v =$ constant on the surface, that is, $\mathbf{P}(u, c)$.

Figure 22.12b shows the geometry required to divide one of these curves into line segments. Assume that a point P_i on the curve is given. The unit normal and unit tangent vectors at this point are $\hat{\mathbf{n}}$ and $\mathbf{t}$, respectively. The point and the unit vectors are obtained from the surface representation. The tolerance e is known from the surface machining data. The step length s of a line segment that a tool traverses during machining must be consistent with the tolerance requirement.

For a small enough step, it is reasonable to approximate the curve around P_i by its osculating circle (circle with the greatest contact with the curve at P_i). Its center O is on the normal to the curve, and its radius r is equal to the radius of curvature at P_i.

If the tool moves the step length s, point Q on the osculating circle results, which may not exactly coincide with point P_{i+1} on the curve $\mathbf{P}(u, c)$. However, for a small step size s, point Q is very close to point P_{i+1}. In this algorithm, they are both assumed to be the same.

The step length s can be written in terms of the specified tolerance, e, by considering the triangle P_iOH, that is,

$$r^2 = \left(\frac{s}{2}\right)^2 + (r - e)^2 \tag{22.1}$$

or

$$s^2 = 4e(2r - e) \tag{22.2}$$

The unit vector $\hat{\mathbf{m}}$ in the direction of the step length s is given by:

$$\frac{s}{2}\hat{\mathbf{m}} = h_1\hat{\mathbf{t}} - \frac{h_2}{h_2 + r - e}(r\hat{\mathbf{n}} + h_1\hat{\mathbf{t}}) \tag{22.3}$$

or

$$s\hat{\mathbf{m}} = \frac{2h_1(r-e)}{h_2 + r - e}\hat{\mathbf{t}} - \frac{2h_2 r}{h_2 + r - e}\hat{\mathbf{n}} \tag{22.4}$$

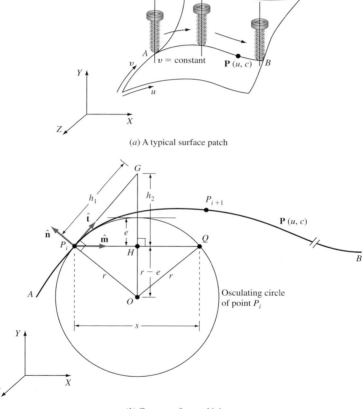

(a) A typical surface patch

(b) Geometry for machining

Figure 22.12   Toolpath geometry.

Using the right triangle $P_i OG$ and the perpendicular line $P_i H$, we can write:

$$h_1 = \frac{rs}{2(r-e)} \tag{22.5}$$

and

$$h_2 = \frac{s^2}{4(r-e)} \tag{22.6}$$

Substituting Eqs. (22.5) and (22.6) into Eq. (22.4), and using Eq. (22.2) to reduce the result, we obtain:

$$\hat{\mathbf{m}} = \frac{r-e}{r}\hat{\mathbf{t}} - \frac{s}{2r}\hat{\mathbf{n}} \tag{22.7}$$

The position vector $\mathbf{Q}$ of point Q on the osculating circle is given by:

$$\mathbf{Q} = \mathbf{P}_i + s\hat{\mathbf{m}} \tag{22.8}$$

or

$$\mathbf{Q} = \mathbf{P}_i + \frac{s(r-e)}{r}\hat{\mathbf{t}} - \frac{s^2}{2r}\hat{\mathbf{n}} \tag{22.9}$$

The parametric value u_{i+1} which corresponds to point P_{i+1} must now be evaluated so that the unit normal and tangent vectors as well as the radius of curvature r at point P_{i+1} can be evaluated using the surface equation.

Figure 22.13 shows the points needed for the calculation. We use the second divided difference method to approximate u_{i+1}. We choose an intermediate point D between point P_i and P_{i+1} (e.g., choose the midpoint) and calculate its corresponding parameter u_D. Let us assume that point C, and its u_C, is the previous intermediate point used to calculated u_i at point P_i.

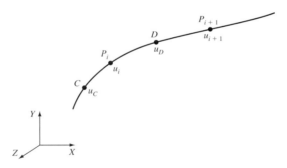

Figure 22.13   Points needed to calculate u_{i+1} at point P_{i+1}.

Using the four points C, P_i, D, and P_{i+1} and their corresponding parameter values, first calculate the distance between the two points C and D in each of the coordinate directions, and find the maximum absolute value and its direction. Then select the corresponding value in that direction (could be x, y, or z) as variable w. Using the second divided difference method, we can write:

$$u_{i+1} = u_C + (w_{i+1} - w_C)[K_1 + (w_{i+1} - w_i)K_3] \tag{22.10}$$

where

$$K_1 = \frac{u_i - u_C}{w_i - w_C} \tag{22.11}$$

$$K_2 = \frac{u_D - u_i}{w_D - w_i} \tag{22.12}$$

and

$$K_3 = \frac{K_2 - K_1}{w_D - w_C} \tag{22.13}$$

Having u_{i+1}, we can now correct the point P_{i+1}, which was assumed to be the same as point Q, by calculating its coordinates from the surface equations using u_{i+1}. In addition, the unit tangent vector $\hat{t}$, the unit normal vector $\hat{n}$, and the radius of curvature r can be evaluated at P_{i+1} to compute the next point P_{i+2} and its parameter value u_{i+2}.

To machine a part contour, the points as calculated previously must be offset by the radius of the cutting tool to produce the correct toolpath. Assuming a ball-end cutter with diameter d, the offset geometry is shown in Figure 22.14.

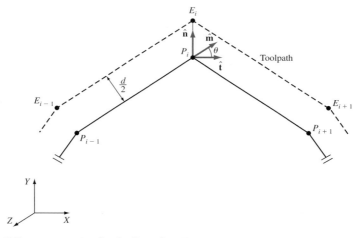

Figure 22.14   Offset geometry for ball-end cutter.

To calculate the end points E_i of the toolpath line segments, we can write the following equations using Figure 22.14:

$$\cos\theta = \hat{t} \cdot \hat{m} \tag{22.14}$$

$$\frac{d}{2} = \left|E_i - P_i\right| \cos\theta \tag{22.15}$$

$$P_i = E_i - \left|E_i - P_i\right|\hat{n} \tag{22.16}$$

Reducing Eq. (22.16) using Eqs. (22.14) and (22.15) gives:

$$E_i = P_i + \frac{d}{2t \cdot \hat{m}} \hat{n} \qquad (22.17)$$

Equation (22.17) gives the cutter offset points E_i, which determine the line segments (toolpath) of the cutter center path. Figure 22.15 shows sample toolpaths that may be obtained using Eq. (22.17). The toolpath takes the same form as the curve. The toolpath may not be a simple curve for relatively large cutter radii and complex curves (Figure 22.15b).

A toolpath with loops and cusps results when the radius of curvature at a point on the curve is less than the cutter radius and the normal vector points away from the center of curvature. In practice, this problem can be rectified by selecting cutters with proper radii.

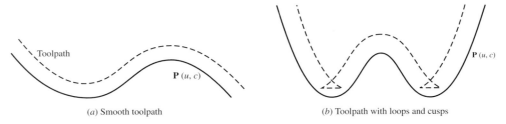

(a) Smooth toolpath   (b) Toolpath with loops and cusps

Figure 22.15 Sample toolpaths.

The toolpath generation algorithm described previously can be summarized as follows:

1. Select the machining direction of the surface. Assuming the u direction is selected, set $v = 0$. The surface representation is assumed to include the tolerance on the surface.
2. Set $u_C = 0$, and evaluate the position, unit tangent, and unit normal vectors and the radius of curvature at $u_A = 0$.
3. Evaluate an extra point D at a parameter value u_D.
4. Select the variable w.
5. Update the parameter value u using Eq. (22.10).
6. If $u = u_{max}$, then let $u = u_{max}$, otherwise go to step 7.
7. Evaluate the position, unit tangent, and unit normal vectors and the radius of curvature at the new u value.
8. Calculate the cutter path point at u using Eq. (22.17).
9. If $u < u_{max}$, set $u_C = u_D$ and go to step 3, otherwise go to step 10.
10. If $v < v_{max}$, set $v = v + D_v$ and go to step 2, otherwise go to step 11.
11. Stop.

Further enhancements and extensions to the preceding algorithm are possible. A collision detection mechanism between the tool, the workpiece, and the machine can be developed. The algorithm can also be extended to calculate multiple passes where a number of cutting passes may be required to machine an object. In addition, the algorithm can be extended to calculate paths using other types of milling cutters.

The preceding algorithm is based on the fact that the surface patch to be machined occupies the range described by $0 < u < u_{max}$ and $0 < v < v_{max}$. If the patch range is not same as this range, the algorithm must be modified significantly. This case is shown in Figure 22.16. In practice, we refer to this problem as machining trimmed or untrimmed surfaces. The use of the same algorithm would require the programmer to subdivide the trimmed surface into a few untrimmed ones.

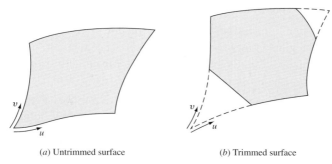

(a) Untrimmed surface   (b) Trimmed surface

Figure 22.16   Types of surfaces for machining.

22.11 Toolpath Verification

NC part programs and the toolpaths that are supposed to guide tools during actual machining usually include a lot of coordinate values that are impossible to verify manually. Graphical and visual display of the toolpaths are beneficial in the verification process. NC verification software can simulate the actual machining process by displaying the cutting tool moving, following its toolpath, relative to the stock and jigs and fixtures. Shaded images of the tool and stock are used to enhance the visualization process. This enables NC programmers to spot any potential errors in NC programs.

The animated toolpath is generated by displaying the tool position and orientation, using the NC program data, at various points on the toolpath, creating frames, and storing the frames for playback.

The advantages of toolpath verification are numerous. We can check whether:

1. The cutter removes the necessary material from the stock.
2. The cutter hits any clamps or fixtures on approach.
3. The cutter passes through the floor or side of a pocket, or through a rib.
4. The toolpaths are as efficient as they could be.

Other advantages include rapid turnaround of program development, rapid training for potential NC programmers without danger, and freeing machine tools to only cut real parts (no testing of programs or training individuals). In addition, toolpath verification by software reduces the wear on machine tools. When polyurethane foam is cut for verification purposes, it rises in the air and then falls into the machine gears, where it acts as a grinding paste.

Even with toolpath verification, some machining problems go undetected. For example, problems such as tool chatter and stock warpage (due to heat stresses during machining) are simply beyond the scope of verification. However, tool verification detects the majority of errors in general.

22.12 CAD/CAM NC Programming

CAD/CAM systems have NC packages (modules) which can be used for NC programming in an integrated CAD/CAM environment. The common functions of such an NC package are features recognition, toolpath generation, toolpath verification, collision detection, APT program generation, and postprocessing.

The NC package provides programmers with tool libraries that they can use to define various tools with the proper tool geometric parameters. Once a tool is defined, a toolpath can be generated using the defined tool and the part geometry. When a toolpath is generated, it can be verified on the graphics display.

Toolpath verification involves the display of an animation sequence that shows the tool moving along its generated path, which is superimposed on the part geometry. If mistakes are spotted, the toolpath can be modified accordingly and reverified. Once accepted, the time and cost estimates can be calculated.

In addition, the NC package can generate the APT program automatically from the toolpath information stored in the part database.

Finally, the NC package can postprocess the APT program to translate it to the low-level machine language, including the G-code and M-code, suitable for the MCU.

22.13 Tutorials

22.13.1 Machine a Part

In this tutorial we use a CAD/CAM system to machine the part shown in Figure 22.17. The part is to be machined from a 1.2 inch–thick low-carbon steel. We use the NC package of the CAD/CAM system to generate the APT program to mill the top face of the part, mill the pocket out, and drill the four holes. The speed and feed for milling are 400 rpm and 0.1 in/min, respectively, and for drilling they are 900 rpm and 3.0 in/min, respectively. The cutting tools for the three machining operations are: 1 in diameter flat-end mill to mill the top face, 0.25 in diameter flat-end mill to mill the pocket, and 0.5 in diameter carbide drill to drill the holes. The general tolerance on the part dimensions is 0.02 in.

Follow the following steps to generate the APT program and the CL-data file:

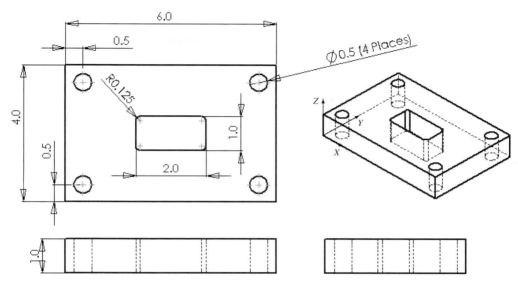

Figure 22.17 Part geometry that results from machining.

1. **Define the finished (final) part.** Create the part model as shown in Figure 22.17. Start the NC package, and open the part file. This part is the final machined part.

2. **Define the workpiece blank.** This is the stock to be machined. Use the model from step 1 to define the blank. The size of the blank is 0.2 in larger than the finished part. The following screenshot shows the blank. We also show, on the right, as an illustration, the user interface of the NC package that creates the blank.

Workpiece blank

User interface of NC package

3. **Define the operation setup.** Specify the name (of choice) of the operation, the name of the NC machine tool to be used in machining, and other setup parameters such as machine zero and tolerance requirements. The screenshot on the right shows the setup.

4. **Define the machine tool setup.** Specify the name of the machine tool (same name as in Step 3), the type of the machine tool, its number of axes, and so forth as shown in the following screenshot.

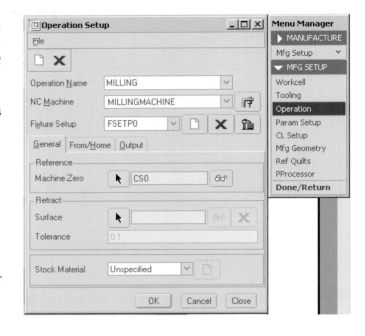

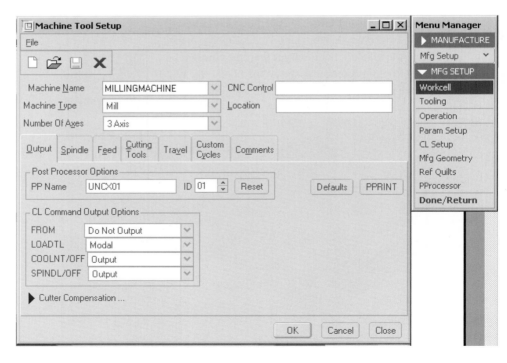

5. **Select/define the cutting tools.** We can define all the cutting tools in this step and select them in Step 6 when we select the operations, or we can skip this step and proceed to Step 6. We use the former approach here. We define three tools: two for milling and one for drilling. The following screenshots show one milling tool and the drill bit.

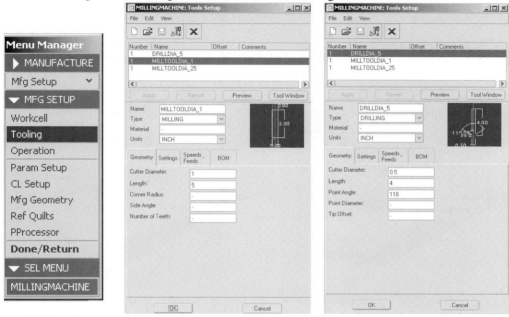

6. **Select the operations.** The three operations we select are face milling for the top face, pocket (volume) milling for the pocket, and drilling for the holes.

7. **Set the operations' parameters.** For each of the three operations of Step 6, we specify the cutting tool (defined in Step 5), the operation parameters such as spindle speed, feedrate, cutting depth, and so forth. The following screenshot shows some parameters.

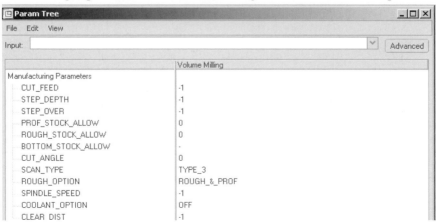

8. **View/verify the toolpath.** The NC package generates an animated toolpath that the machinist can view for verification purposes. The machinist can view the toolpath step by step at will, or jump to any tool position during the animation. The following screenshots show the three toolpaths.

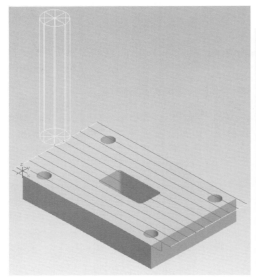

Tool path to mill the top face

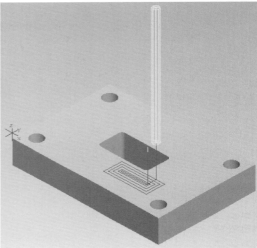

Tool path to mill the pocket

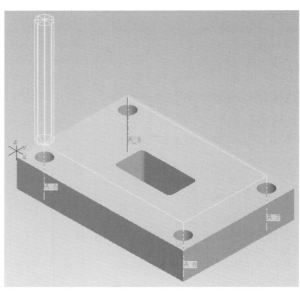

Tool path to drill the holes

9. **Generate the CL-data file.** Generate the APT program for each operation, and output it to a file. These are ASCII files. These programs are shown after step 10 annotated in C-like style.

10. **Postprocess the APT files.** Use the postprocessor of the NC package to convert the APT files to the required format (G-code and M-code) for the particular NC controller of the machine tool.

```
//toolpath to mill top face (this is not an APT statement)
$$* Pro/CLfile  Version Wildfire - 2003210
$$-> MFGNO / TUT22_1
PARTNO / TUT22_1
$$-> FEATNO / 12
MACHIN / UNCX01, 1
$$-> CUTCOM_GEOMETRY_TYPE / OUTPUT_ON_CENTER
UNITS / INCHES
LOADTL / 1
$$-> CUTTER / 1.000000
$$-> CSYS / 1.0000000000, 0.0000000000, 0.0000000000, 0.0000000000,
$
          0.0000000000, 1.0000000000, 0.0000000000, 0.0000000000,
$
            0.0000000000, 0.0000000000, 1.0000000000, 0.0000000000
SPINDL / RPM, 400.000000,   CLW
RAPID
GOTO / 6.5000000000, 0.0000000000, 0.2000000000
FEDRAT / 0.100000,   IPM
GOTO / 6.5000000000, 0.0000000000, -0.2000000000
GOTO / -0.5000000000, 0.0000000000, -0.2000000000
RAPID
GOTO / -0.5000000000, 4.0000000000, -0.2000000000
FEDRAT / 0.100000,   IPM
GOTO / 6.5000000000, 4.0000000000, -0.2000000000
RAPID
GOTO / 6.5000000000, 0.4000000000, -0.2000000000
FEDRAT / 0.100000,   IPM
GOTO / -0.5000000000, 0.4000000000, -0.2000000000
RAPID
GOTO / -0.5000000000, 3.6000000000, -0.2000000000
FEDRAT / 0.100000,   IPM
GOTO / 6.5000000000, 3.6000000000, -0.2000000000
RAPID
GOTO / 6.5000000000, 0.8000000000, -0.2000000000
FEDRAT / 0.100000,   IPM
GOTO / -0.5000000000, 0.8000000000, -0.2000000000
RAPID
GOTO / -0.5000000000, 3.2000000000, -0.2000000000
FEDRAT / 0.100000,   IPM
```

```
GOTO / 6.5000000000, 3.2000000000, -0.2000000000
RAPID
GOTO / 6.5000000000, 1.2000000000, -0.2000000000
FEDRAT / 0.100000,  IPM
GOTO / -0.5000000000, 1.2000000000, -0.2000000000
RAPID
GOTO / -0.5000000000, 2.8000000000, -0.2000000000
FEDRAT / 0.100000,  IPM
GOTO / 6.5000000000, 2.8000000000, -0.2000000000
RAPID
GOTO / 6.5000000000, 1.6000000000, -0.2000000000
FEDRAT / 0.100000,  IPM
GOTO / -0.5000000000, 1.6000000000, -0.2000000000
RAPID
GOTO / -0.5000000000, 2.4000000000, -0.2000000000
FEDRAT / 0.100000,  IPM
GOTO / 6.5000000000, 2.4000000000, -0.2000000000
RAPID
GOTO / 6.5000000000, 2.0000000000, -0.2000000000
FEDRAT / 0.100000,  IPM
GOTO / -0.5000000000, 2.0000000000, -0.2000000000
GOTO / -0.5000000000, 2.0000000000, 0.2000000000
SPINDL / OFF
$$-> END /
FINI


//toolpath to mill pocket (this is not an APT statement)
$$*          Pro/CLfile  Version Wildfire - 2003210
$$-> MFGNO / TUT22_1
PARTNO / TUT22_1
$$-> FEATNO / 22
MACHIN / UNCX01, 1
$$-> CUTCOM_GEOMETRY_TYPE / OUTPUT_ON_CENTER
UNITS / INCHES
LOADTL / 1
$$-> CUTTER / 0.250000
$$-> CSYS / 1.0000000000, 0.0000000000, 0.0000000000, 0.0000000000,
$
        0.0000000000, 1.0000000000, 0.0000000000, 0.0000000000,
$
          0.0000000000, 0.0000000000, 1.0000000000, 0.0000000000
SPINDL / RPM, 400.000000,  CLW
RAPID
GOTO / 3.5250000000, 2.0250000000, 0.2000000000
RAPID
GOTO / 3.5250000000, 2.0250000000, 0.0000000000
FEDRAT / 0.100000,  IPM
GOTO / 3.5250000000, 2.0250000000, -1.2000000000
```

```
GOTO / 2.4750000011, 2.0250000000, -1.2000000000
GOTO / 2.4750000011, 1.9749999970, -1.2000000000
GOTO / 3.5250000000, 1.9749999970, -1.2000000000
GOTO / 3.5250000000, 2.0750000030, -1.2000000000
GOTO / 2.4249999971, 2.0750000030, -1.2000000000
GOTO / 2.4249999971, 1.9250000010, -1.2000000000
GOTO / 3.5749999969, 1.9250000010, -1.2000000000
GOTO / 3.5749999969, 2.0750000030, -1.2000000000
GOTO / 3.5250000000, 2.0750000030, -1.2000000000
GOTO / 3.5250000000, 2.1750000020, -1.2000000000
GOTO / 2.3250000033, 2.1750000020, -1.2000000000
GOTO / 2.3250000033, 1.8250000020, -1.2000000000
GOTO / 3.6749999978, 1.8250000020, -1.2000000000
GOTO / 3.6749999978, 2.1750000020, -1.2000000000
GOTO / 3.5250000000, 2.1750000020, -1.2000000000
GOTO / 3.5250000000, 2.2750000010, -1.2000000000
GOTO / 2.2250000024, 2.2750000010, -1.2000000000
GOTO / 2.2250000024, 1.7250000030, -1.2000000000
GOTO / 3.7749999987, 1.7250000030, -1.2000000000
GOTO / 3.7749999987, 2.2750000010, -1.2000000000
GOTO / 3.5250000000, 2.2750000010, -1.2000000000
GOTO / 3.5250000000, 2.3750000000, -1.2000000000
GOTO / 2.1250000015, 2.3750000000, -1.2000000000
GOTO / 2.1250000015, 1.6249999970, -1.2000000000
GOTO / 3.8749999996, 1.6249999970, -1.2000000000
GOTO / 3.8749999996, 2.3750000000, -1.2000000000
GOTO / 3.5250000000, 2.3750000000, -1.2000000000
GOTO / 3.5250000000, 2.3750000000, 0.2000000000
SPINDL / OFF
$$-> END /
FINI


//toolpath to drill holes (this is not an APT statement)
$$*           Pro/CLfile  Version Wildfire - 2003210
$$-> MFGNO / TUT22_1
PARTNO / TUT22_1
$$-> FEATNO / 528
MACHIN / UNCX01, 1
$$-> CUTCOM_GEOMETRY_TYPE / OUTPUT_ON_CENTER
UNITS / INCHES
LOADTL / 1
$$-> CUTTER / 0.500000
$$-> CSYS / 1.0000000000, 0.0000000000, 0.0000000000, 0.0000000000,
$
           0.0000000000, 1.0000000000, 0.0000000000, 0.0000000000,
$
              0.0000000000, 0.0000000000, 1.0000000000, 0.0000000000
SPINDL / RPM, 900.000000,  CLW
```

```
RAPID
GOTO / 0.5000000000, 3.5000000000, 0.2000000000
CYCLE / DRILL, DEPTH, 1.550215, IPM, 3.000000, CLEAR, 0.200000,
RETURN, $
 0.200000
GOTO / 0.5000000000, 3.5000000000, 0.0000000000
GOTO / 5.5000000000, 3.5000000000, 0.0000000000
GOTO / 5.5000000000, 0.5000000000, 0.0000000000
GOTO / 0.5000000000, 0.5000000000, 0.0000000000
CYCLE / OFF
SPINDL / OFF
$$-> END /
FINI
```

Tutorial 22.13.1 discussion:

The three APT programs follow the syntax of the APT language. Use the *XYZ* coordinate system shown with each toolpath and follow each program, statement by statement. You should be able to understand the toolpath shown in the screenshot for each program.

Each program positions its cutting tool, in rapid movement, at a distance of 0.2 in above the top face of the workpiece. The milling programs use the workpiece dimensions defined in step 2—they shoot 0.2 in beyond the dimensions of the finished part. While the toolpaths for the face milling and the hole drillings are obvious, the toolpath for the pocket milling is less obvious. We start milling the pocket at its center, and cut outward until we finish.

The three operations finish the cutting in one pass. The face milling shaves off 0.2 in. The pocket milling shaves off the full depth of the workpiece — 1 in. The hole drillings create through holes.

Tutorial 22.13.1 hands-on exercise:

Change the three NC programs. Use surface milling, instead of face milling, to generate the NC program. Compare the new program with the one listed here. Is it any longer?

Change the pocket milling program so that it mills the perimeter of the pocket in one pass. This forces the part of the workpiece material that corresponds to the pocket to fall off. This is faster to do than milling all material out.

Change the drilling program to use a different sequence of drilling.

PROBLEMS

Part I: Theory

22.1 What should you do during geometric modeling to ensure the proper linking between CAD and NC?

22.2 List and discuss the guidelines to follow when entering CAD data for successful use in NC programming.

22.3 What are the two units of the MCU of a machine tool? What does each unit do?

22.4 Describe the axes of motion for $2\frac{1}{2}$-axis, 3-axis, and 5-axis machine tools.

22.5 What are the two types of machining that a machine tool can perform?

22.6 What is the difference between NC, CNC, and DNC?

22.7 Sketch and discuss the coordinate system of a machine tool.

22.8 What are the two methods of cutter programming? Which is a better method? Why?

22.9 What is the benefit of using tool offset in NC programming?

22.10 List and describe the different groups of statements of a part programming language.

22.11 List and describe the low-level code words that NC controllers use.

22.12 List and describe the types of statements of the APT language.

22.13 Redo Example 22.1, but for your state code—for example, MA for Massachusetts.

22.14 Using the APT language, generate the NC part programs (CL-files) for the parts shown in the following figure. For each part, write two programs, one to drill the holes, and the other to perform contour milling. The part material is low-carbon steel, and the cutters are HSS. Part thicknesses are 0.500 in.

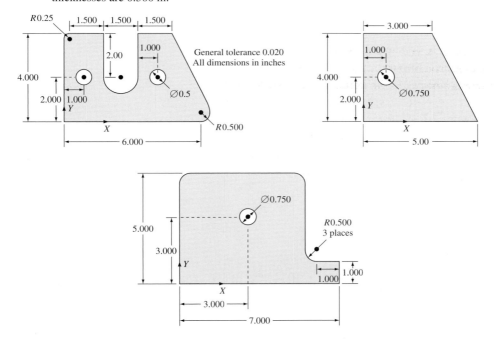

Part II: Laboratory

Use your in-house CAD/CAM system to answer the questions in this part.

22.15 Using an NC package, generate the APT programs for the parts of Problem 22.14. Compare with the manual procedure you followed in Problem 22.14.

22.16 Using an NC package, generate the APT programs for the parts shown at the end of Chapters 6, 7, 9, and 10.

Part III: Programming

Use OpenGL with C or C++, or use Java 3D.

22.17 Write a computer program that implements the algorithm described in Section 22.10 to generate toolpaths for surfaces.

Product Lifecycle Management

OAL

Understand PLM, why it is crucial for companies to implement, what a PLM system offers, what PDM is and its relationship to PLM, and analyze some PLM and PDM case studies.

BJECTIVES

After reading this chapter, you should understand the following concepts:

- The current engineering and design practice

- Types of product information

- PLM goal

- PLM benefits

- PLM systems

- PLM enabling technologies

- PDM

- Case studies of implementing PLM and PDM

CHAPTER HEADLINES

23.1   Introduction

Historically, different computer applications evolve to handle separate parts of the product lifecycle. CAD focuses on product design. CAE focuses on product analysis. CAM focuses on product manufacturing. PDM (product data management) handles the management of design and drafting files from conceptualization through detailed design. ERM (enterprise resource management) handles the release of manufacturing data. SCM (supply chain management) deals with synchronizing orders of supplies and material required to make the product, and coordinates production logistics. CRM (customer relations management) brings customer voices and feedback to product design and development.

These application tools and the engineers who use them have always been, to a large extent, standalone and isolated from each other. While this engineering attitude worked in the past, it does not work anymore. The engineering landscape has changed dramatically because the way in which companies work has changed in many ways:

- **Operations occur in distributed facilities.** An organization may have design and manufacturing facilities in different geographical locations.
- **Product design occurs around the clock and around the world.** This prompts the use of the concepts of collaborative design and Web-based communication.
- **Design responsibilities have been decentralized.** This results in concurrency problems. How can designers of a team know and access the latest design changes?

Companies have to change because of the change in business climate and markets where they compete. Markets are fragmented and change rapidly. One strategy for responding to these ever-changing markets is mass customization which attempts to deliver a product of one at or near mass production cost. Mass customization attempts to deliver customized products, one per customer. Interested readers should consult the literature for more coverage.

Companies realize that CAD, CAE, CAM, PDM, ERP, SCM, CRM, and other systems must work together and exchange information at multiple points during the product lifecycle. For example, PDM is the key to maintaining an accurate product definition for complex products, especially those with long lifecycles. ERP can provide valuable manufacturing production and procurement information for designers in the early conceptual stages of product design.

This highlights the critical importance of a continuous exchange of information and the significant value of individual systems working together throughout the entire lifecycle. **PLM** (product lifecycle management) is a framework that integrates all the different facets and activities of a product cycle together.

23.2   Product Information

A product is a complex entity that unifies an organization and its operations. It also generates a wealth of information at different phases of its lifecycle. The organization must handle and manage product information effectively. At an abstract level, we identify three types of product information: product definition, product production, and operations support. The three types of information must interact with each other continuously to provide effective and current knowledge to all involved teams and individuals in all roles.

Product definition is primarily concerned with managing knowledge as an intellectual asset that is comprised of the product's total definition, including product specifications, conceptual design, part geometry, analysis results, engineering drawings, assembly drawings, and so forth. Thus, product definition encompasses all CAD, CAE, and CAM information.

Activities related to product definition information include:

- **Product configuration management.** As the product design and configuration evolve, the resulting information must be managed in an effective way to ensure its availability on a timely basis to whoever needs it.
- **Change management.** This is a crucial activity. Any change that is made by a member of the product team must be documented and made available immediately. If an organization has offices in different locations, no office should work with obsolete information.
- **Product design and design optimization.** This is a major activity of product definition. All engineering, design, and analysis tools are used here to ensure the proper design and product dimensions.
- **Material selection.** Engineers choose material based on design calculations.

Product production is primarily concerned with the physical operations that are performed to produce the product itself. Activities related to product production information include:

- **Material purchasing.** This activity requires an awareness of suppliers and what they offer. Here, SCM tools should come in handy.
- **Production planning and scheduling.** This is an outcome of the process planning.
- **Equipment design.** Some organizations design their own production equipment in-house. This equipment is highly specialized and is not available off-the-shelf in mass markets.
- **Equipment ordering.** This is off-the-shelf equipment. SCM tools should help decision makers in ordering this equipment.
- **Facilities planning.** Facilities planning must be done before equipment can be installed and operated.
- **Equipment installation and operations.** This signals the beginning of production. This is the ultimate outcome of all of the preceding activities.
- **Maintenance.** Production facilities must be maintained on a scheduled basis to prevent unpredictable shutdowns and failures during critical production times.
- **Manufacturing and quality assurance (QA).** QA tests the accuracy of the production and manufacturing processes. It is an important part of customer satisfaction. QA uses various statistical methods and tools, including Six Sigma.

- **Delivery.** At last, the product can be shipped to customers. Delivery is part of the material handling and logistics problems.
- **Support and customer service.** This activity is the beginning of product revisions and future changes. Customer feedback, field service, and service data often provide a valuable measure of the success of product design and manufacturing.

Operations support includes activities directed at managing finances, human resources, and organizational structure design and management.

The conclusion that we can make now is that a product is synonymous with generating information and managing information. Without information, organizations cannot innovate or create new products; they cannot grow. Therefore their demise and collapse becomes a matter of time. Without information management, organizations cannot compete as their time-to-market lead time becomes excessive. Therefore they cannot make profit nor survive in the marketplace, although they may have excellent products.

Without successful information generation and management, the productivity of individuals and their organizations suffers and declines. Productivity leads to growth, growth leads to profitability, and profitability leads to survival in the marketplace.

Having realized the importance, the complexity, and the heterogeneity of product information, and having realized the importance of its management, the central question can be stated as follows: How can we manage this information effectively? We now can appreciate that the scope of the sum of the product information is much bigger and more complex than the scope of any individual area, including CAD, CAE, CAM, and so forth. One effective answer to the question is PLM.

23.3 PLM Framework

PLM is a framework that attempts to manage product information electronically in a timely manner. It is a vision and philosophy that is built around a digital form of product information. If we can create a complete digital product definition, we have an effective solution to the product information management problem described in Section 23.2.

The main goal of PLM is the creation of a timely communication among all entities and personnel of an organization who are responsible for a product, regardless of their geographical location. Figure 23.1 shows this spirit of communication. The dashed lines in Figure 23.1*a* indicates no or little communication. Without PLM, only suppliers and manufacturing communicate directly, by the virtue of necessity. Figure 23.1*b* shows that PLM fosters and encourages communication at both the physical and intellectual levels among engineering, manufacturing, human resources, suppliers, and customers.

PLM unifies disparate applications that enterprises depend on for daily decision making into one holistic viewpoint. These applications are related to the three types of product information discussed in Section 23.2; they include PDM, ERM, CRM, and SCM. Suppliers are an integral part of PLM. Even customer feedback from existing product lines can be incorporated into the PLM system. This is the "lifecycle" part of PLM.

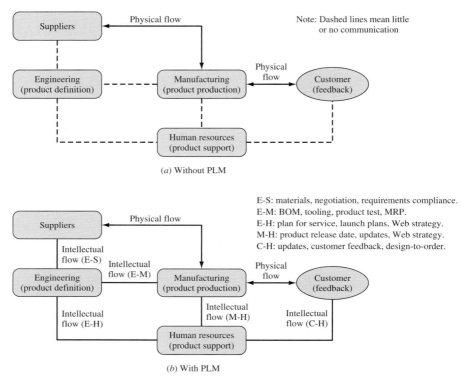

Figure 23.1   PLM framework.

Thus, PLM allows a company to design, analyze, and manage its products from initial conception to retirement. The focus is on the product and its delivery and management from conception to whatever the end of its life is; all done electronically, and in a digital form. In other words, PLM creates a digital product value chain that allows product features, costs, materials, and so forth to be easily and efficiently communicated among all parties involved in a part lifecycle.

Engineering computer applications are morphing today into full-blown PLM tools used by decision makers to increase profits and improve efficiencies across the extended enterprise. PLM enables companies to make better business decisions and deliver greater value to customers, by improving the efficiency of product development processes and the company's capacity to use product-related information.

PLM provides companies with the missing product development link: the ability to truly integrate existing ERM and CRM capabilities with integrated and real-time product information.

By doing so, it brings the informed input of every relevant constituent into a product's planning, definition, design, development, manufacture, sale, movement, support, and even retirement. This kind of rigorous product lifecycle management means that companies are better able to rationalize the design, development, and overall management of their product portfolios.

PLM systems are geared to give enterprises a 360-degree view of all phases of a product's lifecycle, from cradle to grave. From the very first iteration of an idea through its design and development to the manufacturing floor and beyond, PLM systems attempt to incorporate all aspects of a product's costs, features, and functionality into a unified view by combining all manner of data from across the enterprise into the development process as early as possible.

PLM systems work in an enterprise by:

- Providing a single, virtual workspace in which product content, process information, and program status information are fully integrated and under configuration and version control.
- Enabling the precise status of a product at any stage of its life to be accessed and modified through appropriate security protocols, both inside and outside the company.
- Integrating resource information with content creation processes through the connection of workflow and cost information with the product's developmental state.

Unlike so many other enterprise applications, PLM is not a single solution installed from a few CD-ROMs, even though it does involve software and integration tools. Rather, it is a way to reorganize and rethink the entire process of how a company brings products to market. It alters established workflow and integrates business processes. It's a conglomeration of multiple solutions that a company implements and integrates through a common architecture to make all this possible.

23.4 Benefits

PLM fundamentally changes the nature of product development. All parties work collaboratively, particularly while the product's design is still in a position to be influenced. PLM helps companies achieve:

- **Repeatability,** by leveraging intellectual property assets and enforcing standardized, repeatable, and dependable work methods.
- **Lower and steeper learning curves,** through access to the right information for a multitude of enterprise-wide participants, not just engineering.
- **Concurrent work efforts** on different product components, by synchronizing development around object configurations and ongoing changes and updates, something that is impossible to do at any scale without technology enablement.
- **Improved product quality,** by adhering to standards and improving collaboration with customers.

- **Reduced time-to-market,** by reducing the number of engineering change orders, increasing product modularity, and fundamentally altering the method and speed with which information is shared and accessed by employees, from serial/sequential mode to a concurrent one.

- **Reduced total product cost,** by enabling customer product configuration through design-to-order, and by utilizing an integrated approach to product development.

- **Increased product standardization,** by increasing product modularity and encouraging customer input.

- **Improved information exchange,** by increasing collaboration with all parties involved, including customers and suppliers.

- **Increased product customization,** by increasing the generation of product ideas and improving collaboration with customers.

- **Improved predictability,** by reducing the variability in manufacturing processes.

- **Decreased customer response time,** by generating rapid customer quotes through Web technology.

As these benefits point out, PLM is ultimately about even more than getting better products to market faster. PLM can deliver cost-management and revenue-generating benefits as well. PLM can lead to productivity, which leads to growth which, in turn, leads to profitability.

23.5 Implementation

Although many companies like IBM are using PLM systems, few companies outside of the auto and aerospace industries have adopted it across the enterprise. More often than not in today's tight budgetary climate, companies start with one aspect or module of PLM, like a document collaboration suite, on the way to a broader, more comprehensive solution.

In this way, the massive changes an organization may undergo because of PLM can be phased in gradually over time. This lets companies meander toward a solution rather than making a headlong rush toward PLM. Also, with the cost spread out over a longer period, ROI (return on investment) can be more easily measured before the next phase is implemented.

While PLM is aimed at hard-product manufacturers, like automakers, other industries, such as petrochemicals and major pharmaceuticals, also are adopting specialized PLM solutions from niche providers. For the most part, though, PLM is aimed at organizations that produce hard products. PLM is designed to turn the entire extended enterprise into a shared data repository with a common purpose: make a better product, make it faster, and make it for less money.

The four particular areas in which PLM can help resolve long-standing hurdles to innovation, productivity, and profitability are

1. **Knowing which product to pursue.** Companies often expend resources on dead-end ideas before they get the right product completed. By aggregating key insights and capturing known facts, dramatic new approaches to what to do and when to do it are possible.

2. **Long product cycle times.** Margin erosion, excessive discounting, and erratic materials management and inventory profiles are common signs that product delivery performance is too far behind a market window. To do it right the first time and get it done faster requires an unprecedented degree of enterprise-wide synchronization, concurrent development, work management, and data and configuration control.

3. **High product development and launch costs.** High recurring and nonrecurring product costs often indicate an over-reliance on internal solutions, a squandering of resources on developmental dead ends, and a failure to manage evolving insights in a way that allows the true enterprise costs of certain decisions to be determined. Appropriate responses include earlier and more effective collaboration with third parties, integration with enterprise resource information, and the enhanced integration of product-engineering tools. PLM provides a foundation for building these vital links.

4. **Substandard product quality.** Excessive failure rates often point to a breakdown in requirements definition at the front end and weak engineering change-order mechanisms after the problem has surfaced. PLM provides access to key insights early, facilitates bringing the right parties to the table at the right time, and accelerates the ability to process a change when it is required.

23.6 Enabling Technologies

PLM is a framework, vision, and philosophy that encompasses and integrates existing design, manufacturing, and communication technologies. These technologies include CAD, CAE, CAM, PDM, and Web tools. PLM is viewed as an innovation in product development technology. As such, it can be viewed as an extension to current CAD/CAE/CAM/PDM technologies. This view is actually beneficial because it allows us to understand PLM better, and it helps us to put it in context relative to past, present, and future technological developments.

Figure 23.2 shows such a context. It shows the evolution of innovation in product development technology. Four generations are identified over the past four decades. The first generation is 2D CAD and drafting. These early systems were slow, limited by both hardware and software issues at the time.

The second generation is 3D wireframe and surface modeling with limited CAD and CAM applications. These systems were proprietary and closed architecture.

The third generation is what is currently available. Currently, we have standalone systems that support CAD/CAM/CAE, parametric design, associative drafting, and digital prototyping.

The fourth generation has been evolving. Its goal is develop a suite of integrated applications and systems that allow the development and support of a digital product value chain. Fourth-generation systems are based on the PLM framework. An example of a PLM system is Teamcenter from EDS.

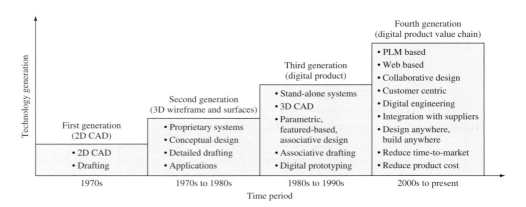

Figure 23.2   Evolution of product development technologies.

23.7 Examples of Business Problems

PLM industrial examples point to reaping substantial benefits and cost savings from the integration of different product technologies. For example, a German automaker uses IBM Thinkpads plugged into automobile dashboards to help dealership mechanics diagnose problems. In the process of diagnosis, the application captures useful service information about the car. That information is transferred back to the car design and manufacturing departments. The integration between the dealership and the car manufacturer saves the manufacturer money on warranty replacement parts, while making for happier and more loyal customers. This results in a win-win situation.

Another example is related to RFQ (request for quotation) in the automotive industry. An automaker has a slow and inconsistent response to customer RFQs. It can take longer than 45 days for a customer to receive a quotation. The reason behind the slow response is a 100-step manual, arcane, nonmeasurable process. The PLM solution is to automate the manual process through workflow, change process, and document management capability. Moreover, the solution provides measurable RFQ metrics via the Web to allow continuous feedback from customers in a timely fashion.

A third example is related to inefficient engineering and manufacturing operations in a company. The company has inefficient engineering throughput, operational deficiencies in design engineering, and lack of availability of the latest, most accurate design data to manufacturing. The root of the problem is the use of multiple CAD/CAM systems and an inefficient engineering change process. The PLM solution is to use only one CAD/CAM system and to use workflow and document management techniques to document, deploy, and control the engineering change process.

23.8 Product Data Management

PDM is an important, if not the most important, PLM enabling technology. Some may think of PLM as the new and more encompassing PDM. CAD/CAM vendors recognize the importance of PDM and offer PDM systems with extensive capabilities. Existing PDM systems include ENOVIA and SmartTeam from Dassault Systems (maker of CATIA CAD/CAM software), iMAN from EDS (maker of Unigraphics), Metaphase from EDS (maker of IDEAS), ProductCenter from AutoCAD, PDMWorks and SmartTeam (from SolidWorks partners), and Windchill from PTC (maker of Pro/E).

PDM software manages, organizes, and controls engineering design information through the transformation from prototype design to a product in full production and then to product obsolescence. Good PDM systems provide flexible product assembly structures, supporting complex relationships between parts, drawings, supporting documents, metadata, and teams.

Current PDM applications employ sophisticated configuration management technology which automatically organizes fragmented product data derived from many sources. With an emphasis on content, meaning, status, and relationships, PDM applications transform product data into product information that can be used reliably within other applications or by users making product data queries.

23.8.1 Motivation

PDM is here out of necessity. There is a lot of engineering data with more being created each day by different groups and teams for a given product. We need PDM because product data

- is subject to different interpretations by different people;
- exists in many different versions;
- is used by many computer programs, running on different computers;
- is stored on many different types of media;
- is used by different people in different functions, at different geographical locations;
- has multiple relationships;
- has to be maintained for years.

23.8.2 Evolution of PDM Systems

PDM systems have been evolving over the years. Current systems combine PDM and collaborative techniques, producing what is known as cPDM, collaborative PDM.

The conception of PDM started when CAD/CAM companies began to see their customers' data management problems. In the late 1980s, they introduced the first-generation commercial PDM systems. Typically, they focused on selling their PDM systems to their existing customers and often used them to leverage sales of CAD/CAM/CAE.

The basis for these PDM systems is the database engine—typically a commercially available relational database management system (RDBMS). The database is used to keep records of parts and related files. The PDM systems then provide the security, file storage, revision control, classification, notification, and application integration, in addition to facilitating the engineering process.

A common theme among first-generation PDM systems is the focus on the downstream engineering process. These systems have two key capabilities: release management and change management. They manage the initial release of design data to manufacturing and the engineering change order process. These PDM systems enforce very strict engineering procedures to control the product process effectively.

The main focus of first-generation PDM systems is on manufacturing. However, other areas of focus are also needed. The second-generation PDM systems support the entire product lifecycle—from initial concept to product obsolescence, including purchasing and product support. They can support concurrent engineering practices in both a flexible, conceptual design phase and in the better-defined engineering change process. They can also manage several levels of release of a design.

Some systems go beyond managing engineering procedures and manage the engineering process (workflow) as well. Users can share up-to-date, prereleased engineering data, and the flow of that data can be intelligently, yet flexibly, controlled according to a set of rules.

23.8.3 Scope of PDM Systems

A PDM system works at every level and across all divisions within a company, even in facilities located around the globe. It integrates data from different types of hardware and software applications to enable members of a multiperson design team to locate and access project data quickly.

A PDM system also permits project leaders to oversee data better throughout the entire design effort, to ensure that team members are working on the most current version of the design, making only permissible modifications to the design, and are not given access to designs that are outdated, unapproved, or "frozen" for review.

A PDM system provides the project team or work groups with a flexible data management tool for meeting the engineering requirements for productivity and quality. At the same time, management of the entire design effort is improved by providing an advanced query capability for accessing data via common attributes and increasing data security.

By overseeing application-level tasks such as checking and tracking design revisions, authorizations, and drawings, a PDM system provides procedural control over the design methodology by facilitating approvals and notifying team members of a project's status. As a result, it provides the improved internal coordination required to allow companies to implement concurrent engineering and compress the overall product development cycle. It also facilitates a team-oriented approach to product development, therefore compressing the time required to bring new products to market.

23.8.4 Benefits of PDM Systems

PDM allows team members to work collaboratively and to have access to current documents at any time, anywhere. PDM systems help companies to achieve

- **Improved design productivity,** due to immediate access to design documents and changes.
- **Improved design and manufacturing accuracy**, because all changes are approved timely.
- **Fewer design changes,** due to collaborative efforts and instant communications.
- **Better management of engineering change,** because project leaders have electronic control.
- **Reduced development time,** thus reducing time-to-market.
- **Better audit trails,** because all changes are well documented electronically in a centralized database.
- **Improved communications between team members,** because they have instant access to information.
- **Faster customer response,** because customers are part of the product team through CRM.
- **Reduced overheads,** because PDM streamlines operations and eliminates redundancies.
- **A major step toward total quality management (TQM),** because all product data is under the control of one system.
- **Better use of creative team skills,** because PDM makes team members feel like one family.
- **Comfortable and easy use of product information,** because every employee knows where information is and how to access it at any time.
- **Data integrity safeguarding,** because it is all stored in electronic vaults.
- **Better control of projects,** because of the timely feedback from all team members.

23.8.5 Implementing a PDM System

The implementation of a PDM system by companies requires careful planning to ensure the success of their use in the engineering process. We offer the following implementation guidelines:

1. **Expectations.** Identify the needs and requirements. Be realistic about them, as well as being practical about the plans for how to meet them.
2. **Analysis.** Do not overanalyze the product data management problem before starting the project.
3. **Team building.** Get the right people together who understand how information flows throughout the company and know where the biggest bottleneck areas exist.
4. **Requirements.** Determine requirements in a few meetings.
5. **Business comes first.** Focus on business requirements first, not technology.

6. **Who offers what.** Develop a list of vendors that offer potential solutions.

7. **Evaluate.** Choose an evaluation methodology that fits the company style, needs, and time frame.

8. **Decide.** Evaluate how a vendor and its products match up with the company business and technology requirements.

9. **Acquire.** Execute the decision. Ask for references, move on, and buy the PDM system.

10. **Use.** When implementing, focus on incremental projects that scale up.

23.8.6 Software Capabilities

PDM systems range from simple, off-the-shelf packages to complex, tailorable systems that can be further developed to fit a company's requirements exactly. Most PDM systems are custom-built to user specifications.

PDM software integrates with, and partially substitutes for, other software used by engineering organizations such as CAD, document management, engineering document management (EDM), free-text databases, ERP, manufacturing resource planning (MRP), and workflow/groupware.

The engineering content of CAD files, scanned images, and word-processed documents are stored in a secure vault, and free-text indexes may be available for the text documents. Check-in and check-out of documents and drawings is part of a simple change control and document distribution process, which can be elaborated by custom programming, either by using the PDM software's own scripting language or by making calls to its published API.

The software functionality and capabilities a company requires depend on many things, including the type of product, its organization, the systems that are in use, the skills of the people who will support the PDM system, and the progress that has been made toward an effective engineering environment.

Some of the features that the software must provide are flexibility, ease of use, powerful access and viewing tools, strong data vaulting technologies, open architecture, full scalability, rules-based and event-driven nature, availability on a wide variety of platforms, and functionality across heterogeneous networks and client/server configurations.

23.8.7 Software Functions

PDM software can be viewed from three perspectives. From a manufacturer's perspective, it permits management and control of the engineering process information. From a product perspective, it can help organize design revisions, track versions of an evolving design concept, and retrieve archived data and other product-specific information. From a process perspective, PDM software can orchestrate procedural events such as design reviews, approvals, product releases, and so forth.

To satisfy these three perspectives, PDM software must provide the following functions:

1. User functions

- *Design release management* provides security and access control, check-in and check-out, establishment of data relationships, global release definitions, user lists, and metadata management.
- *Change management* specifies who approves what, and when. This also includes revision and version relationships and control, baselines, and other configuration management functions.
- *Product structure management* provides parts list and bill of material functions, parts definitions, and parts relationship attributes.
- *Classification* provides tools to search for and retrieve standard parts and existing design data.
- *Program management* creates work breakdown structures and schedules resources.

2. Utility functions

- *Communications and notifications functions* handle interactions within the context of the PDM system and provide interfaces to external e-mail systems.
- *Data transport* provides mechanisms to move data among users, as well as applications and PDM functions to and from other products.
- *Data translation* provides access to tools that translate data between applications such as CAD and CAM.
- *Image services* provide a "viewing" capability for reviewing graphical images and may provide redline mark up.
- *Administration* provides functions which enable PDM users to set up, customize, and manage the PDM system.

23.8.8 A Case Study

Some companies such as NEC, Xerox, Texas Instruments, Groupe Schneider, Honeywell, and General Electric have already invested in PDM technology. We offer a case study as an example to see PDM in action.

NEC Mobile Communications Division (NEC MCD) is a supplier of cellular telephones, pagers, and other infrastructure equipment, such as base stations and terminals. It outsources its design and manufacturing operations to subsidiaries who are far from the headquarters.

To improve its ability to track and share product data with its subsidiaries, NEC MCD implemented the Obbligato PDM system, with the goals of creating a more effective design process, better management of engineering information, and enabling work within a distributed environment.

The Obbligato design environment allows for concurrent engineering. It allows, at the same time, NEC MCD mechanical and electrical engineers to refer to each other's designs and documents, manufacturing engineers to test for design for manufacturability (DFM), and purchasing to readily identify available parts and suppliers. Moreover, Obbligato is also used for managing engineering documents, parts information, drawing information, bill of materials, specification sheets, and customer information.

Some of the outcomes and benefits that NEC MDC has enjoyed from implementing Obbligato include automated functionality, easy access to older files for reference, document management facility that allows data reuse, easy document change by only one person at a time, concurrent viewing of a document by multiple authorized personnel, and the replacement of the inter-company manual delivery system by an electronic distribution system.

PROBLEMS

Part I: Theory

23.1 What is the current business and engineering practice for product lifecycle management?

23.2 What are the three types of product information?

23.3 Describe the PLM framework.

23.4 How does a PLM system work in an enterprise?

23.5 What are the benefits that PLM helps companies achieve?

23.6 How does PLM help innovation, productivity, and profitability?

23.7 List the PLM enabling technologies.

23.8 Find one or two case studies of implementing PLM in industry. Submit a typed report that documents each case as follows: describe the problems that prompted the company to consider PLM, the PLM solution and the system that the company implemented, and the benefits that the company gained.

23.9 Why do we need PDM?

23.10 What are the benefits of PDM?

23.11 List and explain the implementation guidelines for PLM systems.

23.12 List and explain the functions that PLM software provides.

Part II: Laboratory

Use your in-house CAD/CAM system to answer the questions in this part.

23.13 If you have access to a PLM or PDM system, find what it does and how to use it. Apply the chapter concepts to each system to better understand them.

Bibliography

This appendix lists some resources for additional readings and research on the book topics. We list the names of the journals and the conferences, rather than specific articles. The listings are organized by the major topics.

A.1 CAD

1. Advances in Engineering Software (Elsevier Science)
2. American Society of Mechanical Engineers, Design Engineering Division (ASME)
3. *Artificial Intelligence in Engineering* (Computational Mechanics Publications)
4. ASME Database Symposium, Computers in Engineering (ASME)
5. *CAD, Computer Aided Design* (Elsevier Science)
6. *CAE, Computer-Aided Engineering* (Penton Publishing Co.)
7. *Computer Aided Engineering* (Kluwer Academic Publishers)
8. *Computer Aided Geometric Design* (Elsevier Science)
9. Computers in Engineering, Proceedings of the International Computers in Engineering Conference and Exhibit (ASME)
10. *Computers in Industry* (Elsevier Science)
11. *Design Engineering* (Rogers Media)
12. *Engineered Systems* (Business News Publishing Co.)
13. Engineering Data Management: The Technology for Integration (ASCE)
14. *Engineering with Computers* (Springer-Verlag London Ltd.)
15. IEEE CAD-Based Vision Workshop – Proceedings (IEEE)
16. IEEE International Symposium on Intelligent Control – Proceedings (IEEE)
17. IEEE Symposium on Human-Centric Computing (IEEE)

18. *IEEE Transactions on Engineering Management* (IEEE)
19. *IFIP Transactions B: Computer Applications in Technology* (Elsevier Science)
20. *International Journal of Computer Applications in Technology* (Inderscience Enterprises Ltd.)
21. *International Journal of Production Research* (Taylor and Francis Ltd.)
22. *Journal of Computer-Aided Design and Computer Graphics* (Institute of Computing Technology)
23. *Journal of Engineering Technology* (ASEE)
24. *Journal of Materials Processing Technology* (Elsevier Science)
25. *Journal of Mechanical Design* (ASME)
26. Machine Design (Penton Publishing Co.)
27. Proceedings – Design Automation Conference (IEEE)
28. Proceedings of the 5th International Conference on Frontiers of Design and Manufacturing (ICFDM'2002 Program and Organizing Committee)
29. Proceedings of the ASME Design Engineering Technical Conference (ASME)
30. Proceedings of the Symposium on Solid Modeling and Applications (ACM)
31. SAE Technical Paper Series (SAE)
32. Southern Graphics
33. TAPPI Fall Technical Conference and Trade Fair

A.2 CAM

34. American Machinist (McGraw-Hill)
35. American Society of Mechanical Engineers, Design Engineering Division (ASME)
36. Computer Integrated Manufacturing Systems (Butterworth-Heinemann Ltd.)
37. Integrated Manufacturing Systems (Emerald)
38. *International Journal of Advanced Manufacturing Technology* (Springer-Verlag London Ltd.)
39. *International Journal of Computer Integrated Manufacturing* (Taylor and Francis Ltd.)
40. *International Journal of Machine Tools and Manufacture* (Pergamon)
41. *Journal of Engineering Manufacture* (Mechanical Engineering Publications Ltd.)
42. *Journal of Intelligent Manufacturing* (Kluwer Academic Publishers)
43. *Journal of Manufacturing Science and Engineering*, Transactions of the ASME (ASME)
44. *Journal of Manufacturing Systems* (SME)
45. Manufacturing Computer Solutions (Findlay Publications Ltd.)
46. Mechanical Engineering (ASME)
47. Proceedings of the 5th International Conference on Frontiers of Design and Manufacturing (ICFDM'2002 Program and Organizing Committee)
48. Robotics and Computer-Integrated Manufacturing (Elsevier Science)

A.3 Computer Graphics

49. ACM Transactions on Graphics (ACM)
50. Advances in Engineering Software (Elsevier Science)
51. Computer Graphics – A Publication of ACM SIGGRAPH (ACM)
52. Computer Graphics Forum (Blackwell Publishing Ltd.)
53. Computer Graphics World (PennWell Publishing Co.)
54. *Computers and Graphics* (Pergamon)
55. *Computing in Science and Engineering* (IEEE Computer Society)
56. EDN, Engineering Design News (Cahners Publishing Co.)
57. IEEE Computer Graphics and Applications (IEEE)
58. PC World (PCW Communications Inc.)
59. Proceedings – Graphics Interface (Canadian Information Processing Society)
60. Proceedings of SPIE – The International Society for Optical Engineering (SPIE)
61. Proceedings of the ACM SIGGRAPH Conference on Computer Graphics (ACM)
62. *The Engineering Design Graphics* Journal (ASEE)
63. *The Visual Computer* – International Journal of Computer Graphics (Springer-Verlag)

A.4 CAPP

64. Advances in Engineering Software (Elsevier Science)
65. *Computers & Industrial Engineering* (Pergamon Press Inc.)
66. *Computers & Operations Research* (Pergamon Press)
67. *International Journal of Production Research* (Taylor & Francis Ltd.)
68. Machinery and Production Engineering
69. Proceedings of the Industrial Engineering Research Conference (IIE)
70. Proceedings of the Japan/USA Symposium on Flexible Automation (ASME)
71. Technical Paper – Society of Manufacturing Engineers (SME)

A.5 Collaborative Design

72. *Artificial Intelligence for Engineering Design, Analysis and Manufacturing* (Cambridge University Press)
73. *BT Technology Journal* (BT Laboratories plc)
74. Concurrency Practice and Experience (John Wiley & Sons Ltd)
75. Concurrent Engineering Research and Applications (SAGE Publications Ltd.)
76. Design Studies (Elsevier Science)
77. Future Generation Computer Systems (Elsevier Science B.V.)
78. *Journal of Supercomputing* (Kluwer Academic Publishers)
79. Proceedings of the 1st ACM International Conference on Multimedia (ACM)
80. Proceedings of Asian Simulation Conference; System Simulation and Scientific Computing (International Academic Publishers)

81. Proceedings of the 3rd International Conference on Collaborative Virtual Environments (ACM)

82. Proceedings of the 4th International Conference on Collaborative Virtual Environments (ACM)

83. Proceedings of the Annual Symposium on the Virtual Reality Modeling Language, VRML (ACM)

84. Proceedings of the Conference on Designing Interactive Systems: Processes, Practices, Methods, and Techniques, DIS (ACM)

85. Proceedings of the Hawaii International Conference on System Sciences (IEEE Comp Soc)

86. Proceedings of the International Conference on Computer Supported Cooperative Work in Design (National Research Council Canada)

87. Proceedings of the Workshop on Enabling Technologies: Infrastructure for Collaborative Enterprises (IEEE)

88. Research in Engineering Design – Theory, Applications, and Concurrent Engineering (Springer-Verlag GmbH & Company KG)

A.6 PLM

89. AFE Facilities Engineering Journal (Association for Facilities Engineering)

90. ASEE Annual Conference Proceedings (ASEE)

91. *CAE, Computer-Aided Engineering* (Penton Publishing Co.)

92. Control Solutions (PennWell Publishing Co.)

93. Decision Support Systems (Elsevier Science)

94. Design Engineering (Miller Freeman plc)

95. Electronic Engineering (Miller Freeman plc)

96. Engineering Designer

97. Industrial Management and Data Systems (Emerald)

98. Industry Week (Penton Publishing Co.)

99. InTech (ISA – Instrumentation, Systems, and Automation Society)

100. International Journal of Advanced Manufacturing Technology (Springer-Verlag London Ltd.)

101. Managing Automation (Thomas Publishing Co.)

102. Manufacturing Computer Solutions (Findlay Publications Ltd.)

103. Manufacturing Engineer (IEE)

104. Mechanical Engineering (ASME)

105. Mechanical Science and Technology (Northwestern Polytechnical University)

106. Naval Architect (Royal Institution of Naval Architects)

107. Performance Evaluation Review (ACM)

108. Proceedings of the 5th International Conference on Frontiers of Design and Manufacturing (ICFDM'2002 Program and Organizing Committee)

109. Proceedings of the Conference on Technology of Object-Oriented Languages and Systems (IEEE Comp Soc)
110. Proceedings of the Hawaii International Conference on System Sciences (IEEE Comp Soc)
111. Proceedings of the IEEE International Conference on Computer Vision (IEEE)
112. Production Planning and Control (Taylor & Francis)
113. Professional Engineering (Institution of Mechanical Engineers)
114. Technical Papers – Society of Manufacturing Engineers (SME)

Linear Algebra

T his appendix serves as a review of vector and matrix operations. The appendix should enable readers to follow the mathematical developments and derivations presented throughout the book.

Each of the vectors that we use has three components only—x, y, and z because they represent positions, tangents, or curvatures. Moreover, the matrices we cover here are square with a size of 3×3.

B.1 Vectors

We define three vectors:

$$\mathbf{A} = [A_x \ \ A_y \ \ A_z]^T, \ \mathbf{B} = [B_x \ \ B_y \ \ B_z]^T, \ \mathbf{C} = [C_x \ \ C_y \ \ C_z]^T \tag{B.1}$$

B.1.1   Vector Magnitude

$$|\mathbf{A}| \ = \ \sqrt{A_x^2 + A_y^2 + A_z^2} \tag{B.2}$$

where A_x, A_y, and A_z are the cartesian components of the vector $\mathbf{A}$.

B.1.2   Unit Vector in Direction of A

$$\hat{\mathbf{n}}_A \ = \ \frac{\mathbf{A}}{|\mathbf{A}|} \ = \ \mathbf{n}_{Ax}\hat{\mathbf{i}} + \mathbf{n}_{Ay}\hat{\mathbf{j}} + \mathbf{n}_{Az}\hat{\mathbf{k}} \tag{B.3}$$

The components of $\hat{\mathbf{n}}_A$ are also the direction cosines of the vector $\mathbf{A}$.

B.1.3 Vector Equality

If two vectors $\mathbf{A}$ and $\mathbf{B}$ are equal, then

$$A_x = B_y \qquad A_x = B_y \qquad \text{and} \qquad A_z = B_z \tag{B.4}$$

B.1.4 Add or Subtract Two Vectors

Adding or subtracting the two vectors $\mathbf{A}$ and $\mathbf{B}$ results in a vector $\mathbf{R}$ such that

$$\mathbf{R} = \mathbf{A} \pm \mathbf{B} = [A_x \pm B_x \quad A_y \pm B_y \quad A_z \pm B_z] \tag{B.5}$$

The properties of adding or subtracting vectors are

$$\mathbf{A} \pm \mathbf{B} = \mathbf{B} \pm \mathbf{A} \qquad \text{(commutative law)} \tag{B.6}$$

$$(\mathbf{A} \pm \mathbf{B}) \pm \mathbf{C} = \mathbf{A} \pm (\mathbf{B} \pm \mathbf{C}) \qquad \text{(associative law)} \tag{B.7}$$

$$\mathbf{A} - \mathbf{A} = \mathbf{0} \qquad \text{a null (zero) vector} \tag{B.8}$$

B.1.5 Multiply a Vector by a Scalar, s

$$s\mathbf{A} = [sA_x \quad sA_y \quad sA_z]^T \tag{B.9}$$

$$s(\mathbf{A} \pm \mathbf{B}) = s\mathbf{B} \pm s\mathbf{A} \qquad \text{(commutative law)} \tag{B.10}$$

$$s_1(s_2\mathbf{A}) = (s_1 s_2)\mathbf{A} \qquad \text{(associative law)} \tag{B.11}$$

B.1.6 Scalar Product of Two Vectors

The scalar product is also known as the dot or inner product. The scalar product is a scalar value given by

$$\mathbf{A} \cdot \mathbf{B} = \mathbf{B} \cdot \mathbf{A} = A_x B_x + A_y B_y + A_z B_z = |\mathbf{A}||\mathbf{B}|\cos\theta \tag{B.12}$$

where θ is the angle between $\mathbf{A}$ and $\mathbf{B}$. Equation (B.12) gives

$$\cos\theta = \frac{\mathbf{A} \cdot \mathbf{B}}{|\mathbf{A}||\mathbf{B}|} \tag{B.13}$$

The scalar product can give the component of a vector $\mathbf{A}$ in the direction of another vector $\mathbf{B}$ as

$$\mathbf{A} \cdot \hat{\mathbf{n}}_B = |\mathbf{A}|\cos\theta \tag{B.14}$$

Other properties of the scalar product of vectors are

$$\mathbf{A} \cdot \mathbf{A} = |\mathbf{A}|^2 \tag{B.15}$$

$$\mathbf{A} \cdot \mathbf{B} = \mathbf{B} \cdot \mathbf{A} \tag{B.16}$$

$$\mathbf{A} \cdot (\mathbf{B} + \mathbf{C}) = \mathbf{A} \cdot \mathbf{B} + \mathbf{A} \cdot \mathbf{C} \tag{B.17}$$

$$(K\mathbf{A}) \cdot \mathbf{B} = \mathbf{A} \cdot (K\mathbf{B}) = K(\mathbf{A} \cdot \mathbf{B}) \tag{B.18}$$

$$\hat{\mathbf{i}} \cdot \hat{\mathbf{i}} = \hat{\mathbf{j}} \cdot \hat{\mathbf{j}} = \hat{\mathbf{k}} \cdot \hat{\mathbf{k}} = 1 \qquad \hat{\mathbf{i}} \cdot \hat{\mathbf{j}} = \hat{\mathbf{j}} \cdot \hat{\mathbf{k}} = \hat{\mathbf{k}} \cdot \hat{\mathbf{i}} = 0 \tag{B.19}$$

B.1.7   Cross Product of Two Vectors

The cross product is also known as the vector product. It is a vector perpendicular to the plane formed by $\mathbf{A}$ and $\mathbf{B}$, and is given by

$$\mathbf{A} \times \mathbf{B} = \begin{bmatrix} \hat{\mathbf{i}} & \hat{\mathbf{j}} & \hat{\mathbf{k}} \\ A_x & A_y & A_z \\ B_x & B_y & B_z \end{bmatrix} = (A_y B_z - A_z B_y)\hat{\mathbf{i}} + (A_z B_x - A_x B_z)\hat{\mathbf{j}} + (A_x B_y - A_y B_x)\hat{\mathbf{k}} = (|\mathbf{A}||\mathbf{B}|\sin\theta)\hat{\mathbf{n}} \tag{B.20}$$

where $\hat{\mathbf{n}}$ is the perpendicular unit vector to the plane of $\mathbf{A}$ and $\mathbf{B}$. Its sense is determined by the right-hand rule—it is positive in the direction of the advancement of the tip of a right-hand screw when it rotates from $\mathbf{A}$ to $\mathbf{B}$.

Utilizing both the scalar and cross products, the angle between two vectors $\mathbf{A}$ and $\mathbf{B}$ is written as

$$\tan\theta = \frac{|\mathbf{A} \times \mathbf{B}|}{\mathbf{A} \cdot \mathbf{B}} \tag{B.21}$$

The cross product can give the component of a vector $\mathbf{A}$ in the perpendicular direction to another vector $\mathbf{B}$, defined by the unit vector $\hat{\mathbf{n}}_B$, as

$$|\mathbf{A} \times \hat{\mathbf{n}}_B| = |\mathbf{A}|\sin\theta \tag{B.22}$$

Other properties of the cross product of vectors are

$$\mathbf{A} \times \mathbf{B} = -\mathbf{B} \times \mathbf{A} \tag{B.23}$$

$$|\mathbf{A} \times \mathbf{B}|^2 = |\mathbf{A}|^2 |\mathbf{B}|^2 \sin^2\theta \tag{B.24}$$

$$\mathbf{A} \times (\mathbf{B} + \mathbf{C}) = \mathbf{A} \times \mathbf{B} + \mathbf{A} \times \mathbf{C} \tag{B.25}$$

$$(K\mathbf{A}) \times \mathbf{B} = \mathbf{A} \times (K\mathbf{B}) = K(\mathbf{A} \times \mathbf{B}) \tag{B.26}$$

$$\hat{\mathbf{i}} \times \hat{\mathbf{i}} = \hat{\mathbf{j}} \times \hat{\mathbf{j}} = \hat{\mathbf{k}} \times \hat{\mathbf{k}} = 0 \tag{B.27}$$

$$\hat{\mathbf{i}} \times \hat{\mathbf{j}} = \hat{\mathbf{k}} \tag{B.28}$$

$$\hat{\mathbf{j}} \times \hat{\mathbf{k}} = \hat{\mathbf{i}} \tag{B.29}$$

$$\hat{\mathbf{k}} \times \hat{\mathbf{i}} = \hat{\mathbf{j}} \tag{B.30}$$

$$\mathbf{A} \times (\mathbf{B} \times \mathbf{C}) = \mathbf{B}(\mathbf{C} \cdot \mathbf{A}) - \mathbf{C}(\mathbf{A} \cdot \mathbf{B}) \qquad \text{(vector triple product)} \tag{B.31}$$

$$\mathbf{A} \cdot (\mathbf{B} \times \mathbf{C}) = \mathbf{B} \cdot (\mathbf{C} \times \mathbf{A}) = \mathbf{C} \cdot (\mathbf{A} \times \mathbf{B}) = \begin{vmatrix} A_x & A_y & A_z \\ B_x & B_y & B_z \\ C_x & C_y & C_z \end{vmatrix} \quad \text{(scalar triple product)} \quad \text{(B.32)}$$

B.1.8 Parallel Vectors

Two vectors $\mathbf{A}$ and $\mathbf{B}$ are parallel if and only if

$$\hat{\mathbf{n}}_A \cdot \hat{\mathbf{n}}_B = 1 \quad \text{or} \quad \left| \hat{\mathbf{n}}_A \times \hat{\mathbf{n}}_B \right| = 0 \quad\quad \text{(B.33)}$$

B.1.9 Perpendicular Vectors

Two vectors $\mathbf{A}$ and $\mathbf{B}$ are perpendicular if and only if

$$\hat{\mathbf{n}}_A \cdot \hat{\mathbf{n}}_B = 0 \quad \text{or} \quad \left| \hat{\mathbf{n}}_A \times \hat{\mathbf{n}}_B \right| = 1 \quad\quad \text{(B.34)}$$

The conditions for parallelism and perpendicularity are written as shown in Eqs. (B.33) and (B.34) to reflect the expected skew symmetry of the two properties. For computation purposes, it might be useful to replace $\left| \hat{\mathbf{n}}_A \times \hat{\mathbf{n}}_B \right| = 0$ with $\hat{\mathbf{n}}_A \times \hat{\mathbf{n}}_B = \mathbf{0}$ or with $\mathbf{A} \times \mathbf{B} = \mathbf{0}$ for parallel lines, and replace $\hat{\mathbf{n}}_A \cdot \hat{\mathbf{n}}_B = 0$ with $\mathbf{A} \cdot \mathbf{B} = 0$ for perependicular lines.

B.1.10 Linear Independence of Vectors

Vectors $\mathbf{P}_1, \mathbf{P}_2, ...,$ and $\mathbf{P}_n$ are said to be linearly independent if and only if

$$\sum_{i=1}^{n} c_i \mathbf{P}_i = \mathbf{0} \quad \text{for all } c_i = 0 \quad\quad \text{(B.35)}$$

where c_i are scalar quantities.

The vectors are said to be linearly dependent if and only if

$$\sum_{i=1}^{n} c_i \mathbf{P}_i = \mathbf{0} \quad \text{for some } c_i \neq 0 \quad\quad \text{(B.36)}$$

For example, the two vectors $\mathbf{A} = [1 \;\; 2]^T$ and $\mathbf{B} = [3 \;\; 6]^T$ are linearly dependent because there exists $c_1 = 3$ and $c_2 = -1$ such that $c_1 \mathbf{A} + c_2 \mathbf{B} = \mathbf{0}$.

B.2 Matrices

We define three matrices:

$$[A] = \begin{bmatrix} a_{11} & a_{12} & a_{13} \\ a_{21} & a_{22} & a_{23} \\ a_{31} & a_{32} & a_{33} \end{bmatrix}, \; [B] = \begin{bmatrix} b_{11} & b_{12} & b_{13} \\ b_{21} & b_{22} & b_{23} \\ b_{31} & b_{32} & b_{33} \end{bmatrix}, \; [C] = \begin{bmatrix} c_{11} & c_{12} & c_{13} \\ c_{21} & c_{22} & c_{23} \\ c_{31} & c_{32} & c_{33} \end{bmatrix} \quad \text{(B.37)}$$

B.2.1   Identity Matrix

The identity matrix $[I]$ is a matrix whose diagonal elements are ones and off-diagonal elements are zeros:

$$[AI] = \begin{bmatrix} 1 & 0 & 0 \\ 0 & 1 & 0 \\ 0 & 0 & 1 \end{bmatrix} \tag{B.38}$$

B.2.2   Transpose of a Matrix

The matrix $[A]^T$ is the transpose of $[A]$ if the element a_{ij} in $[A]$ is equal to the element a_{ji} in $[A]^T$ for all i and j. For example, if

$$[A] = \begin{bmatrix} 2 & 5 \\ 3 & 7 \\ 9 & 1 \end{bmatrix} \tag{B.39}$$

then,

$$[A]^T = \begin{bmatrix} 2 & 3 & 9 \\ 5 & 7 & 1 \end{bmatrix} \tag{B.40}$$

In general, $[A]^T$ is obtained by interchanging the rows and columns of $[A]$. Thus, if $[A]$ is of the order $m \times n$, then $[A]^T$ is of the order $n \times m$.

B.2.3   Zero Matrix

A matrix $[A] = 0$ is a zero matrix if every element of $[A]$ is zero.

B.2.4   Equal Matrices

Two matrices $[A]$ and $[B]$ are equal (i.e., $[A] = [B]$) if and only if they have the same order and if $a_{ij} = b_{ij}$ for all i and j.

B.2.5   Add or Subtract Two Matrices

Adding or subtracting the two matrices $[A]$ and $[B]$ results in a matrix $[R]$ such that

$$[R] = [A] \pm [B] = \begin{bmatrix} a_{11} \pm b_{11} & a_{12} \pm b_{12} & a_{13} \pm b_{13} \\ a_{21} \pm b_{21} & a_{22} \pm b_{22} & a_{23} \pm b_{23} \\ a_{31} \pm b_{31} & a_{32} \pm b_{32} & a_{33} \pm b_{33} \end{bmatrix} \tag{B.41}$$

The properties of adding or subtracting vectors are

$$[A] \pm [B] = [B] \pm [A] \qquad \text{(commutative law)} \tag{B.42}$$

$$([A]\pm[B])\pm[C]=[A]\pm([B]\pm[C]) \quad \text{(associative law)} \tag{B.43}$$

$$[A]\pm[B])^T = [A]^T\pm[B]^T \tag{B.44}$$

B.2.6   Multiply a Matrix by a Scalar, s

$$s\mathbf{A} = \begin{bmatrix} sa_{11} & sa_{12} & sa_{13} \\ sa_{21} & sa_{22} & sa_{23} \\ sa_{31} & sa_{32} & sa_{33} \end{bmatrix} \tag{B.45}$$

$$s(\mathbf{A}\pm\mathbf{B}) = s\mathbf{B}\pm s\mathbf{A} \qquad \text{(commutative law)} \tag{B.46}$$

$$s_1(s_2\mathbf{A}) = (s_1s_2)\mathbf{A} \qquad \text{(associative law)} \tag{B.47}$$

B.2.7   Matrix Multiplication

Two matrices $[A]$ and $[B]$ can be multiplied in the order $[A][B]$ if and olny if the number of columns of $[A]$ is equal to the number of rows of $[B]$; that is, if $[A]$ is of the order $m \times n$, then $[B]$ must be of the order $n \times r$. If $[C] = [A][B]$, then $[C]$ is of the order $m \times r$. The elements c_{ij} of $[C]$ are given by

$$c_{ij} = \sum_{k=1}^{n} a_{ik}b_{kj} \qquad \text{for all } i \text{ and } j \tag{B.48}$$

For example, if

$$[A] = \begin{bmatrix} 2 & 3 \\ 1 & 0 \\ 6 & 4 \end{bmatrix} \quad \text{and} \quad [B] = \begin{bmatrix} 1 & 9 \\ 8 & 7 \end{bmatrix} \tag{B.49}$$

then

$$C] = [A][B] = \begin{bmatrix} (2)(1)+(3)(8) & (2)(9)+(3)(7) \\ (1)(1)+(0)(8) & (1)(9)+(0)(7) \\ (6)(1)+(4)(8) & (6)(9)+(4)7 \end{bmatrix} = \begin{bmatrix} 26 & 39 \\ 1 & 9 \\ 38 & 82 \end{bmatrix} \tag{B.50}$$

Observe that, in general, matrix multiplication is not commutative; that is, $[A][B] \neq [B][A]$, assuming that $[B][A]$ exists (it does if both matrices are square).

Other properties of matrix multiplication are

$$[I][A] = [A][I] \tag{B.51}$$

$$([A][B])[C] = [A]([B][C]) \tag{B.52}$$

$$[A]([B] + [C]) = [A][B] + [A][C] \tag{B.53}$$

$$([A] + [B])[C]) = [A][C] + [B][C] \tag{B.54}$$

$$s([A][B]) = (s[A])[B] = [A](s[B]) \tag{B.55}$$

B.2.8   Multiplication of Partitioned Matrices

If $[A]$ and $[B]$ are partitioned into the following submatrices

$$\left[\begin{array}{c|c|c} [A_{11}] & [A_{12}] & [A_{13}] \\ \hline [A_{21}] & [A_{22}] & [A_{23}] \end{array}\right] \quad \text{and} \quad \left[\begin{array}{c|c} [B_{11}] & [B_{12}] \\ \hline [B_{21}] & [B_{22}] \\ \hline [B_{31}] & [B_{32}] \end{array}\right] \tag{B.56}$$

then (assuming we can multiply the partitions)

$$[A][B] = \left[\begin{array}{c|c} [A_{11}][B_{11}] + [A_{12}][B_{21}] + [A_{13}][B_{31}] & [A_{11}][B_{12}] + [A_{12}][B_{22}] + [A_{13}][B_{32}] \\ \hline [A_{21}][B_{11}] + [A_{22}][B_{21}] + [A_{23}][B_{31}] & [A_{21}][B_{12}] + [A_{22}][B_{22}] + [A_{23}][B_{32}] \end{array}\right] \tag{B.57}$$

Matrix partitioning is useful in matrix inversion.

B.2.9   Determinant of a Square Matrix

The determinant of a matrix $[A]$ is written as $|A|$. The determinant is a scalar value. If $[A]$ is a 3×3 matrix given by

$$[A] = \begin{bmatrix} a_{11} & a_{12} & a_{13} \\ a_{21} & a_{22} & a_{23} \\ a_{31} & a_{32} & a_{33} \end{bmatrix}, \tag{B.58}$$

then its determinant is given by

$$|A| = a_{11}(a_{22}a_{33} - a_{23}a_{32}) - a_{12}(a_{21}a_{33} - a_{31}a_{23}) + a_{13}(a_{21}a_{32} - a_{22}a_{31}) \tag{B.59}$$

The following properties of determinants are useful:

1. If all the elements of a row or a column of a matrix are zeros, its determinant is zero.
2. If two rows (or columns) of a matrix $[A]$ are interchanged, its determinant becomes $-|A|$.
3. The value of the determinant of a matrix remains unchanged if its rows and columns are interchanged.
4. If two rows (columns) of a matrix $[A]$ are identical, $|A| = 0$.
5. If $[A]$ and $[B]$ are two square matrices of the same size, then

$$|AB| = |A||B| \tag{B.60}$$

B.2.10   Inverse of a Matrix

If $[A]$ and $[B]$ are two square matrices of the same size (order) such that $[A][B] = [I]$, then $[B]$ is the inverse of $[A]$, written as $[A]^{-1}$. Thus, $[A][A]^{-1} = [I]$, or $[A]^{-1}[A] = [I]$.

A matrix $[A]$ is singular if its inverse does not exist—this implies that $|A| = 0$. If $[A]$ and $[B]$ are nonsingular square matrices, then

$$([A][B])^{-1} = [B]^{-1}[A]^{-1} \tag{B.61}$$

If $[A]$ is nonsingular, $[A][B] = [A][C]$ implies that $[B] = [C]$.

Three methods exist to compute the inverse of a matrix. They are the adjoint matrix method, the Gauss-Jordan (row operations) method, and the partitioned matrix method. We cover the first method only.

Utilizing the adjoint matrix method, the inverse of a nonsingular matrix $[A]$ is given by

$$[A]^{-1} = \frac{1}{|A|}\text{adj}([A]) = \frac{1}{|A|}[C]^T \tag{B.62}$$

where $[C]^T$ is the adjoint matrix of $[A]$. $[C]^T$ is the transpose of $[C]$. $[C]$ is the cofactor matrix that is formed by the cofactors C_{ij} of the elements a_{ij} of $[A]$. The cofactors C_{ij} are given by

$$C_{ij} = (-1)^{i+j}M_{ij} \tag{B.63}$$

where M_{ij} is a unique scalar associated with the element a_{ij}. M_{ij} is defined as the determinant of the $(n-1) \times (n-1)$ matrix obtained from the $n \times n$ matrix $[A]$ by crossing out the ith row and the jth column. If $[A]$ is a 3×3, its cofactor matrix becomes

$$[C] = \begin{bmatrix} C_{11} & C_{12} & C_{13} \\ C_{21} & C_{22} & C_{23} \\ C_{31} & C_{32} & C_{33} \end{bmatrix} \tag{B.64}$$

For example, for

$$[A] = \begin{bmatrix} 3 & 4 & 3 \\ 0 & 2 & 4 \\ 2 & 1 & 0 \end{bmatrix}, \tag{B.65}$$

$$|A| = 8 \quad \text{and} \quad \text{adj}([A]) = \begin{bmatrix} -4 & 3 & 10 \\ 8 & -6 & -12 \\ -4 & 5 & 6 \end{bmatrix} \tag{B.66}$$

Thus,

$$[A]^{-1} = \frac{1}{8}\begin{bmatrix} -4 & 3 & 10 \\ 8 & -6 & -12 \\ -4 & 5 & 6 \end{bmatrix} = \begin{bmatrix} -1/2 & 3/8 & 5/4 \\ 1 & -3/4 & -3/2 \\ -1/2 & 5/8 & 3/4 \end{bmatrix} \tag{B.67}$$

ANSI and ISO
Tolerance Tables

T his appendix lists the ANSI and ISO tolerance tables. Use these tables with the conventional tolerances covered in Chapter 20. Use the tables to calculate the tolerances and limits on dimensions that have specific fits assigned to them. Refer to Chapter 20 for more discussions on tolerance calculations.

Table C.1   Clearance running fits.[a]

Basic size	RC1			RC2			RC3		
	Limits of clearance	Standard limits		Limits of clearance	Standard limits		Limits of clearance	Standard limits	
Over - To		Hole H5	Shaft g4		Hole H6	Shaft g5		Hole H7	Shaft f6
0 – 0.12	0.1	+0.2	−0.1	0.1	+0.25	−0.1	0.3	+0.4	−0.3
	0.45	0	−0.25	0.55	0	−0.3	0.95	0	−0.55
0.12 – 0.24	0.15	+0.2	−0.15	0.15	+0.3	−0.15	0.4	+0.5	−0.4
	0.5	0	−0.3	0.65	0	−0.35	1.12	0	−0.7
0.24 – 0.40	0.2	+0.25	−0.2	0.2	+0.4	−0.2	0.5	+0.6	−0.5
	0.6	0	−0.35	0.85	0	−0.45	1.5	0	−0.9
0.40 – 0.71	0.25	+0.3	−0.25	0.25	+0.4	−0.25	0.6	+0.7	−0.6
	0.75	0	−0.45	0.95	0	−0.55	1.7	0	−1.0
0.71 – 1.19	0.3	+0.4	−0.3	0.3	+0.5	−0.3	0.8	+0.8	−0.8
	0.95	0	−0.55	1.2	0	−0.7	2.1	0	−1.3
1.19 – 1.97	0.4	+0.4	−0.4	0.4	+0.6	−0.4	1.0	+1.0	−1.0
	1.1	0	−0.7	1.4	0	−0.8	2.6	0	−1.6
1.97 – 3.15	0.4	+0.5	−0.4	0.4	+0.7	−0.4	1.2	1.2	−1.2
	1.2	0	−0.7	1.6	0	−0.9	3.1	0	−1.9
3.15 – 4.73	0.5	+0.6	−0.5	0.5	+0.9	−0.5	1.4	+1.4	−1.4
	1.5	0	−0.9	2.0	0	−1.1	3.7	0	−2.3
4.73 – 7.09	0.6	+0.7	−0.6	0.6	+1.0	−0.6	1.6	1.6	−1.6
	1.8	0	−1.1	2.3	0	−1.3	4.2	0	−2.6
7.09 – 9.85	0.6	+0.8	−0.6	0.6	+1.2	−0.6	2.0	1.8	−2.0
	2.0	0	−1.2	2.6	0	−1.4	5.0	0	−3.2
9.85 – 12.41	0.8	+0.9	−0.8	0.8	+1.2	−0.8	2.5	2.0	−2.5
	2.3	0	−1.4	2.9	0	−1.7	5.7	0	−3.7

a.   Table entries are in inches. Multiply the entries in *Limits of clearance* and *Standard limits* columns by 10^{-3}.

Follow these steps to use the table (and the other tables in this appendix):

1. Locate the *Basic size* row that includes your basic size. For example, if your basic size is 1.000 in, the *Basic size* row would be 0.71 – 1.19.
2. Locate your fit column—for example, RC1.
3. Go right across the row and down along the column as shown in the table. The intersection (shaded cells) of the two directions provides the tolerance values.

Table C.1   Clearance runing fits.[a] (Continued)

Basic size	RC4			RC5			RC6		
	Limits of clearance	Standard limits		Limits of clearance	Standard limits		Limits of clearance	Standard limits	
Over - To		Hole H8	Shaft f7		Hole H8	Shaft e7		Hole H9	Shaft e8
0 – 0.12	0.3	+0.6	–0.3	0.6	+0.6	–0.6	0.6	+1.0	–0.6
	1.3	0	–0.7	1.6	0	–1.0	2.2	0	–1.2
0.12 – 0.24	0.4	+0.7	–0.4	0.8	+0.7	–0.8	0.8	+1.2	–0.8
	1.6	0	–0.9	2.0	0	–1.3	2.7	0	–1.5
0.24 – 0.40	0.5	+0.9	–0.5	1.0	+0.9	–1.0	1.0	+1.4	–1.0
	2.0	0	–1.1	2.5	0	–1.6	3.3	0	–1.9
0.40 - 0.71	0.6	+1.0	–0.6	1.2	+1.0	–1.2	1.2	+1.6	–1.2
	2.3	0	–1.3	2.9	0	–1.9	3.8	0	–2.2
0.71 – 1.19	0.8	+1.2	–0.8	1.6	+1.2	–1.6	1.6	+2.0	–1.6
	2.8	0	–1.6	3.6	0	–2.4	4.8	0	–2.8
1.19 – 1.97	1.0	+1.6	–1.0	2.0	+1.6	–2.0	2.0	+2.5	–2.0
	3.6	0	–2.0	4.6	0	–3.0	6.1	0	–3.6
1.97 – 3.15	1.2	+1.8	–1.2	2.5	+1.8	–2.5	2.5	+3.0	–2.5
	4.2	0	–2.4	5.5	0	–3.7	7.3	0	–4.3
3.15 – 4.73	1.4	+2.2	–1.4	3.0	+2.2	–3.0	3.0	+3.5	–3.0
	5.0	0	–2.8	6.6	0	–4.4	8.7	0	–5.2
4.73 – 7.09	1.6	+2.5	–1.6	3.5	+2.5	–2.5	3.5	+4.0	–3.5
	5.7	0	–3.2	7.6	0	–5.1	10.0	0	–6.0
7.09 – 9.85	2.0	+2.8	–2.0	4.0	+2.8	–4.0	4.0	+4.5	–4.0
	6.6	0	–3.8	8.6	0	–5.8	11.3	0	–6.8
9.85 – 12.41	2.5	+3.0	–2.5	5.0	+3.0	–5.0	5.0	+5.0	–5.0
	7.5	0	–4.5	10.0	0	–7.0	13.0	0	–8.0

a.   Table entries are in inches. Multiply the entries in *Limits of clearance* and *Standard limits* columns by 10^{-3}.

Table C.1   Clearance running fits.[a] (Continued)

Basic size Over - To	RC7 Limits of clearance	RC7 Standard limits Hole H9	RC7 Standard limits Shaft d8	RC8 Limits of clearance	RC8 Standard limits Hole H10	RC8 Standard limits Shaft c9	RC9 Limits of clearance	RC9 Standard limits Hole H11	RC9 Standard limits Shaft c11
0 – 0.12	1.0 2.6	+1.0 0	−1.0 −1.6	2.5 5.1	+1.6 0	−2.5 −3.5	4.0 8.1	+ 2.5 0	−4.0 −5.6
0.12 – 0.24	1.2 3.1	+1.2 0	−1.2 −1.9	2.8 5.8	+1.8 0	−2.8 −4.0	4.5 9.0	+ 3.0 0	−4.5 −6.0
0.24 – 0.40	1.6 3.9	+1.4 0	−1.6 −2.5	3.0 6.6	+2.2 0	−3.0 −4.4	5.0 10.7	+ 3.5 0	−5.0 −7.2
0.40 – 0.71	2.0 4.6	+1.6 0	−2.0 −3.0	3.5 7.9	+2.8 0	−3.5 −5.1	6.0 12.8	+ 4.0 0	−6.0 −8.8
0.71 – 1.19	2.5 5.7	+2.0 0	−2.5 −3.7	4.5 10.0	+3.5 0	−4.5 −6.5	7.0 15.5	+ 5.0 0	−7.0 −10.5
1.19 – 1.97	3.0 7.1	+2.5 0	−3.0 −4.6	5.0 11.5	+4.0 0	−5.0 −7.5	8.0 18.0	+ 6.0 0	−8.0 −12.0
1.97 – 3.15	4.0 8.8	+3.0 0	−4.0 −5.8	6.0 13.5	+4.5 0	−6.0 −9.0	9.0 20.5	+ 7.0 0	−9.0 −13.5
3.15 – 4.73	5.0 10.7	+3.5 0	−5.0 −7.2	7.0 15.5	+5.0 0	−7.0 −10.5	10.0 24.0	+ 9.0 0	−10.0 −15.0
4.73 – 7.09	6.0 12.5	+4.0 0	−6.0 −8.5	8.0 18.0	+6.0 0	−8.0 −12.0	12.0 28.0	+10.0 0	−12.0 −18.0
7.09 – 9.85	7.0 14.3	+4.5 0	−7.0 −9.8	10.0 21.5	+7.0 0	−10.0 −14.5	15.0 34.0	+ 12.0 0	−15.0 −22.0
9.85 – 12.41	8.0 16.0	+5.0 0	−8.0 −11.0	12.0 25.0	+8.0 0	−12.0 −17.0	18.0 38.0	+12.0 0	−18.0 −26.0

a.   Table entries are in inches. Multiply the entries in *Limits of clearance* and *Standard limits* columns by 10^{-3}.

Table C.2   Clearance location fits.[a]

Basic size	LC1			LC2			LC3		
	Limits of clearance	Standard limits		Limits of clearance	Standard limits		Limits of clearance	Standard limits	
Over - To		Hole H6	Shaft h5		Hole H7	Shaft h6		Hole H8	Shaft h7
0 – 0.12	0	+0.25	0	0	+0.4	0	0	+0.6	0
	0.45	0	−0.2	0.65	0	−0.25	1	0	−0.4
0.12 – 0.24	0	+0.3	0	0	+0.5	0	0	+0.7	0
	0.5	0	−0.2	0.8	0	−0.3	1.2	0	−0.5
0.24 – 0.40	0	+0.4	0	0	+6	0	0	+0.9	0
	0.65	0	−0.25	1.0	0	−0.4	1.5	0	−0.6
0.40 – 0.71	0	+0.4	0	0	+0.7	0	0	+1.0	0
	0.7	0	−0.3	1.1	0	−0.4	1.7	0	−0.7
0.71 – 1.19	0	+0.5	0	0	+0.8	0	0	+1.2	0
	0.9	0	−0.4	1.3	0	−0.5	2	0	−0.8
1.19 – 1.97	0	+0.6	0	0	+1.0	0	0	+1.6	0
	1.0	0	−0.4	1.6	0	−0.6	2.6	0	−1
1.97 – 3.15	0	+0.7	0	0	+1.2	0	0	+1.8	0
	1.2	0	−0.5	1.9	0	−0.7	3	0	−1.2
3.15 – 4.73	0	+0.9	0	0	+1.4	0	0	+2.2	0
	1.5	0	−0.6	2.3	0	−0.9	3.6	0	−1.4
4.73 – 7.09	0	+1.0	0	0	+1.6	0	0	+2.5	0
	1.7	0	−0.7	2.6	0	−1.0	4.1	0	−1.6
7.09 – 9.85	0	+1.2	0	0	+1.8	0	0	+2.8	0
	2.0	0	−0.8	3.0	0	−1.2	4.6	0	−1.8
9.85 – 12.41	0	+1.2	0	0	+2.0	0	0	+3.0	0
	2.1	0	−0.9	3.2	0	−1.2	5	0	−2.0

a.   Table entries are in inches. Multiply the entries in *Limits of clearance* and *Standard limits* columns by 10^{-3}.

Table C.2   Clearance location fits.[a] (Continued)

Basic size	LC4			LC5			LC6		
	Limits of clearance	Standard limits		Limits of clearance	Standard limits		Limits of clearance	Standard limits	
Over - To		Hole H10	Shaft h9		Hole H7	Shaft g6		Hole H9	Shaft f8
0 – 0.12	0	+1.6	0	0.1	+0.4	−0.1	0.3	+1.0	−0.3
	2.6	0	−1.0	0.75	0	−0.35	1.9	0	−0.9
0.12 – 0.24	0	+1.8	0	0.15	+0.5	−0.15	0.4	+1.2	−0.4
	3.0	0	−1.2	0.95	0	−0.45	2.3	0	−1.1
0.24 – 0.40	0	+2.2	0	0.2	+0.6	−0.2	0.5	+1.4	−0.5
	3.6	0	−1.4	1.2	0	−0.6	2.8	0	−1.4
0.40 – 0.71	0	+2.8	0	0.25	+0.7	−0.25	0.6	+1.6	−0.6
	4.4	0	−1.6	1.35	0	−0.65	3.2	0	−1.6
0.71 – 1.19	0	+3.5	0	0.3	+0.8	−0.3	0.8	+2.0	−0.8
	5.5	0	−2.0	1.6	0	−0.8	4.0	0	−2.0
1.19 – 1.97	0	+4.0	0	0.4	+1.0	−0.4	1.0	+2.5	−1.0
	6.5	0	−2.5	2.0	0	−1.0	5.1	0	−2.6
1.97 – 3.15	0	+4.5	0	0.4	+1.2	−0.4	1.2	+3.0	−1.2
	7.5	-0	−3	2.3	0	−1.1	6.0	0	−3.0
3.15 – 4.73	0	+5.0	0	0.5	+1.4	−0.5	1.4	+3.5	−1.4
	8.5	0	−3.5	2.8	0	−1.4	7.1	0	−3.6
4.73 – 7.09	0	+6.0	0	0.6	+1.6	−0.6	1.6	+4.0	−1.6
	10	0	−4	3.2	0	−1.6	8.1	0	−4.1
7.09 – 9.85	0	+7.0	0	0.6	+1.8	−0.6	2.0	+4.5	−2.0
	11.5	0	−4.5	3.6	0	−1.8	9.3	0	−4.8
9.85 – 12.41	0	+8.0	0	0.7	+2.0	−0.7	2.2	+5.0	−2.2
	13	0	−5	3.9	0	−1.9	10.2	0	−5.2

a.   Table entries are in inches. Multiply the entries in *Limits of clearance* and *Standard limits* columns by 10^{-3}.

Table C.2 Clearance location fits.[a] (Continued)

Basic size	LC7			LC8			LC9		
	Limits of clearance	Standard limits		Limits of clearance	Standard limits		Limits of clearance	Standard limits	
Over - To		Hole H10	Shaft e9		Hole H10	Shaft d9		Hole H11	Shaft c10
0 – 0.12	0.6	+1.6	−0.6	1.0	+1.6	−1.0	2.5	+2.5	−2.5
	3.2	0	−1.6	3.6	0	−2.0	6.6	0	−4.1
0.12 – 0.24	0.8	+1.8	−0.8	1.2	+1.8	−1.2	2.8	+3.0	−2.8
	3.8	0	−2.0	4.2	0	−2.4	7.6	0	−4.6
0.24 – 0.40	1.0	+2.2	−1.0	1.6	+2.2	−1.6	3.0	+3.5	−3.0
	4.6	0	−2.4	5.2	0	−3.0	8.7	0	−5.2
0.40 – 0.71	1.2	+2.8	−1.2	2.0	+2.8	−2.0	3.5	+4.0	−3.5
	5.6	0	−2.8	6.4	0	−3.6	10.3	0	−6.3
0.71 – 1.19	1.6	+3.5	−1.6	2.5	+3.5	−2.5	4.5	+5.0	−4.5
	7.1	0	−3.6	8.0	0	−4.5	13.0	0	−8.0
1.19 – 1.97	2.0	+4.0	−2.0	3.0	+4.0	−3.0	5.0	+6.0	−5.0
	8.5	0	−4.5	9.5	0	−5.5	15.0	0	−9.0
1.97 – 3.15	2.5	+4.5	−2.5	4.0	+4.5	−4.0	6.0	+7.0	−6.0
	10.0	0	−5.5	11.5	0	−7.0	17.5	0	−10.5
3.15 – 4.73	3.0	+5.0	−3.0	5.0	+5.0	−5.0	7.0	+9.0	−7.0
	11.5	0	−6.5	13.5	0	−8.5	21.0	0	−12.0
4.73 – 7.09	3.5	+6.0	−3.5	6.0	+6.0	−6.0	8.0	+10.0	−8.0
	13.5	0	−7.5	16.0	0	−10.0	24.0	0	−14.0
7.09 – 9.85	4.0	+7.0	−4.0	7.0	+7.0	−7.0	10.0	+12.0	−10.0
	15.5	0	−8.5	18.5	0	−11.5	29.0	0	−17.0
9.85 – 12.41	4.5	+8.0	−4.5	7.0	+8.0	−7.0	12.0	+12.0	−12.0
	17.5	0	−9.5	20.0	0	−12.0	32.0	0	−20.0

a. Table entries are in inches. Multiply the entries in *Limits of clearance* and *Standard limits* columns by 10^{-3}.

Table C.2   Clearance location fits.[a] (Continued)

Basic size	LC10			LC11		
	Limits of clearance	Standard limits		Limits of clearance	Standard limits	
Over - To		Hole H12	Shaft c12		Hole H13	Shaft c13
0 – 0.12	4.0 12	+4.0 0	−4.0 −8.0	5.0 17.0	+6.0 0	−5.0 −11.0
0.12 – 0.24	4.5 14.5	+5.0 0	−4.5 −9.5	6.0 20.0	+7.0 0	−6.0 −13.0
0.24 – 0.40	5.0 17	+6.0 0	−5.0 −11.0	7.0 25.0	+9.0 0	−7.0 −16.0
0.40 – 0.71	6.0 20.0	+7.0 0	−6.0 −13.0	8.0 28.0	+10.0 0	−8.0 −18.0
0.71 – 1.19	7.0 23.0	+8.0 0	−7.0 −15.0	10.0 34.0	+12.0 0	−10.0 −22.0
1.19 – 1.97	8.0 28.0	+10.0 0	−8.0 −18.0	12.0 44.0	+16.0 0	−12.0 −28.0
1.97 – 3.15	10.0 34.0	+12.0 0	−10.0 −22.0	14.0 50.0	+18.0 0	−14.0 −32.0
3.15 – 4.73	11.0 39.0	+14.0 0	−11.0 −25.0	16.0 60.0	+22.0 0	−16.0 −38.0
4.73 – 7.09	12.0 44.0	+16.0 0	−12.0 −28.0	18.0 68.0	+25.0 0	−18.0 −43.0
7.09 – 9.85	16.0 52.0	+18.0 0	−16.0 −34.0	22.0 78.0	+28.0 0	−22.0 −50.0
9.85 – 12.41	20.0 60.0	+20.0 0	−20.0 −40.0	28.0 88.0	+30.0 0	−28.0 −58.0

a.   Table entries are in inches. Multiply the entries in *Limits of clearance* and *Standard limits* columns by 10^{-3}.

Table C.3   Transition fits.[a]

Basic size Over - To	LT1			LT2			LT3		
	Limits of fit	Standard limits		Limits of fit	Standard limits		Limits of fit	Standard limits	
		Hole H7	Shaft js6		Hole H8	Shaft js7		Hole H7	Shaft k6
0 – 0.12	−0.10 +0.50	+0.4 0	+0.10 −0.10	−0.2 +0.8	+0.6 0	+0.2 −0.2			
0.12 – 0.24	−0.15 +0.65	+0.5 0	+0.15 −0.15	−0.25 +0.95	+0.7 0	+0.25 −0.25			
0.24 – 0.40	−0.2 +0.8	+0.6 0	+0.2 −0.2	−0.3 +1.2	+0.9 0	+0.3 −0.3	−0.5 +0.5	+0.6 0	+0.5 +0.1
0.40 – 0.71	−0.2 +0.9	+0.7 0	+0.2 −0.2	−0.35 + 1.35	+1.0 0	+0.35 −0.35	−0.5 +0.6	+0.7 0	+0.5 +0.1
0.71 – 1.19	−0.25 +1.05	+0.8 0	+0.25 −0.25	−0.4 +1.6	+1.2 0	+0.4 −0.4	−0.6 +0.7	+0.8 0	+0.6 +0.1
1.19 – 1.97	−0.3 +1.3	+1.0 0	+0.3 −0.3	−0.5 +2.1	+1.6 0	+0.5 −0.5	−0.7 +0.9	+1.0 0	+0.7 +0.1
1.97 – 3.15	−0.3 +1.5	+1.2 0	+0.3 −0.3	−0.6 +2.4	+1.8 0	+0.6 −0.6	−0.8 + 1.1	+1.2 0	+0.8 +0.1
3.15 – 4.73	−0.4 +1.8	+1.4 0	+0.4 −0.4	−0.7 +2.9	+2.2 0	+0.7 −0.7	−1.0 + 1.3	+1.4 0	+1.0 +0.1
4.73 – 7.09	−0.5 +2.1	+1.6 0	+0.5 −0.5	−0.8 +3.3	+2.5 0	+0.8 −0.8	−1.1 +1.5	+1.6 0	+1.1 +0.1
7.09 – 9.85	−0.6 +2.4	+1.8 0	+0.6 −0.6	−0.9 +3.7	+2.8 0	+0.9 −0.9	−1.4 +1.6	+1.8 0	+1.4 +0.2
9.85 – 12.41	−0.6 +2.6	+2.0 0	+0.6 −0.6	−1.0 +4.0	+3.0 0	+1.0 −1.0	−1.4 + 1.8	+2.0 0	+1.4 +0.2
12.41 – 15.75	−0.7 +2.9	+2.2 0	+0.7 −0.7	−1.0 +4.5	+3.5 0	+1.0 −1.0	−1.6 +2.0	+2.2 0	+1.6 +0.2
15.75 – 19.69	−0.8 +3.3	+2.5 0	+0.8 −0.8	−1.2 +5.2	+4.0 0	+1.2 −1.2	−1.8 +2.3	+2.5 0	+1.8 +0.2

a.   Table entries are in inches. Multiply the entries in *Limits of fit* and *Standard limits* columns by 10^{-3}.

Table C.3 Transition fits.[a] (Continued)

Basic size Over - To	LT4			LT5			LT6		
	Limits of fit	Standard limits		Limits of fit	Standard limits		Limits of fit	Standard limits	
		Hole H8	Shaft k7		Hole H7	Shaft n6		Hole H7	Shaft n7
0 – 0.12				−0.5 +0.15	+0.4 0	+0.5 +0.25	−0.65 +0.15	+0.4 0	+0.65 +0.25
0.12 – 0.24				−0.6 +0.2	+0.5 0	+0.6 +0.3	−0.8 +0.2	+0.5 0	+0.8 +0.3
0.24 – 0.40	−0.7 +0.8	+0.9 0	+0.7 +0.1	−0.8 +0.2	+0.6 0	+0.8 +0.4	−1.0 +0.2	+0.6 0	+1.0 +0.4
0.40 – 0.71	−0.8 +0.9	+1.0 0	+0.8 +0.1	−0.9 +0.2	+0.7 0	+0.9 +0.5	−1.2 +0.2	+0.7 0	+1.2 +0.5
0.71 – 1.19	−0.9 +1.1	+1.2 0	+0.9 +0.1	−1.1 +0.2	+0.8 0	+1.1 +0.6	−1.4 +0.2	+0.8 0	+1.4 +0.6
1.19 – 1.97	−1.1 +1.5	+1.6 0	+1.1 +0.1	−1.3 +0.3	+1.0 0	+1.3 +0.7	−1.7 +0.3	+1.0 0	+1.7 +0.7
1.97 – 3.15	−1.3 +1.7	+1.8 0	+1.3 +0.1	−1.5 +0.4	+1.2 0	+1.5 +0.8	−2.0 +0.4	+ 1.2 0	+2.0 +0.8
3.15 – 4.73	−1.5 +2.1	+2.2 0	+1.5 +0.1	−1.9 +0.4	+1.4 0	+1.9 +1.0	−2.4 +0.4	+ 1.4 0	+2.4 +1.0
4.73 – 7.09	−1.7 +2.4	+2.5 0	+1.7 +0.1	−2.2 +0.4	+1.6 0	+2.2 +1.2	−2.8 +0.4	+1.6 0	+2.8 +1.2
7.09 – 9.85	−2.0 +2.6	+2.8 0	+2.0 +0.2	−2.6 +0.4	+1.8 0	+2.6 +1.4	−3.2 +0.4	+1.8 0	+3.2 +1.4
9.85 – 12.41	−2.2 +2.8	+3.0 0	+2.2 +0.2	−2.6 +0.6	+2.0 0	+2.6 +1.4	−3.4 +0.6	+2.0 0	+3.4 +1.4
12.41 – 15.75	+2.4 +3.3	+3.5 0	+2.4 +0.2	−3.0 +0.6	+2.2 0	+3.0 +1.6	−3.8 +0.6	+2.2 0	+3.8 +1.6
15.75 – 19.69	−2.7 +3.8	+4.0 0	+2.7 +0.2	−3.4 +0.6	+2.5 0	+3.4 +1.8	−4.3 +0.7	+2.5 0	+4.3 +1.8

a.   Table entries are in inches. Multiply the entries in *Limits of fit* and *Standard limits* columns by 10^{-3}.

Table C.4   Interference fits.[a]

Basic size	LN1			LN2			LN3		
	Limits of interference	Standard limits		Limits of interference	Standard limits		Limits of interference	Standard limits	
Over - To		Hole H6	Shaft n5		Hole H7	Shaft p6		Hole H7	Shaft r6
0 – 0.12	0	+0.25	+0.45	0	+0.4	+0.65	0.1	+0.4	+0.75
	0.45	0	+0.25	0.65	0	+0.4	0.75	0	+0.5
0.12 – 0.24	0	+0.3	+0.5	0	+0.5	+0.8	0.1	+0.5	+0.9
	0.5	0	+0.3	0.8	0	+0.5	0.9	0	+0.6
0.24 – 0.40	0	+0.4	+0.65	0	+0.6	+1.0	0.2	+0.6	+1.2
	0.65	0	+0.4	1.0	0	+0.6	1.2	0	+0.8
0.40 – 0.71	0	+0.4	+0.8	0	+0.7	+1.1	0.3	+0.7	+1.4
	0.8	0	+0.4	1.1	0	+0.7	1.4	0	+1.0
0.71 – 1.19	0	+0.5	+1.0	0	+0.8	+1.3	0.4	+0.8	+1.7
	1.0	0	+0.5	1.3	0	+0.8	1.7	0	+1.2
1.19 – 1.97	0	+0.6	+1.1	0	+1.0	+1.6	0.4	+1.0	+2.0
	1.1	0	+0.6	1.6	0	+1.0	2.0	0	+1.4
1.97 – 3.15	0.1	+0.7	+1.3	0.2	+1.2	+2.1	0.4	+1.2	+2.3
	1.3	0	+0.7	2.1	0	+1.4	2.3	0	+1.6
3.15 – 4.73	0.1	+0.9	+1.6	0.2	+ 1.4	+2.4	0.6	+1.4	+2.9
	1.6	0	+1.0	2.5	0	+1.6	2.9	0	+2.0
4.73 – 7.09	0.2	+1.0	+1.9	0.2	+1.6	+2.8	0.9	+1.6	+3.5
	1.9	0	+1.2	2.8	0	+1.8	3.5	0	+2.5
7.09 – 9.85	0.2	+1.2	+2.2	0.2	+1.8	+3.2	1.2	+1.8	+4.2
	2.2	0	+1.4	3.2	0	+2.0	4.2	0	+3.0
9.85 – 12.41	0.2	+1.2	+2.3	0.2	+2.0	+3.4	1.5	+2.0	+4.7
	2.3	0	+1.4	3.4	0	+2.2	4.7	0	+3.5

a.   Table entries are in inches. Multiply the entries in *Limits of interference* and *Standard limits* columns by 10^{-3}.

Table C.5   Interference force fits.[a]

Basic size	FN1			FN2			FN3		
	Limits of interference	Standard limits		Limits of interference	Standard limits		Limits of interference	Standard limits	
Over - To		H6	n6		H7	s6		H7	t6
0 – 0.12	0.05	+0.25	+0.5	0.2	+0.4	+0.85			
	0.5	0	+0.3	0.85	0	+0.6			
0.12 – 0.24	0.1	+0.3	+0.6	0.2	+0.5	+1.0			
	0.6	0	+0.4	1.0	0	+0.7			
0.24 – 0.40	0.1	+0.4	+0.75	0.4	+0.6	+1.4			
	0.75	0	+0.5	1.4	0	+1.0			
0.40 – 0.56	0.1	+0.4	+0.8	0.5	+0.7	+1.6			
	0.8	0	+0.5	1.6	0	+1.2			
0.56 – 0.71	0.2	+0.4	+0.9	0.5	+0.7	+1.6			
	0.9	0	+0.6	1.6	0	+1.2			
0.71 – 0.95	0.2	+0.5	+1.1	0.6	+0.8	+1.9			
	1.1	0	+0.7	1.9	0	+1.4			
0.95 – 1.19	0.3	+0.5	+1.2	0.6	+0.8	+1.9	0.8	+0.8	+2.1
	1.2	0	+0.8	1.9	0	+1.4	2.1	0	+1.6
1.19 – 1.58	0.3	+0.6	+1.3	0.8	+1.0	+2.4	1.0	+1.0	+2.6
	1.3	0	+0.9	2.4	0	+1.8	2.6	0	+2.0
1.58 – 1.97	0.4	+0.6	+1.4	0.8	+1.0	+2.4	1.2	+1.0	+2.8
	1.4	0	+1.0	2.4	0	+1.8	2.8	0	+2.2
1.97 – 2.56	0.6	+0.7	+1.8	0.8	+1.2	+2.7	1.3	+1.2	+3.2
	1.8	0	+1.3	2.7	0	+2.0	3.2	0	+2.5
2.56 – 3.15	0.7	+0.7	+1.9	1.0	+1.2	+2.9	1.8	+1.2	+3.7
	1.9	0	+1.4	2.9	0	+2.2	3.7	0	+3.8
3.15 – 3.94	0.9	+0.9	+2.4	1.4	+1.4	+3.7	2.1	+1.4	+4.4
	2.4	0	+1.8	3.7	0	+2.8	4.4	0	+3.5
3.94 – 4.73	1.1	+0.9	+2.6	1.6	+1.4	+3.9	2.6	+1.4	+4.9
	2.6	0	+2.0	3.9	0	+3.0	4.9	0	+4.0
4.73 – 5.52	1.2	+1.0	+2.9	1.9	+1.6	+4.5	3.4	+1.6	+6.0
	2.9	0	+2.2	4.5	0	+3.5	6.0	0	+5.0
5.52 – 6.30	1.5	+1.0	+3.2	2.4	+1.6	+5.0	3.4	+1.6	+6.0
	3.2	0	+2.5	5.0	0	+4.0	6.0	0	+5.0
6.30 – 7.09	1.8	+1.0	+3.5	2.9	+1.6	+5.5	4.4	+1.6	+7.0
	3.5	0	+2.8	5.5	0	+4.5	7.0	0	+6.0
7.09 – 7.88	1.8	+1.2	+3.8	3.2	+1.8	+6.2	5.2	+1.8	+8.2
	3.8	0	+3.0	6.2	0	+5.0	8.2	0	+7.0
7.88 – 8.86	2.3	+1.2	+4.3	3.2	+1.8	+6.2	5.2	+1.8	+8.2
	4.3	0	+3.5	6.2	0	+5.0	8.2	0	+7.0
8.86 – 9.85	2.3	+1.2	+4.3	4.2	+1.8	+7.2	6.2	+1.8	+9.2
	4.3	0	+3.5	7.2	0	+6.0	9.2	0	+8.0
9.85 – 11.03	2.8	+1.2	+4.9	4.0	+2.0	+7.2	7.0	+2.0	+10.2
	4.9	0	+4.0	7.2	0	+6.0	10.2	0	+9.0
11.03 – 12.41	2.8	+1.2	+4.9	5.0	+2.0	+8.2	7.0	+2.0	+10.2
	4.9	0	+4.0	8.2	0	+7.0	10.2	0	+9.0

a.   Table entries are in inches. Multiply the entries in *Limits of interference* and *Standard limits* columns by 10^{-3}.

Table C.5   Interference force fits.[a] (Continued)

Basic size	FN4			FN5		
	Limits of interference	Standard limits		Limits of interference	Standard limits	
Over - To		H7	u6		H8	x7
0 – 0.12	0.3	+0.4	+0.95	0.3	+0.6	+1.3
	0.95	0	+0.7	1.3	0	+0.9
0.12 – 0.24	0.4	+0.5	+1.2	0.5	+0.7	+1.7
	1.2	0	+0.9	1.7	0	+1.2
0.24 – 0.40	0.6	+0.6	+1.6	0.5	+0.9	+2.0
	1.6	0	+1.2	2.0	0	+1.4
0.40 – 0.56	0.7	+0.7	+1.8	0.6	+1.0	+2.3
	1.8	0	+1.4	2.3	0	+1.6
0.56 – 0.71	0.7	+0.7	+1.8	0.8	+1.0	+2.5
	1.8	0	+1.4	2.5	0	+1.8
0.71 – 0.95	0.8	+0.8	+2.1	1.0	+1.2	+3.0
	2.1	0	+1.6	3.0	0	+2.2
0.95 – 1.19	1.0	+0.8	+2.3	1.3	+1.2	+3.3
	2.3	0	+1.8	3.3	0	+2.5
1.19 – 1.58	1.5	+1.0	+3.1	1.4	+1.6	+4.0
	3.1	0	+2.5	4.0	0	+3.0
1.58 – 1.97	1.8	+1.0	+3.4	2.4	+1.6	+5.0
	3.4	0	+2.8	5.0	0	+4.0
1.97 – 2.56	2.3	+1.2	+4.2	3.2	+1.8	+6.2
	4.2	0	+3.5	6.2	0	+5.0
2.56 – 3.15	2.8	+1.2	+4.7	4.2	+1.8	+7.2
	4.7	0	+4.0	7.2	0	+6.0
3.15 – 3.94	3.6	+1.4	+5.9	4.8	+2.2	+8.4
	5.9	0	+5.0	8.4	0	+7.0
3.94 – 4.73	4.6	+1.4	+6.9	5.8	+2.2	+9.4
	6.9	0	+6.0	9.4	0	+8.0
4.73 – 5.52	5.4	+1.6	+8.0	7.5	+2.5	+11.6
	8.0	0	+7.0	11.6	0	+10.0
5.52 – 6.30	5.4	+1.6	+8.0	9.5	+2.5	+13.6
	8.0	0	+7.0	13.6	0	+12.0
6.30 – 7.09	6.4	+1.6	+9.0	9.5	+2.5	+13.6
	9.0	0	+8.0	13.6	0	+12.0
7.09 – 7.88	7.2	+1.8	+10.2	11.2	+2.8	+15.8
	10.2	0	+9.0	15.8	0	+14.0
7.88 – 8.86	8.2	+1.8	+11.2	13.2	+2.8	+17.8
	11.2	0	+10.0	17.8	0	+16.0
8.86 – 9.85	10.2	+1.8	+13.2	13.2	+2.8	+17.8
	13.2	0	+12.0	17.8	0	+16.0
9.85 – 11.03	10.2	+2.0	+13.2	15.0	+3.0	+20.0
	13.2	0	+12.0	20.0	0	+18.0
11.03 – 12.41	12.0	+2.0	+15.2	17.0	+3.0	+22.0
	15.2	0	+14.0	22.0	0	+20.0

a.   Table entries are in inches. Multiply the entries in *Limits of interference* and *Standard limits* columns by 10^{-3}.

Table C.6   Tolerance grades — US (English) units.[a]

Basic size	Tolerance grade					
Over - To	IT6	IT7	IT8	IT9	IT10	IT11
0 – 0.12	0.2	0.4	0.6	1.0	1.6	2.4
0.12 – 0.24	0.3	0.5	0.7	1.2	1.9	3.0
0.24 – 0.40	0.4	0.6	0.9	1.4	2.3	3.5
0.40 – 0.72	0.4	0.7	1.1	1.7	2.8	4.3
0.72 – 1.20	0.5	0.8	1.3	2.0	3.3	5.1
1.20 – 2.00	0.6	1.0	1.5	2.4	3.9	6.3
2.00 – 3.20	0.7	1.2	1.8	2.9	4.7	7.5
3.20 – 4.80	0.9	1.4	2.1	3.4	5.5	8.7
4.80 – 7.20	1.0	1.6	2.5	3.9	6.3	9.8
7.20 – 10.00	1.1	1.8	2.8	4.5	7.3	11.4
10.00 – 12.60	1.3	2.0	3.2	5.1	8.3	12.6
12.60 – 16.00	1.4	2.2	3.5	5.5	9.1	14.2

a.   Table entries are in inches. Multiply the entries in *Tolerance grade* columns by 10^{-3}.

Table C.7   Fundamental deviation — US (English) units.[a]

Basic size	Shaft tolerance symbol									
Over - To	c	d	f	g	h	k	n	p	s	u
0 – 0.12	−2.4	−0.8	−0.2	−0.1	0	0	+0.2	+0.2	+0.6	+0.7
0.12 – 0.24	−2.8	−1.2	−0.4	−0.2	0	0	+0.3	+0.5	+0.7	+0.9
0.24 – 0.40	−3.1	−1.6	−0.5	−0.2	0	0	+0.4	+0.6	+0.9	+1.1
0.40 – 0.72	−3.7	−2.0	−0.6	−0.2	0	0	+0.5	+0.7	+1.1	+1.3
0.72 – 0.96	−4.3	−2.6	−0.8	−0.3	0	+0.1	+0.6	+0.9	+1.4	+1.6
0.96 – 1.20	−4.3	−2.6	−0.8	−0.3	0	+0.1	+0.6	+0.9	+1.4	+1.9
1.20 – 1.60	−4.7	−3.1	−1.0	−0.4	0	+0.1	+0.7	+1.0	+1.7	+2.4
1.60 – 2.00	−5.1	−3.1	−1.0	−0.4	0	+0.1	+0.7	+1.0	+1.7	+2.8
2.00 – 2.60	−5.5	−3.9	−1.2	−0.4	0	+0.1	+0.8	+1.3	+2.1	+3.4
2.60 – 3.20	−5.9	−3.9	−1.2	−0.4	0	+0.1	+0.8	+1.3	2.3	+4.0
3.20 – 4.00	−6.7	−4.7	−1.4	−0.5	0	+0.1	+0.9	+1.5	+2.8	+4.9
4.00 – 4.80	−7.1	−4.7	−1.4	−0.5	0	+0.1	+0.9	+1.5	+3.1	+5.7
4.80 – 5.60	−7.9	−5.7	−1.7	−0.6	0	+0.1	+1.1	+1.7	+3.6	+6.7
5.60 – 6.40	−8.3	−5.7	−1.7	−0.6	0	+0.1	+1.1	+1.7	+3.9	+7.5
6.40 – 7.20	−9.1	−5.7	−1.7	−0.6	0	+0.1	+1.1	+1.7	+4.3	+8.3
7.20 – 8.00	−9.4	−6.7	−2.0	−0.6	0	+0.2	+1.2	+2.0	+4.8	+9.3
8.00 – 9.00	−10.2	−6.7	−2.0	−0.6	0	+0.2	+1.2	+2.0	+5.1	+10.2
9.00 – 10.00	−11.0	−6.7	−2.0	−0.6	0	+0.2	+1.2	+2.0	+5.5	+11.2
10.00 – 11.20	−11.8	−7.5	−2.2	−0.7	0	+0.2	+1.3	+2.2	+6.2	+12.4
11.20 – 12.60	−13.0	−7.5	−2.2	−0.7	0	+0.2	+1.3	+2.2	+6.7	+13.0
12.60 – 14.20	−14.2	−8.3	−2.4	−0.7	0	+0.2	+1.5	+2.4	+7.5	+15.4
14.20 – 16.00	−15.7	−8.3	−2.4	−0.7	0	+0.2	+1.5	+2.4	+8.2	+17.1

a.   Table entries are in inches. Multiply the entries in *Shaft tolerance symbol* columns by 10^{-3}.

Table C.8   Tolerance grades — Metric (SI) units.[a]

Basic size	Tolerance grade					
Over - To	IT6	IT7	IT8	IT9	IT10	IT11
0 – 3	6	10	14	25	40	60
3 – 6	8	12	18	30	48	75
6 – 10	9	15	22	36	58	90
10 – 18	11	18	27	43	70	110
18 – 30	13	21	33	52	84	130
30 – 50	16	25	39	62	100	160
50 – 80	19	30	46	74	120	190
80 – 120	22	35	54	87	140	220
120 – 180	25	40	63	100	160	250
180 – 250	29	46	72	115	185	290
250 – 315	32	52	81	130	210	320
315 – 400	36	57	89	140	230	360

a.   Table entries are in millimeters. Multiply the entries in *Tolerance grade* columns by 10^{-3}.

Table C.9   Fundamental deviation — Metric (SI) units.[a]

Basic size	Shaft tolerance symbol									
Over - To	c	d	f	g	h	k	n	p	s	u
0 – 3	−60	−20	−6	−2	0	0	+4	+6	+14	+18
3 – 6	−70	−30	−10	−4	0	+1	+8	+12	+19	+23
6 – 10	−80	−40	−13	−5	0	+1	+10	+15	+23	+28
10 – 14	−95	−50	−16	−6	0	+1	+12	+18	+28	+33
14 – 18	−95	−50	−16	−6	0	+1	+12	+18	+28	+33
18 – 24	−110	−65	−20	−7	0	+2	+15	+22	+35	+41
24 – 30	−110	−65	−20	−7	0	+2	+15	+22	+35	+48
30 – 40	−120	−80	−25	−9	0	+2	+17	+26	+43	+60
40 – 50	−130	−80	−25	−9	0	+2	+17	+26	+43	+70
50 – 65	−140	−100	−30	−10	0	+2	+20	+32	+53	+87
65 – 80	−150	−100	−30	−10	0	+2	+20	+32	+59	+102
80 – 100	−170	−120	−36	−12	0	+3	+23	+37	+71	+124
100 – 120	−180	−120	−36	−12	0	+3	+23	+37	+79	+144
120 – 140	−200	−145	−43	−14	0	+3	+27	+43	+92	+170
140 – 160	−210	−145	−43	−14	0	+3	+27	+43	+100	+190
160 – 180	−230	−145	−43	−14	0	+3	+27	+43	+108	+210
180 – 200	−240	−170	−50	−15	0	+4	+31	+50	+122	+236
200 – 225	−260	−170	−50	−15	0	+4	+31	+50	+130	+258
225 – 250	−280	−170	−50	−15	0	+4	+31	+50	+140	+284
250 – 280	−300	−190	−56	−17	0	+4	+34	+56	+158	+315
280 – 315	−330	−190	−56	−17	0	+4	+34	+56	+170	+350
315 – 355	−360	−210	−62	−18	0	+4	+37	+62	+190	+390
355 – 400	−400	−210	−62	−18	0	+4	+37	+62	+208	+435

a.   Table entries are in millimeters. Multiply the entries in *Shaft tolerance symbol* columns by 10^{-3}.

Index